Sampling

Design and Analysis

Third Edition

CHAPMAN & HALL/CRC
Texts in Statistical Science Series

Recently Published Titles

Beyond Multiple Linear Regression
Applied Generalized Linear Models and Multilevel Models in R
Paul Roback, Julie Legler

Bayesian Thinking in Biostatistics
Gary L. Rosner, Purushottam W. Laud, and Wesley O. Johnson

Linear Models with Python
Julian J. Faraway

Modern Data Science with R, Second Edition
Benjamin S. Baumer, Daniel T. Kaplan, and Nicholas J. Horton

Probability and Statistical Inference
From Basic Principles to Advanced Models
Miltiadis Mavrakakis and Jeremy Penzer

Bayesian Networks
With Examples in R, Second Edition
Marco Scutari and Jean-Baptiste Denis

Time Series
Modeling, Computation, and Inference, Second Edition
Raquel Prado, Marco A. R. Ferreira and Mike West

A First Course in Linear Model Theory, Second Edition
Nalini Ravishanker, Zhiyi Chi, Dipak K. Dey

Foundations of Statistics for Data Scientists
With R and Python
Alan Agresti and Maria Kateri

Fundamentals of Causal Inference
With R
Babette A. Brumback

Sampling: Design and Analysis, Third Edition
Sharon L. Lohr

For more information about this series, please visit: https://www.crcpress.com/Chapman--Hall/CRC-Texts-in-Statistical-Science/book-series/CHTEXSTASCI

Sampling

Design and Analysis

Third Edition

Sharon L. Lohr

CRC Press is an imprint of the
Taylor & Francis Group, an **informa** business
A CHAPMAN & HALL BOOK

Third edition published 2022
by CRC Press
6000 Broken Sound Parkway NW, Suite 300, Boca Raton, FL 33487-2742

and by CRC Press
2 Park Square, Milton Park, Abingdon, Oxon, OX14 4RN

Second edition published by CRC Press 2019

CRC Press is an imprint of Taylor & Francis Group, LLC

Library of Congress Cataloging-in-Publication Data

Names: Lohr, Sharon L., author.
Title: Sampling : design and analysis / Sharon L. Lohr.
Description: Third edition. | Boca Raton : CRC Press, 2022. | Series: Chapman & Hall CRC texts in statistical science | Includes index. | Summary: ""The level is appropriate for an upper-level undergraduate or graduate-level statistics major. Sampling: Design and Analysis (SDA) will also benefit a non-statistics major with a desire to understand the concepts of sampling from a finite population. A student with patience to delve into the rigor of survey statistics will gain even more from the content that SDA offers. The updates to SDA have potential to enrich traditional survey sampling classes at both the undergraduate and graduate levels. The new discussions of low response rates, non-probability surveys, and internet as a data collection mode hold particular value, as these statistical issues have become increasingly important in survey practice in recent years... I would eagerly adopt the new edition of SDA as the required textbook." (Emily Berg, Iowa State University) What is the unemployment rate? What is the total area of land planted with soybeans? How many persons have antibodies to the virus causing COVID-19? Sampling: Design and Analysis, Third Edition shows you how to design and analyze surveys to answer these and other questions. This authoritative text, used as a standard reference by numerous survey organizations, teaches the principles of sampling with examples from social sciences, public opinion research, public health, business, agriculture, and ecology. Readers should be familiar with concepts from an introductory statistics class including probability and linear regression; optional sections contain statistical theory for readers familiar with mathematical statistics. The third edition, thoroughly revised to incorporate recent research and applications, includes a new chapter on nonprobability samples-when to use them and how to evaluate their quality. More than 200 new examples and exercises have been added to the already extensive sets in the second edition. SDA's companion website contains data sets, computer code, and links to two free downloadable supplementary books (also available in paperback) that provide step-by-step guides-with code, annotated output, and helpful tips-for working through the SDA examples. Instructors can use either R or SAS® software. SAS® Software Companion for Sampling: Design and Analysis, Third Edition by Sharon L. Lohr (2022, CRC Press) R Companion for Sampling: Design and Analysis, Third Edition by Yan Lu and Sharon L. Lohr (2022, CRC Press)"-- Provided by publisher.
Identifiers: LCCN 2021025531 (print) | LCCN 2021025532 (ebook) | ISBN 9780367279509 (hardback) | ISBN 9781032130590 (paperback) | ISBN 9780429298899 (ebook)
Subjects: LCSH: Sampling (Statistics)
Classification: LCC HA31.2 .L64 2022 (print) | LCC HA31.2 (ebook) | DDC 001.4/33--dc23
LC record available at https://lccn.loc.gov/2021025531
LC ebook record available at https://lccn.loc.gov/2021025532

ISBN: 978-0-367-27950-9 (hbk)
ISBN: 978-1-032-13059-0 (pbk)
ISBN: 978-0-429-29889-9 (ebk)

DOI: 10.1201/9780429298899

Typeset in LM Roman
by KnowledgeWorks Global Ltd.

Access the Support Material: http://routledge.com/9780367279509

To Doug

Contents

Preface

We rarely have complete information in life. Instead, we make decisions from partial information, often in the form of a sample from the population we are interested in. *Sampling: Design and Analysis* teaches the statistical principles for selecting samples and analyzing data from a sample survey. It shows you how to evaluate the quality of estimates from a survey, and how to design and analyze many different forms of sample surveys.

The third edition has been expanded and updated to incorporate recent research on theoretical and applied aspects of survey sampling, and to reflect developments related to the increasing availability of massive data sets ("big data") and samples selected via the internet. The new chapter on nonprobability sampling tells how to analyze and evaluate information from samples that are not selected randomly (including big data), and contrasts nonprobability samples with low-response-rate probability samples. The chapters on nonsampling errors have been extensively revised to include recent developments on treating nonresponse and measurement errors. Material in other chapters has been revised where there has been new research or I felt I could clarify the presentation of results. Examples retained from the second edition have been updated when needed, and new examples have been added throughout the book to illustrate recent applications of survey sampling.

The third edition has also been revised to be compatible with multiple statistical software packages. Two supplementary books, available for FREE download from the book's companion website (see page xviii for how to obtain the books), provide step-by-step guides of how to use SAS® and R software to analyze the examples in *Sampling: Design and Analysis.* Both books are also available for purchase in paperback form, for readers who prefer a hard copy.

Lohr, S. (2022). *SAS® Software Companion for Sampling: Design and Analysis, Third Edition.* Boca Raton, FL: Chapman & Hall/CRC Press.

Lu, Y. and Lohr, S. (2022). *R Companion for Sampling: Design and Analysis, Third Edition.* Boca Raton, FL: Chapman & Hall/CRC Press.

Instructors can choose which software package to use in the class (SAS software alone, R software alone, or, if desired, both software packages) and have students download the appropriate supplementary book. See the Computing section on page xvi for more information about the supplementary books and about choice of statistical software.

Features of *Sampling: Design and Analysis, Third Edition*

- The book is accessible to students with a wide range of statistical backgrounds, and is flexible for content and level. By appropriate choice of sections, this book can be used for an upper-level undergraduate class in statistics, a first- or second-year graduate class for statistics students, or a class with students from business, sociology, psychology, or biology who want to learn about designing and analyzing data from sample surveys. It is also useful for persons who analyze survey data and want to learn more about the statistical aspects of surveys and recent developments. The book is intended for anyone who is interested in using sampling methods to learn about a population, or who wants to understand how data from surveys are collected, analyzed, and interpreted.

Chapters 1–8 can be read by students who are familiar with basic concepts of probability and statistics from an introductory statistics course, including independence and expectation, confidence intervals, and straight-line regression. Appendix A reviews the probability concepts needed to understand probability sampling. Parts of Chapters 9 to 16 require more advanced knowledge of mathematical and statistical concepts. Section 9.1, on linearization methods for variance estimation, assumes knowledge of calculus. Chapter 10, on categorical data analysis, assumes the reader is familiar with chi-square tests and odds ratios. Chapter 11, on regression analysis for complex survey data, presupposes knowledge of matrices and the theory of multiple regression for independent observations.

Each chapter concludes with a chapter summary, including a glossary of key terms and references for further exploration.

- The examples and exercises feature real data sets from the social sciences, engineering, agriculture, ecology, medicine, business, and a variety of other disciplines. Many of the data sets contain other variables not specifically referenced in the text; an instructor can use these for additional exercises and activities.

  The data sets are available for download from the book's companion website. Full descriptions of the variables in the data sets are given in Appendix A of the supplementary books described above (Lohr, 2022; Lu and Lohr, 2022).

  The exercises also give the instructor much flexibility for course level (see page xv). Some emphasize mastering the mechanics, but many encourage the student to think about the sampling issues involved and to understand the structure of sample designs at a deeper level. Other exercises are open-ended and encourage further exploration of the ideas.

  In the exercises, students are asked to design and analyze data from real surveys. Many of the data examples and exercises carry over from chapter to chapter, so students can deepen their knowledge of the statistical concepts and see how different analyses are performed with the sample. Data sets that are featured in multiple chapters are listed in the "Data sets" entry of the Index so you can follow them across chapters.

- *Sampling: Design and Analysis, Third Edition* includes many topics not found in other textbooks at this level. Chapters 7–11 discuss how to analyze complex surveys such as those administered by federal statistical agencies, how to assess the effects of nonresponse and weight the data to adjust for it, how to use computer-intensive methods for estimating variances in complex surveys, and how to perform chi-square tests and regression analyses using data from complex surveys. Chapters 12–14 present methods for two-phase sampling, using a survey to estimate population size, and designing a survey to study a subpopulation that is hard to identify or locate. Chapter 15, new for the third edition, contrasts probability and nonprobability samples, and provides guidance on how to evaluate the quality of nonprobability samples. Chapter 16 discusses a total quality framework for survey design, and presents some thoughts on the future of sampling.

- Design of surveys is emphasized throughout, and is related to methods for analyzing the data from a survey. The book presents the philosophy that the design is by far the most important aspect of any survey: No amount of statistical analysis can compensate for a badly designed survey.

- *Sampling: Design and Analysis, Third Edition* emphasizes the importance of graphing the data. Graphical analysis of survey data is challenging because of the large sizes and complexity of survey data sets but graphs can provide insight into the data structure.

- While most of the book adopts a randomization-based perspective, I have also included sections that approach sampling from a model-based perspective, with the goal of placing sampling methods within the framework used in other areas of statistics. Many important results in survey research have involved models, and an understanding of both approaches is essential for the survey practitioner. All methods for dealing with nonresponse are model-based. The model-based approach is introduced in Section 2.10 and further developed in successive chapters; those sections can be covered while those chapters are taught or discussed at any time later in the course.

Exercises. The book contains more than 550 exercises, organized into four types. More than 150 of the exercises are new to the third edition.

A. **Introductory** exercises are intended to develop skills on the basic ideas in the book.

B. **Working with Survey Data** exercises ask students to analyze data from real surveys. Most require the use of statistical software; see section on Computing below.

C. **Working with Theory** exercises are intended for a more mathematically oriented class, allowing students to work through proofs of results in a step-by-step manner and explore the theory of sampling in more depth. They also include presentations of additional results about survey sampling that may be of interest to more advanced students. Many of these exercises require students to know calculus, probability theory, or mathematical statistics.

D. **Projects and Activities** exercises contain activities suitable for classroom use or for assignment as a project. Many of these activities ask the student to design, collect, and analyze a sample selected from a population. The activities continue from chapter to chapter, allowing students to build on their knowledge and compare various sampling designs. I always assigned Exercise 35 from Chapter 7 and its continuation in subsequent chapters as a course project, and asked students to write a report with their findings. These exercises ask students to download data from a survey on a topic of their choice and analyze the data. Along the way, the students read and translate the survey design descriptions into the design features studied in class, develop skills in analyzing survey data, and gain experience in dealing with nonresponse or other challenges.

Suggested chapters for sampling classes. Chapters 1–6 treat the building blocks of simple random, stratified, and cluster sampling, as well as ratio and regression estimation. To read them requires familiarity with basic ideas of expectation, sampling distributions, confidence intervals, and linear regression—material covered in most introductory statistics classes. Along with Chapters 7 and 8, these chapters form the foundation of a one-quarter or one-semester course. Sections on the statistical theory in these chapters are marked with asterisks—these require more familiarity with probability theory and mathematical statistics. The material in Chapters 9–16 can be covered in almost any order, with topics chosen to fit the needs of the students.

Sampling: Design and Analysis, Third Edition can be used for many different types of classes, and the choice of chapters to cover can be tailored to meet the needs of the students in that class. Here are suggestions of chapters to cover for four types of sampling classes.

Undergraduate class of statistics students: Chapters 1–8, skipping sections with asterisks; Chapters 15 and 16.

One-semester graduate class of statistics students: Chapters 1–9, with topics chosen from the remaining chapters according to the desired emphasis of the class.

Two-semester graduate class of statistics students: All chapters, with in-depth coverage of Chapters 1–8 in the first term and Chapters 9–16 in the second term. The exercises contain many additional theoretical results for survey sampling; these can be presented in class or assigned for students to work on.

Students from social sciences, biology, business, or other subjects: Chapters 1–7 should be covered for all classes, skipping sections with asterisks. Choice of other material depends on how the students will be using surveys in the future. Persons teaching classes for social scientists may want to include Chapters 8 (nonresponse), 10 (chi-square tests), and 11 (regression analyses of survey data). Persons teaching classes for biology students may want to cover Chapter 11 and Chapter 13 on using surveys to estimate population sizes. Students who will be analyzing data from large government surveys would want to learn about replication-based variance estimation methods in Chapter 9. Students who may be using nonprobability samples should read Chapter 15.

Any of these can be taught as activity-based classes, and that is how I structured my sampling classes. Students were asked to read the relevant sections of the book at home before class. During class, after I gave a ten-minute review of the concepts, students worked in small groups with their laptops on designing or analyzing survey data from the chapter examples or the "Projects and Activities" section, and I gave help and suggestions as needed. We ended each class with a group discussion of the issues and a preview of the next session's activities.

Computing. You need to use a statistical software package to analyze most of the data sets provided with this book. I wrote *Sampling: Design and Analysis, Third Edition* for use with either SAS or R software. You can choose which software package to use for computations: SAS software alone, R alone, or both, according to your preference. Both software packages are available at no cost for students and independent learners, and the supplementary books tell how to obtain them.

The supplementary books, *SAS® Software Companion for Sampling: Design and Analysis, Third Edition* by Sharon L. Lohr, and *R Companion for Sampling: Design and Analysis, Third Edition* by Yan Lu and Sharon L. Lohr, available for FREE download from the book's companion website, demonstrate how to use SAS and R software, respectively, to analyze the examples in *Sampling: Design and Analysis, Third Edition.* Both books are also available for purchase in paperback form, for readers who prefer hard copies. The two supplementary books are written in parallel format, making it easy to find how a particular example is coded in each software package. They thus would also be useful for a reader who is familiar with one of the software packages but would like to learn how to use the other.

The supplementary books provide the code used to select, or produce estimates or graphs from, the samples used for the examples in Chapters 1–13 of this book. They display and interpret the output produced by the code, and discuss special features of the procedure or function used to produce the output. Each chapter concludes with tips and warnings on how to avoid common errors when designing or analyzing surveys.

Which software package should you use? If you are already familiar with R or SAS software, you may want to consider adopting that package when working through *Sampling: Design and Analysis, Third Edition.* You may also want to consider the following features of the two software packages for survey data.

Features of SAS software for survey data:

- Students and independent learners anywhere in the world can access a FREE, cloud-based version of the software: SAS® OnDemand for Academics (`https://www.sas.com/en_us/software/on-demand-for-academics.html`) contains all of the programs needed to select samples, compute estimates, and graph data for surveys. Short online videos for instructors show how to create a course site online, upload data that can be accessed by all students, and give students access to the course material. Additional short video tutorials help students become acquainted with the basic features of the system; other videos, and online help materials, introduce students to basic concepts of programming in SAS software.

- Most of the data analyses or sample selections for this book's examples and exercises can be done with short programs (usually containing five or fewer lines) that follow a standard syntax.

  The survey analysis procedures in SAS/STAT® software, which at this writing include the SURVEYMEANS, SURVEYREG, SURVEYFREQ, SURVEYLOGISTIC, and SURVEYPHREG procedures, are specifically designed to produce estimates from complex surveys. The procedures can calculate either linearization-based variance estimates (used in Chapters 1–8) or the replication variance estimates described in Chapter 9, and they will construct replicate weights for a survey design that you specify. They will also produce appropriate survey-weighted plots of the data. The output provides the statistics you request as well as additional information that allows you to verify the design and weighting information used in the analysis. The procedures also print warnings if you have written code that is associated with some common survey data analysis errors.

  The SURVEYSELECT procedure will draw every type of probability sample discussed in this book, again with output that confirms the procedure used to draw the sample.

- SAS software is designed to allow you to manipulate and manage large data sets (some survey data sets contain tens of thousands or even millions of records), and compute estimates for those data sets using numerically stable and efficient algorithms. Many large survey data sets (such as the National Health and Nutrition Examination Survey data discussed in Chapter 7) are distributed as SAS data sets; you can also import files from spreadsheet programs, comma- or tab-delimited files, and other formats.

- The software is backward compatible—that is, code written for previous versions of the software will continue to work with newer versions. All programs are thoroughly tested before release, and the customer support team resolves any problems with the software that users might discover after release (they do not answer questions about how to do homework problems, though!). Appendix 5 of SAS Institute Inc. (2020) describes the methods used to quality-check and validate statistical procedures in SAS software.

- You do not need to learn computer programming to perform standard survey data analyses with SAS software. But for advanced users, the software offers the capability to write programs in SAS/IML® software or use macros. In addition, many user-contributed macros that perform specialized analyses of survey data have been published.

Features of the R statistical software environment for survey data:

- The software is available FREE from `https://www.r-project.org/`. It is open-source software, which means anyone can use it without a license or fee. Many tutorials on how to use R are available online; these tell you how to use the software to compute statistics and to create customized graphics.

- Base R contains functions that will select and analyze data from simple random samples. To select and analyze data from other types of samples, however—those discussed after Chapter 2 of this book—R users must either (1) write their own R functions or (2) use functions that have been developed by other R users and made available through a contributed package. As of September 2020, the Comprehensive R Archive Network (CRAN) contained more than 16,000 contributed packages. If a statistical method has been published, there is a good chance that someone has developed a contributed package for R that performs the computations.

Contributed packages for R are not peer-reviewed or quality-checked unless the package authors arrange for such review. Functions in base R and contributed packages can change at any time, and are not always backward compatible.

But the open-source nature of R means that other users can view and test the functions in the packages. The book by Lu and Lohr (2022) makes use of functions in two popular contributed packages that have been developed for survey data by Lumley (2020) and by Tillé and Matei (2021). These functions will compute estimates and select samples for every type of probability sampling design discussed in *Sampling: Design and Analysis, Third Edition.*

- You need to learn how to work with functions in R in order to use it to analyze or select surveys. After you have gained experience with R, however, you can write functions to produce estimates for new statistical methods or to conduct simulation studies such as that requested in Exercise 21 of Chapter 4.

Software packages other than SAS and R can also be used with the book, as long as they have programs that correctly calculate estimates from complex survey data. Brogan (2015) illustrated the errors that result when non-survey software is used to analyze data from a complex survey. Software packages with survey data capabilities include SUDAAN® (RTI International, 2012), Stata® (Kolenikov, 2010), SPSS® (Zou et al., 2020), Mplus® (Muthén and Muthén, 2017), WesVar® (Westat, 2015), and IVEware (Raghunathan et al., 2016). See West et al. (2018) for reviews of these and other packages. New computer programs for analyzing survey data are developed all the time; the newsletter of the International Association of Survey Statisticians (`http://isi-iass.org`) is a good resource for updated information.

Website for the book. The book's website can be reached from either of the following addresses:

`https://www.sharonlohr.com`
`https://www.routledge.com/9780367279509.`

It contains links to:

- Downloadable pdf files for the supplementary books *SAS® Software Companion for Sampling: Design and Analysis, Third Edition* and *R Companion for Sampling: Design and Analysis, Third Edition.* The pdf files are identical to the published paperback versions of the books.

- All data sets referenced in the book. These are available in comma-delimited (.csv), SAS, or R format. The data sets in R format are also available in the R contributed package *SDAResources* (Lu and Lohr, 2021).

- Other resources related to the book.

A solutions manual for the book is available (for instructors only) from the publisher at `https://www.routledge.com/9780367279509.`

Acknowledgments. I have been fortunate to receive comments and advice from many people who have used or reviewed one or more of the editions of this book. Serge Alalouf, David Bellhouse, Emily Berg, Paul Biemer, Mike Brick, Trent Buskirk, Ted Chang, Ron Christensen, Mark Conaway, Dale Everson, Andrew Gelman, James Gentle, Burke Grandjean, Michael Hamada, David Haziza, Nancy Heckman, Mike Hidiroglou, Norma Hubele, Tim Johnson, Jae-Kwang Kim, Stas Kolenikov, Partha Lahiri, Yan Lu, Steve MacEachern, David Marker, Ruth Mickey, Sarah Nusser, N. G. N. Prasad, Minsun Riddles, Deborah Rumsey, Thomas P. Ryan, Fritz Scheuren, Samantha Seals, Elizabeth Stasny, Imbi Traat, Shap Wolf, Tommy Wright, Wesley Yung, and Elaine Zanutto have all provided suggestions that resulted in substantial improvements in the exposition. I am profoundly grateful that these extraordinary statisticians were willing to take the time to share their insights about how the book could better meet the needs of students and sampling professionals.

I'd like to thank Sandra Clark, Mark Asiala, and Jason Fields for providing helpful suggestions and references for the material on the American Community Survey and Household Pulse Survey. Kinsey Dinan, Isaac McGinn, Arianna Fishman, and Jayme Day answered questions and pointed me to websites with information about the procedures for the annual point-in-time count described in Example 3.13. Pierre Lavallée, Dave Chapman, Jason Rivera, Marina Pollán, Roberto Pastor-Barriuso, Sunghee Lee, Mark Duda, and Matt Hayat generously helped me with questions about various examples in the book.

J. N. K. Rao has provided encouragement, advice, and suggestions for this book since the first edition. I began collaborating with Jon on research shortly after receiving tenure, and have always been awed at his ability to identify and solve the important problems in survey sampling—often years before anyone else realizes how crucial the topics will be. I can think of no one who has done more to develop the field of survey sampling, not only through his research contributions but also through his strong support for young statisticians from all over the world. Thank you, Jon, for all your friendship and wise counsel over the years.

John Kimmel, editor extraordinaire at CRC Press, encouraged me to write this third edition, and it was his idea to have supplemental books showing how to use SAS and R software with the book examples. I feel immensely privileged to have had the opportunity to work with him and to benefit from his amazing knowledge of all things publishing.

Sharon L. Lohr
April 2021

Symbols and Acronyms

The number in parentheses is the page where the notation is introduced.

ACS	American Community Survey. (4)
ASA	American Statistical Association. (91)
ANOVA	Analysis of variance. (90)
B	Ratio t_y/t_x or, more generally, a regression coefficient. (122)
BMI	Body mass index (variable measured in NHANES). (291)
χ^2	Chi-square. (349)
$\mathcal{C}$	Set of units in a convenience (or other nonprobability) sample. (528)
cdf	Cumulative distribution function. (281)
CI	Confidence interval. (46)
Cov	Covariance. (57)
CV	Coefficient of variation. (42)
deff	Design effect. (286)
df	Degrees of freedom. (48)
D_i	Random variable indicating inclusion in phase II of a two-phase sample. (460)
E	Expected value. (36)
f	Probability density or mass function. (281)
F	Cumulative distribution function. (281) In other contexts, F represents the F distribution. (404)
fpc	Finite population correction, $= (1 - n/N)$ for a simple random sample. (41)
GREG	Generalized regression. (444)
GVF	Generalized variance function. (379)
HT	Horvitz-Thompson estimator or variance estimator. (236)
ICC	Intraclass correlation coefficient. (176)
IPUMS	Integrated Public Use Microdata Series. (78)
ln	Natural logarithm. (338)
logit	$\text{Logit}(p) = \ln[p/(1-p)]$. (441)
M_i	Number of ssus in the population from psu i. (170)
m_i	Number of ssus in the sample from psu i. (171)
M_0	Total number of ssus in the population, in all psus. (170)
MAR	Missing at random given covariates, a mechanism for missing data. (322)
MCAR	Missing completely at random, a mechanism for missing data. (321)

MICE Multivariate imputation by chained equations. (338)

MSE Mean squared error. (37)

μ Theoretical value of mean in an infinite population, used in model-based inference. (56)

NHANES National Health and Nutrition Examination Survey. (273)

NMAR Not missing at random, a mechanism for missing data. (323)

N Number of units in the population. (34)

n Number of units in the sample. (32)

OLS Ordinary least squares. (420)

P Probability operator. (34)

p Proportion of units in the population having a characteristic. (38)

$\hat{p}$ Estimated proportion of units in the population having a characteristic. (39)

PES Post-enumeration survey. (487)

π_i Probability that unit i is in the sample. (34)

π_{ik} Probability that units i and k are both in the sample (joint inclusion probability). (235)

ϕ_i Probability that unit i responds to a survey after being selected for the sample, called the response propensity. (321)

ψ_i Probability that unit i is selected on the first draw in a with-replacement sample. (220)

pps Probability proportional to size. (229)

psu Primary sampling unit. (167)

Q_i Random variable indicating the number of times unit i appears in a with-replacement sample. (73)

$\mathcal{R}$ Set of respondents to the survey. (323)

R_i Random variable indicating whether unit i responds to a survey after being selected for the sample. (321) In Chapter 15, R_i is the random variable indicating participation in a non-probability sample. (525)

R^2 Coefficient of determination for a regression analysis. (421)

R_a^2 Adjusted R^2. (177)

$\mathcal{S}$ Set of units in a probability sample. (34)

$\mathcal{S}_h$ Set of units sampled from stratum h in a stratified sample. (84)

$\mathcal{S}_i$ Set of ssus sampled from psu i in a cluster sample. (171)

$\mathcal{S}^{(1)}$ Phase I sample. (459)

$\mathcal{S}^{(2)}$ Phase II sample. (460)

S^2 Population variance of y. (38)

s^2 Sample variance of y in a simple random sample. (42)

S Population standard deviation of y, $= \sqrt{S^2}$. (38)

S_h^2 Population variance in stratum h. (84)

s_h^2	Sample variance in stratum h, in a stratified random sample. (84)
σ	Theoretical value of standard deviation for an infinite population, used in model-based theory. (59)
SE	Standard error. (42)
SRS	Simple random sample without replacement. (39)
SRSWR	Simple random sample with replacement. (39)
ssu	Secondary sampling unit. (167)
SYG	Sen-Yates-Grundy, specifying an estimator of the variance. (236)
t	Population total, with $t = t_y = \sum_{i=1}^{N} y_i$. (35)
T	Population total in model-based approach. (59) When used as superscript on a vector or matrix, as in $\mathbf{x}^T$, T denotes transpose. (404)
$\hat{t}$	Estimator of population total. (35)
$\hat{t}_{\text{HT}}$	Horvitz–Thompson estimator of the population total. (236)
$t_{\alpha/2,k}$	The $100(1-\alpha/2)$th percentile of a t distribution with k degrees of freedom. (48)
tsu	Tertiary (third-level) sampling unit. (243)
$\mathcal{U}$	Set of units in the population, also called the universe. (34)
V	Variance. (37)
$\mathcal{W}$	Set of units in a with-replacement probability sample, including the repeated units multiple times. (226)
w_i	Weight associated with unit i in the sample. (44)
WLS	Weighted least squares. (432)
x_i	An auxiliary variable for unit i in the population. This symbol is in boldface when a vector of auxiliary variables is considered. (121)
y_i	A characteristic of interest observed for sampled unit i. (35)
Y_i	A random variable used in model-based inference; y_i is the realization of Y_i in the sample. (59)
$\bar{y}_{\mathcal{U}}$	Population mean, $= \dfrac{1}{N}\sum_{i=1}^{N} y_i$. (38)
$\bar{y}$	Sample mean, $= \dfrac{1}{n}\sum_{i\in\mathcal{S}} y_i$. (35)
$\hat{\bar{y}}$	An estimator of the population mean. (122)
$\bar{Y}_{\mathcal{S}}$	Sample mean, in model-based approach. (59)
$\bar{y}_{\mathcal{C}}$	Sample mean from a convenience or other nonprobability sample of size n, $= \dfrac{1}{n}\sum_{i\in\mathcal{C}} y_i$. (528)
$z_{\alpha/2}$	The $100(1-\alpha/2)$th percentile of the standard normal distribution. (48)
Z_i	Random variable indicating inclusion in a without-replacement probability sample. $Z_i = 1$ if unit i is in the sample and 0 otherwise. (56)

1

Introduction

When statistics are not based on strictly accurate calculations, they mislead instead of guide. The mind easily lets itself be taken in by the false appearance of exactitude which statistics retain in their mistakes, and confidently adopts errors clothed in the form of mathematical truth.

—Alexis de Tocqueville, *Democracy in America*

1.1 Guidance from Samples

We all use data from samples to make decisions. When tasting soup to correct the seasoning, deciding to buy a book after reading the first page, choosing a major after taking first-year college classes, or buying a car following a test drive, we rely on partial information to judge the whole.

External data used to help with those decisions come from samples, too. Statistics such as the average rating for a book in online reviews, the median salary of psychology majors, the percentage of persons with an undergraduate mathematics degree who are working in a mathematics-related job, or the number of injuries resulting from automobile accidents in 2018 are all derived from samples. So are statistics about unemployment and poverty rates, inflation, number and characteristics of persons with diabetes, medical expenditures of persons aged 65 and over, persons experiencing food insecurity, criminal victimizations not reported to the police, reading proficiency among fourth-grade children, household expenditures on energy, public opinion of political candidates, land area under cultivation for rice, livestock owned by farmers, contaminants in drinking water, size of the Antarctic population of emperor penguins—I could go on, but you get the idea. Samples, and statistics calculated from samples, surround us.

But statistics from some samples are more trustworthy than those from others. What distinguishes, using Tocqueville's words beginning this page, statistics that "mislead" from those that "guide"?

This book sets out the statistical principles that tell you how to design a sample survey, and analyze data from a sample, so that statistics calculated from a sample accurately describe the population from which the sample was drawn. These principles also help you evaluate the quality of any statistic you encounter that originated from a sample survey.

Before embarking on our journey, let's look at how a statistic from a now-infamous survey misled readers in 1936.

Example 1.1. *The Survey That Killed a Magazine.* Any time a pollster predicts the wrong winner of an election, some commentator is sure to mention the *Literary Digest* Poll of 1936. It has been called "one of the worst political predictions in history" (Little, 2016) and is regularly cited as the classic example of poor survey practice. What went wrong with the poll, and was it really as flawed as it has been portrayed?

DOI: 10.1201/9780429298899-1

In the first three decades of the twentieth century, *The Literary Digest*, a weekly news magazine founded in 1890, was one of the most respected news sources in the United States. In presidential election years, it, like many other newspapers and magazines, devoted page after page to speculation about who would win the election. For the 1916 election, however, the editors wrote that "[p]olitical forecasters are in the dark" and asked subscribers in five states to mail in a ballot indicating their preferred candidate (Literary Digest, 1916).

The 1916 poll predicted the correct winner in four of the five states, and the magazine continued polling subsequent presidential elections, with a larger sample each time. In each of the next four election years—1920, 1924 (the first year the poll collected data from all states), 1928, and 1932—the person predicted to win the presidency did so, and the magazine accurately predicted the margin of victory. In 1932, for example, the poll predicted that Franklin Roosevelt would receive 56% of the popular vote and 474 votes in the Electoral College; in the actual election, Roosevelt received 57% of the popular vote and 472 votes in the Electoral College.

With such a strong record of accuracy, it is not surprising that the editors of *The Literary Digest* gained confidence in their polling methods. Launching the 1936 poll, they wrote:

> The Poll represents thirty years' constant evolution and perfection. Based on the "commercial sampling" methods used for more than a century by publishing houses to push book sales, the present mailing list is drawn from every telephone book in the United States, from the rosters of clubs and associations, from city directories, lists of registered voters, classified mail-order and occupational data. (Literary Digest, 1936b, p. 3)

On October 31, 1936, the poll predicted that Republican Alf Landon would receive 54% of the popular vote, compared with 41% for Democrat Franklin Roosevelt. The final article on polling before the election contained the statement, "We make no claim to infallibility. We did not coin the phrase 'uncanny accuracy' which has been so freely applied to our Polls" (Literary Digest, 1936a). It is a good thing *The Literary Digest* made no claim to infallibility. In the election, Roosevelt received 61% of the vote; Landon, 37%. It is widely thought that this polling debacle contributed to the demise of the magazine in 1938.

What went wrong? One problem may have been that names of persons to be polled were compiled from sources such as telephone directories and automobile registration lists. Households with a telephone or automobile in 1936 were generally more affluent than other households, and opinion of Roosevelt's economic policies was generally related to the economic class of the respondent. But the mailing list's deficiencies do not explain all of the difference. Postmortem analyses of the poll (Squire, 1988; Calahan, 1989; Lusinchi, 2012) indicated that even persons with both a car and a telephone tended to favor Roosevelt, though not to the degree that persons with neither car nor telephone supported him.

Nonresponse—the failure of persons selected for the sample to provide data—was likely the source of much of the error. *Ten million* questionnaires were mailed out, and more than 2.3 million were returned—an enormous sample, but fewer than one-quarter of those solicited. In Allentown, Pennsylvania, for example, the survey was mailed to every registered voter, but the poll results for Allentown were still incorrect because only one-third of the ballots were returned (Literary Digest, 1936c). Squire (1988) reported that persons supporting Landon were much more likely to have returned the survey; in fact, many Roosevelt supporters did not remember receiving a survey even though they were on the mailing list.

One lesson to be learned from *The Literary Digest* poll is that the sheer size of a sample is no guarantee of its accuracy. The *Digest* editors became complacent because they sent out questionnaires to more than one-quarter of all registered voters and obtained a huge sample of more than 2.3 million people. But large unrepresentative samples can perform as badly as small unrepresentative samples. A large unrepresentative sample may even do more harm than a small one because many people think that large samples are always superior to small ones. In reality, as we shall discuss in this book, the design of the sample survey—how units are selected to be in the sample—is far more important than its size.

Another lesson is that past accuracy of a flawed sampling procedure does not guarantee future results. The *Literary Digest* poll was accurate for five successive elections—until suddenly, in 1936, it wasn't. Reliable statistics result from using statistically sound sampling and estimation procedures. With good procedures, statisticians can provide a measure of a statistic's accuracy; without good procedures, a sampling disaster can happen at any time even if previous statistics appeared to be accurate. ■

Some of today's data sets make the size of the *Literary Digest's* sample seem tiny by comparison, and some types of data can be gathered almost instantaneously from all over the world. But the challenges of inferring the characteristics of a population when we observe only part of it remain the same. The statistical principles underlying sampling apply to any sample, of any size, at any time or place in the universe.

Chapters 2 through 7 of this book show you how to design a sample so that its data can be used to estimate characteristics of unobserved parts of the population; Chapters 9 through 14 show how to use survey data to estimate population sizes, relationships among variables, and other characteristics of interest. But even though you might design and select your sample in accordance with statistical principles, in many cases, you cannot guarantee that everyone selected for the sample will agree to participate in it. A typical election poll in 2021 has a much lower response rate than the *Literary Digest* poll, but modern survey samplers use statistical models, described in Chapters 8 and 15, to adjust for the nonresponse. We'll return to the *Literary Digest* poll in Chapter 15 and see if a nonresponse model would have improved the poll's forecast (and perhaps have saved the magazine).

1.2 Populations and Representative Samples

In the 1947 movie "Magic Town," the public opinion researcher played by James Stewart discovered a town that had exactly the same characteristics as the whole United States: Grandview had exactly the same proportion of people who voted Republican, the same proportion of people under the poverty line, the same proportion of auto mechanics, and so on, as the United States taken as a whole. All that Stewart's character had to do was to interview the people of Grandview, and he would know public opinion in the United States.

Grandview is a "scaled-down" version of the population, mirroring every characteristic of the whole population. In that sense, it is **representative** of the population of the United States because any numerical quantity that could be calculated from the population can be inferred from the sample.

But a sample does not necessarily have to be a small-scale replica of the population to be representative. As we shall discuss in Chapters 2 and 3, a sample is representative if it can be used to "reconstruct" what the population looks like—and if we can provide an accurate assessment of how good that reconstruction is.

Some definitions are needed to make the notions of a "population" and a "representative sample" more precise.

Observation unit An object on which a measurement is taken, sometimes called an **element**. In surveys of human populations, observation units are often individual persons;

in agriculture or ecology surveys, they may be small areas of land; in audit surveys, they may be financial records.

Target population The complete collection of observations we want to study. Defining the target population is an important and often difficult part of the study. For example, in a political poll, should the target population be all adults eligible to vote? All registered voters? All persons who voted in the last election? The choice of target population will profoundly affect the statistics that result.

Sample A subset of a population.

Sampled population The collection of all possible observation units that might have been chosen in a sample; the population from which the sample was taken.

Sampling unit A unit that can be selected for a sample. We may want to study individuals but do not have a list of all individuals in the target population. Instead, households serve as the sampling units, and the observation units are the individuals living in the households.

Sampling frame A list, map, or other specification of sampling units in the population from which a sample may be selected. For a telephone survey, the sampling frame might be a list of telephone numbers of registered voters, or simply the collection of all possible telephone numbers. For a survey using in-person interviews, the sampling frame might be a list of all street addresses. For an agricultural survey, a sampling frame might be a list of all farms, or a map of areas containing farms.

In an ideal survey, the sampled population will be identical to the target population, but this ideal is rarely met exactly. In surveys of people, the sampled population is usually smaller than the target population. As illustrated in Figure 1.1, some persons in the target population are missing from the sampling frame, and some will not respond to the survey. It is also possible for the sampled population to include units that are not in the target population, for example, if the target population consists of persons at least 18 years old, but some persons who complete the survey are younger than that.

The target population for the American Community Survey (ACS), an annual survey conducted by the U.S. Census Bureau, is the resident population of the United States (U.S. Census Bureau, 2020e). The sampling frame comes from the Census Bureau's lists of residential housing units (for example, houses, apartments, and mobile homes) and group quarters (for example, prisons, skilled nursing facilities, and college dormitories). These lists are regularly updated to include new construction. A sample of about 3.5 million housing unit addresses is selected randomly from the housing unit list; an adult at each sampled address is asked to fill out the questionnaire and provide information about all household members. Approximately 2% of the group quarters population is also sampled. The sampled population consists of persons who reside at one of the places on the lists, can be contacted, and are willing to answer the survey questions. Some U.S. residents, such as persons experiencing homelessness or residing at an unlisted location, may be missing from the sampling frame; others cannot be contacted or refuse or are unable to participate in the survey (U.S. Census Bureau, 2014).

In an agricultural survey taken to estimate crop acreages and livestock inventories, the target population may be all areas of land that are used for agriculture. Area frames are often used for agricultural surveys, particularly when there is no list of all farm operators or of households that engage in agriculture, or when lists of farm operators or land under agricultural production may be outdated. The land area of a country is divided into smaller areas that form the sampling units. The sampling frame is the list of all of the areas, which

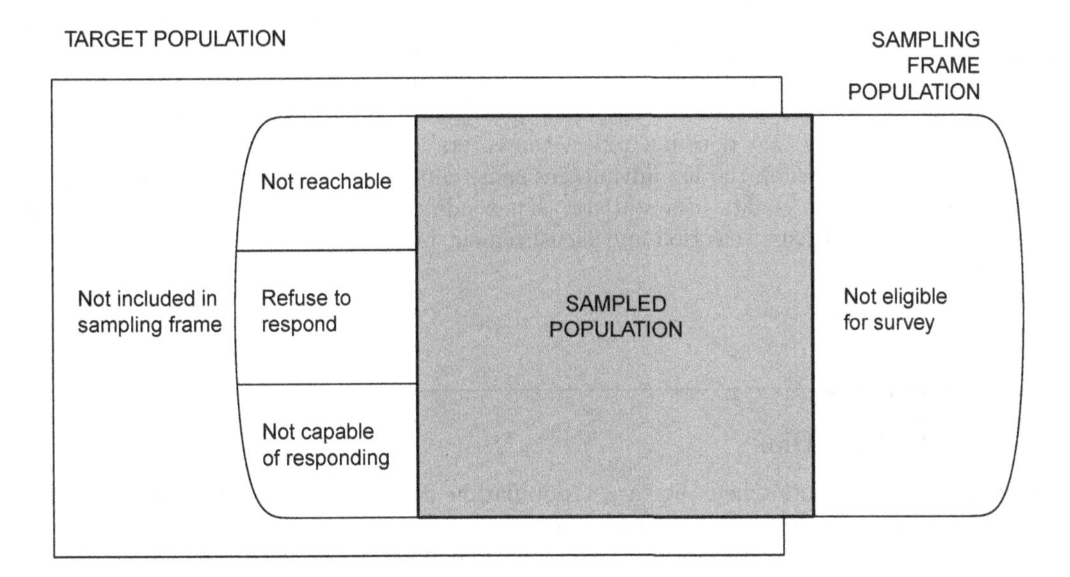

FIGURE 1.1
Target population and sampled population in a telephone survey of registered voters. Some persons in the target population do not have a telephone or will not be associated with a telephone number in the sampling frame. In some households with telephones, the residents are not registered to vote and hence are not eligible for the survey. Some eligible persons in the sampling frame population do not respond because they cannot be contacted, some refuse to respond to the survey, and some may be ill and incapable of responding.

together comprise the target population of all land that could be used for agriculture in the country. A sample of land areas is randomly selected. In some agricultural surveys, the sampler directly observes the acreage devoted to different crops and counts the livestock in the sampled areas. In others, the sampler conducts interviews with all farm operators operating within the boundaries of the sampled areas; in this case, the sampling unit is the area, and the observation unit is the farm operator.

In the *Literary Digest* poll, the characteristic of interest was the percentage of 1936 election-day voters who would support Roosevelt. An individual person was an element. The target population was all persons who would vote on election day in the United States. The sampled population was persons on the lists used by the *Literary Digest* who would return the sample ballot.

Election polls conducted in the 21st century have the same target population (persons who will vote in the election) and elements (individual voters) as the *Literary Digest* poll, but the sampled populations differ from poll to poll. In some polls, the sampled population consists of persons who can be reached by telephone and who are judged to be likely to vote in the next election (see Figure 1.1); in other polls, the sampled population consists of persons who are recruited over the internet and meet screening criteria for participation; in still others, the sampled population consists of anyone who clicks on a website and expresses a preference for one of the election candidates.

Mismatches between the target population and sampled population can cause the sample to be unrepresentative and statistics calculated from it to be **biased**. **Bias** is a systematic error in the sampling, measurement, or estimation procedures that results in a statistic being consistently larger (or consistently smaller) than the population characteristic that it estimates. In an election poll, bias can occur if, unknown to the pollster, the sample selection procedure systematically excludes or underrepresents voters supporting one of the candidates (as occurred in the *Literary Digest* poll); or if support for one or more candidates is measured in a way that does not reflect the voters' actual opinions (for example, if the ordering of candidates on the list advantages some candidates relative to others); or if the estimation procedure results in a statistic that tends to be too small (or too large). The next two sections discuss selection and measurement bias; estimation bias is considered in Chapter 2.

1.3 Selection Bias

Selection bias occurs when the target population does not coincide with the sampled population or, more generally, when some population units are sampled at a different rate than intended by the investigator. If a survey designed to study household income has fewer poor households than would be obtained in a representative sample, the survey estimates of the average or median household income will be too large.

The following examples indicate some ways in which selection bias can occur.

1.3.1 Convenience Samples

Some persons who are conducting surveys use the first set of population units they encounter as the sample. The problem is that the population units that are easiest to locate or collect may differ from other units in the population on the measures being studied. The sample selection may, unknown to the investigators, depend on some characteristic associated with the properties of interest.

For example, a group of investigators took a convenience sample of adolescents to study how frequently adolescents talk to their parents and teachers about AIDS. But adolescents willing to talk to the investigators about AIDS are probably also more likely to talk to other authority figures about AIDS. The investigators, who simply averaged the amounts of time that adolescents in the sample said they spent talking with their parents and teachers, probably overestimated the amount of communication occurring between parents and adolescents in the population.

1.3.2 Purposive or Judgment Samples

Some survey conductors deliberately or purposively select a "representative" sample. If we want to estimate the average amount a shopper spends at the Mall of America in a shopping trip, and we sample shoppers who look like they have spent an "average" amount, we have deliberately selected a sample to confirm our prior opinion. This type of sample is sometimes called a **judgment sample**—the investigators use their judgment to select the specific units to be included in the sample.

1.3.3 Self-Selected Samples

A self-selected sample consists entirely of volunteers—persons who select themselves to be in the sample. Such is the case in radio and television call-in polls, and in many surveys conducted over the internet. The statistics from such surveys cannot be trusted. At best, they are entertainment; at worst, they mislead.

Yet statistics from call-in polls or internet surveys of volunteers are cited as supporting evidence by independent research institutes, policy organizations, news organizations, and scholarly journals. For example, Maher (2008) reported that about 20 percent of the 1,427 people responding to an internet poll (described in the article as an "informal survey" that solicited readers to take the survey on a website) said they had used one of the cognitive-enhancing drugs methylphenidate (Ritalin), modafinil, or beta blockers for non-medical reasons in order to "stimulate their focus, concentration or memory." As of 2020, the statistic had been cited in more than 200 scientific journal articles, but few of the citing articles mentioned the volunteer nature of the original sample or the fact that the statistic applies only to the 1,427 persons who responded to the survey and not to a more general population. In fact, all that can be concluded from the poll is that about 280 people who visited a website said they had used one of the three drugs; nothing can be inferred about the rest of the population without making heroic assumptions.

An additional problem with volunteer samples is that some individuals or organizations may respond multiple times to the survey, skewing the results. This occurred with an internet poll conducted by *Parade* magazine that asked readers whether they blamed actor Tom Cruise, or whether they blamed the media, for his "disastrous public relations year" (United Press International, 2006, reporting on the poll, mentioned an incident in which Cruise had jumped on the couch during Oprah Winfrey's television show). The editors grew suspicious, however, when 84 percent of respondents said the media—not Cruise—was to blame. The magazine's publicist wrote: "We did some investigating and found out that more than 14,000 (of the 18,000-plus votes) that came in were cast from only 10 computers. One computer was responsible for nearly 8,400 votes alone, all blaming the media for Tom's troubles. We also discovered that at least two other machines were the sources of inordinate numbers of votes .... It seems these folks (whoever they may be) resorted to extraordinary measures to try to portray Tom in a positive light for the Parade.com survey."

Example 1.2. Many researchers collect samples from persons who sign up to take surveys on the internet and are paid for their efforts. How well do such samples represent the population for which inference is desired?

Ellis et al. (2018) asked a sample of 1,339 U.S. adults to take a survey about eating behavior. The study participants were recruited from Amazon's Mechanical Turk, a crowd-sourcing website that allows persons or businesses to temporarily hire persons who are registered on the site as "Workers." Workers who expressed interest in the study were directed to the online survey and paid 50 cents upon completing it. The sample was thus self-selected—participants first chose to register with Mechanical Turk and then chose to take and complete the survey.

Do the survey participants have the same eating behavior patterns as U.S. adults as a whole? We can't tell from this survey, but the participants differed from the population of U.S. adults on other characteristics. According to the 2017 ACS, about 51 percent of the U.S. population aged 18 and over was female; 63 percent was white non-Hispanic; and 29 percent had a bachelor's degree or higher (U.S. Census Bureau, 2020b). The sample of Mechanical Turk Workers was 60 percent female and 80 percent white non-Hispanic, and 52 percent had a bachelor's degree or higher. As found in other research (see, for example, Hitlin, 2016; Mortensen et al., 2018), the persons recruited from the Mechanical Turk website were more likely to be female, highly educated, white, and non-Hispanic than persons not in the sample. In addition, all of the persons who took the survey had access to—and used—the internet and were willing to take a 15-minute survey in exchange for a tiny remuneration.

The study authors made no claims that their sample represents the U.S. population.

Their purpose was to explore potential relationships between picky eating and outcomes such as social eating anxiety, body mass index, and depressive symptoms. As the authors stated, further research would be needed to determine whether the relationships found in this study apply more generally.

Because the sample was self-selected, statistics calculated from it describe *only* the 1,339 adults who provided answers, *not* the adult population as a whole. About 18 percent of the persons in the sample fit the "picky eater" profile, but we cannot conclude from the study that 18 percent of all adults in the United States are picky eaters. Even if the sample resembled the population with respect to all demographic characteristics, picky eaters could have chosen to participate in the survey at a higher (or lower) rate than non-picky eaters. ■

1.3.4 Undercoverage

Undercoverage occurs when the sampling frame fails to include some members of the target population. Population units that are not in the sampling frame have no chance of being in the sample; if they differ systematically from population units that are in the frame, statistics calculated from the sample may be biased.

Undercoverage occurs in telephone surveys because some households and persons do not have telephones. In 2020, nearly all telephone surveys in the United States used sampling frames that included both cellular telephones and landline telephones. In earlier years, however, many telephone surveys excluded cellular telephones, which meant that persons in households with no landline were not covered.

A mail survey has undercoverage of persons whose addresses are missing from the address list or who have no fixed address. An online or e-mail survey fails to cover persons who lack internet access. A survey of anglers that uses a state's list of persons with fishing licenses as a sampling frame has undercoverage of unlicensed anglers or anglers from out-of-state.

1.3.5 Overcoverage

Overcoverage occurs when units not in the target population can end up in the sample.

It is not always easy to construct a sampling frame that corresponds exactly with the target population. There might be no list of all households with children under age 5, persons who are employed in science or engineering fields, or businesses that sell food products to consumers. To survey those populations, samplers often use a too-large sampling frame, then screen out ineligible units. For example, the sampling frame might consist of all household addresses in the area, and interviewers visiting sampled addresses would exclude households with no children under age 5. But overcoverage can occur when persons not in the target population are not screened out of the sample, or when data collectors are not given clear instructions on sample eligibility. In some surveys, particularly when payment is offered for taking the survey, overcoverage may occur when persons not eligible for the survey falsely claim to meet the eligibility criteria (Kan and Drummey, 2018).

Another form of overcoverage occurs when individual units appear multiple times in the sampling frame, and thus have multiple chances to be included in the sample, but the multiplicity is not adjusted for in the analysis. In its simplest form, random digit dialing prescribes selecting a random sample of 10-digit telephone numbers. Households with more than one telephone line have a higher chance of being selected in the sample. This multiplicity can be compensated in the estimation (we'll discuss this in Section 6.5); if it is ignored, bias can result. One might expect households with more telephone lines to be larger or more affluent, so if no adjustment is made for those households having a higher probability of being selected for the sample, estimates of average income or household size may be too large. Similarly, a person with multiple e-mail addresses has a higher chance of being selected in an e-mail survey.

Some surveys have both undercoverage and overcoverage. Political polls attempt to predict election results from a sample of likely voters. But defining the set of persons who will vote in the election is difficult. Pollsters use a variety of different methods and models to predict who will vote in the election, but the predictions can exclude some voters and include some nonvoters.

To assess undercoverage and overcoverage, you need information that is external to the survey. In the ACS, for example, coverage errors are assessed for the population by comparing survey estimates with independent population estimates that are calculated from data on housing, births, deaths, and immigration (U.S. Census Bureau, 2014).

1.3.6 Nonresponse

Nonresponse—failing to obtain responses from some members of the chosen sample—distorts the results of many surveys, even surveys that are carefully designed to minimize other sources of selection bias. Many surveys reported in newspapers or research journals have dismal response rates—in some, fewer than one percent of the households or persons selected to be in the sample agree to participate.

Numerous studies comparing respondents and nonrespondents have found differences between the two groups. Although survey samplers attempt to adjust for the nonresponse using methods we'll discuss in Chapter 8, systematic differences between the respondents and nonrespondents may persist even after the adjustments. Typically knowledge from an external source is needed to assess effects of nonresponse—you cannot tell the effects of nonresponse by examining data from the respondents alone.

Example 1.3. Response rates for the U.S. National Health Interview Survey, an annual survey conducted in person at respondents' residences, have been declining since the early 1990s. The survey achieved household response rates exceeding 90% in the 1990s, but by 2015 only about 70% of the households selected to participate did so. The goal of the survey is to provide information about the health status of U.S. residents and their access to health care. If the nonrespondents are less healthy than the persons who answer the survey, however, then estimates from the survey may overstate the health of the nation.

Evaluating effects of nonresponse can be challenging: the nonrespondents' health status is, in general, unknown to the survey conductor because nonrespondents do not provide answers to the survey. Sometimes, though, information about the nonrespondents can be obtained from another source. By matching National Health Interview Survey respondents from 1990 through 2009 with a centralized database of death record information, Keyes et al. (2018) were able to determine which of the survey respondents had died as of 2011. They found that the mortality rates for survey respondents were lower than those for the general population, indicating that respondents may be healthier, on average, than persons who are not in the sampling frame or who do not respond to the survey. ■

1.3.7 What Good Are Samples with Selection Bias?

Selection bias is of concern when it is desired to use estimates from a sample to describe the population. If we want to estimate the total number of violent crime victims in the United States, or the percentage of likely voters in the United Kingdom who intend to vote for the Labour Party in the next election, selection bias can cause estimates from the sample to be far from the corresponding population quantities.

But samples with selection bias may provide valuable information for other purposes, particularly in the early stages of an investigation. Such was the case for a convenience sample taken in fall 2019.

Example 1.4. As of October 2019, more than 1,600 cases of lung injuries associated with use of electronic cigarettes (e-cigarettes) had occurred, including 34 deaths (Moritz et al., 2019), but the cause of the injuries was unknown. Lewis et al. (2019) conducted interviews with 53 patients in Utah who had used e-cigarette products within three months of experiencing lung injury. Forty-nine of them (92 percent) reported using cartridges containing tetrahydrocannabinol (THC is the psychoactive ingredient in marijuana). Most of the THC-containing products were acquired from friends or from illicit dealers.

The study authors identified possible sources of selection bias in their report. Although they attempted to interview all 83 patients who were reported to have lung injuries following use of e-cigarettes, only 53 participated, and the nonresponse might cause estimates to be biased. Additional bias might occur because physicians may have reported only the more serious cases, or because THC was illegal in Utah and patients might have underreported its use. Persons with lung injuries who did not seek medical care were excluded from the study. The sample used in the study was likely not representative of e-cigarette users with lung injuries in the United States as a whole, or even in Utah.

But even with the selection bias, the sample provided new information about the lung injuries. The majority of the persons with lung injury in the sample had been using e-cigarettes containing THC, and this finding led the authors to recommend that the public stop using these products, pending further research. The purpose of the sample was to provide timely information for improving public health, not to produce statistics describing the entire population of e-cigarette users, and the data in the sample provided a basis for further investigations. ■

1.4 Measurement Error

A good sample has accurate responses to the items of interest. When a response in the survey differs from the true value, **measurement error** has occurred. **Measurement bias** occurs when the response has a tendency to differ from the true value in one direction. As with selection bias, measurement error and bias must be considered and minimized in the design stage of the survey; no amount of statistical analysis will disclose that the scale erroneously added 5 kilograms to the weight of every person in the health survey.

Measurement error is a concern in all surveys and can be insidious. In many surveys of vegetation, for example, areas to be sampled are divided into smaller plots. A sample of plots is selected, and the number of plants in each plot is recorded. When a plant is near the boundary of the region, the field researcher needs to decide whether to include the plant in the tally. A person who includes all plants near or on the boundary in the count is likely to produce an estimate of the total number of plants in the area that is too high because some plants may be counted twice. High-quality ecological surveys have clearly defined protocols for counting plants near the boundaries of the sampled plots.

Example 1.5. Measurement errors may arise for reasons that are not immediately obvious. More than 20,000 households participated in a survey conducted in Afghanistan in 2018. Because the survey asked several hundred questions, the questions were divided among several modules. Two modules, however, gave very different estimates of the percentage of children who had recently had a fever, and the investigators struggled to understand why. After all, the same question—"Has your child had fever in the past two weeks"—was asked of the same set of sampled households about the same children. Why was the estimated percentage of children who recently had fever twice as high for Module 2 as Module 1?

Alba et al. (2019) found potential reasons for the discrepancy. Questions in the two modules were answered by different persons in the household and had different contexts. Men were asked the questions in Module 1, which concerned medical expenditures. Women were asked the questions in Module 2, which concerned treatment practices for childhood illnesses. The context of medical expenditures in Module 1 may have focused recall on fevers requiring professional medical treatment, and respondents may have neglected to mention less serious fevers. In addition, women, more likely to be the children's primary caregivers, may have been aware of more fever episodes than men. ■

Sometimes measurement bias is unavoidable. In the North American Breeding Bird Survey, observers stop every one-half mile on designated routes and count all birds heard or seen during a 3-minute period within a quarter-mile radius (Ziolkowski et al., 2010; Sauer et al., 2017). The count of birds at a stop is almost always smaller than the true number of birds in the area because some birds are silent and unseen during the 3-minute count; scientists use statistical models and information about the detectability of different bird species to obtain population estimates. If data are collected with the same procedure and with similarly skilled observers from year to year, however, the survey counts can be used to estimate trends in the population of different species—the biases from different years are expected to be similar, and may cancel when year-to-year differences are calculated.

Obtaining accurate responses is challenging in all types of surveys, but particularly so in surveys of people:

- People sometimes do not tell the truth. In an agricultural survey, farmers in an area with food-aid programs may underreport crop yields, hoping for more food aid. Obtaining truthful responses is a particular challenge in surveys involving sensitive subject matter, such as surveys about drug use.

- People forget. A victimization survey may ask respondents to describe criminal victimizations that occurred to them within the past year. Some persons, however, may forget to mention an incident that occurred; others may include a memorable incident that occurred more than a year ago.

- People do not always understand the questions. Confusing questions elicit confused responses. A question such as "Are you concerned about housing conditions in your neighborhood?" has multiple sources of potential confusion. What is meant by "concern," "housing conditions," or "neighborhood"? Even the pronoun "you" may be ambiguous in this question. Is it a singular pronoun referring to the individual survey respondent or a collective pronoun referring to the entire neighborhood?

- People may give different answers to surveys conducted by different modes (Dillman, 2006; de Leeuw, 2008; Hox et al., 2017). The **survey mode** is the method used to distribute and collect answers to the survey. Some surveys are conducted using a single mode—in-person, internet, telephone, or mail—while others allow participants to choose their mode when responding. Respondents may perceive questions differently when they hear them than when they read them.

  Respondents may also give different answers to a self-administered survey (for example, an internet or mail survey where respondents enter answers directly) than to a survey in which questions are asked by interviewers. This is particularly true for questions on sensitive topics such as drug use, criminal activity, or health risk behaviors—people may be more willing to disclose information that puts them in a bad light in a self-administered survey than to an interviewer (Kreuter et al., 2008; Lind et al., 2013).

Conversely, people may be more likely to provide "socially desirable" answers that portray them in a positive light to an interviewer. Dillman and Christian (2005) found that people are more likely to rate their health as excellent when in a face-to-face interview than when they fill out a questionnaire sent by mail. In another experiment, Keeter (2015) randomly assigned persons taking a survey to telephone mode (with an interviewer) or internet mode (with no interviewer). Among those taking the survey by telephone, 62 percent said they were "very satisfied" with their family life; among those taking the survey over the internet, 44 percent said they were "very satisfied."

- People may say what they think an interviewer wants to hear or what they think will impress, or not offend, the interviewer. West and Blom (2017) reviewed studies finding that the race or gender of an interviewer may influence survey responses. Eisinga et al. (2011) reported that survey respondents were more likely to report dietary behavior such as fasting or skipping meals to an interviewer who was overweight.

  In experiments done with questions beginning "Do you agree or disagree with the following statement," it has been found that a subset of the population tends to agree with any statement regardless of its content (Krosnick and Presser, 2010). Lenski and Leggett (1960) found that about one-tenth of their sample agreed with both of the following statements:

  > It is hardly fair to bring children into the world, the way things look for the future.
  >
  > Children born today have a wonderful future to look forward to.

  Schuman and Presser (1981) reported on an experiment conducted with two national samples. In the first, 60 percent agreed with the statement: "Individuals are more to blame than social conditions for crime and lawlessness in this country." In the second, 57 percent agreed with the statement: "Social conditions are more to blame than individuals for crime and lawlessness in this country"—a statement that is the exact opposite of the first. Later, the respondents to each survey were recontacted and given the opposite statement of the one they had answered in the initial survey. About one-quarter of the respondents agreed with both the original statement and its opposite.

- Question wording and question order can have a large effect on the responses obtained. Two surveys were taken in late 1993/early 1994 about Elvis Presley, who had died in 1977. One survey asked, "In the past few years, there have been a lot of rumors and stories about whether Elvis Presley is really dead. How do you feel about this? Do you think there is any possibility that these rumors are true and that Elvis Presley is still alive, or don't you think so?" The other survey asked, "A recent television show examined various theories about Elvis Presley's death. Do you think it is possible that Elvis is alive or not?" Eight percent of the respondents to the first question said it is possible that Elvis is still alive; 16% of respondents to the second question said it is possible that Elvis is still alive.

In some cases, accuracy can be increased by careful questionnaire design.

1.5 Questionnaire Design

This section gives a very brief introduction to writing and testing questions. It provides some general guidelines and examples, but if you are writing a questionnaire, you should consult one of the more comprehensive references on questionnaire design listed at the end of this chapter.

The most important step in writing a questionnaire is to decide what you want to find out. Write down the goals of your survey, and be precise. "I want to learn something about persons experiencing homelessness" won't do. Instead, you should write down specific research questions, such as "What percentage of persons using homeless shelters in Chicago between January and March 2021 are under 16 years old?" Then, write or select survey questions that will elicit accurate answers to the research questions.

When reporting results from a survey, always give the text of the actual question asked of the survey respondents, so your reader knows how the concept was measured. With some topics, answers change dramatically when questions are asked differently. Estimated rates of sexual assault, for example, can be ten times as large in one survey as in another survey, simply because different questions are asked (Lohr, 2019a, Chapter 8).

- *Test the questions before taking the survey.* Ideally, the questions would be tested on a small sample of members of the target population. Try different versions for the questions, and ask respondents in your pretest how they interpret the questions.

  Any question considered for the ACS is extensively tested for several years before being included in the survey (U.S. Census Bureau, 2017). Testing methods include interviewing people about how they understand the question and about any parts that may be unclear or ambiguous, using expert review, and trying out the questions in the field with designed experiments.

  You will not necessarily catch misinterpretations of questions by trying them out on friends or colleagues; your friends and colleagues may have backgrounds similar to yours, and may not have the same understanding of words as persons in your target population.

- *Keep it simple and clear.* Strunk and White advised writers to "Prefer the specific to the general, the definite to the vague, the concrete to the abstract" (1959, p. 15). Good questions result from good writing.

  Questions that seem clear to you may not be clear to someone listening over the telephone or to a person with a different native language. Belson (1981, p. 240) tested the question "What proportion of your evening viewing time do you spend watching news programmes?" on 53 people. Only 14 people correctly interpreted the word "proportion" as "percentage," "part," or "fraction." Others interpreted it as "how long do you watch" or "which news programs do you watch."

- *Define your terms.* Some words mean different things to different people. A simple question such as "Do you own a car?" may be answered yes or no depending on the respondent's interpretation of "you" (does it refer to just the individual, or to the household?), "own" (does it count as ownership if you are making payments to a finance company?), or "car" (are pickup trucks included?). Suessbrick et al. (2000) found that the concepts in a seemingly clear question such as "Have you smoked at least 100 cigarettes in your entire life?" were commonly interpreted in a different way than the authors intended: Some respondents included marijuana cigarettes or cigars, while others excluded partially smoked or hand-rolled cigarettes.

Instead of asking "How many rooms are in this home," the ACS carefully defines what counts as a room, with the question

> How many separate rooms are in this house, apartment, or mobile home? Rooms must be separated by built-in archways or walls that extend out at least 6 inches and go from floor to ceiling.
>
> —Include bedrooms, kitchens, etc.
>
> —Exclude bathrooms, porches, balconies, foyers, halls, or unfinished basements. (U.S. Census Bureau, 2019a)

- *Relate your questions to the concept of interest.* This seems obvious but is forgotten or ignored in many surveys. In some disciplines, a standard set of questions has been developed and tested, and these are then used by subsequent researchers. Often, use of a common survey instrument allows results from different studies to be compared. In some cases, however, the standard questions are inappropriate for addressing the research hypotheses.

  Pincus (1993) criticized early research that concluded that persons with arthritis were more likely to have psychological problems than persons without arthritis. In those studies, persons with arthritis were given the Minnesota Multiphasic Personality Inventory, a test of 566 true/false questions commonly used in psychological research. Patients with rheumatoid arthritis tended to have high scores on the scales of hypochondriasis, depression, and hysteria. Part of the reason they scored highly on those scales is clear when the actual questions are examined. A person with arthritis can truthfully answer false to questions such as "I am about as able to work as I ever was," "I am in just as good physical health as most of my friends," and "I have few or no pains" without being either hysterical or a hypochondriac.

  In the 2019 National Student Survey, final-year undergraduates in the United Kingdom were asked about their experiences and satisfaction at university. For each of 27 statements, the student was asked to choose one of the following options: definitely agree, mostly agree, neither agree nor disagree, mostly disagree, definitely disagree, or not applicable. One of the items stated, "It is clear how students' feedback on the course has been acted on." A student might "definitely agree" with this statement if it is clear that the students' feedback has been completely ignored. A high percentage of agreement with the statement does not necessarily say anything about the quality of the course (Fisher, 2019).

- *Decide whether to use open or closed questions.* An **open question** allows respondents to form their own response categories; in a **closed question** (multiple choice), the respondent chooses from a set of categories read or displayed. Each has advantages in specific situations.

  A well-written closed question may prompt the respondent to remember response options that might otherwise be forgotten. For example, the closed version of the question "what is your marital status" is usually accompanied by five response options: now married, widowed, divorced, separated, or never married. With an open question, unaccompanied by response options, some persons in each of the last four categories may describe their status as "single."

  If the survey is exploratory or questions are sensitive, however, an open question may be better: Bradburn and Sudman (1979) noted that respondents reported a higher frequency of drinking alcoholic beverages when asked an open question than a closed question with categories "never" through "daily."

TABLE 1.1
Response options and percentages from two surveys about satisfaction with life.

Question	Response Option	Percentage
Survey 1: To what extent are you satisfied with the life you currently lead?	Extraordinarily satisfied	8.4
	Very satisfied	35.5
	Satisfied	45.1
	Fairly satisfied	7.6
	Not very satisfied	3.4
Survey 2: On the whole how satisfied are you with the life you lead?	Very satisfied	51.5
	Fairly satisfied	44.8
	Not very satisfied	3.1
	Not at all satisfied	0.6

Source: de Jonge et al. (2017, p. 22).

Schuman and Scott (1987) concluded that depending on the context, either open or closed questions can limit the types of responses received. In one experiment, the most common responses to the open question "What do you think is the most important problem facing this country today?" were "unemployment" (17%) and "general economic problems" (17%). The closed version asked, "Which of the following do you think is the most important problem facing this country today—the energy shortage, the quality of public schools, legalized abortion, or pollution—or if you prefer, you may name a different problem as most important"; 32% or respondents chose "the quality of public schools." In this case, the limited options in the closed question guided respondents to one of the listed responses. In another experiment, Schuman and Scott (1987) asked respondents to name one or two of the most important national world events or changes during the last 50 years. Persons asked the open question most frequently gave responses such as World War II or the Vietnam War; they typically did not mention events such as the invention of the computer, which was the most common response to the closed question including this option.

If using a closed-ended question, always have an "other" category. In one study of sexual activity among adolescents, adolescents were asked from whom they felt the most pressure to have sex. Categories for the closed question were "friends of same sex," "boyfriend/girlfriend," "friends of opposite sex," "TV or radio," "don't feel pressure," and "other." The response "parents" or "father" was written in by a number of the adolescents taking the survey, a response that had not been anticipated by the researchers.

- *Pay attention to response option scales.* Estimates from two surveys taken in the Netherlands in 2008, given in Table 1.1, were both reported to the public as measuring satisfaction with life. But the surveys had widely divergent estimates of the percentage who are "fairly satisfied." This difference may occur partly because respondents were prompted about different time frames (the first asks about the "life you currently lead" and the second asks about life "on the whole") or because the questions were asked in different contexts. De Jonge et al. (2017) ascribed much of the difference to the scales used for the response options. The first survey had an asymmetric scale with five options, where "fairly satisfied" was second from the bottom. In the second survey, "fairly satisfied" was the second-most-positive option. Thus, in the context of the response options given, many people would view "fairly satisfied" as a negative response in the first survey but as a positive response in the second survey.

  This example also underscores the importance of reporting the actual question that respondents were asked, along with the response options.

- *Avoid questions that prompt or motivate the respondent to answer in a particular way.* These are often called **leading** or **loaded questions**.

  In the first season of the television show *Parks and Recreation*, fictional Parks Department deputy director Leslie Knope wanted her survey to show that the neighborhood supported her proposed park project. She deliberately asked leading questions intended to obtain an affirmative response: "Wouldn't you rather have a park than a storage facility for nuclear waste?" and "Wouldn't you agree, like most decent Americans, that it would be a good idea to turn the abandoned lot on Sullivan Street into a beautiful community park?"

- *Consider the social desirability of responses to questions, and write questions and use modes that elicit accurate responses.* Some behaviors are considered more socially acceptable than others. People tend to overreport behavior such as contributing to charity, exercising, and eating healthful foods, and underreport behavior such as drinking, drug use, cheating, and gambling (Krumpal, 2013).

  There is a large body of empirical research studying how to elicit accurate answers to sensitive questions, or questions in which some answers are more socially desirable than others. Techniques include using self-administered questionnaires, using special statistical methods that have been developed to keep information private (a method we'll discuss in Section 16.3.4 prevents even the interviewer from knowing the true response but still allows population percentages to be estimated), and testing alternative question wordings.

- *Avoid double negatives.* Double negatives needlessly confuse the respondent. A question such as "Do you favor or oppose not allowing drivers to use cell phones while driving?" might elicit either "favor" or "oppose" from a respondent who thinks persons should not use cell phones while driving.

- *Use forced-choice rather than agree/disagree questions.* As noted earlier, some persons will agree with almost any statement. Saris et al. (2010) reviewed 50 years of research on the "acquiescence bias" that can be prompted by agree/disagree questions.

  Schuman and Presser (1981, p. 223) reported the following differences from an experiment comparing agree/disagree with forced-choice versions:

  Q1: Do you agree or disagree with this statement: Most men are better suited emotionally for politics than are most women.

  Q2: Would you say that most men are better suited emotionally for politics than are most women, that men and women are equally suited, or that women are better suited than men in this area?

	Years of schooling		
	0–11	12	13+
Q1: Percent "agree"	57	44	39
Q2: Percent "men better suited"	33	38	28

- *Ask only one concept per question.* In particular, avoid what are sometimes called **compound** or **double-barreled** questions, so named because if one barrel of the shotgun does not get you, the other one will.

  A question such as "Do you agree or disagree: The president should reduce taxes and government spending" asks about two separate concepts: reducing taxes and lowering government spending. The question wording may confuse a respondent who thinks the president should reduce government spending but not taxes. A better option is to ask separate questions for each concept (and also avoid the agree/disagree format and define the terms).

- *Pay attention to question order effects.* Questions asked early in the survey can affect answers to later questions. If you ask more than one question on a topic, it is usually (but not always) better to ask the more general question first and follow it by the specific questions (Cialdini, 2009; Garbarski et al., 2015).

  McFarland (1981) conducted an experiment in which half of the respondents were given general questions (for example, "How interested would you say you are in religion: very interested, somewhat interested, or not very interested") first, followed by specific questions on the subject ("Did you, yourself, happen to attend church in the last seven days?"); the other half were asked the specific questions first and then asked the general questions. When the general question was asked first, 56% reported that they were "very interested in religion"; the percentage rose to 64% when the specific question was asked first.

  Brick and Lohr (2019) reported on experiments performed to evaluate question order effects in a mail survey on crime and attitudes about community safety. Each questionnaire tested asked nine questions on attitudes about community safety and local police, as well as questions about crimes that occurred to household residents during the past year. The questionnaires differed in the placement of those questions. In one version, the safety and police questions were at the beginning of the questionnaire, before the crime questions; in the other version, they were at the end, after the crime questions. Questionnaires that started with the safety and police questions yielded higher percentages of persons who reported a place within a mile radius where they would be afraid to walk alone at night, and lower percentages who said that the police department is doing an excellent or good job.

1.6 Sampling and Nonsampling Errors

Most surveys report a "margin of error." Many merely say that the margin of error is 3 percentage points. Others give more detail, as in this excerpt from a *New York Times* poll: "In theory, in 19 cases out of 20 the results based on such samples will differ by no more than three percentage points in either direction from what would have been obtained by interviewing all Americans."

The margin of error describes **sampling error**, the error that results from taking one randomly selected sample instead of examining the whole population. If we randomly selected a different sample, we would most likely obtain a different sample percentage of persons who visited the public library last week. Sampling errors are usually reported in probabilistic terms. We discuss the calculation of sampling errors for different survey designs in Chapters 2 through 7.

Selection bias and measurement error are examples of **nonsampling errors**, which are any errors that cannot be attributed to the sample-to-sample variability. Sometimes the sampling error that is reported for the survey is negligible compared to the nonsampling errors; you often see surveys with a 5% response rate proudly proclaiming their 3% margin of error, while ignoring the tremendous potential selection bias in their results.

The goal of this chapter is to sensitize you to various forms of selection bias and inaccurate responses. We can reduce some forms of selection bias by using probability sampling methods, as described in the next chapter. Measurement error can often be reduced through careful design and testing of the survey instrument, thorough training of interviewers, and pretesting the survey. We shall return to nonsampling errors in Chapter 8, where we discuss methods that have been proposed for trying to reduce nonresponse error after the survey has been collected (sneak preview: none of the methods is as good as obtaining a high response rate to begin with), and Chapter 16, where we present a unified approach to survey design that attempts to minimize both sampling and nonsampling errors.

1.7 Why Use Sampling?

With the abundance of poorly done surveys, it is not surprising that some people are skeptical of *all* surveys. "After all," some say, "my opinion has never been asked, so how can the survey results claim to represent me?" Public questioning of the validity of surveys intensifies after pollsters predict the wrong winner of an election, as occurred after the *Literary Digest* poll of 1936, or after the 2016 U.S. presidential election in which most pollsters predicted that Hillary Clinton would defeat Donald Trump. After a polling misfire in the United Kingdom, one member of Parliament expressed his opinion that "Extrapolating what tens of millions are thinking from a tiny sample of opinions affronts human intelligence and negates true freedom of thought."

Some people insist that only a complete census, in which every element of the population is measured, will be satisfactory. This objection to sampling has a long history. After Anders Kiaer, director of the Norwegian Bureau of Statistics, proposed using sampling to collect official governmental statistics during the 1895 meeting of the International Institute of Statistics (Kiaer, 1896), some statisticians at the meeting recommended further investigation of his idea. Others called the idea of taking a sample instead of a census "very dangerous," stated that no inductive procedure can replace a census, and cautioned against "calling such an evaluation a true statistic." Like the member of Parliament a century later, they wondered how one can possibly infer the characteristics of a population of 10 million people from a sample of 1,000 or even 100,000 people.

Kiaer, however, did not propose that one could take just any sample at hand to represent the population. He said that it is essential to use "good methods to obtain a representative sample; one must avoid the pitfall of easy-to-collect data since the data that are easy to collect often give an incorrect representation."

Researchers who took samples to investigate social trends during the early part of the twentieth century found they gave useful information more quickly and with less expense than a complete enumeration. During the 1930s and 1940s, agricultural demands, the Great Depression, and World War II spurred a need for more and faster statistics. Survey samples were taken to measure unemployment, agricultural production, family spending patterns during wartime, retail trade, industrial output, and other measures needed for economic and social planning. Although some still thought of sampling as a "cheap substitute for a complete count," it proved its practical worth for obtaining reliable information. Between 1920 and 1960, statisticians produced a series of mathematical results that prove why a **probability sample**—that is, a sample chosen using random selection methods—produces reliable statistics that apply to the population from which the sample was drawn. Chapter 2 shows how that works for the simplest form of probability sampling.

In Kiaer's day, and in the 1930s and 1940s, collecting or processing almost any kind of data was difficult and expensive. Even if all of the records were available in filing cabinets, someone had to go through the files, transcribe the data of interest, and calculate the statistics by hand. It made good sense to substitute a representative sample for this laborious exercise. In the 21st century, some types of data can be collected easily and inexpensively, and sometimes it is just as easy to take a census as a sample. In other situations, a sample may yield higher quality information than a census. The next sections list advantages and disadvantages of each type of data collection.

1.7.1 Advantages of Taking a Census

When data are collected through an automated process—for example, when all medical records are collected electronically—or the population is small, it makes sense to obtain data for every member of the population. If you can easily collect the information you need on the entire target population, there are advantages to doing so.

The primary advantage of a census is that you have sampled everyone. You do not have to infer characteristics of the entire population from a sample measuring only part of it, because you have the data for the entire population. A statistic calculated from a census has no sampling error.

A census also provides information on every subpopulation, no matter how small. Suppose that in a country with 3 million deaths per year, 900 of those deaths (three-hundredths of one percent) are pregnancy-related. With a census of every death certificate, containing detailed and accurate information on the cause of death and characteristics of each decedent, you have information on all 900 of those pregnancy-related deaths. If you did not have a census and instead had a random sample of 30,000 death certificates, you would expect to have about 9 pregnancy-related deaths in the sample—much less information for studying this subpopulation.

Most countries that have censuses of birth and death records have gone through enormous efforts to establish accurate large-scale data collections (see, for example, Hetzel, 1997; Wallgren and Wallgren, 2014). Other types of censuses might not be taken with such care: they may miss parts of the target population, or may not contain the data needed to answer the question of interest, or may have inaccurate data. In such cases, a sample may produce more accurate results and save money.

1.7.2 Advantages of Taking a Sample Instead of a Census

In many situations, taking a complete census of a population uses a great deal of time and money, and does not eliminate error. The biggest causes of error in a survey are often undercoverage, nonresponse, and sloppiness in data collection. It is often much better to take a high-quality sample and allocate resources elsewhere, for instance, by being more careful in collecting or recording data, or doing follow-up studies, or measuring more variables.

After all, the *Literary Digest* poll discussed in Example 1.1 predicted the vote wrong even in counties where ballots were mailed to every registered voter. The U.S. Decennial Census, which attempts to enumerate every resident of the country, misses some of the population. Assessments of the coverage of the 2010 Census found that it missed about 2% of the Black/African American population, about 5% of the American Indian and Alaska Native population living on reservations, and about 5% of children between the ages of 0 and 4 (Hogan et al., 2013; O'Hare, 2014).

There are three main justifications for using sampling:

- Sampling can provide reliable information at far less cost than a census. You only need to collect data on part of the population, not every member. With probability samples (described in the next chapter), you can quantify the sampling error from a survey.

In some situations, an observation unit must be destroyed to be measured, as when a cookie must be pulverized to determine the fat content. Yes, you could evaluate quality by testing every cookie in the batch, but then you would have no cookies left to sell. A sample of cookies provides reliable information about the population; a census destroys the population and, with it, the need for information about it.

- When it is expensive or time-consuming to collect the data, as when data are collected through interviews, estimates from a sample can often be produced more quickly and inexpensively. An estimate of the unemployment rate for 2020 is not very helpful if it takes until 2030 to interview every household.
- Finally, and less well known, estimates based on sample surveys are often more accurate than those based on a census because investigators can be more careful when collecting data. A complete census often requires a large administrative organization and involves many persons in the data collection. With the administrative complexity and the pressure to produce timely estimates, many types of errors can be injected into the census. In a sample, more attention can be devoted to data quality through training personnel and following up on nonrespondents. It is far better to have good measurements on a representative sample than unreliable or biased measurements on the whole population.

Deming wrote: "Sampling is not mere substitution of a partial coverage for a total coverage. Sampling is the science and art of controlling and measuring the reliability of useful statistical information through the theory of probability" (Deming, 1950, p. 2). In the remaining chapters of this book, we explore this science and art in detail.

1.8 Chapter Summary

Survey sampling is used in every area of life, but samples must be collected and analyzed carefully for their results to apply to the target population. Statistics from a representative sample can be generalized to the population from which the population was drawn, and can be accompanied by a measure of how accurate the statistics are.

Bias is a major concern for samples intended to generalize to the population. Selection bias occurs when some population members are, without the intention of the sampler, overrepresented or underrepresented in the sample. Sources of selection bias include undercoverage (when the sampling frame fails to include part of the target population), overcoverage (when the sample can include units not in the target population), and nonresponse (when units designated for the sample fail to provide data). Self-selected samples (in which persons volunteer for the survey) and convenience samples (in which the sample consists of easy-to-locate or easy-to-collect units) often have selection bias.

Measurement error occurs when the data recorded for observation units differ from the true values. In surveys of people, question wording and ordering are common sources of measurement error, and these can sometimes be mitigated by careful questionnaire design.

Selection bias and measurement error are examples of nonsampling errors. Nonsampling errors can affect any type of data collection, even those that are intended to be a census of the population. It may be possible to reduce nonsampling errors by taking a carefully collected sample, even though estimates from the sample have sampling error because only part of the population is measured.

Key Terms

Bias: A systematic error in the sampling, measurement, or estimation procedures that results in a statistic being consistently larger (or consistently smaller) than the population characteristic that it estimates.

Census: A data collection procedure in which an attempt is made to enumerate or obtain information on all members of the population of interest.

Convenience sample: A sample in which the primary consideration for sample selection is the ease with which units can be recruited or located.

Coverage: The percentage of the population of interest that is included in the sampling frame.

Measurement error: The difference between the response coded and the true value of the characteristic being studied for a respondent.

Mode: The method used to distribute surveys and collect answers.

Nonresponse: Failure of some units in the sample to provide data for the survey.

Nonsampling error: An error from any source other than sampling error. Examples include undercoverage, nonresponse, and measurement error.

Observation unit: An object on which a measurement is taken in the survey. This is not always the same as the sampling unit.

Purposive sample: A sample selected based on the knowledge or judgment of the investigator.

Representative sample: A sample that can be used to estimate quantities of interest in a population and provide measures of accuracy about the estimates.

Respondent: An observation unit that provides data for the sample survey.

Sample: A subset of a population.

Sampled population: The collection of all possible observation units that might have been chosen using the sampling procedure. This may, or may not, coincide with the target population.

Sampling error: Error in estimation due to taking a sample instead of measuring every unit in the population.

Sampling frame: A list, map, or other specification of units in the population from which a sample may be selected. Examples include a list of all university students, a telephone directory, or a map of geographic areas.

Sampling unit: An object that can be sampled from the population.

Selection bias: Bias that occurs because the actual probabilities with which units are sampled differ from the selection probabilities specified by the investigator.

Social desirability bias: The tendency of some respondents to report an answer that they consider to be more socially acceptable than the "true" answer.

Target population: The set of units that the researcher wants to study. It is desired to generalize results from the sample to the target population.

For Further Reading

Dillman et al. (2014) give clear, practical, research-supported guidance on everything from writing questions to choosing the survey mode to the optimal timing of follow-ups with nonrespondents. Groves et al. (2011) discuss statistical and non-statistical issues in survey sampling, with examples from large-scale surveys. For a shorter read (less than 200 pages), Fowler (2014) walks you through issues and best practices concerning survey questions, survey mode, interviewing, data coding, and ethical issues. Toepoel (2016) provides a concise guide for conducting online surveys; Couper (2008) gives a longer, detailed treatment.

The article by Oberski (2018) and the books written or edited by Saris and Gallhofer (2014), Madans et al. (2011), Krosnick and Presser (2010), Bradburn et al. (2004), and Presser et al. (2004) review the scientific research on writing survey questions. They describe best practices as well as methods for testing questions; all are clearly written and list additional references. The concise books by Converse and Presser (1986) and Fowler (1995), and the book chapter by Fowler and Cosenza (2009), give practical advice for improving survey questions, with numerous examples. Lenzner and Menold (2016) give a 14-page summary of Dos and Don'ts for writing questions.

Tourangeau and Bradburn (2010) discuss the psychology of survey response—the cognitive processes that respondents use when interpreting and answering survey questions. Schaeffer et al. (2010) and West and Blom (2017) review research about how interviewers affect answers given to surveys.

The American Association of Public Opinion Research website, `www.aapor.org`, contains many resources for the sampling practitioner, including a guide to Standards and Best Practices. The journal *Public Opinion Quarterly* regularly publishes articles about survey quality, nonsampling errors, and questionnaire design.

1.9 Exercises

A. Introductory Exercises

For each survey in Exercises 1–31, describe the target population, sampling frame, sampling unit, and observation unit. Discuss any possible sources of selection bias or inaccuracy of responses.

1. Every year since 1927, *Time* magazine has chosen a Person of the Year to recognize a person or group who has had the most influence on the world that year. Although the final choice is made by the magazine's editors, readers can vote for a Person of the Year on the poll website. In 2018, the editors chose The Guardians—journalists who were murdered in 2018 for "speaking up and speaking out"—as the Person of the Year. The winner of the readers' poll was BTS, a K-pop band with a large online following.

2. A student wants to estimate the percentage of mutual funds that invest in a particular stock. She selects every twentieth fund in a list of all funds, and calculates the percentage that invest in the stock.

3. An online bookseller summarizes readers' reviews of the books it sells. Persons who want to rate a book can submit a review online; the bookseller reports the average rating from all reviews on its website.

4. Potential jurors in most Pennsylvania counties are sampled randomly from a list of county residents who are registered voters or licensed drivers over age 18. In 2017, 1,224,826 jury summons were mailed to residents of Pennsylvania on the list. About 91,000 of those were returned from the post office as undeliverable; for another 284,000 of the summons, no one responded. Approximately 153,000 persons were unqualified for service because they were not citizens, were under 18, were convicted felons, or had other reasons that disqualified them from serving on a jury. An additional 187,000 were exempted or excused from jury service because of age, illness, financial hardship, military service, or another acceptable reason. The final sample consisted of 246,000 persons who were told to report for service and reported (Nieves et al., 2018).

5. Many scholars and policy-makers are interested in the proportion of persons experiencing homelessness who are mentally ill. Wright (1988) estimated that 33% of all homeless people are mentally ill, by sampling homeless persons who received medical attention from one of the clinics in the Health Care for the Homeless (HCH) project. He argued that selection bias is not a serious problem because the clinics were easily accessible to the homeless and because the demographic profiles of HCH clients were close to those of the general homeless population in each city in the sample. Do you agree?

6. In July 2019, the Iowa State Fair Straw Poll asked fair-goers to indicate their choices for candidates for U.S. president and Congress in the 2020 election. Fair-goers indicated their choices by stopping at the Iowa Secretary of State's booth and filling out the poll on an electronic tablet. More than 4,000 persons participated.

7. A survey is conducted to find the average weight of cows in a region. A list of all farms is available for the region, and 50 farms are selected at random. Then the weight of each cow at the 50 selected farms is recorded.

8. To study nutrient content of menus in boarding homes for the elderly in Washington State, Goren et al. (1993) mailed surveys to all 184 licensed homes in Washington State, directed to the administrator and food service manager. Of those, 43 were returned by the deadline and included menus.

9. Entries in the online encyclopedia *Wikipedia* can be written or edited by anyone with internet access. This has given rise to concern about the accuracy of the information. Giles (2005) reported on a *Nature* study assessing the accuracy of *Wikipedia* science articles. Fifty subjects were chosen "on a broad range of scientific disciplines." For each subject, the entries from *Wikipedia* and *Encyclopaedia Brittanica* were sent to a relevant expert; 42 sets of usable reviews were returned. The editors of *Nature* then tallied the number of errors reported for each encyclopedia.

10. The December 2003 issue of *PC World* reported the results from a survey of over 32,000 subscribers asking about reliability and service for personal computers and other electronic equipment. The magazine "invited subscribers to take the Web-based survey from April 1 through June 30, 2003" and received 32,051 responses. Survey respondents were entered in a drawing to win prizes. They reported that 46% of desktop PC's had at least one significant malfunction.

11. Karras (2008) reported on a survey conducted by *SELF* magazine on the prevalence of eating disorders in women. The survey, posted online at `self.com`, obtained responses from 4,000 women. Based on these responses, the article reported that 27% of women in the survey "say they would be 'extremely upset' if they gained just 5 pounds"; it was estimated that 10% of women have eating disorders such as anorexia or bulimia.

12. Shen and Hsieh (1999) took a purposive sample of 29 higher education institutions; the institutions were "representative in terms of institutional type, geographic and demographic diversity, religious/nonreligious affiliation, and the public/private dimension" (p. 318). They then mailed a survey about improving the professional status of teaching to 2,042 faculty members in the institutions, of whom 1,219 returned the survey.

13. In October 2018, each of the 15,769 persons in the current membership list of the American Statistical Association (ASA) was sent an e-mail message with the subject line "American Statistical Association Study — Please Participate." The invitation included a link to a survey about members' experiences of sexual or gender-based misconduct at ASA events, and stated: "It's important for you to participate whether or not you have personally witnessed or experienced any such misconduct." A total of 3,191 persons completed the survey.

    Respondents who had attended an ASA event (86% of female respondents and 85% of male respondents) were asked, "While attending an ASA event, have you ever witnessed a sexually oriented conversation that you felt was inappropriate?" Thirteen percent of women and 5 percent of men said yes (Langer Research Associates, 2019).

14. Deal and Olson (2015) reported results from the National Sample of Couples and Remarriage, including the statistic that 63 percent of couples in the study (55 percent of "happy couples" and 81 percent of "unhappy couples") fear a relationship breakup. The sample consisted of 50,575 couples who took a marital assessment survey as part of couples' therapy or premarital counseling with a professional counselor or trained clergyperson. Eighteen percent of the couples in the sample lived in rural areas, 58 percent lived in the suburbs or a small city, and 24 percent lived in a large city.

15. Kripke et al. (2002) claimed that persons who sleep 8 or more hours per night have a higher mortality risk than persons who sleep 6 or 7 hours. They analyzed data from the 1982 Cancer Prevention Study II of the American Cancer Society, a national survey taken by about 1.1 million people. The survival or date of death was determined for about 98% of the sample six years later. Most of the respondents were friends and relatives of American Cancer Society volunteers; the purpose of the original survey was to explore factors associated with the development of cancer, but the survey also contained a few questions about sleep and insomnia.

16. In lawsuits about trademarks, a plaintiff claiming that another company is infringing on its trademarks must often show that the marks have a "secondary meaning" in the marketplace—that is, potential users of the product associate the trademarks with the plaintiff even when the company's name is missing. In the court case *Harlequin Enterprises Ltd v. Gulf & Western Corporation* (503 F. Supp. 647, 1980), the publisher of Harlequin Romances persuaded the court that the cover design for "Harlequin Presents" novels had acquired secondary meaning. Part of the evidence presented was a survey of 500 women from three cities who identified themselves as readers of romance fiction. They were shown copies of unpublished "Harlequin Presents" novels with the Harlequin name hidden; over 50% identified the novel as a Harlequin product.

17. Theoharakis and Skordia (2003) asked statisticians who responded to their survey to rank statistics journals in terms of prestige, importance, and usefulness. They gathered e-mail addresses for 12,053 statisticians from the online directories of statistical organizations and sent an e-mail invitation to each to participate in the online survey. A total of 2,190 responses were obtained. The authors suggested that the results of their survey could help universities make promotion and tenure decisions about statistics faculty by providing information about the perceived quality of statistics journals.

18. The U.S. National Intimate Partner and Sexual Violence Survey was launched in 2010 to assess sexual violence, stalking, and intimate partner violence victimization among adult men and women. The survey was conducted by telephone, and two sampling frames were used. The landline telephone frame consisted of hundred-banks of telephone numbers in which at least one of the numbers in the bank was known to be residential. (A hundred-bank is the set of hundred numbers in which the area code and first five digits are fixed, and the last two digits take on any value between 00 and 99.) Numbers known to belong to businesses were excluded. The cellular telephone phone frame consisted of telephone banks known to be in use for cell phones.

    In 2010, of the 201,881 landline and cell telephone numbers sampled, approximately 31% were ineligible (for example, they belonged to a business or were nonworking numbers). Another 53% were of unknown eligibility (usually because no one answered the telephone). When someone from an eligible household answered, the adult with the most recent birthday was asked to take the survey. From the 31,241 households determined to be eligible, a total of 18,049 persons were interviewed, of whom 16,507 completed the survey (U.S. Department of Health and Human Services, Centers for Disease Control and Prevention, 2014).

19. The November 2019 issue of *Consumer Reports* contained satisfaction scores for 53 automobile insurance providers from a survey of 90,352 *Consumer Reports* members. Answer the general questions about the target population, sampling frame, and units for this survey. The organization has more than 6 million members. Under what circumstances will the magazine's statement that 54 percent of its members "have been with their current company for 15 years or more" be accurate?

20. Ebersole (2000) studied how students in selected public schools describe their use of the internet. Five school districts in a Western state were selected to give a cross-section of urban and rural schools that have internet access. A survey, administered electronically, was installed as the home page in middle and high school media centers in the districts "for a period of time to gather approximately 100 responses from each school." Students who had parental permission to access the internet were permitted to access the computer-administered survey. Participation in the survey was voluntary.

21. A news report with the headline "CTE found in 99 percent of former NFL players studied" began with the sentence: "A new study suggests that chronic traumatic encephalopathy (CTE), a progressive, degenerative brain disease found in people with a history of repeated head trauma, may be more common among football players than previously thought" (Moran, 2017). The 202 brains of former football players used in the study were obtained from the Concussion Legacy Foundation Brain Bank; most had been donated by family members in order to further research on repetitive brain trauma and CTE. Autopsies were conducted to determine the presence or absence of CTE: 110 of the 111 former National Football League (NFL) players, and 67 of the other 91 former football players, were diagnosed with CTE.

22. A newspaper columnist asked readers to e-mail her their opinions about stores being open on Thanksgiving. Ninety-five percent of the respondents said that stores should be closed.

23. A medical clinic, wishing to assess patients' satisfaction with the healthcare services received, conducts in-person exit interviews with a sample of patients. The survey takes 15 minutes to complete. Every 20th person entering the clinic is selected to be interviewed.

24. For the situation described in Exercise 23, instead of sampling every 20th person entering the clinic, the interviewer tries to conduct as many interviews as possible. Upon completing an interview with a patient, the interviewer conducts the survey with the next patient to leave a consultation room.

25. Half of the scoring on the television show *Dancing with the Stars* depends on votes cast by the audience. Persons can vote for their favorite dance couple online or by text message during the live broadcast of the show in the eastern and central time zones. The vote score for a couple is (total number of votes received for the couple) divided by (total number of votes received for all couples).

26. A television station in Phoenix, Arizona, asked viewers to visit its website to answer the survey question "Should Arizona participate in daylight savings time?" Options were "yes—love the extra sunshine" or "no—we have enough daylight." The poll showed 10% yes and 90% no.

27. The League of Women Voters of Texas (2019) evaluated the county election websites for all 254 counties in Texas. They found that 201 of the counties had an insecure website—that is, the internet address began with "http" rather than "https." Only nine used the ".gov" domain that indicates a bona fide U.S.-based government organization.

28. Schuitemaker (2019) surveyed 502 people around the world about their experiences as non-parenting adults. Survey participants were recruited through social media and on the author's website. The sample had 171 persons from the Netherlands, 78 from the United States, 61 from the United Kingdom, 57 from Italy, and the remainder from 26 other countries.

29. The *Survey of America's Physicians* was e-mailed to nearly every physician in the United States having an e-mail address in the American Medical Association's Physician Master File—more than 700,000 in all. Responses from 8,774 physicians were received (The Physicians Foundation, 2018).

30. The Sustainability Office of Little Rock, Arkansas, conducted an online survey to assess residents' and visitors' views on waste-reduction proposals. The questionnaire was on the city's website and distributed on social media platforms for about a month; 2,479 of 3,631 respondents said they supported a ban or fee on single-use plastic bags in the city (Herzog, 2020).

31. Van Patter et al. (2019) reported on a survey undertaken by the Cat Population Taskforce of Guelph, Ontario, to study residents' attitudes about the management of community cats, defined as cats "that spend all of their time outdoors and do not currently have an owner who is actively caring for them." The survey was administered two ways: in-person and online. For the in-person part of the survey, seven volunteers recruited 116 survey participants at six public locations in the city. An additional 333 persons visited the web link and took the online survey, which was publicized through articles in the local newspaper; through pamphlets distributed to veterinary offices, pet stores, and cafés; and through social media. About two-thirds of the participants were female, and 84% had some level of education beyond a high school degree.

32. Evaluate each of the following survey questions (most came from real surveys) according to the criteria in Section 1.5. Which (if any) problem or problems (for example, leading or double-barreled) does the question exhibit? What is the concept of interest, and how would you rewrite the question to measure it?

(a) How many potatoes did you eat last year?

(b) According to the American Film Institute, *Citizen Kane* is the greatest American movie ever made. Have you seen this movie?

(c) Do you favor or oppose changing environmental regulations so that while they still protect the public, they cost American businesses less and lower product costs?

(d) Agree or disagree: We should not oppose policies that reduce carbon emissions.

(e) How many of your close friends use marijuana products?

(f) Do you want a candidate for president who would make history based on race, gender, sexual orientation, or religion?

(g) Who was or is the worst president in the history of the United States?

(h) Agree or disagree: Health care is not a right, it's a privilege.

(i) The British Geological Survey has estimated that the UK has 1,300 trillion cubic feet of natural gas from shale. If just 10% of this could be recovered, it would be enough to meet the UK's demand for natural gas for nearly 50 years or to heat the UK's homes for over 100 years. From what you know, do you think the UK should produce natural gas from shale?

(j) What hours do you usually work each day?

(k) Do you support or oppose gun control?

(l) Looking back, do you regret having moved so rapidly to be divorced, and do you now feel that had you waited, the marriage might have been salvaged?

(m) Are the Patriots and the Packers good football teams?

(n) Antarctic ice is now melting three times faster than it was before 2012. And the Arctic is warming at twice the rate of the rest of the world. On a scale from 1 to 5, with 5 being the highest, how alarmed are you by melting ice, the rise of global sea levels, and increased flooding of coastal cities and communities?

(o) Overall, how strong was your emotional reaction to the performance of *La Bohème*? (No emotional response, Weak response, Moderate response, Strong response, or Not applicable)

(p) How often do you wish you hadn't gotten into a relationship with your current partner? (All the time, Most of the time, More often than not, Occasionally, Rarely, Never)

(q) Do you believe that for every dollar of tax increase there should be $2.00 in spending cuts with the savings earmarked for deficit and debt reduction?

(r) There probably won't be a woman president of the United States for a long time and that's probably just as well.

(s) How satisfied are you that your health insurance plan covers what you expect it to cover? (Very satisfied, Satisfied, Dissatisfied, Very dissatisfied, or Not applicable)

(t) Given the world situation, does the government protect too many documents by classifying them as secret and top secret?

D. Projects and Activities

33. Many universities ask students to fill out a course evaluation at the end of the term. Obtain a copy of the questionnaire that is used by the department offering your sampling class.

    (a) Evaluate the questions on this survey. How could they be improved?

    (b) What procedure is used to distribute and collect the surveys? What are some possible sources of selection bias?

    (c) Explain why you think the student evaluations do, or do not, give an accurate measurement of student satisfaction with the course.

34. *For students of U.S. history.* Eighty-five letters appeared in New York City newspapers in 1787 and 1788, with the purpose of drawing support in the state for the newly drafted Constitution. Collectively, these letters are known as *The Federalist.* Read Number 54 of *The Federalist*, in which the author (thought to be James Madison) discusses using a population census to apportion elected representatives and taxes among the states. This article explains part of Article I, Section II of the United States Constitution.

    Write a short paper discussing Madison's view of a population census. What is the target population and sampling frame? What sources of bias does Madison mention, and how does he propose to reduce bias? What is your reaction to Madison's plan, from a statistical point of view?

35. Statistics from an informal poll taken in 1975–1976 by syndicated advice columnist Ann Landers created a furor and continued to be cited long after the original survey. In 1975, Landers published a letter from a young couple who were about to be married and debated whether to have children, and she asked readers to respond to the question posed in the last line of the letter.

    > "So many of our friends," the letter said, "seem to resent their children. They envy us our freedom to go and come as we please. Then there's the matter of money. They say their kids keep them broke. One couple we know had their second child in January. Last week, she had her tubes tied and he had a vasectomy—just to make sure. All this makes me wonder, Ann Landers. Is parenthood worth the trouble? Jim and I are very much in love. Our relationship is beautiful. We don't want anything to spoil it. All around us we see couples who were so much happier before they were tied down with a family. Will you please ask your readers the question: If you had it to do over again, would you have children?" Landers (1976)

    Landers received more than 10,000 responses, 80% of those from women. About 70% of the readers who responded said "No."

    (a) Discuss the features of Ann Landers' survey that might cause the estimated percentage to be biased.

    (b) In a sidebar that accompanied Landers (1976), *Good Housekeeping* magazine editors wrote: "All of us at *Good Housekeeping* know that no mother will be able to read Ann Landers' report without passionately agreeing or disagreeing. We would like to know what your reaction is. Won't you therefore, take a minute or two to let us know how you would answer the question: if you it had to do over again, would you have children?" 95% of the respondents to the *Good Housekeeping* survey said yes, they would have children if doing it over. How do you explain the discrepancy between this result and the overwhelming negative response to the Landers survey?

(c) In 1976, *Newsday* magazine (1976) asked the question "If you had it to do over again, would you or would you not have children?" of a "representative nationwide sample of 1,373 parents aged 16 and older"; 91 percent of respondents said they would and 7 percent said they would not. The magazine did not say, however, how the sample was selected or contacted. What questions would you ask about the survey to see if it was, indeed, representative?

36. Find a recent survey reported in a newspaper, academic journal, or popular magazine. Describe the survey. What are the target population and sampled population? What conclusions are drawn about the survey in the article? Do you think those conclusions are justified? What are possible sources of bias for the survey?

37. Write two questions about a topic you are interested in. Write the first question with the intention of eliciting an honest response, and write the second question so that it leads the respondent to give a particular answer. Now, from a set of 20 persons, choose 10 randomly to be asked question 1 and the other 10 to be asked question 2. How do the statistics calculated from the two groups differ? Note that in some universities, Institutional Review Board approval may be needed before doing this exercise.

38. Find a survey being conducted on the internet. Write a paragraph or two describing the survey and its results (most online surveys allow you to see the statistics from all the persons who have taken the survey). What are the target population and sampled population? What biases do you think might occur in the results?

39. Some polling organizations recruit volunteers for an internet panel and then take samples from the panel to measure public opinion. Find a survey organization that takes samples from an internet panel. Report how the organization recruits participants and produces estimates. What are possible sources of selection bias?

2

Simple Probability Samples

[Kennedy] read every fiftieth letter of the thirty thousand coming weekly to the White House, as well as a statistical summary of the entire batch, but he knew that these were often as organized and unrepresentative as the pickets on Pennsylvania Avenue.

—Theodore Sorensen, *Kennedy*

The examples of misleading surveys in Chapter 1—for example, the *Literary Digest* poll in Section 1.1—had major flaws that resulted in unrepresentative samples. In this chapter, we discuss how to use **probability sampling** to conduct surveys. In a probability sample, each unit in the population has a known probability of selection, and a random number table or other randomization mechanism is used to choose the specific units to be included in the sample. If a probability sampling design is implemented well, an investigator can use a relatively small sample to make inferences about an arbitrarily large population.

In Chapters 2 through 6, we explore survey design and properties of estimators for the three major design components used in a probability sample: simple random sampling, stratified sampling, and cluster sampling. We shall integrate all these ideas, and show how they are combined in complex surveys such as the U.S. National Health and Nutrition Examination Survey, in Chapter 7. To simplify presentation of the concepts, we assume for now that the sampled population is the target population, that the sampling frame is complete, that there is no nonresponse or missing data, and that there is no measurement error. We return to nonsampling errors in Chapter 8.

As you might suppose, you need to know some probability to be able to understand probability sampling. You may want to review the material in Sections A.1 and A.2 of Appendix A while reading this chapter.

Statistical software for survey data. Calculations for the examples in this book can be done with multiple software packages. Two supplementary books, available for free download from the book website and also available commercially in paperback, show how to do the calculations for the examples in this book using SAS® and R software.

Lohr, S. L. (2022). SAS® Software Companion for *Sampling: Design and Analysis, Third Edition.* Boca Raton, FL: CRC Press.

Lu, Y. and Lohr, S. L. (2022). R Companion for *Sampling: Design and Analysis, Third Edition.* Boca Raton, FL: CRC Press.

These books walk the reader through code and output for the examples, and discuss tips and programming practices for using these two software packages to select probability samples and analyze survey data; see Section 2.6 for examples. They also contain detailed descriptions of the data sets and variables referenced in this book. Page xviii of the Preface tells how to download the supplementary books and the data sets. You may want to obtain one of these books now (if you have not done so already) if you want to follow along with the software for the examples.

DOI: 10.1201/9780429298899-2

2.1 Types of Probability Samples

The terms *simple random sample*, *stratified sample*, *cluster sample*, and *systematic sample* are basic to any discussion of sample surveys, so let's define them now.

- A **simple random sample** (SRS) is the simplest form of probability sample. An SRS of size n is taken when every possible subset of n units in the population has the same chance of being the sample. SRSs are the focus of this chapter and the foundation for more complex sampling designs. In taking a random sample, the investigator is in effect mixing up the population before grabbing n units. The investigator does not need to examine every member of the population for the same reason that a medical technician does not need to drain you of blood to measure your red blood cell count: Your blood is sufficiently well mixed that any sample should be representative. SRSs are discussed in Section 2.3, after we present the basic framework for probability samples in Section 2.2.

- In a **stratified random sample**, the population is divided into subgroups called **strata**. Then an SRS is selected from each stratum, and the SRSs in the strata are selected independently. The strata are often subgroups of interest to the investigator—for example, the strata might be defined by regions of the country in a survey of people, types of terrain in an ecological survey, or sizes of firms in a business survey. Elements in the same stratum often tend to be more similar than randomly selected elements from the whole population, so stratification often increases precision, as we shall see in Chapter 3.

- In a **cluster sample**, observation units in the population are aggregated into larger sampling units, called **clusters**. Suppose you want to survey Lutheran church members in Minneapolis but do not have a list of all church members in the city, so you cannot take an SRS of Lutheran church members. However, you do have a list of all the Lutheran churches. You can then take an SRS of the churches and then subsample all or some church members in the selected churches. In this case, the churches form the clusters, and the church members are the observation units. It is more convenient to sample at the church level; however, members of the same church may have more similarities than Lutherans selected at random in Minneapolis, so a cluster sample of 500 Lutherans may not provide as much information as an SRS of 500 Lutherans. We shall explore this idea further in Chapter 5.

- In a **systematic sample**, a starting point is chosen from a list of population members using a random number. That unit, and every kth unit thereafter, is chosen to be in the sample. A systematic sample thus consists of units that are equally spaced in the list. Systematic samples will be discussed in more detail in Sections 2.8 and 5.5.

Suppose you want to estimate the average amount of time that professors at your university say they spent grading homework in a specific week. To take an SRS, construct a list of all professors and randomly select n of them to be your sample. Now ask each professor in your sample how much time he or she spent grading homework that week—you would, of course, have to define the words *homework* and *grading* carefully in your questionnaire. In a stratified sample, you might classify faculty by college: engineering, liberal arts and sciences, business, nursing, and fine arts. You would then take an SRS of faculty in the engineering college, a separate SRS of faculty in liberal arts and sciences, and so on. For a cluster sample, you might randomly select 10 of the 60 academic departments in the university and ask each faculty member in those departments how much time he or she spent grading homework. A systematic sample could be chosen by selecting an integer at random between 1 and 20; if the random integer is 16, say, then you would include professors in positions 16, 36, 56, and so on, in the list.

Example 2.1. Figure 2.1 illustrates the differences among simple random, stratified, cluster, and systematic sampling for selecting a sample of 20 integers from the population $\{1, 2, \ldots, 100\}$. For the stratified sample, the population was divided into the 10 strata $\{1, 2, \ldots, 10\}$, $\{11, 12, \ldots, 20\}$, ..., $\{91, 92, \ldots, 100\}$ and an SRS of 2 numbers was drawn from each of the 10 strata. This ensures that each stratum is represented in the sample. For the cluster sample, the population was divided into 20 clusters $\{1, 2, 3, 4, 5\}$, $\{6, 7, 8, 9, 10\}$, ..., $\{96, 97, 98, 99, 100\}$; an SRS of 4 of these clusters was selected. For the systematic sample, the random starting point was 3, so the sampled units are 3, 8, 13, ..., 93, 98. ■

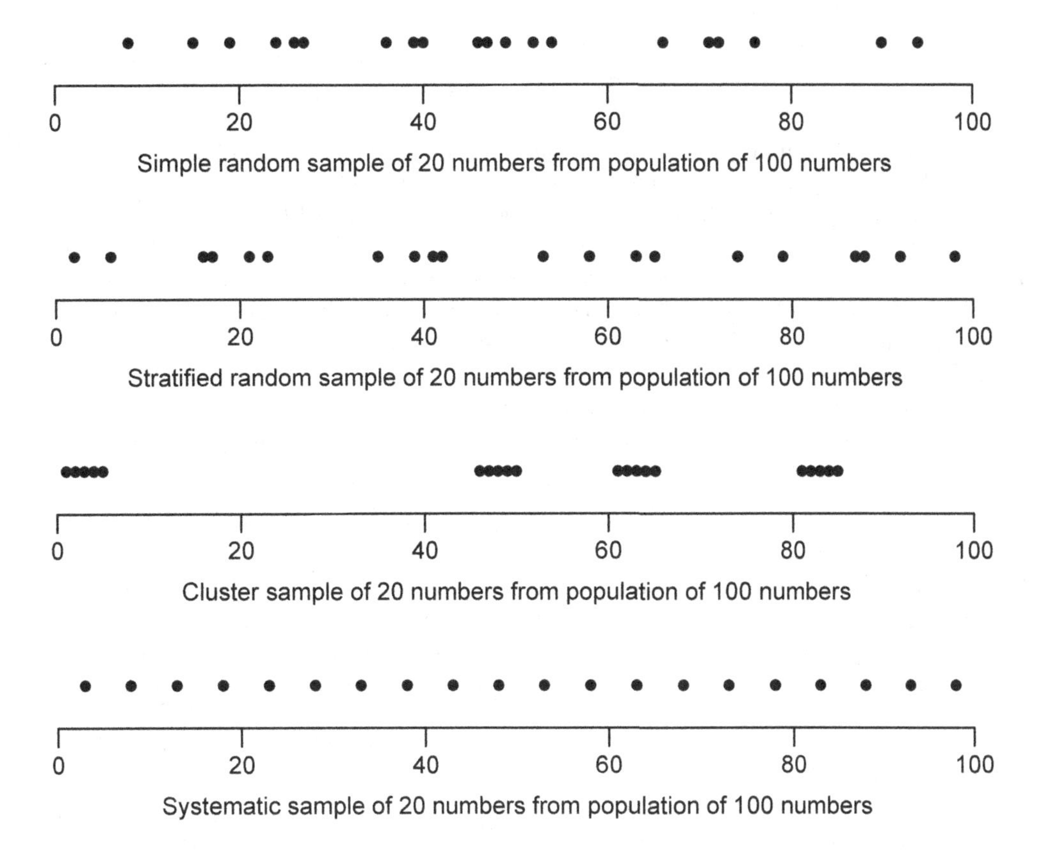

FIGURE 2.1
Examples of a simple random sample, stratified random sample, cluster sample, and systematic sample of 20 integers from the population $\{1, 2, \ldots, 100\}$.

All of these methods—simple random sampling, stratified random sampling, cluster sampling, and systematic sampling—involve random selection of units to be in the sample. In an SRS, the observation units themselves are selected at random from the population of observation units; in a stratified random sample, observation units within each stratum are randomly selected; in a cluster sample, the clusters are randomly selected from the population of all clusters. Each method is a form of probability sampling, which we discuss in the next section.

2.2 Framework for Probability Sampling

To show how probability sampling works, let us assume there is a list of the N units in the finite population. The finite **population,** or **universe,** of N units is denoted by the index set

$$\mathcal{U} = \{1, 2, \ldots, N\}. \tag{2.1}$$

Out of this population we can choose various samples, which are subsets of $\mathcal{U}$. The particular sample chosen is denoted by $\mathcal{S}$, a subset consisting of n of the units in $\mathcal{U}$.

Suppose the population has four units: $\mathcal{U} = \{1, 2, 3, 4\}$. Six different samples of size 2 could be chosen from this population:

$$\begin{array}{lll} \mathcal{S}_1 = \{1,2\} & \mathcal{S}_2 = \{1,3\} & \mathcal{S}_3 = \{1,4\} \\ \mathcal{S}_4 = \{2,3\} & \mathcal{S}_5 = \{2,4\} & \mathcal{S}_6 = \{3,4\} \end{array}$$

In probability sampling, each possible sample $\mathcal{S}$ from the population has a known probability $P(\mathcal{S})$ of being chosen, and the probabilities of the possible samples sum to 1. One possible sample design for a probability sample of size 2 would have $P(\mathcal{S}_1) = 1/3$, $P(\mathcal{S}_2) = 1/6$, and $P(\mathcal{S}_6) = 1/2$, and $P(\mathcal{S}_3) = P(\mathcal{S}_4) = P(\mathcal{S}_5) = 0$. The probabilities $P(\mathcal{S}_1)$, $P(\mathcal{S}_2)$, and $P(\mathcal{S}_6)$ of the possible samples are known before the sample is drawn. One way to select the sample would be to place six labeled balls in a box; two of the balls are labeled 1, one is labeled 2, and three are labeled 6. Now choose one at random; if a ball labeled 6 is chosen, then $\mathcal{S}_6$ is the sample.

In a probability sample, since each possible sample has a known probability of being the chosen sample, each unit in the population has a known probability of appearing in our selected sample. We calculate

$$\pi_i = P(\text{unit } i \text{ in sample}) \tag{2.2}$$

by summing the probabilities of all possible samples that contain unit i. In probability sampling, the π_i are known before the survey commences, and we assume that $\pi_i > 0$ for every unit in the population. For the sample design described above, $\pi_1 = P(\mathcal{S}_1) + P(\mathcal{S}_2) + P(\mathcal{S}_3) = 1/2$, $\pi_2 = P(\mathcal{S}_1) + P(\mathcal{S}_4) + P(\mathcal{S}_5) = 1/3$, $\pi_3 = P(\mathcal{S}_2) + P(\mathcal{S}_4) + P(\mathcal{S}_6) = 2/3$, and $\pi_4 = P(\mathcal{S}_3) + P(\mathcal{S}_5) + P(\mathcal{S}_6) = 1/2$.

Of course, we rarely write all possible samples down and calculate the probability with which we would choose each possible sample—this would take far too long. But such enumeration tacitly underlies all of probability sampling. Investigators using a probability sample have much less discretion about which units are included in the sample, so using probability samples helps us avoid some of the selection biases described in Chapter 1. In a probability sample, the interviewer cannot choose to substitute a friendly looking person for the grumpy person selected to be in the sample by the random selection method. A forester taking a probability sample of trees cannot simply measure the trees near the road but must measure the trees designated for inclusion in the sample. Taking a probability sample is usually more difficult than taking a convenience sample, but a probability sampling procedure guarantees that each unit in the population could appear in the sample and provides information that can be used to assess the precision of statistics calculated from the sample.

It is this quality that makes a probability sample **representative** of the population. Within the framework of probability sampling, we can quantify how likely it is that our sample is a "good" one. A single probability sample is not guaranteed to mirror the population with regard to the characteristics of interest, but we can quantify what proportion of the possible samples will give a statistic that is within a fixed degree of tolerance from the population quantity. The notion is the same as that of confidence intervals: We do not know whether the particular 95% confidence interval we construct for the mean contains the true value of the population mean. We do know, however, that if the assumptions for the confidence interval procedure are valid and if we repeat the procedure over and over again, we can expect 95% of the resulting confidence intervals to contain the true value of the mean.

Let y_i be a characteristic associated with the ith unit in the population. We consider y_i to be a fixed quantity; if Farm 723 is included in the sample, then the amount of corn produced on Farm 723, y_{723}, is known exactly.

Example 2.2. To illustrate these concepts, let's look at an artificial situation in which we know the value of y_i for each of the $N = 8$ units in the whole population. The index set for the population is

$$\mathcal{U} = \{1, 2, 3, 4, 5, 6, 7, 8\}.$$

The values of y_i are

i	1	2	3	4	5	6	7	8
y_i	1	2	4	4	7	7	7	8

There are 70 possible samples of size 4 that may be drawn without replacement from this population; the samples are listed in file `sample70.csv`. If the sample consisting of units $\{1, 2, 3, 4\}$ were chosen, the corresponding values of y_i would be 1, 2, 4, and 4. The values of y_i for the sample $\{2, 3, 6, 7\}$ are 2, 4, 7, and 7. Define $P(\mathcal{S}) = 1/70$ for each distinct subset of size four from $\mathcal{U}$. As you will see after you read Section 2.3, this design is an SRS without replacement. Each unit is in exactly 35 of the possible samples, so $\pi_i = 1/2$ for $i = 1, 2, \ldots, 8$.

A random mechanism is used to select one of the 70 possible samples. One possible mechanism for this example, because we have listed all possible samples, is to generate a random number between 1 and 70 and select the corresponding sample. With large populations, however, the number of samples is so great that it is impractical to list all possible samples—instead, another method is used to select the sample. Methods that will give an SRS will be described in Section 2.3. ■

Sampling distributions. Most results in sampling rely on the **sampling distribution** of a statistic, which is the distribution of different values of the statistic obtained by the process of taking all possible samples from the population. A sampling distribution is an example of a discrete probability distribution.

Suppose we want to use a sample to estimate a population quantity, say the population total $t = \sum_{i=1}^{N} y_i$. One estimator we might use for t is $\hat{t}_{\mathcal{S}} = N\bar{y}_{\mathcal{S}}$, where $\bar{y}_{\mathcal{S}}$ is the average of the y_is in $\mathcal{S}$, the chosen sample. In Example 2.2, $t = 40$. If the sample $\mathcal{S}$ consists of units 1, 3, 5, and 6, then $\hat{t}_{\mathcal{S}} = 8 \times (1 + 4 + 7 + 7)/4 = 38$. Since we know the whole population here, we can find $\hat{t}_{\mathcal{S}}$ for each of the 70 possible samples. The probabilities of selection for the samples give the sampling distribution of $\hat{t}$:

$$P\{\hat{t} = k\} = \sum_{\mathcal{S}:\hat{t}_{\mathcal{S}}=k} P(\mathcal{S}).$$

The summation is over all samples $\mathcal{S}$ for which $\hat{t}_{\mathcal{S}} = k$. We know the probability $P(\mathcal{S})$ with which we select a sample $\mathcal{S}$ because we take a probability sample.

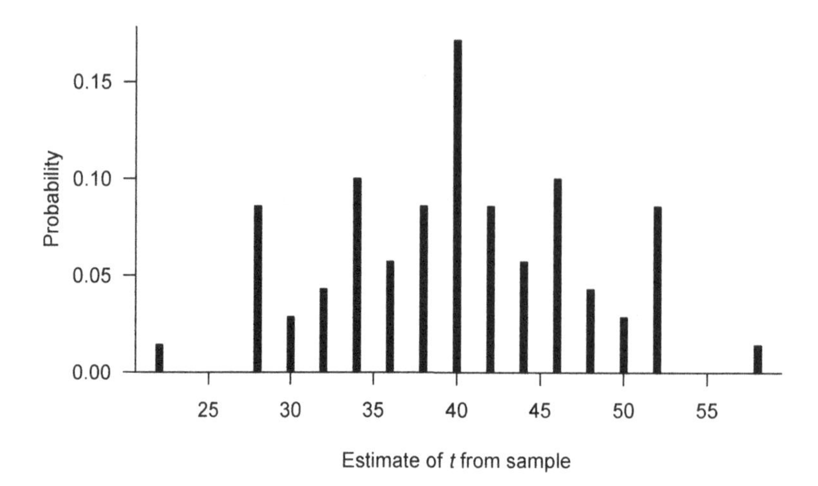

FIGURE 2.2
Sampling distribution of the estimated sample total $\hat{t}$ for Example 2.3.

Example 2.3. The sampling distribution of $\hat{t}$ for the population and sampling design in Example 2.2 derives entirely from the probabilities of selection for the various samples. Four samples ({3,4,5,6}, {3,4,5,7}, {3,4,6,7}, and {1,5,6,7}) result in the estimate $\hat{t} = 44$, so $P\{\hat{t} = 44\} = 4/70$. For this example, we can write out the sampling distribution of $\hat{t}$ because we know the values for the entire population.

k	22	28	30	32	34	36	38	40	42	44	46	48	50	52	58
$P\{\hat{t} = k\}$	$\frac{1}{70}$	$\frac{6}{70}$	$\frac{2}{70}$	$\frac{3}{70}$	$\frac{7}{70}$	$\frac{4}{70}$	$\frac{6}{70}$	$\frac{12}{70}$	$\frac{6}{70}$	$\frac{4}{70}$	$\frac{7}{70}$	$\frac{3}{70}$	$\frac{2}{70}$	$\frac{6}{70}$	$\frac{1}{70}$

Figure 2.2 displays the sampling distribution of $\hat{t}$. ■

The **expected value** of $\hat{t}$, $E[\hat{t}\,]$, is the mean of the sampling distribution of $\hat{t}$:

$$E[\hat{t}\,] = \sum_{\mathcal{S}} \hat{t}_{\mathcal{S}}\, P(\mathcal{S}) = \sum_{k} k\, P(\hat{t} = k). \tag{2.3}$$

The expected value of the statistic is the weighted average of the possible sample values of the statistic, weighted by the probability that particular value of the statistic would occur.

The **estimation bias** of the estimator $\hat{t}$ is

$$\text{Bias}\,[\hat{t}\,] = E[\hat{t}\,] - t. \tag{2.4}$$

If $\text{Bias}[\hat{t}\,] = 0$, we say that the estimator $\hat{t}$ is **unbiased** for t. For the data in Example 2.2 the expected value of $\hat{t}$ is

$$E[\hat{t}] = \frac{1}{70}(22) + \frac{6}{70}(28) + \ldots + \frac{1}{70}(58) = 40.$$

Thus, the estimator is unbiased.

Note that the mathematical definition of estimation bias in (2.4) is *not* the same thing as the selection or measurement bias described in Chapter 1. All indicate a systematic deviation from the population value, but from different sources. Selection bias is due to the method of selecting the sample—the investigator may act as though every possible sample $\mathcal{S}$ has the same probability of being selected, but some subsets of the population actually have a different probability of selection. With undercoverage, for example, the probability of including a unit not in the sampling frame is zero. Measurement bias means that the y_is are measured inaccurately, so although $\hat{t}$ may be unbiased in the sense of (2.4) for $t = \sum_{i=1}^N y_i$, t itself would not be the true total of interest. Estimation bias means that the estimator chosen results in bias—for example, if we used $\hat{t}_{\mathcal{S}} = \sum_{i \in \mathcal{S}} y_i$ and did not take a census, $\hat{t}$ would be biased.

To illustrate these distinctions, suppose you wanted to estimate the average height of male actors belonging to the Screen Actors Guild. Selection bias would occur if you took a convenience sample of actors on the set—perhaps taller actors are more or less likely to be working. Measurement bias would occur if your tape measure inaccurately added 3 centimeters (cm) to each actor's height. Estimation bias would occur if you took an SRS from the list of all actors in the Guild, but estimated mean height by the average height of the six shortest men in the sample—the sampling procedure is good, but the estimator is bad.

The **variance** of the sampling distribution of $\hat{t}$ is

$$V(\hat{t}) = E\left[\left(\hat{t} - E[\hat{t}]\right)^2\right] = \sum_{\substack{\text{all possible} \\ \text{samples } \mathcal{S}}} P(\mathcal{S}) \left(\hat{t}_{\mathcal{S}} - E[\hat{t}]\right)^2. \tag{2.5}$$

For the data in Example 2.2,

$$V\left(\hat{t}\right) = \frac{1}{70}(22 - 40)^2 + \ldots + \frac{1}{70}(58 - 40)^2 = \frac{3840}{70} = 54.86.$$

Because we sometimes use biased estimators, we often use the **mean squared error** (MSE) rather than variance to measure the accuracy of an estimator:

$$\begin{aligned}
\text{MSE}[\hat{t}] &= E\left[\left(\hat{t} - t\right)^2\right] \\
&= E\left[(\hat{t} - E[\hat{t}] + E[\hat{t}] - t)^2\right] \\
&= E\left[(\hat{t} - E[\hat{t}])^2\right] + \left(E\left[\hat{t}\right] - t\right)^2 + 2E\left[(\hat{t} - E[\hat{t}])\left(E[\hat{t}] - t\right)\right] \\
&= V\left(\hat{t}\right) + \left[\text{Bias}(\hat{t})\right]^2.
\end{aligned} \tag{2.6}$$

Thus an estimator $\hat{t}$ of t is **unbiased** if $E(\hat{t}) = t$, **precise** if $V(\hat{t}) = E[(\hat{t} - E[\hat{t}])^2]$ is small, and **accurate** if $\text{MSE}[\hat{t}] = E[(\hat{t} - t)^2]$ is small. A badly biased estimator may be precise but it will not be accurate; accuracy (MSE) is how close the estimate is to the true value, while precision (variance) measures how close estimates from different samples are to each other. Figure 2.3 illustrates these concepts.

In summary, the finite population $\mathcal{U}$ consists of units $\{1, 2, \ldots, N\}$ whose measured values are $\{y_1, y_2, \ldots, y_N\}$. We select a sample $\mathcal{S}$ of n units from $\mathcal{U}$ using the probabilities of selection that define the sampling design. The y_is are fixed but unknown quantities—unknown unless that unit happens to appear in our sample $\mathcal{S}$. Unless we make additional assumptions, the only information we have about the set of y_is in the population is in the set $\{y_i : i \in \mathcal{S}\}$.

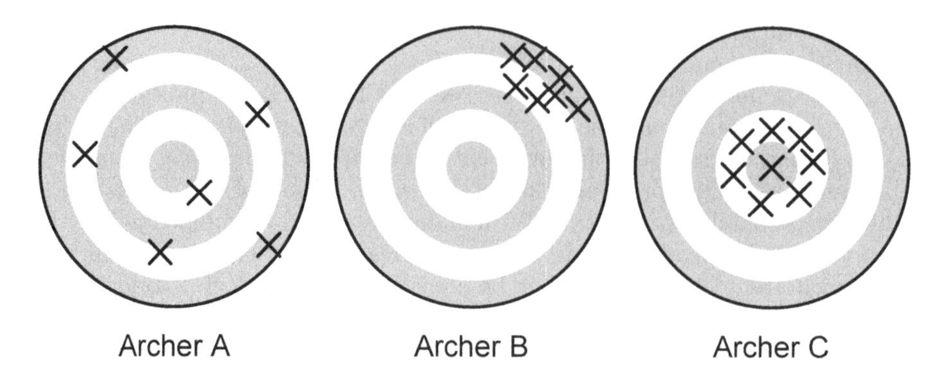

FIGURE 2.3
Unbiased, precise, and accurate archers. Archer A is unbiased—the average position of all arrows is at the bull's-eye. Archer B is precise but not unbiased—all arrows are close together but systematically away from the bull's-eye. Archer C is accurate—all arrows are close together and near the center of the target.

You may be interested in many different population quantities from your population. Historically, however, the main impetus for developing theory for sample surveys has been estimating population means and totals. Suppose we want to estimate the total number of persons in Canada who have diabetes, or the average number of oranges produced per orange tree in Florida. The population total is

$$t = \sum_{i=1}^{N} y_i, \tag{2.7}$$

and the mean of the population (the subscript $\mathcal{U}$ always denotes a population quantity) is

$$\bar{y}_{\mathcal{U}} = \frac{1}{N} \sum_{i=1}^{N} y_i. \tag{2.8}$$

Almost all populations exhibit some variability; for example, households have different incomes and trees have different diameters. Define the **variance** of the population values about the mean as

$$S^2 = \frac{1}{N-1} \sum_{i=1}^{N} (y_i - \bar{y}_{\mathcal{U}})^2. \tag{2.9}$$

The population **standard deviation** is $S = \sqrt{S^2}$.

It is sometimes helpful to have a special notation for proportions. The proportion of units having a characteristic is simply a special case of the mean, obtained by letting $y_i = 1$ if unit i has the characteristic of interest, and $y_i = 0$ if unit i does not have the characteristic. Let

$$p = \frac{\text{number of units with the characteristic in the population}}{N}.$$

Example 2.4. For the population in Example 2.2, let

$$y_i = \begin{cases} 1 \text{ if unit } i \text{ has the value 7} \\ 0 \text{ if unit } i \text{ does not have the value 7} \end{cases}$$

Let $\hat{p}_{\mathcal{S}} = \sum_{i \in \mathcal{S}} y_i/4$, the proportion of 7s in the sample. The list of all possible samples in the data file `sample70.csv` has 5 samples with no 7s, 30 samples with exactly one 7, 30 samples with exactly two 7s, and 5 samples with three 7s. Since each of the possible samples is selected with probability 1/70, the sampling distribution of $\hat{p}$ is:

k	0	$\frac{1}{4}$	$\frac{1}{2}$	$\frac{3}{4}$
$P\{\hat{p} = k\}$	$\frac{5}{70}$	$\frac{30}{70}$	$\frac{30}{70}$	$\frac{5}{70}$

■

2.3 Simple Random Sampling

Simple random sampling is the most basic form of probability sampling, and provides the theoretical basis for the more complicated forms. There are two ways of taking a simple random sample: with replacement, in which the same unit may be included more than once in the sample, and without replacement, in which all units in the sample are distinct.

A **simple random sample with replacement** (SRSWR) of size n from a population of N units can be thought of as drawing n independent samples of size 1. One unit is randomly selected from the population to be the first sampled unit, with probability $1/N$. Then the sampled unit is replaced in the population, and a second unit is randomly selected with probability $1/N$. This procedure is repeated until the sample has n units, which may include duplicates from the population.

In finite population sampling, however, sampling the same person twice provides no additional information. We usually prefer to sample without replacement, so that the sample contains no duplicates. A **simple random sample without replacement** (SRS) of size n is selected so that every possible subset of n distinct units in the population has the same probability of being selected as the sample. There are $\binom{N}{n}$ possible samples (see Appendix A), and each is equally likely, so the probability of selecting any individual sample $\mathcal{S}$ of n units is

$$P(\mathcal{S}) = \frac{1}{\binom{N}{n}} = \frac{n!(N-n)!}{N!}. \tag{2.10}$$

As a consequence of this definition, the probability that population unit i appears in the sample is $\pi_i = n/N$, as shown in Section 2.9.

To take an SRS, you need a list of all observation units in the population; this list is the sampling frame. In an SRS, the sampling unit and observation unit coincide. Each unit is assigned a number, and a sample is selected so that each possible sample of size n has the same chance of being the sample actually selected. This can be thought of as drawing numbers out of a hat; in practice, computer-generated pseudo-random numbers are usually used to select a sample.

Example 2.5. *Selecting an SRS from a population.* One method for selecting an SRS of size n from a population of size N is to generate N random numbers between 0 and 1, then select the units corresponding to the n smallest random numbers to be the sample. For example, if $N = 10$ and $n = 4$, we generate 10 numbers between 0 and 1:

Unit i	1	2	3	4	5	6	7	8	9	10
Random number	0.837	0.636	0.465	0.609	0.154	0.766	0.821	0.713	0.987	0.469

The smallest 4 of the random numbers are 0.154, 0.465, 0.469, and 0.609, leading to the sample with units $\{3, 4, 5, 10\}$. Other methods that might be used to select an SRS are described in Example 2.6 and Exercises 31, 32, and 33. Lohr (2022) and Lu and Lohr (2022) show how to select an SRS in SAS and R software, respectively. ■

Example 2.6. The U.S. government conducts a Census of Agriculture every five years, collecting data on all farms (defined as any place from which \$1000 or more of agricultural products were produced and sold). The file `agpop.csv` contains historical information from 1982, 1987, and 1992 on the number of farms, total acreage devoted to farms, number of farms with fewer than 9 acres, and number of farms with more than 1000 acres for the population consisting of the $N = 3078$ counties and county-equivalents in the United States (U.S. Bureau of the Census, 1995).

To take an SRS of size 300 from this population, I generated 300 random numbers between 0 and 1 on the computer, multiplied each by 3078, and rounded the result up to the next highest integer. This procedure generates an SRSWR. If the population is large relative to the sample, it is likely that each unit in the sample only occurs once in the list. In this case, however, 13 of the 300 numbers were duplicates. The duplicates were discarded and replaced with new randomly generated numbers between 1 and 3078 until all 300 numbers were distinct; file `agsrs.csv` contains the data from the SRS.

The counties selected to be in the sample may not "feel" very random at first glance. For example, counties 2840, 2841, and 2842 are all in the sample while none of the counties between 2740 and 2787 appear. The sample contains 18% of Virginia counties but no counties in Alaska, Arizona, Connecticut, Delaware, Hawaii, Rhode Island, Utah, or Wyoming. There is a quite natural temptation to want to "adjust" the random number list, to spread it out a bit more. If you want a random sample, you must resist this temptation. Research, dating back to Neyman (1934), has repeatedly demonstrated that purposive samples often do not represent the population on key variables. If you deliberately substitute other counties for those in the randomly generated sample, you may be able to match the population on one particular characteristic such as geographic distribution; however, you will likely fail to match the population on characteristics of interest such as the number of farms or average farm size. If you want to ensure that all states are represented, do not adjust your randomly selected sample purposively but take a stratified sample (to be discussed in Chapter 3).

Let's look at the variable *acres92*, the number of acres devoted to farms in 1992. A small number of counties in the population are missing the value of *acres92* to prevent disclosing data on individual farms. Thus we first check to see the extent of the missing data in our sample. Fortunately, our sample has no missing data. Figure 2.4 displays a histogram of the acreage devoted to farms in each of the 300 counties. ■

Estimating a population mean. For estimating the population mean $\bar{y}_{\mathcal{U}}$ from an SRS, we use the sample mean

$$\bar{y}_{\mathcal{S}} = \frac{1}{n} \sum_{i \in \mathcal{S}} y_i. \tag{2.11}$$

In the following, we use $\bar{y}$ to refer to the sample mean and drop the subscript $\mathcal{S}$ unless it is needed for clarity. As will be shown in Section 2.9, $\bar{y}$ is an unbiased estimator of the

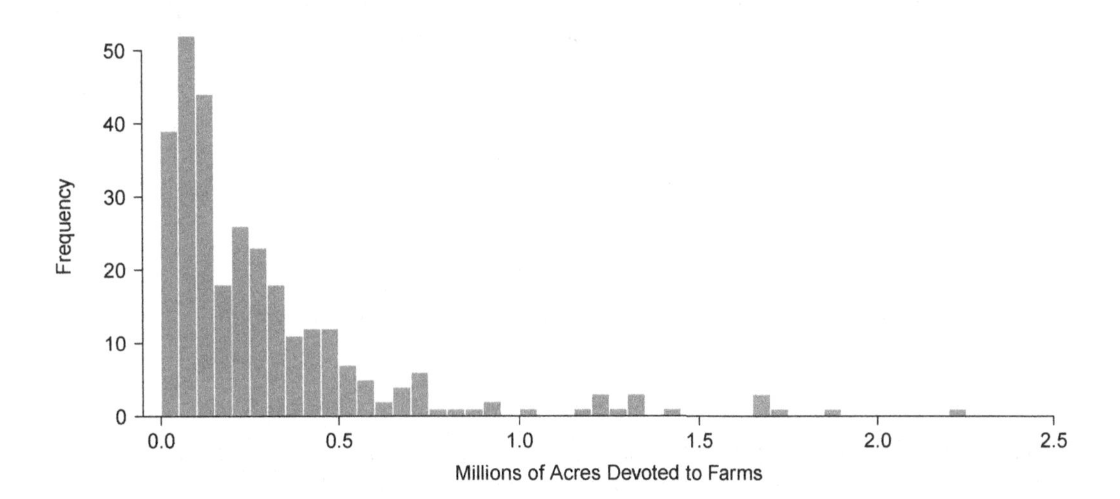

FIGURE 2.4
Histogram of number of acres devoted to farms in 1992, for an SRS of 300 counties. Note the skewness of the data. Most of the counties have fewer than 500,000 acres in farms; some counties, however, have more than 1.5 million acres in farms.

population mean $\bar{y}_{\mathcal{U}}$, and the variance of $\bar{y}$ is

$$V(\bar{y}) = \frac{S^2}{n}\left(1 - \frac{n}{N}\right) \tag{2.12}$$

for S^2 defined in (2.9). The variance $V(\bar{y})$ measures the variability among estimates of $\bar{y}_{\mathcal{U}}$ from different samples.

The factor $(1 - n/N)$ is called the **finite population correction** (fpc). Intuitively, we make this correction because with small populations the greater the **sampling fraction** n/N, the more information we have about the population and thus the smaller the variance. If $N = 10$ and we sample all 10 observations, we would expect the variance of $\bar{y}$ to be 0 (which it is). If $N = 10$, there is only one possible sample $\mathcal{S}$ of size 10 without replacement, with $\bar{y}_{\mathcal{S}} = \bar{y}_{\mathcal{U}}$, so there is no variability due to taking a sample. For a census, the fpc, and hence $V(\bar{y})$, is 0. When the sampling fraction n/N is large in an SRS without replacement, the sample is closer to a census, which has no sampling variability.

For most samples that are taken from extremely large populations, the fpc is approximately 1. For large populations it is the size of the sample taken, not the percentage of the population sampled, that determines the precision of the estimator: If your soup is well stirred, you need to taste only one or two spoonfuls to check the seasoning, whether you have made 1 liter or 20 liters of soup. A sample of size 100 from a population of 100,000 units has almost the same precision as a sample of size 100 from a population of 100 million units:

$$V[\bar{y}] = \frac{S^2}{100}\,\frac{99{,}900}{100{,}000} = \frac{S^2}{100}(0.999) \quad \text{for } N = 100{,}000$$

$$V[\bar{y}] = \frac{S^2}{100}\,\frac{99{,}999{,}900}{100{,}000{,}000} = \frac{S^2}{100}(0.999999) \quad \text{for } N = 100{,}000{,}000$$

The population variance S^2, which depends on the values for the entire population, is in general unknown. We estimate it by the sample variance:

$$s^2 = \frac{1}{n-1} \sum_{i \in \mathcal{S}} (y_i - \bar{y})^2. \tag{2.13}$$

An unbiased estimator of the variance of $\bar{y}$ is (see Section 2.9):

$$\hat{V}(\bar{y}) = \left(1 - \frac{n}{N}\right) \frac{s^2}{n}. \tag{2.14}$$

The **standard error** (SE) is the square root of the estimated variance of $\bar{y}$:

$$\text{SE}\ (\bar{y}) = \sqrt{\left(1 - \frac{n}{N}\right) \frac{s^2}{n}}. \tag{2.15}$$

The population standard deviation is often related to the mean. A population of trees might have a mean height of 10 meters and standard deviation of one meter. A population of small cacti, however, with a mean height of 10 cm, might have a standard deviation of 1 cm. The **coefficient of variation** (CV) of the estimator $\bar{y}$ in an SRS is a measure of relative variability, which may be defined when $\bar{y}_{\mathcal{U}} \neq 0$ as:

$$\text{CV}(\bar{y}) = \frac{\sqrt{V(\bar{y})}}{E(\bar{y})} = \sqrt{1 - \frac{n}{N}} \frac{S}{\sqrt{n}} \frac{1}{\bar{y}_{\mathcal{U}}}. \tag{2.16}$$

If tree height is measured in meters, then $\bar{y}_{\mathcal{U}}$ and S are also in meters. The CV does not depend on the unit of measurement. Our trees and cacti have the same CV for a sample of size n.

We can estimate $\text{CV}(\bar{y})$ using the standard error divided by the mean (defined only when the mean is nonzero). In an SRS,

$$\widehat{\text{CV}}(\bar{y}) = \frac{\text{SE}\ (\bar{y})}{\bar{y}} = \sqrt{1 - \frac{n}{N}} \frac{s}{\sqrt{n}} \frac{1}{\bar{y}}. \tag{2.17}$$

The estimated CV is thus the standard error expressed as a fraction of the mean.

Estimating a population total. A population total, t, can be expressed as the population mean multiplied by N:

$$t = \sum_{i=1}^{N} y_i = N \bar{y}_{\mathcal{U}}.$$

To estimate t, we use the unbiased estimator

$$\hat{t} = N \bar{y}. \tag{2.18}$$

Then, from (2.12),

$$V(\hat{t}) = N^2 V(\bar{y}) = N^2 \left(1 - \frac{n}{N}\right) \frac{S^2}{n} \tag{2.19}$$

and

$$\hat{V}(\hat{t}) = N^2 \left(1 - \frac{n}{N}\right) \frac{s^2}{n}. \tag{2.20}$$

Note that $\text{CV}(\hat{t}) = \sqrt{V(\hat{t})}/E(\hat{t})$ is the same as $\text{CV}(\bar{y})$ for an SRS.

Example 2.7. For the data in Example 2.6, $N = 3078$ and $n = 300$, so the sampling fraction is $300/3078 = 0.097$. The sample statistics are $\bar{y} = 297{,}897$, $s = 344{,}551.9$, and $\hat{t} = N\bar{y} = 916{,}927{,}110$. Standard errors are

$$\text{SE}\,(\bar{y}) = \sqrt{\frac{s^2}{n}\left(1 - \frac{300}{3078}\right)} = 18{,}898.434428$$

and

$$\text{SE}\,(\hat{t}) = (3078)(18{,}898.434428) = 58{,}169{,}381.$$

The estimated coefficient of variation is

$$\widehat{\text{CV}}\,(\bar{y}) = \frac{\text{SE}\,(\bar{y})}{\bar{y}} = \frac{18{,}898.434428}{297{,}897} = 0.06344.$$

Since these data are so highly skewed, we should also report the median number of farm acres in a county, which is 196,717. ■

Estimating a population proportion. We might also want to estimate the proportion of counties in Example 2.6 with fewer than 200,000 acres in farms. Since estimating a proportion is a special case of estimating a mean, the results in (2.11)–(2.17) hold for proportions as well, and they take a simple form. Suppose we want to estimate the proportion of units in the population that have some characteristic—call this proportion p. Define y_i to be 1 if the unit has the characteristic and to be 0 if the unit does not have that characteristic. Then $p = \sum_{i=1}^{N} y_i/N = \bar{y}_{\mathcal{U}}$, and p is estimated by $\hat{p} = \bar{y}$. Consequently, $\hat{p}$ is an unbiased estimator of p. For the response y_i, taking on values 0 or 1,

$$S^2 = \frac{\sum_{i=1}^{N}(y_i - p)^2}{N-1} = \frac{\sum_{i=1}^{N} y_i^2 - 2p\sum_{i=1}^{N} y_i + Np^2}{N-1} = \frac{N}{N-1}p(1-p).$$

Thus, (2.12) implies that

$$V(\hat{p}) = \left(\frac{N-n}{N-1}\right)\frac{p(1-p)}{n}. \tag{2.21}$$

Also,

$$s^2 = \frac{1}{n-1}\sum_{i\in\mathcal{S}}(y_i - \hat{p})^2 = \frac{n}{n-1}\hat{p}(1-\hat{p}),$$

so from (2.14),

$$\hat{V}(\hat{p}) = \left(1 - \frac{n}{N}\right)\frac{\hat{p}(1-\hat{p})}{n-1}. \tag{2.22}$$

Example 2.8. For the sample described in Example 2.6, the estimated proportion of counties with fewer than 200,000 acres in farms is

$$\hat{p} = \frac{153}{300} = 0.51$$

with standard error

$$\text{SE}(\hat{p}) = \sqrt{\left(1 - \frac{300}{3078}\right)\frac{(0.51)(0.49)}{299}} = 0.0275. \quad ■$$

2.4 Sampling Weights

In (2.2), we defined π_i to be the probability that unit i is included in the sample. In probability sampling, these inclusion probabilities are used to calculate point estimates such as $\hat{t}$ and $\bar{y}$. Define the **sampling weight**, sometimes called the design weight, to be the reciprocal of the inclusion probability:

$$w_i = \frac{1}{\pi_i}. \tag{2.23}$$

The sampling weight of sampled unit i can be interpreted as the number of population units represented by unit i.

In an SRS, each unit has inclusion probability $\pi_i = n/N$; consequently, all sampling weights are the same with $w_i = 1/\pi_i = N/n$. We can thus think of every unit in the sample as representing the same number of units, N/n, in the population—itself plus $N/n - 1$ of the unsampled units. We call such a sample, in which every unit has the same sampling weight, a **self-weighting** sample.

Recall from Chapter 1 that a sample is representative if it can be used to "reconstruct" what the population looks like, and provide an accurate measure of how close statistics calculated from the reconstructed population are likely to be to statistics calculated from the true, but unknown, population. Since we do not know the original population, we estimate it, conceptually, from the sample by creating w_i copies of unit i, for each unit in $\mathcal{S}$. In a self-weighting sample such as an SRS, we can think of the estimated population as consisting of N/n copies of the sample.

Of course, we do not create a physical data file with N entries. Instead, we use the sampling weights directly for all calculations. The population size is estimated by the sum of the weights for the sample,

$$\hat{N} = \sum_{i \in \mathcal{S}} w_i. \tag{2.24}$$

The population total for y is estimated by

$$\hat{t} = \sum_{i \in \mathcal{S}} w_i y_i, \tag{2.25}$$

and the population mean is estimated by

$$\bar{y} = \frac{\displaystyle\sum_{i \in \mathcal{S}} w_i y_i}{\displaystyle\sum_{i \in \mathcal{S}} w_i} = \frac{\hat{t}}{\hat{N}}. \tag{2.26}$$

Survey software packages use (2.25) and (2.26) to calculate $\hat{t}$ and $\bar{y}$ for all probability sampling designs.

For an SRS, where $w_i = N/n$, the expressions in (2.24) to (2.26) simplify to $\hat{N} = \sum_{i \in \mathcal{S}}(N/n) = N$, $\hat{t} = (N/n)\sum_{i \in \mathcal{S}} y_i$, and $\bar{y} = (1/n)\sum_{i \in \mathcal{S}} y_i$, the same expressions we had in Section 2.3.

Example 2.9. Let's look at the sampling weights for the sample described in Example 2.6. Here, $N = 3078$ and $n = 300$, so the sampling weight is $w_i = 3078/300 = 10.26$ for each

TABLE 2.1
Sampling weights and calculations for Example 2.9.

County	State	weight	acres92	lt200k	weight*acres92	weight*lt200k
Coffee	AL	10.26	175209	1	1797644.34	10.26
Colbert	AL	10.26	138135	1	1417265.10	10.26
Lamar	AL	10.26	56102	1	575606.52	10.26
Marengo	AL	10.26	199117	1	2042940.42	10.26
Marion	AL	10.26	89228	1	915479.28	10.26
Tuscaloosa	AL	10.26	96194	1	986950.44	10.26
Columbia	AR	10.26	57253	1	587415.78	10.26
Faulkner	AR	10.26	210692	0	2161699.92	0.00
⋮	⋮	⋮	⋮	⋮	⋮	⋮
Pleasants	WV	10.26	15650	1	160569.00	10.26
Putnam	WV	10.26	55827	1	572785.02	10.26
Sum		3078.00			916927109.60	1569.78

unit in the sample. The first county in the data file `agsrs.csv`, Coffee County, Alabama, thus represents itself and 9.26 counties from the 2778 counties not included in the sample.

All estimates of means and totals can be calculated using the sampling weights, as shown in Table 2.1. The sampling weight for each sampled county is in Column 3, with header *weight*. Column 6 is formed by multiplying columns *weight* and *acres92*, so the entries of Column 6 are $w_i y_i$ for y_i the value of *acres92*. We see that the sum of the values in Column 6 is $\sum_{i \in \mathcal{S}} w_i y_i$ = 916,927,110, which is the same value we obtained for the estimated population total in Example 2.6.

To calculate the estimated proportion from Example 2.8, form variable *lt200k*, which takes on value 1 if *acres92* $<$ 200,000 and takes on value 0 if *acres92* $\geq$ 200,000. Then multiply each value in Column 5 (*lt200k*) by the weight in Column 3 to form Column 7. The sum of the values in the *weight*lt200k* column, 1569.78, estimates the total number of counties with fewer than 200,000 acres in farms. We divide this by the sum of the weights to obtain $\hat{p} = 1569.78/3078 = 0.51$. ■

All SRSs are self-weighting, but not all self-weighting samples are SRSs. In an SRS, $\pi_i = n/N$ and $w_i = N/n$ for all units $i = 1, \ldots, N$. However, many other probability sampling designs also have $\pi_i = n/N$ and $w_i = N/n$ for all units but are not SRSs. To have an SRS, it is not sufficient for every individual to have the same probability of being in the sample; in addition, *every possible sample* of size n must have the same probability $1/\binom{N}{n}$ of being the sample selected, as defined in (2.10).

The cluster sampling design in Example 2.1, in which the population of 100 integers is divided into 20 clusters $\{1, 2, 3, 4, 5\}$, $\{6, 7, 8, 9, 10\}$, ..., $\{96, 97, 98, 99, 100\}$ and an SRS of 4 of these clusters selected, has $\pi_i = 20/100$ for each unit in the population but is *not* an SRS of size 20 because different possible samples of size 20 have different probabilities of being selected. To see this, let's look at two particular subsets of $\{1, 2, \ldots, 100\}$. Let $\mathcal{S}_1$ be the cluster sample depicted in the third panel of Figure 2.1, with

$$\mathcal{S}_1 = \{1, 2, 3, 4, 5, 46, 47, 48, 49, 50, 61, 62, 63, 64, 65, 81, 82, 83, 84, 85\},$$

and let

$$\mathcal{S}_2 = \{1, 6, 11, 16, 21, 26, 31, 36, 41, 46, 51, 56, 61, 66, 71, 76, 81, 86, 91, 96\}.$$

The cluster sampling design specifies taking an SRS of 4 of the 20 clusters, so

$$P(\mathcal{S}_1) = \frac{1}{\binom{20}{4}} = \frac{4!(20-4)!}{20!} = \frac{1}{4845}.$$

Sample $\mathcal{S}_2$ cannot occur under this design, however, so $P(\mathcal{S}_2) = 0$. An SRS with $n = 20$ from a population with $N = 100$ would have

$$P(\mathcal{S}) = \frac{1}{\binom{100}{20}} = \frac{20!(100-20)!}{100!} = \frac{1}{5.359834 \times 10^{20}}$$

for *every* subset $\mathcal{S}$ of size 20 from the population $\{1, 2, \ldots, 100\}$.

2.5 Confidence Intervals

When you take a sample survey, it is not sufficient to simply report the average height of trees or the sample proportion of voters who intend to vote for Candidate B in the next election. You also need to give an indication of how accurate your estimates are. **Confidence intervals** (CIs) indicate the accuracy of an estimate.

A 95% confidence interval is often explained heuristically: If we take samples from our population over and over again, and construct a confidence interval using our procedure for each possible sample, we expect 95% of the resulting intervals to include the true value of the population parameter. The expected proportion of intervals that include the true population value is called the **coverage probability** of the interval.

In probability sampling from a finite population, only a finite number of possible samples exist and we know the probability with which each will be chosen; if we were able to generate all possible samples from the population, we would be able to calculate the exact coverage probability (confidence level) for a confidence interval procedure.

Example 2.10. *Repeated sampling interpretation of a confidence interval.* Return to Example 2.2, in which the entire population is known. Let's choose an arbitrary procedure for calculating a confidence interval, constructing interval estimates for t as

$$\text{CI}(\mathcal{S}) = [\hat{t}_{\mathcal{S}} - 4s_{\mathcal{S}}, \hat{t}_{\mathcal{S}} + 4s_{\mathcal{S}}].$$

There is no theoretical reason to choose this procedure, but it will illustrate the concept of a confidence interval. Define $u(\mathcal{S})$ to be 1 if $\text{CI}(\mathcal{S})$ contains the true population value 40, and 0 if $\text{CI}(\mathcal{S})$ does not contain 40. Since we know the population, we can calculate the confidence interval $\text{CI}(\mathcal{S})$ and the value of $u(\mathcal{S})$ for each possible sample $\mathcal{S}$. Some of the 70 confidence intervals are shown in Table 2.2 (all entries in the table are rounded to two decimals).

Each individual confidence interval either does or does not contain the population total 40. The probability statement in the confidence interval is made about the collection of all possible samples; for this confidence interval procedure and population, the coverage probability is

$$\sum_{\mathcal{S}} P(\mathcal{S})u(\mathcal{S}) = 0.77.$$

TABLE 2.2
Confidence intervals for possible samples from a small population. The procedure in Example 2.10 produces 77% confidence intervals.

Sample $\mathcal{S}$	$y_i, i \in \mathcal{S}$	$\hat{t}_\mathcal{S}$	$s_\mathcal{S}$	CI($\mathcal{S}$)	$u(\mathcal{S})$
$\{1,2,3,4\}$	1,2,4,4	22	1.50	[16.00, 28.00]	0
$\{1,2,3,5\}$	1,2,4,7	28	2.65	[17.42, 38.58]	0
$\{1,2,3,6\}$	1,2,4,7	28	2.65	[17.42, 38.58]	0
$\{1,2,3,7\}$	1,2,4,7	28	2.65	[17.42, 38.58]	0
$\{1,2,3,8\}$	1,2,4,8	30	3.10	[17.62, 42.38]	1
$\{1,2,4,5\}$	1,2,4,7	28	2.65	[17.42, 38.58]	0
$\{1,2,4,6\}$	1,2,4,7	28	2.65	[17.42, 38.58]	0
$\{1,2,4,7\}$	1,2,4,7	28	2.65	[17.42, 38.58]	0
$\{1,2,4,8\}$	1,2,4,8	30	3.10	[17.62, 42.38]	1
$\{1,2,5,6\}$	1,2,7,7	34	3.20	[21.19, 46.81]	1
⋮	⋮	⋮	⋮	⋮	⋮
$\{2,3,4,8\}$	2,4,4,8	36	2.52	[25.93, 46.07]	1
$\{2,3,5,6\}$	2,4,7,7	40	2.45	[30.20, 49.80]	1
$\{2,3,5,7\}$	2,4,7,7	40	2.45	[30.20, 49.80]	1
$\{2,3,5,8\}$	2,4,7,8	42	2.75	[30.98, 53.02]	1
$\{2,3,6,7\}$	2,4,7,7	40	2.45	[30.20, 49.80]	1
⋮	⋮	⋮	⋮	⋮	⋮
$\{4,5,6,8\}$	4,7,7,8	52	1.73	[45.07, 58.93]	0
$\{4,5,7,8\}$	4,7,7,8	52	1.73	[45.07, 58.93]	0
$\{4,6,7,8\}$	4,7,7,8	52	1.73	[45.07, 58.93]	0
$\{5,6,7,8\}$	7,7,7,8	58	0.50	[56.00, 60.00]	0

That means that if we take an SRS of four elements without replacement from this population of eight elements, there is a 77% chance that our sample is one of the "good" ones whose confidence interval contains the true value 40. This procedure thus creates a 77% confidence interval.

Of course, in real life, we only take one sample and do not know the value of the population total t. Without further investigation, we have no way of knowing whether the sample we obtained is one of the "good" ones, such as $\mathcal{S} = \{2,3,5,6\}$, or one of the "bad" ones such as $\mathcal{S} = \{4,6,7,8\}$. The confidence interval gives us only a probabilistic statement of how often we expect to be right. ■

In practice, we do not know the values of statistics from all possible samples, so we cannot calculate the exact confidence coefficient for a procedure as done in Example 2.10. In your introductory statistics class, you relied largely on **asymptotic** (as the sample size goes to infinity) results to construct confidence intervals for an unknown mean μ. The central limit theorem says that if we have a simple random sample with replacement, then the probability distribution of $\sqrt{n}(\bar{y} - \mu)$ converges to a normal distribution as the sample size n approaches infinity.

In most sample surveys, though, we only have a finite population. To use asymptotic results in finite population sampling, we pretend that our population is itself part of a larger **superpopulation**; the superpopulation is itself a subset of a larger superpopulation, and so on until the superpopulations are as large as we could wish. Our population is embedded in a series of increasing finite populations. This embedding can give us properties such as consistency and asymptotic normality. One can imagine the superpopulations as "alternative universes" in which circumstances were slightly different (see Section 15.2).

Central limit theorem for an SRS. Hájek (1960) proved a central limit theorem for simple random sampling without replacement (also see Sections 2.8 and 4.4 of Lehmann, 1999). In practical terms, Hájek's theorem says that if certain technical conditions hold and if n, N, and $N-n$ are all "sufficiently large," then the sampling distribution of

$$\frac{\bar{y}-\bar{y}_{\mathcal{U}}}{\sqrt{\left(1-\frac{n}{N}\right)}\frac{S}{\sqrt{n}}}$$

is approximately normal (Gaussian) with mean 0 and variance 1. A large-sample $100(1-\alpha)\%$ CI for the population mean is

$$\left[\bar{y}-z_{\alpha/2}\sqrt{1-\frac{n}{N}}\frac{S}{\sqrt{n}},\ \bar{y}+z_{\alpha/2}\sqrt{1-\frac{n}{N}}\frac{S}{\sqrt{n}}\right], \tag{2.27}$$

where $z_{\alpha/2}$ is the $(1-\alpha/2)$th percentile of the standard normal distribution. In simple random sampling without replacement, 95% of the possible samples that could be chosen will give a 95% CI for $\bar{y}_{\mathcal{U}}$ that contains the true value of $\bar{y}_{\mathcal{U}}$. Usually, S is unknown, so in large samples s is substituted for S, giving the large-sample CI:

$$\left[\bar{y}-z_{\alpha/2}\text{SE}(\bar{y}),\ \bar{y}+z_{\alpha/2}\text{SE}(\bar{y})\right].$$

In practice, we often substitute $t_{\alpha/2,n-1}$, the $(1-\alpha/2)^{th}$ percentile of a t distribution with $n-1$ degrees of freedom (df), for $z_{\alpha/2}$. For large samples, $t_{\alpha/2,n-1}\approx z_{\alpha/2}$. In smaller samples, using $t_{\alpha/2,n-1}$ instead of $z_{\alpha/2}$ produces a wider CI. Most software packages use the following CI for the population mean from an SRS:

$$\left[\bar{y}-t_{\alpha/2,n-1}\sqrt{1-\frac{n}{N}}\frac{s}{\sqrt{n}}, \bar{y}+t_{\alpha/2,n-1}\sqrt{1-\frac{n}{N}}\frac{s}{\sqrt{n}}\right], \tag{2.28}$$

How large is "sufficiently large"? The imprecise term "sufficiently large" occurs in the central limit theorem because the adequacy of the normal approximation depends on n and on how closely the population $\{y_i, i=1,\ldots,N\}$ resembles a population generated from the normal distribution. The "magic number" of $n=30$, often cited in introductory statistics books as a sample size that is "sufficiently large" for the central limit theorem to apply, often does not suffice in finite population sampling problems. Many populations we sample are highly skewed—we may measure income, number of acres on a farm that are devoted to corn, or the concentration of mercury in Minnesota lakes. For all of these examples, we expect most of the observations to be relatively small, but a few to be very, very large, so that a smoothed histogram of the entire population would look like this:

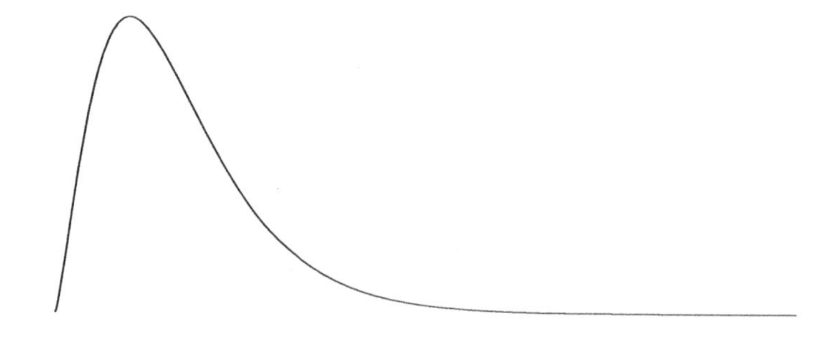

Thinking of observations as generated from some distribution is useful in deciding whether it is safe to use the central limit theorem. If you can think of the generating distribution as being somewhat close to normal, it is probably safe to use the central limit theorem with a sample size as small as 50. If the sample size is too small and the sampling distribution of $\bar{y}$ is not approximately normal, we need to use another method, relying on distributional assumptions, to obtain a CI for $\bar{y}_{\mathcal{U}}$. Such methods fit in with a model-based perspective for sampling (Section 2.10).

Sugden et al. (2000) recommended a minimum sample size of

$$n_{\min} = 28 + 25\left(\frac{\sum_{i=1}^{N}(y_i - \bar{y}_{\mathcal{U}})^3}{NS^3}\right)^2 \tag{2.29}$$

for the CI in (2.28) to have confidence level approximately equal to $1 - \alpha$. The quantity $\sum_{i=1}^{N}(y_i - \bar{y}_{\mathcal{U}})^3/(NS^3)$ is the skewness of the population; if the skewness is large, a large sample size is needed for the normal approximation to be valid. Another approach for considering whether the sample size is adequate for a normal approximation to be used is to look at a bootstrap approximation to the sampling distribution (see Exercise 27).

Example 2.11. The histogram in Figure 2.4 exhibited an underlying distribution for farm acreage that was far from normal. Is the sample size large enough to apply the central limit theorem? We substitute the sample values $s = 344{,}551.9$ and $\sum_{i\in\mathcal{S}}(y_i - \bar{y})^3/n = 1.05036 \times 10^{17}$ for the population quantities S and $\sum_{i=1}^{N}(y_i - \bar{y}_{\mathcal{U}})^3/N$ in (2.29), giving an estimated minimum sample size of

$$n_{\min} = 28 + 25\left[\frac{1.05036 \times 10^{17}}{(344{,}551.9)^3}\right]^2 \approx 193.$$

For this example, our sample of size 300 appears to be sufficiently large for the sampling distribution of $\bar{y}$ to be approximately normal.

For the data in Example 2.6, an approximate 95% CI for $\bar{y}_{\mathcal{U}}$, using $t_{\alpha/2,299} = 1.968$, is

$$[297{,}897 - (1.968)(18{,}898.434),\ 297{,}897 + (1.968)(18{,}898.434)] = [260{,}706,\ 335{,}088].$$

For the population total t, an approximate 95% CI is

$$\begin{aligned}[916{,}927{,}110 - 1.968(58{,}169{,}381),\ & 916{,}927{,}110 + 1.968(58{,}169{,}381)] \\ &= \left[8.02 \times 10^8, 1.03 \times 10^9\right].\end{aligned}$$

For estimating proportions, the usual criterion that the sample size is large enough to use the normal distribution if both $np \geq 5$ and $n(1-p) \geq 5$ is a useful guideline. (When that guideline is not met you may want to use the Clopper-Pearson CI discussed in Exercise 34.) An approximate 95% CI for the proportion of counties with fewer than 200,000 acres in farms is

$$0.51 \pm 1.968(0.0275), \text{ or } [0.456, 0.564].$$

To find a 95% CI for the total number of counties with fewer than 200,000 acres in farms, we simply multiply all quantities by N, so the point estimate is $3078(0.51) = 1570$, with standard error $3078 \times \text{SE}(\hat{p}) = 84.54$ and 95% CI [1403, 1736]. ■

2.6 Using Statistical Software to Analyze Survey Data

The examples in this book show the formulas, but in practice you should use a statistical software package to calculate estimates for data collected from a probability sample. The software performs the calculations for you, and is often written so as to avoid roundoff errors and to use numerically efficient and stable algorithms for the calculations.

Statistical software packages use the weight variable with formulas (2.25) and (2.26) to calculate $\hat{t}$ and $\bar{y}$, and they use the formula in (2.28), or a slight variation of that formula, to calculate CIs for an SRS. The format of the output varies across statistical packages, but for most packages it is easy to identify the estimates from the output.

Figure 2.5 shows the output from SAS software for the calculations in Examples 2.7 and 2.11 (see the supplementary book by Lohr, 2022, for the code that produces this output, and output for subsequent examples in this book). Here, *acres92* is the number of acres devoted to farms in 1992, and *lt200k* equals 1 if the county has fewer than 200,000 acres in farms and 0 if the county has 200,000 or more acres in farms. The output shows the estimated mean and population total for *acres92*, as well as estimated proportions (and total numbers) of population units having *lt200k* = 0 and *lt200k* = 1. These statistics are accompanied by their standard errors and 95% CIs.

The SURVEYMEANS Procedure

Data Summary	
Number of Observations	300
Sum of Weights	3078

Class Level Information		
Variable	Levels	Values
lt200k	2	0 1

Statistics

Variable	Level	Mean	Std Error of Mean	95% CL for Mean		Coeff of Variation	Sum	Std Error of Sum	95% CL for Sum	
acres92		297897	18898	260706.257	335087.836	0.063439	916927110	58169381	802453859	1031400361
lt200k	0	0.490000	0.027465	0.436	0.544	0.056051	1508.220000	84.537220	1342	1675
	1	0.510000	0.027465	0.456	0.564	0.053853	1569.780000	84.537220	1403	1736

FIGURE 2.5
Partial output from SAS software for calculations in Examples 2.7 and 2.11 (Lohr, 2022).

Figure 2.6 shows the analogous statistics that are calculated by the *survey* package (Lumley, 2020) in the R software environment. Lu and Lohr (2022) show how to produce this output and how to perform calculations in R for subsequent examples in this book.

2.7 Determining the Sample Size

An investigator often measures multiple variables and has a number of goals for a survey. Anyone designing an SRS must decide what amount of sampling error in the estimates is tolerable and must balance the precision of the estimates with the cost of the survey. Even

```
> dsrs <- svydesign(id = ~1, weights = ~sampwt, fpc = ~rep(3078,300),
                    data = agsrs)
> smean <- svymean(~acres92,dsrs)
> smean
          mean    SE
acres92 297897 18898

> confint(smean, df=299)
           2.5 %   97.5 %
acres92 260706.3 335087.8

> svymean(~factor(lt200k),dsrs)
                 mean     SE
factor(lt200k)0 0.49 0.0275
factor(lt200k)1 0.51 0.0275
```

FIGURE 2.6
Partial output from R for calculations in Examples 2.7 and 2.11 (Lu and Lohr, 2022).

though many variables may be measured, an investigator can often focus on one or two responses that are of primary interest in the survey, and use these for estimating a sample size.

For a single response, follow these steps to estimate the sample size:

1. Ask "What is expected of the sample, and how much precision do I need?" What are the consequences of the sample results? How much error is tolerable? If your survey measures the unemployment rate every month, you would like your estimates to have high precision so that you can detect changes in unemployment rates from month to month. A preliminary investigation, however, often needs less precision than an ongoing survey.

   Instead of asking about required precision, many people ask, "What percentage of the population should I include in my sample?" This is usually the wrong question to be asking. Except in very small populations, precision is obtained through the absolute size of the sample, not the proportion of the population covered. We saw in Section 2.3 that the fpc, which is the only place that the population size N occurs in the variance formula, has little effect on the variance of the estimator in large populations.

2. Find an equation relating the sample size n and your expectations of the sample.

3. Estimate any unknown quantities and solve for n.

4. If you are relatively new at designing surveys, you will find at this point that the sample size you calculated in Step 3 is much larger than you can afford. Go back and adjust some of your expectations for the survey and try again. In some cases, you will find that you cannot even come close to the precision you need with the resources you have available; in that case, perhaps you should consider whether you should even conduct your study.

Specify the tolerable error. Only the investigators in the study can say how much precision is needed. The desired precision is often expressed in absolute terms, as

$$P\left(|\bar{y} - \bar{y}_{\mathcal{U}}| \leq e\right) = 1 - \alpha.$$

The investigator must decide on reasonable values for α and e; e is called the **margin of error**. The margin of error is one-half of the width of a 95% CI. For many surveys of people in which a proportion is measured, the margin of error is set at $e = 0.03$ and $\alpha = 0.05$.

Sometimes you would like to achieve a desired relative precision, controlling the CV in (2.16) rather than the absolute error. In that case, if $\bar{y}_{\mathcal{U}} \neq 0$ the precision may be expressed as

$$P\left(\left|\frac{\bar{y} - \bar{y}_{\mathcal{U}}}{\bar{y}_{\mathcal{U}}}\right| \leq r\right) = 1 - \alpha.$$

Find an equation. The simplest equation relating the precision and sample size comes from the confidence intervals in the previous section. To obtain absolute precision e, find a value of n that satisfies

$$e = z_{\alpha/2}\sqrt{\left(1 - \frac{n}{N}\right)}\frac{S}{\sqrt{n}}.$$

To solve this equation for n, we first find the sample size n_0 that we would use for an SRSWR:

$$n_0 = \left(\frac{z_{\alpha/2}S}{e}\right)^2. \tag{2.30}$$

Then (see Exercise 9) the desired sample size is

$$n = \frac{n_0}{1 + \dfrac{n_0}{N}} = \frac{z^2_{\alpha/2}S^2}{e^2 + \dfrac{z^2_{\alpha/2}S^2}{N}}. \tag{2.31}$$

Of course, if $n_0 \geq N$ we simply take a census with $n = N$.

In surveys in which one of the main responses of interest is a proportion, it is often easiest to use that response when setting the sample size. For large populations, $S^2 \approx p(1-p)$, which attains its maximal value when $p = 1/2$. Thus, using $n_0 = 1.96^2/(4e^2)$ will result in a 95% CI with width at most $2e$.

To calculate a sample size to obtain a specified relative precision, substitute $r\bar{y}_{\mathcal{U}}$ for e in (2.30) and (2.31). This results in sample size

$$n = \frac{z^2_{\alpha/2}S^2}{(r\bar{y}_{\mathcal{U}})^2 + \dfrac{z^2_{\alpha/2}S^2}{N}} = \frac{z^2_{\alpha/2}(S/\bar{y}_{\mathcal{U}})^2}{r^2 + \dfrac{z^2_{\alpha/2}(S/\bar{y}_{\mathcal{U}})^2}{N}}. \tag{2.32}$$

To achieve a specified relative precision, the sample size may be determined using only the ratio $S/\bar{y}_{\mathcal{U}}$, the CV for a sample of size 1.

Example 2.12. Suppose we want to estimate the proportion of recipes in a cookbook that do not involve animal products. We plan to take an SRS of the $N = 1251$ recipes in the cookbook, and want to use a 95% CI with margin of error 0.03. Then,

$$n_0 = \frac{(1.96)^2\left(\frac{1}{2}\right)\left(1 - \frac{1}{2}\right)}{(0.03)^2} \approx 1067.$$

The sample size ignoring the fpc is large compared with the population size, so in this case we would make the fpc adjustment and use

$$n = \frac{n_0}{1 + \dfrac{n_0}{N}} = \frac{1067}{1 + \dfrac{1067}{1251}} = 576. \quad \blacksquare$$

In Example 2.12, the fpc makes a difference in the sample size because N is only 1251. If N is large, however, typically n_0/N will be very small so that for large populations we usually have $n \approx n_0$. Thus, we need approximately the same sample size for any large population—whether that population has 10 million or 1 billion or 100 billion units.

Example 2.13. Many public opinion polls specify using a sample size of about 1100. That number comes from rounding the value of n_0 in Example 2.12 up to the next hundred, and then noting that the population size is so large relative to the sample that the fpc should be ignored. For large populations, it is the size of the sample, not the proportion of the population that is sampled, that determines the precision. ■

Example 2.14. Why does the American Community Survey (ACS) need a sample of 3.5 million households, when the margin of error for an estimated proportion from a much smaller SRS of size 40,000 is less than 0.005, half of one percentage point? If the survey's only goal were to produce statistics for the country as a whole, a much smaller sample would suffice. But the survey also produces detailed estimates for subareas of the country, and the sample size in each subarea must be large enough to produce estimates of a specified precision for that subarea. ■

Estimate unknown quantities. When interested in a proportion, we can use 1/4 as an upper bound for S^2. For other quantities, S^2 must be estimated or guessed at. Some methods for estimating S^2 include:

1. Use sample quantities obtained when pretesting your survey. This is probably the best method, as your pretest should be similar to the survey you take. A **pilot sample**, a small sample taken to provide information and guidance for the design of the main survey, can be used to estimate quantities needed for setting the sample size.

2. Use previous studies or data available in the literature. You are rarely the first person in the world to study anything related to your investigation. You may be able to find estimates of variances that have been published in related studies, and use these as a starting point for estimating your sample size. But you have no control over the quality or design of those studies, and their estimates may be unreliable or may not apply for your study. In addition, estimates may change over time and vary in different geographic locations.

   Sometimes you can use the CV for a sample of size 1, the ratio of the standard deviation to the mean, to obtain an estimate of variability. The CV of a quantity is a measure of relative error, and tends to be more stable over time and location than the variance. If we take a random sample of houses for sale in the United States today, we will find that the variability in price will be much greater than if we had taken a similar survey in 1930. But the average price of a house has also increased from 1930 to today. We would probably find that the CV today is close to the CV in 1930.

3. If nothing else is available, guess the variance. Sometimes a hypothesized distribution of the data will give us information about the variance. For example, if you believe the population to be normally distributed, you may not know what the variance is, but you may have an idea of the range of the data. You could then estimate S by range/4 or range/6, as approximately 95% of values from a normal population are within 2 standard deviations of the mean, and 99.7% of the values are within 3 standard deviations of the mean.

Example 2.15. Before taking the sample of size 300 in Example 2.6, we looked at a pilot sample of size 30 from the population. One county in the pilot sample of size 30 was missing the value of *acres92*; the sample standard deviation of the remaining 29 observations was 519,085. Using this value, and a desired margin of error of 60,000,

$$n_0 = (1.96)^2 \frac{519{,}085^2}{60{,}000^2} = 288.$$

We took a sample of size 300 in case the estimated standard deviation from the pilot sample is too low. Also, we ignored the fpc in the sample size calculations; in most populations, the fpc will have little effect on the sample size.

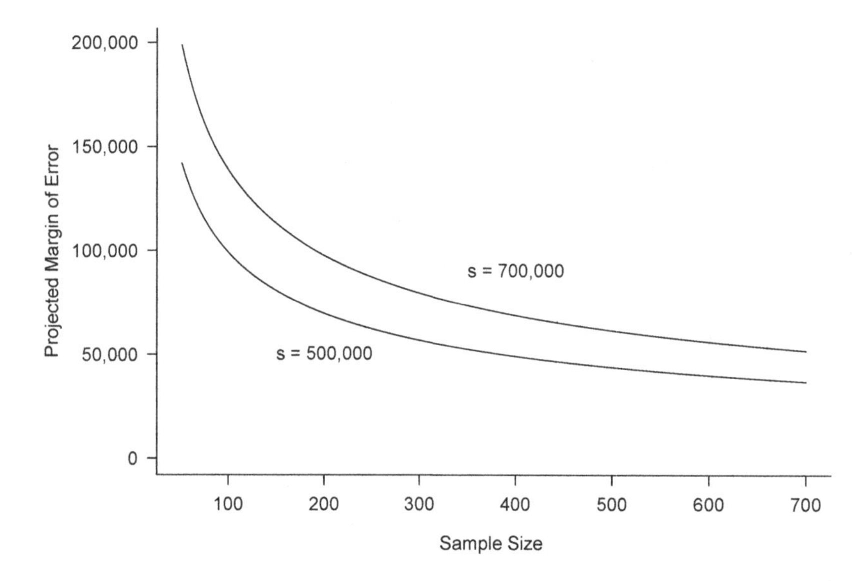

FIGURE 2.7
Plot of $t_{0.025,n-1}s/\sqrt{n}$ vs. n, for two possible values of the standard deviation s.

You may also view possible consequences of different sample sizes graphically. Figure 2.7 shows the value of $t_{0.025,n-1}\, s/\sqrt{n}$, for a range of sample sizes between 50 and 700, and for two possible values of the standard deviation s. The plot shows that if we ignore the fpc and if the standard deviation is about 500,000, a sample of size 300 will give a margin of error of about 60,000. ■

Determining the sample size is one of the early steps that must be taken in an investigation, and no magic formula will tell you the perfect sample size for your investigation (you only know that in hindsight, after you have completed the study!). Choosing a sample size is somewhat like deciding how much food to take on a picnic. You have a rough idea of how many people will attend, but do not know how much food you should have brought until after the picnic is over. You also need to bring extra food to allow for unexpected happenings, such as 2-year-old Freddie feeding a bowl of potato salad to the ducks or cousin Ted bringing along some extra guests. But you do not want to bring too much extra food, or it will spoil and you will have wasted money. Of course, the more picnics you have organized, and the better acquainted you are with the picnic guests, the better you become at bringing the right amount of food. It is comforting to know that the same is true of determining sample sizes—experience and knowledge about the population make you much better at designing surveys.

The results in this section can give you some guidance in choosing the size of the sample, but the final decision is up to you. In general, the larger the sample, the smaller the sampling error. Remember, though, that in most surveys you also need to worry about nonsampling errors, and need to budget resources to control selection and measurement bias. In many cases, nonsampling errors are greater when a larger sample is taken—with a large sample, it is easy to introduce additional sources of error (for example, it becomes more difficult to control the quality of the interviewers or to follow up on nonrespondents) or to become more relaxed about selection bias. In Chapter 16, we shall revisit the issue of designing a sample with the aim of reducing nonsampling as well as sampling error.

2.8 Systematic Sampling

Sometimes **systematic sampling** is used as a proxy for simple random sampling, when no list of the population exists or when the list is in roughly random order. To obtain a systematic sample, choose a sample size n. If N/n is an integer, let $k = N/n$; otherwise, let k be the next integer after N/n. Then find a random integer R between 1 and k, which determines the sample to be the units numbered $R, R + k, R + 2k, \ldots, R + (n - 1)k$. For example, to select a systematic sample of 45 students from the list of 45,000 students at a university, the sampling interval k is 1000. Suppose the random integer we choose is 597. Then the students numbered 597, 1597, 2597, ..., 44,597 would be in the sample.

If the list of students is ordered by randomly generated student identification numbers, we shall probably obtain a sample that will behave much like an SRS—it is unlikely that a person's position in the list is associated with the characteristic of interest. However, systematic sampling is not the same as simple random sampling; it does not have the property that every possible group of n units has the same probability of being the sample. In the example above, it is impossible to have students 345 and 346 both appear in the sample. Systematic sampling is technically a form of cluster sampling, as will be discussed in Chapter 5.

If the population is in random order, the systematic sample will behave much like an SRS, and SRS methods can be used in the analysis. The population itself can be thought of as being mixed. In the quote at the beginning of this chapter, Sorensen reported that President Kennedy used to read a systematic sample of letters written to him at the White House. This systematic sample most likely behaved much like a random sample. Note that Kennedy was well aware that the letters he read, while representative of letters written to the White House, were not at all representative of public opinion.

Systematic sampling does not necessarily give a representative sample, though, if the listing of population units is in some periodic or cyclical order. If male and female names alternate in the list, for example, and k is even, the systematic sample will contain either all men or all women—this cannot be considered a representative sample. In ecological surveys done on agricultural land, a ridge-and-furrow topography may be present that would lead to a periodic pattern of vegetation. If a systematic sampling scheme follows the same cycle, the sample will not behave like an SRS.

On the other hand, some populations are in increasing or decreasing order. A list of accounts receivable may be ordered from largest amount to smallest amount. In this case, estimates from the systematic sample may have smaller variance than comparable estimates from the SRS. A systematic sample from an ordered list of accounts receivable is forced to contain some large amounts and some small amounts. It is possible for an SRS to contain all small amounts or all large amounts, so there may be more variability among the sample means of all possible SRSs than there is among the sample means of all possible systematic samples.

In systematic sampling, we must still have a sampling frame and be careful when defining the target population. Sampling every 20th student to enter the library will not give a representative sample of the student body. Sampling every 10th person exiting an airplane, though, will probably give a representative sample of the persons on that flight. The sampling frame for the airplane passengers is not written down, but it exists all the same.

2.9 Randomization Theory for Simple Random Sampling*

In this section, we show that $\bar{y}$ is an unbiased estimator of $\bar{y}_{\mathcal{U}}$. We also calculate the variance of $\bar{y}$ given in (2.12), and show that the variance estimator in (2.14) is unbiased over repeated sampling. [Sections with asterisks involve the theory of sampling, and require more familiarity with probability theory and mathematical statistics than other sections.]

No distributional assumptions are made about the y_is in order to ascertain that $\bar{y}$ is unbiased for estimating $\bar{y}_{\mathcal{U}}$. We do not, for instance, assume that the y_is are normally distributed with mean μ. In the **randomization theory** (also called **design-based**) approach to sampling, the y_is are considered to be fixed but unknown numbers—the random variables used in randomization theory inference indicate which population units are in the sample.

Let's see how the randomization theory works for deriving properties of the sample mean in simple random sampling. As in Cornfield (1944), define

$$Z_i = \begin{cases} 1 & \text{if unit } i \text{ is in the sample} \\ 0 & \text{otherwise} \end{cases} . \tag{2.33}$$

Then

$$\bar{y} = \sum_{i \in \mathcal{S}} \frac{y_i}{n} = \sum_{i=1}^{N} Z_i \frac{y_i}{n}. \tag{2.34}$$

The Z_is are the *only* random variables in (2.34) because, according to randomization theory, the y_is are fixed quantities. When we choose an SRS of n units out of the N units in the population, $\{Z_1, \ldots, Z_N\}$ are identically distributed Bernoulli random variables with

$$\pi_i = P(Z_i = 1) = P(\text{select unit } i \text{ in sample}) = \frac{n}{N} \tag{2.35}$$

and

$$P(Z_i = 0) = 1 - \pi_i = 1 - \frac{n}{N}.$$

The probability in (2.35) follows from the definition of an SRS. To see this, note that if unit i is in the sample, then the other $n-1$ units in the sample must be chosen from the other $N-1$ units in the population. A total of $\binom{N-1}{n-1}$ possible samples of size $n-1$ may be drawn from a population of size $N-1$, so

$$P(Z_i = 1) = \frac{\text{number of samples including unit } i}{\text{number of possible samples}} = \frac{\binom{N-1}{n-1}}{\binom{N}{n}} = \frac{n}{N}.$$

As a consequence of (2.35),

$$E\left[Z_i\right] = E\left[Z_i^2\right] = \frac{n}{N}$$

and

$$E[\bar{y}] = E\left[\sum_{i=1}^{N} Z_i \frac{y_i}{n}\right] = \sum_{i=1}^{N} E[Z_i]\frac{y_i}{n} = \sum_{i=1}^{N} \frac{n}{N}\frac{y_i}{n} = \sum_{i=1}^{N} \frac{y_i}{N} = \bar{y}_{\mathcal{U}}. \quad (2.36)$$

This shows that $\bar{y}$ is an unbiased estimator of $\bar{y}_{\mathcal{U}}$. Note that in (2.36), the random variables are $Z_1, \ldots, Z_N$; $y_1, \ldots, y_N$ are treated as constants.

The variance of $\bar{y}$ is also calculated using properties of the random variables $Z_1, \ldots Z_N$. Note that

$$\mathrm{V}(Z_i) = E[Z_i^2] - (E[Z_i])^2 = \frac{n}{N} - \left(\frac{n}{N}\right)^2 = \frac{n}{N}\left(1 - \frac{n}{N}\right).$$

For $i \neq j$,

$$\begin{aligned} E[Z_i Z_j] &= P(Z_i = 1 \text{ and } Z_j = 1) \\ &= P(Z_j = 1 \mid Z_i = 1)P(Z_i = 1) \\ &= \left(\frac{n-1}{N-1}\right)\left(\frac{n}{N}\right). \end{aligned}$$

Because the population is finite, the Z_is are not quite independent—if we know that unit i is in the sample, we do have a small amount of information about whether unit j is in the sample, reflected in the conditional probability $P(Z_j = 1 \mid Z_i = 1)$. Consequently, for $i \neq j$, the covariance of Z_i and Z_j is:

$$\begin{aligned} \mathrm{Cov}(Z_i, Z_j) &= E[Z_i Z_j] - E[Z_i]E[Z_j] \\ &= \frac{n-1}{N-1}\frac{n}{N} - \left(\frac{n}{N}\right)^2 \\ &= -\frac{1}{N-1}\left(1 - \frac{n}{N}\right)\left(\frac{n}{N}\right). \end{aligned}$$

The negative covariance of Z_i and Z_j is the source of the fpc. The following derivation shows how we can use the random variables $Z_1, \ldots, Z_N$ and the properties of covariances given in Appendix A to find $V(\bar{y})$:

$$\begin{aligned} V(\bar{y}) &= \frac{1}{n^2} V\left(\sum_{i=1}^{N} Z_i y_i\right) \\ &= \frac{1}{n^2}\mathrm{Cov}\left(\sum_{i=1}^{N} Z_i y_i, \sum_{j=1}^{N} Z_j y_j\right) \\ &= \frac{1}{n^2}\sum_{i=1}^{N}\sum_{j=1}^{N} y_i y_j \,\mathrm{Cov}(Z_i, Z_j) \\ &= \frac{1}{n^2}\left[\sum_{i=1}^{N} y_i^2 V(Z_i) + \sum_{i=1}^{N}\sum_{j \neq i}^{N} y_i y_j \,\mathrm{Cov}(Z_i, Z_j)\right] \\ &= \frac{1}{n^2}\left[\frac{n}{N}\left(1 - \frac{n}{N}\right)\sum_{i=1}^{N} y_i^2 - \frac{1}{N-1}\left(1 - \frac{n}{N}\right)\left(\frac{n}{N}\right)\sum_{i=1}^{N}\sum_{j \neq i}^{N} y_i y_j\right] \\ &= \frac{1}{n^2}\frac{n}{N}\left(1 - \frac{n}{N}\right)\left[\sum_{i=1}^{N} y_i^2 - \frac{1}{N-1}\sum_{i=1}^{N}\sum_{j \neq i}^{N} y_i y_j\right] \end{aligned}$$

$$= \frac{1}{n}\left(1-\frac{n}{N}\right)\frac{1}{N(N-1)}\left[(N-1)\sum_{i=1}^{N} y_i^2 - \left(\sum_{i=1}^{N} y_i\right)^2 + \sum_{i=1}^{N} y_i^2\right]$$

$$= \frac{1}{n}\left(1-\frac{n}{N}\right)\frac{1}{N(N-1)}\left[N\sum_{i=1}^{N} y_i^2 - \left(\sum_{i=1}^{N} y_i\right)^2\right]$$

$$= \left(1-\frac{n}{N}\right)\frac{S^2}{n}.$$

To show that the estimator in (2.14) is an unbiased estimator of the variance, we need to show that $E[s^2] = S^2$. The argument proceeds much like the previous one. Since $S^2 = \sum_{i=1}^{N}(y_i - \bar{y}_{\mathcal{U}})^2/(N-1)$, it makes sense when trying to find an unbiased estimator to find the expected value of $\sum_{i \in \mathcal{S}}(y_i - \bar{y})^2$, and then find the multiplicative constant that will give the unbiasedness:

$$\begin{aligned}
E\left[\sum_{i\in\mathcal{S}}(y_i-\bar{y})^2\right] &= E\left[\sum_{i\in\mathcal{S}}\left\{(y_i-\bar{y}_{\mathcal{U}})-(\bar{y}-\bar{y}_{\mathcal{U}})\right\}^2\right] \\
&= E\left[\sum_{i\in\mathcal{S}}(y_i-\bar{y}_{\mathcal{U}})^2 - n\left(\bar{y}-\bar{y}_{\mathcal{U}}\right)^2\right] \\
&= E\left[\sum_{i=1}^{N} Z_i(y_i-\bar{y}_{\mathcal{U}})^2\right] - n\,\mathrm{V}(\bar{y}) \\
&= \frac{n}{N}\sum_{i=1}^{N}(y_i-\bar{y}_{\mathcal{U}})^2 - \left(1-\frac{n}{N}\right)S^2 \\
&= \frac{n(N-1)}{N}S^2 - \frac{N-n}{N}S^2 \\
&= (n-1)S^2.
\end{aligned}$$

Thus,

$$E\left[\frac{1}{n-1}\sum_{i\in\mathcal{S}}(y_i-\bar{y})^2\right] = E\left[s^2\right] = S^2.$$

2.10 Model-Based Theory for Simple Random Sampling*

Unless you have studied randomization theory for the design of experiments (see, for example, Kempthorne, 1952; Box et al., 1978; Oehlert, 2000), the proofs in the preceding section probably seemed strange to you. The observations y_i in randomization theory are fixed constants, not random variables. The random variables $Z_1, \ldots, Z_N$ are indicator variables that tell us whether the ith unit is in the sample or not. In a design-based, or randomization-theory, approach to sampling inference, the only relationship between units sampled and units not sampled is that the nonsampled units could have been sampled had we used a different starting value for the random number generator.

In Section 2.9 we found properties of the sample mean $\bar{y}$ using randomization theory: $y_1, y_2, \ldots, y_N$ were considered to be fixed values, and $\bar{y}$ is unbiased because $\bar{y} =$

$(1/n)\sum_{i=1}^{N} Z_i y_i$ and $E[Z_i] = P(Z_i = 1) = n/N$. The only probabilities used in finding the expected value and variance of $\bar{y}$ are the probabilities that units are included in the sample. The quantity measured on unit i, y_i can be anything: Whether y_i is number of television sets owned, systolic blood pressure, or acreage devoted to soybeans, the properties of estimators depend exclusively on the joint distribution of the random variables $\{Z_1, \ldots, Z_N\}$.

Model-based inference. In your other statistics classes, you most likely learned a different approach to inference. There, you had random variables $\{Y_i\}$ that followed some probability distribution, and the actual sample values were realizations of those random variables. Thus you assumed, for example, that $Y_1, Y_2, \ldots, Y_n$ were independent and identically distributed from a normal distribution with mean μ and variance σ^2, and used properties of independent random variables and the normal distribution to find expected values of various statistics.

We can extend this approach to sampling by thinking of random variables $Y_1, Y_2, \ldots, Y_N$ generated from some model. The actual values for the finite population, $y_1, y_2, \ldots, y_N$, are one realization of the random variables. The joint probability distribution of $Y_1, Y_2, \ldots, Y_N$ supplies the link between units in the sample and units not in the sample in this **model-based** approach—a link that is missing in the randomization approach. Here, we sample $\{y_i, i \in \mathcal{S}\}$ and use these data to predict the unobserved values $\{y_i, i \notin \mathcal{S}\}$. Thus, problems in finite population sampling may be thought of as prediction problems.

In an SRS, a simple model to adopt is:

$$Y_1, Y_2, \ldots, Y_N \text{ independent with } E_M[Y_j] = \mu \text{ and } V_M[Y_j] = \sigma^2. \tag{2.37}$$

The subscript M indicates that the expectation uses the model, not the randomization distribution used in Section 2.9. Here, μ and σ^2 represent unknown infinite population parameters, not the finite population quantities in Section 2.9. This model makes assumptions about the observations not in the sample; namely, that they have the same mean and variance as observations that are in the sample. We take a sample $\mathcal{S}$ and observe the values y_i for $i \in \mathcal{S}$. That is, we see realizations of the random variables Y_i for $i \in \mathcal{S}$. The other observations in the population $\{y_i, i \notin \mathcal{S}\}$ are also realizations of random variables, but we do not see those. The finite population total t for our sample can be written as

$$t = \sum_{i=1}^{N} y_i = \sum_{i \in \mathcal{S}} y_i + \sum_{i \notin \mathcal{S}} y_i \tag{2.38}$$

and is one possible value that can be taken on by the random variable

$$T = \sum_{i=1}^{N} Y_i = \sum_{i \in \mathcal{S}} Y_i + \sum_{i \notin \mathcal{S}} Y_i.$$

We observe the values in the sample, $\{y_i, i \in \mathcal{S}\}$. To estimate t, we need to predict values for the y_is not in the sample. This is where our model of the common mean μ comes in. The least squares estimator of μ from the sample is $\bar{Y}_{\mathcal{S}} = \sum_{i \in \mathcal{S}} Y_i/n$, and this is the best linear unbiased predictor (under the model) of each unobserved random variable, so that $\hat{Y}_i = \bar{Y}_{\mathcal{S}}$ and

$$\hat{T} = \sum_{i \in \mathcal{S}} Y_i + \sum_{i \notin \mathcal{S}} \hat{Y}_i = \sum_{i \in \mathcal{S}} Y_i + \frac{N-n}{n} \sum_{i \in \mathcal{S}} Y_i = \frac{N}{n} \sum_{i \in \mathcal{S}} Y_i.$$

The estimator $\hat{T}$ is **model-unbiased**: if the model is correct, then the average of $\hat{T} - T$ over repeated realizations of the population is

$$E_M[\hat{T} - T] = \frac{N}{n} \sum_{i \in \mathcal{S}} E_M[Y_i] - \sum_{i=1}^{N} E_M[Y_i] = 0.$$

Notice the difference between finding expectations under the model-based approach and under the design-based approach. In the model-based approach, the Y_is are the random variables, and the sample has no information for calculating expected values, so we can take the sum $\sum_{i \in \mathcal{S}}$ outside of the expected value. In the design-based approach, the random variables are implicitly contained in the sample $\mathcal{S}$ because $\sum_{i \in \mathcal{S}} y_i = \sum_{i=1}^{N} Z_i y_i$.

The mean squared error is also calculated as the average squared deviation between the estimate and the finite population total. For any given realization of the random variables, the squared error is

$$(\hat{t} - t)^2 = \left[\frac{N}{n} \sum_{i \in \mathcal{S}} y_i - \sum_{i=1}^{N} y_i\right]^2.$$

Averaging this quantity over all possible realizations of the random variables gives the mean squared error under the model assumptions:

$$\begin{aligned}
E_M[(\hat{T} - T)^2] &= E_M\left[\left(\frac{N}{n} \sum_{i \in \mathcal{S}} Y_i - \sum_{i=1}^{N} Y_i\right)^2\right] \\
&= E_M\left[\left\{\left(\frac{N}{n} - 1\right) \sum_{i \in \mathcal{S}} Y_i - \sum_{i \notin \mathcal{S}} Y_i\right\}^2\right] \\
&= E_M\left[\left\{\left(\frac{N}{n} - 1\right) \sum_{i \in \mathcal{S}} Y_i - \sum_{i \notin \mathcal{S}} Y_i - \left(\frac{N}{n} - 1\right) n\mu + (N - n)\mu\right\}^2\right] \\
&= E_M\left[\left(\frac{N}{n} - 1\right)^2 \left(\sum_{i \in \mathcal{S}} Y_i - n\mu\right)^2 + \left(\sum_{i \notin \mathcal{S}} Y_i - (N - n)\mu\right)^2\right] \\
&= \left(\frac{N}{n} - 1\right)^2 n\sigma^2 + (N - n)\sigma^2 \\
&= N^2 \left(1 - \frac{n}{N}\right) \frac{\sigma^2}{n}.
\end{aligned}$$

In practice, if the model in (2.37) were adopted, you would estimate σ^2 by the sample variance s^2. Thus the design-based approach and the model-based approach—with the model in (2.37)—lead to the same estimator of the population total and the same variance estimator. If a different model were adopted, however, the estimators might differ. We shall see in Chapter 4 how a design-based approach and a model-based approach can lead to different inferences.

The design-based approach and the model-based approach with model (2.37) also lead to the same CI for the mean. These CIs have different interpretations, however. The design-based CI for $\bar{y}_{\mathcal{U}}$ may be interpreted as follows: If we take all possible SRSs of size n from the finite population of size N, and construct a 95% CI for each sample, we expect 95% of all the CIs constructed to include the true population value $\bar{y}_{\mathcal{U}}$. Thus, the design-based CI has a *repeated sampling* interpretation. Statistical inference is based on repeated sampling from the finite population.

To construct CIs in the model-based approach, we rely on a central limit theorem that states that if n/N is small, the standardized prediction error

$$\frac{\hat{T} - T}{\sqrt{E_M[(\hat{T} - T)^2]}}$$

converges to a standard normal distribution (Valliant et al., 2000, Section 2.5). For the model in (2.37), with $E_M[(\hat{T} - T)^2] = N^2 \left(1 - n/N\right) \sigma^2/n$, this central limit theorem says that for sufficiently large sample sizes,

$$P\left[\hat{T} - z_{\alpha/2} N \sqrt{\left(1 - \frac{n}{N}\right)\frac{\sigma^2}{n}} \leq T \leq \hat{T} + z_{\alpha/2} N \sqrt{\left(1 - \frac{n}{N}\right)\frac{\sigma^2}{n}}\right] \approx 1 - \alpha.$$

Substituting the sample standard deviation s for σ, we get the large-sample CI

$$\hat{T} \pm z_{\alpha/2} N \sqrt{\left(1 - \frac{n}{N}\right)\frac{s^2}{n}},$$

which has the same form as the design-based CI for the population total t. This model-based CI is also interpreted using repeated sampling ideas, but in a different way than in Section 2.9. The design-based confidence level gives the expected proportion of CIs that will include the true finite population total $t = \sum_{i=1}^{N} y_i$, from the set of all CIs that could be constructed by taking an SRS of size n from the finite population of fixed values $\{y_1, y_2, \ldots, y_N\}$. The model-based confidence level gives the expected proportion of CIs that will include the realization of the population total, from the set of all sample values that could be generated from the model in (2.37).

Model assumptions. In the model-based approach, the probability model is proposed for all population units, whether in the sample or not. If the model assumptions are valid, model-based inference does not require random sampling—it is assumed that *all* units in the population follow the assumed model, so it makes no difference which ones are chosen for the sample. Thus, model-based analyses can be—and indeed must be, as we shall see in Chapter 15—used for nonprobability samples.

The assumptions for model-based analysis are strong—for the model in (2.37), it is assumed that all random variables for the response of interest in the population are independent and have mean μ and variance σ^2. But we only observe the units in the sample, and cannot examine the assumption of whether the model holds for units not in the sample. If you take a sample of your friends to estimate the average amount of time students at your university spend studying, there is no reason to believe that the students not in your sample spend the same average amount of time studying as your friends do. If the model is deficient, inferences made using a model-based analysis may be seriously flawed.

A note on notation. Many books (Cochran, 1977, for example) and journal articles use Y to represent the population total (t in this book) and $\bar{Y}$ to represent the finite population mean (our $\bar{y}_{\mathcal{U}}$). In this book, we reserve Y and T to represent random variables in a model-based approach. Our usage is consistent with other areas of statistics, in which capital letters near the end of the alphabet represent random variables. However, you should be aware that notation in the survey sampling literature is not uniform.

2.11 When Should a Simple Random Sample Be Used?

Simple random samples are usually easy to design and easy to analyze. But they are not the best design to use in the following situations:

- Before taking an SRS, consider whether a survey sample is the best method for studying your research question. If you want to study whether a certain brand of bath oil is an effective mosquito repellent, you should perform a controlled experiment, not take a survey. You should take a survey if you want to estimate how many people use the bath oil as a mosquito repellent, or if you want to estimate how many mosquitoes are in an area.

- You may not have a list of the observation units, or it may be expensive in terms of travel time to take an SRS. If interested in the proportion of mosquitoes in southwestern Wisconsin that carry an encephalitis virus, you cannot construct a sampling frame of the individual mosquitoes. You would need to sample different areas, and then examine some or all of the mosquitoes found in those areas, using a form of cluster sampling. Cluster sampling will be discussed in Chapters 5 and 6.

- You may have additional information that can be used to design a more cost-effective sampling scheme. In a survey to estimate the total number of mosquitoes in an area, an entomologist would know what terrain would be likely to have high mosquito density, and what areas would be likely to have low mosquito density, before any samples were taken. You would save effort in sampling by dividing the area into *strata*, groups of similar units, and then sampling plots within each stratum. Stratified sampling will be discussed in Chapter 3.

You may want to use an SRS in these situations:

- Little extra information is available that can be used when designing the survey. If your sampling frame is merely a list of university students' names in alphabetical order and you have no additional information such as major or year, simple random or systematic sampling is probably the best probability sampling strategy.

- Persons using the data will insist on using SRS formulas, whether they are appropriate or not. Some persons will not be swayed from the belief that one should only estimate the mean by taking the average of the sample values—in that case, you should design a sample in which averaging the sample values is the right thing to do. SRSs are often recommended when sample evidence is used in legal actions; sometimes, when a more complicated sampling scheme is used, an opposing counsel may try to persuade the jury that the sample results are not valid.

- You wish to perform analyses intended for independent observations. Although it is possible to create a complex-survey-design analog of almost any analysis you would care to perform, this might entail developing new statistical theory or software. If there are no compelling reasons to take a stratified or cluster sample, it might be easier to take an SRS so you can exploit the theory and programs that already exist for data assumed to be independent. An SRS may be more convenient when you wish to perform computationally intensive analyses.

Sampling and data mining. Credit card transactions, sensors, surveillance cameras, stock market trades, internet searches, transportation or energy usage, social media, medical devices attached to patients, air quality monitors, and other sources of "big data" may generate terabytes or even petabytes of data.

Data mining methods (see, for example, Hastie et al., 2009) are sometimes used to identify patterns in these massive data sets. But there are many challenges to working with the whole data set. The data may contain miscoded values or outliers that may be difficult to detect but may unduly affect statistics or models. Data cleaning and editing are time-consuming and may be impractical for the entire data set. Statistical analyses on massive data sets can take a great deal of time, even on the fastest computers. The data set may even be too large to store or process on a single machine.

An alternative is to take a probability sample from the massive data set. Analysts can then verify accuracy, examine outliers, and clean the data in the sample. If the data are too large to store or the population size is unknown, an SRS can be selected using the procedure in Exercise 33. If the population does not have a periodic or cyclical ordering, another option is to take a systematic sample.

The sample size n can be as large as feasible to work with and store. If n is sufficiently large, statistics calculated from the entire SRS will have negligible sampling variability. If measurement errors have been rectified in the sample, the statistics calculated from the SRS will be more accurate than statistics from the entire population, because, as discussed in Section 1.7, they will have fewer nonsampling errors.

What if you are interested in small subpopulations of the data, such as suspected fraudulent transactions in credit card data, or persons with a rare disease? An SRS—even an enormous SRS—might contain few observations from a rare subpopulation. But you can take a probability sample that has a pre-specified sample size from each subpopulation and is still representative of the population as a whole. The next chapter shows how to do this.

2.12 Chapter Summary

In probability sampling, every possible subset from the population has a known probability of being selected as the sample. These probabilities provide a basis for inference to the finite population.

Simple random sampling without replacement is the simplest of all probability sampling methods. In an SRS, each subset of the population of size n has the same probability of being chosen as the sample. The probability that unit i of the population appears in the sample is

$$\pi_i = \frac{n}{N}.$$

The sampling weight for each unit in the sample is

$$w_i = \frac{1}{\pi_i} = \frac{N}{n};$$

each unit in the sample can be thought of as representing N/n units in the population.

Estimators for an SRS, displayed in Table 2.3, are similar to those from your introductory statistics class, using $\bar{y} = \sum_{i \in \mathcal{S}} y_i/n$ and $s^2 = \sum_{i \in \mathcal{S}}(y_i - \bar{y})^2/(n-1)$.

Standard errors for without-replacement random samples contain a finite population correction, $(1-n/N)$, which does not appear in standard errors for with-replacement random samples. The fpc reduces the standard error when the sample size is large relative to the population size. In most surveys done in practice, the fpc is so close to one that it can be ignored.

TABLE 2.3
Estimators for an SRS.

Population Quantity	Estimator	Standard Error
Population total, $t = \sum_{i=1}^{N} y_i$	$\hat{t} = \sum_{i \in \mathcal{S}} w_i y_i = N\bar{y}$	$N\sqrt{\left(1 - \frac{n}{N}\right)\frac{s^2}{n}}$
Population mean, $\bar{y}_{\mathcal{U}} = \frac{t}{N}$	$\frac{\hat{t}}{N} = \frac{\sum_{i \in \mathcal{S}} w_i y_i}{\sum_{i \in \mathcal{S}} w_i} = \bar{y}$	$\sqrt{\left(1 - \frac{n}{N}\right)\frac{s^2}{n}}$
Population proportion, p	$\hat{p}$	$\sqrt{\left(1 - \frac{n}{N}\right)\frac{\hat{p}(1-\hat{p})}{n-1}}$

For "sufficiently large" sample sizes, an approximate 95% CI is given by

$$\text{estimate } \pm z_{\alpha/2} \text{ SE (estimate)}.$$

In a smaller sample, substitute the t critical value with $n - 1$ df, $t_{\alpha/2,n-1}$, for $z_{\alpha/2}$. The margin of error of an estimate is $z_{\alpha/2} \times$ SE (estimate). This is half of the width of a 95% CI.

Key Terms

Cluster sample: A probability sample in which each population unit belongs to a group, or cluster, and the clusters are sampled according to the sampling design.

Coefficient of variation (CV): The CV of a statistic $\hat{\theta}$, where $E(\hat{\theta}) > 0$, is $\text{CV}(\hat{\theta}) = \sqrt{V(\hat{\theta})}/E(\hat{\theta})$. The estimated CV is $\text{SE}(\hat{\theta})/\hat{\theta}$.

Confidence interval (CI): An interval estimate for a population quantity, where the probability that the random interval contains the true value of the population quantity is known.

Coverage probability: The expected proportion of CIs from repeated samples that include the true value of the population quantity.

Design-based inference: Inference for finite population characteristics based on the survey design, also called **randomization inference**.

Finite population correction (fpc): A correction factor which, when multiplied by the with-replacement variance, gives the without-replacement variance. For an SRS of size n from a population of size N, the fpc is $1 - n/N$.

Inclusion probability: $\pi_i = P(i \in \mathcal{S})$, the probability that unit i is in the sample.

Margin of error: Half of the width of a symmetric 95% CI.

Model-based inference: Inference for finite population characteristics based on a model for the population, also called **prediction inference**.

Probability sampling: Method of sampling in which every unit in the population has a known probability of being included in the sample.

Representative sample: A sample that can be used to estimate quantities of interest in a population and provide measures of accuracy about the estimates.

Sampling distribution: The probability distribution of a statistic generated by the sampling design.

Sampling fraction: The fraction of the population that is included in the sample ($= n/N$).

Sampling weight: Reciprocal of the inclusion probability; $w_i = 1/\pi_i$.

Self-weighting sample: A sample in which all probabilities of inclusion π_i are equal, so that the sampling weights w_i are the same for all units.

Simple random sample with replacement (SRSWR): A probability sample in which the first unit is selected from the population with probability $1/N$; then the unit is replaced and the second unit is selected from the set of N units with probability $1/N$, and so on until n units are selected.

Simple random sample without replacement (SRS): An SRS of size n is a probability sample in which any possible subset of n units from the population has the same probability ($= n!(N-n)!/N!$) of being the sample selected.

Standard error (SE): The square root of the estimated variance of a statistic.

Stratified sample: A probability sample in which population units are partitioned into strata, and then a probability sample of units is taken from each stratum.

Systematic sample: A probability sample in which every kth unit in the population is selected to be in the sample, starting with a randomly chosen starting value. Systematic sampling is a special case of cluster sampling.

For Further Reading

For readers with a background in mathematical statistics, the theory of probability sampling is explored in depth by Raj (1968), Särndal et al. (1992), Thompson (1997), Fuller (2009), and Wu and Thompson (2020). Ardilly and Tillé (2006) present exercises and worked solutions for problems from different probability sampling designs, including simple random sampling. Valliant et al. (2000), Brewer (2002), Chambers and Clark (2012), and Chambers et al. (2012) develop the theory of survey sampling from a model-based perspective.

Six books published around 1950 by the pioneers of survey sampling are now considered classics. Many of the issues faced then are similar to those faced today, and these books describe the innovations developed to meet the challenge of how to make inferences about the unseen parts of a population from a sample. Yates's *Sampling Methods for Censuses and Surveys* was the first of this set to appear, in 1949. Parten (1950) set out the principles of survey methodology, emphasizing the importance of designing surveys to control sampling and nonsampling errors. Cochran's book *Sampling Techniques* was first published in 1953, then revised in 1963 and 1977. The first edition of Sukhatme's book *Sampling Theory of Surveys with Applications* appeared in 1954. The books by Deming (1950) and Hansen, Hurwitz, and Madow (1953) were written by researchers who developed survey sampling theory and methods at the U.S. Census Bureau.

2.13 Exercises

Data files referenced in the exercises are available on the book website (see page xviii), and the variables are described in Appendix A of each of the supplementary books (Lohr, 2022; Lu and Lohr, 2022).

A. Introductory Exercises

1. Let $N = 6$ and $n = 3$. For purposes of studying sampling distributions, assume that all population values are known.

$$y_1 = 98 \qquad y_2 = 102 \qquad y_3 = 154$$
$$y_4 = 133 \qquad y_5 = 190 \qquad y_6 = 175$$

We are interested in $\bar{y}_{\mathcal{U}}$, the population mean. Two sampling plans are proposed.

- Plan 1. Eight possible samples may be chosen.

Sample Number	Sample, $\mathcal{S}$	$P(\mathcal{S})$
1	{1,3,5}	1/8
2	{1,3,6}	1/8
3	{1,4,5}	1/8
4	{1,4,6}	1/8
5	{2,3,5}	1/8
6	{2,3,6}	1/8
7	{2,4,5}	1/8
8	{2,4,6}	1/8

- Plan 2. Three possible samples may be chosen.

Sample Number	Sample, $\mathcal{S}$	$P(\mathcal{S})$
1	{1,4,6}	1/4
2	{2,3,6}	1/2
3	{1,3,5}	1/4

(a) What is the value of $\bar{y}_{\mathcal{U}}$?

(b) Let $\bar{y}$ be the mean of the sample values. For each sampling plan, find

$$\text{(i) } E[\bar{y}]; \quad \text{(ii) } V[\bar{y}]; \quad \text{(iii) Bias}(\bar{y}); \quad \text{(iv) MSE}(\bar{y}).$$

(c) Which sampling plan do you think is better? Why?

2. For the population in Example 2.2, consider the following sampling scheme:

$\mathcal{S}$	$P(\mathcal{S})$
{1,3,5,6}	1/8
{2,3,7,8}	1/4
{1,4,6,8}	1/8
{2,4,6,8}	3/8
{4,5,7,8}	1/8

   (a) Find the probability of selection π_i for each unit i.

   (b) What is the sampling distribution of $\hat{t} = 8\bar{y}$?

3. Each of the 10,000 shelves in a certain library is 300 cm long. To estimate how many books in the library need rebinding, a librarian takes a sample of 50 books using the following procedure: He first generates a random integer between 1 and 10,000 to select a shelf, and then generates a random number between 0 and 300 to select a location on that shelf. Thus, the pair of random numbers (2531, 25.4) would tell the librarian to include the book that is above the location 25.4 cm from the left end of shelf number 2531 in the sample. Does this procedure generate an SRS of the books in the library? Explain why or why not.

4. For the population in Example 2.2, find the sampling distribution of $\bar{y}$ for

   (a) an SRS of size 3 (without replacement)

   (b) an SRSWR of size 3 (with replacement).

   For each, draw the histogram of the sampling distribution of $\bar{y}$. Which sampling distribution has the smaller variance, and why?

5. An SRS of size 30 is taken from a population of size 100. The sample values are given below, and in the data file `srs30.csv`.

   8 5 2 6 6 3 8 6 10 7 15 9 15 3 5 6 7 10 14 3 4 17 10 6 14 12 7 8 12 9

   (a) What is the sampling weight for each unit in the sample?

   (b) Use the sampling weights to estimate the population total, t.

   (c) Give a 95% CI for t. Does the fpc make a difference for this sample?

6. A university has 807 faculty members. For each faculty member, the number of refereed publications was recorded. This number is not directly available on the database, so requires the investigator to examine each record separately. A frequency table for number of refereed publications is given below for an SRS of 50 faculty members.

Refereed Publications	0	1	2	3	4	5	6	7	8	9	10
Faculty Members	28	4	3	4	4	2	1	0	2	1	1

   (a) Plot the data using a histogram. Describe the shape of the data.

   (b) Estimate the mean number of publications per faculty member, and give the SE for your estimate.

   (c) Do you think that $\bar{y}$ from (b) will be approximately normally distributed? Why or why not?

(d) Estimate the proportion of faculty members with no publications and give a 95% CI.

7. A letter in the December 1995 issue of *Dell Champion Variety Puzzles* stated: "I've noticed over the last several issues there have been no winners from the South in your contests. You always say that winners are picked at random, so does this mean you're getting fewer entries from the South?" In response, the editors took a random sample of 1,000 entries from the last few contests, and found that 175 of those came from the South.

   (a) Find a 95% CI for the percentage of entries that come from the South.

   (b) According to *Statistical Abstract of the United States*, 30.9% of the U.S. population lived in states that the editors considered to be in the South. Is there evidence from your CI that the percentage of entries from the South differs from the percentage of persons living in the South?

8. Discuss whether an SRS would be a good design choice for the following situations. What other designs might be used?

   (a) For an e-mail survey of students, a sampling frame is available that contains a list of e-mail addresses for all students.

   (b) A researcher wants to take a sample of patients of board-certified allergists.

   (c) A researcher wants to estimate the percentage of topics in a medical website that have errors.

   (d) A county election official wants to assess the accuracy of the machine that counts the ballots by taking a sample of the paper ballots and comparing the estimated vote tallies for candidates from the sample to the machine counts.

9. Show that if $n_0/N \leq 1$, the value of n in (2.31) satisfies

$$e = z_{\alpha/2}\sqrt{\left(1 - \frac{n}{N}\right)}\frac{S}{\sqrt{n}}.$$

10. Which of the following SRS designs will give the most precision for estimating a population mean? Assume each population has the same value of the population variance S^2.

    1. An SRS of size 400 from a population of size 4,000
    2. An SRS of size 30 from a population of size 300
    3. An SRS of size 3,000 from a population of size 300,000,000

11. As stated in Example 2.13, many public opinion polls set the sample size to achieve a predetermined margin of error for proportions estimated from the poll.

    (a) Show that for a fixed value of n, the margin of error is largest when $\hat{p} = 0.5$. That value is sometimes reported as the margin of error for all proportions calculated from a survey since that is the largest possible margin of error for that sample size.

    (b) Suppose the population has $N = 50{,}000{,}000{,}000$ members, so that an fpc will make little difference in the precision. Construct a table showing the maximum margin of error for estimating a proportion when the sample size is: (i) 50 (ii) 100 (iii) 500 (iv) 1,000 (v) 5,000 (vi) 10,000 (vii) 50,000 (viii) 100,000 (ix) 1,000,000.

(c) Repeat part (b) for a population of size N = 1,000,000. For which sample sizes does the fpc make a difference?

12. The 2010 U.S. census listed the following populations for these Arizona cities:

City	Population
Casa Grande	48,571
Gila Bend	1,922
Jerome	444
Phoenix	1,445,632
Tempe	161,719

Suppose that you are interested in estimating the percentage of persons who have been immunized against measles in each city and can take an SRS of persons. What should your sample size be in each of the 5 cities if you want the estimate from each city to have margin of error of 4 percentage points? For which cities does the finite population correction make a difference?

13. *Objects in bin.* A large bin contains 15,000 squares and 5,000 circles. Some are black, the rest are gray, and all have different sizes.

(a) Select an SRS of size 200 from the population in file `shapespop.csv`. What is the sampling weight for each sampled object? Save your sample for exercises in later chapters.

(b) Draw a histogram of the areas for the objects in the sample.

(c) Using the sample, estimate the average area for objects in the bin. Give a 95% CI. What is the estimate of the total area covered by all objects in the bin?

(d) Using the sample, estimate the total number of gray objects in the bin. Give a 95% CI.

(e) Using the sample, estimate the total number of circles in the population, along with a 95% CI. Does the CI include the population value of 5,000 circles?

B. Working with Survey Data

14. To study how many people could be identified from relatives who have contributed their DNA to a genetic database, Erlich et al. (2018) took an SRS of 30 persons whose records were in a genetic database containing 1.28 million records. For each member of the SRS, they found the closest DNA match from a different person in the database. They found that for 23 of the persons in the sample, the closest DNA match corresponded to a third cousin or closer relative.

(a) Estimate the proportion of persons in the database who have a third cousin or closer relative also in the database, and give a 95% CI for the proportion.

(b) The proportion from (a) was reported in newspapers as the probability that a typical person in the U.S. could be matched in the DNA database. What assumptions are needed for this to be true?

15. Mayr et al. (1994) took an SRS of 240 children who visited their pediatric outpatient clinic. They found the following frequency distribution for the age (in months) of free (unassisted) walking among the children:

Age (Months)	9	10	11	12	13	14	15	16	17	18	19	20
Number of Children	13	35	44	69	36	24	7	3	2	5	1	1

(a) Construct a histogram of the distribution of age at walking. Is the shape normally distributed? Do you think the sampling distribution of the sample average will be normally distributed? Why, or why not?

(b) Find the mean, SE, and a 95% CI for the average age for onset of free walking.

(c) Suppose the researchers wanted to do another study in a different region, and wanted a 95% CI for the mean age of onset of walking to have margin of error 0.5. Using the estimated standard deviation for these data, what sample size would they need to take?

16. Cullen (1994) took a sample of the 580 children served by an Auckland family practice to estimate the percentage of children overdue for a vaccination.

    (a) What sample size in an SRS (without replacement) would be necessary to estimate the proportion with 95% confidence and margin of error 0.10?

    (b) Cullen actually took an SRS *with* replacement of size 120, of whom 93 were overdue for vaccination. Give a 95% CI for the proportion of children served by the practice who were overdue for vaccination.

17. Einarsen et al. (1998) selected an SRS of 935 assistant nurses from a Norwegian county with 2700 assistant nurses. A total of 745 assistant nurses (80%) responded to the survey.

    (a) 20% of the 745 respondents reported that bullying occurred in their department. Using these respondents as the sample, give a 95% CI for the total number of nurses in the county who would report bullying in their department.

    (b) What assumptions must you make about the nonrespondents for the analysis in (a)?

18. In 2005, the Statistical Society of Canada (SSC) had 864 members listed in the online directory. An SRS of 150 of the members was selected; the sex and employment category (industry, academic, government) was ascertained for each person in the SRS, with results in file `ssc.csv`.

    (a) What are the possible sources of selection bias in this sample?

    (b) Estimate the percentage of members who are female, and give a 95% CI for your estimate.

    (c) Assuming that all members are listed in the online directory, estimate the total number of SSC members who are female, along with a 95% CI.

19. The data set `agsrs.csv` also contains information on other variables. For each of the following quantities, plot the data, and estimate the population mean for that variable along with a 95% CI.

    (a) Number of acres devoted to farms in 1987

    (b) Number of farms, 1992

    (c) Number of farms with 1000 acres or more, 1992 (use variable *largef92*)

    (d) Number of farms with 9 acres or fewer, 1992 (use variable *smallf92*)

20. Data from an SRS of 120 golf courses, selected from a list of 14,938 golf courses in the U.S., are in file `golfsrs.csv`.

(a) Display the data in a histogram for the weekday greens fees for nine holes of golf (variable *wkday9*). How would you describe the shape of the data?

(b) Find the average weekday greens fee to play nine holes of golf, and give the SE for your estimate.

21. Repeat Exercise 20 for the back tee yardage (variable *backtee*).

22. For the data in `golfsrs.csv`, estimate the proportion of golf courses that have 18 holes, and give a 95% CI for the population proportion.

23. The Chicago database of crimes known to police between 2001 and 2019 at `https://data.cityofchicago.org/` contained 7,048,107 records. To reduce the size of the data for performing analyses, an SRS of 5,000 records was drawn from the database and is in file `crimes.csv`.

    (a) What is the sampling weight for each record in the sample?

    (b) What is the estimated percentage of crimes in which an arrest was made? Give a 95% CI.

    (c) Estimate the total number of burglaries known to police between 2001 and 2019 (use variable *crimetype* to identify the burglaries), along with a 95% CI.

    (d) What percentage of crimes were domestic-related (variable *domestic*)? What is the coefficient of variation (CV) for the estimate?

C. Working with Theory

24. Using the method illustrated in Example 2.10, find the exact confidence level for a CI based on an SRS (without replacement) of size 4 from the population in Example 2.2, where the confidence interval procedure is defined by

$$\text{CI}(\mathcal{S}) = \left[\hat{t}_{\mathcal{S}} - 1.96 \text{ SE}(\hat{t}_{\mathcal{S}}), \, \hat{t}_{\mathcal{S}} + 1.96 \text{ SE}(\hat{t}_{\mathcal{S}}\right].$$

Does your confidence level equal 95%?

25. Suppose we are interested in estimating the proportion p of a population that has a certain disease. As in Section 2.3 let $y_i = 1$ if person i has the disease, and $y_i = 0$ if person i does not have the disease. Then $\hat{p} = \bar{y}$.

    (a) Show, using the definition in (2.16), that

$$\text{CV}(\hat{p}) = \sqrt{\frac{N-n}{N-1}\frac{1-p}{np}}.$$

If the population is large and the sampling fraction is small, so that $(N-n)/(N-1) \approx 1$, write (2.32) in terms of the CV for a sample of size 1.

    (b) Suppose that the fpc is approximately equal to 1. Consider populations with p taking the successive values

$$0.001, 0.005, 0.01, 0.05, 0.10, 0.30, 0.50, 0.70, 0.90, 0.95, 0.99, 0.995, 0.999.$$

For each value of p, find the sample size needed to estimate the population proportion (a) with fixed margin of error 0.03, using (2.31), and (b) with relative error $0.03p$, using (2.32). What happens to the sample sizes for small values of p?

26. *Decision theoretic approach for sample size estimation* (requires calculus). In a decision theoretic approach, two functions are specified:

$$L(n) = \text{Loss or "cost" of a bad estimate}$$
$$C(n) = \text{Cost of taking the sample}$$

Suppose that for some constants c_0, c_1, and k,

$$L(n) = k\,V(\bar{y}_{\mathcal{S}}) = k\left(1 - \frac{n}{N}\right)\frac{S^2}{n}$$
$$C(n) = c_0 + c_1 n.$$

What sample size n minimizes the total cost $L(n) + C(n)$?

27. (Requires computing.) If you have a large SRS, you can estimate the sampling distribution of $\bar{y}_{\mathcal{S}}$ by repeatedly taking samples of size n with replacement from the list of sample values. A histogram of the means from 1,000 samples of size 300 with replacement from the data in Example 2.6 is displayed in Figure 2.8; the shape may be slightly skewed, but still appears approximately normal.

Would a sample of size 100 from this population be sufficiently large to use the central limit theorem? Take 500 samples with replacement of size 100 from the variable *acres92* in `agsrs.csv`, and draw a histogram of the 500 means. The approach described in this exercise is known as the **bootstrap** (see Efron and Tibshirani, 1993); we discuss the bootstrap further in Section 9.3.

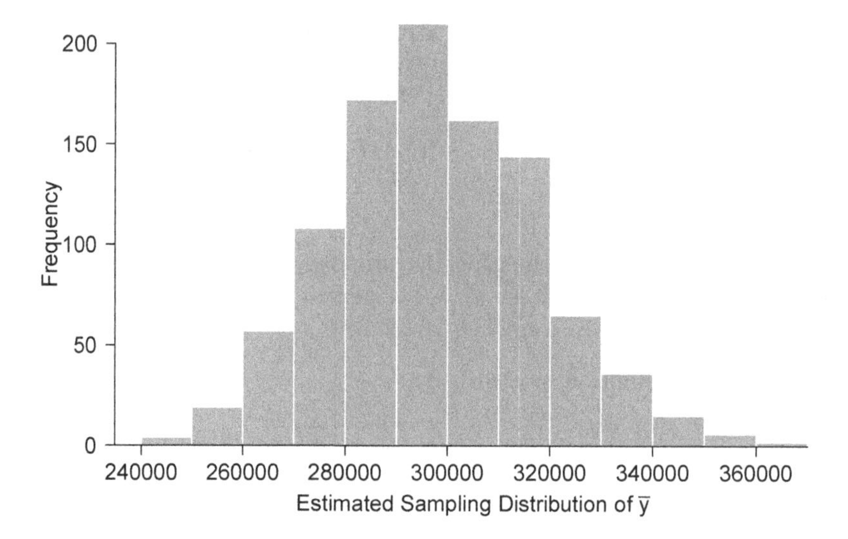

FIGURE 2.8
Histogram of the means of 1,000 samples of size 300, taken with replacement from the data in Example 2.6.

28. (Requires probability.) In an SRS, each possible subset of n units has probability $1/\binom{N}{n}$ of being chosen as the sample; in this chapter, we showed that this definition implies that each unit has probability n/N of appearing in the sample. The

converse is not true, however. Show that the inclusion probability π_i for each unit in a systematic sample is n/N, but that condition (2.10) is not met.

29. (Requires probability.) Many opinion polls survey about 1,000 adults. Suppose that the sampling frame contains 100 million adults including yourself, and that an SRS of 1,000 adults is chosen from the frame.

    (a) What is the probability that you are selected to be in the sample?

    (b) Now suppose that 2,000 such samples are selected, each sample selected independently of the others. What is the probability that you will *not* be in any of the samples?

    (c) How many samples must be selected for you to have a 0.5 probability of being in at least one sample?

30. (Requires probability.) In an SRSWR, a population unit can appear in the sample anywhere between 0 and n times. Let

$$Q_i = \text{number of times unit } i \text{ appears in the sample,}$$

and

$$\hat{t} = \frac{N}{n}\sum_{i=1}^{N} Q_i y_i.$$

    (a) Argue that the joint distribution of $Q_1, Q_2, \ldots, Q_N$ is multinomial with n trials and $p_1 = p_2 = \cdots = p_N = 1/N$.

    (b) Using (a) and properties of the multinomial distribution, show that $E[\hat{t}] = t$.

    (c) Using (a) and properties of the multinomial distribution, find $V[\hat{t}]$.

31. One way of selecting an SRS is to assign a number to every unit in the population, then use a random number table to select units from the list. A page from a random number table is given in file `rnt.csv`. Explain why each of the following methods will or will not result in a simple random sample.

    (a) The population has 742 units, and we want to take an SRS of size 30. Divide the random digits into segments of size 3 and throw out any sequences of three digits not between 001 and 742. If a number occurs that has already been included in the sample, ignore it. If we used this method with the first line of random numbers in `rnt.csv`, the sequence of three-digit numbers would be

    749 700 699 611 136 ...

    We would include units 700, 699, 611, and 136 in the sample.

    (b) For the situation in (a), when a random three-digit number is larger than 742, eliminate only the first digit and start the sequence with the next digit. With this procedure, the first five numbers would be 497, 006, 611, 136, and 264.

    (c) Now suppose the population has 170 items. If we used the procedures described in (a) or (b), we would throw away many of the numbers from the list. To avoid this waste, divide every random three-digit number by 170 and use the rounded remainder as the unit in the sample. If the remainder is 0, use unit 170. For the sequence in the first row of the random number table, the numbers generated would be

$$69 \quad 20 \quad 19 \quad 101 \quad 136 \quad \ldots$$

(d) Suppose the population has 200 items. Take two-digit sequences of random numbers and put a decimal point in front of each to obtain the sequence

$$0.74 \quad 0.97 \quad 0.00 \quad 0.69 \quad 0.96 \quad \ldots$$

Then multiply each decimal by 200 to get the units for the sample (convert 0.00 to 200):

$$148 \quad 194 \quad 200 \quad 138 \quad 192 \quad \ldots$$

(e) A school has 20 homeroom classes; each homeroom class contains between 20 and 40 students. To select a student for the sample, draw a random number between 1 and 20; then select a student at random from the chosen class. Do not include duplicates in your sample.

(f) For the situation in the preceding question, select a random number between 1 and 20 to choose a class. Then select a second random number between 1 and 40. If the number corresponds to a student in the class then select that student; if the second random number is larger than the class size, then ignore this pair of random numbers and start again. As usual, eliminate duplicates from your list.

32. *Sampling from a data stream, population size known* (requires probability). Suppose you would like to take an SRS of size n from a population of known size N, whose units arrive sequentially. As each unit in the population arrives, you can consider it as a candidate for the final sample, but you cannot revisit units already excluded from the sample. Prove that the following procedure results in an SRS of size n.

    Set $j := 0$ as the initial value of the accumulated sample size. For $k = 1, \ldots, N$, do the following:

    (a) Generate a random number u_k between 0 and 1.

    (b) If $u_k \leq (n-j)/(N-k+1)$ then select unit k for the sample and increment $j := j+1$. If $j = n$, then stop; you have a sample of size n.

    (c) If $u_k > (n-j)/(N-k+1)$ then skip unit k and move on to the next unit.

33. *Sampling from a data stream, population size unknown* (requires probability). Suppose you would like to take an SRS of size n from a population of unknown size, whose units arrive sequentially. This is like the situation in Exercise 32 except that N is unknown. Consider the following procedure:

    (a) Set $\mathcal{S}_n = \{1, 2, \ldots, n\}$, so that the initial sample for consideration consists of the first n units on the list.

    (b) For $k = n+1, n+2, \ldots$, generate a random number u_k between 0 and 1. If $u_k > n/k$, then set $\mathcal{S}_k$ equal to $\mathcal{S}_{k-1}$. If $u_k \leq n/k$, then select one of the units in $\mathcal{S}_{k-1}$ at random and replace it by unit k to form $\mathcal{S}_k$.

    Show that $\mathcal{S}_N$ from this procedure is an SRS of size n. HINT: Use induction.

34. *Clopper-Pearson (1934) CI for a proportion.* The CIs for a proportion in Section 2.5 assume that $\hat{p}$ is approximately normally distributed. When $\hat{p}$ is close to 0 or 1, however, its distribution may be skewed, and a CI based on a normal approximation may have incorrect coverage probability. Leemis and Trivedi (1996) gave the following formula for the Clopper-Pearson $100(1-\alpha)\%$ CI for p:

$$\left[\frac{n\hat{p}F_1}{n\hat{p}F_1 + n(1-\hat{p}) + 1}, \frac{(n\hat{p}+1)F_2}{(n\hat{p}+1)F_2 + n(1-\hat{p})}\right], \tag{2.39}$$

where F_1 is the $(\alpha/2)$th percentile of the F distribution with $2n\hat{p}$ and $2(n - n\hat{p} + 1)$ degrees of freedom, and F_2 is the $(1-\alpha/2)$th percentile of the F distribution with $2(n\hat{p}+1)$ and $2n(1-\hat{p})$ degrees of freedom.

(a) The CI based on the normal approximation is $\hat{p} \pm 1.96\sqrt{\hat{V}(\hat{p})}$, with $\hat{V}(\hat{p})$ given in (2.22). Compute the normal-approximation CI when (i) $n = 10$ and $\hat{p} = 0.5$, (ii) $n = 10$ and $\hat{p} = 0.9$, (iii) $n = 100$ and $\hat{p} = 0.5$, (iv) $n = 100$ and $\hat{p} = 0.9$, (v) $n = 100$ and $\hat{p} = 0.99$, (vi) $n = 1000$ and $\hat{p} = 0.99$, and (vii) $n = 1000$ and $\hat{p} = 0.999$. Assume that N is very large so you do not need to use the fpc.

(b) Compute the Clopper-Pearson CI for each of situations (i)–(vii) in part (a).

(c) For which situations is the normal-approximation CI similar to the exact Clopper-Pearson CI? What do you notice about the normal-approximation CI for situations (ii), (v), and (vii)?

D. Projects and Activities

35. *Rectangles.* This activity was suggested by Gnanadesikan et al. (1997). Figure 2.9 contains a population of 100 rectangles. Your goal is to estimate the total area of all the rectangles by taking a sample of 10 rectangles. Keep your results from this exercise; you will use them again in later chapters.

(a) Select a purposive sample of 10 rectangles that you think will be representative of the population of 100 rectangles. Record the area (number of small squares) for each rectangle in your sample. Use your sample to estimate the total area. How did you choose the rectangles for your sample?

(b) Find the sample variance for your purposive sample of 10 rectangles from part (a), and use (2.28) to form an interval estimate for the total area t.

(c) Now take an SRS of 10 rectangles. Use your SRS to estimate the total area of all 100 rectangles, and find a 95% CI for the total area.

(d) Compare your intervals with those of other students in the class. What percentage of the intervals from part (b) include the true total area of 3079? What about the CIs from part (c)?

36. *Mutual funds.* The websites of companies such as Fidelity (`www.fidelity.com`), Vanguard (`www.vanguard.com`), and T. Rowe Price (`www.troweprice.com`) list the mutual funds of those companies, along with some statistics about the performance of those funds. Take an SRS of 25 mutual funds from one of these companies. Describe how you selected the SRS. Find the mean and a 95% CI for the mean of a variable you are interested in, such as daily percentage change, or 1-year performance, or length of time the fund has existed.

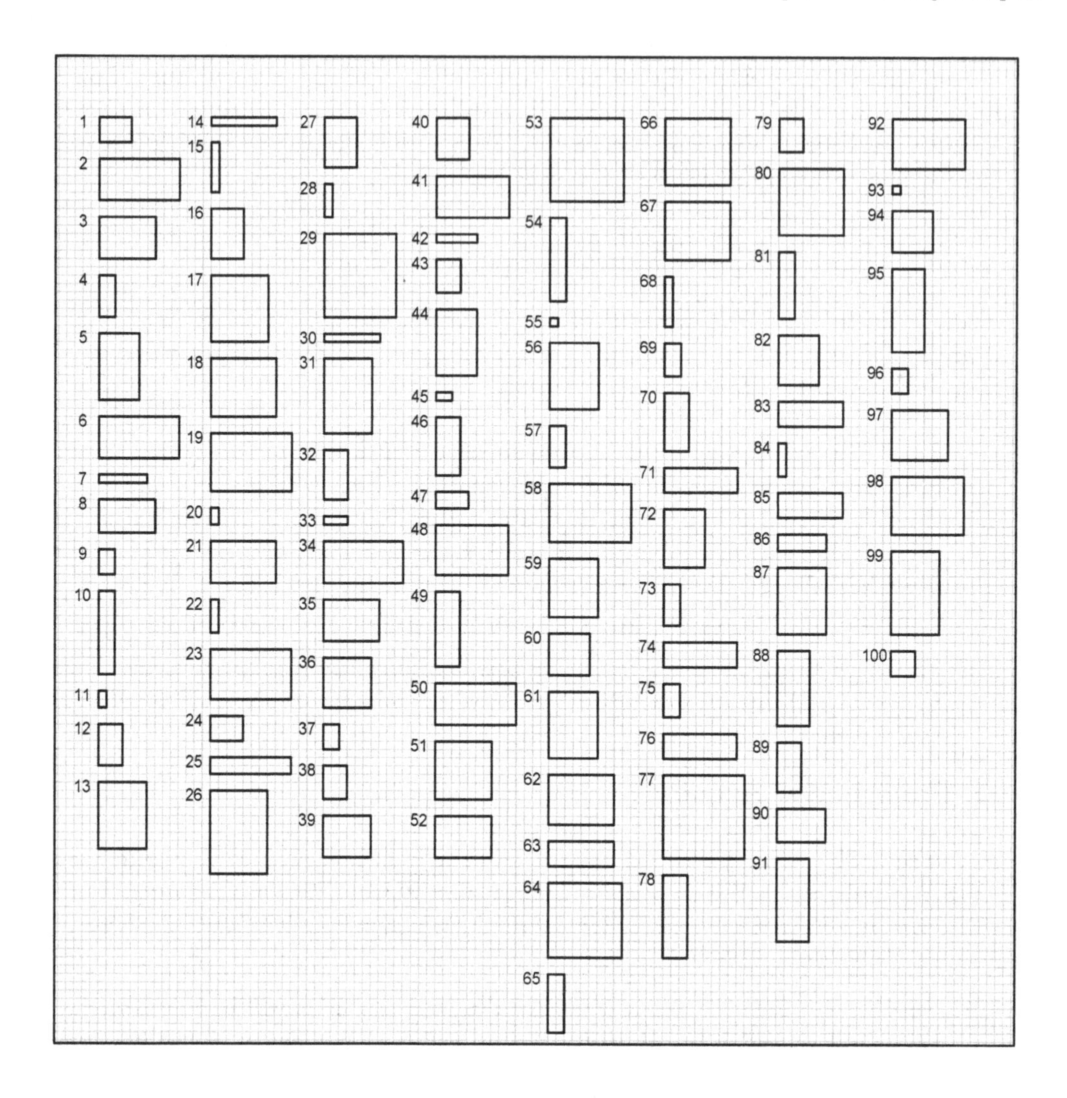

FIGURE 2.9
Population of 100 rectangles for Exercise 35.

37. *Baseball data.* This activity was created by Jenifer Boshes, who also compiled the data from Forman (2004) and publicly available salary information. The data file `baseball.csv` contains statistics on 797 baseball players from the rosters of all major league teams in November, 2004. In this exercise (which will be continued in later chapters), you will treat the file `baseball.csv` as a population and draw samples from it using different sampling designs.

    (a) Take an SRS of 150 players from the file. Describe how you selected the SRS. Save your data set for use in future exercises.

    (b) Calculate logsal = ln(salary). Construct a histogram of the variables *salary* and *logsal* from your SRS. Does the distribution of *salary* appear approximately normal? What about *logsal*?

    (c) Find the mean of the variable *logsal*, and give a 95% CI.

(d) Estimate the proportion of players in the data set who are pitchers, and give a 95% CI.

(e) Since you have the full data file for the population, you can find the true mean and proportion for the population. Do your CIs in (c) and (d) contain the true population values?

38. *Online bookstore.* Find a website for an online bookstore.

    (a) In the books search window, type in a genre you like, such as mystery or sports; you may want to narrow your search by selecting a subcategory if an upper bound is placed on the number of books that can be displayed. Choose a genre with at least 20 pages of listings. The list of books forms your population.

    (b) What is your target population? What is the population size, N?

    (c) Take an SRS of 50 books from your population. Describe how you selected the SRS, and record the amount of time you spent taking the sample and collecting the data.

    (d) Record the following information for each book in your SRS: price, number of pages, and whether the book is paperback or hardback.

    (e) Give a point estimate and a 95% CI for the mean price of books in the genre you selected.

    (f) Give a point estimate and a 95% CI for the mean number of pages for books in the genre you selected.

    (g) Explain, to a person who knows no statistics, what your estimates and CIs mean.

39. Take a small SRS of something you are interested in. Explain what it is you decide to study and carefully describe how you chose your SRS (give the random numbers generated and explain how you translated them into observations), report your data, and give a point estimate and the SE for the quantity or quantities of interest.

    The data collection for this exercise should not take a great deal of effort, as you are surrounded by things waiting to be sampled. Some examples: web pages that result from an internet search about a topic, actual weights of 1-pound bags of carrots at the supermarket, or the cost of a used dining room table from an online classified advertisement site.

40. *Estimating the size of an audience.* A common method for estimating the size of an audience is to take an SRS of n of the N rows in an auditorium, count the number of people in each of the selected rows, then multiply the total number of people in your sample by N/n.

    (a) Why is it important to take an SRS instead of a convenience sample of the first 10 rows?

    (b) Go to a performance or a lecture, and count the number of rows in the auditorium. Take an SRS of 10 or 20 rows, count the number of people in each row, and estimate the number of people in the audience using this method. Give a 95% CI.

41. *Estimating crowd size.* A method similar to that in Exercise 40 can be used to estimate the size of a crowd from a high-resolution photograph. Obtain an aerial photograph of a large outdoor gathering. Divide the photograph into a grid of equal-sized squares, and take an SRS of the squares. Use the counts of people in the sampled squares to estimate the crowd size along with a 95% CI. Explain how you decided on the grid size and the number of squares to sample. What rule did you use for counting people on the boundary of two squares?

42. *IPUMS data.* This exercise is designed for the Integrated Public Use Microdata Series (IPUMS), a collection of data from the U.S. decennial census and ACS (Ruggles et al., 2004). In the following exercises, we treat a self-weighting sample from the 1980 census data, in file `ipums.csv`, as a population.

    (a) The variable *inctot* is total personal income from all sources. To protect the confidentiality of the respondents, the variable is "topcoded" at $75,000—that is, anyone with an income greater than $75,000 is listed as having income $75,000. What effect does the topcoding have on estimates from the file?

    (b) Draw a pilot sample (SRS) of size 50 from the IPUMS population. Use the sample variance you get for *inctot* to determine the sample size that is needed to estimate the average of *inctot* with a margin of error of 700 or less.

    (c) Take an SRS of your desired sample size from the population. Estimate the total income for the population, and give a 95% CI. Make sure you save the sample (or the random seed used to generate it) for use in later chapters.

43. *Large public database.* Many cities post data sets about city operations on the internet. For example, the sample for Exercise 23 was drawn from a large database made public by the city of Chicago.

    Find a large data set from a major city near you (do an internet search for city name and "open data"). Select an SRS from the data set and analyze variables of interest using your SRS. What considerations should you make when determining the sample size?

3

Stratified Sampling

> One of the things she [Mama] taught me should be obvious to everyone, but I still find a lot of cooks who haven't figured it out yet. Put the food on first that takes the longest to cook.
>
> —Pearl Bailey, *Pearl's Kitchen*

3.1 What Is Stratified Sampling?

Example 3.1. The Federal Deposit Insurance Corporation (FDIC) was created in 1933 by the U.S. Congress to supervise banks; it insures deposits at member banks up to a specified limit. When a bank fails, the FDIC acquires the assets from that bank and uses them to help pay the insured depositors. Valuing the assets is time-consuming, and Chapman (2005) described a procedure used to estimate the total amount recovered from financial institutions from a sample. The assets from failed institutions fall into several types: (1) consumer loans, (2) commercial loans, (3) securities, (4) real estate mortgages, (5) other owned real estate, (6) other assets, and (7) net investments in subsidiaries. A simple random sample (SRS) of assets may result in an imprecise estimate of the total amount recovered. Consumer loans tend to be much smaller on average than assets in the other classes, so the sample variance from an SRS can be very large. In addition, an SRS might contain no assets from one or more of the types; if category (2) assets tend to have the most monetary value and the sample chosen has no assets from category (2), that sample may result in an estimate of total assets that is too small. It would be desirable to have a method for sampling that prevents samples that we know would produce bad estimates, and that increases the precision of the estimators. **Stratified sampling** can accomplish these goals. ■

Often, we have supplementary information that can help us design our sample. For example, we would know before undertaking a survey that men generally earn more than women, that New York City residents pay more for housing than residents of Des Moines, or that rural residents shop for groceries less frequently than urban residents. The FDIC has information on the type of each asset, which is related to the value of the asset.

If the variable we are interested in takes on different mean values in different subpopulations, we may be able to obtain more precise estimates of population quantities by taking a **stratified** random sample. The word *stratify* means "to arrange or deposit in layers"; we divide the population into H subpopulations, called **strata**. The strata do not overlap, and they constitute the whole population so that each sampling unit belongs to exactly one stratum. We draw an independent probability sample from each stratum, then pool the information to obtain overall population estimates.

We use stratified sampling for one or more of the following reasons:

1. We want to be protected from the possibility of obtaining a really bad sample. When taking an SRS of size 100 from a population of 1,000 male and 1,000 female students,

DOI: 10.1201/9780429298899-3

obtaining a sample with no or very few males is theoretically possible, although such a sample is not likely to occur. Most people would not consider such a sample to be representative of the population and would worry that men and women might respond differently on the item of interest. In a stratified sample, you can take an SRS of 50 males and an independent SRS of 50 females, guaranteeing that the proportion of males in the sample is the same as that in the population. With this design, a sample with no or few males cannot be selected.

2. We may want data of known precision for subgroups of the population. These subgroups should be the strata. McIlwee and Robinson (1992) sampled graduates from electrical and mechanical engineering programs at public universities in southern California. They were interested in comparing the educational and workforce experiences of male and female graduates, so they stratified their sampling frame by gender and took separate random samples of male and female graduates. Because there were many more male than female graduates, they sampled a higher fraction of female graduates than male graduates in order to obtain comparable precisions for the two groups.

3. A stratified sample may be more convenient to administer and may result in a lower cost for the survey. For example, sampling frames may be constructed differently in different strata, or different sampling designs or field procedures may be used. In a survey of businesses, an internet survey might be used for large firms while a mail or telephone survey is used for small firms. In other surveys, a different procedure may be used for sampling households in urban strata than in rural strata.

4. Stratified sampling often gives more precise (having lower variance) estimates for population means and totals. Persons of different ages tend to have different blood pressures, so in a blood pressure study it would be helpful to stratify by age groups. If studying the concentration of plants in an area, one would stratify by type of terrain; marshes would have different plants than woodlands. Stratification works for lowering the variance because the variance within each stratum is often lower than the variance in the whole population. Prior knowledge can be used to save money in the sampling procedure.

Example 3.2. Refer to Example 2.6, in which we took an SRS to estimate the average number of farm acres per county. Even though we scrupulously generated an SRS, some areas of the country were overrepresented, and others were not represented at all. Taking a stratified sample can provide some balance in the sample on the stratifying variable.

The SRS in Example 2.6 exhibited a wide range of values for y_i, the number of acres devoted to farms in county i in 1992. You might conjecture that part of the large variability arises because counties in the western United States are larger, and thus tend to have larger values of y, than counties in the eastern United States.

For this example, we use the four census regions of the United States—Northeast, North Central, South, and West—as strata. The SRS in Example 2.6 sampled about 10% of the population; to be able to compare the results of the stratified sample with the SRS, we also sample about 10% of the counties in each stratum, as shown in column 3 of Table 3.1. (We discuss other stratified sampling designs later in the chapter.)

We select four separate SRSs, one from each of the four strata. To select the SRS from the Northeast stratum, we number the counties in that stratum from 1 to 220, and select 21 numbers randomly from $\{1, \ldots, 220\}$. We follow a similar procedure for the other three strata, selecting 103 counties at random from the 1054 in the North Central region, 135 counties from the 1,382 in the South, and 41 counties from the 422 in the West. The four SRSs are selected independently: Knowing which counties are in the sample from the Northeast tells us nothing about which counties are in the sample from the South.

TABLE 3.1
Population size, sample size, and summary statistics for each stratum in Example 3.2.

Region	Number of Counties in Population	Number of Counties in Sample	Sample Average in Region	Sample Variance in Region
Northeast	220	21	97,629.8	7,647,472,708
North Central	1054	103	300,504.2	29,618,183,543
South	1382	135	211,315.0	53,587,487,856
West	422	41	662,295.5	396,185,950,266
Total	3078	300		

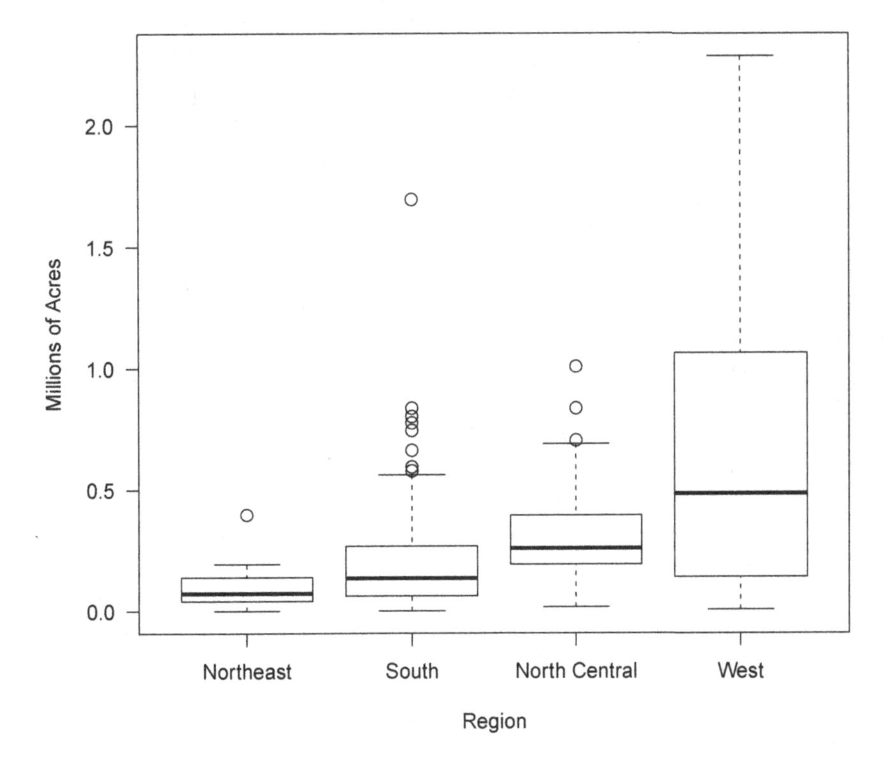

FIGURE 3.1
Boxplot of data from Example 3.2. The thick line for each region is the median of the sample data from that region; the other horizontal lines in the boxes are the 25th and 75th percentiles. The Northeast region has a relatively small median and small variance; the West region, however, has a much higher median and variance. The distribution of farm acreage appears to be positively skewed in each of the regions.

The data sampled from all four strata are in file `agstrat.csv`. The last two columns of Table 3.1 give summary statistics for the sampled counties in each stratum: the sample average and sample variance of the number of acres devoted to farms. Figure 3.1 displays boxplots showing the median and quartiles for the strata.

Since we took an SRS in each stratum, we can use (2.18) and (2.20) to estimate the population quantities for each stratum. We use

$$(220)(97{,}629.81) = 21{,}478{,}558.2$$

to estimate the total number of acres devoted to farms in the Northeast, with estimated variance

$$(220)^2\left(1-\frac{21}{220}\right)\frac{7{,}647{,}472{,}708}{21}=1.594316\times 10^{13}.$$

Table 3.2 gives estimates of the total number of farm acres and estimated variance of the total for each of the four strata.

TABLE 3.2
Estimated total number of farm acres, along with its estimated variance, for each stratum.

Stratum	Estimated Total of Farm Acres	Estimated Variance of Estimated Total
Northeast	21,478,558	1.59432×10^{13}
North Central	316,731,380	2.88232×10^{14}
South	292,037,391	6.84076×10^{14}
West	279,488,706	1.55365×10^{15}
Total	909,736,035	2.54190×10^{15}

We can estimate the total number of acres devoted to farming in the United States in 1992 by summing the estimated totals for each stratum in Table 3.2. Because sampling was done independently in each stratum, the variance of the total is the sum of the variances of the estimated stratum totals. Thus we estimate the total number of acres devoted to farming as 909,736,035, with standard error (SE) $\sqrt{2.5419 \times 10^{15}} = 50{,}417{,}248$. We estimate the average number of acres devoted to farming per county as 909,736,035/3078 = 295,561, with standard error 50,417,248/3078 = 16,380.

For comparison, the estimate of the population total in Example 2.6, using an SRS of size 300, was 916,927,110, with standard error 58,169,381. For this example, stratified sampling ensures that each region of the United States is represented in the sample, and produces an estimate with smaller standard error than an SRS with the same number of observations. The sample variance in Example 2.6 was $s^2 = 1.1872 \times 10^{11}$. Only the West had sample variance larger than s^2; the sample variance in the Northeast was only 7.647×10^9.

Observations within a stratum are often more homogeneous than observations in the population as a whole, and the reduction in variance in the individual strata can lead to a reduced variance for the population estimate. In this example, the relative gain from stratification can be estimated by the ratio

$$\frac{\text{estimated variance from stratified sample, with } n=300}{\text{estimated variance from SRS, with } n=300}=\frac{2.5419\times 10^{15}}{3.3837\times 10^{15}}=0.75.$$

If these figures were the population variances, we would expect that we would need only (300)(0.75) = 225 observations with a stratified sample to obtain the same precision as from an SRS of 300 observations.

Of course, no law says that you must sample the same fraction of observations in every stratum. In this example, there is far more variability from county to county in the western region. If acres devoted to farming were the primary variable of interest, you would reduce the variance of the estimated total even further by taking a higher sampling fraction in the western region than in the other regions. You will explore an alternative sampling design in Exercise 25. ■

3.2 Theory of Stratified Sampling

We divide the population of N sampling units into H "layers" or strata, with N_h sampling units in stratum h. For stratified sampling to work, we must know the stratum membership for every unit in the population. Thus, we know the values of $N_1, N_2, \ldots, N_H$, and, because each population unit is in exactly one stratum, $N_1 + N_2 + \ldots N_H = N$.

In **stratified random sampling**, the simplest form of stratified sampling, we independently take an SRS from each stratum, so that n_h observations are randomly selected from the N_h population units in stratum h. The total sample size is $n = n_1 + n_2 + \ldots + n_H$.

Table 3.3 defines the notation used for stratified random sampling. Statistics and sample sizes for stratum h have subscript h, so that $\bar{y}_{h\mathcal{U}}$ is the population mean in stratum h, $\bar{y}_h$ is the sample mean in stratum h, and similarly for other quantities.

Suppose we sampled only the hth stratum. In effect, we have a population of N_h units and take an SRS of n_h units. We estimate $\bar{y}_{h\mathcal{U}}$ by $\bar{y}_h$, and t_h by $\hat{t}_h = N_h \bar{y}_h$. The total for the entire population is $t = \sum_{h=1}^{H} t_h$, so we estimate t by

$$\hat{t}_{\text{str}} = \sum_{h=1}^{H} \hat{t}_h = \sum_{h=1}^{H} N_h \bar{y}_h. \tag{3.1}$$

To estimate $\bar{y}_{\mathcal{U}}$, then, we use

$$\bar{y}_{\text{str}} = \frac{\hat{t}_{\text{str}}}{N} = \sum_{h=1}^{H} \frac{N_h}{N} \bar{y}_h. \tag{3.2}$$

This is a weighted average of the sample stratum means; $\bar{y}_h$ is multiplied by N_h/N, the proportion of the population units in stratum h. Note that (3.1) requires knowing the population stratum sizes N_h and (3.2) requires knowing the relative sizes N_h/N.

The properties of these estimators follow directly from the properties of SRS estimators:

- **Unbiasedness.** $\bar{y}_{\text{str}}$ and $\hat{t}_{\text{str}}$ are unbiased estimators of $\bar{y}_{\mathcal{U}}$ and t. An SRS is taken in each stratum, so (2.36) implies that $E[\bar{y}_h] = \bar{y}_{h\mathcal{U}}$ and consequently

$$E\left[\sum_{h=1}^{H} \frac{N_h}{N} \bar{y}_h\right] = \sum_{h=1}^{H} \frac{N_h}{N} E[\bar{y}_h] = \sum_{h=1}^{H} \frac{N_h}{N} \bar{y}_{h\mathcal{U}} = \bar{y}_{\mathcal{U}}.$$

- **Variance of the estimators.** The key to calculating the variance is that the samples from different strata are selected independently: Knowing which units are sampled in stratum h gives no information about which units are sampled from the other strata. Since we know $V(\hat{t}_h)$ from the SRS theory, and all $\hat{t}_h$ are mutually independent, (2.19) and Property 8 in Table A.1 of Appendix A imply that

$$V\left(\hat{t}_{\text{str}}\right) = \sum_{h=1}^{H} V\left(\hat{t}_h\right) = \sum_{h=1}^{H} \left(1 - \frac{n_h}{N_h}\right) N_h^2 \frac{S_h^2}{n_h} \tag{3.3}$$

and

$$V(\bar{y}_{\text{str}}) = \frac{1}{N^2} V(\hat{t}_{\text{str}}) = \sum_{h=1}^{H} \left(1 - \frac{n_h}{N_h}\right) \left(\frac{N_h}{N}\right)^2 \frac{S_h^2}{n_h}. \tag{3.4}$$

TABLE 3.3
Notation for stratified random sampling.

Symbol	Formula	Description
	Population Quantities	
N_h		number of population units in stratum h
y_{hj}		value of jth unit in stratum h
t_h	$\sum_{j=1}^{N_h} y_{hj}$	population total in stratum h
t	$\sum_{h=1}^{H} t_h$	overall population total
$\bar{y}_{h\mathcal{U}}$	$\frac{t_h}{N_h} = \frac{1}{N_h}\sum_{j=1}^{N_h} y_{hj}$	population mean in stratum h
$\bar{y}_{\mathcal{U}}$	$\frac{t}{N} = \frac{1}{N}\sum_{h=1}^{H}\sum_{j=1}^{N_h} y_{hj}$	overall population mean
S_h^2	$\frac{1}{N_h - 1}\sum_{j=1}^{N_h} (y_{hj} - \bar{y}_{h\mathcal{U}})^2$	population variance in stratum h
	Sample Quantities	
Symbol	**Formula**	**Description**
$\mathcal{S}_h$		set of units from stratum h included in sample
n_h		sample size taken in stratum h
n	$\sum_{h=1}^{H} n_h$	total sample size, from all strata
$\bar{y}_h$	$\frac{1}{n_h}\sum_{j\in\mathcal{S}_h} y_{hj}$	sample mean in stratum h
$\hat{t}_h$	$\frac{N_h}{n_h}\sum_{j\in\mathcal{S}_h} y_{hj} = N_h\bar{y}_h$	estimated population total in stratum h
s_h^2	$\frac{1}{n_h - 1}\sum_{j\in\mathcal{S}_h} (y_{hj} - \bar{y}_h)^2$	sample variance in stratum h

- **Estimated variances and standard errors for stratified samples.** We can obtain unbiased estimators of $V(\hat{t}_{\text{str}})$ and $V(\bar{y}_{\text{str}})$ by substituting the sample variances s_h^2 for the population variances S_h^2 in (3.3) and (3.4). Note that we must sample at least two units from stratum h to be able to calculate $s_h^2 = \sum_{j\in\mathcal{S}_h} (y_{hj} - \bar{y}_h)^2 / (n_h - 1)$. The estimated variances are:

$$\hat{V}\left(\hat{t}_{\text{str}}\right) = \sum_{h=1}^{H}\left(1 - \frac{n_h}{N_h}\right) N_h^2 \frac{s_h^2}{n_h}, \tag{3.5}$$

$$\hat{V}(\bar{y}_{\text{str}}) = \frac{1}{N^2}\hat{V}(\hat{t}_{\text{str}}) = \sum_{h=1}^{H}\left(1 - \frac{n_h}{N_h}\right)\left(\frac{N_h}{N}\right)^2 \frac{s_h^2}{n_h}. \tag{3.6}$$

As always, the standard error of an estimator is the square root of the estimated variance: $\text{SE}(\bar{y}_{\text{str}}) = \sqrt{\hat{V}(\bar{y}_{\text{str}})}$.

- **Confidence intervals for stratified samples.** If either (1) the sample sizes within each stratum are large, or (2) the sampling design has a large number of strata, an approximate $100(1-\alpha)\%$ confidence interval (CI) for the population mean $\bar{y}_{\mathcal{U}}$ is

$$\bar{y}_{\text{str}} \pm z_{\alpha/2}\ \text{SE}\ (\bar{y}_{\text{str}}).$$

The central limit theorem used for constructing this CI is stated in Krewski and Rao (1981). Most survey software packages use the percentile of a t distribution with $n - H$ degrees of freedom (df) rather than the percentile of the normal distribution.

Example 3.3. Siniff and Skoog (1964) used stratified random sampling to estimate the size of the Nelchina herd of Alaska caribou in February of 1962. In January and early February, several sampling techniques were field-tested. The field tests told the investigators that several of the proposed sampling units, such as equal-flying-time sampling units, were difficult to implement in practice, and that an equal-area sampling unit of 4 square miles would work well for the survey. The biologists used preliminary estimates of caribou densities to divide the area of interest into six strata; each stratum was then divided into a grid of 4-square-mile sampling units. Stratum A, for example, contained $N_1 = 400$ sampling units; $n_1 = 98$ of these were randomly selected to be in the survey.

TABLE 3.4
Data and calculations for Example 3.3.

Stratum	N_h	n_h	$\bar{y}_h$	s_h^2	$\hat{t}_h = N_h\bar{y}_h$	$\left(1 - \frac{n_h}{N_h}\right) N_h^2 \frac{s_h^2}{n_h}$
A	400	98	24.1	5,575	9,640	6,872,040.82
B	30	10	25.6	4,064	768	243,840.00
C	61	37	267.6	347,556	16,323.6	13,751,945.51
D	18	6	179.0	22,798	3,222	820,728.00
E	70	39	293.7	123,578	20,559	6,876,006.67
F	120	21	33.2	9,795	3984	5,541,171.43
Total	699	211			54,496.6	34,105,732.43

Table 3.4 displays the data and the calculations for finding the stratified sampling estimates. The estimated total number of caribou is 54,497 with standard error $\sqrt{34{,}105{,}732.43} = 5{,}840$. An approximate 95% CI for the total number of caribou is

$$54{,}497 \pm 1.96(5840) = [43{,}051,\ 65{,}943].$$

Of course, this CI reflects only the uncertainty due to sampling error; if the field procedure for counting caribou tends to miss animals, then the entire CI will be too low. ■

Stratified sampling for proportions. As we observed in Section 2.3, a proportion is a mean of a variable that takes on values 0 and 1. To make inferences about proportions, we simply use the results in (3.1)–(3.6), with $\bar{y}_h = \hat{p}_h$ and $s_h^2 = \dfrac{n_h}{n_h - 1}\hat{p}_h(1 - \hat{p}_h)$. Then

$$\hat{p}_{\text{str}} = \sum_{h=1}^{H} \frac{N_h}{N}\hat{p}_h \tag{3.7}$$

and

$$\hat{V}(\hat{p}_{\text{str}}) = \sum_{h=1}^{H} \left(1 - \frac{n_h}{N_h}\right)\left(\frac{N_h}{N}\right)^2 \frac{\hat{p}_h(1 - \hat{p}_h)}{n_h - 1}. \tag{3.8}$$

Estimating the total number of population units having a specified characteristic is similar:

$$\hat{t}_{\text{str}} = \sum_{h=1}^{H} N_h \hat{p}_h,$$

so the estimated total number of population units with the characteristic is the sum of the estimated totals in each stratum. Similarly, $\hat{V}(\hat{t}_{\text{str}}) = N^2\hat{V}(\hat{p}_{\text{str}})$.

Example 3.4. The American Council of Learned Societies (ACLS) used a stratified random sample of selected ACLS societies in seven disciplines to study publication patterns and computer and library use among scholars who belong to one of the member organizations of the ACLS (Morton and Price, 1989). The data are shown in Table 3.5.

TABLE 3.5
Data from ACLS survey.

Discipline	Membership	Number Mailed	Valid Returns	Female Members (%)
Literature	9,100	915	636	38
Classics	1,950	633	451	27
Philosophy	5,500	658	481	18
History	10,850	855	611	19
Linguistics	2,100	667	493	36
Political Science	5,500	833	575	13
Sociology	9,000	824	588	26
Total	44,000	5,385	3,835	

Ignoring the nonresponse for now (we'll return to the nonresponse in Exercise 12 of Chapter 8) and supposing there are no duplicate memberships, let's use the stratified sample to estimate the percentage and number of respondents of the major societies in those seven disciplines that are female. Here, let N_h be the membership figures, and let n_h be the number of valid surveys. Thus,

$$\hat{p}_{\text{str}} = \sum_{h=1}^{7} \frac{N_h}{N}\hat{p}_h = \frac{9100}{44{,}000}0.38 + \ldots + \frac{9000}{44{,}000}0.26 = 0.2465$$

and

$$\text{SE}(\hat{p}_{\text{str}}) = \sqrt{\sum_{h=1}^{7} \left(1 - \frac{n_h}{N_h}\right)\left(\frac{N_h}{N}\right)^2 \frac{\hat{p}_h(1 - \hat{p}_h)}{n_h - 1}} = 0.0071.$$

The estimated total number of female members in the societies is $\hat{t}_{\text{str}} = 44{,}000 \times (0.2465) = 10{,}847$, with $\text{SE}(\hat{t}_{\text{str}}) = 44{,}000 \times (0.0071) = 312$. ■

3.3 Sampling Weights in Stratified Random Sampling

We introduced the notion of sampling weight, $w_i = 1/\pi_i$, in Section 2.4. For an SRS, the sampling weight for each observation is the same since all of the inclusion probabilities π_i are the same. In stratified sampling, however, we may have different inclusion probabilities in different strata so that the weights may be unequal for some stratified sampling designs.

The stratified sampling estimator $\hat{t}_{\text{str}}$ can be expressed as a weighted sum of the individual sampling units. Using (3.1),

$$\hat{t}_{\text{str}} = \sum_{h=1}^{H} N_h \bar{y}_h = \sum_{h=1}^{H} \sum_{j \in \mathcal{S}_h} \frac{N_h}{n_h} y_{hj}.$$

The estimator of the population total in stratified sampling may thus be written as

$$\hat{t}_{\text{str}} = \sum_{h=1}^{H} \sum_{j \in \mathcal{S}_h} w_{hj} y_{hj}, \tag{3.9}$$

where the sampling weight for unit j of stratum h is $w_{hj} = (N_h/n_h)$. The sampling weight can again be thought of as the number of units in the population represented by the sample member y_{hj}. If the population has 1600 men and 400 women, and the stratified sample design specifies sampling 200 men and 200 women, then each man in the sample has weight 8 and each woman has weight 2. Each woman in the sample represents herself and another woman not selected to be in the sample, and each man represents himself and seven other men not in the sample. Note that the probability of including unit j of stratum h in the sample is $\pi_{hj} = n_h/N_h$, the sampling fraction in stratum h. Thus, as before, the sampling weight is simply the reciprocal of the inclusion probability:

$$w_{hj} = \frac{1}{\pi_{hj}}. \tag{3.10}$$

The sum of the sampling weights in stratified random sampling equals the population size N; each sampled unit in stratum h "represents" N_h/n_h the population, so the whole sample "represents" the whole population. In a stratified random sample, the population mean is thus estimated by:

$$\bar{y}_{\text{str}} = \frac{\displaystyle\sum_{h=1}^{H} \sum_{j \in \mathcal{S}_h} w_{hj} y_{hj}}{\displaystyle\sum_{h=1}^{H} \sum_{j \in \mathcal{S}_h} w_{hj}}. \tag{3.11}$$

Figure 3.2 illustrates a stratified random sample for a population with 3 strata. The sampling weights are smallest in Stratum 1, where half of the stratum population units are sampled.

A stratified sample is self-weighting if the sampling fraction n_h/N_h is the same for each stratum. In that case, the sampling weight for each observation is N/n, exactly the same as in an SRS. The variance of a stratified random sample, however, depends on the stratification, so (3.5) must be used to estimate the variance of $\hat{t}_{\text{str}}$. Equation (3.5) requires that you calculate the variance separately within each stratum; the weights do not tell you the stratum membership of the observations.

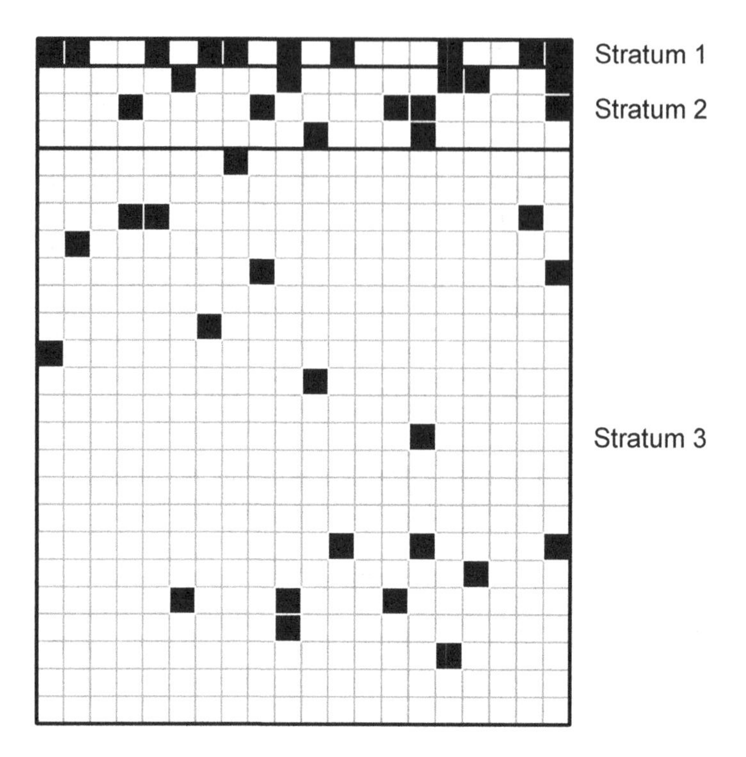

FIGURE 3.2
A stratified random sample from a population with $N = 500$. The top row is Stratum 1; rows 2–4 are Stratum 2; the bottom 21 rows are Stratum 3. Units in the sample are shaded. Stratum 1 has $N_1 = 20$ and $n_1 = 10$, so the sampling weight for each unit in Stratum 1 is 2. Each unit in Stratum 2, with $N_2 = 60$ and $n_2 = 12$, has sampling weight 5. Each unit in Stratum 3, with $N_3 = 420$ and $n_3 = 20$, has sampling weight 21.

Example 3.5. For the caribou survey in Example 3.3, the weights are

Stratum	N_h	n_h	w_{hj}
A	400	98	4.08
B	30	10	3.00
C	61	37	1.65
D	18	6	3.00
E	70	39	1.79
F	120	21	5.71

In stratum A, each sampling unit represents 4.08 sampling units in the stratum (including itself); in stratum B, a sampling unit in the sample represents itself and 2 other sampling units that are not in the sample. The population total is estimated using (3.9) with a weight variable that contains the value 4.08 for each of the 98 sampled units in stratum A, 3.00 for each of the 10 sampled units in stratum B, and so on. ■

Example 3.6. The sample in Example 3.2 was designed so that each county in the United States would have approximately the same probability of appearing in the sample. To estimate the total number of acres devoted to agriculture, we created the variable *strwt* in file `agstrat.csv` with the sampling weights; *strwt* equals 220/21 for counties in the Northeast stratum, 1054/103 for the North Central counties, 1382/135 for the South counties, and 422/41 for the West counties. We can use (3.9) to estimate the population total by forming a new variable containing the product of variables *strwt* and *acres92*, then calculating the sum of the new variable (this same procedure was used in Table 2.1 for an SRS). In doing so, we calculate $\hat{t}_{\text{str}} = 909{,}736{,}035$, the same estimate as obtained in Example 3.2. Note that even though this sample is approximately self-weighting, it is not exactly self-weighting because the stratum sample sizes must be integers. When calculating estimates, use the actual weights from each stratum.

The variable *strwt* can be used to estimate population means or totals for every variable measured in the sample, and most computer software packages for surveys use the weight variable to calculate point estimates. Note, however, that you cannot calculate the standard error of $\hat{t}_{\text{str}}$ with the weights alone—you need to use (3.5), or specify the stratification design information in the software, to estimate the variance. ■

3.4 Allocating Observations to Strata

So far, we have simply analyzed data from a survey that someone else has designed. Designing the survey is the most important part of using a survey in research: If the survey is badly designed, then no amount of analysis will yield the needed information. Survey design includes methods for controlling nonsampling as well as sampling error. We discuss design issues for nonsampling error in Chapters 8 and 16. In this chapter, we discuss design features that affect the sampling error.

Simple random sampling involved one design feature: the sample size (Section 2.7). For stratified random sampling, we need to define the strata, then decide how many observations to sample in each stratum. It is somewhat easier to look at these steps in reverse order, so let's start with methods of allocating observations to strata that have already been defined.

3.4.1 Proportional Allocation

If you are taking a stratified sample in order to ensure that the sample reflects the population with respect to the stratification variable and you would like your sample to be a miniature version of the population, you should use proportional allocation.

In **proportional allocation**, so called because the number of sampled units in each stratum is proportional to the size of the stratum, the inclusion probability for unit j in stratum h, $\pi_{hj} = n_h/N_h$, is the same ($= n/N$) for all strata. In a population of 2,400 men and 1,600 women, proportional allocation with a 10% sample specifies sampling 240 men and 160 women. Thus the probability that an individual will be selected to be in the sample, n/N, is the same as in an SRS, but many of the "bad" samples that could occur in an SRS (for example, a sample in which all 400 persons are men) cannot be selected in a stratified sample with proportional allocation.

With proportional allocation, each unit in the sample represents the same number of units in the population. In our example, each man in the sample represents 10 men in the population, and each woman represents 10 women in the population. The sampling weight for every unit in the sample thus equals 10, and the stratified sampling estimator of the population mean is simply the average of all of the observations. Proportional allocation thus results in a self-weighting sample, where $\bar{y}_{\text{str}}$ is the average of all observations in the sample.

When the strata are large enough, the variance of $\bar{y}_{\text{str}}$ under proportional allocation is usually at most as large as the variance of the sample mean from an SRS with the same number of observations. This is true no matter how silly the stratification scheme may be. To see why, let's display the variances between strata and within strata, for proportional allocation, in an Analysis of Variance (ANOVA) table for the population (Table 3.6).

TABLE 3.6
Population ANOVA table.

Source	df	Sum of Squares
Between strata	$H-1$	$\text{SSB} = \sum_{h=1}^{H}\sum_{j=1}^{N_h}(\bar{y}_{h\mathcal{U}} - \bar{y}_{\mathcal{U}})^2 = \sum_{h=1}^{H} N_h(\bar{y}_{h\mathcal{U}} - \bar{y}_{\mathcal{U}})^2$
Within strata	$N-H$	$\text{SSW} = \sum_{h=1}^{H}\sum_{j=1}^{N_h}(y_{hj} - \bar{y}_{h\mathcal{U}})^2 = \sum_{h=1}^{H}(N_h-1)S_h^2$
Total, about $\bar{y}_{\mathcal{U}}$	$N-1$	$\text{SSTO} = \sum_{h=1}^{H}\sum_{j=1}^{N_h}(y_{hj} - \bar{y}_{\mathcal{U}})^2 = (N-1)\,S^2$

In a stratified sample of size n with proportional allocation, $n_h/N_h = n/N$. Equation (3.3) and Table 3.6 then imply that

$$
\begin{aligned}
V_{\text{prop}}(\hat{t}_{\text{str}}) &= \sum_{h=1}^{H}\left(1 - \frac{n_h}{N_h}\right) N_h^2 \frac{S_h^2}{n_h} \\
&= \left(1 - \frac{n}{N}\right)\frac{N}{n}\sum_{h=1}^{H} N_h S_h^2 \\
&= \left(1 - \frac{n}{N}\right)\frac{N}{n}\left(\text{SSW} + \sum_{h=1}^{H} S_h^2\right).
\end{aligned}
$$

The sums of squares add up, with SSTO = SSW + SSB, so the variance of the estimated population total from an SRS of size n is:

$$
\begin{aligned}
V_{\text{SRS}}(\hat{t}_{\text{SRS}}) &= \left(1 - \frac{n}{N}\right) N^2 \frac{S^2}{n} \\
&= \left(1 - \frac{n}{N}\right)\frac{N^2}{n}\frac{\text{SSTO}}{N-1} \\
&= \left(1 - \frac{n}{N}\right)\frac{N^2}{n(N-1)}(\text{SSW} + \text{SSB}) \\
&= V_{\text{prop}}(\hat{t}_{\text{str}}) + \left(1 - \frac{n}{N}\right)\frac{N}{n(N-1)}\left[N(\text{SSB}) - \sum_{h=1}^{H}(N - N_h)S_h^2\right]. \qquad (3.12)
\end{aligned}
$$

Equation (3.12) shows that proportional allocation with stratification always gives smaller variance than an SRS *unless*

$$
\text{SSB} < \sum_{h=1}^{H}\left(1 - \frac{N_h}{N}\right) S_h^2. \qquad (3.13)
$$

This rarely happens when the N_h are large and the stratum means differ; generally, the large population sizes of the strata will force $N_h(\bar{y}_{h\mathcal{U}} - \bar{y}_{\mathcal{U}})^2 > S_h^2$. In general, the variance of the estimator of t from a stratified sample with proportional allocation will be smaller than the variance of the estimator of t from an SRS with the same number of observations.

The more unequal the stratum means $\bar{y}_{h\mathcal{U}}$, the more precision you will gain by using proportional allocation. The variance of $\hat{t}_{\text{str}}$ depends primarily on SSW; since SSTO is a fixed value for the finite population, SSW is smaller when SSB is larger. Of course, this result only holds for population variances; it is possible for a variance estimate from proportional allocation to be larger than that from an SRS merely because the particular sample selected had large within-stratum sample variances.

Example 3.7. Lohr (2019b) selected a stratified random sample from the set of 529 statisticians who had been awarded Fellow of the American Statistical Association (ASA) between 2000 and 2018. One of the goals was to estimate the percentage of Fellows who had majored in mathematics as undergraduate students (see Exercise 14). She stratified the list by gender and year of award, and sampled approximately 10% of the Fellows in each stratum.

This yielded an approximately self-weighting sample that was guaranteed to have about the same percentage of women, and the same percentage of persons awarded Fellow from each specific year, as the population. The sample design was easy to explain to a general audience, since each Fellow had the same chance of appearing in the sample, and sample estimates of population percentages could be described as the percentage of the sample having the characteristic. ■

3.4.2 Optimal Allocation

If the variances S_h^2 are more or less equal across all the strata, proportional allocation is probably the best allocation for increasing precision. In cases where the S_h^2 vary greatly among the strata, **optimal allocation** can result in higher precision or lower costs. In practice, when we are sampling units of different sizes, the larger units are likely to be more variable than the smaller units, and we would sample them with a higher sampling fraction. For example, if we were to take a sample of American corporations with the goal of estimating the amount of trade with Europe, the variation among large corporations would be greater than the variation among small ones. As a result, we would sample a higher percentage of the large corporations. Optimal allocation works well for sampling units such as corporations, cities, and hospitals, which vary greatly in size. It is also effective when some strata are much more expensive to sample than others.

Example 3.8. One goal of the Canadian Survey of Employment, Payrolls and Hours (Statistics Canada, 2020) is to estimate the weekly payroll and the number of hours worked for each industry sector across Canada. Statistics Canada selects a stratified random sample of about 15,000 businesses from the Business Register, a continually updated listing of the businesses operating in Canada. The sample is stratified by province, industry sector, and number of employees (establishment size). One would, in general, expect businesses in the large-establishment-size strata to have higher payroll and total number of hours worked, and to have higher variability in these quantities, than businesses with smaller establishment sizes. Taking disproportionally larger sample sizes in the large-establishment-size strata will therefore reduce the variance in Equation (3.3) for the estimate of total payroll.

About 800 businesses are considered so important to the quality of the estimates that they are permanently in the sample. These businesses are in *certainty strata*, where units are selected with probability 1 and thus have sampling weight 1. Units in certainty strata represent themselves alone, and do not contribute to the variance $V(\hat{t}_{\text{str}})$ in a stratified random sample because the fpc $(1 - n_h/N_h)$ for the stratum equals 0. ■

Example 3.9. How are musicians paid when their compositions are performed? In the U.S., many composers are affiliated with the American Society of Composers, Authors and Publishers (ASCAP, 2015). Television networks, local television and radio stations, background music services, internet streaming services, symphony orchestras, restaurants, nightclubs, and other operations pay ASCAP an annual license fee, based largely on the size of the audience, that allows them to play compositions in the ASCAP catalog. ASCAP then distributes royalties to composers whose works are played.

Theoretically, ASCAP members should get royalties every time one of their compositions is played. When feasible, ASCAP takes a census of every piece of music played on a medium. For example, ASCAP relies on television producers' cue sheets, which provide details on the music used in a program, to identify and tabulate musical pieces played on major network television and cable channels.

Stratified sampling is used to sample radio stations for the survey (Massarsky, 2013). Radio stations are grouped into strata based on the license fee paid to ASCAP, the type of community the station is in, and the geographic region. As stations paying higher license fees contribute more money for royalties, they are more likely to be sampled. Experts identify the musical compositions aired in subsampled time periods from the sampled stations. ASCAP thus uses a form of optimal allocation. Strata with the highest radio fees, and thus with the highest variability in royalty amounts, have larger sampling fractions than strata containing radio stations that pay small fees. ■

The objective in optimal allocation is to gain the most information for the least cost. Let c_0 represent overhead costs such as maintaining an office and c_h represent the cost of taking an observation in stratum h. The costs $c_0, c_1, c_2, \ldots, c_H$ are assumed to be known. Then a simple function specifying the total cost C for conducting the survey is:

$$C = c_0 + \sum_{h=1}^{H} c_h n_h. \tag{3.14}$$

Optimal allocation assigns units to strata so as to minimize $V(\bar{y}_{\text{str}})$ in (3.6) for a fixed total cost C, or, equivalently, to minimize C for a fixed $V(\bar{y}_{\text{str}})$. For either minimization, the optimal allocation has n_h proportional to $N_h S_h/\sqrt{c_h}$ for each h (see Exercise 35). Thus, the optimal sample size in stratum h is

$$n_h = \left(\frac{\dfrac{N_h S_h}{\sqrt{c_h}}}{\displaystyle\sum_{l=1}^{H} \frac{N_l S_l}{\sqrt{c_l}}} \right) n. \tag{3.15}$$

Choice of n is discussed in Section 3.4.5 and Exercise 35.

If $S_h/\sqrt{c_h} = S_l/\sqrt{c_l}$ for all strata h and l, then (3.15) simplifies to proportional allocation with $n_h = nN_h/N$. Otherwise, the allocation is **disproportional**—one or more of the strata is **oversampled** relative to its share of the population. For oversampled strata, $n_h/n > N_h/N$.

Optimal allocation specifies taking a larger sample size in a stratum if:

- The stratum accounts for a large part of the population (N_h is large).
- The variance within the stratum (S_h^2) is large; we sample more units to compensate for the heterogeneity.
- Sampling in the stratum is inexpensive (c_h is small).

Sometimes applying the optimal allocation formula in Equation (3.15) results in one or more of the "optimal" n_h's being larger than the population size N_h in those strata. In that case, take a sample size of N_h in those strata, and then apply (3.15) again with the remaining strata.

Neyman allocation is a special case of optimal allocation, used when the costs in the strata (but not the variances) are approximately equal. Under Neyman allocation, n_h is proportional to $N_h S_h$. If the variances S_h^2 are specified correctly, Neyman allocation will give an estimator with smaller variance than proportional allocation (see Exercise 36).

Example 3.10. The caribou survey in Example 3.3 used Neyman allocation to determine the n_h. Before taking the survey, the investigators obtained approximations of the caribou densities and distribution, and constructed strata to be relatively homogeneous in terms of population density. They set the total sample size as $n = 225$. They then used the estimated count in each stratum as a rough estimate of the standard deviation, with the result shown in Table 3.7. The investigators wanted the sampling fraction to be at least 1/3 in smaller strata, so they used the Neyman allocations in the n_h column as a guideline for determining the final sample sizes in the last column. ■

TABLE 3.7
Quantities used for designing the caribou survey in Example 3.10.

Stratum	N_h	s_h	$N_h s_h$	$n_h = 225\frac{N_h s_h}{\sum_l N_l s_l}$	**Sample size**
A	400	3,000	1,200,000	96.26	98
B	30	2,000	60,000	4.81	10
C	61	9,000	549,000	44.04	37
D	18	2,000	36,000	2.89	6
E	70	12,000	840,000	67.38	39
F	120	1,000	120,000	9.63	21
Total	699		2,805,000	225.00	211

3.4.3 Allocation for Specified Precision within Strata

Sometimes you are less interested in the precision of the estimate of the population total or mean for the whole population than in comparing means or totals among different strata. In that case, you would determine the sample size needed for the individual strata using the guidelines in Section 2.7.

Example 3.11. The U.S. Postal Service has conducted surveys asking postal customers about their perceptions of the quality of mail service. The population of residential postal service customers is stratified by geographic area, and it is desired that the precision be ± 3 percentage points, at a 95% confidence level, within each area. If there were no nonresponse, such a requirement would lead to sampling at least 1067 households in each stratum, as calculated in Example 2.12. Such an allocation is neither proportional, as the number of residential households in the population varies a great deal from stratum to stratum, nor optimal in the sense of providing the greatest efficiency for estimating percentages for the whole population. It does, however, provide the desired precision within each stratum. ■

3.4.4 Which Allocation to Use?

Proportional allocation is a good choice if you want the sample to be a miniature version of the population. Every unit in the sample has weight of approximately N/n, just as in an SRS, and represents approximately the same number of units in the population. Estimates from a proportionally allocated stratified random sample usually have smaller variances than corresponding estimates from an SRS of the same size. The stratified sample is forced to mirror the population exactly with respect to the stratification variables. The sampling distributions of variables that are associated with the stratification variables are carried along, and they become closer to their population distributions as well.

A proportionally allocated sample often "feels" more representative than an SRS. The sample in Example 3.2 is distributed among the regions proportionally to their population sizes. In an SRS, the number of sampled counties from each region is a random variable, and varies from sample to sample. The SRS in Example 2.6, for example, has 24 counties from the Northeast. A different SRS selected from the population may well have a different number of counties from the Northeast. It is even possible (though unlikely) that an SRS would have *no* counties from the Northeast. By contrast, *every* possible sample from this proportionally allocated design will have exactly 21 counties from the Northeast.

If y is related to the stratification variables (that is, the population means $\bar{y}_{h\mathcal{U}}$ differ among the strata), then the proportionally allocated stratified sample is expected to resemble the population more closely than an SRS of the same size. The proportional allocation does not improve the precision of estimates for variables that are unrelated to the stratification variables, but it does no harm.

Disproportional allocations specify oversampling one or more of the strata. There are often very good reasons to oversample some strata. You may want to guarantee that estimates calculated from individual strata will achieve a specified degree of precision, as in Example 3.11. Or you may want to use Neyman allocation to oversample strata thought to have high variances.

But a disproportional allocation is more complicated, and you want to make sure that the efficiency gains are worth giving up the simplicity and self-weighting property of proportional allocation. A stratified sample with disproportional allocation is representative, in the sense that it can be used to estimate any numerical quantity that could be calculated from the population, but it is not a small-scale replica of the population. You *must* use the sampling weights when graphing data and computing estimates; otherwise, your estimates will be biased. If you sample 200 men and 200 women from a population of 20,000 men and 5,000 women, each sampled man represents 100 men in the population and each sampled woman represents 25 women in the population. If men tend to be taller than women and you report the sample average of the 400 sampled persons, ignoring the weights, your estimate of the average height in the population will be too small.

Neyman allocation specifies oversampling strata that have higher variances S_h^2. This reduces the variance of $\bar{y}_{\text{str}}$ for the particular response y considered for the allocation, but not necessarily for other response variables.

Example 3.12. The file `college.csv`, abstracted from online data published by the U.S. Department of Education (2020), contains selected variables on each institution in a population of 1,372 U.S. colleges and universities. Suppose that the primary variables to be studied in a proposed survey from this population, such as total instructional budget or number of students who have full-time employment five years after graduation, are thought to be related to the size of the institution. Table 3.8 shows ten strata defined by size and residential status (variable *ccsizset*). The variances of the variables to be measured in the survey are unknown, but may be thought to be roughly proportional to the variance of *ugds*, the number of undergraduate students at the institution. The value of S_h in Table 3.8 is the population standard deviation of *ugds* for the colleges in stratum h.

TABLE 3.8
Allocations for a stratified sample of size 200 in Example 3.12.

Stratum	N_h	S_h	**Proportional**	**Neyman**
Very small	195	251	28	3
Small, primarily nonresidential	45	784	7	2
Small, primarily residential	123	515	18	3
Small, highly residential	347	508	51	10
Medium, primarily nonresidential	80	2490	12	11
Medium, primarily residential	160	2150	23	19
Medium, highly residential	158	1473	23	13
Large, primarily nonresidential	95	11273	14	59
Large, primarily residential	126	9178	18	64
Large, highly residential	43	6844	6	16
Total	1372		200	200

The Neyman allocation has much larger sample sizes in the last three strata than the proportional allocation. This results in much lower variances for the estimated mean of variable *ugds*, which provided the population variances for the table. The anticipated variance for the mean of *ugds*, calculated using (3.4), is 82,937 for proportional allocation and 20,814 for Neyman allocation. Either allocation gives a huge reduction from the variance of the mean of *ugds* from an SRS of size 200, which is 270,689.

The gains in precision from the Neyman allocation are not likely to be as dramatic for variables that were not used for the optimization, since these variables have different relative values of S_h^2. But one would expect the Neyman allocation to reduce variances for variables that are correlated with *ugds* such as total instructional budget.

The Neyman allocation would be expected to yield higher variances than proportional allocation, however, for variables that are not related to *ugds*. Consider variable *majwomen*, which equals 1 if the majority of the undergraduate students in the college are women and 0 otherwise. The correlation between *majwomen* and *ugds* is about -0.07. The anticipated margin of error for estimating the percentage of colleges with *majwomen* $= 1$ under proportional allocation with $n = 200$ is about 5.1 percentage points, slightly less than the anticipated margin of error for an SRS (5.2 percentage points). But the anticipated margin of error for Neyman allocation is about 11 percentage points. For *majwomen*, proportional allocation results in a slightly lower variance than an SRS, but the variance from Neyman allocation is much higher than that from an SRS.

This example illustrates some of the trade-offs that must be considered when deciding on an allocation. An allocation that is optimal for one variable may be sub-optimal for another. For this survey, where most of the variables of interest are thought to be correlated with *ugds* but a few variables such as *majwomen* are not, one might want to consider an allocation that is between the proportional and Neyman (for *ugds*) allocations. ■

Allocation methods provide guidelines, not prescriptions. You can explore as many designs as you like before choosing one. And you may want to mix features of different allocations so that your sample will be suitable for multiple objectives. For example, in a stratified random sample of persons, you may want to assign an initial sample size of 400 persons to each stratum, so that estimates of percentages in each stratum have margin of error at most 5 percentage points, and then assign the remaining sample size proportionally or optimally to the strata.

3.4.5 Determining the Total Sample Size

The total sample size n for a stratified random sample should be large enough to achieve the desired precision for key variables. After strata are constructed (see Section 3.5) and observations allocated to strata, Equation (3.4) can be used to determine the sample size necessary to achieve a prespecified margin of error e.

As we saw in (3.15), proportional and optimal allocation methods specify the relative allocations n_h/n, so let's write the margin of error e for stratified sampling as a function of the relative allocations. Define

$$v = \sum_{h=1}^{H} \frac{n}{n_h}\left(\frac{N_h S_h}{N}\right)^2 \tag{3.16}$$

and write

$$e = z_{\alpha/2}\sqrt{V(\bar{y}_{\text{str}})} = z_{\alpha/2}\sqrt{\sum_{h=1}^{H}\left(1-\frac{n_h}{N_h}\right)\left(\frac{N_h}{N}\right)^2\frac{S_h^2}{n_h}} = z_{\alpha/2}\sqrt{\frac{v}{n} - \sum_{h=1}^{H}\frac{N_h S_h^2}{N^2}}. \tag{3.17}$$

Solving for n in (3.17) gives

$$n = \frac{v z_{\alpha/2}^2}{e^2 + z_{\alpha/2}^2 \displaystyle\sum_{h=1}^{H}\frac{N_h S_h^2}{N^2}} = \frac{n_0}{1 + \dfrac{n_0}{N}\displaystyle\sum_{h=1}^{H}\frac{N_h}{N}\frac{S_h^2}{v}}, \tag{3.18}$$

where $n_0 = z_{\alpha/2}^2 v/e^2$ is the sample size that would be used if all stratum fpcs could be ignored. The sample size for an SRS in (2.31) is a special case of (3.18) when $H = 1$.

The quantity v in (3.16) can be thought of as an "average" variability per unit in a stratified random sample with the specified allocation, just as S^2 in (2.9) is the variability per unit in an SRS. If the stratum fpcs are close to 1, then $V(\bar{y}_{\text{str}}) \approx v/n$, while the variance of the mean from an SRS of the same size is $V(\bar{y}_{\text{SRS}}) \approx S^2/n$. With proportional allocation, $v = \sum_{h=1}^{H}(N_h/N)S_h^2$ is a weighted average of the within-stratum variances and thus will usually be smaller than the overall population variance S^2.

When the fpcs can be ignored, the sample size needed to achieve margin of error e for stratified sampling can be calculated as $(n_{\text{SRS}})v/S^2$, where n_{SRS} is the sample size calculated for an SRS in (2.31). If $v < S^2$, as it almost always will be for proportional allocation (and for some variables with optimal allocation), then stratified sampling allows you to achieve a desired precision with a smaller sample size.

As we saw in Example 3.12, the variance reduction from stratified sampling may be greater for some variables than for others. The final sample size should be determined so as to meet the precision requirements for all (or most) of the key survey variables.

3.5 Defining Strata

As argued in Section 3.4, stratified sampling with proportional allocation improves (or does not harm) the precision for every variable y measured in the survey. If you have information in the sampling frame that can be used to form the strata, and you know the stratum membership of every population unit, it is almost always better, from a variance standpoint, to take a proportionally allocated stratified random sample than an SRS. Sometimes the efficiency can be improved even more by using a disproportional allocation.

Remember, stratification is most efficient when the stratum means differ widely; then the between sum of squares in Table 3.6 is large, and the variability within strata will be smaller. Consequently, when constructing strata, we want the stratum means to be as different as possible. Ideally, we would stratify by the values of y; if our survey is to estimate total business expenditures on advertising, we would like to put businesses that spent the most on advertising in stratum 1, businesses with the next highest level of advertising expenditures in stratum 2, and so on, until the last stratum contained businesses that spent nothing on advertising. The problem with this scheme is that we do not know the advertising expenditures for all the businesses while designing the survey—if we did, we would not need to do a survey at all!

Instead, we try to find some variables closely related to y. For estimating total business expenditures on advertising, we might stratify by number of employees or size of the business and by the type of product or service. For farm income, we might use the size of the farm as a stratifying variable, since we expect larger farms to have higher incomes. The sample in Example 3.12 was stratified by size and residential status, since we expect those variables to be related to variables of interest such as the college's instructional budget, the number of students who are working full time while attending college, the number of students who are employed five years after graduation, and similar quantities.

Most surveys measure more than one variable, so any stratification variable should be related to many characteristics of interest. In the Canadian Survey of Employment, Payrolls, and Hours, business establishments are stratified by industry, province, and estimated number of employees. If several stratification variables are available, use the variables associated with the most important responses.

How many strata? The number of strata you choose depends upon many factors, such as the difficulty in constructing a sampling frame with stratifying information, and the cost of stratifying. A general rule to keep in mind is: The more information you have, the more strata you should use. Thus, if little prior information about the target population is available, you may want to take an SRS. If you have a great deal of prior information on variables thought to be associated with variables to be measured in the survey, you may want to use a highly stratified design—many surveys are stratified to the point that only two sampling units are observed in each stratum.

You can often collect preliminary data that can be used to stratify your design. If you are taking a survey to estimate the number of fish in a region, you can use physical features of the area that are related to fish density, such as depth, salinity, and water temperature. Or you can use survey information from previous years, or data from a preliminary cruise to aid in constructing strata.

The information used to construct the strata does not have to be perfect. Example 3.12 stratified the population of colleges by institution size. But even if some colleges were misclassified—for example, if a large residential college was mistakenly placed in one of the other strata—the misclassification does not affect the validity of estimates from the survey, whose properties depend only on the stratification that was used.

For many surveys, stratification can increase precision dramatically and often well repays the effort used to construct the strata. Example 3.13 describes how strata were constructed for estimating the number of persons experiencing homelessness in New York City.

Example 3.13. In the United States, state and local agencies responsible for coordinating homeless services produce annual estimates of the number and characteristics of persons experiencing homelessness in their regions. Each agency chooses one of the last ten nights in January to conduct the count of the sheltered (persons in emergency shelters or transitional housing) and unsheltered (persons staying in a place not intended for human habitation, such as on the streets, under bridges, or in cars, tents, bus or subway stations, or abandoned buildings) populations at that particular point in time. The survey is conducted on one night to reduce possible double-counting, and to reduce the effort required for the teams of volunteers who canvass areas to survey the unsheltered population (statistics about the sheltered population are obtained directly from the shelters). Most agencies also use the survey as an opportunity to disseminate information about resources and social services that are available for persons experiencing homelessness, and to collect data about needs and demographics that can be used in subsequent community outreach efforts. Thus, it is desired to have a sampling plan that (1) produces as accurate an estimate of the size of the unsheltered population as possible and (2) allows for contact with as many unsheltered persons in the area as possible.

Many cities have large land area, making it impossible for volunteers to visit every location in the city during one night. New York City, with a land area of about 800 square kilometers, takes a carefully designed stratified random sample to conduct its annual point-in-time count (Schneider et al., 2016; New York City Department of Homeless Services, 2019).

The city is partitioned into approximately 7,000 areas. Using previous years' point-in-time counts as well as information from service providers and community organizations, each area is classified in the autumn before the count as "high-density" or "low-density." An area in Manhattan or in the subway system is high-density if expected to contain at least two persons experiencing homelessness; an area in the Bronx, Queens, Staten Island, or Brooklyn is high-density if expected to contain at least one person experiencing homelessness. Areas not classified as high-density are considered to be low-density.

This classification results in 12 strata for the city: one high-density stratum and one low-density stratum for each of the 5 boroughs and for the subway system. Most of the areas in the low-density strata are expected to have low counts of persons experiencing homelessness. If teams of volunteers were sent to all of the low-density areas, many would end up with a count of zero. On the other hand, if none of the low-density areas were visited, the city's estimated count would be too small because it is likely that some of the low-density areas contain persons experiencing homelessness, and a few may, despite the classification as low-density, have a large number.

New York City solves this problem by using different sampling fractions for the strata. The sampling fraction for each of the six high-density strata is 1; all of the areas in high-density strata are canvassed. The sampling fractions for the low-density strata vary from borough to borough and from year to year, and are designed to give an estimate with high precision for each borough and for the subway system (see Exercise 30). In 2019, about 21 percent of the 7,000 areas in the city were canvassed.

The stratified sampling design allows New York City to attain its objectives for the one-night count with the number of volunteers available (typically between 2,000 and 2,500). The design deploys most of the volunteers to areas thought to contain unsheltered persons and thus obtains a large number of unsheltered persons in the sample (and, with the larger sample size, more information about their characteristics), yet still produces statistically valid estimates because a random sample of areas is visited from each low-density stratum.

Even with the careful design, however, the point-in-time count has measurement error for estimating the number of persons experiencing homelessness on the night of the count. Some persons remain out of sight or in locations deemed unsafe for the volunteers to visit. In other instances, the volunteers may mistakenly decide that persons encountered are not homeless when they are—or are homeless when they are not. Volunteers may also fail to survey one or more individuals on their routes; Exercise 14 of Chapter 13 describes one method the city uses to estimate how many persons are missed. ■

3.6 Model-Based Theory for Stratified Sampling*

The one-way ANOVA model with fixed effects provides an underlying structure for stratified sampling. Here,

$$Y_{hj} = \mu_h + \varepsilon_{hj}, \tag{3.19}$$

where the ε_{hj}s are independent with mean 0 and variance σ_h^2. Then the least squares estimator of μ_h is $\bar{Y}_h$, the average in stratum h.

Let the random variable

$$T_h = \sum_{j=1}^{N_h} Y_{hj}$$

represent the total in stratum h and the random variable

$$T = \sum_{h=1}^{H} T_h$$

represent the overall total. From Section 2.10, the best linear unbiased estimator for T_h is

$$\hat{T}_h = \frac{N_h}{n_h} \sum_{j \in \mathcal{S}_h} Y_{hj}.$$

Then, from the results shown for simple random sampling in Section 2.10,

$$E_M[\hat{T}_h - T_h] = 0$$

and

$$E_M[(\hat{T}_h - T_h)^2] = N_h^2 \left(1 - \frac{n_h}{N_h}\right) \frac{\sigma_h^2}{n_h}.$$

Since observations in different strata are independent under the model in (3.19),

$$\begin{aligned}
E_M[(\hat{T} - T)^2] &= E_M\left[\left\{\sum_{h=1}^{H} (\hat{T}_h - T_h)\right\}^2\right] \\
&= E_M\left[\sum_{h=1}^{H} (\hat{T}_h - T_h)^2 + \sum_{h=1}^{H} \sum_{k \neq h} (\hat{T}_h - T_h)(\hat{T}_k - T_k)\right] \\
&= E_M\left[\sum_{h=1}^{H} (\hat{T}_h - T_h)^2\right] \\
&= \sum_{h=1}^{H} N_h^2 \left(1 - \frac{n_h}{N_h}\right) \frac{\sigma_h^2}{n_h}.
\end{aligned}$$

The theoretical variance σ_h^2 can be estimated by s_h^2. Adopting the model in (3.19) results in the same estimators for t and its standard error as found under randomization theory in (3.6). If a different model is used, however, then different estimators are obtained.

3.7 Chapter Summary

Stratification uses additional information about a population in the survey design. In the simplest form, stratified random sampling, we take an SRS of size n_h from each stratum h in the population. The samples from the H strata are selected independently. To use stratification, we must know the relative population size N_h/N for each stratum; we must also know the stratum membership for every unit in the population. The inclusion probability for unit i in stratum h is $\pi_{hi} = n_h/N_h$; consequently, the sampling weight for that unit is $w_{hi} = N_h/n_h$.

To estimate the population total t using a stratified random sample, let $\hat{t}_h$ estimate the population total in stratum h. Then

$$\hat{t}_{\text{str}} = \sum_{h=1}^{H} \hat{t}_h = \sum_{h=1}^{H} \sum_{j \in \mathcal{S}_h} w_{hj} y_{hj}$$

and

$$\hat{V}(\hat{t}_{\text{str}}) = \sum_{h=1}^{H} \hat{V}(\hat{t}_h) = \sum_{h=1}^{H} \left(1 - \frac{n_h}{N_h}\right) N_h^2 \frac{s_h^2}{n_h}.$$

The population mean $\bar{y}_{\mathcal{U}} = t/N$ is estimated by:

$$\bar{y}_{\text{str}} = \frac{\hat{t}_{\text{str}}}{N} = \sum_{h=1}^{H} \frac{N_h}{N} \bar{y}_h = \frac{\sum_{h=1}^{H} \sum_{j \in \mathcal{S}_h} w_{hj} y_{hj}}{\sum_{h=1}^{H} \sum_{j \in \mathcal{S}_h} w_{hj}}$$

with $\hat{V}(\bar{y}_{\text{str}}) = \hat{V}(\hat{t}_{\text{str}})/N^2$.

Stratified sampling has three major design issues: defining the strata, choosing the total sample size, and allocating the observations to the defined strata. With proportional allocation, the same sampling fraction is used in each stratum. Proportional allocation almost always results in smaller variances for estimated means and totals than simple random sampling. Disproportional allocation may be preferred if some strata should have higher sampling fractions than others, for example, if it is desired to have larger sample sizes for strata with minority populations or for strata with large companies. Optimal allocation specifies taking larger sampling fractions in strata that have larger variances or lower sampling costs.

Key Terms

Certainty stratum: A stratum in which all units are selected to be in the sample; the inclusion probability for each unit in the stratum equals 1.

Disproportional allocation: Allocation of sampling units to strata so that the sampling fractions n_h/N_h are unequal.

Optimal allocation: Allocation of sampling units to strata so that the variance of the estimator is minimized for a given total cost.

Oversampling: Selecting more observations in a stratum than would be called for in proportional allocation. An oversampled stratum has $n_h/N_h > n/N$.

Proportional allocation: Allocation of sampling units to strata so that $n_h/N_h = n/N$ for each stratum. Proportional allocation results in a self-weighting sample.

Stratified random sampling: Probability sampling method in which population units are partitioned into strata and then an SRS is taken from each stratum.

Stratum: A subpopulation in which an independent probability sample is taken. Every unit in the population is in exactly one stratum. The strata together comprise the entire population.

For Further Reading

The references in Chapter 2 also describe stratified sampling. Chapter 4 of Raj (1968) gives a rigorous and concise treatment of stratified sampling theory. Cochran (1977) has further results on allocation and construction of strata, and uses ANOVA tables to compare precisions of sampling designs (first described in Cochran, 1939).

Allocation of stratified samples with multiple variables is often best done computationally. Bethel (1989) and Stokes and Plummer (2004) described algorithms for allocating a stratified sample when multiple variables are of interest. Brito et al. (2015) developed an integer programming method for optimal allocation.

Neyman (1934) wrote one of the most important papers in the historical development of survey sampling. He presented a framework for stratified sampling and demonstrated its superiority to purposive selection methods. Those interested in the history of survey sampling should read this paper, which shows how he arrived at the ideas of stratified sampling and confidence intervals.

3.8 Exercises

A. Introductory Exercises

1. What stratification variable(s) would you use for each of the following situations:

   (a) A political poll to estimate the percentage of registered voters in Arizona who approve of the governor's performance.

   (b) An e-mail survey of students at your university, to estimate the total amount of money students spend on textbooks in a term.

   (c) A sample of high schools in New York City to estimate the number of schools that offer one or more classes in computer programming.

   (d) A sample of public libraries in California to study the availability and usage of computer resources.

   (e) A survey of anglers visiting a freshwater lake, to learn about which species of fish are preferred.

(f) An aerial survey to estimate the number of walrus in the pack ice near Alaska between 173° East and 154° West longitude.

(g) A survey of businesses to learn about policies for paid leave.

2. Consider the hypothetical population and stratification below (this population is also used in Example 2.2), with $N_1 = N_2 = 4$.

Unit number	Stratum	y
1	1	1
2	1	2
3	1	4
8	1	8
4	2	4
5	2	7
6	2	7
7	2	7

Consider the stratified sampling design in which $n_1 = n_2 = 2$.

(a) Write out all possible SRSs of size 2 from stratum 1, and find the probability of each sample. Do the same for stratum 2.

(b) Using your work in (a), find the sampling distribution of $\hat{t}_{\text{str}}$.

(c) Find the mean and variance of the sampling distribution of $\hat{t}_{\text{str}}$. How do these compare to the mean and variance in Example 2.2?

3. Consider a population of 6 students. Suppose we know the test scores of the students to be

Student	1	2	3	4	5	6
Score	66	59	70	83	82	71

(a) Find the mean $\bar{y}_{\mathcal{U}}$ and variance S^2 of the population.

(b) How many SRS's of size 4 are possible?

(c) List the possible SRS's. For each, find the sample mean. Using Equation (2.12), find $V(\bar{y})$.

(d) Now let stratum 1 consist of students 1–3 and stratum 2 consist of students 4–6. How many stratified random samples of size 4 are possible in which 2 students are selected from each stratum?

(e) List the possible stratified random samples. Which of the samples from (c) cannot occur with the stratified design?

(f) Find $\bar{y}_{\text{str}}$ for each possible stratified random sample. Find $V(\bar{y}_{\text{str}})$, and compare it to $V(\bar{y})$.

4. For Example 3.4, construct a data set with 3,835 observations. Include three columns: column 1 is the stratum number (from 1 to 7), column 2 contains the response variable of gender (0 for males and 1 for females), and column 3 contains the sampling weight N_h/n_h for each observation. Using columns 2 and 3 along with (3.11), calculate $\hat{t}_{\text{str}}$ and $\hat{p}_{\text{str}}$, the total number and proportion of females. Is it possible to calculate $\text{SE}(\hat{p}_{\text{str}})$ by using only columns 2 and 3, with no additional information?

5. The survey in Example 3.4 collected additional information. Another of the survey's questions asked whether the respondent agreed with the following statement: "When I look at a new issue of my discipline's major journal, I rarely find an article that interests me." The results are as follows:

Discipline	**Agree (%)**
Literature	37
Classics	23
Philosophy	23
History	29
Linguistics	19
Political Science	43
Sociology	41

   (a) What is the sampled population in this survey?

   (b) Find an estimate of the population percentage of persons who agree with the statement, and give the standard error of your estimate.

6. Suppose that a city has 90,000 dwelling units, of which 35,000 are houses, 45,000 are apartments, and 10,000 are condominiums.

   (a) You believe that the mean electricity usage is about twice as much for houses as for apartments or condominiums, and that the standard deviation is proportional to the mean so that $S_1 = 2S_2 = 2S_3$. How would you allocate a stratified sample of 900 observations if you wanted to estimate the mean electricity consumption for all households in the city?

   (b) Now suppose that you take a stratified random sample with proportional allocation and want to estimate the overall proportion of households in which energy conservation is practiced. If 45% of house dwellers, 25% of apartment dwellers, and 3% of condominium residents practice energy conservation, what is p for the population? What gain would the stratified sample with proportional allocation offer over an SRS, that is, what is $V_{\text{prop}}(\hat{p}_{\text{str}})/V_{\text{SRS}}(\hat{p}_{\text{SRS}})$?

7. In Exercise 6 of Chapter 2, data on numbers of publications were given for an SRS of 50 faculty members. Not all departments were represented, however, in the SRS. The SRS contained several faculty members from psychology and from chemistry, but none from foreign languages. The following data are from a stratified sample of faculty, using the areas biological sciences, physical sciences, social sciences, and humanities as the strata.

Stratum	**Number of Faculty Members in Stratum**	**Number of Faculty Members in Sample**
Biological Sciences	102	7
Physical Sciences	310	19
Social Sciences	217	13
Humanities	178	11
Total	807	50

The frequency table for number of publications in the strata is given below.

Number of Refereed Publications	Number of Faculty Members			
	Biological	Physical	Social	Humanities
0	1	10	9	8
1	2	2	0	2
2	0	0	1	0
3	1	1	0	1
4	0	2	2	0
5	2	1	0	0
6	0	1	1	0
7	1	0	0	0
8	0	2	0	0

(a) Estimate the total number of refereed publications by faculty members in the college, and give the standard error.

(b) How does your result from (a) compare with the result from the SRS in Exercise 6 of Chapter 2?

(c) Estimate the proportion of faculty with no refereed publications, and give the standard error.

(d) Did stratification increase precision in this example? Explain why you think it did or did not.

8. A public opinion researcher has a budget of \$150,000 for interviewing households in a survey. She knows that 90% of all households have telephones. Telephone interviews cost \$100 each; in-person interviews cost \$300 each if all interviews are conducted in person, and \$400 each if only nonphone households are interviewed in person (because there will be extra travel costs). Assume that the variances in the phone and nonphone groups are similar. How many households should be interviewed in each group if

   (a) All households are interviewed in person.

   (b) Households with a phone are contacted by telephone and households without a phone are contacted in person.

9. *Objects in bin.* In Exercise 13 of Chapter 2, you selected an SRS of 200 objects was drawn from the population in file `shapespop.csv`. But an SRS does not take advantage of the knowledge that the population contains 15,000 squares and 5,000 circles.

   Draw a stratified sample of 200 objects from the population with proportional allocation, stratified by *shape*. Answer parts (c)–(e) of Exercise 13 of Chapter 2 for the stratified sample. How do the CIs for the stratified sample compare with those for the SRS? Why does the CI for total number of circles have zero width?

10. Repeat Exercise 9 for a stratified sample of 100 objects from each stratum.

11. *Jury selection in ancient Athens.* Aristotle (350 BCE) and Orlandini (2018) described how a kleroterion, a stone slab containing a grid of slots (see Figure 3.3), was used to select the Athenian city council and juries from the citizenry, which consisted of men who had been born free in Athens. Each citizen belonged to exactly one of ten tribes, and had a bronze ticket containing his name and information about tribe membership. The tickets for citizens from a tribe who were eligible and willing to serve were placed in a basket corresponding to that tribe. The presiding magistrate shook and then drew one ticket from each basket; this person, randomly selected to be the tribe's ticket-inserter,

drew tickets one by one from the tribe's basket and placed them in the kleroterion column corresponding to the tribe, starting with the slots in the first row. After the insertions, the first column of the kleroterion contained the tickets of citizens from tribe 1, the second column contained the tickets of citizens from tribe 2, and so on.

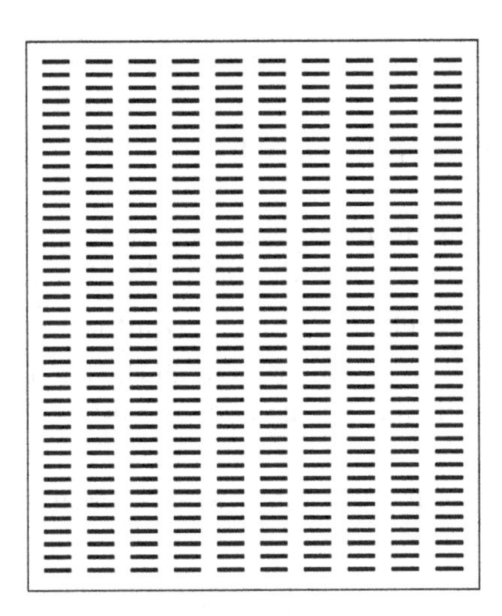

FIGURE 3.3
Schematic of a kleroterion. Each citizen's bronze identification ticket was placed in a slot in the column corresponding to his tribe.

Suppose that a jury of $10 \times J$ men was desired, and the shortest column in the kleroterion had K tickets. To select the jury, the magistrate drew marbles one at a time from a set of K marbles, of which J were white and the other $K - J$ were black. If the first marble drawn was white, then all ten citizens in the first row were selected for the jury; if it was black, all of the citizens in the first row were excused from service. Similarly, the ten citizens in the second row were selected if the second marble was white and excused if it was black, and so on until the prescribed number of jurors was selected.

Under what conditions will this procedure result in a set of jurors that is a stratified random sample of Athenian citizens? What are the strata and selection probabilities? When will the procedure produce a self-weighting sample?

B. Working with Survey Data

12. *Public health articles.* Probability sampling (random selection) allows researchers to generalize from the sample to the population, and random assignment in experiments allows researchers to infer causation, but how often are these methods employed in public health research? Hayat and Knapp (2017) drew a stratified random sample of 198 articles from the 547 research articles published in 2013 by three leading public health journals. For each article, they determined the number of authors, the type of statistical inference used (confidence intervals, hypothesis tests, both, or neither), and whether random selection or random assignment was used. The data are in file `healthjournals.csv`.

    The three journals served as the strata for this study, with population and sample sizes in the following table.

Journal	Population Size	Sample Size
American Journal of Public Health	280	100
American Journal of Preventive Medicine	103	38
Preventive Medicine	164	60

(a) What is the sampling weight for each stratum? What type of allocation was used in the sample design?

(b) This sample comprises more than one-third of the population. When should an fpc be used for analyses with this data set?

(c) Estimate the percentage of the 547 articles that used: (i) confidence intervals only, (ii) hypothesis tests only, (iii) both, and (iv) neither, along with 95% CIs.

(d) Give a point estimate and 95% CI for the mean number of authors per article.

(e) Estimate the percentage and total number of articles (among the 547 in the population) that use *neither* random selection nor random assignment, along with 95% CIs.

(f) Do the statistics calculated in this exercise describe *all* public health research articles? To what population do they apply?

13. An avid mystery reader selected a stratified random sample of 60 books from the list of 655 books that had been nominated through 2020 for an Edgar® award from the Mystery Writers of America. Twelve strata were formed using the cross-classification of when the book was nominated (1946–1980, 1981–2000, 2001–2020), type of nomination (best novel or best first novel), and whether the book won the award (yes or no). Four books were selected randomly from each of the three strata with award winners, and six books were sampled randomly from each stratum with books that were nominated but did not win; this procedure oversampled the strata with winning books.

    The file `mysteries.csv` contains selected information about each book, as well as the information on N_h and n_h used to calculate the sampling weights in variable *p1weight*. Give 95% CIs for your estimates in (b)–(d).

    (a) Should you use an fpc when calculating variances? Why, or why not?

    (b) What percentage of the books had at least one female author?

    (c) What percentage of the books had at least one female detective?

    (d) Estimate the total number of female detectives in the 655 books.

14. File `asafellow.csv` contains the data described in Example 3.7. There are 18 strata, defined by the cross-classification of variables *awardyr* and *gender*.

    (a) Calculate the sampling weight for each member of the sample. Is the sample exactly self-weighting?

    (b) Estimate the proportion of Fellows who work in academia. In 2020, approximately 48% of the population of ASA members worked in academia. Does your 95% CI include the value 0.48?

    (c) Estimate the proportion of Fellows who majored in mathematics as undergraduates, along with its standard error. Use the same method as in Example 3.4 to deal with the missing data.

15. The data file `agstrat.csv` also contains information on other variables. For each of the following quantities, plot the data, and estimate the population mean for that variable along with its standard error and a 95% CI. Compare your answers with those from the SRS in Exercise 19 of Chapter 2.

    (a) Number of acres devoted to farms, 1987

    (b) Number of farms, 1992

    (c) Number of farms with 1000 acres or more, 1992

    (d) Number of farms with 9 acres or fewer, 1992

16. Hard shell clams may be sampled by using a dredge. Clams do not tend to be uniformly distributed in a body of water, however, as some areas provide better habitat than others. Thus, taking a simple random sample is likely to result in a large estimated variance for the number of clams in an area. Russell (1972) used stratified random sampling to estimate the total number of bushels of hard shell clams (*Mercenaria mercenaria*) in Narragansett Bay, Rhode Island. The area of interest was divided into four strata based on preliminary surveys that identified areas in which clams were abundant. Then n_h dredge tows were made in stratum h, for $h = 1, 2, 3, 4$. The acreage for each stratum was known, and Russell calculated that the area fished during a standard dredge tow was 0.039 acres, so that we may use $N_h = 25.6 \times \text{Area}_h$.

    (a) Here are the results from the survey taken before the commercial season. Estimate the total number of bushels of clams in the area, and give the standard error of your estimate.

Stratum	Area (Acres)	Number of Tows Made	Average Number of Bushels per Tow	Sample Variance for Stratum
1	222.81	4	0.44	0.068
2	49.61	6	1.17	0.042
3	50.25	3	3.92	2.146
4	197.81	5	1.80	0.794

    (b) Another survey was performed at the end of the commercial season. In this survey, strata 1, 2, and 3 were collapsed into a single stratum, called stratum 1 below. Estimate the total number of bushels of clams (with standard error) at the end of the season.

Stratum	Area (Acres)	Number of Tows Made	Average Number of Bushels per Tow	Sample Variance for Stratum
1	322.67	8	0.63	0.083
4	197.81	5	0.40	0.046

17. Lydersen and Ryg (1991) used stratification techniques to estimate ringed seal populations in Svalbard fjords. The 200 km^2 study area was divided into three zones: Zone 1, outer Sassenfjorden, was covered with relatively new ice during the study period in March, 1990, and had little snow cover; Zone 3, Tempelfjorden, had a stable ice cover throughout the year; Zone 2, inner Sassenfjorden, was intermediate between the stable Zone 3 and the unstable Zone 1. Ringed seals need good ice to establish territories with breathing holes, and snow cover enables females to dig out birth lairs. Thus, it was thought that the three zones would have different seal densities. The investigators took

a stratified random sample of 20% of the 200 1-km^2 areas. The following table gives the number of plots, and the number of plots sampled, in each zone:

Zone	Number of Plots	Plots Sampled
1	68	17
2	84	12
3	48	11
Total	200	40

In each sampled area, Imjak the Siberian husky tracked seal structures by scent; the number of breathing holes in sampled square was recorded. A total of 199 breathing holes were located in zones 1, 2, and 3 altogether. The data (reconstructed from information given in the paper) are in the file `seals.csv`.

(a) Estimate the total number of breathing holes in the study region, along with its standard error.

(b) If you were designing this survey, how would you allocate observations to strata if the goal is to estimate the total number of breathing holes? If the goal is to compare the density of breathing holes in the three zones?

18. Burnard (1992) sent a questionnaire to a stratified sample of nursing tutors and students in Wales, to study what the tutors and students understood by the term *experiential learning.* The population size and sample size obtained for each of the four strata are given below:

Stratum	Population Size	Sample Size
General nursing tutors (GT)	150	109
Psychiatric nursing tutors (PT)	34	26
General nursing students (GS)	2680	222
Psychiatric nursing students (PS)	570	40
Total	3434	397

Respondents were asked which of the following techniques could be identified as experiential learning methods; the numbers of students in each group who identified the method as an experiential learning method are given below:

Method	GS	PS	PT	GT
Role play	213	38	26	104
Problem solving activities	182	33	22	95
Simulations	95	20	22	64
Empathy-building exercises	89	25	20	54
Gestalt exercises	24	4	5	12

Estimate the overall percentage of nursing students and tutors who identify each of these techniques as "experiential learning." Be sure to give standard errors for your estimates.

19. Hayes (2000) took a stratified sample of New York City food stores. The sampling frame consisted of 1,408 food stores with at least 4,000 square feet of retail space. The population of stores was stratified into three strata using median household income

within the zip code. The prices of a "market basket" of goods were determined for each store; the goal of the survey was to investigate whether prices differ among the three strata. Hayes used the logarithm of total price for the basket as the response y. Results are given in the following table:

Stratum, h	N_h	n_h	$\bar{y}_h$	s_h
1 Low income	190	21	3.925	0.037
2 Middle income	407	14	3.938	0.052
3 Upper income	811	22	3.942	0.070

(a) The planned sample size was 30 in each stratum; this was not achieved because some stores went out of business while the data were being collected. What are the advantages and disadvantages of sampling the same number of stores in each stratum?

(b) Estimate $\bar{y}_{\mathcal{U}}$ for these data and give a 95% CI.

(c) Is there evidence that prices are different in the three strata?

20. Kruuk et al. (1989) used a stratified sample to estimate the number of otter (*Lutra lutra*) dens along the 1400-km coastline of Shetland, UK. The coastline was divided into 242 (237 that were not predominantly buildings) 5-km sections, and each section was assigned to the stratum whose terrain type predominated. Then sections were chosen randomly from the sections in each stratum. In each section chosen, the investigators counted the total number of dens in a 110-m-wide strip along the coast. The data are in file `otters.csv`. The population sizes for the strata are as follows:

Stratum	**Total Sections**	**Sections Counted**
1 Cliffs over 10 m	89	19
2 Agriculture	61	20
3 Not 1 or 2, peat	40	22
4 Not 1 or 2, non-peat	47	21

(a) Estimate the total number of otter dens in Shetland, along with a standard error for your estimate.

(b) Discuss possible sources of bias in this study. Do you think it is possible to avoid all selection and measurement bias?

21. Until 1996, the National Center for Health Statistics collected detailed data about marriages and divorces from the states. State and local officials provided annual counts of marriages and divorces in each county. In addition, some states sent computer tapes of additional data, or microfilm copies of marriage or divorce certificates. These additional data were used to calculate statistics about age at marriage or divorce, previous marital status of marrying couples, and children involved in divorce. In 1987, if a state sent a computer tape, all records were included in the divorce statistics; if a state sent microfilm copies, a specified fraction of the divorce certificates was randomly sampled, and data recorded. The sampling rates (probabilities of selection) and number of records sampled in each state in the divorce registration area for 1987 are in file `divorce.csv`.

(a) How many divorces were there in the divorce registration area in 1987? HINT: Construct and use the sampling weights.

(b) Why were different sampling rates used in different states?

(c) Estimate the total number of divorces granted to men aged 24 or less. To women aged 24 or less. Give 95% CIs for your estimates.

(d) In what proportion of all divorces was the husband between 40 and 50 years old? In what proportion was the wife between 40 and 50? Give 95% CIs for your estimates.

22. Wilk et al. (1977) reported data on the number and types of fishes and environmental data for the area of the Atlantic continental shelf between eastern Long Island, New York and Cape May, New Jersey. The ocean survey area was divided into strata based on depth. Sampling was done at a higher rate close to shore than farther away from shore: "Inshore strata (0–28 m) were sampled at a rate of approximately one station per 515 km^2 and off-shore strata (29–366 m) were sampled at a rate of approximately one station per 1,030 km^2" (p. 1). Thus each record in strata 3–6 represents twice as much area as each record in strata 1 and 2. In calculating average numbers of fish caught and numbers of species, we may use a relative sampling weight of 1 for strata 1 and 2, and weight 2 for strata 3–6.

Stratum	**Depth (m)**	**Relative Sampling Weight**
1	0–19	1
2	20–28	1
3	29–55	2
4	56–100	2
5	111–183	2
6	184–366	2

The data file `nybight.csv` contains data on the total catch for sampling stations visited in June 1974 and June 1975.

(a) Construct side-by-side boxplots of the number of fish caught in the trawls in June, 1974. Does there appear to be a large variation among the strata?

(b) Calculate estimates of the average number and average weight of fish caught per haul in June 1974, along with the standard error.

(c) Calculate estimates of the average number and average weight of fish caught per haul in June 1975, along with the standard error.

(d) Is there evidence that the average weight of fish caught per haul differs between June 1974 and June 1975? Answer using an appropriate hypothesis test.

23. In January 1995, the Office of University Evaluation at Arizona State University surveyed faculty and staff members to find out their reaction to the closure of the university during Winter Break, 1994. Faculty and staff in academic units that were closed during the winter break were divided into four strata and subsampled.

Stratum	**Employee Type**	N_h	n_h
1	Faculty	1374	500
2	Classified staff	1960	653
3	Administrative staff	252	98
4	Academic professional	95	95

Questionnaires were sent through campus mail to persons in strata 1 through 4; the sample size in the above table is the number of questionnaires mailed in each stratum.

We'll come back to the issue of nonresponse in this survey in Chapter 8; for now, just analyze the respondents in the stratified sample of employees in closed units; the data are in the file `winter.csv`. For this exercise, look at the answers to the question "Would you want to have Winter Break Closure again?" (variable *breakaga*).

(a) Not all persons in the survey responded to the question. Find the number of persons that responded to the question in each of the four strata. For this exercise, use these values as the n_h.

(b) Use (3.7) and (3.8) to estimate the proportion of faculty and staff that would answer yes to the question "Would you want to have Winter Break Closure again" and give the standard error.

(c) Create a new variable, in which persons who respond "yes" to the question take on the value 1, persons who respond "no" to the question take on the value 0, and persons who do not respond are either left blank (if you are using a spreadsheet) or assigned the missing value code (if you are using statistical software). Construct a column of sampling weights N_h/n_h for the observations in the sample. (The sampling weight will be 0 or missing for nonrespondents.) Now use (3.11) to estimate the proportion of faculty and staff that would answer yes to the question "Would you want to have Winter Break Closure again."

(d) Using the column of 0s and 1s you constructed in the previous question, find s_h^2 for each stratum by calculating the sample variance of the observations in that stratum. Now use (3.6) to calculate the standard error of your estimate of the proportion. Why is your answer the same as you calculated in (b)?

(e) Stratification is sometimes used as a method of dealing with nonresponse. Calculate the response rates (the number of persons who responded divided by the number of questionnaires mailed) for each stratum. Which stratum has the lowest response rate for this question? How does stratification treat the nonrespondents?

24. The data in the file `radon.csv` were collected from 1,003 homes in Minnesota in 1987 (Tate, 1988) in order to estimate the prevalence and distribution of households with high indoor radon concentrations. Since the investigators were interested in how radon levels varied across counties, each of the 87 counties in Minnesota served as a stratum. An SRS of telephone numbers from county telephone directories was selected in each county. When a household could not be contacted or was unwilling to participate in the study, an alternate telephone number was used, until the desired sample size in the stratum was reached.

(a) Discuss possible sources of nonsampling error in this survey.

(b) Calculate the sampling weight for each observation, using the values for N_h and n_h in the data file.

(c) Treating the sample as a stratified random sample, estimate the average radon level for Minnesota homes, along with a 95% CI. Do the same for the response log(radon).

(d) Estimate the total number of Minnesota homes that have radon level of 4 picocuries per liter or higher, with a 95% CI.

25. Proportional allocation was used in the stratified sample in Example 3.2. It was noted, however, that variability was much higher in the West than in the other regions. Using the estimated variances in Example 3.2, and assuming that the sampling costs are the same in each stratum, find an optimal allocation for a stratified sample of size 300.

26. Select a stratified random sample of size 300 from the data in the file `agpop.csv`, using your allocation in Exercise 25. Estimate the total number of acres devoted to farming in the United States, and give the standard error of your estimate. How does this standard error compare with that found in Example 3.2?

27. Suppose that, instead of e-mailing all 15,769 persons in the current membership list of the ASA as described in Exercise 13 of Chapter 1, the researchers wanted to take a stratified random sample of 1,100 of the 5,803 female and 9,966 male ASA members. What should the sample sizes for men and women be if:

    (a) Proportional allocation is used?

    (b) Neyman allocation is used, and it is assumed that $S^2_{\text{women}} = (11/5) \times S^2_{\text{men}}$?

28. *College scorecard data.* For the data in Example 3.12, consider a design for stratified random sample of 200 colleges that is stratified by the cross-classification of *control* (whether the institution is public or private) and *ccbasic* (type of institution).

    (a) Find the number of colleges in each of the 16 strata.

    (b) What sample size would be taken in each of the strata under proportional allocation?

    (c) It is thought that the variance of many of the quantities of interest will be proportional to the variance of the number of undergraduates (variable *ugds*). Calculate the variance of *ugds* for each stratum, and use these variances to find the Neyman allocation.

    (d) Compare the variance of the estimated mean of *ugds* for this design with the variance for the allocation in Example 3.12. Which stratification performs better for this variable?

29. Example 3.13 described the stratified sampling plan used for New York City's annual point-in-time count. The summary statistics in Table 3.9 are from a fictional survey conducted by a fictional city with 4 geographic divisions that uses the same type of design (the full data set is in file `pitcount.csv`). In each sampled area, y_{hj} represents the number of persons encountered in area (hj) who were counted as unsheltered.

TABLE 3.9
Summary statistics for Exercise 29.

Stratum, h	1	2	3	4	5	6	7	8
Division	1	2	3	4	1	2	3	4
Density	High	High	High	High	Low	Low	Low	Low
N_h	12	8	4	16	384	192	240	144
n_h	12	8	4	16	24	12	15	9
$\bar{y}_h$	2.167	3.500	5.750	9.188	0.375	0.167	0.133	0.444
s_h^2	3.424	2.571	0.250	12.829	0.332	0.152	0.124	1.028

    (a) Estimate the total number of persons experiencing homelessness in the city, along with a 95% CI.

    (b) How much difference does the fpc make for this estimate?

(c) The CI captures the sampling error in the estimate. What are some of the sources of nonsampling error?

30. Use the data in Exercise 29 to explore the relative efficiency of four designs for meeting the two goals of the survey: (1) obtain an accurate estimate of the population total, and (2) include as many unsheltered persons in the sample as possible.

    Design 1: Proportional allocation

    Design 2: Neyman allocation for estimating $\hat{t}$

    Design 3: Sample all areas in high-density strata, then allocate the remaining sample size to the low-density strata with Neyman allocation.

    Design 4: Allocate two observations to each stratum. Then allocate the remainder of the sample as follows. First, take all areas in the stratum with the highest value of the mean $\bar{y}_{h\mathcal{U}}$, then take all areas in the stratum with the second highest value of $\bar{y}_{h\mathcal{U}}$, and so on until the entire sample size has been allocated.

    (a) What is the allocation of a sample of 100 areas for each design, assuming that at least two areas are to be sampled in each stratum? You may use the sample quantities $\bar{y}_h$ and s_h^2 as estimates of the population quantities $\bar{y}_{h\mathcal{U}}$ and S_h^2.

    (b) What do you anticipate $V(\hat{t})$ and $\sum_{h=1}^{H} \sum_{i \in \mathcal{S}_h} y_{hi}$ to be for each allocation in (a)?

    (c) What design would you use if your primary consideration is goal (1)? Goal (2)? If the two goals are equally important?

31. The guidelines for conducting the count described in Example 3.13 (U.S. Department of Housing and Urban Development, 2014, p. 14) stated that it is preferable for organizations "to conduct a census count when practicable, as it is by definition the most complete and accurate information available."

    Will a census always give more complete and accurate information than a sample? Under what conditions would a census be preferred? When would a stratified sample produce a more accurate count?

C. Working with Theory

32. Construct a small population and stratification for which $V(\hat{t}_{\text{str}})$ using proportional allocation is larger than the variance that would be obtained by taking an SRS with the same number of observations. HINT: Use (3.13).

33. A stratified sample is being designed to estimate the prevalence p of a rare characteristic, say the proportion of residents in Milwaukee, Wisconsin, who have Lyme disease. Stratum 1, with N_1 units, has a high prevalence of the characteristic; stratum 2, with N_2 units, has low prevalence. Assume that the cost to sample a unit (for example, the cost to select a person for the sample and determine whether he or she has Lyme disease) is the same for each stratum, and that at most 2000 persons are to be sampled.

    (a) Let p_1 and p_2 be the proportions in stratum 1 and stratum 2 with the rare characteristic. If $p_1 = 0.10$, $p_2 = 0.03$, and $N_1/N = 0.4$, what are n_1 and n_2 under optimal allocation?

    (b) If $p_1 = 0.10$, $p_2 = 0.03$, and $N_1/N = 0.4$, what is $V(\hat{p}_{\text{str}})$ under proportional allocation? Under optimal allocation? What is the variance if you take an SRS of 2000 units from the population?

(c) (Requires computing.) Now fix $p = 0.05$. Let p_1 range from 0.05 to 0.50, and N_1/N range from 0.01 to 0.50 (these two values then determine the value of p_2). For each combination of p_1 and N_1/N, find the optimal allocation, and the variance under both proportional allocation and optimal allocation. Also, find the variance from an SRS of 2000 units. When does the optimal allocation give a substantial increase in precision when compared to proportional allocation? When compared to an SRS?

34. (Requires probability.) We know from Section 2.3 that there are $\binom{N}{n} = \dfrac{N!}{n!(N-n)!}$ possible SRSs of size n from a population of size N. Suppose we stratify the population into H strata, where each stratum contains $N_h = N/H$ units. A stratified sample is to be selected using proportional allocation, so that $n_h = n/H$.

    (a) How many possible stratified samples are there?

    (b) Stirling's formula approximates $k!$, when k is large, by $k! \approx \sqrt{2\pi k}\,(k/e)^k$, where $e = \exp(1) \approx 2.718282$. Use Stirling's formula to approximate

$$\frac{\text{number of possible stratified samples of size } n}{\text{number of possible SRSs of size } n}.$$

35. *Optimal allocation* (requires calculus).

    (a) Show that the allocation in (3.15) minimizes $V(\bar{y}_{\text{str}})$ for a fixed cost C with the cost function in (3.14). (Or, if you prefer, show that the allocation minimizes the total sampling cost $\sum_{h=1}^{H} c_h n_h$ if the variance is set equal to $V(\bar{y}_{\text{str}})$; the allocation is the same either way.) HINT: Use Lagrange multipliers.

    (b) If the variance is minimized for a fixed cost C, show by substituting the value of n_h from (3.15) into (3.14) that

$$n = \frac{(C - c_0)\sum_{l=1}^{H} N_l S_l/\sqrt{c_l}}{\sum_{h=1}^{H} N_h S_h \sqrt{c_h}}.$$

    (c) If minimizing C for a fixed value V_0 of $V(\bar{y}_{\text{str}})$, show that

$$n = \frac{\sum_{h=1}^{H}\left[N_h S_h \sqrt{c_h}\right]\sum_{l=1}^{H}\left[N_l S_l/\sqrt{c_l}\right]}{N^2 V_0 + \sum_{h=1}^{H} N_h S_h^2}.$$

36. Under Neyman allocation, discussed in Section 3.4.2, the optimal sample size in stratum h is

$$n_{h,\text{Neyman}} = n\frac{N_h S_h}{\sum_{l=1}^{H} N_l S_l}.$$

    (a) Show that the variance of $\hat{t}_{\text{str}}$ if Neyman allocation is used is

$$V_{\text{Neyman}}(\hat{t}_{\text{str}}) = \frac{1}{n}\left(\sum_{h=1}^{H} N_h S_h\right)^2 - \sum_{h=1}^{H} N_h S_h^2.$$

    (b) We showed in Section 3.4.1 that the variance of $\hat{t}_{\text{str}}$ if proportional allocation is used is

$$V_{\text{prop}}(\hat{t}_{\text{str}}) = \frac{N}{n}\sum_{h=1}^{H} N_h S_h^2 - \sum_{h=1}^{H} N_h S_h^2.$$

Prove that the theoretical variance from Neyman allocation (for the variable studied, assuming that the values of S_h^2 are accurate) is always less than or equal to the theoretical variance from proportional allocation by showing that

$$V_{\text{prop}}(\hat{t}_{\text{str}}) - V_{\text{Neyman}}(\hat{t}_{\text{str}}) = \frac{N^2}{n} \sum_{h=1}^{H} \frac{N_h}{N} \left(S_h - \sum_{l=1}^{H} \frac{N_l}{N} S_l \right)^2 \geq 0.$$

(c) From (b), we see that the gain in precision from using Neyman allocation relative to using proportional allocation is higher if the stratum standard deviations S_h vary widely. When $H = 2$, show that

$$V_{\text{prop}}(\hat{t}_{\text{str}}) - V_{\text{Neyman}}(\hat{t}_{\text{str}}) = \frac{N_1 N_2}{n}(S_1 - S_2)^2.$$

37. *Optimal allocation for multiple variables.* Suppose there are K variables of interest, and variable k has relative importance $a_k > 0$, where $\sum_{k=1}^{K} a_k = 1$. Let $\hat{t}_{y_k}$ be the estimated population total for variable k, and let S_{kh}^2 be the population variance for variable k in stratum h. Then the optimal allocation problem is to minimize

$$\sum_{k=1}^{K} a_k V(\hat{t}_{y_k})$$

subject to the constraint $C = c_0 + \sum_{h=1}^{H} n_h c_h$. Show that the optimal allocation for fixed total sample size n gives

$$n_h = n \left(N_h \sqrt{\frac{\sum_{k=1}^{K} a_k S_{kh}^2}{c_h}} \right) \Bigg/ \left(\sum_{j=1}^{H} N_j \sqrt{\frac{\sum_{k=1}^{K} a_k S_{kj}^2}{c_j}} \right).$$

38. *Power allocation.* Bankier (1988) described power allocations that can be used when it is desired to have accurate estimates for (1) the population as a whole and (2) each individual stratum. A power allocation minimizes

$$f(n_1, n_2, \ldots, n_H) = \sum_{h=1}^{H} [X_h^q \, \text{CV}(\bar{y}_h)]^2$$

subject to the constraint $\sum_{h=1}^{H} n_h = n$, where X_h is a measure of the size or importance of stratum h, the power q is a constant between 0 and 1, and $\text{CV}(\bar{y}_h) = \sqrt{V(\bar{y}_h)}/\bar{y}_{h\mathcal{U}}$.

(a) Show that $f(n_1, n_2, \ldots, n_H)$ is minimized subject to $\sum_{h=1}^{H} n_h = n$ when

$$n_h = n \frac{X_h^q S_h / \bar{y}_{h\mathcal{U}}}{\sum_{l=1}^{H} X_l^q S_l / \bar{y}_{l\mathcal{U}}}.$$

(b) Show that when $X_h = t_h$ and $q = 1$, the power allocation is equivalent to Neyman allocation.

(c) Show that when $q = 0$ and the values of $S_h/\bar{y}_{h\mathcal{U}}$ are roughly equal among the strata, the power allocation is $n_h = n/H$.

39. Using the stratification in Example 3.12 and variable *ugds* of `college.csv`, find the power allocations (see Exercise 38) for $q = 0, 0.25, 0.5,$ and 1. What happens as q moves from 0 to 1?

40. *Integer allocation.* The stratum sample sizes n_h from the optimal allocation in (3.15) are often fractions. If the budget is flexible, these can be rounded up to the next integer. Sometimes, however, the total sample size n is fixed exactly and cannot be exceeded, as in the allocation problem described in Exercise 47.

    Wright (2012) described an allocation method that may be used when (i) the total sample size must exactly equal n, (ii) each stratum must have an integer number of observations, and (iii) each stratum must have at least k observations. Often, k will be set equal to 1 (because any stratified sample must have at least 1 observation per stratum) or 2 (because a stratified sample needs at least 2 observations per stratum to be able to estimate the variance from the sample).

    Create a matrix with H rows and $n-k$ columns. Row h (for $h = 1, \ldots, H$) has values

$$\frac{N_h^2 S_h^2}{(k)(k+1)}, \frac{N_h^2 S_h^2}{(k+1)(k+2)}, \frac{N_h^2 S_h^2}{(k+2)(k+3)}, \ldots, \frac{N_h^2 S_h^2}{(n-1)n}.$$

    Select the largest $n - kH$ values from the matrix. Let a_h denote the number of values selected from row h. Set $n_h = k + a_h$.

    Show that the values of n_h from this procedure minimize $V(\hat{t}_{\text{str}})$ in (3.3) subject to the constraints in (i), (ii), and (iii). HINT: First show that when $k \geq 2$ and $n_h \geq k$,

$$\frac{1}{n_h} = \frac{1}{k} - \frac{1}{(k)(k+1)} - \frac{1}{(k+1)(k+2)} - \frac{1}{(k+2)(k+3)} - \cdots - \frac{1}{(n_h-1)(n_h)}.$$

41. *Sampling from a data stream* (requires computing). Exercise 33 of Chapter 2 gave a method for selecting an SRS from a population that arrives as a data stream. If you know the sample size n_h to be taken from each stratum, you can apply the SRS procedure to each stratum to draw a stratified random sample. But suppose you want to take a proportionally or optimally allocated stratified random sample of size n and do not know $N_1, \ldots, N_H$. Collins and Lu (2019) showed that, when N_h and n_h are large, the following procedure results in a stratified random sample with approximate optimal allocation.

    Take the first n units as the initial sample $\mathcal{S}_n = \{1, 2, \ldots, n\}$. Determine the stratum membership for each unit in $\mathcal{S}_n$. For each stratum h ($h = 1, \ldots, H$), calculate

$$\begin{aligned} n_h &= \text{number of observations in } \mathcal{S}_n \text{ falling in stratum } h, \\ S_h &= \text{standard deviation of the units in } \mathcal{S}_n \text{ falling in stratum } h. \end{aligned}$$

    Initially set $N_h := n_h$ for $h = 1, \ldots, H$; N_h is the population size in stratum h for the population consisting of the first n units in the stream. Then carry out the following steps for $k = n+1, n+2, \ldots$, until the entire data stream has been examined.

    i Determine the stratum membership of unit k in the stream. If unit k is in stratum h, set $N_h := N_h + 1$ (increment N_h by 1) and update S_h to be the population standard deviation for stratum h from units $1, \ldots, k$. For large populations, you may want to use a numerically stable one-pass algorithm for updating variances such as that in Welford (1962).

    ii Calculate the sample size for optimal allocation for a population of size k as $n_l^* = na_l / \sum_{h=1}^{H} a_h$, for $l = 1, \ldots, H$, where $a_l = N_l S_l / \sqrt{c_l}$.

    iii Generate a random number u_k between 0 and 1. Use this number to determine $\mathcal{S}_k$, the working sample at Step k.

a. If $u_k > n_h^*/N_h$, then set $\mathcal{S}_k$ equal to $\mathcal{S}_{k-1}$.

b. If $u_k \leq n_h^*/N_h$ and $n_h < n_h^*$, then find the stratum with the largest oversampling $n_l - n_l^*$ (call it stratum j). Randomly select a unit from $\mathcal{S}_{k-1}$ that is in stratum j, and remove it from the sample. Add unit k (which is in stratum h) to the sample. Set $n_h := n_h + 1$ and $n_j := n_j - 1$ to give the current stratum sample sizes in $\mathcal{S}_k$.

c. If $u_k \leq n_h^*/N_h$ and $n_h \geq n_h^*$, then select one of the stratum-h units in $\mathcal{S}_{k-1}$ at random and replace it by unit k to form sample $\mathcal{S}_k$.

(a) Show that the algorithm gives the correct population stratum sizes N_h.

(b) Show that if there is one stratum, this algorithm simplifies to the procedure in Exercise 33 of Chapter 2.

(c) Use this procedure to draw a stratified random sample of size 300 with proportional allocation from `agpop.csv`.

D. Projects and Activities

42. *Rectangles.* This activity continues Exercise 35 of Chapter 2. Divide the rectangles in the population of Figure 2.9 into two strata, based on your judgment of size. Now take a stratified sample of 10 rectangles. State how you decided on the sample size in each stratum. Estimate the total area of all the rectangles in the population, and give a 95% CI, based on your sample. How does your CI compare with that from the SRS in Chapter 2?

43. *Mutual funds.* In Exercise 36 of Chapter 2, you took an SRS of funds from a mutual fund company. Most companies have mutual funds in a number of different categories, for example, domestic stock funds, foreign stock funds, and bond funds, and the returns in these categories differ.

    (a) Divide the funds from the company into strata. You may use categories provided by the fund company, or other categories such as market capitalization. Create a table of the strata, with the number of mutual funds in each stratum.

    (b) Using proportional allocation, take a stratified random sample of size 25 from your population.

    (c) Find the mean and a 95% CI for the mean of the variable you studied in Exercise 36 of Chapter 2. How does your estimate from the stratified sample compare with the estimate from the SRS you found earlier?

44. *Index mutual funds* attempt to mimic the performance of one of the indices of overall stock or bond market performance. The Wilshire 5000 Index^{SM}, for example, is a market-capitalization-weighted index of the market value of all U.S. equity securities with readily available price information. A mutual fund would track the index's performance if, at each time point, it held all of the stocks in the index in the proportions indicated by the weighting. But buying all of the stocks in the index and adjusting the holdings every time the index is revised can lead to excessive transaction costs and tax consequences. As a result, some mutual fund companies use stratified sampling to select stocks to include in their index funds.

    (a) Find a mutual or exchange-traded fund that attempts to replicate an index. Summarize how they use stratified sampling.

(b) Suppose you were asked to devise a stratified sampling plan to represent the Wilshire 5000 Index using market capitalization classes as strata. Using a list of the stocks in the index, construct strata and develop a stratified sampling design.

45. The U.S. Annual Retail Trade Survey, described at `www.census.gov/programs-surveys/arts.html`, provides estimates of sales at retail and food service stores. Read the documentation on Survey Design. How is stratification used?

46. An actor wants to know what proportion of the online articles written about him contain incorrect information. Devise a stratified sampling plan to do this.

47. *Congressional Apportionment.* The U.S. House of Representatives is reapportioned every ten years following the decennial census. The 435 representatives are to be allocated proportionally to each state's population, with each state having at least one representative. This can be thought of as an allocation problem in which the total sample size is $n = 435$ and the strata are the 50 states.

    The U.S. has used the *method of equal proportions* to allocate representatives to states since 1940 (Wright, 2012). This is a special case of the method described in Exercise 40 in which $S_1 = S_2 = \ldots = S_H$ since proportional allocation is desired. The file `census2010.csv` contains the state populations from the 2010 census.

    (a) Calculate nN_h/N, the (possibly fractional) sample size that would be assigned to each state under proportional allocation.

    (b) The Hamilton/Vinton method of apportionment begins by assigning state h the largest integer less than or equal to nN_h/N. The remaining seats are assigned to the states with the largest fractional remainders. Find the number of representatives allocated to each state using this method.

    (c) Find the number of representatives allocated to each state under the method of equal proportions.

    (d) Debates about apportionment were so contentious during the 1920s that no reapportionment was done during that decade (Anderson, 2015). Repeat (a), (b), and (c) for the 1920 census data in `census1920.csv`. Which states in 1920 benefit from the Hamilton/Vinton method? From the equal proportions method?

48. *College scorecard data.* The stratification in Example 3.12 employed 10 strata to make it easy to see the difference between the proportional and Neyman allocations. The file `college.csv`, however, has a great deal of detailed information about the colleges, and a better design for sampling this population would form as many strata as feasible. Use variables *control* and *region* to subdivide the large-institution and medium-institution strata in Example 3.12 so that you end up with 50 strata altogether, each with at least 2 units. Using approximately proportional allocation (it will likely not be exactly proportional since you may have some certainty strata), compare the anticipated variance for *ugds* and *majwomen* for this stratification with that from Example 3.12.

49. *Trucks.* The Vehicle Inventory and Use Survey (VIUS) used a stratified random sampling design to provide information on the number of private and commercial trucks in each state (U.S. Census Bureau, 2006). For the 2002 survey, 255 strata were formed from the sampling frame of truck registrations using stratification variables *state* and *trucktype*. The 50 states plus the District of Columbia formed 51 geographic classes; in each, the truck registrations were partitioned into one of five classes:

1. Pickups
2. Minivans, other light vans, and sport utility vehicles
3. Light single-unit trucks with gross vehicle weight less than 26,000 pounds
4. Heavy single-unit trucks with gross vehicle weight greater than or equal to 26,000 pounds
5. Truck-tractors

Consequently, the full data set has $51 \times 5 = 255$ strata. Selected variables from the data are in the data file `vius.csv`. For each question, give a point estimate and a 95% CI.

(a) The sampling weights are found in variable *tabtrucks* and the stratification is given by variable *stratum*. Estimate the total number of trucks in the United States. (HINT: What should your response variable be?) Why is the standard deviation of your estimator essentially zero?

(b) Estimate the total number of truck miles driven in 2002 (variable *miles_annl*).

(c) Estimate the total number of truck miles driven in 2002 for each of the five *trucktype* classes.

(d) Estimate the total number of vehicles that have gross vehicle weight exceeding 10,000 pounds (variable *vius_gvw*).

50. *Baseball data.* Exercise 37 of Chapter 2 described the population of baseball players in data file `baseball.csv`.

(a) Take a stratified random sample of 150 players from the file, using proportional allocation with the different teams as strata. Describe how you selected the sample.

(b) Estimate the proportion of players in the data set who are pitchers and give a 95% CI.

(c) Find the mean of the variable *logsal*, using your stratified sample, and give a 95% CI.

(d) How do your estimates compare with those of Exercise 37 from Chapter 2?

(e) Examine the sample variances in each stratum. Do you think optimal allocation would be worthwhile for this problem?

(f) Using the sample variances from (e) to estimate the population stratum variances, determine the optimal allocation for a sample in which the cost is the same in each stratum and the total sample size is 150. How much does the optimal allocation differ from the proportional allocation?

51. *Online bookstore.* In Exercise 38 from Chapter 2 you took an SRS of book titles from amazon.com. Use the same book genre for this problem.

(a) Stratify the population into two categories: hardcover and paperback. You can obtain the population counts in the paperback category by refining your search to include the word paperback.

(b) Take a stratified random sample of 40 books from your population using proportional allocation. Record the price and number of pages for each book.

(c) Give a point estimate and a 95% CI for the mean price of books and the mean number of pages for books in the population.

(d) Compare your CI's to those from Exercise 38 of Chapter 2. Does stratification appear to increase the precision of your estimate?

(e) Use your SRS from Chapter 2 to estimate the within-stratum variance of book price for each stratum. In this case, you are using the SRS as a pilot sample to help design a subsequent sample. Find the optimal allocation for a stratified random sample of 40 books. How does the optimal allocation differ from the proportional allocation?

52. *IPUMS exercises.* Exercise 42 of Chapter 2 described the IPUMS data.

(a) Using one or more of the following variables: *age*, *sex*, *race*, or *marstat*, divide the population into strata. Explain how you decided upon your stratification variable and how you chose the number of strata to use. (Note: It is NOT FAIR to use the values of *inctot* in the population to choose your strata! However, you may draw a pilot sample of size 200 using an SRS to aid you in constructing your strata.)

(b) Using the strata you constructed, draw a stratified random sample using proportional allocation. Use the same overall sample size you used for your SRS in Exercise 42 of Chapter 2. Explain how you calculated the sample size to be drawn from each stratum.

(c) Using the stratified sample you selected with proportional allocation, estimate the total income for the population, along with a 95% CI.

(d) Using the pilot sample of size 200 to estimate the within-stratum variances, use optimal allocation to determine sample stratum sizes. Use the same value of n as in part 52b, which is the same n from the SRS in Exercise 42 of Chapter 2. Draw a stratified random sample from the population along with a 95% CI.

(e) Under what conditions can optimal allocation be expected to perform much better than proportional allocation? Do these conditions exist for this population? Comment on the relative performance you observed between these two allocations.

(f) Overall, do you think your stratification was worthwhile for sampling from this population? How did your stratified estimates compare with the estimate from the SRS you took in Chapter 2? If you were to start over on the stratification, what would you do differently?

53. *Large public database.* In Exercise 43, you selected an SRS from a large publicly available data set. Now draw a stratified random sample from the database, and analyze variables of interest.

If your primary interest is in estimating statistics for the population as a whole, you may want to use proportional allocation. In other situations, you may want to oversample certain strata. For example, if sampling from the Chicago crime database as in Exercise 23 of Chapter 2, you may want to oversample certain types of crime, or oversample crimes in which no arrest was made. Discuss why you chose the sample size, stratification, and allocation that you used.

4

Ratio and Regression Estimation

> The registers of births, which are kept with care in order to assure the condition of the citizens, can serve to determine the population of a great empire without resorting to a census of its inhabitants, an operation which is laborious and difficult to do with exactness. But for this it is necessary to know the ratio of the population to the annual births. The most precise means for this consists of, first, choosing subdivisions in the empire that are distributed in a nearly equal manner on its whole surface so as to render the general result independent of local circumstances; second, carefully enumerating the inhabitants of several communes in each of the subdivisions, for a specified time period; third, determining the corresponding mean number of annual births, by using the accounts of births during several years preceding and following this time period. This number, divided by that of the inhabitants, will give the ratio of the annual births to the population, in a manner that will be more reliable as the enumeration becomes larger.
>
> —Pierre-Simon Laplace, *Essai Philosophique sur les Probabilités* (trans. S. Lohr)

France had no population census in 1802, and Laplace wanted to estimate the number of persons living there (Laplace, 1814; Cochran, 1978). He obtained a sample of 30 communes spread throughout the country. These communes had a total of 2,037,615 inhabitants on September 23, 1802. In the 3 years preceding September 23, 1802, a total of 215,599 births were registered in the 30 communes. Laplace determined the annual number of registered births in the 30 communes to be 215,599/3 = 71,866.33. Dividing 2,037,615 by 71,866.33, Laplace estimated that each year there was one registered birth for every 28.352845 persons. Reasoning that communes with large populations are also likely to have large numbers of registered births, and judging that the ratio of population to annual births in his sample would likely be similar to that throughout France, he concluded that one could estimate the total population of France by multiplying the total number of annual births in all of France by 28.352845. (For some reason, Laplace decided not to use the actual number of registered births in France in the year prior to September 22, 1802 in his calculation but instead multiplied the ratio by 1 million.)

Laplace was not interested in the total number of registered births for its own sake but used it as auxiliary information for estimating the total population of France. We often have auxiliary information in surveys. In Chapter 3, we used such auxiliary information in designing a survey. In this chapter, we use auxiliary information in the estimators. Ratio and regression estimation use variables that are correlated with the variable of interest to improve the precision of estimators of the mean and total of a population.

4.1 Ratio Estimation in Simple Random Sampling

For ratio estimation to apply, two quantities y_i and x_i must be measured on each sample unit; x_i is often called an **auxiliary variable** or **subsidiary variable**. In the population

DOI: 10.1201/9780429298899-4

of size N

$$t_y = \sum_{i=1}^{N} y_i, \quad t_x = \sum_{i=1}^{N} x_i$$

and their ratio[1] is

$$B = \frac{t_y}{t_x} = \frac{\bar{y}_\mathcal{U}}{\bar{x}_\mathcal{U}}.$$

In the simplest use of ratio estimation, a simple random sample (SRS) of size n is taken, and the information in both x and y is used to estimate B, t_y, or $\bar{y}_\mathcal{U}$.

Ratio and regression estimation both take advantage of the correlation of x and y in the population; the higher the correlation, the better they work. Define the **population correlation coefficient** of x and y to be

$$R = \frac{\sum_{i=1}^{N}(x_i - \bar{x}_\mathcal{U})(y_i - \bar{y}_\mathcal{U})}{(N-1)S_x S_y}. \tag{4.1}$$

Here, S_x is the population standard deviation of the x_is, S_y is the population standard deviation of the y_is, and R is simply the Pearson correlation coefficient of x and y for the N units in the population.

Example 4.1. Suppose the population consists of agricultural fields of different sizes. Let

$$\begin{aligned} y_i &= \text{bushels of grain harvested in field } i \\ x_i &= \text{acreage of field } i \end{aligned}$$

Then

$$\begin{aligned} B &= \text{average yield in bushels per acre} \\ \bar{y}_\mathcal{U} &= \text{average yield in bushels per field} \\ t_y &= \text{total yield in bushels.} \quad \blacksquare \end{aligned}$$

If an SRS is taken, natural estimators for B, t_y, and $\bar{y}_\mathcal{U}$ are:

$$\hat{B} = \frac{\bar{y}}{\bar{x}} = \frac{\hat{t}_y}{\hat{t}_x} \tag{4.2}$$

$$\hat{t}_{yr} = \hat{B} t_x \tag{4.3}$$

$$\hat{\bar{y}}_r = \hat{B} \bar{x}_\mathcal{U}, \tag{4.4}$$

where t_x and $\bar{x}_\mathcal{U}$ are assumed known.

4.1.1 Why Use Ratio Estimation?

1. Sometimes, we simply want to estimate a ratio. In Example 4.1, B—the average yield per acre—is of interest and is estimated by the ratio of the sample means $\hat{B} = \bar{y}/\bar{x}$. If the fields differ in size, both numerator and denominator are random quantities; if a different sample is selected, both $\bar{y}$ and $\bar{x}$ are likely to change. In other survey situations,

[1] Why use the letter B to represent the ratio? As we shall see in Section 4.6, ratio estimation is motivated by a regression model: $Y_i = \beta x_i + \varepsilon_i$, with $E[\varepsilon_i] = 0$ and $V[\varepsilon_i] = \sigma^2 x_i$. Thus the ratio of t_y and t_x is actually a regression coefficient.

ratios of interest might be the ratio of liabilities to assets, the ratio of the number of fish caught to the number of hours spent fishing, or the per-capita income of household members in Australia.

Some ratio estimates appear disguised because the denominator looks like it is just a regular sample size. To determine whether you need to use ratio estimation for a quantity, ask yourself, "If I took a different sample, would the denominator be a different number?" If yes, then you are using ratio estimation. Suppose you are interested in the percentage of pages in your favorite magazine that contain at least one advertisement. You might take an SRS of 10 issues from the most recent 60 issues of the magazine, and measure the following for each issue:

$$x_i = \text{total number of pages in issue } i$$
$$y_i = \text{total number of pages in issue } i \text{ that contain at least one advertisement.}$$

The proportion of interest can be estimated as

$$\hat{B} = \frac{\hat{t}_y}{\hat{t}_x}.$$

The denominator is the estimated total number of pages in the 60 issues. If a different sample of 10 issues is selected, the denominator will likely be different. In this example, we have an SRS of magazine issues; we have a **cluster sample** (we briefly discussed cluster samples in Section 2.1) of pages from the most recent 60 issues of the magazine. In Chapter 5, we shall see that ratio estimation is commonly used to estimate means in cluster sampling.

Technically, we are using ratio estimation every time we take an SRS and estimate a mean or proportion for a subpopulation, as will be discussed in Section 4.3.

2. Sometimes we want to estimate a population total, but the population size N is unknown. Then we cannot use the estimator $\hat{t}_y = N\bar{y}$ from Chapter 2. But we know that $N = t_x/\bar{x}_\mathcal{U}$ and can estimate N by $t_x/\bar{x}$. We thus use another measure of size, t_x, instead of the population count N.

   To estimate the total number of fish in a haul that are longer than 12 cm, you could take a random sample of fish, estimate the proportion that are larger than 12 cm, and multiply that proportion by the total number of fish, N. Such a procedure cannot be used if N is unknown. You can, however, weigh the total haul of fish, and use the fact that having a length of more than 12 cm (y) is related to weight (x), so

$$\hat{t}_{yr} = \bar{y}\,\frac{t_x}{\bar{x}}.$$

   The total weight of the haul, t_x, is easily measured, and $t_x/\bar{x}$ estimates the total number of fish in the haul.

3. Ratio estimation is often used to increase the precision of estimated means and totals. Laplace used ratio estimation for this purpose in the example at the beginning of this chapter, and increasing precision will be the main use discussed in the chapter.

   In Laplace's use of ratio estimation,

$$y_i = \text{number of persons in commune } i$$
$$x_i = \text{number of registered births in commune } i.$$

Laplace could have estimated the total population of France by multiplying the average number of persons in the 30 communes ($\bar{y}$) by the total number of communes in France (N). He reasoned that the ratio estimator would attain more precision: on average, the larger the population of a commune, the higher the number of registered births. Thus the population correlation coefficient R, defined in (4.1), is likely to be positive. Since $\bar{y}$ and $\bar{x}$ are then also positively correlated [see (A.20) in Appendix A], the sampling distribution of $\bar{y}/\bar{x}$ will have less variability than the sampling distribution of $\bar{y}/\bar{x}_{\mathcal{U}}$. So if

$$t_x = \text{ total number of registered births}$$

is known, the mean squared error (MSE) of $\hat{t}_{yr} = \hat{B}t_x$ is likely to be smaller than the MSE of $N\bar{y}$, an estimator that does not use the auxiliary information of registered births.

4. Ratio estimation is used to adjust estimates from the sample so that they reflect demographic totals. An SRS of 400 students taken at a university with 4,000 students may contain 240 women and 160 men, with 84 of the sampled women and 40 of the sampled men planning to follow careers in teaching. Using only the information from the SRS, you would estimate that

$$\frac{4000}{400} \times 124 = 1240$$

students plan to be teachers. Knowing that the college has 2,700 women and 1,300 men, a better estimate of the number of students planning teaching careers might be

$$\frac{84}{240} \times 2700 + \frac{40}{160} \times 1300 = 1270.$$

Ratio estimation is used within each gender: In the sample, 60% are women, but 67.5% of the population are women, so we adjust the estimate of the total number of students planning a career in teaching accordingly. To estimate the total number of women who plan to follow a career in teaching, let

$$y_i = \begin{cases} 1 \text{ if woman and plans career in teaching} \\ 0 \text{ otherwise.} \end{cases}$$

$$x_i = \begin{cases} 1 \text{ if woman} \\ 0 \text{ otherwise.} \end{cases}$$

Then $(84/240) \times 2700 = (\bar{y}/\bar{x})t_x$ is a ratio estimate of the total number of women planning a career in teaching. Similarly, $(40/160) \times 1300$ is a ratio estimate of the total number of men planning a teaching career.

This use of ratio estimation, called **poststratification**, will be discussed in Sections 4.4, 8.6, and 15.3.

5. Ratio estimation may be used to adjust for nonresponse, as will be discussed in Chapter 8. Suppose a sample of businesses is taken; let y_i be the amount spent on health insurance by business i and x_i be the number of employees in business i. Assume that t_x, the total number of employees in all businesses in the population, is known. We expect that the amount a business spends on health insurance will be related to the number of employees. Some businesses may not respond to the survey, however. One method of adjusting for the nonresponse when estimating total insurance expenditures

is to multiply the ratio $\bar{y}/\bar{x}$ (using data only from the respondents) by the population total t_x. If companies with few employees are less likely to respond to the survey, and if y_i is proportional to x_i, then we would expect the estimate $N\bar{y}$ to overestimate the population total t_y. In the ratio estimate $t_x\bar{y}/\bar{x}$, $t_x/\bar{x}$ is likely to be smaller than N because companies with many employees are more likely to respond to the survey. Thus a ratio estimate of total health care insurance expenditures may help to compensate for the nonresponse of companies with few employees.

Example 4.2. Let's return to the SRS from the U.S. Census of Agriculture, described in Example 2.6. The file `agsrs.csv` contains data from an SRS of 300 of the 3,078 counties.

For this example, suppose we know the population totals for 1987, but have 1992 information only for the SRS of 300 counties. When the same quantity is measured at different times, the response of interest at an earlier time often makes an excellent auxiliary variable. Let

$$\begin{aligned} y_i &= \text{total acreage of farms in county } i \text{ in 1992} \\ x_i &= \text{total acreage of farms in county } i \text{ in 1987.} \end{aligned}$$

In 1987 a total of $t_x = 964{,}470{,}625$ acres were devoted to farms in the United States. The average acres of farms per county for the population is $\bar{x}_{\mathcal{U}} = 964{,}470{,}625/3078 = 313{,}343.3$. The data, and the line through the origin with slope $\hat{B}$, are plotted in Figure 4.1.

A portion of a spreadsheet with the 300 values of x_i and y_i is given in Table 4.1. The estimated ratio is

$$\hat{B} = \frac{\bar{y}}{\bar{x}} = \frac{297897.0467}{301953.7233} = 0.986565,$$

and the ratio estimators of $\bar{y}_{\mathcal{U}}$ and t_y are:

$$\hat{\bar{y}}_r = \hat{B}\bar{x}_{\mathcal{U}} = (\hat{B})(313{,}343.283) = 309{,}133.6,$$

and

$$\hat{t}_{yr} = \hat{B}t_x = (\hat{B})(964{,}470{,}625) = 951{,}513{,}191.$$

Note that $\bar{y}$ for these data is 297,897.0, so $\hat{t}_{y\text{SRS}} = (3078)(\bar{y}) = 916{,}927{,}110$.

In this example, $\bar{x}_{\mathcal{S}} = 301{,}953.7$ is smaller than $\bar{x}_{\mathcal{U}} = 313{,}343.3$. This means that our SRS of size 300 slightly underestimates the true population mean of the x's. Since the x's and y's are positively correlated, we have reason to believe that $\bar{y}_{\mathcal{S}}$ may also underestimate the population value $\bar{y}_{\mathcal{U}}$. Ratio estimation gives a more precise estimate of $\bar{y}_{\mathcal{U}}$ because $\bar{y}_{\mathcal{S}}$ is multiplied by the factor $\bar{x}_{\mathcal{U}}/\bar{x}_{\mathcal{S}}$, which corrects for the underestimation of $\bar{x}_{\mathcal{U}}$ in the particular sample that was drawn. Figure 4.2 shows the ratio and SRS estimates of $\bar{y}_{\mathcal{U}}$ on a graph of the center part of the data. ■

4.1.2 Bias and Mean Squared Error of Ratio Estimators

Unlike the estimators $\bar{y}$ and $N\bar{y}$ in an SRS, ratio estimators are usually *biased* for estimating $\bar{y}_{\mathcal{U}}$ and t_y. We start with the unbiased estimator $\bar{y}$—if we calculate $\bar{y}_{\mathcal{S}}$ for each possible SRS $\mathcal{S}$, then the average of all of the sample means from the possible samples is the population mean $\bar{y}_{\mathcal{U}}$. The estimation bias in ratio estimation arises because $\bar{y}$ is multiplied by $\bar{x}_{\mathcal{U}}/\bar{x}$; if we calculate $\hat{\bar{y}}_r$ for all possible SRSs $\mathcal{S}$, then the average of all the values of $\hat{\bar{y}}_r$ from the different samples will be close to $\bar{y}_{\mathcal{U}}$, but will usually not equal $\bar{y}_{\mathcal{U}}$ exactly.

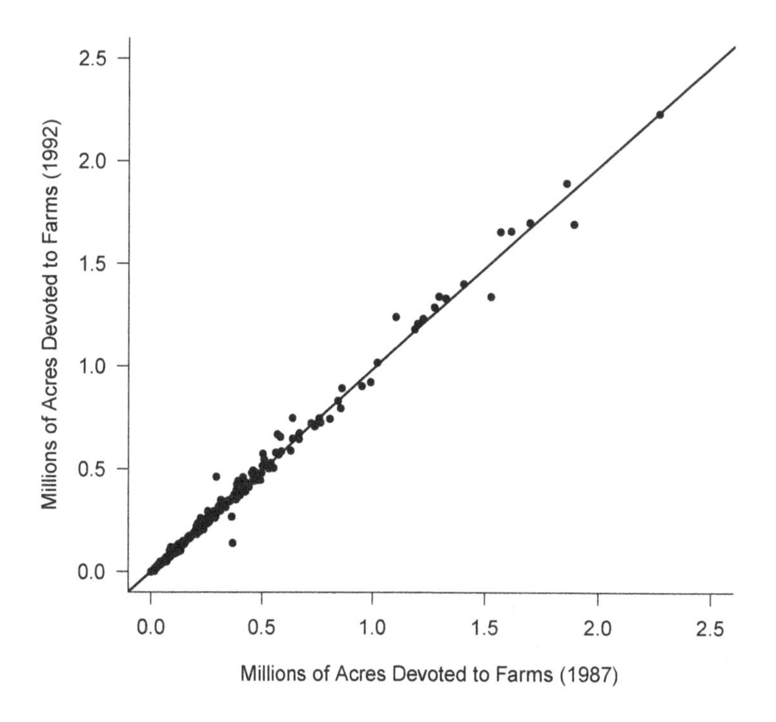

FIGURE 4.1
The plot of acreage, 1992 vs. 1987, for an SRS of 300 counties. The line in the plot goes through the origin and has slope $\hat{B}= 0.9866$. Note that the variability about the line increases with x.

TABLE 4.1
Part of the spreadsheet for the Census of Agriculture data.

County	State	*acres92* (y)	*acres87* (x)	Residual ($y - \hat{B}x$)
Coffee County	AL	175209	179311	−1693.00
Colbert County	AL	138135	145104	−5019.56
Lamar County	AL	56102	59861	−2954.78
Marengo County	AL	199117	220526	−18446.29
⋮	⋮	⋮	⋮	⋮
Rock County	WI	343115	357751	−9829.70
Kanawha County	WV	19956	21369	−1125.91
Pleasants County	WV	15650	15716	145.14
Putnam County	WV	55827	55635	939.44
Column sum		89369114	90586117	0
Column average		297897.0467	301953.7233	0
Column standard deviation		344551.8948	344829.5964	31657.21817

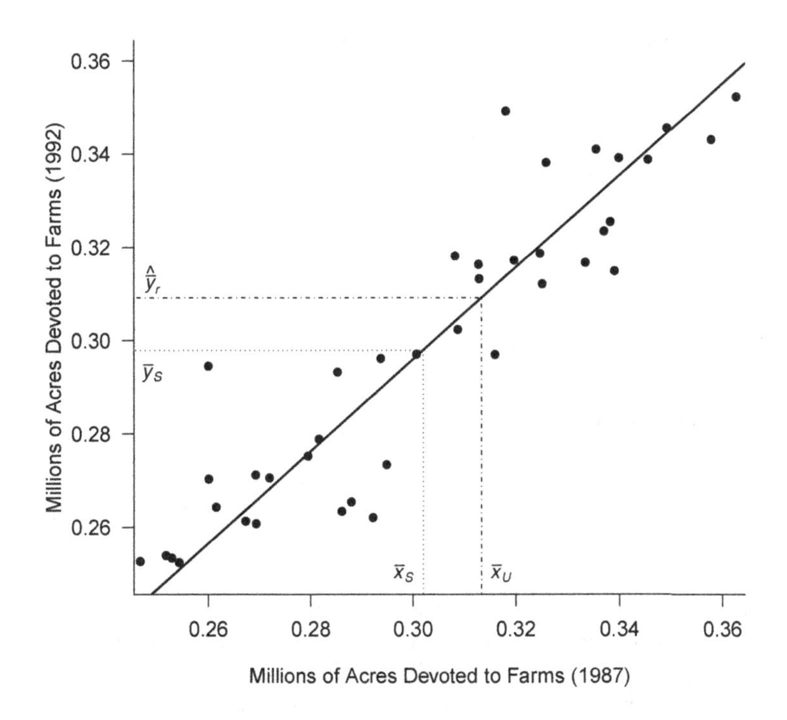

FIGURE 4.2
Detail of the center portion of Figure 4.1. Here, $\bar{x}_{\mathcal{U}}$ is larger than $\bar{x}_{\mathcal{S}}$, so $\hat{\bar{y}}_r$ is larger than $\bar{y}_{\mathcal{S}}$.

The reduced variance of the ratio estimator usually compensates for the presence of bias—although $E[\hat{\bar{y}}_r] \neq \bar{y}_{\mathcal{U}}$, the value of $\hat{\bar{y}}_r$ for any individual sample is likely to be closer to $\bar{y}_{\mathcal{U}}$ than is the sample mean $\bar{y}_{\mathcal{S}}$. After all, we take only one sample in practice; most people would prefer to be able to say that their particular estimate from the sample is likely to be close to the true value, rather than that their particular value of $\bar{y}_{\mathcal{S}}$ may be quite far from $\bar{y}_{\mathcal{U}}$, but that the average deviation $\bar{y}_{\mathcal{S}} - \bar{y}_{\mathcal{U}}$, averaged over all possible samples $\mathcal{S}$ that could be obtained, is zero. For large samples, the sampling distributions of both $\bar{y}$ and $\hat{\bar{y}}_r$ will be approximately normal; if x and y are highly positively correlated, the following pictures illustrate the relative bias and variance of the two estimators:

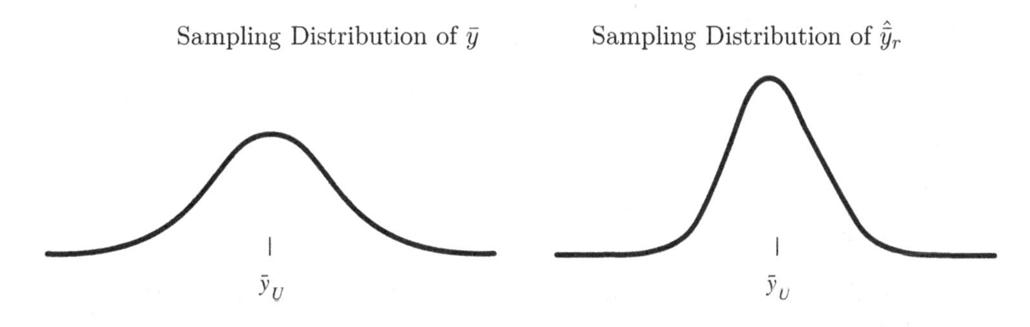

Bias of ratio estimator. To calculate the bias of the ratio estimator of $\bar{y}_{\mathcal{U}}$, note that

$$\hat{\bar{y}}_r - \bar{y}_{\mathcal{U}} = \frac{\bar{y}}{\bar{x}}\bar{x}_{\mathcal{U}} - \bar{y}_{\mathcal{U}} = \bar{y}\left(1 - \frac{\bar{x} - \bar{x}_{\mathcal{U}}}{\bar{x}}\right) - \bar{y}_{\mathcal{U}}.$$

Since $E[\bar{y}] = \bar{y}_{\mathcal{U}}$,

$$\begin{aligned} \text{Bias}\left(\hat{\bar{y}}_r\right) &= E\left[\hat{\bar{y}}_r - \bar{y}_{\mathcal{U}}\right] \\ &= E\left[\bar{y}\right] - \bar{y}_{\mathcal{U}} - E\left[\frac{\bar{y}}{\bar{x}}\left(\bar{x} - \bar{x}_{\mathcal{U}}\right)\right] \\ &= -E\left[\hat{B}\left(\bar{x} - \bar{x}_{\mathcal{U}}\right)\right] \\ &= -\text{Cov}\left(\hat{B}, \bar{x}\right). \end{aligned} \tag{4.5}$$

Consequently, as shown by Hartley and Ross (1954),

$$\frac{|\text{Bias}(\hat{\bar{y}}_r)|}{\sqrt{V(\hat{\bar{y}}_r)}} = \frac{|\text{Cov}\,(\hat{B}, \bar{x})|}{\bar{x}_{\mathcal{U}}\sqrt{V(\hat{B})}} = \frac{|\text{Corr}\,(\hat{B}, \bar{x})|\sqrt{V(\hat{B})V(\bar{x})}}{\bar{x}_{\mathcal{U}}\sqrt{V(\hat{B})}} \leq \frac{\sqrt{V(\bar{x})}}{\bar{x}_{\mathcal{U}}} = \text{CV}(\bar{x}). \tag{4.6}$$

The absolute value of the bias of the ratio estimator is small relative to the standard deviation of the estimator if the coefficient of variation (CV) of $\bar{x}$ is small. For an SRS, $\text{CV}(\bar{x}) \leq S_x^2/(n\bar{x}_{\mathcal{U}})$, so that $\text{CV}(\bar{x})$ decreases as the sample size increases.

The result in Equation (4.5) is exact, but not necessarily easy to use with data. We now find approximations for the bias and variance of the ratio estimator that rely on the variances and covariance of $\bar{x}$ and $\bar{y}$. These approximations are an example of the linearization approach to approximating variances, to be discussed in Section 9.1. Write

$$\hat{\bar{y}}_r - \bar{y}_{\mathcal{U}} = \frac{\bar{x}_{\mathcal{U}}(\bar{y} - B\bar{x})}{\bar{x}} = (\bar{y} - B\bar{x})\left(1 - \frac{\bar{x} - \bar{x}_{\mathcal{U}}}{\bar{x}}\right). \tag{4.7}$$

We can then show that (see Exercise 24)

$$\begin{aligned} \text{Bias}\,(\hat{\bar{y}}_r) = E[\hat{\bar{y}}_r - \bar{y}_{\mathcal{U}}] &\approx \frac{1}{\bar{x}_{\mathcal{U}}}\left[B\,V(\bar{x}) - \text{Cov}\,(\bar{x}, \bar{y})\right] \\ &= \left(1 - \frac{n}{N}\right)\frac{1}{n\bar{x}_{\mathcal{U}}}(BS_x^2 - RS_xS_y), \end{aligned} \tag{4.8}$$

with R the correlation between x and y. The bias of $\hat{\bar{y}}_r$ is thus small if:

- the sample size n is large
- the sampling fraction n/N is large
- $\bar{x}_{\mathcal{U}}$ is large
- S_x is small
- the correlation R is close to 1.

Note that if all x's are the same value ($S_x = 0$), then the ratio estimator is the same as the SRS estimator $\bar{y}$ and the bias is zero.

Mean squared error of ratio estimator. For estimating the MSE of $\hat{\bar{y}}_r$, (4.7) gives:

$$E\left[\left(\hat{\bar{y}}_r - \bar{y}_{\mathcal{U}}\right)^2\right] = E\left[\left\{(\bar{y} - B\bar{x})\left(1 - \frac{\bar{x} - \bar{x}_{\mathcal{U}}}{\bar{x}}\right)\right\}^2\right]$$
$$= E\left[(\bar{y} - B\bar{x})^2 + (\bar{y} - B\bar{x})^2\left\{\left(\frac{\bar{x} - \bar{x}_{\mathcal{U}}}{\bar{x}}\right)^2 - 2\frac{\bar{x} - \bar{x}_{\mathcal{U}}}{\bar{x}}\right\}\right].$$

It can be shown (David and Sukhatme, 1974; Eltinge, 1994) that the second term is generally small compared with the first term, so the variance and MSE are approximated by:

$$\text{MSE}\left(\hat{\bar{y}}_r\right) = E\left[(\hat{\bar{y}}_r - \bar{y}_{\mathcal{U}})^2\right] \approx E\left[(\bar{y} - B\bar{x})^2\right]. \tag{4.9}$$

The term $E\left[(\bar{y} - B\bar{x})^2\right]$ can also be written as (see Exercise 20):

$$E\left[(\bar{y} - B\bar{x})^2\right] = V\left[\frac{1}{n}\sum_{i \in \mathcal{S}}(y_i - Bx_i)\right] = \left(1 - \frac{n}{N}\right)\frac{S_y^2 - 2BRS_xS_y + B^2S_x^2}{n}. \tag{4.10}$$

From (4.9) and (4.10), the approximated MSE of $\hat{\bar{y}}_r$ will be small when

- the sample size n is large
- the sampling fraction n/N is large
- the deviations $y_i - Bx_i$ are small
- the correlation R is close to $+1$.

In large samples, the bias of $\hat{\bar{y}}_r$ is typically small relative to $V(\hat{\bar{y}}_r)$, so that $\text{MSE}\left(\hat{\bar{y}}_r\right) \approx V(\hat{\bar{y}}_r)$ (see Exercise 23). Thus, in the following, we use $\hat{V}(\hat{\bar{y}}_r)$ to estimate both the variance and the MSE.

Note from (4.10) that $E\left[(\bar{y} - B\bar{x})^2\right] = V(\bar{d})$, where $d_i = y_i - Bx_i$ and $\bar{d}_{\mathcal{U}} = 0$. Since the deviations d_i depend on the unknown value B, define the new variable

$$e_i = y_i - \hat{B}x_i,$$

which is the ith residual from fitting the line $y = \hat{B}x$. Estimate $V(\hat{\bar{y}}_r)$ by

$$\hat{V}\left(\hat{\bar{y}}_r\right) = \left(1 - \frac{n}{N}\right)\left(\frac{\bar{x}_{\mathcal{U}}}{\bar{x}}\right)^2\frac{s_e^2}{n}, \tag{4.11}$$

where s_e^2 is the sample variance of the residuals e_i:

$$s_e^2 = \frac{1}{n-1}\sum_{i \in \mathcal{S}} e_i^2.$$

[Exercise 21 explains why we include the factor $\bar{x}_{\mathcal{U}}/\bar{x}$ in (4.11). In large samples, we expect $\bar{x}_{\mathcal{U}}/\bar{x}$ to be approximately equal to 1.] Similarly,

$$\hat{V}\left(\hat{B}\right) = \left(1 - \frac{n}{N}\right)\frac{s_e^2}{n\bar{x}^2} \tag{4.12}$$

and

$$\hat{V}\left(\hat{t}_{yr}\right) = \hat{V}\left(t_x\hat{B}\right) = \left(1 - \frac{n}{N}\right)\left(\frac{t_x}{\bar{x}}\right)^2\frac{s_e^2}{n}. \tag{4.13}$$

If the sample size n is sufficiently large, approximate 95% confidence intervals (CIs) can be constructed using the standard errors (SEs)—the square roots of (4.11), (4.12), and (4.13)—as

$$\hat{B} \pm 1.96\,\text{SE}(\hat{B}), \quad \hat{\bar{y}}_r \pm 1.96\,\text{SE}(\hat{\bar{y}}_r), \quad \text{or} \quad \hat{t}_{yr} \pm 1.96\,\text{SE}(\hat{t}_{yr}). \tag{4.14}$$

Most statistical software packages substitute a t percentile with $n - 1$ degrees of freedom (df) for the normal percentile 1.96.

Example 4.3. For the SRS in Example 4.2, the last column of Table 4.1 contains the residuals $e_i = y_i - \hat{B}x_i$. The sample standard deviation of that column is $s_e = 31{,}657.218$. Thus, using (4.13),

$$\text{SE}(\hat{t}_{yr}) = 3078\sqrt{1 - \frac{300}{3078}}\left(\frac{313{,}343.283}{301{,}953.723}\right)\frac{31{,}657.218}{\sqrt{300}} = 5{,}546{,}162.$$

An approximate 95% CI for the total farm acreage, using the ratio estimator, is

$$951{,}513{,}191 \pm 1.96(5{,}546{,}162) = [940{,}642{,}713, \;\; 962{,}383{,}669].$$

Did using a ratio estimator for the population total improve the precision in this example? The standard error of $\hat{t}_y = N\bar{y}$ is more than 10 times as large:

$$\text{SE}(N\bar{y}) = 3078\sqrt{\left(1 - \frac{150}{3078}\right)}\frac{s_y}{\sqrt{150}} = 58{,}169{,}381.$$

The estimated CV for the ratio estimator is 5,546,162/951,513,191 = 0.0058, as compared with an estimated CV of 0.0634 for the SRS estimator $N\bar{y}$ which does not use the auxiliary information. Including the 1987 information through the ratio estimator has greatly increased the precision. If all quantities to be estimated were highly correlated with the 1987 acreage, we could dramatically reduce the sample size and still obtain high precision by using ratio estimators rather than $N\bar{y}$. ■

Example 4.4. Let's take another look at the hypothetical population used in Example 2.2 and Exercise 2 of Chapter 3. Now, though, instead of using x as a stratification variable in stratified sampling, we use it as auxiliary information for ratio estimation. The population values are the following:

Unit Number	x	y
1	4	1
2	5	2
3	5	4
4	6	4
5	8	7
6	7	7
7	7	7
8	5	8

Note that x and y are positively correlated. We can calculate population quantities since we know the entire population and sampling distribution:

$$t_x = 47 \qquad t_y = 40$$
$$S_x = 1.3562027 \qquad S_y = 2.618615$$
$$R = 0.6838403 \qquad B = 0.8510638$$

TABLE 4.2
Sampling distribution for $\hat{t}_{yr}$.

Sample Number	Sample, $\mathcal{S}$	$\bar{x}_{\mathcal{S}}$	$\bar{y}_{\mathcal{S}}$	$\hat{B}$	$\hat{t}_{\text{SRS}}$	$\hat{t}_{yr}$
1	{1,2,3,4}	5.00	2.75	0.55	22.00	25.85
2	{1,2,3,5}	5.50	3.50	0.64	28.00	29.91
3	{1,2,3,6}	5.25	3.50	0.67	28.00	31.33
4	{1,2,3,7}	5.25	3.50	0.67	28.00	31.33
⋮	⋮	⋮	⋮	⋮	⋮	⋮
67	{4,5,6,8}	6.50	6.50	1.00	52.00	47.00
68	{4,5,7,8}	6.50	6.50	1.00	52.00	47.00
69	{4,6,7,8}	6.25	6.50	1.04	52.00	48.88
70	{5,6,7,8}	6.75	7.25	1.07	58.00	50.48

Part of the sampling distribution for $\hat{t}_{yr}$ for a sample of size $n = 4$ is given in Table 4.2; the full listing with all possible samples is in file `artifratio.csv`. Figure 4.3 displays histograms for the sampling distributions of two estimators of t_y: $\hat{t}_{\text{SRS}} = N\bar{y}$, the estimator used in Chapter 2, and $\hat{t}_{yr}$. The sampling distribution for the ratio estimator is not spread out as much as the sampling distribution for $N\bar{y}$; it is also skewed rather than symmetric. The skewness leads to the slight estimation bias of the ratio estimator. The population total is $t_y = 40$; the mean value of the sampling distribution of $\hat{t}_{yr}$ is 39.85063.

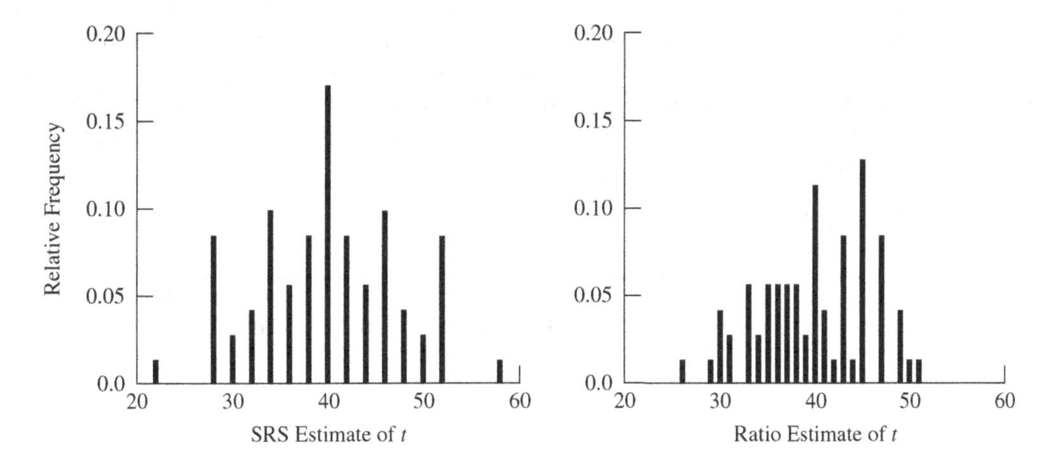

FIGURE 4.3
Sampling distributions for (a) $\hat{t}_{\text{SRS}}$ and (b) $\hat{t}_{yr}$.

The mean value of the sampling distribution of $\hat{B}$ is 0.8478857, so Bias$(\hat{B}) = -0.003178$. The approximate bias of $\hat{B}$, calculated by substituting the population quantities into (4.8) and noting from (4.2) that $\hat{B} = \hat{\bar{y}}_r/\bar{x}_{\mathcal{U}}$, is

$$\left(1 - \frac{n}{N}\right) \frac{1}{n\,\bar{x}_{\mathcal{U}}^2}(BS_x^2 - RS_xS_y) = -0.003126.$$

The variance of the sampling distribution of $\hat{B}$, calculated using the definition of variance in (2.5), is 0.015186446; the approximation in (4.10) is

$$\frac{4}{8}\frac{1}{(4)(5.875)^2}\left(S_y^2 - 2BRS_xS_y + B^2S_x^2\right) = 0.01468762. \blacksquare$$

Example 4.4 demonstrates that the approximation to the MSE in (4.10) is in fact only an approximation; it happens to be a good approximation in that example even though the population and sample are both small.

For (4.9) to be a good approximation to the MSE, the bias should be small and the terms discarded in the approximation of the variance should be small. If the CV of $\bar{x}$ is small—that is, if $\bar{x}_{\mathcal{U}}$ is estimated with high relative precision, the bias is small relative to the square root of the variance. If we form a CI using $\hat{t}_{yr} \pm z_{\alpha/2}\,\text{SE}[\hat{t}_{yr}]$, using (4.13) as the standard error, then the bias will not have a great effect on the coverage probability of the CI. A small $\text{CV}(\bar{x})$ also means that $\bar{x}$ is stable from sample to sample. In more complex sampling designs, though, the bias may be a matter of concern—we return to this issue in Section 4.5 and Chapter 9. For the approximation in (4.9) to work well, we want the sample size to be large, and to have $\text{CV}(\bar{x}) \leq 0.1$ and $\text{CV}(\bar{y}) \leq 0.1$. If these conditions are not met, then (4.9) may underestimate the true MSE.

4.1.3 Ratio Estimation with Proportions

Ratio estimation works exactly the same way when the quantity of interest is a proportion.

Example 4.5. Peart (1994) collected the data shown in Table 4.3 as part of a study evaluating the effects of feral pig activity and drought on the native vegetation on Santa Cruz Island, California. She counted the number of woody seedlings in pig-protected areas under each of ten sampled oak trees in March 1992, following the drought-ending rains of 1991. She put a flag by each seedling, then determined how many were still alive in February 1994. The data are plotted in Figure 4.4.

TABLE 4.3
Santa Cruz Island seedling data.

Tree	Seedlings Alive in March 1992 (x)	Seedlings Alive in February 1994 (y)
1	1	0
2	0	0
3	8	1
4	2	2
5	76	10
6	60	15
7	25	3
8	2	2
9	1	1
10	31	27
Total	206	61
Average	20.6	6.1
Standard deviation	27.4720	8.8248

When most people who have had one introductory statistics course see data like these, they want to find the sample proportion of the 1992 seedlings that are still alive in 1994, and then calculate the standard error as though they had an SRS of 206 seedlings, obtaining a value of $\sqrt{(0.2961)(0.7039)/206} = 0.0318$. This calculation is *incorrect* for these data since

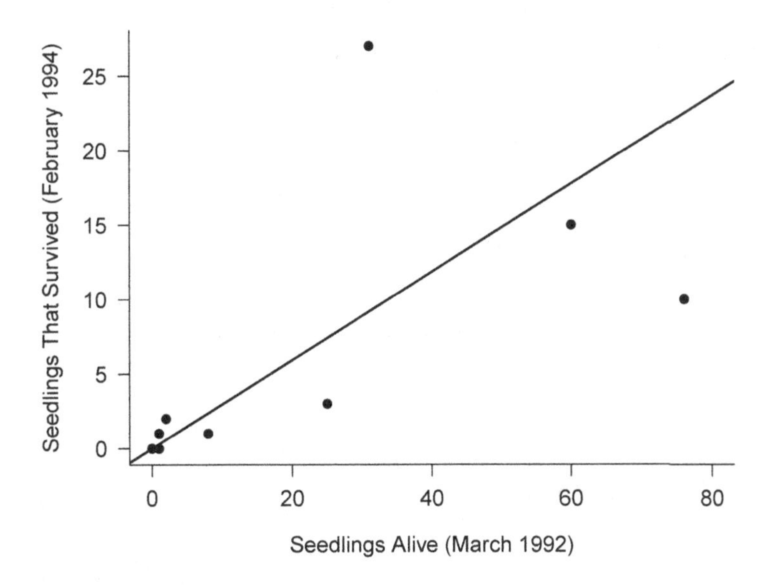

FIGURE 4.4
Plot of the number of seedlings that survived to February 1994 versus seedlings that were alive in March 1992, for ten oak trees.

plots, not individual seedlings, are the sampling units. Seedling survival depends on many factors such as local rainfall, amount of light, and predation. Such factors are likely to affect seedlings in the same plot to a similar degree, leading different plots to have, in general, different survival rates. The sample size in this example is 10, not 206.

The design is actually a **cluster sample**; the clusters are the plots associated with each tree, and the observation units are individual seedlings in those plots. To look at this example from the framework of ratio estimation, let

$$y_i = \text{ number of seedlings near tree } i \text{ that are alive in 1994,}$$

$$x_i = \text{ number of seedlings near tree } i \text{ that are alive in 1992.}$$

Then the ratio estimate of the proportion of seedlings still alive in 1994 is

$$\hat{B} = \hat{p} = \frac{\bar{y}}{\bar{x}} = \frac{6.1}{20.6} = 0.2961.$$

Using (4.12) and ignoring the finite population correction (fpc),

$$\text{SE}(\hat{B}) = \sqrt{\frac{1}{(10)\,(20.6)^2} \frac{\sum_{i \in \mathcal{S}}(y_i - 0.2961165x_i)^2}{9}} = \sqrt{\frac{56.3778}{(10)\,(20.6)^2}} = 0.115.$$

The approximation to the variance of $\hat{B}$ in this example may be a little biased because the sample size is small. Note, however, the difference between the correctly calculated standard error of 0.115, and the incorrect value 0.0318 that would be obtained if one erroneously pretended that the seedlings were selected in an SRS. ■

4.1.4 Ratio Estimation Using Weight Adjustments

In Section 2.4, we defined the sampling weight to be $w_i = 1/\pi_i$, and wrote the estimated population total as a function of the observations y_i and weights w_i:

$$\hat{t}_y = \sum_{i \in \mathcal{S}} w_i y_i.$$

Note that

$$\hat{t}_{yr} = \frac{t_x}{\hat{t}_x}\hat{t}_y = \frac{t_x}{\hat{t}_x}\sum_{i \in \mathcal{S}} w_i y_i.$$

We can think of the modification used in ratio estimation as an adjustment to each weight. Define

$$g_i = \frac{t_x}{\hat{t}_x}.$$

Then

$$\hat{t}_{yr} = \sum_{i \in \mathcal{S}} w_i g_i y_i. \tag{4.15}$$

The estimator $\hat{t}_{yr}$ is a weighted sum of the observations, with weights $w_i^* = w_i g_i$. Unlike the original weights w_i, however, the adjusted weights w_i^* depend upon values from the sample: If a different sample is taken, the weight adjustment $g_i = t_x/\hat{t}_x$ will be different.

The weight adjustments g_i **calibrate** the estimates on the x variable. Since

$$\sum_{i \in \mathcal{S}} w_i g_i x_i = t_x,$$

the adjusted weights force the estimated total for the x variable to equal the known population total t_x. The factors g_i are called the calibration factors.

The variance estimators in (4.11) and (4.13) can be calculated by forming the new variable $u_i = g_i e_i$. Then, for an SRS,

$$\hat{V}(\bar{u}) = \left(1 - \frac{n}{N}\right)\frac{1}{n(n-1)}\sum_{i \in \mathcal{S}}(u_i - \bar{u})^2 = \left(1 - \frac{n}{N}\right)\frac{s_e^2}{n}\left(\frac{t_x}{\hat{t}_x}\right)^2 = \hat{V}(\hat{\bar{y}}_r)$$

and, similarly, $\hat{V}(\hat{t}_u) = \hat{V}(\hat{t}_{yr})$.

Example 4.6. For the Census of Agriculture data used in Examples 4.2 and 4.3, $g_i = t_x/\hat{t}_x = 964{,}470{,}625/929{,}413{,}560 = 1.037719554$ for each observation. Since the sample has $\hat{t}_x < t_x$, each observation's sampling weight is increased by a small amount. The sampling weight for the SRS design is $w_i = 3078/300 = 10.26$, so the ratio-adjusted weight for each observation is

$$w_i^* = w_i g_i = (10.26)(1.037719554) = 10.64700262.$$

Then

$$\sum_{i \in \mathcal{S}} w_i g_i x_i = \sum_{i \in \mathcal{S}} 10.64700262\, x_i = 964{,}470{,}625 = t_x$$

and

$$\sum_{i \in \mathcal{S}} w_i g_i y_i = \sum_{i \in \mathcal{S}} 10.64700262\, y_i = 951{,}513{,}191 = \hat{t}_{yr}.$$

The adjusted weights, however, no longer sum to $N = 3078$:

$$\sum_{i \in \mathcal{S}} w_i^* = \sum_{i \in \mathcal{S}} w_i g_i = (300)(10.64700262) = 3194.$$

Thus, this ratio estimator is calibrated to the population total t_x of the x variable, but is no longer calibrated to the population size N. In Chapter 11, we shall see how to perform calibration with multiple calibrating variables so that the calibrated weights w_i^* have the property $\sum_{i \in \mathcal{S}} w_i^* x_i = t_x$ for multiple variables x. ■

4.1.5 Advantages of Ratio Estimation

Ratio estimation is motivated by the desire to use information about a known auxiliary quantity x to obtain a more accurate estimator of t_y or $\bar{y}_{\mathcal{U}}$. If x and y are perfectly correlated, that is, $y_i = Bx_i$ for all $i = 1, \ldots, N$, then $\hat{t}_{yr} = t_y$ and there is no estimation error. In general, if y_i is roughly proportional to x_i, the MSE will be small.

When does ratio estimation help? If the deviations of y_i from $\hat{B}x_i$ are smaller than the deviations of y_i from $\bar{y}$, then $\hat{V}(\hat{\bar{y}}_r) \leq \hat{V}(\bar{y})$. Recall from Chapter 2 that

$$\text{MSE}(\bar{y}) = V(\bar{y}) = \left(1 - \frac{n}{N}\right)\frac{S_y^2}{n}.$$

Using the approximation in (4.10),

$$\text{MSE}(\hat{\bar{y}}_r) \approx \left(1 - \frac{n}{N}\right)\frac{1}{n}\left(S_y^2 - 2BRS_xS_y + B^2S_x^2\right).$$

Thus,

$$\begin{aligned}\text{MSE}(\hat{\bar{y}}_r) - \text{MSE}(\bar{y}) &\approx \left(1 - \frac{n}{N}\right)\frac{1}{n}\left(S_y^2 - 2BRS_xS_y + B^2S_x^2 - S_y^2\right)\\ &= \left(1 - \frac{n}{N}\right)\frac{1}{n}S_xB\left(-2RS_y + BS_x\right)\end{aligned}$$

so to the accuracy of the approximation,

$$\text{MSE}\left(\hat{\bar{y}}_r\right) \leq \text{MSE}\left(\bar{y}\right) \text{ if and only if } R \geq \frac{BS_x}{2S_y} = \frac{\text{CV}(x)}{2\text{CV}(y)}.$$

If the CVs are approximately equal, then it pays to use ratio estimation when the correlation between x and y is larger than $1/2$.

The ratio estimator is motivated by a regression model, discussed in Section 4.6, in which the relationship between y and x is a straight line through the origin and the variance of y_i about the line is proportional to x_i. Ratio estimation is most efficient, decreasing the variance of estimated means and totals, if that model fits the data well. But the standard errors for ratio estimation are correct even if the model is inappropriate. The variance estimates in (4.11) to (4.13) are derived from properties of the sampling design, so even if the model is a poor fit, the confidence intervals in (4.14) have approximately the correct coverage probability.

4.2 Regression Estimation in Simple Random Sampling

A straight-line regression model. Ratio estimation works best if the data are well fit by a straight line through the origin. Sometimes, data appear to be evenly scattered about a straight line that does not go through the origin—that is, the data look as though the straight-line regression model

$$y = B_0 + B_1 x$$

would provide a good fit.

Let $\hat{B}_1$ and $\hat{B}_0$ be the ordinary least squares regression coefficients of the slope and intercept. For the straight line regression model,

$$\hat{B}_1 = \frac{\sum_{i \in \mathcal{S}} (x_i - \bar{x})(y_i - \bar{y})}{\sum_{i \in \mathcal{S}} (x_i - \bar{x})^2} = \frac{r s_y}{s_x},$$

$$\hat{B}_0 = \bar{y} - \hat{B}_1 \bar{x},$$

and r is the sample correlation coefficient of x and y.

In regression estimation, like ratio estimation, we use the correlation between x and y to obtain an estimator for $\bar{y}_{\mathcal{U}}$ with (we hope) increased precision. Suppose we know $\bar{x}_{\mathcal{U}}$, the population mean for the x's. Then the regression estimator of $\bar{y}_{\mathcal{U}}$ is the predicted value of y from the fitted regression equation when $x = \bar{x}_{\mathcal{U}}$:

$$\hat{\bar{y}}_{\text{reg}} = \hat{B}_0 + \hat{B}_1 \bar{x}_{\mathcal{U}} = \bar{y} + \hat{B}_1 (\bar{x}_{\mathcal{U}} - \bar{x}). \tag{4.16}$$

If $\bar{x}$ from the sample is smaller than the population mean $\bar{x}_{\mathcal{U}}$ and x and y are positively correlated, then we would expect $\bar{y}$ to also be smaller than $\bar{y}_{\mathcal{U}}$. The regression estimator adjusts $\bar{y}$ by the quantity $\hat{B}_1(\bar{x}_{\mathcal{U}} - \bar{x})$.

Bias and mean squared error for regression estimation. Like the ratio estimator, the regression estimator is biased. Let B_1 and B_0 be the least squares regression slope and intercept calculated from all the data in the population:

$$B_1 = \frac{\sum_{i=1}^{N} (x_i - \bar{x}_{\mathcal{U}})(y_i - \bar{y}_{\mathcal{U}})}{\sum_{i=1}^{N} (x_i - \bar{x}_{\mathcal{U}})^2} = \frac{R S_y}{S_x},$$

$$B_0 = \bar{y}_{\mathcal{U}} - B_1 \bar{x}_{\mathcal{U}}.$$

Then, using (4.16), the bias of $\hat{\bar{y}}_{\text{reg}}$ is given by

$$E\left[\hat{\bar{y}}_{\text{reg}} - \bar{y}_{\mathcal{U}}\right] = E\left[\bar{y} - \bar{y}_{\mathcal{U}}\right] + E\left[\hat{B}_1(\bar{x}_{\mathcal{U}} - \bar{x})\right] = -\operatorname{Cov}\left(\hat{B}_1, \bar{x}\right). \tag{4.17}$$

If the regression line goes through all of the points (x_i, y_i) in the population, then the bias is zero: in that situation, $\hat{B}_1 = B_1$ for every sample, so $\operatorname{Cov}(\hat{B}_1, \bar{x}) = 0$. As with ratio estimation, for large SRSs the MSE for regression estimation is approximately equal to the variance (see Exercise 31); the bias is often negligible in large samples.

The method used for approximating the MSE in ratio estimation can also be applied to regression estimation. Let

$$d_i = y_i - B_0 - B_1 x_i = y_i - \bar{y}_{\mathcal{U}} - B_1(x_i - \bar{x}_{\mathcal{U}})$$

denote the ith residual from the regression model fit to the entire population. Then,

$$\begin{aligned}
\text{MSE}(\hat{\bar{y}}_{\text{reg}}) &= E\left[\{\bar{y} + \hat{B}_1(\bar{x}_{\mathcal{U}} - \bar{x}) - \bar{y}_{\mathcal{U}}\}^2\right] \\
&= E\left[\left\{\bar{y} - \bar{y}_{\mathcal{U}} - B_1(\bar{x} - \bar{x}_{\mathcal{U}}) - \left(\hat{B}_1 - B_1\right)(\bar{x} - \bar{x}_{\mathcal{U}})\right\}^2\right] \\
&\approx V(\bar{d}) \\
&= \left(1 - \frac{n}{N}\right)\frac{S_d^2}{n}. \qquad (4.18)
\end{aligned}$$

Using the relation $B_1 = RS_y/S_x$, it may be shown (see Exercise 30) that

$$\begin{aligned}
\left(1 - \frac{n}{N}\right)\frac{S_d^2}{n} &= \left(1 - \frac{n}{N}\right)\frac{1}{n}\sum_{i=1}^{N}\frac{(y_i - \bar{y}_{\mathcal{U}} - B_1[x_i - \bar{x}_{\mathcal{U}}])^2}{N-1} \\
&= \left(1 - \frac{n}{N}\right)\frac{1}{n}S_y^2(1 - R^2). \qquad (4.19)
\end{aligned}$$

Thus, the approximate MSE is small when

- n is large
- n/N is large
- S_y is small
- the correlation R is close to -1 or $+1$.

The standard error may be calculated by substituting estimates for the population quantities in (4.18) or (4.19). We can estimate S_d^2 in (4.18) by using the residuals

$$e_i = y_i - (\hat{B}_0 + \hat{B}_1 x_i);$$

then $s_e^2 = \sum_{i\in\mathcal{S}} e_i^2/(n-1)$ estimates S_d^2 and

$$\text{SE}(\hat{\bar{y}}_{\text{reg}}) = \sqrt{\left(1 - \frac{n}{N}\right)\frac{s_e^2}{n}}. \qquad (4.20)$$

In small samples, we may alternatively calculate s_e^2 using the MSE from a regression analysis: $s_e^2 = \sum_{i\in\mathcal{S}} e_i^2/(n-2)$. This adjusts the estimator for the degrees of freedom in the regression.

To estimate the variance using the formulation in (4.19), substitute the sample variance s_y^2 and the sample correlation r for the population quantities S_y^2 and R, obtaining

$$\text{SE}(\hat{\bar{y}}_{\text{reg}}) = \sqrt{\left(1 - \frac{n}{N}\right)\frac{1}{n}s_y^2(1 - r^2)}. \qquad (4.21)$$

Example 4.7. To estimate the number of dead trees in an area, we divide the area into 100 square plots and count the number of dead trees on a photograph of each plot. Photo counts can be made quickly, but sometimes a tree is misclassified or not detected. So we select an SRS of 25 of the plots for field counts of dead trees. We know that the population mean number of dead trees per plot from the photo count is 11.3. The data—plotted in Figure 4.5—are given below and in file `deadtrees.csv`.

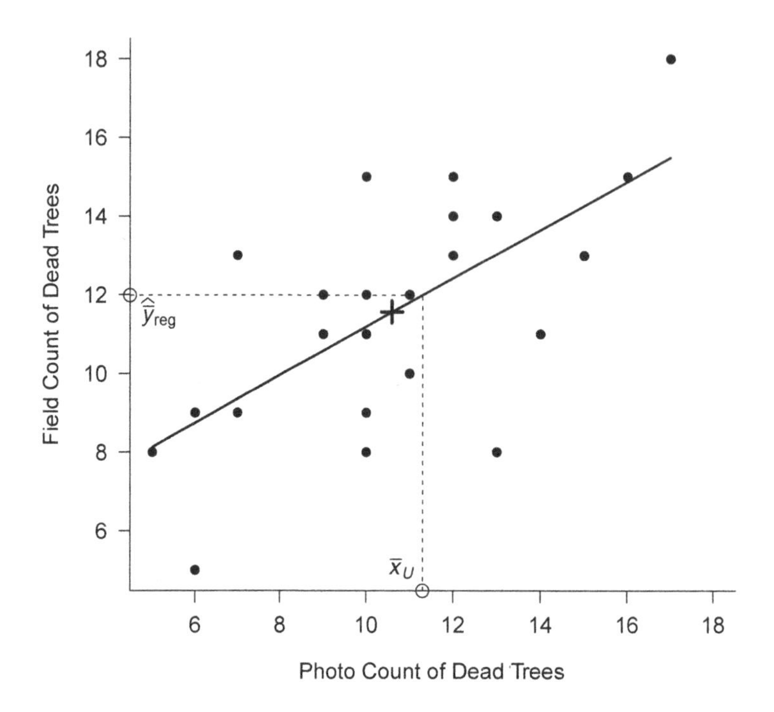

FIGURE 4.5
Scatterplot of photo and field tree-count data, along with the regression line. Note that $\hat{\bar{y}}_{\text{reg}}$ is the predicted value from the regression equation when $x = \bar{x}_{\mathcal{U}}$. The point $(\bar{x}, \bar{y})$ is marked by "+" on the graph.

Photo (x)	10	12	7	13	13	6	17	16	15	10	14	12	10
Field (y)	15	14	9	14	8	5	18	15	13	15	11	15	12

Photo (x)	5	12	10	10	9	6	11	7	9	11	10	10
Field (y)	8	13	9	11	12	9	12	13	11	10	9	8

For these data, $\bar{x} = 10.6$, $\bar{y} = 11.56$, $s_y^2 = 9.09$, and the sample correlation between x and y is $r = 0.62420$. Fitting a straight line regression model gives

$$\hat{y} = 5.059292 + 0.613274x$$

with $\hat{B}_0 = 5.059292$ and $\hat{B}_1 = 0.613274$. In this example, x and y are positively correlated so that $\bar{x}$ and $\bar{y}$ are also positively correlated. Since $\bar{x} < \bar{x}_{\mathcal{U}}$, we expect that the sample mean $\bar{y}$ is also too small; the regression estimate adds the quantity $\hat{B}_1(\bar{x}_{\mathcal{U}} - \bar{x}) = 0.613(11.3-10.6) = 0.43$ to $\bar{y}$ to compensate.

Using (4.16), the regression estimate of the mean is

$$\hat{\bar{y}}_{\text{reg}} = 5.059292 + 0.613274\,(11.3) = 11.99.$$

From (4.21), the standard error is

$$\text{SE}[\hat{\bar{y}}_{\text{reg}}] = \sqrt{\left(1 - \frac{25}{100}\right)\frac{1}{25}(9.09)(1 - 0.62420^2)} = 0.408.$$

Note that the standard error for the sample mean $\bar{y}$ is $\text{SE}(\bar{y}) = \sqrt{(1 - 25/100)\, s_y^2/25} = 0.522$; using regression estimation has increased the precision in this example because the variables photo and field are positively correlated.

To estimate the total number of dead trees, use

$$\hat{t}_{y\text{reg}} = (100)(11.99) = 1199,$$

$$\text{SE}[\hat{t}_{y\text{reg}}] = (100)(0.408) = 40.8.$$

An approximate 95% confidence interval for the total number of dead trees is given by

$$1199 \pm (2.07)(40.8) = [1114, 1283];$$

because of the relatively small sample size, we used the t distribution percentile (with $n - 2 = 23$ degrees of freedom) of 2.07 in the CI rather than the normal distribution percentile of 1.96.

In practice, we recommend using survey regression software to produce regression estimates; Lohr (2022) and Lu and Lohr (2022) provide code and output. These packages may use a slightly different formula for the standard error than the one used in this example, but for large samples the results will be similar (see Section 11.6). ■

4.3 Estimation in Domains

Often we want separate estimates for subpopulations; the subpopulations are called **domains** or **subdomains**. We may want to take an SRS of 1000 people from a population of 50,000 people and estimate the average salary for men and the average salary for women. There are two domains: men and women. We do not know which persons in the population belong to which domain until they are sampled, though. Thus, the number of persons in an SRS who fall into each domain is a random variable, with value unknown at the time the survey is designed.

Estimating domain means is a special case of ratio estimation. Suppose there are D domains. Let $\mathcal{U}_d$ be the index set of the units in the population that are in domain d, and let $\mathcal{S}_d$ be the index set of the units in the sample that are in domain d, for $d = 1, 2, \ldots, D$. Let N_d be the number of population units in $\mathcal{U}_d$, and n_d be the number of sample units in $\mathcal{S}_d$. Suppose we want to estimate the mean salary for the domain of women,

$$\bar{y}_{\mathcal{U}_d} = \frac{\sum_{i \in \mathcal{U}_d} y_i}{N_d} = \frac{\text{total salary for all women in population}}{\text{number of women in population}}.$$

A natural estimator of $\bar{y}_{\mathcal{U}_d}$ is

$$\bar{y}_d = \frac{\sum_{i \in \mathcal{S}_d} y_i}{n_d} = \frac{\text{total salary for women in sample}}{\text{number of women in sample}},$$

which looks at first just like the sample means studied in Chapter 2.

The quantity n_d is a random variable, however: If a different SRS is taken, we will very likely have a different value for n_d, the number of women in the sample. To see that $\bar{y}_d$ uses ratio estimation, let

$$x_i = \begin{cases} 1 & \text{if } i \in \mathcal{U}_d \\ 0 & \text{if } i \notin \mathcal{U}_d, \end{cases}$$

$$u_i = y_i x_i = \begin{cases} y_i & \text{if } i \in \mathcal{U}_d \\ 0 & \text{if } i \notin \mathcal{U}_d. \end{cases}$$

Then $t_x = \sum_{i=1}^N x_i = N_d$, $\bar{x}_{\mathcal{U}} = N_d/N$, $t_u = \sum_{i=1}^N u_i$, $\bar{y}_{\mathcal{U}_d} = t_u/t_x = B$, $\bar{x} = n_d/n$, and

$$\bar{y}_d = \hat{B} = \frac{\bar{u}}{\bar{x}} = \frac{\hat{t}_u}{\hat{t}_x}.$$

Because we are estimating a ratio, we use (4.12) to calculate the standard error:

$$\text{SE}(\bar{y}_d) = \sqrt{\left(1 - \frac{n}{N}\right) \frac{1}{n\bar{x}^2} \frac{\sum_{i \in \mathcal{S}} (u_i - \hat{B}x_i)^2}{n-1}} = \sqrt{\left(1 - \frac{n}{N}\right) \frac{n}{n_d^2} \frac{(n_d - 1)s_{yd}^2}{n-1}}, \tag{4.22}$$

where

$$s_{yd}^2 = \frac{1}{n_d - 1} \sum_{i \in \mathcal{S}_d} (y_i - \bar{y}_d)^2$$

is the sample variance of the sample observations in domain d. If $E(n_d)$ is large, then $(n_d - 1)/n_d \approx 1$ and

$$\text{SE}\,(\bar{y}_d) \approx \sqrt{\left(1 - \frac{n}{N}\right) \frac{s_{yd}^2}{n_d}}. \tag{4.23}$$

In a sufficiently large SRS, the standard error of $\bar{y}_d$ is approximately the same as if we used formula (2.15).

The situation is a little more complicated when estimating a domain total. If N_d is known, we can estimate t_u by $N_d \bar{y}_d$. If N_d is unknown, though, we need to estimate t_u by

$$\hat{t}_{yd} = \hat{t}_u = N\bar{u}.$$

The standard error is

$$\text{SE}(\hat{t}_{yd}) = N\,\text{SE}(\bar{u}) = N\sqrt{\left(1 - \frac{n}{N}\right) \frac{s_u^2}{n}}. \tag{4.24}$$

This standard error incorporates the uncertainty about the number of population members in the domain through the sample variance s_u^2.

Example 4.8. In the SRS of size 300 from the Census of Agriculture (see Examples 2.6, 4.2, and 4.3), 129 counties have at least 600 farms and the remaining 171 counties have fewer than 600 farms. What are the estimated average and total number of acres devoted to farming in each of these domains?

Summary statistics for the two domains are:

Domain, d	n_d	$\bar{y}_d$	s_d
1. At least 600 farms	129	316,565.65	258,249.74
2. Fewer than 600 farms	171	283,813.71	397,643.92

Thus the standard error for the estimated mean in domain 1 is:

$$\mathrm{SE}\,(\bar{y}_1) = \sqrt{\left(1 - \frac{300}{3078}\right)\left(\frac{300}{299}\right)\left(\frac{128}{129}\right)}\frac{258{,}249.74}{\sqrt{129}} = 21{,}553.$$

An approximate 95% CI for the mean farm acreage for counties in domain 1, using the t critical value with 128 df, is $316{,}565.65 \pm 1.979\,(21{,}553)$, or [273,919, 359,212]. A similar calculation for domain 2 yields $\mathrm{SE}\,(\bar{y}_2) = 28{,}852.24$ and an approximate 95% CI of [226,859, 340,769].

Suppose that we do not know how many counties in the population are in each domain. To estimate the total in domain 1, define

$$x_i = \begin{cases} 1, & \text{if county } i \text{ is in domain 1} \\ 0, & \text{otherwise} \end{cases}$$

and $u_i = y_i x_i$. Then

$$\hat{t}_{y1} = \hat{t}_u = \sum_{i \in \mathcal{S}} \frac{3078}{300} u_i = 418{,}987{,}302. \tag{4.25}$$

The standard error is

$$\mathrm{SE}(\hat{t}_{y1}) = N\sqrt{1 - \frac{n}{N}}\frac{s_u}{\sqrt{n}} = 3078\sqrt{\left(1 - \frac{300}{3078}\right)}\frac{230{,}641.22}{\sqrt{300}} = 38{,}938{,}277$$

and a 95% CI for the population total in domain 1, using a t critical value with 299 df, is $418{,}987{,}302 \pm 1.968(38{,}938{,}277) = [342{,}359{,}512,\ 495{,}615{,}092]$. Similarly, a 95% CI for the population total in domain 2 is

$$497{,}939{,}808 \pm 1.968(3078)\sqrt{\left(1 - \frac{300}{3078}\right)}\frac{331{,}225.43}{\sqrt{300}} = [387{,}894{,}116,\ 607{,}985{,}499]\,. \blacksquare$$

In this section, we have shown that estimating domain means is a special case of ratio estimation because the sample size in the domain varies from sample to sample. If the sample size for the domain in an SRS is sufficiently large, we can use SRS formulas for inference about the domain mean.

Inference about totals depends on whether the population size of the domain, N_d, is known. If N_d is known, then the estimated total is $N_d\bar{y}_d$. If N_d is unknown, then define a new variable u_i that equals y_i for observations in the domain and 0 for observations not in the domain; then use $\hat{t}_u$ to estimate the domain total.

The results of this section are only for SRSs, and the approximations depend on having a sufficiently large sample so that $E(n_d)$ is large. Domain estimates for general survey designs are discussed in Section 11.3 and the problem of domain estimates when n_d is small is discussed in Section 14.2.

Domains and strata. Domains and strata are both subsets of the population. What is the difference between them?

Strata are used in the sample design. A stratified random sampling design specifies selecting a predetermined number of observations, n_h, from stratum h. Every sample that could possibly be drawn using that sampling design will have n_h observations from stratum h.

A domain is any population subset of interest, and calculating estimates for domains of interest is part of the analysis. Sometimes it is desired to calculate separate estimates for a stratum; in that case, the stratum is also a domain and the sample size is fixed. But for domains that are not strata, such as those in Example 4.8, the domain sample sizes will vary from sample to sample. It is even possible that a sample will have no members in some of the domains.

Consider the stratified sample of colleges considered in Example 3.12. If we select a sample using the proportional allocation in Table 3.8, that sample will have exactly 28 colleges in the "very small" stratum. If we want to estimate the total number of mathematics faculty members in very small colleges, then that stratum also plays the role of a domain, because it is a subset of the population for which we want a separate estimate.

If we want to estimate the total number of mathematics faculty members in colleges located in Ohio, the set of colleges from Ohio forms a domain (since we want to calculate estimates for that subset of the population) but is not a stratum (since it was not used as a stratum in the sampling design). The number of colleges from Ohio will vary from sample to sample, and it is even possible that a particular sample will contain no Ohio colleges.

4.4 Poststratification

Suppose a sampling frame lists all households in an area, and you would like to estimate the average amount spent on food in a month. One desirable stratification variable might be household size, because large households might be expected to have higher food bills than smaller households. The distribution of household size in the region is known from census data. It is known that 28.37% of the 175,000 households in the region have one person, 34.51% have two persons, 15.07% have three persons, and the remaining 22.05% have four or more persons.

The sampling frame, however, contains no information on household size—it only lists the households. Thus, although you know the population number of households in each subgroup, you cannot take a stratified sample because you do not know the stratum membership of the units in your sampling frame. You can, however, take an SRS and record the amount spent on food as well as the household size for each household in your sample. If n, the size of the SRS, is large enough, then the sample is likely to resemble a stratified sample with proportional allocation: We would expect about 28% of the sample to be one-person households, about 34% to be two-person households, and so on.

But, because of sampling variability, we do not expect the sample percentages in each household-size category to exactly accord with the census percentages. Poststratification adjusts the sample weights so that the estimates of the poststratum counts from the sample equal the population poststratum counts, which are known for this example from the census data.

Considering the different household-size groups to be different domains, we can use the methods from Section 4.3 to estimate the average amount spent on groceries for each domain. Take an SRS of size n. Let $n_1, n_2, \ldots, n_H$ be the numbers of units in the H household-size groups (poststrata) and let $\bar{y}_1, \ldots, \bar{y}_H$ be the sample means for the groups. Since the poststrata are formed *after* the sample is taken, the sample domain sizes $n_1, n_2, \ldots, n_H$ are random quantities. If we selected another SRS from the population, the poststratum sizes in the sample would change. Since the poststratum sizes in the population, N_h, are known, however, we can use the values of N_h in the estimation.

To see how poststratification fits in the framework of ratio estimation, define $x_{ih} = 1$ if observation i is in poststratum h and 0 otherwise. Then

$$\hat{t}_{xh} = \hat{N}_h = \sum_{i \in \mathcal{S}} w_i x_{ih}$$

and

$$\hat{t}_{yh} = \sum_{i \in \mathcal{S}} w_i x_{ih} y_i.$$

Because $t_{xh} = \sum_{i=1}^{N} x_{ih} = N_h$ is known, (4.3) gives the ratio estimator of the population total of y in poststratum h as

$$\hat{t}_{yhr} = \frac{\hat{t}_{yh}}{\hat{t}_{xh}} t_{xh} = \frac{N_h}{\hat{N}_h} \hat{t}_{yh}.$$

The poststratified estimator of the population total is

$$\hat{t}_{y\text{post}} = \sum_{h=1}^{H} \hat{t}_{yhr} = \sum_{h=1}^{H} \frac{N_h}{\hat{N}_h} \hat{t}_{yh} = \sum_{h=1}^{H} N_h \bar{y}_h;$$

ratio estimation is used within each poststratum to estimate the population total in that poststratum.

The poststratified estimator of the population mean $\bar{y}_{\mathcal{U}}$ is

$$\bar{y}_{\text{post}} = \sum_{h=1}^{H} \frac{N_h}{N} \bar{y}_h. \tag{4.26}$$

If N_h/N is known, n_h is reasonably large (≥ 30 or so), and n is large, then we can use the variance for proportional allocation as an approximation to the poststratified variance:

$$\hat{V}(\bar{y}_{\text{post}}) = \left(1 - \frac{n}{N}\right) \sum_{h=1}^{H} \frac{N_h}{N} \frac{s_h^2}{n}. \tag{4.27}$$

The approximation in (4.27) is valid only for an SRS and when the expected sample sizes in each poststratum are large (see Exercise 40). Section 11.6 will look at poststratification for general survey designs.

Poststratification using weights. Poststratification modifies the weights so that they sum to N_h in poststratum h. Let w_i represent the sampling weight for unit i. For an SRS, for example, $w_i = N/n$ for each unit in the sample. Define

$$w_i^* = \sum_{h=1}^{H} \frac{N_h}{\hat{N}_h} w_i x_{ih}$$

so that $w_i^* = (N_h/\hat{N}_h) w_i$ when unit i is in poststratum h. Then estimate the population total t_y by

$$\hat{t}_{y\text{post}} = \sum_{i \in \mathcal{S}} w_i^* y_i \tag{4.28}$$

and the population mean $\bar{y}_{\mathcal{U}}$ by

$$\bar{y}_{\text{post}} = \frac{\sum_{i \in \mathcal{S}} w_i^* y_i}{\sum_{i \in \mathcal{S}} w_i^*}. \tag{4.29}$$

For an SRS, $\sum_{i \in \mathcal{S}} w_i^* = N$ and (4.29) reduces to (4.26).

Example 4.9. Example 3.2 displayed estimates for a stratified random sample from the Census of Agriculture population. The stratified sample, taken with proportional allocation, produced estimates with smaller variances than the SRS in Example 2.6.

But what if you took an SRS and only later realized that you should have taken a stratified sample? Or if you did not have region membership available for the counties in the sampling frame? Let's poststratify the SRS from Example 2.6 and find out. The quantities needed for the calculation are given in Table 4.4.

TABLE 4.4
Weight adjustments for poststratification in Example 4.9. The last two columns, $\bar{y}_h$ and s_h, give the poststratum mean and standard deviation, respectively.

Region	N_h	n_h	$\hat{N}_h$	w_i	w_i^*	$\bar{y}_h$	s_h
Northeast	220	24	246.24	10.26	9.1667	71970.83	65000.06
North Central	1054	107	1097.82	10.26	9.8505	350292.01	294715.13
South	1382	130	1333.80	10.26	10.6308	206246.35	277433.61
West	422	39	400.14	10.26	10.8205	598680.59	516157.67
Total	3078	300	3078.00				

The poststratification-adjusted weights w_i^* differ from the original sampling weights $w_i = 3078/300 = 10.26$. The poststratified weight for every county in the Northeast, where the SRS contained more units than would have been drawn in a stratified sample with proportional allocation, is

$$w_i^* = \frac{N_{\text{Northeast}}}{\hat{N}_{\text{Northeast}}} w_i = \frac{220}{246.24}(10.26) = 9.1667.$$

The poststratified weight for Northeast counties is smaller than 10.26 to account for the fact that the randomly selected sample contained, by chance, more counties in the poststratum than its share of the population. The poststratified weights in the West poststratum are larger than 10.26 to correct for the sample having fewer Western counties than it would under proportional allocation.

The weight adjustments force the estimated counts from each poststratum to equal the true poststratum count, N_h. Thus, $\sum_{i \in \mathcal{S}} w_i^* x_{i1} = (24)(9.1667) = 220$, $\sum_{i \in \mathcal{S}} w_i^* x_{i2} = 1054$, $\sum_{i \in \mathcal{S}} w_i^* x_{i3} = 1382$, and $\sum_{i \in \mathcal{S}} w_i^* x_{i4} = 422$.

The poststratified estimate of the population mean is

$$\bar{y}_{\text{post}} = \frac{\sum_{i \in \mathcal{S}} w_i^* y_i}{\sum_{i \in \mathcal{S}} w_i^*} = \frac{15833583 + 369207778.5 + 285032455.7 + 252643209}{3078} = 299{,}778.$$

From (4.27), the standard error of $\bar{y}_{\text{post}}$ is

$$\text{SE}\,(\bar{y}_{\text{post}}) = \sqrt{\left(1 - \frac{300}{3078}\right) \sum_{h=1}^{H} \frac{N_h}{3078} \frac{s_h^2}{300}} = 17{,}443.$$

By contrast, the standard error of the sample mean, $\bar{y}$, from Example 2.7, is 18,898. The poststratification reduces the standard error because the weighted average of the within-poststratum variances, $\sum_{h=1}^{H}(N_h/N)s_h^2$, is smaller than s^2. ■

Difference between stratification and poststratification. In both stratification and poststratification, each population member belongs to exactly one of H possible groups. In stratified random sampling, independent SRSs are selected from each of the H groups (strata) with fixed sample sizes n_h. Every possible stratified sample has n_h observations from stratum h. Stratification is part of the survey design.

Poststratification is used in the estimation. The sample is selected without stratifying on the H poststrata. The sample size in poststratum h, n_h, will vary from sample to sample. The poststratified weights are calculated so that the sum of the weights in poststratum h equals the known population size N_h. It is possible that some poststrata in the population will not appear in a particular SRS; in that case, one may want to combine poststrata before adjusting the weights.

Many large surveys use poststratification to improve efficiency of the estimators or to correct for the effects of nonresponse or undercoverage (see Chapter 8). We discuss post-stratification for general survey designs in Section 11.6.

4.5 Ratio Estimation with Stratified Sampling

The previous sections proposed ratio and regression estimators for use with SRSs. The concept of ratio estimation, however, is completely general and is easily extended to other sampling designs. In stratified sampling, for example, we can use estimators of the population totals for x and y from a stratified sample to give the **combined ratio estimator**

$$\hat{t}_{yrc} = \hat{B} t_x,$$

where

$$\hat{B} = \frac{\hat{t}_{y,\text{str}}}{\hat{t}_{x,\text{str}}}.$$

As in Sections 3.2 and 3.3,

$$\hat{t}_{y,\text{str}} = \sum_{h=1}^{H} N_h \bar{y}_h = \sum_{h=1}^{H} \sum_{j \in \mathcal{S}_h} w_{hj} y_{hj},$$

where the sampling weight is $w_{hj} = (N_h/n_h)$, and

$$\hat{t}_{x,\text{str}} = \sum_{h=1}^{H} N_h \bar{x}_h = \sum_{h=1}^{H} \sum_{j \in \mathcal{S}_h} w_{hj} x_{hj}.$$

Then, using the arguments in Section 4.1.2,

$$\text{MSE}\left(\hat{t}_{yrc}\right) \approx V\left(\hat{t}_{y,\text{str}} - B\hat{t}_{x,\text{str}}\right) = V\left[\sum_{h=1}^{H} \sum_{j \in \mathcal{S}_h} w_{hj}(y_{hj} - Bx_{hj})\right];$$

we estimate the MSE by

$$\begin{aligned}
\hat{V}(\hat{t}_{yrc}) &= \left(\frac{t_{x,\text{str}}}{\hat{t}_{x,\text{str}}}\right)^2 \hat{V}\left(\sum_{h=1}^{H} \sum_{j \in \mathcal{S}_h} w_{hj} e_{hj}\right) \\
&= \left(\frac{t_{x,\text{str}}}{\hat{t}_{x,\text{str}}}\right)^2 \hat{V}(\hat{t}_{e,\text{str}}) \\
&= \left(\frac{t_{x,\text{str}}}{\hat{t}_{x,\text{str}}}\right)^2 \left[\hat{V}(\hat{t}_{y,\text{str}}) + \hat{B}^2 \hat{V}(\hat{t}_{x,\text{str}}) - 2\hat{B}\widehat{\text{Cov}}\left(\hat{t}_{y,\text{str}}, \hat{t}_{x,\text{str}}\right)\right],
\end{aligned}$$

where $e_{hj} = y_{hj} - \hat{B}x_{hj}$. In the combined ratio estimator, first the strata are combined to estimate t_x and t_y, then ratio estimation is applied.

For the **separate ratio estimator**, ratio estimation is applied first, then the strata are combined. The estimator

$$\hat{t}_{yrs} = \sum_{h=1}^{H} \hat{t}_{yhr} = \sum_{h=1}^{H} t_{xh} \frac{\hat{t}_{yh}}{\hat{t}_{xh}},$$

uses ratio estimation separately in each stratum, with

$$\hat{V}(\hat{t}_{yrs}) = \sum_{h=1}^{H} \hat{V}(\hat{t}_{yhr}).$$

The separate ratio estimator can improve efficiency if the $\hat{t}_{yh}/\hat{t}_{xh}$ vary from stratum to stratum, but should not be used when strata sample sizes are small because each ratio is biased, and the bias can propagate through the strata. Note that poststratification (Section 4.4) is a special case of the separate ratio estimator.

The combined estimator has less bias when the sample sizes in some of the strata are small. When the ratios vary greatly from stratum to stratum, however, the combined estimator does not take as much advantage of the extra efficiency afforded by stratification as does the separate ratio estimator. Many survey software packages, such as SAS software, calculate the combined ratio estimator by default.

Example 4.10. Steffey et al. (2006) described the use of combined ratio estimation in the legal case *Labor Ready v. Gates McDonald.* The plaintiff alleged that the defendant had not thoroughly investigated claims for worker's compensation, resulting in overpayments for these claims by the plaintiff. A total of $N = 940$ claims were considered. For each of these, the incurred cost of the claim (x_i) was known, and consequently, the total amount of incurred costs was known to be $t_x = \$9.407$ million. But the plaintiff contended that the incurred value amounts were unjustified, and that the assessed value (y_i) of some claims after a thorough review would differ from the incurred value.

A sampling plan was devised for estimating the total assessed value of all 940 claims. Since it was expected that the assessed value would be highly correlated with the incurred costs, ratio estimation is desirable here. Two strata were sampled: Stratum 1 consisted of the claims in which the incurred cost exceeded \$25,000, and stratum 2 consisted of the smaller claims (incurred cost less than \$25,000). Summary statistics for the strata are given in the following table, with r_h the sample correlation in stratum h:

Stratum	N_h	n_h	$\bar{x}_h$	s_{xh}	$\bar{y}_h$	s_{yh}	r_h
1	102	70	\$59,549.55	\$64,047.95	\$38,247.80	\$32,470.78	0.62
2	838	101	\$5,718.84	\$5,982.34	\$3,833.16	\$5,169.72	0.77

The sampling fraction was set much higher in stratum 1 than in stratum 2 because the variability is much higher in stratum 1 (the investigators used a modified form of the optimal allocation described in Section 3.4.2). We estimate

$$\hat{t}_{x,\text{str}} = \sum_{h=1}^{2} \hat{t}_{xh} = (102)(59{,}549.55) + (838)(5{,}718.84) = 10{,}866{,}442.02,$$

$$\hat{t}_{y,\text{str}} = \sum_{h=1}^{2} \hat{t}_{yh} = (102)(38{,}247.80) + (838)(3{,}833.16) = 7{,}113{,}463.68,$$

and

$$\hat{B} = \frac{\hat{t}_{y,\text{str}}}{\hat{t}_{x,\text{str}}} = \frac{7{,}113{,}463.68}{10{,}866{,}442.02} = 0.654626755.$$

Using formulas for variances of stratified samples,

$$\hat{V}(\hat{t}_{x,\text{str}}) = \left(1 - \frac{70}{102}\right)(102)^2 \frac{(64{,}047.95)^2}{70} + \left(1 - \frac{101}{838}\right)(838)^2 \frac{(5982.34)^2}{101} = 410{,}119{,}750{,}555,$$

$$\hat{V}(\hat{t}_{y,\text{str}}) = \left(1 - \frac{70}{102}\right)(102)^2 \frac{(32{,}470.78)^2}{70} + \left(1 - \frac{101}{838}\right)(838)^2 \frac{(5169.72)^2}{101} = 212{,}590{,}045{,}044,$$

and

$$\begin{aligned}\widehat{\text{Cov}}(\hat{t}_{x,\text{str}}, \hat{t}_{y,\text{str}}) &= \left(1 - \frac{70}{102}\right)(102)^2 \frac{(32{,}470.78)(64{,}047.95)(0.62)}{70} \\ &\quad + \left(1 - \frac{101}{838}\right)(838)^2 \frac{(5169.72)(5982.34)(0.77)}{101} \\ &= 205{,}742{,}464{,}829.\end{aligned}$$

Using the combined ratio estimator, the total assessed value of the claims is estimated by

$$\hat{t}_{yrc} = (9.407 \times 10^6)(0.654626755) = \$6.158 \text{ million}$$

with standard error

$$\text{SE}(\hat{t}_{yrc}) = \frac{10.866}{9.407}\sqrt{[2.126 + (0.6546)^2(4.101) - 2(0.6546)(2.057)] \times 10^{11}} = \$0.371 \text{ million}.$$

We use 169 = (number of observations) − (number of strata) degrees of freedom for the CI. An approximate 95% CI for the total assessed value of the claims is $6.158 \pm 1.97(0.371)$, or between \$5.43 and \$6.89 million. Note that the CI for t_y does not contain the total incurred value (t_x) of \$9.407 million. This supported the plaintiff's case that the total incurred value was too high. ■

4.6 Model-Based Theory for Ratio and Regression Estimation*

In the design-based theory presented in Sections 4.1 and 4.2, the forms of the estimators $\hat{\bar{y}}_r$ and $\hat{\bar{y}}_{\text{reg}}$ are motivated by regression models. Properties of the estimators, however, depend only on the sampling design. Thus, we found in (4.10) that $V(\hat{\bar{y}}_r) \approx \frac{1}{n}\left(1 - \frac{n}{N}\right)\sum_{i=1}^{N} \frac{(y_i - Bx_i)^2}{N-1}$ for an SRS.

This variance approximation is derived from the simple random sampling formulas in Chapter 2 and does not rely on any assumptions about the model. If the model does not fit the data well, ratio or regression estimation might not increase precision for estimated means and totals, but in large samples CIs for the means or totals will be correct in the sense that a 95% CI will have coverage probability close to 0.95. Inferences about finite population quantities using ratio or regression estimation are correct even if the model does not fit the data well.

The ratio and regression estimators presented in Sections 4.1 and 4.2 are examples of **model-assisted estimators**—a model motivates the form of the estimator, but inference depends on the sampling design. Särndal et al. (1992) presented the theory of model-assisted estimation, in which inference is based on randomization theory.

If you have studied regression analysis, you likely learned a different approach to model-fitting in which you make assumptions about the regression model, find the least squares estimators of the regression parameters under the model, and plot residuals and explore regression diagnostics to check how well the model fits the data. A model-based approach for survey data was pioneered by Brewer (1963) and Royall (1970). As in the model-based approach outlined in Section 2.10, the model is used to predict population values for units not in the sample.

This section presents models that give the point estimators in (4.2) and (4.16) for ratio and regression estimation. The variances under a model-based approach, however, differ from variances obtained under a design-based approach, as we shall see.

4.6.1 A Model for Ratio Estimation

Section 4.1.4 mentioned that ratio estimation works well for estimating population totals in an SRS when a straight line through the origin fits well and when the variance of the observations about the line is proportional to x. We can state these conditions as a linear regression model: Assume that $x_1, x_2, \ldots, x_N$ are known (and all are greater than zero) and that $Y_1, Y_2, \ldots, Y_N$ are independent and follow the model

$$Y_i = \beta x_i + \varepsilon_i, \tag{4.30}$$

where $E_M(\varepsilon_i) = 0$ and $V_M(\varepsilon_i) = \sigma^2 x_i$ (the subscript M indicates that the expectation is computed using the probability distribution associated with the model). This model has the assumptions that (1) a straight line through the origin fits the data, since the model has no intercept term, (2) the variance of the observations about the regression line is proportional to x, and (3) all observations are independent.

Under the model in (4.30), $T_y = \sum_{i=1}^{N} Y_i$ is a random variable and the population total of interest, t_y, is one realization of the random variable T_y (this is in contrast to the randomization approach, in which t_y is considered to be a fixed but unknown quantity and the only random variables are the sample indicators Z_i).

The population total t_y for the particular sample $\mathcal{S}$ that was drawn is:

$$t_y = \sum_{i \in \mathcal{S}} y_i + \sum_{i \notin \mathcal{S}} y_i.$$

We observe the values of y_i for units in the sample, and predict those for units not in the sample as $\hat{y}_i = \hat{\beta} x_i$, where $\hat{\beta} = \bar{y}/\bar{x}$ is the weighted least squares estimate of β under the model in (4.30) (see Exercise 35). Then t_y is estimated by

$$\hat{t}_{yM} = \sum_{i \in \mathcal{S}} y_i + \hat{\beta} \sum_{i \notin \mathcal{S}} x_i = n\bar{y} + \frac{\bar{y}}{\bar{x}} \sum_{i \notin \mathcal{S}} x_i = \frac{\bar{y}}{\bar{x}} \sum_{i=1}^{N} x_i = \frac{\bar{y}}{\bar{x}} t_x.$$

This model results in the same ratio estimator of t_y as in Section 4.1. Indeed, the finite population ratio $B = t_y/t_x$ is the weighted least squares estimate of β, applied to the entire population.

In many common sampling designs, we find that if we adopt a model consistent with the reasons we would adopt a certain sampling design or method of estimation, the point estimators obtained using the model are very close to the design-based estimators. The model-based variance, though, usually differs from the variance from the randomization theory. In design-based sampling, the *sampling design* determines how sampling variability is estimated. In model-based sampling, the *model* determines how variability is estimated, and the sampling design is irrelevant—as long as the model holds for all population units, you could choose any n units you want as the sample from the population.

The model-based estimator

$$\hat{T}_y = \sum_{i \in \mathcal{S}} Y_i + \hat{\beta} \sum_{i \notin \mathcal{S}} x_i$$

is unbiased under the assumed model in (4.30) since

$$E_M\left[\hat{T}_y - T\right] = E_M\left[\hat{\beta}\sum_{i\notin\mathcal{S}} x_i - \sum_{i\notin\mathcal{S}} Y_i\right] = \beta \sum_{i\notin\mathcal{S}} x_i - E_M\left[\sum_{i\notin\mathcal{S}}(\beta x_i + \varepsilon_i)\right] = 0.$$

The model-based variance is

$$V_M[\hat{T}_y - T] = V_M\left[\hat{\beta}\sum_{i\notin\mathcal{S}} x_i - \sum_{i\notin\mathcal{S}} Y_i\right] = V_M\left[\hat{\beta}\sum_{i\notin\mathcal{S}} x_i\right] + V_M\left[\sum_{i\notin\mathcal{S}} Y_i\right];$$

the last equality follows because $\hat{\beta}$ and $\sum_{i\notin\mathcal{S}} Y_i$ are independent under the model assumptions. The model in (4.30) is assumed to apply to all population units and does not depend on which population units are selected to be the sample $\mathcal{S}$. Consequently, under the model, the set of units in $\mathcal{S}$ is a fixed quantity and

$$V_M\left[\sum_{i\notin\mathcal{S}} Y_i\right] = V_M\left[\sum_{i\notin\mathcal{S}} (\beta x_i + \varepsilon_i)\right] = V_M\left[\sum_{i\notin\mathcal{S}} \varepsilon_i\right] = \sigma^2 \sum_{i\notin\mathcal{S}} x_i,$$

and, similarly,

$$V_M\left[\hat{\beta}\sum_{i\notin\mathcal{S}} x_i\right] = \left(\sum_{i\notin\mathcal{S}} x_i\right)^2 V_M\left[\frac{\sum_{i\in\mathcal{S}} Y_i}{\sum_{i\in\mathcal{S}} x_i}\right] = \left(\frac{\sum_{i\notin\mathcal{S}} x_i}{\sum_{i\in\mathcal{S}} x_i}\right)^2 \sum_{i\in\mathcal{S}} V_M[Y_i] = \frac{\sigma^2\left(\sum_{i\notin\mathcal{S}} x_i\right)^2}{\sum_{i\in\mathcal{S}} x_i}.$$

Combining the two terms,

$$V_M[\hat{T}_y - T] = \frac{\sigma^2 \sum_{i\notin\mathcal{S}} x_i}{\sum_{i\in\mathcal{S}} x_i}\left(\sum_{i\in\mathcal{S}} x_i + \sum_{i\notin\mathcal{S}} x_i\right) = \frac{\sigma^2 \sum_{i\notin\mathcal{S}} x_i}{\sum_{i\in\mathcal{S}} x_i} t_x = \left(1 - \frac{\sum_{i\in\mathcal{S}} x_i}{t_x}\right)\frac{\sigma^2 t_x^2}{\sum_{i\in\mathcal{S}} x_i}. \quad (4.31)$$

Note that if the sample size is small relative to the population size, then

$$V_M[\hat{T}_y - T] \approx \frac{\sigma^2 t_x^2}{\sum_{i\in\mathcal{S}} x_i};$$

the quantity $(1 - \sum_{i\in\mathcal{S}} x_i / t_x)$ in (4.31) serves as an fpc in the model-based approach to ratio estimation.

The model-based standard error of the estimated total, using (4.31), is

$$\mathrm{SE}\left(\hat{t}_{yM}\right) = \sqrt{\left(1 - \frac{\sum_{i \in \mathcal{S}} x_i}{t_x}\right) \frac{\hat{\sigma}^2 t_x^2}{\sum_{i \in \mathcal{S}} x_i}}, \tag{4.32}$$

where $\hat{\sigma}^2 = \sum e_{i,wt}^2/(n-2)$ is calculated using the sum of squares of the weighted residuals

$$e_{i,wt} = \frac{y_i - \hat{\beta} x_i}{\sqrt{x_i}}. \tag{4.33}$$

In software packages that compute weighted least squares estimates, $\hat{\sigma}^2$ is the MSE from the ANOVA table.

Example 4.11. Let's perform a model-based analysis of the SRS from the Census of Agriculture, used in Examples 4.2 and 4.3. We already plotted the data in Figure 4.1, and it looked as though a straight line through the origin would fit well and that the variability about the line was greater for observations with larger values of x. For the data points with x positive, we can fit a regression model with no intercept and with weight variable $1/x$. Only 299 observations are used in this analysis since observation 179, Hudson County, New Jersey, has $x_{179} = 0$. The regression coefficient $\hat{\beta}$, which can be computed from any statistical software package that performs weighted least squares analyses, is shown in the following output:

Variable	DF	Parameter Estimate	Standard Error	t Value	Pr > |t|
acres87	1	0.986565	0.00484	203.68	<.0001

The slope, 0.986565, and the model-based estimate of the total, (0.986565)(964,470,625) = 951,513,191, are the same as the design-based estimates obtained in Example 4.2. Using (4.32), and calculating $\hat{\sigma}^2 = 2125.19126$ from the weighted residuals,

$$\mathrm{SE}_M[\hat{T}_y] = \sqrt{\left(1 - \frac{90{,}586{,}117}{964{,}470{,}625}\right) \frac{(2125.19126)(964{,}470{,}625)^2}{90{,}586{,}117}} = 4{,}446{,}719.$$

Note that for this example, the model-based standard error is smaller than the standard error we calculated using randomization inference in Example 4.3, which was 5,344,568. The model-based analysis assumes that the model, a straight line through the origin with $V_M(\varepsilon_i) = \sigma^2 x_i$, describes the data; the design-based analysis does not require such an assumption. ■

Evaluating model assumptions. The model for estimating a population total in (4.30) has the following assumptions:

1. The model is correct, that is, $E_M(Y_i) = x_i \beta$.
2. The variance structure is correct, that is, $V_M(Y_i) = \sigma^2 x_i$.
3. The observations are independent.
4. The regression model applies to population units that are not in the sample.

The assumptions say nothing about how the data were collected, which is the foundation of inference from a randomization-based perspective. Model-based inference can be applied to convenience samples and other types of nonprobability samples—indeed, as we shall see in Chapter 15, model-based inference *must* be used when analyzing nonprobability samples. But the model has strong assumptions, and anyone performing a model-based analysis should attempt to verify that the assumptions hold for the population of interest. If the assumptions from the model are not met, the inference from the analysis is flawed.

In a typical regression analysis, assumptions 1 and 2 are evaluated by plotting the data and examining and plotting residuals from the model. Regression books such as Chatterjee and Hadi (2012) and Montgomery et al. (2012) tell you how to plot residuals and assess the fit of the model. Since the ratio model in (4.30) assumes that $V(\varepsilon_i)$ is proportional to x_i, the regression diagnostics should be performed with the weighted residuals in (4.33), $e_{i,wt}$. If the variance of y_i about the line is proportional to x_i, then a plot of the weighted residuals $e_{i,wt}$ against x_i or the predicted values $\hat{y}_i$ should not exhibit any patterns.

Assumption 3 is more difficult to evaluate, and requires knowledge of how the data were collected. Generally, if you take an SRS, then you may act as though the observations are independent. For other types of data collection—in particular, for convenience samples—you may be able to check for some types of dependence such as autocorrelation, but other types of dependence (for example, clustering) may be undetectable.

Assumption 4 is the most important in the list. When estimating the population mean or total with a model-based approach, you use the model to predict the value of y_i for the population units that are not in the sample. If the model does not apply to observations outside of the sample, then those predictions will be wrong and your estimates of population characteristics may be badly biased. Residual analyses and other model diagnostics can help you evaluate how well the model fits the data in the sample, but you do not know y_i for the observations not in the sample and thus cannot evaluate how well the model predicts them unless you have additional information.

An SRS is representative of the population from which it is drawn, so if your model fits the data in an SRS it likely is also appropriate for the full population. You usually cannot evaluate Assumption 4 for a convenience sample.

4.6.2 A Model for Regression Estimation

For regression estimation, a straight-line regression model is adopted:

$$Y_i = \beta_0 + \beta_1 x_i + \varepsilon_i,$$

where the ε_is are independent and identically distributed with mean 0 and constant variance σ^2. The least squares estimators of β_0 and β_1 in this model are

$$\hat{\beta}_1 = \frac{\sum_{i \in \mathcal{S}} (x_i - \bar{x}_{\mathcal{S}})(Y_i - \bar{Y}_{\mathcal{S}})}{\sum_{i \in \mathcal{S}} (x_i - \bar{x}_{\mathcal{S}})^2}$$

and

$$\hat{\beta}_0 = \bar{Y}_{\mathcal{S}} - \hat{\beta}_1 \bar{x}_{\mathcal{S}}.$$

Then, substituting the predicted values for the units not sampled,

$$\begin{aligned}
\hat{T}_y &= \sum_{i\in\mathcal{S}} Y_i + \sum_{i\notin\mathcal{S}} \left(\hat{\beta}_0 + \hat{\beta}_1 x_i\right) \\
&= n\bar{Y}_{\mathcal{S}} + \sum_{i\notin\mathcal{S}} \left(\hat{\beta}_0 + \hat{\beta}_1 x_i\right) \\
&= n\left(\hat{\beta}_0 + \hat{\beta}_1 \bar{x}_{\mathcal{S}}\right) + \sum_{i\notin\mathcal{S}} \left(\hat{\beta}_0 + \hat{\beta}_1 x_i\right) \\
&= \sum_{i=1}^{N} \left(\hat{\beta}_0 + \hat{\beta}_1 x_i\right) \\
&= N\left(\hat{\beta}_0 + \hat{\beta}_1 \bar{x}_{\mathcal{U}}\right).
\end{aligned}$$

The regression estimator of T_y is thus $N\times$ (predicted value under the model at $\bar{x}_{\mathcal{U}}$).

In practice, if the sample size is small relative to the population size and we have an SRS, we can simply ignore the fpc and use the standard error for estimating the mean value of a response. From regression theory (see one of the regression books listed in the For Further Reading section of Chapter 11), the variance of $(\hat{\beta}_0 + \hat{\beta}_1 \bar{x}_{\mathcal{U}})$ is

$$\sigma^2\left[\frac{1}{n} + \frac{(\bar{x}_{\mathcal{U}} - \bar{x})^2}{\sum_{i\in\mathcal{S}}(x_i - \bar{x})^2}\right].$$

Thus if n/N is small,

$$V_M[\hat{T}_y - T] \approx N^2\sigma^2\left[\frac{1}{n} + \frac{(\bar{x}_{\mathcal{U}} - \bar{x}_{\mathcal{S}})^2}{\sum_{i\in\mathcal{S}}(x_i - \bar{x}_{\mathcal{S}})^2}\right]. \tag{4.34}$$

Example 4.12. In Example 4.7, the predicted value from the model when $x = 11.3$ is the regression estimator for $\bar{y}_{\mathcal{U}}$. The predicted value is $(5.05929 + 0.61327 \times 11.3) = 11.9893$. The model-based standard error is obtained from (4.34):

$$\text{SE}_M(\hat{\bar{Y}}_{\text{reg}}) = \sqrt{\hat{\sigma}^2\left[\frac{1}{n} + \frac{(\bar{x}_{\mathcal{U}} - \bar{x}_{\mathcal{S}})^2}{\sum_{i\in\mathcal{S}}(x_i - \bar{x}_{\mathcal{S}})^2}\right]} = \sqrt{5.79\left[\frac{1}{25} + \frac{(11.3 - 10.6)^2}{226.006}\right]} = 0.494.$$

These values can be calculated directly using any statistical software package. The standard error in (4.34) does not incorporate an fpc, and Exercise 37 examines an fpc for model-based regression. ■

4.6.3 Differences between Model-Based and Design-Based Estimators

Under the ratio and regression models presented in this section, the point estimator for the population total is the same as in the design-based approach, but the variance differs from that for the design-based estimator. Why aren't the standard errors the same as in randomization theory? That is, how can we have two different variances for the same estimator?

The discrepancy is due to the different definitions of *variance*: In design-based sampling, the variance is the average squared deviation of the estimate from its expected value, averaged over all samples that could be obtained using a given design. If we are using a model, the variance is again the average squared deviation of the estimate from its expected value, but here the average is over all possible samples that could be generated from the population model.

The model-based estimator uses a prediction approach, in which the values of y_i not in the sample are predicted using the model. In the model-based approach,

$$\hat{T}_y = \sum_{i \in \mathcal{S}} Y_i + \sum_{i \notin \mathcal{S}} \hat{Y}_i = \sum_{i=1}^{N} \hat{Y}_i + \sum_{i \in \mathcal{S}} (Y_i - \hat{Y}_i),$$

where $\hat{Y}_i$ is a predictor of the random variable Y_i under the model adopted.

If you were absolutely certain that your model was correct, you could minimize the model-based variance of the regression estimator in Section 4.6.2 by including only the members of the population with the largest and smallest values of x to be in the sample, and excluding units with values of x between those extremes. No one would recommend such a design in practice, of course, because one never has that much assurance in a model. However, nothing in the model says that you should take an SRS or any other type of probability sample, or that the sample needs to be representative of the population—*as long as the model is correct.*

What if the model is wrong? The model-based estimates are model-unbiased—that is, they are unbiased only within the structure of that particular model. If the model is wrong, the model-based estimators will be biased, but, from within the model, we will not necessarily be able to tell how big the bias is. Thus, if the model is wrong, the model-based estimator of the variance generally underestimates the MSE. When using model-based inference in sampling, then, you need to be very careful to check the assumptions of the model by examining residuals and using other diagnostic tools. The assumption of independence is typically the most difficult to check. You can (and should!) perform diagnostics to check some of the assumptions of the model for the sampled data, but need to realize that you are making a strong, untestable assumption that the model applies to population units you did not observe.

In design-based inference, the standard errors and confidence intervals are valid even when the model is inappropriate for the data. The inference depends only on how the sample was selected. If a regression model fits the data well, then the design-based variance for the regression estimator will usually be smaller than the design-based variance for $\bar{y}$. If the regression model is a poor choice for describing the relationship between x and y, for example, if x and y are uncorrelated, then using the regression estimator will not reduce the SRS variance [see Equation (4.19)], but the design-based estimator of the variance is an approximately unbiased estimator of the true variance. In that sense, design-based estimates are robust to the particular model adopted—the inference is valid whether the model describes the data or not.

Hansen et al. (1983) pointed out that generally, randomization-theory adherents have a model in mind when designing the survey and take that model into account to improve efficiency—through stratification in the design and through ratio or regression estimation in the analysis. And every survey data analyst, regardless of philosophy, relies on models for nonresponse and for nonprobability samples, as we shall see in Chapters 8 and 15.

We shall return to the issue of inference for regression models in Chapter 11.

4.7 Chapter Summary

Ratio and regression estimation use an auxiliary variable that is highly correlated with the variable of interest to reduce the MSE of estimated population means or totals. We "know" that y is correlated with x, and we know how far $\bar{x}$ is from $\bar{x}_{\mathcal{U}}$, so we use this information to adjust $\bar{y}$ and (we hope) increase the precision of our estimate. The estimators in ratio and regression estimation come from models that we hope describe the data, but the randomization-theory properties of the estimators do not depend on these models.

As will be seen in Chapter 11, the ratio and regression estimators discussed in this chapter are special cases of a generalized regression estimator. All three estimators of the population total discussed so far—$\hat{t}_y$, $\hat{t}_{yr}$, and $\hat{t}_{y\text{reg}}$—can be expressed in terms of regression coefficients. For an SRS of size n, the estimators are given in Table 4.5. For each, the estimated variance depends on s_e^2, the sample variance of the residuals e_i.

TABLE 4.5
Estimators and residuals for ratio and regression estimators of t.

Estimator	**Formula**	**Residual,** e_i
SRS	$\hat{t}_y$	$y_i - \bar{y}$
Ratio	$\hat{t}_{yr} = \hat{t}_y \left(\dfrac{t_x}{\hat{t}_x}\right)$	$y_i - \hat{B}x_i$
Regression	$\hat{t}_{y\text{reg}} = N\left[\bar{y} + \hat{B}_1(\bar{x}_{\mathcal{U}} - \bar{x})\right]$	$y_i - \hat{B}_0 - \hat{B}_1 x_i$

In an SRS, ratio or regression estimators give greater precision than $\hat{t}_y$ when $\sum_{i\in\mathcal{S}} e_i^2$ for the method is smaller than $\sum_{i\in\mathcal{S}}(y_i - \bar{y})^2$. Ratio estimation is especially useful in cluster sampling, as we shall see in chapters 5 and 6.

We often want to find estimates for subpopulations of interest, for example, different age groups. If the sampling frame contains information on the age group for units in the population, a stratified sample can be designed as in Chapter 3. If the sampling frame does not contain this information but we know the population sizes of the subpopulations, then the subpopulations are poststrata and we can estimate the population total in group h by $N_h\bar{y}_h$, where $\bar{y}_h$ is the mean of the sampled observations in the subpopulation. If we do not know the population sizes of the subpopulations, then they are domains and we estimate the population total in group h by $N(n_h/n)\bar{y}_h$.

In this chapter, we discussed ratio and regression estimation using just one auxiliary variable x. In practice, you may want to use several auxiliary variables. The principles for using multiple regression models will be the same; we shall present the theory for general surveys in Section 11.6.

Key Terms

Auxiliary variable: A variable that is used to improve the sampling design or to help with the estimation of variables of interest. The auxiliary variable is often known for every unit in the population.

Calibration: A procedure in which weights are adjusted so that estimated population totals of auxiliary variables agree with the known population totals of those variables.

Domain: A subpopulation for which estimates are desired. The domain sample sizes are generally random variables.

Poststratification: A form of ratio estimation in which sampled units are divided into subgroups based on characteristics measured in the sample. The weights in each subgroup are adjusted so that they sum to the population size of that subgroup (assumed to be known).

Ratio estimator: An estimator of the population mean or total based on a ratio with an auxiliary quantity for which the population mean or total is known.

Regression estimator: An estimator of the population mean or total based on a regression model using an auxiliary quantity for which the population mean or total is known.

For Further Reading

Raj (1968) and Cochran (1977) have good treatments of ratio and regression estimation in SRSs. For regression models in a general framework, discussed in Chapter 11 of this book, see Särndal et al. (1992) and Levy and Lemeshow (2008). The overview paper by Särndal (2007) summarizes the use of ratio and regression estimation for calibrating survey estimates to known population totals. Chapter 2 of Fuller (2009) provides a rigorous treatment of the theory underlying ratio and regression estimation. Valliant (2002) compares variance estimators for ratio and regression estimation, and Henry and Valliant (2009) explore how auxiliary information can be used with establishment surveys. Deville and Särndal (1992), Särndal (2007), Kott (2016) Devaud and Tillé (2019) discuss general methods used for calibration.

The books by Thompson (1997), Valliant et al. (2000), Brewer (2002), and Chambers and Clark (2012) describe differences between design-based and model-based approaches to survey inference, as do the articles by Hansen et al. (1983), Rao (1997), Lohr (2001), and Little (2004). Welsh and Wiens (2013) explore designs that are robust to choice of model.

4.8 Exercises

A. Introductory Exercises

1. For each of the following situations, indicate how you might use ratio or regression estimation.

   (a) Estimate the proportion of time in television news broadcasts in your city that is devoted to sports.

   (b) Estimate the average number of fish caught per hour for anglers visiting a lake in August.

   (c) Estimate the average amount that undergraduate students at your university spent on textbooks in the fall semester.

(d) Estimate the total weight of usable meat (discarding bones, fat, and skin) in a shipment of chickens.

2. Consider the hypothetical population below, with population values:

Unit number	x	y
1	13	10
2	7	7
3	11	13
4	12	17
5	1	8
6	3	1
7	11	15
8	3	7
9	5	4

(a) Find the values of the population quantities t_x, t_y, S_x, S_y, R, and B.

(b) Construct a table like that in Table 4.2, giving the sampling distribution of $N\bar{y}$ and of $\hat{t}_{yr}$, for a sample of size $n = 3$.

(c) Draw a histogram of the sampling distribution of $\hat{t}_{yr}$. Compare this histogram to a histogram of the sampling distribution of $N\bar{y}$.

(d) Find the mean and variance of the sampling distribution of $\hat{t}_{yr}$. How do these compare to the mean and variance of $N\bar{y}$? What is the bias of $\hat{t}_{yr}$?

(e) Use Equation (4.8), together with the population quantities you calculated in (a) to find an approximation to Bias $(\hat{t}_{yr}) = N\text{Bias}\,(\hat{\bar{y}}_r)$. How close is the approximation to the true bias in (c)?

3. Foresters want to estimate the average age of trees in a stand. Determining age is cumbersome, because one needs to count the tree rings on a core taken from the tree. In general, though, the older the tree, the larger the diameter, and diameter is easy to measure. The foresters measure the diameter of all 1132 trees and find that the population mean equals 10.3. They then randomly select 20 trees for age measurement.

Tree Number	Diameter, x	Age, y	Tree Number	Diameter, x	Age, y
1	12.0	125	11	5.7	61
2	11.4	119	12	8.0	80
3	7.9	83	13	10.3	114
4	9.0	85	14	12.0	147
5	10.5	99	15	9.2	122
6	7.9	117	16	8.5	106
7	7.3	69	17	7.0	82
8	10.2	133	18	10.7	88
9	11.7	154	19	9.3	97
10	11.3	168	20	8.2	99

(a) Draw a scatterplot of y vs. x.

(b) Estimate the population mean age of trees in the stand using ratio estimation and give an approximate standard error for your estimate.

(c) Repeat (b) using regression estimation.

(d) Label your estimates on your graph. How do they compare?

4. An SRS of 1,500 licensed boat owners in a state was sampled from a list of 400,000 names with currently licensed boats. Define two domains: domain 1 consists of the 472 licensed boat owners who own large motorboats (defined as open motorboats longer than 16 feet) and domain 2 consists of the 1,028 licensed boat owners who do not own large motorboats. The following frequency table gives the number of respondents in each domain who reported having that number of children.

Number of Children	**Domain 1**	**Domain 2**
0	76	528
1	139	189
2	166	225
3	63	76
4	19	8
5	5	2
6	3	0
8	1	0
Total	472	1028

Estimate, along with a 95% CI,

(a) The percentage of boat owners who have children in each domain.

(b) The average number of children per household among boat owners in domain 1.

(c) The total number of children who live in households where someone owns a large motorboat.

B. Working with Survey Data

5. Use the data in `ssc.csv`, described in Exercise 18 of Chapter 2, for this problem.

(a) Estimate the proportion of female members who are in academia. Note that this is a domain mean, with $x_i = 1$ if person i is female and 0 otherwise, and $y_i = 1$ if person i is female and in academia and 0 otherwise. Give a 95% CI.

(b) Estimate the total number of female members in academia, along with a 95% CI.

6. Use the data in file `golfsrs.csv` for this problem. Estimate the average greens fee on a weekend to play 18 holes (variable *wkend18*) for the domain of 18-hole courses. Give a 95% CI.

7. Using the data in file `golfsrs.csv`, plot the weekend 9-hole greens fee versus the backtee yardage. Estimate the regression parameters for predicting weekend 9-hole greens fees from backtee yardage. Is there a strong relationship between the two variables? Use regression estimation to estimate the mean weekend 9-hole greens fee with its standard error, assuming that the population mean backtee yardage is 5,415.

8. Use the data in file `golfsrs.csv` for this problem.

(a) Estimate the mean weekday greens fee to play 9 holes, for courses with a golf professional available.

(b) Now estimate the mean weekday greens fee to play 9 holes, for courses without a golf professional.

9. Use the data set in `crimes.csv` (see Exercise 23 of Chapter 2) to estimate, along with a 95% CI:

   (a) The percentage of crimes of each type (variable *crimetype*) for which an arrest was made.

   (b) The total number of aggravated assaults that are domestic-related.

10. The data set `agsrs.csv` also contains information on the number of farms in 1987 for the SRS of $n = 300$ counties from the population of the $N = 3078$ counties in the United States (see Example 2.6). In 1987, the United States had a total of 2,087,759 farms.

    (a) Plot the data.

    (b) Use ratio estimation to estimate the total number of acres devoted to farming in 1992 (variable *acres92*), using the number of farms in 1987 (variable *farms87*) as the auxiliary variable.

    (c) Repeat (b), using regression estimation.

    (d) Which method gives the most precision: ratio estimation with auxiliary variable *acres87*, ratio estimation with auxiliary variable *farms87*, or regression estimation with auxiliary variable *farms87*? Why?

11. The data set `cherry.csv`, from Hand et al. (1994), contains measurements of diameter (inches), height (feet), and timber volume (cubic feet) for a sample of 31 black cherry trees. Diameter and height of trees are easily measured, but volume is more difficult to measure.

    (a) Plot volume vs. diameter for the 31 trees.

    (b) Suppose that these trees are an SRS from a forest of $N = 2967$ trees and that the sum of the diameters for all trees in the forest is $t_x = 41{,}835$ inches. Use ratio estimation to estimate the total volume for all trees in the forest. Give a 95% CI.

    (c) Use regression estimation to estimate the total volume for all trees in the forest. Give a 95% CI.

12. *Public health articles* (continuation of Exercise 12 in Chapter 3).

    (a) Some researchers who carry out hypothesis tests report the exact p-value; others report categories using asterisks, where *, **, and *** denote p-value < 0.05, p-value < 0.01, and p-value < 0.001, respectively. Of the articles in the stratified random sample in `healthjournals.csv` that use hypothesis tests, what percentage report p-value categories using asterisks (variable *asterisk*)? Give a 95% CI.

    (b) Hayat and Knapp (2017) argued that the legitimacy of inferences can be questioned when randomness is not used. Of the articles that *do not* use random selection or assignment (see Exercise 12) in Chapter 3), what percentage include statistical inferences (using a confidence interval, hypothesis test, or both)? Give a 95% CI.

13. The data file `counties.csv` contains information on land area, population, number of physicians, unemployment, and other characteristics in 1993 for an SRS of 100 of the 3141 counties in the United States (U.S. Census Bureau, 1994). The total land area for the United States in 1993 was 3,536,278 square miles and the total population was 255,077,536.

    (a) Draw a histogram of the number of physicians for the 100 counties.

(b) Estimate the total number of physicians in the United States, along with its standard error, using $N\bar{y}$.

(c) Plot the number of physicians vs. population for each county. Which method do you think is more appropriate for these data: ratio estimation or regression estimation?

(d) Using the method you chose in (c), use the auxiliary variable population to estimate the total number of physicians in the United States, along with the standard error.

(e) The "true" value for the total number of physicians in the population is 532,638. Which method of estimation came closer?

14. Repeat parts (a)–(d) of Exercise 13 with y = farm population and x = land area.

15. Repeat parts (a)–(d) of Exercise 13 with y = number of veterans and x = population.

16. (Model-based analysis; requires material in Section 4.6.) Refer to the situation in Exercise 13. Use a model-based analysis to estimate the total number of physicians in the United States. Which model did you choose, and why? What are the assumptions for the model? Do you think they are met? Be sure to examine the residual plots for evidence of the inadequacy of the model. How do your results differ from those you obtained in Exercise 13?

17. Jackson et al. (1987) compared the precision of systematic and stratified sampling for estimating the average concentration of lead and copper in the soil, in milligrams per kilogram. The one-square-kilometer area was divided into 100-meter squares, and a soil sample was collected at each of the resulting 121 grid intersections. Summary statistics from this systematic sample are given below.

Element	n	Average	Range	Standard Deviation
Lead	121	127	22-942	146
Copper	121	35	15-90	16

The investigators also poststratified the same region. Stratum A consisted of farmland away from roads, villages, and woodlands. Stratum B contained areas within 50 meters of roads, and was expected to have larger concentrations of lead. Stratum C contained the woodlands, which were also expected to have larger concentrations of lead because the foliage would capture airborne particles. The data on concentration of lead and copper were not used in determining the strata. The data from the grid points falling in each stratum are in the following table:

Element	Stratum	n_h	Average	Range	Standard Deviation
Lead	A	82	71	22–201	28
Lead	B	31	259	36–942	232
Lead	C	8	189	88–308	79
Copper	A	82	28	15–68	9
Copper	B	31	50	22–90	18
Copper	C	8	45	31–69	15

(a) Calculate a 95% CI for the average concentration of lead in the area, using the systematic sample. (You may assume that this sample behaves like an SRS.) Repeat for the average concentration of copper.

(b) Now use the poststratified sample, and find 95% CIs for the average concentration of lead and copper. How do these compare with the CIs in (a)? Do you think that using stratification in future surveys would increase precision?

18. For the data in `mysteries.csv` (see Exercise 13 of Chapter 3), estimate (a) the ratio of total number of male authors to total number of female authors and (b) the ratio of total number of male detectives to total number of female detectives, along with 95% CIs. Use the combined ratio estimator with weight variable *p1weight*.

19. For the data in `mysteries.csv` (see Exercise 13 of Chapter 3), consider the two domains defined by variable *authorgender*: books with female authors and books with male authors. For each domain, estimate (with 95% CI) the proportion of books that have at least one female detective. Use weight variable *p1weight*.

C. Working with Theory

20. (Requires probability.) Use covariances derived in Appendix A to show the result in (4.10).

21. (Requires computing.) In Equation (4.11), we used

$$\hat{V}_1[\hat{\bar{y}}_r] = \left(1 - \frac{n}{N}\right)\left(\frac{\bar{x}_{\mathcal{U}}}{\bar{x}}\right)^2 \frac{s_e^2}{n}$$

to estimate $\left(1 - \frac{n}{N}\right)\frac{S_d^2}{n}$. An alternative estimator that has been proposed is

$$\hat{V}_2[\hat{\bar{y}}_r] = \left(1 - \frac{n}{N}\right)\frac{s_e^2}{n}.$$

Generate a population of size 1,000 from the model $y_i = \beta x_i + \varepsilon_i$ where $\varepsilon_i \sim N(0, \sigma^2 x_i)$. Now take $R = 2000$ different samples, each with $n = 50$, from the population. Calculate $a_j = \hat{\bar{y}}_r$, $b_j = \hat{V}_1[\hat{\bar{y}}_r]$, and $c_j = \hat{V}_2[\hat{\bar{y}}_r]$ from sample j, for $j = 1, \ldots, 2{,}000$. Compare $\sum_{j=1}^R b_j/R$ and $\sum_{j=1}^R c_j/R$ with the empirical estimate of the variance $\hat{V}_3 = \sum_{j=1}^R (a_j - \bar{a})^2/(R-1)$. Which variance estimator, $\hat{V}_1$ or $\hat{V}_2$, comes closer to $\hat{V}_3$?

If the variability about the line $y = \beta x$ increases as x increases, as is the case for the data generated above, then, if $\bar{x} < \bar{x}_{\mathcal{U}}$ we would expect $\frac{1}{n-1}\sum_{i\in\mathcal{S}}(y_i - \hat{B}x_i)^2$ to be smaller than $\frac{1}{N-1}\sum_{i=1}^N (y_i - Bx_i)^2 = S_d^2$. Using $\hat{V}_1$ instead of $\hat{V}_2$ partially compensates for this. See Valliant (2002) for a discussion of why $\hat{V}_1$ is preferred to $\hat{V}_2$ from a conditional inference perspective.

22. Some books use the formula

$$\hat{V}[\hat{B}] = \left(1 - \frac{n}{N}\right)\frac{1}{n\bar{x}_{\mathcal{U}}^2}(s_y^2 - 2\hat{B}rs_xs_y + \hat{B}^2s_x^2),$$

where r is the sample correlation coefficient of x and y for the values in the sample, to estimate the variance of a ratio.

(a) Show that this formula is algebraically equivalent to (4.12).

(b) It often does not work as well as (4.12) in practice, however: If s_x and s_y are large, many computer packages will truncate some of the significant digits so that the subtraction will be inaccurate. For the data in Example 4.2, calculate the values of s_y^2, s_x^2, r, and $\hat{B}$. Use the formula above to calculate the estimated variance of $\hat{B}$. Is it exactly the same as the value from (4.12)?

23. (Requires probability.) Recall from Section 2.2 that MSE = variance + (Bias)^2. Using (4.8) and other approximations in that section, show that $[E(\hat{\bar{y}}_r - \bar{y}_{\mathcal{U}})]^2$ is small compared to $V(\hat{\bar{y}}_r)$, when n is large.

24. (Requires probability.) Prove (4.8). HINT: Use (4.7) and the derivation of the covariance of $\bar{x}$ and $\bar{y}$ in (A.11) of Appendix A.

25. Use Equation (4.8) to find the approximate bias of $\hat{t}_{yr}$ and of $\hat{B}$.

26. *Comparing two domain means in an SRS.* Suppose there are two domains, defined by indicator variable

$$x_i = \begin{cases} 1 \text{ if unit } i \text{ is in domain 1} \\ 0 \text{ if unit } i \text{ is in domain 2} \end{cases}.$$

Then, letting $u_i = x_i y_i$, the population values of the two domain means are

$$\bar{y}_{\mathcal{U}1} = \frac{\sum_{i=1}^N x_i y_i}{\sum_{i=1}^N x_i} = \frac{t_u}{t_x} = \frac{\bar{u}_{\mathcal{U}}}{\bar{x}_{\mathcal{U}}}$$

and

$$\bar{y}_{\mathcal{U}2} = \frac{\sum_{i=1}^N (1-x_i) y_i}{\sum_{i=1}^N (1-x_i)} = \frac{t_y - t_u}{N - t_x} = \frac{\bar{y}_{\mathcal{U}} - \bar{u}_{\mathcal{U}}}{1 - \bar{x}_{\mathcal{U}}}.$$

If an SRS of size n is taken from a population of size N, the population domain means may be estimated by

$$\bar{y}_1 = \frac{\hat{t}_u}{\hat{t}_x} = \frac{\bar{u}}{\bar{x}}, \qquad \bar{y}_2 = \frac{\hat{t}_y - \hat{t}_u}{N - \hat{t}_x} = \frac{\bar{y} - \bar{u}}{1 - \bar{x}}.$$

(a) Use an argument similar to that in the discussion following (4.7) to show that

$$\text{Cov}(\bar{y}_1, \bar{y}_2) \approx \frac{1}{\bar{x}_{\mathcal{U}}(1 - \bar{x}_{\mathcal{U}})} \text{Cov}\left[\left(\bar{u} - \frac{t_u}{t_x}\bar{x}\right), \left\{\bar{y} - \bar{u} - \frac{t_y - t_u}{N - t_x}(1 - \bar{x})\right\}\right]. \tag{4.35}$$

(b) For an SRS, show using (A.11) that

$$\text{Cov}\left[\left(\bar{u} - \frac{t_u}{t_x}\bar{x}\right), \left\{\bar{y} - \bar{u} - \frac{t_y - t_u}{N - t_x}(1 - \bar{x})\right\}\right] = 0.$$

[Consequently, since Property 7 of Expected Value in Section A.2 implies that $V(\bar{y}_1 - \bar{y}_2) = V(\bar{y}_1) + V(\bar{y}_2) - 2\,\text{Cov}(\bar{y}_1, \bar{y}_2)$, in an SRS $V(\bar{y}_1 - \bar{y}_2) \approx V(\bar{y}_1) + V(\bar{y}_2)$ and an approximate 95% CI for $\bar{y}_{\mathcal{U}1} - \bar{y}_{\mathcal{U}2}$ is given by

$$\bar{y}_1 - \bar{y}_2 \pm 1.96\sqrt{\hat{V}(\bar{y}_1) + \hat{V}(\bar{y}_2)}.$$

Thus, for an SRS, the large-sample CI for the difference of two domain means is the same (if we ignore the fpc) as you learned in your introductory statistics class. Note, though, that this result holds only for an SRS. For more complex sampling designs, the covariance of the estimated domain means may be nonzero (see Exercise 25 of Chapter 6), so more general methods discussed in Section 11.3 must be used.]

27. *Showing (4.10)* (requires mathematical statistics). Suppose that $n/N \to 0$ as $n \to \infty$, so that the fpc can be ignored. The central limit theorem tells us that under regularity conditions,
$$\sqrt{n}\begin{bmatrix} \bar{x} - \bar{x}_{\mathcal{U}} \\ \bar{y} - \bar{y}_{\mathcal{U}} \end{bmatrix} \xrightarrow{\mathcal{L}} N\left(\mathbf{0}, \begin{bmatrix} S_x^2 & RS_xS_y \\ RS_xS_y & S_y^2 \end{bmatrix}\right),$$
where $\mathcal{L}$ denotes convergence in distribution. Show that the limiting distribution of $\sqrt{n}(\hat{\bar{y}}_r - \bar{y}_{\mathcal{U}})$ has mean 0 and variance $S_y^2 - 2BRS_xS_y + B^2S_x^2$.

28. Show that if we consider approximations to the MSE in (4.10) and (4.18) to be accurate, then the variance of $\hat{\bar{y}}_r$ from ratio estimation is at least as large as the variance of $\hat{\bar{y}}_{\text{reg}}$ from regression estimation. HINT: Look at $V(\hat{\bar{y}}_r) - V(\hat{\bar{y}}_{\text{reg}})$ using the formulas in (4.10) and (4.18), and show that the difference is non-negative.

29. (Requires probability.) Prove (4.17).

30. (Requires probability.) Prove (4.19).

31. (Requires probability.) Let $d_i = y_i - [\bar{y}_{\mathcal{U}} + B_1(x_i - \bar{x}_{\mathcal{U}})]$. Show that for regression estimation,
$$E[\hat{\bar{y}}_{\text{reg}} - \bar{y}_{\mathcal{U}}] \approx -\frac{1 - n/N}{n\, S_x^2} \sum_{i=1}^{N} \frac{d_i(x_i - \bar{x}_{\mathcal{U}})^2}{N - 1}.$$
As in Exercise 23, show that $(E[\hat{\bar{y}}_{\text{reg}} - \bar{y}_{\mathcal{U}}])^2$ is small compared to $\text{MSE}[\hat{\bar{y}}_{\text{reg}}]$, when n is large.

32. *Difference estimation* is a special case of regression estimation where the slope is fixed at $B_1 = 1$ rather than estimated from the data. In accounting, for example, an auditor knows the book value x_i for each of the N accounts in a list of accounts receivable, then determines the audited value y_i (the actual amount owed) for each account in an SRS of size n drawn from the population. The difference estimator of the population mean $\bar{y}_{\mathcal{U}}$ is
$$\hat{\bar{y}}_{\text{diff}} = \bar{y} + (\bar{x}_{\mathcal{U}} - \bar{x}).$$
This is similar to (4.16), but with $\hat{B}_1$ replaced by $B_1 = 1$. The sample mean is adjusted by the difference between the population and sample mean of x, $(\bar{x}_{\mathcal{U}} - \bar{x})$.

   (a) Show that $E\left[\hat{\bar{y}}_{\text{diff}}\right] = \bar{y}_{\mathcal{U}}$.

   (b) What is $V(\hat{\bar{y}}_{\text{diff}})$?

   (c) Compare $V(\hat{\bar{y}}_{\text{diff}})$ with the variance of the ratio estimator in (4.10). When will the difference estimator have smaller variance than the ratio estimator?

33. (Requires probability.) Consider the combined ratio estimator of the population total, $\hat{t}_{yrc}$, from Section 4.5.

   (a) Show that
$$\frac{\text{Bias}[\hat{t}_{yrc}]}{\sqrt{V(\hat{t}_{yrc})}} \leq \text{CV}\ (\hat{t}_x).$$
   HINT: See (4.6).

   (b) In a stratified random sample, find the approximate bias and MSE of $\hat{t}_{yrc}$.

34. (Requires probability.) Consider the separate ratio estimator of the population total, $\hat{t}_{yrs}$, from Section 4.5. Find the bias and an approximation to the MSE of $\hat{t}_{yrs}$ in a stratified random sample. Allow different ratios, B_h, in each stratum. When will the bias be small?

35. (Requires linear model theory.) Suppose we have a stochastic model

$$Y_i = \beta x_i + \varepsilon_i$$

where the ε_is are independent with mean 0 and variance $\sigma^2 x_i$. Show that the weighted least squares estimator of β is $\hat{\beta} = \bar{Y}/\bar{x}$. Is the standard error for $\hat{\beta}$ that comes from weighted least squares the same as that for $\hat{B}$ in (4.12)?

36. (Requires linear model theory.) Suppose that the model in (4.30) misspecifies the variance structure and that a better model has $V_M[\varepsilon_i] = \sigma^2$.

    (a) What is the weighted least squares estimator of β if $V_M[\varepsilon_i] = \sigma^2$? What is the corresponding estimator of the population total for y?

    (b) Derive $V_M[\hat{T}_y - T_y]$.

    (c) Apply your estimators to the data in `agsrs.csv`. How do these estimates compare with those in Examples 4.2 and 4.11?

37. Equation (4.34) gave the model-based variance for a population total when it is assumed that the sample size is small relative to the population size. Derive the variance incorporating the finite population correction.

38. The quantity $\hat{B}$ used in ratio estimation is sometimes called the *ratio-of-means estimator*. An alternative that has been proposed is the *mean-of-ratios estimator*: Let $b_i = y_i/x_i$ for unit i; then the mean-of-ratios estimator is

$$\bar{b} = \frac{1}{n}\sum_{i \in \mathcal{S}} b_i.$$

    (a) Do you think the mean-of-ratios estimator is appropriate for the data in Example 4.5? Why, or why not?

    (b) Show that, for the ratio-of-means estimator $\hat{B}$, $t_x\hat{B} = t_y$ when the entire population is sampled (that is, $\mathcal{S} = \mathcal{U}$).

    (c) Give an example to show that it is possible to have $t_x\bar{b} \neq t_y$ when the entire population is sampled.

    (d) Define

$$S_{bx} = \frac{1}{N-1}\sum_{i=1}^{N}(b_i - \bar{b}_{\mathcal{U}})(x_i - \bar{x}_{\mathcal{U}}).$$

    Show that for an SRS of size n, the bias of $\bar{b}$ under the sampling design is

$$E[\bar{b} - B] = -\frac{(N-1)S_{bx}}{t_x}.$$

    As a consequence, if $S_{bx} \neq 0$ the bias does not decrease as n increases.

    (e) (Requires linear model theory.) Show that $\bar{b}$ is the weighted least squares estimator of β under the model

$$Y_i = \beta x_i + \varepsilon_i$$

    when ε_i has mean 0 and variance $\sigma^2 x_i^2$.

39. (Requires computing.)

    (a) Generate 500 data sets, each with 30 pairs of observations (x_i, y_i). Use a bivariate normal distribution with means 0, standard deviations 1 and correlation 0.5 to generate each pair (x_i, y_i). For each data set, calculate $\bar{y}$ and $\hat{\bar{y}}_{\text{reg}}$, using $\bar{x}_{\mathcal{U}} = 0$. Graph a histogram of the 500 values of $\bar{y}$ and another histogram of the 500 values of $\hat{\bar{y}}_{\text{reg}}$. What do you see?

    (b) Repeat part (a) for 500 data sets, each with 60 pairs of observations.

40. (Requires computing.) Use the population in `agpop.csv` for this exercise.

    (a) Take 500 independent SRSs, each of size $n = 40$, from the data set. For each SRS, calculate the sample mean $\bar{y}$, the poststratified mean $\bar{y}_{\text{post}}$ from (4.26), the estimated variance of $\bar{y}$, and the estimated variance of $\bar{y}_{\text{post}}$ using (4.27).

    (b) Calculate the sample variance of the 500 values of $\bar{y}$. This gives an estimate of the true value of $V(\bar{y})$. Compare your value with the average of the 500 values of $\hat{V}(\bar{y})$. Since $\hat{V}(\bar{y})$ is an unbiased estimator of $V(\bar{y})$ for any sample size, these values should be close.

    (c) Calculate the sample variance of the 500 values of $\bar{y}_{\text{post}}$. This gives an estimate of the true value of $V(\bar{y}_{\text{post}})$. Compare your value with the average of the 500 values of $\hat{V}(\bar{y}_{\text{post}})$.

D. Projects and Activities

41. Find a dictionary of a language you have studied. Choose 30 pages at random from the dictionary. For each, record

$$x = \text{number of words on the page}$$
$$y = \text{number of words that you know on the page (be honest!)}$$

    How many words do you estimate are in the dictionary? How many do you estimate that you know? What percentage of the words do you know? Give standard errors for all your estimates.

42. Epidemiologists often want to estimate fatality rates for a disease. The population mortality rate for a disease is defined as (number of persons who have died from the disease) divided by (number of persons in the population). The case fatality rate for a disease is defined as (number of persons who have died from the disease) divided by (number of confirmed cases of the disease).

    Investigate how the case fatality rate and population mortality rate are calculated in your country for a disease such as influenza or COVID-19. How is ratio estimation used? What are sources of sampling error, measurement error, and selection bias for each rate?

43. *Online bookstore.* Use your sample from Exercise 38 in Chapter 2 for this exercise.

    (a) Estimate the ratio (average price/average number of pages) and give the standard error.

    (b) Consider two domains: hardcover books and paperback books. Estimate the mean price (with standard error) for books in each domain.

44. *Trucks.* Use the data described in Exercise 49 of Chapter 3 for this exercise.

(a) The variable *business* describes the primary business in which the vehicle was used in 2002. Estimate the total miles driven for each type of business in 2002, along with a 95% CI. How is this a special case of estimating a domain total?

(b) Estimate the average miles per gallon (MPG) for each of the transmission types (*transmssn*), along with a 95% CI.

(c) Estimate the ratio of miles driven in 2002 (*miles_annl*) to lifetime miles driven (*miles_life*), along with a 95% CI.

45. *Baseball data.*

(a) Using your SRS from Exercise 37 of Chapter 2, estimate the mean log salary for players in each position along with the standard errors.

(b) Estimate the ratio (total number of home runs)/(number of runs scored) for the population and give a 95% CI.

46. *IPUMS exercises.*

(a) Using your SRS from Exercise 42 of Chapter 2, try estimating total income *(inctot)* using ratio estimation with *age* as an auxiliary variable. Does it decrease the standard error? Why, or why not? (Include a plot as part of your answer.)

(b) Using one of the following variables: *age*, *sex*, *race*, or *marstat*, use regression estimation to calibrate your estimate of total income to the category totals for the variable you chose.

47. *Large public database.* For the sample you drew in Exercise 43 of Chapter 2, estimate the ratio of two variables.

5

Cluster Sampling with Equal Probabilities

"But averages aren't real," objected Milo; "they're just imaginary."

"That may be so," he agreed, "but they're also very useful at times. For instance, if you didn't have any money at all, but you happened to be with four other people who had ten dollars apiece, then you'd each have an average of eight dollars. Isn't that right?"

"I guess so," said Milo weakly.

"Well, think how much better off you'd be, just because of averages," he explained convincingly. "And think of the poor farmer when it doesn't rain all year: if there wasn't an average yearly rainfall of 37 inches in this part of the country, all his crops would wither and die."

It all sounded terribly confusing to Milo, for he had always had trouble in school with just this subject.

—Norton Juster, *The Phantom Tollbooth*

In all the sampling procedures discussed so far, we have assumed that the population is given and all we must do is reach in and take a suitable sample of units. But units are not necessarily nicely defined, even when the population is. There may be several ways of listing the units, and the unit size we choose may very well contain smaller subunits.

Suppose we want to find out how many bicycles are owned by residents in a community of 10,000 households. We could take a simple random sample (SRS) of 400 households, or we could divide the community into blocks of about 20 households each and sample every household (or subsample some of the households) in each of 20 blocks selected at random from the 500 blocks in the community. The latter plan is an example of cluster sampling. The blocks are the **primary sampling units** (psus), or **clusters**. (In this chapter, we use the terms cluster and psu interchangeably.) The households are the **secondary sampling units** (ssus); often, the ssus are the elements in the population.

The cluster sample of 400 households is likely to give less precision than an SRS of 400 households; some blocks of the community are composed mainly of families (with more bicycles), while the residents of other blocks are mainly retirees (with fewer bicycles). Twenty households in the same block are not as likely to reflect the diversity of the community as well as 20 households chosen at random. Thus, cluster sampling in this situation will probably result in less information per observation than an SRS of the same size. However, if you conduct the survey in person, it is much cheaper and easier to interview all 20 households in a block than 20 households selected at random from the community, so cluster sampling may well result in more information per dollar spent.

In cluster sampling, individual elements of the population are allowed in the sample only if they belong to a cluster (psu) that is included in the sample. The sampling unit (psu) is not the same as the observation unit (ssu), and the two sizes of experimental units must be considered when calculating standard errors from cluster samples.

DOI: 10.1201/9780429298899-5

Why use cluster samples?

1. Constructing a sampling frame list of observation units may be difficult, expensive, or impossible. We cannot list all honeybees in a region or all customers of a store; we may be able to construct a list of all trees in a stand of northern hardwood forest or a list of individuals in a city for which we only have a list of housing units, but constructing the list will be time-consuming and expensive.

2. The population may be widely distributed geographically or may occur in natural clusters such as households or schools, and it is less expensive to take a sample of clusters rather than an SRS of individuals. If the target population is residents of nursing homes in the United States, it is much cheaper to sample nursing homes and interview some or all residents in the selected homes than to interview an SRS of nursing home residents: With an SRS of residents, you might have to travel to a nursing home just to interview one resident. If taking an archaeological survey, you would examine all of the artifacts found in a region—you would not just choose points at random and examine only artifacts occurring at those isolated points.

Clusters bear a superficial resemblance to strata: A cluster, like a stratum, is a grouping of the members of the population. The selection process, though, is quite different in the two methods. Similarities and differences between cluster samples and stratified samples are illustrated in Figure 5.1.

Whereas stratification generally increases precision when compared with simple random sampling, cluster sampling generally decreases it. Members of the same cluster tend to be more similar than elements selected at random from the whole population—members of the same household tend to have similar political views; fish in the same lake tend to have similar concentrations of mercury; residents of the same nursing home tend to have similar opinions of the quality of care. These similarities usually arise because of some underlying factors that may or may not be measurable—residents of the same nursing home may have similar opinions because the care is poor, or the concentration of mercury in the fish may reflect the concentration of mercury in the lake. Thus, we do not obtain as much information about all nursing home residents in the United States by sampling two residents in the same home as by sampling two residents in different homes, because the two residents in the same home are likely to have more similar opinions. By sampling everyone in the cluster, we partially repeat the same information instead of obtaining new information, and that gives us less precision for estimates of population quantities.

Most large surveys that collect data in person or through site visits employ cluster sampling. It is cheaper and more convenient to sample in clusters than randomly in the population. Usually, to increase precision, survey conductors group the population psus into strata and select a probability sample of psus within each stratum. Chapter 7 will show how to combine clustering with stratification. In Chapters 5 and 6, we consider the effects of clustering on its own so you can see how clustering affects the variance of estimates.

When clustering is ignored. One of the biggest mistakes made by researchers using survey data is to analyze a cluster sample as if it were an SRS. Such confusion usually results in the researchers reporting standard errors that are smaller than they should be, and this gives the impression that the statistics from the survey are much more precise than they really are. Exercise 40 presents an activity for exploring what happens to properties of confidence intervals (CIs) when clustered data are analyzed incorrectly.

Example 5.1. Many studies in education involve cluster samples. Instead of taking an SRS of students, investigators select a sample of schools or classrooms, then obtain data from all

Stratified Sampling	**Cluster Sampling**
Each element of the population is in exactly one stratum.	Each element of the population is in exactly one cluster.
Population of H strata 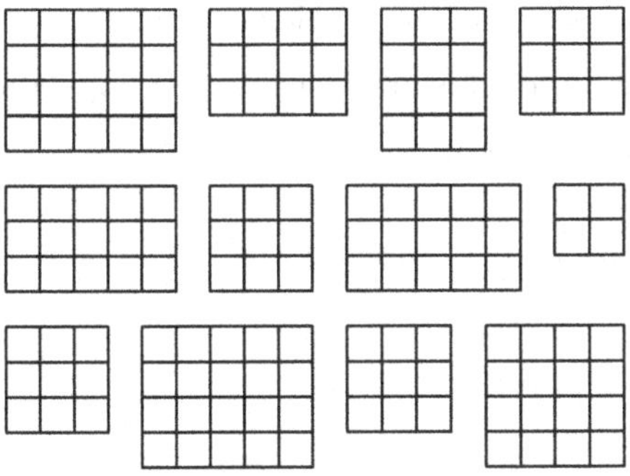	Population of N clusters: 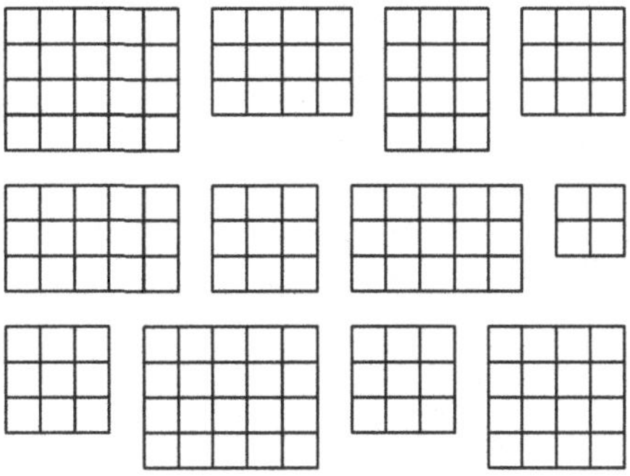
Take an SRS from *every* stratum. 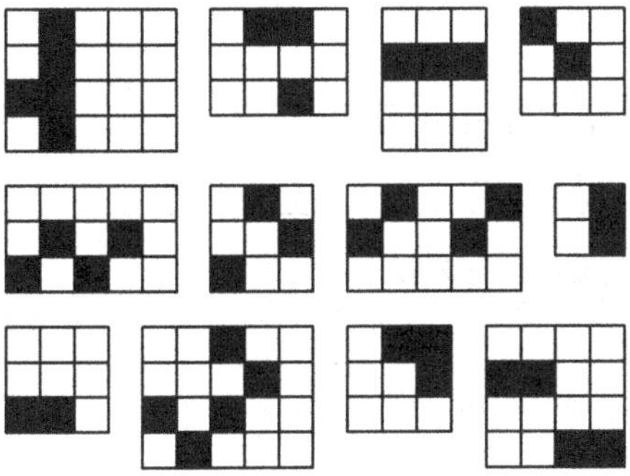	Take an SRS of clusters; observe all elements within the clusters in the sample. 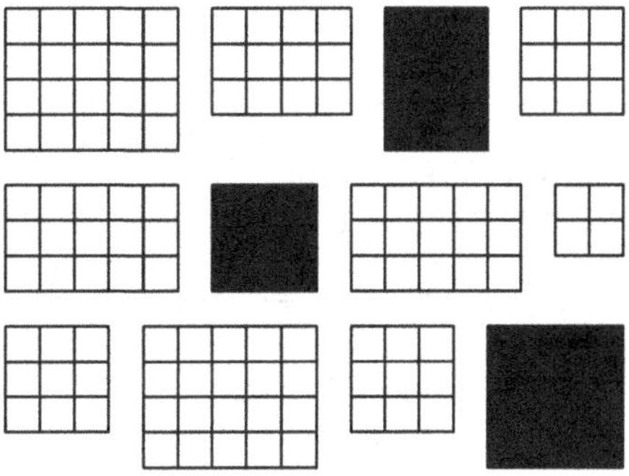
The variance of the estimator of $\bar{y}_{\mathcal{U}}$ depends on the variability of values *within* strata.	The cluster is the sampling unit; the more clusters sampled, the smaller the variance. The variance of the estimator of $\bar{y}_{\mathcal{U}}$ depends primarily on the variability *between* cluster means.
For greatest precision, individual elements within each stratum should have similar values, but stratum means should differ from each other as much as possible.	For greatest precision, individual elements within each cluster should be heterogeneous, and cluster means should be similar to one another.

FIGURE 5.1
Contrasting stratified sampling with one-stage cluster sampling.

or a subsample of the students in those schools or classrooms. Basow and Silberg (1987) and Basow and Martin (2012) presented results of research on whether students evaluate female college professors differently than they evaluate male college professors. In the first study, the authors matched 16 female professors with 16 male professors by subject taught, years of teaching experience, and tenure status, and gave evaluation questionnaires to students in those professors' classes. The sample size for analyzing this study is $n = 32$, the number of faculty studied; it is not 1029, the number of students who returned questionnaires. The 1029 students are from a cluster sample, where the classrooms form the clusters.

TABLE 5.1
Notation for cluster sampling: Population quantities.

psu Level: Population Quantities		
Symbol	**Formula**	**Description**
y_{ij}		measurement for jth element in ith psu
N		number of clusters (psus) in the population
M_i		number of ssus in psu i
M_0	$\sum_{i=1}^{N} M_i$	total number of ssus in the population
t_i	$\sum_{j=1}^{M_i} y_{ij}$	total in psu i
t	$\sum_{i=1}^{N} t_i = \sum_{i=1}^{N}\sum_{j=1}^{M_i} y_{ij}$	population total
S_t^2	$\frac{1}{N-1}\sum_{i=1}^{N}\left(t_i - \frac{t}{N}\right)^2$	population variance of the psu totals
ssu Level: Population Quantities		
Symbol	**Formula**	**Description**
$\bar{y}_{\mathcal{U}}$	$\frac{1}{M_0}\sum_{i=1}^{N}\sum_{j=1}^{M_i} y_{ij}$	population mean
$\bar{y}_{i\mathcal{U}}$	$\frac{1}{M_i}\sum_{j=1}^{M_i} y_{ij} = \frac{t_i}{M_i}$	population mean in psu i
S^2	$\frac{1}{M_0-1}\sum_{i=1}^{N}\sum_{j=1}^{M_i}(y_{ij} - \bar{y}_{\mathcal{U}})^2$	population variance (per ssu)
S_i^2	$\frac{1}{M_i-1}\sum_{j=1}^{M_i}(y_{ij} - \bar{y}_{i\mathcal{U}})^2$	population variance within psu i

Students' evaluations of faculty reflect the different styles of faculty teaching; students within the same class are likely to have some agreement in their rating of the professor and should not be treated as independent observations because their ratings will be positively correlated. If this positive correlation is ignored and the student ratings are treated as independent observations, variance estimates will be too small and differences will be declared statistically significant far more often than they should be. ■

Outline of chapter. This chapter shows you how to design a cluster sample and how to obtain correct inferences when analyzing data from a cluster sample. After a brief journey into "notation land" in Section 5.1, we discuss **one-stage cluster sampling**, in which every element within a sampled cluster is included in the sample, in Section 5.2. We then generalize the results to **two-stage cluster sampling**, in which we subsample only some of the elements of selected clusters, in Section 5.3. In Section 5.4, we discuss design issues

TABLE 5.2
Notation for cluster sampling: Sample quantities.

Symbol	Formula	Description
$\mathcal{S}$		sample of psus from the population of psus
$\mathcal{S}_i$		sample of ssus chosen from the psu i
n		number of clusters (psus) in the sample
m_i		number of ssus in the sample from psu i
$\bar{y}_i$	$\frac{1}{m_i}\sum_{j\in\mathcal{S}_i} y_{ij}$	sample mean (per ssu) in psu i
$\hat{t}_i$	$\frac{M_i}{m_i}\sum_{j\in\mathcal{S}_i} y_{ij}$	estimated total for psu i
s_t^2	$\frac{1}{n-1}\sum_{i\in\mathcal{S}}\left(\hat{t}_i - \frac{\sum_{j\in\mathcal{S}}\hat{t}_j}{n}\right)^2$	sample variance among estimated psu totals
s_i^2	$\frac{1}{m_i-1}\sum_{j\in\mathcal{S}_i}(y_{ij}-\bar{y}_i)^2$	sample variance within psu i
w_{ij}		sampling weight for ssu j in psu i

for cluster sampling, including selection of subsample and sample sizes. In Section 5.5, we return to systematic sampling, which we previously discussed in Section 2.8, and show that it is a special case of cluster sampling. The chapter concludes with theory of cluster sampling from the model-based perspective; we shall derive the design-based theory in the more general setting of Section 6.6.

5.1 Notation for Cluster Sampling

In simple random sampling, the units sampled are also the elements observed. In cluster sampling, the sampling units are the clusters (psus) and the elements observed are the ssus within the clusters. The universe $\mathcal{U}$ is the population of N psus; $\mathcal{S}$ designates the sample of psus chosen from the population of psus, and $\mathcal{S}_i$ is the sample of ssus chosen from the ith psu. The measured quantities are

$$y_{ij} = \text{measurement for } j\text{th element in } i\text{th psu},$$

but in cluster sampling, it is easiest to think at the psu level in terms of cluster totals.

No matter how you define it, the notation for cluster sampling is messy because you need notation for both the psu and the ssu levels. The notation used in this chapter and Chapter 6 for population and sample quantities is presented in Tables 5.1 and 5.2 for easy reference. Note that in Chapters 5 and 6, N is the number of psus, not the number of observation units.

5.2 One-Stage Cluster Sampling

In one-stage cluster sampling, either all or none of the elements from a cluster (= psu) are in the sample. One-stage cluster sampling is used in many surveys in which the cost of sampling ssus is negligible compared with the cost of sampling psus. For education surveys, a natural psu is the classroom; all students in a selected classroom are often included as the ssus since little extra cost is added by handing out a questionnaire to all students in the classroom rather than some.

In the population of N psus, the ith psu contains M_i ssus (elements). In the simplest design, we take an SRS of n psus from the population and measure the variables of interest on *every* element in the sampled psus. Thus, for one-stage cluster sampling, $M_i = m_i$.

5.2.1 Clusters of Equal Sizes: Estimation

Let's consider the simplest case in which each psu has the same number of elements, with $M_i = m_i = M$. Most naturally occurring clusters of people do not fit into this framework, but it can occur in agricultural and industrial sampling. Estimating population means or totals is simple: We treat the psu means or totals as the observations and simply ignore the individual elements.

Thus, we have an SRS of n data points $\{t_i, i \in \mathcal{S}\}$; t_i is the total for all the elements in psu i. Then $\bar{t}_{\mathcal{S}} = \sum_{i\in\mathcal{S}} t_i/n$ estimates the average of the cluster totals. In a household survey to estimate income in two-person households, the individual observations y_{ij} are the incomes of individual persons within the household, t_i is the total income for household i (t_i is *known* for sampled households because both persons are interviewed), $\bar{t}_{\mathcal{U}}$ is the average income per household, and $\bar{y}_{\mathcal{U}}$ is the average income per person. To estimate the total income t, we can use the estimator

$$\hat{t} = \frac{N}{n}\sum_{i\in\mathcal{S}} t_i. \tag{5.1}$$

The results in Sections 2.3 and 2.9 apply to $\hat{t}$ because we have an SRS of n units from a population of N units. As a result, $\hat{t}$ is an unbiased estimator of t, with variance given by

$$V(\hat{t}) = N^2\left(1 - \frac{n}{N}\right)\frac{S_t^2}{n} \tag{5.2}$$

and with standard error

$$\text{SE}(\hat{t}) = N\sqrt{\left(1 - \frac{n}{N}\right)\frac{s_t^2}{n}}, \tag{5.3}$$

where S_t^2 and s_t^2 are the population and sample variance, respectively, of the psu totals:

$$S_t^2 = \frac{1}{N-1}\sum_{i=1}^{N}\left(t_i - \frac{t}{N}\right)^2$$

and

$$s_t^2 = \frac{1}{n-1}\sum_{i\in\mathcal{S}}\left(t_i - \frac{\sum_{j\in\mathcal{S}} t_j}{n}\right)^2 = \frac{1}{n-1}\sum_{i\in\mathcal{S}}\left(t_i - \frac{\hat{t}}{N}\right)^2.$$

To estimate $\bar{y}_{\mathcal{U}}$, divide the estimated total by the number of persons, obtaining

$$\hat{\bar{y}} = \frac{\hat{t}}{NM}, \tag{5.4}$$

with

$$V(\hat{\bar{y}}) = \left(1 - \frac{n}{N}\right) \frac{S_t^2}{nM^2} \tag{5.5}$$

and

$$\text{SE}\,(\hat{\bar{y}}) = \frac{1}{M}\sqrt{\left(1 - \frac{n}{N}\right) \frac{s_t^2}{n}}. \tag{5.6}$$

No new ideas are introduced to carry out one-stage cluster sampling; we simply use the results for simple random sampling with the psu totals as the observations.

Confidence intervals, too, are calculated the same way as for an SRS. But this is an SRS of *n psus*, not an SRS of *nM observation units.* Therefore the degrees of freedom (df) used for the t percentile in a cluster sample is $n-1$, not $nM-1$. A 95% CI for the population mean is of the form $\hat{\bar{y}} \pm t_{\alpha/2,n-1}\text{SE}\,(\hat{\bar{y}})$, for $\text{SE}\,(\hat{\bar{y}})$ in (5.6).

Example 5.2. A student wants to estimate the average grade point average (GPA) in his dormitory. Instead of obtaining a listing of all students in the dorm and conducting an SRS, he notices that the dorm consists of 100 suites, each with four students; he chooses 5 of those suites at random, and obtains the GPA of every person in the 5 suites. The results are as follows:

Person	**Suite (psu)**				
Number	1	2	3	4	5
1	3.08	2.36	2.00	3.00	2.68
2	2.60	3.04	2.56	2.88	1.92
3	3.44	3.28	2.52	3.44	3.28
4	3.04	2.68	1.88	3.64	3.20
Total	12.16	11.36	8.96	12.96	11.08

The psus are the suites, so $N = 100$, $n = 5$, and $M = 4$. The estimate of the population total (the estimated sum of all the GPAs for everyone in the dorm—a meaningless quantity for this example but useful for demonstrating the procedure) is

$$\hat{t} = \frac{100}{5}(12.16 + 11.36 + 8.96 + 12.96 + 11.08) = 1130.4.$$

The average of the suite totals is estimated by $\bar{t} = 1130.4/100 = 11.304$, and

$$s_t^2 = \frac{1}{5-1}\left[(12.16 - 11.304)^2 + \ldots + (11.08 - 11.304)^2\right] = 2.256.$$

Note that s_t^2 is simply the usual sample variance of the 5 suite totals. Thus, using (5.4) and (5.6), $\hat{\bar{y}} = 1130.4/400 = 2.826$, and

$$\text{SE}(\hat{\bar{y}}) = \sqrt{\left(1 - \frac{5}{100}\right)\frac{2.256}{(5)(4)^2}} = 0.164.$$

A 95% CI for the mean is given by

$$2.826 \pm 2.776(0.164) = [2.37, 3.28]\,,$$

where 2.776 is the percentile from a t distribution with $(n-1) = 4$ df.

Note that in these calculations, only the "total" row of the data table is used—the individual GPAs are used only for their contribution to the suite total. ■

Sampling weights. The weight for observation unit j in psu i is

$$w_{ij} = \frac{1}{P\{\text{ssu } j \text{ of psu } i \text{ is in sample}\}} = \frac{N}{n}.$$

One-stage cluster sampling with an SRS of psus produces a self-weighting sample. The sum of the sampling weights is the number of observation units in the population:

$$\sum_{i \in \mathcal{S}} \sum_{j \in \mathcal{S}_i} w_{ij} = \sum_{i \in \mathcal{S}} \sum_{j \in \mathcal{S}_i} \frac{N}{n} = NM.$$

When analyzing data from a cluster sample, always verify your weight calculations by checking that the weights sum to the population number of ssus.

As in simple random and stratified sampling, we estimate the population total and mean using the weights, with

$$\hat{t} = \sum_{i \in \mathcal{S}} \sum_{j \in \mathcal{S}_i} w_{ij} y_{ij} \tag{5.7}$$

and

$$\hat{\bar{y}} = \frac{\displaystyle\sum_{i \in \mathcal{S}} \sum_{j \in \mathcal{S}_i} w_{ij} y_{ij}}{\displaystyle\sum_{i \in \mathcal{S}} \sum_{j \in \mathcal{S}_i} w_{ij}}. \tag{5.8}$$

For the data in Example 5.2,

$$\hat{t} = \sum_{i \in \mathcal{S}} \sum_{j \in \mathcal{S}_i} w_{ij} y_{ij} = \frac{N}{n}(3.08 + 2.60 + \ldots + 3.28 + 3.20) = \frac{100}{5}(56.52) = 1130.4$$

and

$$\hat{\bar{y}} = \frac{\displaystyle\sum_{i \in \mathcal{S}} \sum_{j \in \mathcal{S}_i} w_{ij} y_{ij}}{\displaystyle\sum_{i \in \mathcal{S}} \sum_{j \in \mathcal{S}_i} w_{ij}} = \frac{1130.4}{NM} = 2.826.$$

Survey software packages use the weights to calculate estimated population totals and means (see Lohr, 2022; Lu and Lohr, 2022, for calculations in SAS and R software).

If we had taken an SRS of nM elements, each element in the sample would have been assigned weight $(NM)/(nM) = N/n$—the same weights we obtain for cluster sampling. The precision obtained for the two types of sampling, however, can differ greatly; the difference in precision is explored in the next section.

5.2.2 Clusters of Equal Sizes: Theory

In this section, we compare cluster sampling with simple random sampling: Cluster sampling almost always provides less precision for the estimators than one would obtain by taking an SRS with the same number of elements.

TABLE 5.3
Population ANOVA table—cluster sampling.

Source	df	Sum of Squares	Mean Square
Between psus	$N-1$	$\text{SSB} = \sum_{i=1}^{N}\sum_{j=1}^{M}(\bar{y}_{i\mathcal{U}} - \bar{y}_{\mathcal{U}})^2$	MSB
Within psus	$N(M-1)$	$\text{SSW} = \sum_{i=1}^{N}\sum_{j=1}^{M}(y_{ij} - \bar{y}_{i\mathcal{U}})^2$	MSW
Total, about $\bar{y}_{\mathcal{U}}$	$NM-1$	$\text{SSTO} = \sum_{i=1}^{N}\sum_{j=1}^{M}(y_{ij} - \bar{y}_{\mathcal{U}})^2$	S^2

As in stratified sampling, let's look at the ANOVA table (Table 5.3) for the whole population. In stratified sampling, the variance of the estimator of t depends on the variability *within* the strata; Equation (3.3) and Table 3.6 imply that the variance in stratified sampling is small if SSW is small relative to SSTO, or equivalently, if the within mean square (MSW) is small relative to S^2. In stratified sampling, you have some information about *every* stratum, so you need not worry about variability due to unsampled strata. If MSB/MSW is large—that is, the variability among the strata means is large when compared with the variability within strata—then stratified sampling increases precision.

The opposite situation occurs in cluster sampling. In one-stage cluster sampling when each psu has M ssus, the variability of the unbiased estimator of t depends entirely on the *between*-psu part of the variability, because

$$S_t^2 = \sum_{i=1}^{N} \frac{(t_i - \bar{t}_{\mathcal{U}})^2}{N-1} = \sum_{i=1}^{N} \frac{M^2(\bar{y}_{i\mathcal{U}} - \bar{y}_{\mathcal{U}})^2}{N-1} = M\,(\text{MSB}).$$

Thus, for cluster sampling,

$$V(\hat{t}_{\text{cluster}}) = N^2\left(1 - \frac{n}{N}\right)\frac{M(\text{MSB})}{n}. \tag{5.9}$$

If MSB/MSW is large in cluster sampling, then cluster sampling decreases precision. In that situation, MSB is relatively large because it measures the cluster-to-cluster variability: Elements in different clusters often vary more than elements in the same cluster because different clusters have different means. If we took a cluster sample of classes and sampled all students within the selected classes, we would likely find that average reading scores varied from class to class. An excellent reading teacher might raise the reading scores for the entire class; a class of students from an area with much poverty might tend to be undernourished and not score as highly at reading. Unmeasured factors, such as teaching skill or poverty, can affect the overall mean for a cluster and thus cause the MSB to be large.

Within a class, too, students' reading scores vary. The MSW is the pooled value of the within-cluster variances: the variance from element to element, present for all elements of the population. If the clusters are relatively homogeneous—if, for example, students in the same class have similar scores—the MSW will be small.

Comparing a cluster sample with an SRS of the same size. If, instead of taking a cluster sample of M elements in each of n clusters, we had taken an SRS with nM observations, the variance of the estimated total would have been

$$V(\hat{t}_{\text{SRS}}) = (NM)^2\left(1 - \frac{nM}{NM}\right)\frac{S^2}{nM} = N^2\left(1 - \frac{n}{N}\right)\frac{M\,S^2}{n}.$$

Comparing this with (5.9), we see that if MSB $> S^2$, then cluster sampling is less efficient than simple random sampling.

The **intraclass** (sometimes called **intracluster**) **correlation coefficient** (ICC) tells us how similar elements in the same cluster are. It provides a **measure of homogeneity** within the clusters. The ICC is defined to be the Pearson correlation coefficient for the $NM(M-1)$ pairs (y_{ij}, y_{ik}) for i between 1 and N and $j \neq k$ (see Exercise 25) and can be written in terms of the population ANOVA table quantities as

$$\text{ICC} = 1 - \frac{M}{M-1}\frac{\text{SSW}}{\text{SSTO}}. \tag{5.10}$$

Because $0 \leq \text{SSW}/\text{SSTO} \leq 1$, it follows from (5.10) that

$$-\frac{1}{M-1} \leq \text{ICC} \leq 1.$$

If the clusters are perfectly homogeneous and hence SSW $= 0$, then ICC $= 1$. Equation (5.10) also implies that

$$\text{MSB} = \frac{NM-1}{M(N-1)}S^2[1 + (M-1)\text{ICC}]. \tag{5.11}$$

How much precision do we lose by taking a cluster sample? From (5.9) and (5.11),

$$\frac{V(\hat{t}_{\text{cluster}})}{V(\hat{t}_{\text{SRS}})} = \frac{\text{MSB}}{S^2} = \frac{NM-1}{M(N-1)}[1 + (M-1)\text{ICC}] \tag{5.12}$$

If N, the number of psus in the population, is large so that $NM - 1 \approx M(N-1)$, then the ratio of the variances in (5.12) is approximately $1 + (M-1)$ICC. So $1 + (M-1)$ICC ssus, taken in a one-stage cluster sample, give us approximately the same amount of information as one ssu from an SRS. If ICC $= 1/2$ and $M = 5$, then $1 + (M-1)\text{ICC} = 3$, and we would need to measure 300 elements using a cluster sample to obtain the same precision as an SRS of 100 elements. We hope, though, that because it is often much cheaper and easier to collect data in a cluster sample, that we will have more precision per dollar spent in cluster sampling.

The ICC provides a measure of homogeneity for the clusters. The ICC is positive if elements within a psu tend to be similar; then, SSW will be small relative to SSTO, and the ICC relatively large. When the ICC is positive, cluster sampling is less efficient than simple random sampling of elements.

If the clusters occur naturally in the population, the ICC is usually positive. Elements within the same cluster tend to be more similar than elements selected at random from the population. This may occur because the elements in a cluster share a similar environment—we would expect wells in the same geographic cluster to have similar levels of pesticides, or we would expect one area of a city to have a different incidence of measles than another area of a city. In human populations, personal choice, as well as interactions among household members or neighbors, may cause the ICC to be positive—wealthy households tend to live in similar neighborhoods, and persons in the same neighborhood may share similar opinions.

The ICC is negative if elements within a cluster are dispersed *more* than a randomly chosen group would be. This forces the cluster means to be very nearly equal—because SSTO $=$ SSW $+$ SSB, if SSTO is held fixed and SSW is large, then SSB must be small. If ICC < 0, cluster sampling is more efficient than simple random sampling of elements. The ICC is rarely negative in naturally occurring clusters, but negative values can occur in some systematic samples or artificial clusters, as discussed in Section 5.5.

The ICC is defined for clusters of equal sizes. An alternative measure of homogeneity in general populations is the adjusted R^2, called R_a^2 and defined as

$$R_a^2 = 1 - \frac{\text{MSW}}{S^2}. \tag{5.13}$$

If all psus are of the same size, then the increase in variance due to cluster sampling is

$$\frac{V(\hat{t}_{\text{cluster}})}{V(\hat{t}_{\text{SRS}})} = \frac{\text{MSB}}{S^2} = 1 + \frac{N(M-1)}{N-1} R_a^2;$$

by comparing with (5.12), you can see that for many populations, R_a^2 is close to the ICC. The quantity R_a^2 is a reasonable measure of homogeneity because of its interpretation in linear regression: It is the relative amount of variability in the population explained by the psu means, adjusted for the number of degrees of freedom. If the psus are homogeneous, then the psu means are highly variable relative to the variation within psus, and R_a^2 will be high.

Example 5.3. Consider two artificial populations, each having three psus with three elements per psu. Table 5.4 shows the population values, summary statistics, and population ANOVA tables for Population A and Population B.

TABLE 5.4
Population values, summary statistics and ANOVA tables for two populations.

Population A

	y_{i1}	y_{i2}	y_{i3}	$\bar{y}_{i\mathcal{U}}$	S_i^2
psu 1	10	20	30	20	100
psu 2	11	20	32	21	111
psu 3	9	17	31	19	124

ANOVA Table:

Source	df	SS	MS
Between psus	2	6	3
Within psus	6	670	111.67
Total, about $\bar{y}_{\mathcal{U}}$	8	676	84.5

$$R_a^2 = 1 - \frac{111.67}{84.5} = -0.3215$$

$$\text{ICC} = 1 - \left(\frac{3}{2}\right)\frac{670}{676} = -0.4867$$

Population B

	y_{i1}	y_{i2}	y_{i3}	$\bar{y}_{i\mathcal{U}}$	S_i^2
psu 1	9	10	11	10	1
psu 2	17	20	20	19	3
psu 3	31	32	30	31	1

ANOVA Table:

Source	df	SS	MS
Between psus	2	666	333
Within psus	6	10	1.67
Total, about $\bar{y}_{\mathcal{U}}$	8	676	84.5

$$R_a^2 = 1 - \frac{1.67}{84.5} = 0.9803$$

$$\text{ICC} = 1 - \left(\frac{3}{2}\right)\frac{10}{676} = 0.9778$$

The elements are the same in the two populations, so they share the values $\bar{y}_{\mathcal{U}} = 20$ and $S^2 = 84.5$. In population A, the psu means are similar and most of the variability occurs within psus ($S_i^2 \geq 100$ for each psu). In population B, the within-psu variances are small ($S_i^2 \leq 3$ for each psu) and most of the variability occurs between psus.

Population A has much variation among elements within the psus but little variation among the psu means. This is reflected in the large negative values of the ICC and R_a^2. Elements in the same cluster are actually less similar than randomly selected elements from the whole population. For this situation, cluster sampling is more efficient than simple random sampling.

The opposite situation occurs in population B: Most of the variability occurs between psus, and the psus themselves are relatively homogeneous. The ICC and R_a^2 are very close to 1, indicating that little new information would be gleaned by sampling more than one element per psu. Here, one-stage cluster sampling is much less efficient than simple random sampling. ■

Most real-life populations fall somewhere between these two extremes. The ICC is usually positive, but not overly close to 1. Thus, there is a penalty in efficiency for using cluster sampling, and that decreased efficiency should be offset by cost savings.

Example 5.4. When all psus are the same size, we can estimate the variance of $\hat{t}$ as well as the ICC from the sample ANOVA table. Here is the sample ANOVA table for the GPA data from Example 5.2:

Source	df	SS	MS
Between suites	4	2.2557	0.56392
Within suites	15	2.7756	0.18504
Total, about mean	19	5.0313	0.26480

In one-stage cluster sampling with equal psu sizes, the mean squares for within suites and between suites are unbiased estimators of the corresponding quantities in the population ANOVA table (see Exercise 28). Thus

$$E\left[\widehat{\text{MSB}}\right] = \text{MSB} = \frac{S_t^2}{M}$$

and, using (5.9),

$$\text{SE}\,(\hat{\bar{y}}) = \sqrt{\left(1 - \frac{n}{N}\right)\frac{\widehat{\text{MSB}}}{nM}} = \sqrt{\left(1 - \frac{5}{100}\right)\frac{0.56392}{(5)(4)}} = 0.164,$$

as calculated in Example 5.2.

The sample mean square total is biased for estimating S^2, though (see Exercise 29). Note that we can estimate the sums of squares from the population ANOVA table by $\widehat{\text{SSB}} = (N-1)\widehat{\text{MSB}}$ and $\widehat{\text{SSW}} = N(M-1)\widehat{\text{MSW}}$, so an unbiased estimator of S^2 is

$$\hat{S}^2 = \frac{(N-1)\widehat{\text{MSB}} + N(M-1)\widehat{\text{MSW}}}{NM-1}.$$

For the GPA data, $\widehat{\text{SSB}} = (99)(0.56392) = 55.828$ and $\widehat{\text{SSW}} = (300)(0.18504) = 55.512$. Consequently, $\widehat{\text{SSTO}} = 55.828 + 55.512 = 111.340$. The estimates of the population sums of squares are given in the following table:

Source	df	$\widehat{\text{SS}}$ (estimated)	$\widehat{\text{MS}}$
Between suites	99	55.828	0.56392
Within suites	300	55.512	0.18504
Total, about mean	399	111.340	0.279

Using these estimates, $\hat{S}^2 = 111.340/399 = 0.279$ (note the difference between this estimate and the one from the sample ANOVA table, 0.265). In addition,

$$\widehat{\text{ICC}} = 1 - \frac{M}{M-1}\frac{\widehat{\text{SSW}}}{\widehat{\text{SSB}} + \widehat{\text{SSW}}} = 1 - \left(\frac{4}{3}\right)\frac{55.512}{111.34} = 0.335$$

and

$$\hat{R}_a^2 = 1 - \frac{\widehat{\text{MSW}}}{\hat{S}^2} = 1 - \frac{0.18504}{0.279} = 0.337.$$

The increase in variance for using cluster sampling is estimated to be

$$\frac{\widehat{\text{MSB}}}{\hat{S}^2} = \frac{0.56392}{0.279} = 2.02.$$

This says that we need to sample about $2.02\,n$ elements in a cluster sample to get the same precision as an SRS of size n. There are 4 persons in each psu, so in terms of precision, one psu is worth about $4/2.02 = 1.98$ SRS persons. ■

Example 5.5. When is a cluster not a cluster? When it's the whole population.

Consider the situation of sampling oak trees on Santa Cruz Island, described in Example 4.5. There, the sampling unit was one tree, and an observation unit was a seedling by the tree. The population of interest was seedlings of oak trees on Santa Cruz Island. An SRS of trees was used to estimate quantities of interest about the population of oak trees on the island.

But suppose the investigator had been interested in seedling survival in all of California, had divided the regions with oak trees into equal-sized areas, and had randomly selected five of those areas to be in the study. Then the primary sampling unit is the area, and trees are subsampled in each area. If Santa Cruz Island had been selected as one of the five areas, we could no longer treat the ten trees on Santa Cruz Island as though they were part of a random sample of trees from the population; instead, those trees are part of the Santa Cruz Island cluster. We would expect all ten trees on Santa Cruz Island to experience, as a group, different environmental factors (such as weather conditions and numbers of seedling eaters) than the ten trees selected in the Santa Ynez Valley on the mainland. Thus the ICC within each cluster (area) would likely be positive.

However, suppose we were only interested in the seedlings from Tree #10 on Santa Cruz Island. Then the population is all seedlings from Tree #10, and the primary sampling unit is the seedling. In this situation, the tree is not a cluster but is the entire population. ■

5.2.3 Clusters of Unequal Sizes

Clusters are rarely of equal sizes in social surveys. In one of the early probability samples (Converse, 1987), the Enumerative Check Census of 1937, a 2% sample of postal routes was chosen, and questionnaires were distributed to all households on each chosen postal route with the goal of checking unemployment figures. Since postal routes had different numbers of households, the cluster sizes could vary greatly.

In a one-stage cluster sample of n of the N psus, we know how to estimate population totals and means in two ways: using unbiased estimation and using ratio estimation.

Unbiased estimation. An **unbiased** estimator of t is calculated exactly as in (5.1):

$$\hat{t}_{\text{unb}} = \frac{N}{n} \sum_{i \in \mathcal{S}} t_i, \tag{5.14}$$

and, by (5.3),

$$\text{SE}(\hat{t}_{\text{unb}}) = N\sqrt{\left(1 - \frac{n}{N}\right)\frac{s_t^2}{n}}. \tag{5.15}$$

The difference between unequal- and equal-sized clusters is that the variation among the individual cluster totals t_i is likely to be large when the clusters have different sizes. The investigators conducting the Enumerative Check Census of 1937 were interested in the total number of unemployed persons, and t_i would be the number of unemployed persons on postal route i. One would expect to find more persons, and hence more unemployed persons, on a postal route with a large number of households than on a postal route with a small number of households. So we would expect that t_i would be large when the psu size M_i is large, and small when M_i is small. Often, then, s_t^2 is larger in a cluster sample when the psus have unequal sizes than when the psus all have the same number of ssus.

We can use (5.14) and (5.15) to derive an unbiased estimator for $\bar{y}_{\mathcal{U}}$ and to find its standard error. Define $M_0 = \sum_{i=1}^{N} M_i$ as the total number of ssus in the population; then $\hat{\bar{y}}_{\text{unb}} = \hat{t}_{\text{unb}}/M_0$ and $\text{SE}(\hat{\bar{y}}_{\text{unb}}) = \text{SE}(\hat{t}_{\text{unb}})/M_0$. The unbiased estimator of the mean $\hat{\bar{y}}_{\text{unb}}$ can be inefficient when the values of M_i are unequal since its variance, like that of $\hat{t}_{\text{unb}}$, depends on the variability s_t^2 of the cluster totals t_i. It also requires that M_0 be known; however, we often know M_i only for the sampled clusters. In the Enumerative Check Census, for example, the number of households on a postal route would only be ascertained for the postal routes actually chosen to be in the sample. When the population psu sizes M_i are unequal, we usually use a ratio estimator of the population mean, which often has smaller variance. Let's look at that now.

Ratio estimation. We usually expect t_i to be positively correlated with M_i. If psus are counties, we would expect the total number of households living in poverty in county i (t_i) to be roughly proportional to the total number of households in county i (M_i). The population mean $\bar{y}_{\mathcal{U}}$ is a ratio

$$\bar{y}_{\mathcal{U}} = \frac{\sum_{i=1}^{N} t_i}{\sum_{i=1}^{N} M_i} = \frac{t}{M_0},$$

where t_i and M_i are usually positively correlated. Thus, $\bar{y}_{\mathcal{U}} = B$ as in Section 4.1 (substituting t_i for y_i and using M_i as the auxiliary variable x_i). Define

$$\hat{\bar{y}}_r = \frac{\hat{t}_{\text{unb}}}{\hat{M}_0} = \frac{\dfrac{N}{n}\sum_{i\in\mathcal{S}} t_i}{\dfrac{N}{n}\sum_{i\in\mathcal{S}} M_i} = \frac{\sum_{i\in\mathcal{S}} M_i\bar{y}_i}{\sum_{i\in\mathcal{S}} M_i}. \tag{5.16}$$

The estimator $\hat{\bar{y}}_r$ in (5.16) is the quantity $\hat{B}$ in (4.2): the denominator is a random quantity that depends on which particular psus are included in the sample. If the M_is are unequal and a different cluster sample of size n is taken, the denominator will likely be different. From (4.12), taking $e_i = t_i - M_i\hat{\bar{y}}_r = M_i(\bar{y}_i - \hat{\bar{y}}_r)$, we have

$$\text{SE}(\hat{\bar{y}}_r) = \sqrt{\left(1 - \frac{n}{N}\right)\frac{s_r^2}{n\overline{M}^2}}, \tag{5.17}$$

where

$$s_r^2 = \frac{1}{n-1}\sum_{i\in\mathcal{S}} M_i^2\left(\bar{y}_i - \hat{\bar{y}}_r\right)^2 \tag{5.18}$$

and $\overline{M} = \hat{M}_0/N$ is the average size of the psus in the sample. The variance of the ratio estimator depends on the variability of the means per element in the psus, and can be much smaller than that of the unbiased estimator $\hat{\bar{y}}_{\text{unb}}$.

If the total number of elements in the population, $M_0 = \sum_{i=1}^{N} M_i$, is known, we can also use ratio estimation to estimate the population total: the ratio estimator is $\hat{t}_r = M_0 \hat{\bar{y}}_r$ with $\text{SE}(\hat{t}_r) = M_0 \, \text{SE}(\hat{\bar{y}}_r)$. Note, though, that $\hat{t}_r$ requires that we know the total number of elements in the population, M_0; the unbiased estimator in (5.14) makes no such requirement.

Estimation using weights. The estimators $\hat{t}_{\text{unb}}$ and $\hat{\bar{y}}_r$ are usually calculated using the sampling weights. The probability that a psu is in the sample is n/N, because an SRS is taken of n of the N psus. With one-stage cluster sampling, an ssu is included in the sample whenever its psu is included in the sample. Thus, as in Section 5.2.1,

$$w_{ij} = \frac{1}{P(\text{ssu } j \text{ of psu } i \text{ is in sample})} = \frac{N}{n}.$$

One-stage cluster sampling produces a self-weighting sample when the psus are selected with equal probabilities.

When each psu has the same number of elements, the weights sum to the population size M_0. When the psus have different sizes, however, the sum of the weights *estimates* the population size, with

$$\hat{M}_0 = \sum_{i \in \mathcal{S}} \sum_{j \in \mathcal{S}_i} w_{ij} = \sum_{i \in \mathcal{S}} \sum_{j \in \mathcal{S}_i} \frac{N}{n} = \frac{N}{n} \sum_{i \in \mathcal{S}} M_i.$$

The estimated population size $\hat{M}_0$ is a random variable, and varies from sample to sample when the M_i are unequal.

Using the weights, the estimated population total in (5.14) is calculated as

$$\hat{t}_{\text{unb}} = \sum_{i \in \mathcal{S}} \sum_{j \in \mathcal{S}_i} w_{ij} y_{ij}. \tag{5.19}$$

The estimated population mean in (5.16) is calculated as

$$\hat{\bar{y}}_r = \frac{\hat{t}_{\text{unb}}}{\hat{M}_0} = \frac{\displaystyle\sum_{i \in \mathcal{S}} \sum_{j \in \mathcal{S}_i} w_{ij} y_{ij}}{\displaystyle\sum_{i \in \mathcal{S}} \sum_{j \in \mathcal{S}_i} w_{ij}}. \tag{5.20}$$

Example 5.6. One-stage cluster samples are often used in educational studies, since students are naturally clustered into classrooms or schools. Consider a population of 187 high school algebra classes in a city. An investigator takes an SRS of 12 of those classes and gives each student in the sampled classes a test about function knowledge. The (hypothetical) data are given in the file `algebra.csv`, with summary statistics in Table 5.5.

We can use either (5.16) or (5.20) to estimate the mean score in the population. Using (5.16), we have

$$\hat{\bar{y}}_r = \frac{\displaystyle\sum_{i \in \mathcal{S}} M_i \bar{y}_i}{\displaystyle\sum_{i \in \mathcal{S}} M_i} = \frac{18{,}708}{299} = 62.57.$$

Statistical software packages typically use (5.20) to estimate the population mean $\bar{y}_{\mathcal{U}}$, and this formulation gives the same number. The weight for each observation is $w_{ij} = 187/12 =$

TABLE 5.5
Summary statistics for Example 5.6.

Class Number	M_i	$\bar{y}_i$	t_i	$M_i^2(\bar{y}_i - \hat{\bar{y}}_r)^2$
23	20	61.5	1,230	456.7298
37	26	64.2	1,670	1,867.7428
38	24	58.4	1,402	9,929.2225
39	34	58.0	1,972	24,127.7518
41	26	58.0	1,508	14,109.3082
44	28	64.9	1,816	4,106.2808
46	19	55.2	1,048	19,825.3937
51	32	72.1	2,308	93,517.3218
58	17	58.2	989	5,574.9446
62	21	66.6	1,398	7,066.1174
106	26	62.3	1,621	33.4386
108	26	67.2	1,746	14212.7867
Total	299		18,708	194,827.0387

15.5833, and the sum of the weights for the sample, 4659.41667, estimates the total number of students in the 187 high school algebra classes, resulting in the same estimate

$$\hat{\bar{y}}_r = \frac{\sum_{i \in \mathcal{S}} \sum_{j=1}^{M_i} w_{ij} y_{ij}}{\sum_{i \in \mathcal{S}} \sum_{j=1}^{M_i} w_{ij}} = \frac{291{,}533}{4659.41667} = 62.57.$$

The standard error, from (5.17), is

$$\text{SE}(\hat{\bar{y}}_r) = \sqrt{\left(1 - \frac{12}{187}\right) \frac{1}{(12)(24.92)^2} \frac{194{,}827}{11}} = 1.49.$$

A 95% CI is given by

$$62.57 \pm 2.20\,(1.49) = [59.29, 65.85],$$

where 2.20 is the percentile from a t distribution with 11 df. ■

5.3 Two-Stage Cluster Sampling

In one-stage cluster sampling, we observe all the ssus within the selected psus. In some situations, though, the ssus in a psu may be so similar that measuring all of the ssus within a psu wastes resources; alternatively, it may be expensive to measure ssus relative to the cost of sampling psus. In these situations, it may be much cheaper to take a subsample within each psu that is selected. The stages within a two-stage cluster sample, when we sample the psus and subsample the ssus with equal probabilities, are:

1. Select an SRS $\mathcal{S}$ of n psus from the population of N psus.

2. Select an SRS of ssus from each selected psu. The SRS of m_i elements from the ith psu is denoted $\mathcal{S}_i$.

The difference between one-stage and two-stage cluster sampling is illustrated in Figure 5.2. The extra stage complicates the notation and estimators, as we need to consider variability arising from both stages of data collection. The point estimators of t and $\bar{y}_{\mathcal{U}}$ are analogous to those in one-stage cluster sampling, but the variance formulas become messier.

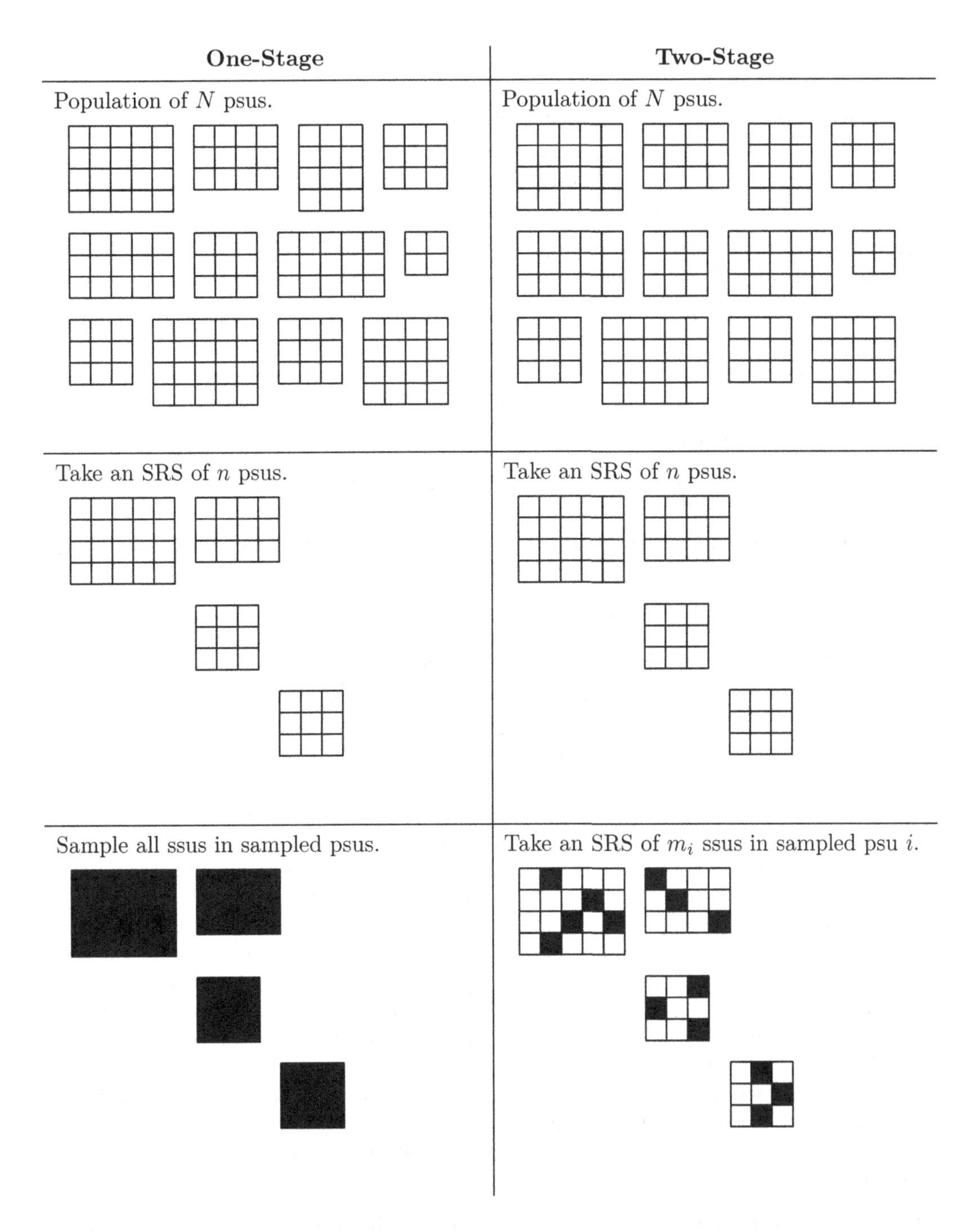

FIGURE 5.2
Differences between one-stage and two-stage cluster sampling.

Unbiased estimator of population total. In one-stage cluster sampling, we could estimate the population total by $\hat{t}_{\text{unb}} = (N/n)\sum_{i\in\mathcal{S}} t_i$; the psu totals t_i were known because we sampled every ssu in the selected psus. In two-stage cluster sampling, however, since we do not observe every ssu in the sampled psus, we need to estimate the individual psu totals by

$$\hat{t}_i = \sum_{j\in\mathcal{S}_i} \frac{M_i}{m_i} y_{ij} = M_i \bar{y}_i$$

and an unbiased estimator of the population total is

$$\hat{t}_{\text{unb}} = \frac{N}{n}\sum_{i\in\mathcal{S}} \hat{t}_i = \frac{N}{n}\sum_{i\in\mathcal{S}} M_i\bar{y}_i = \sum_{i\in\mathcal{S}}\sum_{j\in\mathcal{S}_i} \frac{N}{n}\frac{M_i}{m_i} y_{ij}. \tag{5.21}$$

Weights in two-stage cluster sampling. Most survey statisticians use sampling weights to estimate means and totals in cluster samples. Equation (5.21) suggests that the sampling weight for ssu j of psu i is $\dfrac{N}{n}\dfrac{M_i}{m_i}$, and we can see that this is so by calculating the inclusion probability. For cluster sampling,

$$\begin{aligned}
&P(j\text{th ssu in } i\text{th psu is selected for the sample})\\
&\quad= P(i\text{th psu selected}) \times P(j\text{th ssu selected} \mid i\text{th psu selected})\\
&\quad= \frac{n}{N}\frac{m_i}{M_i}.
\end{aligned}$$

Recall from Section 2.4 that the weight of an element is the reciprocal of the probability of its selection. Thus,

$$w_{ij} = \frac{NM_i}{nm_i}. \tag{5.22}$$

If psus are blocks, for example, and ssus are households, then household j in psu i represents $(NM_i)/(nm_i)$ households in the population: itself, and $(NM_i)/(nm_i) - 1$ households that are not sampled. Then,

$$\hat{t}_{\text{unb}} = \sum_{i\in\mathcal{S}}\sum_{j\in\mathcal{S}_i} w_{ij} y_{ij}. \tag{5.23}$$

In two-stage cluster sampling, a self-weighting design has each ssu representing the same number of ssus in the population. To take a cluster sample of persons in Illinois, we could take an SRS of counties in Illinois and then take an SRS of m_i of the M_i persons from county i in the sample. To have every person in the sample represent the same number of persons in the population, m_i needs to be proportional to M_i, so that m_i/M_i is approximately constant. Thus, we would subsample more persons in the large counties than in the small counties to have a self-weighting sample. For Illinois, this would mean that Cook County, home of Chicago, would have a huge subsample while small rural counties would have small values of m_i. In Chapter 6 we shall study a better way to obtain a self-weighting sample of ssus when the psu sizes vary.

The sampling weights provide a convenient way of calculating point estimates; they do not avoid associated shortcomings such as large variances. Also, the sampling weights give no information on how to find standard errors, and we need to derive the formula for the variance using the sampling design.

Variance estimation for two-stage cluster samples. In two-stage sampling, the $\hat{t}_i$s are random variables. Consequently, the variance of $\hat{t}_{\text{unb}}$ has two components: (1) the variability between psus and (2) the variability of ssus within psus. We do not have to worry about component (2) in one-stage cluster sampling.

The variance of $\hat{t}_{\text{unb}}$ in (5.21) equals the variance of $\hat{t}_{\text{unb}}$ from one-stage cluster sampling plus an extra term to account for the extra variance due to estimating the $\hat{t}_i$s rather than measuring them directly. For two-stage cluster sampling,

$$V\left(\hat{t}_{\text{unb}}\right) = N^2\left(1 - \frac{n}{N}\right)\frac{S_t^2}{n} + \frac{N}{n}\sum_{i=1}^{N}\left(1 - \frac{m_i}{M_i}\right)M_i^2\frac{S_i^2}{m_i}, \tag{5.24}$$

where S_t^2 is the population variance of the psu totals, and S_i^2 is the population variance among the elements within psu i. The first term in (5.24) is the variance from one-stage cluster sampling, and the second term is the additional variance due to subsampling within the psus. If $m_i = M_i$ for each psu i, as occurs in one-stage cluster sampling, then the second term in (5.24) is 0. To prove (5.24), we need to condition on the psus included in the sample. This is more easily done in the general setting of unequal probability sampling, and the variance for general cluster samples is derived in Section 6.6. (If you prefer to see the proof before you use the variance results, read Section 6.6 now.)

To estimate $V(\hat{t}_{\text{unb}})$, let

$$s_t^2 = \frac{1}{n-1}\sum_{i\in\mathcal{S}}\left(\hat{t}_i - \frac{\sum_{j\in\mathcal{S}}\hat{t}_j}{n}\right)^2 = \frac{1}{n-1}\sum_{i\in\mathcal{S}}\left(\hat{t}_i - \frac{\hat{t}_{\text{unb}}}{N}\right)^2 \tag{5.25}$$

be the sample variance among the estimated psu totals and let

$$s_i^2 = \frac{1}{m_i - 1}\sum_{j\in\mathcal{S}_i}\left(y_{ij} - \bar{y}_i\right)^2 \tag{5.26}$$

be the sample variance of the ssus sampled in psu i. As will be shown in Section 6.6, an unbiased estimator of the variance in (5.24) is given by

$$\hat{V}\left(\hat{t}_{\text{unb}}\right) = N^2\left(1 - \frac{n}{N}\right)\frac{s_t^2}{n} + \frac{N}{n}\sum_{i\in\mathcal{S}}\left(1 - \frac{m_i}{M_i}\right)M_i^2\frac{s_i^2}{m_i}. \tag{5.27}$$

The standard error, $\text{SE}(\hat{t}_{\text{unb}})$, is of course the square root of (5.27).

As in one-stage cluster sampling with unequal cluster sizes, s_t^2 is affected both by variations in the unit sizes (the M_i) and by variations in the $\bar{y}_i$. If the cluster sizes are disparate, this component is large even if the cluster means are fairly constant.

Simplified variance estimation. In many situations when N is large, the contribution of the second term in (5.27) to the variance estimator is negligible compared with that of the first term. We show in Section 6.6 that

$$E\left[s_t^2\right] = S_t^2 + \frac{1}{N}\sum_{i=1}^{N}\left(1 - \frac{m_i}{M_i}\right)M_i^2\frac{S_i^2}{m_i} = S_t^2 + \frac{1}{N}\sum_{i=1}^{N}V\left(\hat{t}_i\right). \tag{5.28}$$

We expect the sample variance of the estimated psu totals $\hat{t}_i$ to be larger than the sample variance of the true psu totals t_i because $\hat{t}_i$ will be different if we take a different subsample in psu i. Thus the first term in (5.27) estimates the first term of (5.24) plus part of the second term of (5.24).

The variance estimator in (5.27) requires knowledge of the sample and population sizes at both stages of sampling. If there are more than two stages of cluster sampling, the unbiased variance estimator gets even more complicated, with additional terms for each stage. To simplify calculations, most software packages for analyzing survey data approximate the variance using only the first term of (5.27), after omitting the finite population correction (fpc) of $(1 - n/N)$.

The simpler estimator

$$\hat{V}_{\text{WR}}(\hat{t}_{\text{unb}}) = N^2 \frac{s_t^2}{n} \tag{5.29}$$

estimates the with-replacement variance for a cluster sample, as will be seen in Section 6.3. If the first-stage sampling fraction n/N is small, then the with-replacement variance estimate in (5.29) approximately equals the without-replacement variance estimate in (5.27); you will derive the expected value of (5.29) for a without-replacement sample in Exercise 31. Alternatively, a replication method of variance estimation from Chapter 9 can be used.

Ratio estimator of population mean. As in one-stage cluster sampling, we use a ratio estimator for the population mean. Again, the y's of Chapter 4 are the psu totals (now estimated by $\hat{t}_i$), and the x's are the psu sizes M_i. As in (5.16),

$$\hat{\bar{y}}_r = \frac{\sum_{i \in \mathcal{S}} \hat{t}_i}{\sum_{i \in \mathcal{S}} M_i} = \frac{\sum_{i \in \mathcal{S}} M_i \bar{y}_i}{\sum_{i \in \mathcal{S}} M_i}. \tag{5.30}$$

Using the sampling weights in (5.22) with $w_{ij} = (NM_i)/(nm_i)$, we can rewrite $\hat{\bar{y}}_r$ as

$$\hat{\bar{y}}_r = \frac{\hat{t}_{\text{unb}}}{\hat{M}_0} = \frac{\sum_{i \in \mathcal{S}} \sum_{j \in \mathcal{S}_i} w_{ij} y_{ij}}{\sum_{i \in \mathcal{S}} \sum_{j \in \mathcal{S}_i} w_{ij}}. \tag{5.31}$$

The weights are different, but the form of the estimator is the same as in (5.20).

The variance estimator is again based on the approximation in (4.12):

$$\hat{V}(\hat{\bar{y}}_r) = \frac{1}{\overline{M}^2}\left(1 - \frac{n}{N}\right)\frac{s_r^2}{n} + \frac{1}{nN\overline{M}^2}\sum_{i \in \mathcal{S}} M_i^2 \left(1 - \frac{m_i}{M_i}\right)\frac{s_i^2}{m_i}, \tag{5.32}$$

where s_r^2 is defined in (5.18), s_i^2 is defined in (5.26), and $\overline{M} = \hat{M}_0/N$ is the average size of the psus in the sample.

As with $\hat{t}_{\text{unb}}$, the second term in (5.32) is usually negligible compared with the first term, and most survey software packages calculate the variance using

$$\hat{V}_{\text{WR}}(\hat{\bar{y}}_r) = \frac{s_r^2}{n\overline{M}^2}. \tag{5.33}$$

Degrees of freedom for a two-stage cluster sample are the same as for a one-stage cluster sample: the number of psus minus 1.

Example 5.7. In education surveys where measurements on individual students are relatively easy to obtain once the school or classroom is selected (as is assumed for the SRS of 12 algebra classes in Example 5.6, where students are given a written test), a one-stage cluster sample is often the most practical. But if measuring the responses of interest requires a lot of resources from the research team (for example, if students are interviewed in person by assessment specialists), then a two-stage cluster sample may be less expensive.

File `schools.csv` contains data from a two-stage sample of students. In the first stage of sampling, an SRS of $n = 10$ schools is selected from a population of $N = 75$ schools. At stage 2, an SRS of $m_i = 20$ students is selected from each sampled school and administered assessments for reading and math. These data are fictional, but the summary statistics are consistent with those seen in educational studies.

Side-by-side boxplots for the math scores at the sampled schools are displayed in Figure 5.3, with the schools ordered from lowest median score to highest median score. Note the variation in the medians from school to school. This indicates that math scores within the same school tend to be more similar than the scores of randomly selected students from different schools, and that the cluster sample does not provide as much information per student as would an SRS of 200 students from the population of schools.

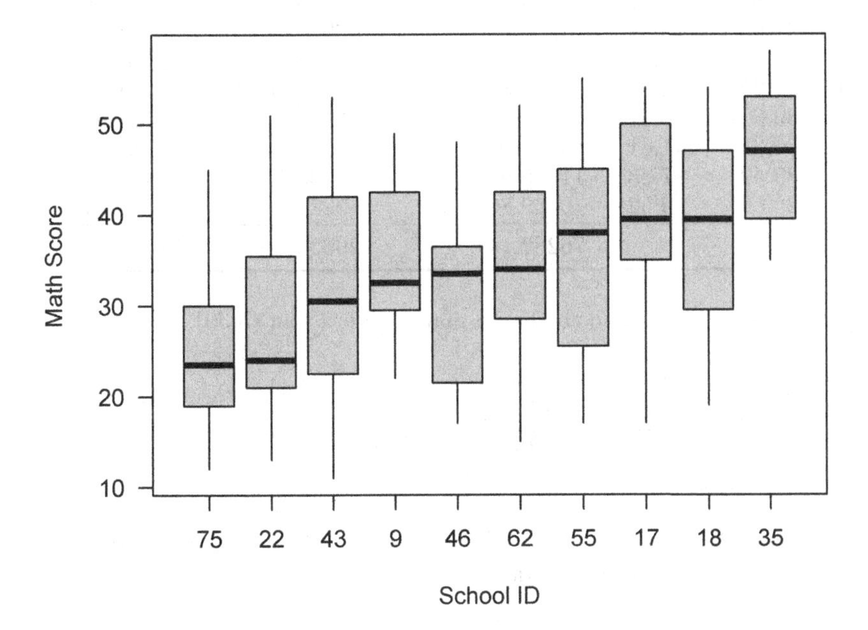

FIGURE 5.3
Side-by-side boxplots of math test scores for Example 5.7.

Statistical survey software gives the output in Figure 5.6 for estimating the mean math score (variable *math*) for the students in the population of 75 schools, as well as the proportion and total number of students who have scores higher than 40 (having variable *mathlevel* = 2). The with-replacement standard error and 95% CI are given for each statistic (see Lohr, 2022; Lu and Lohr, 2022, for the code used to produce output similar to this).

TABLE 5.6
Output from survey software for Example 5.7.

Variable	DF	Mean	Std Error of Mean	95% CI for Mean	
math	9	33.12295	1.759894	29.1417926	37.1041040
mathlevel=2	9	0.28769	0.054168	0.1651535	0.4102264
Variable	**DF**	**Total**	**Std Error of Total**	**95% CI for Total**	
mathlevel=2	9	4969.125	676.25819	3439.32269	6498.9273

Let's see where the numbers in Table 5.6 come from. Table 5.7, showing the calculations needed to use the formulas to compute the estimates, is similar to Table 5.5, but now, instead of being able to calculate the psu total t_i from all students in a sampled class, we estimate the psu total by $\hat{t}_i = M_i \bar{y}_i$. We also calculate the within-psu variances s_i^2.

TABLE 5.7
Calculations using formulas for math scores in Example 5.7.

School	M_i	$\bar{y}_i$	s_i^2	$\hat{t}_i$	$M_i^2\left(1-\dfrac{m_i}{M_i}\right)\dfrac{s_i^2}{m_i}$	$M_i^2(\bar{y}_i - \hat{\bar{y}}_r)^2$
9	163	34.75	74.51	5664.25	86841	70336
17	180	40.80	111.01	7344.00	159855	1909563
18	114	37.85	124.87	4314.90	66906	290396
22	367	27.95	109.31	10257.65	696046	3604196
35	109	46.10	50.31	5024.90	24401	2000806
43	219	32.20	162.80	7051.80	354749	40855
46	318	30.60	86.57	9730.80	410178	643681
55	259	36.35	141.61	9414.65	438284	698572
62	311	35.40	97.83	11009.40	442693	501495
75	263	24.60	69.52	6469.80	222134	5024481
Sum	2303			76282.15	2902087	14784382

We use the ratio estimator to estimate the mean math score. From (5.30),

$$\hat{\bar{y}}_r = \frac{\sum_{i\in\mathcal{S}} \hat{t}_i}{\sum_{i\in\mathcal{S}} M_i} = \frac{76282.15}{2303} = 33.12.$$

Or we can, equivalently, use (5.31) to calculate $\hat{\bar{y}}_r$ using the sampling weights (as is done by most software packages). The weight for student j in school i is:

$$w_{ij} = \frac{N}{n}\frac{M_i}{m_i} = \frac{75}{10}\frac{M_i}{20}.$$

Using (5.31), the estimated mean again is

$$\hat{\bar{y}}_r = \frac{\sum_{i\in\mathcal{S}}\sum_{j\in\mathcal{S}_i} w_{ij}y_{ij}}{\sum_{i\in\mathcal{S}}\sum_{j\in\mathcal{S}_i} w_{ij}} = \frac{572116.1}{17272.5} = 33.12.$$

The weights do not allow us to calculate the standard error, however. We need the clustering information to do that. From Table 5.7,

$$s_r^2 = \frac{1}{n-1}\sum_{i\in\mathcal{S}} M_i^2(\bar{y}_i - \hat{\bar{y}}_r)^2 = \frac{14784382}{9} = 1642709$$

and $\overline{M} = 2303/10 = 230.3$. The output in Table 5.6 uses the with-replacement variance

$$\hat{V}_{\text{WR}}(\hat{\bar{y}}_r) = \frac{s_r^2}{n\overline{M}^2} = \frac{1642709}{(10)(230.3)^2} = 3.097$$

with $\text{SE}_{\text{WR}}(\hat{\bar{y}}_r) = \sqrt{3.097} = 1.76$.

We can calculate the without-replacement variance with some extra work, by plugging into the formula in (5.32):

$$\hat{V}(\hat{\bar{y}}_r) = \frac{1}{(230.3)^2}\left[\left(1 - \frac{10}{75}\right)\frac{1642709}{10} + \frac{1}{75}\frac{2902087}{10}\right] = 2.684 + 0.073 = 2.757.$$

Note that the second term in the estimated variance is small relative to the first term. The standard error, $\text{SE}(\hat{\bar{y}}_r) = \sqrt{2.757} = 1.66$, is slightly smaller than the with-replacement standard error for this example because the first-stage sampling fraction, 10/75, is relatively small. A 95% confidence interval is calculated using the critical value from a t distribution with $n - 1 = 9$ df as $33.12 \pm 2.26(1.66)$, or [29.4, 36.9].

To estimate the proportion and total number of students who have *mathlevel* = 2, define $u_{ij} = 1$ if student j in school i has *mathlevel* = 2 and 0 otherwise. The sum of the weights for sample observations having *mathlevel* = 2 is

$$\hat{t}_{\text{unb}} = \sum_{i \in \mathcal{S}} \sum_{j \in \mathcal{S}_i} w_{ij} u_{ij} = 4969.125$$

so the estimated proportion is $\hat{p} = 4969.125/17272.5 = 0.28769$. Using a spreadsheet similar to that in Table 5.7, we can calculate

$$s_t^2 = \frac{1}{n-1} \sum_{i \in \mathcal{S}} \left(M_i \hat{p}_i - \frac{4969.125}{75}\right)^2 = \frac{1}{9}(7317.2) = 813.02$$

and

$$s_r^2 = \frac{1}{n-1} \sum_{i \in \mathcal{S}} M_i^2 \left(\hat{p}_i - \hat{p}\right)^2 = \frac{1}{9}(14006) = 1556.23,$$

where $\hat{p}_i$ is the proportion of students in school i having *mathlevel*=2.

Using (5.29) and (5.33), the with-replacement variances are

$$\hat{V}_{\text{WR}}\left(\hat{t}_{\text{unb}}\right) = N^2 \frac{s_t^2}{n} = (75)^2 \frac{813.02}{10} = 457325$$

and

$$\hat{V}_{\text{WR}}(\hat{p}) = \frac{s_r^2}{n\overline{M}^2} = \frac{1556.23}{(10)(230.3)^2} = 0.00293,$$

with standard errors $\text{SE}_{\text{WR}}(\hat{t}_{\text{unb}}) = 676.3$ and $\text{SE}_{\text{WR}}(\hat{p}) = 0.054$

If desired, the without-replacement variances can be calculated using (5.27) and (5.32). These require the additional calculation of

$$\sum_{i \in \mathcal{S}} M_i^2 \left(1 - \frac{m_i}{M_i}\right) \frac{s_i^2}{m_i} = 4828.13,$$

where s_i^2 is the variance of the u_{ij}s in psu i, giving

$$\hat{V}\left(\hat{t}_{\text{unb}}\right) = \left(1 - \frac{10}{75}\right) \hat{V}_{\text{WR}}\left(\hat{t}_{\text{unb}}\right) + \frac{75}{10}(4828.13) = 396348 + 36211 = 432599$$

and

$$\hat{V}(\hat{p}) = \left(1 - \frac{10}{75}\right) \hat{V}_{\text{WR}}(\hat{p}) + \frac{1}{(10)(75)(230.3)^2}(4828.13) = 0.00254 + 0.00012 = 0.00266.$$

As before, the without-replacement standard errors (657.7 for the estimated total and 0.052 for the estimated proportion) are smaller than the with-replacement standard errors, but only slightly so. The most important feature of the calculation is to account for the variability induced by the cluster sampling. ■

We can calculate the full sampling weight and the without-replacement variances for the schools data because we know the number of schools in the population, N. But we can calculate means along with their with-replacement variances even when N is unknown, as the following example shows.

Example 5.8. The data in the file `coots.csv` come from Arnold's (1991) work on egg size and volume of American Coot eggs in Minnedosa, Manitoba. In this data set, we look at volumes of a subsample of eggs in clutches (nests of eggs) with at least two eggs available for measurement.

The data are plotted in Figure 5.4. Because we have only two observations per clutch, we can plot the individual data points. Note that the two eggs in each clutch have similar volumes, and that most of the variability in the data comes from the different clutches, indicating that the cluster sample does not provide as much information per egg as would an SRS of eggs. Data from a cluster sample can be plotted in many ways, and you often need to construct more than one type of plot to see features of the data. We shall return to the issue of plotting data from complex surveys in Section 7.6.

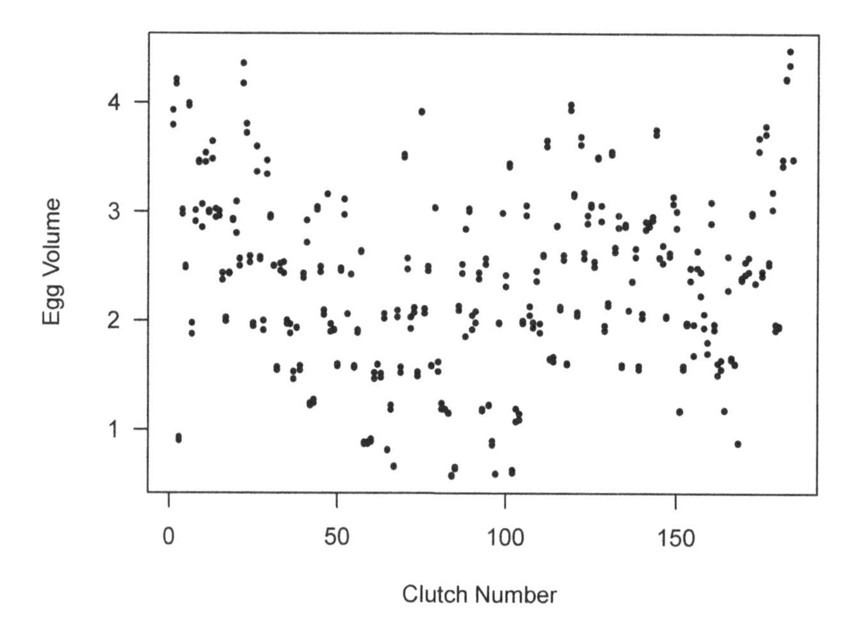

FIGURE 5.4
Plot of egg volume data.

As in Example 5.7, we use the ratio estimator to estimate the mean egg volume, calculating $\hat{\bar{y}}_r$ in (5.31) with the sampling weights. The weight for egg j in clutch i is:

$$w_{ij} = \frac{N}{n}\frac{M_i}{m_i} = \frac{N}{184}\frac{M_i}{2}.$$

Because N is unknown, and because we are calculating a mean where the factor $N/184$ will cancel in numerator and denominator, we use relative weights $w_{ij} = M_i/2$, which gives

$$\hat{\bar{y}}_r = \frac{\sum_{i \in \mathcal{S}} \sum_{j \in \mathcal{S}_i} w_{ij} y_{ij}}{\sum_{i \in \mathcal{S}} \sum_{j \in \mathcal{S}_i} w_{ij}} = \frac{4375.947}{1757} = 2.49.$$

For standard error calculations, we can construct a table similar to Table 5.7 to calculate $M_i^2(\bar{y}_i - \hat{\bar{y}}_r)^2$ for each clutch, giving

$$s_r^2 = \frac{1}{n-1} \sum_{i \in \mathcal{S}} M_i^2(\bar{y}_i - \hat{\bar{y}}_r)^2 = \frac{11{,}438.99}{183} = 62.51$$

and $\bar{M} = \sum_{i \in \mathcal{S}} M_i/n = 1758/184 = 9.554$. From (5.32), then,

$$\hat{V}(\hat{\bar{y}}_r) = \frac{1}{9.554^2} \left[\left(1 - \frac{184}{N}\right) \frac{62.51}{184} + \frac{1}{N} \frac{46.31}{184} \right].$$

Now N, the total number of clutches in the population, is unknown but presumed to be large (and known to be larger than 184). Thus, we may take the psu-level fpc to be 1, and note that the second term in the estimated variance will be very small relative to the first term. We then use

$$V_{\text{WR}}(\hat{\bar{y}}_r) = \frac{1}{(9.554)^2} \frac{62.51}{184} = 0.0037.$$

a 95% confidence interval is calculated using the critical value from a t distribution with 183 df as $2.49 \pm 1.97\sqrt{0.0037}$, or [2.37, 2.61]. The estimated coefficient of variation for $\hat{\bar{y}}_r$ is

$$\frac{\text{SE}_{\text{WR}}(\hat{\bar{y}}_r)}{\hat{\bar{y}}_r} = \frac{0.061}{2.49} = 0.0245. \blacksquare$$

In Example 5.8, we could not use the unbiased estimator for the mean or estimate population totals because we know neither N nor M_0. The M_is, however, did not vary widely, so the unbiased estimator would probably have had similar coefficient of variation. If all the M_is are equal, in fact, the unbiased estimator is the same as the ratio estimator (see Exercise 28); if the M_is vary, the unbiased estimator often performs poorly. The next example illustrates that the unbiased estimator of t may have large variance when the cluster sizes are highly variable.

Example 5.9. *The case of the six-legged puppy.* Suppose we want to estimate the average number of legs on the healthy puppies in Sample City puppy homes. Sample City has two puppy homes: Puppy Palace with 30 puppies, and Dog's Life with 10 puppies. Let's select one puppy home with probability 1/2. After the home is selected, then select 2 puppies at random from the home, and use $\hat{\bar{y}}_{\text{unb}}$ to estimate the average number of legs per puppy.

Suppose we select Puppy Palace. Not surprisingly, each of the two puppies sampled has four legs, so $\hat{t}_{\text{PP}} = 30 \times 4 = 120$. Then, using (5.21), an unbiased estimate for the total number of puppy legs in both homes is

$$\hat{t}_{\text{unb}} = \frac{2}{1}\hat{t}_{\text{PP}} = 240.$$

We divide the estimated total number of legs by the total number of puppies to estimate the mean number of legs per puppy as $240/40 = 6$.

If we select Dog's Life instead, $\hat{t}_{\text{DL}} = 10 \times 4 = 40$, and

$$\hat{t}_{\text{unb}} = \frac{2}{1}\hat{t}_{\text{DL}} = 80.$$

If Dog's Life is selected, the unbiased estimate of the mean number of legs per puppy is $80/40 = 2$.

These are not good estimates of the number of legs per puppy. But the estimator is mathematically unbiased: $(6 + 2)/2 = 4$, so averaging over all possible samples results in the right number. The poor quality of the estimator is reflected in the very large variance of the estimator, calculated using (5.24):

$$\begin{aligned} V(\hat{t}_{\text{unb}}) &= \left(1 - \frac{1}{2}\right) 2^2 \frac{S_t^2}{1} + \frac{2}{1}\sum_{i=1}^{2}\left(1 - \frac{m_i}{M_i}\right) M_i^2 \frac{S_i^2}{m_i} \\ &= \frac{1}{2}(4)(3200) = 6400. \end{aligned}$$

The ratio estimator, on the other hand, is right on target. If Puppy Palace is selected, $\hat{\bar{y}}_r = 120/30 = 4$; if Dog's Life is selected, $\hat{\bar{y}}_r = 40/10 = 4$. Because the estimate is the same for all possible samples, $V(\hat{\bar{y}}_r) = 0$. ■

In general, the unbiased estimator of the population total is inefficient if the cluster sizes are unequal and t_i is roughly proportional to M_i. The variance of $\hat{t}_{\text{unb}}$ depends on the variance of the t_is, and that variance may be large if the M_is are unequal.

The ratio estimator, however, generally performs well when t_i is roughly proportional to M_i. Recall from (4.9) that the approximate mean squared error (MSE) of the estimator $\hat{B}$ is proportional to the variance of the residuals from the model: Using the notation of this chapter, the approximate MSE of $\hat{\bar{y}}_r$ ($= \hat{B}$) is proportional to $\sum_{i=1}^{N}(t_i - \bar{y}_{\mathcal{U}} M_i)^2$. When t_i (the response variable) is highly positively correlated with M_i (the auxiliary variable), the residuals are small. In Example 5.9, the total number of puppy legs in a puppy home (t_i) is exactly four times the total number of puppies in the home (M_i), so the variance of the ratio estimator is zero.

This is an important issue, since many naturally occurring clusters are of unequal sizes, and we expect that the cluster totals will often be proportional to the number of ssus. In a cluster sample of nursing homes, we expect that a larger number of residents will be satisfied with the level of care in a home with 500 residents than in a home with 20 residents, even though the proportions of residents who are satisfied may be the same. The total of the math scores for all students in a class will be much greater for large classes than for small classes. In general, we expect to see more honeybees in a large area than in a small area. For all of these situations, then, while the estimator $\hat{\bar{y}}_r$ can work well, the estimator $\hat{t}_{\text{unb}}$ tends to have large variability. In Chapter 6, we discuss an alternative design and estimator for cluster sampling that result in a much lower variance for the estimated population total when t_i is proportional to M_i.

5.4 Designing a Cluster Sample

Persons and organizations taking an expensive, large-scale survey need to devote a great deal of time to designing the survey; typically, large government surveys take several years to design and test. Even then, the Fundamental Principle of Survey Design often holds true:

You can best design the survey you should have taken after you have finished the survey. After the survey is completed, you can assess the effect of the clustering on the estimates and know where you could have allocated more resources to obtain better information.

The more you know about a population, the better you can design an efficient sampling scheme to study it. If you know the value of y_{ij} for every person in your population, then you can design a flawless (but unnecessary because you already know everything!) survey for studying the population. If you know very little about the population, chances are that you will gain information about it after collecting the survey, but you may not have the most efficient design possible for that survey. You may, however, be able to use your newly gained knowledge to make your next survey more efficient.

When designing a cluster sample, you need to address four major issues:

1. What overall precision is needed?
2. What size should the psus be?
3. How many ssus should be sampled in each psu selected for the sample?
4. How many psus should be sampled?

Question 1 must be faced in any survey design. To answer questions 2 through 4, you need to know the cost of sampling a psu for possible psu sizes, the cost of sampling a ssu, and a measure of homogeneity (R_a^2 or ICC) for the possible sizes of psu.

5.4.1 Choosing the psu Size

The psu is often a naturally occuring unit. In Example 5.8, a clutch of eggs was an obvious cluster unit. A survey to estimate calf mortality might use farms as the psus; a survey of sixth-grade students might use classes or schools as the psus.

In other surveys, however, the investigator may have a wide choice for psu size. In a survey to estimate the sex and age ratios of mule deer in a region of Colorado, psus might be designated areas and ssus might be individual deer or groups of deer in those areas. But should the size of the psus be 1 km^2, 2 km^2, or 100 m^2?

A general principle in area surveys is that the larger the psu size, the more variability you expect to see within a psu. Hence you expect R_a^2 and ICC to be smaller with a large psu than with a small psu. If the psu size is too large, however, you may lose the cost savings of cluster sampling.

The theory provides useful guidance for designing your own survey. There are many ways to "try out" different psu sizes before taking your survey. One way is to postulate a model for the relationship between R_a^2 or MSW and M, and to fit the model using preliminary data or information from other studies. Then use different combinations of R_a^2 and M, and compare the costs. Another way is to perform an experiment and collect data on relative costs and variances with different psu sizes.

Example 5.10. The Colorado potato beetle has long been considered a major pest of potatoes. Zehnder et al. (1990) studied different sizes of sampling units that could be used to estimate potato beetle counts. Ten randomly selected sites were sampled from each of ten fields. The investigators visually inspected each site for small larvae, large larvae, and adults on all foliage from a single stem on each of five adjacent plants.

They then considered different sizes of psu, ranging from one stem per site to five stems per site. To study the efficiency of a one-stem-per-site design, they examined data from stem 1 of each site. Similarly, the data from stems 1 and 2 of each site gave a cluster sample with two ssus per psu, and so on. It takes about 30 minutes to walk among the sites in each field; sampling one stem requires about 10 seconds during the early part of the season. Thus the total cost to sample all ten sites with the one-stem-per-site design is estimated to be 30 + 100/60 = 31.67 minutes. Data for estimating the number of small larvae are given in Table 5.8.

TABLE 5.8
Relative net precision in the potato beetle study.

Number of Stems Sampled per Site	$\hat{\bar{y}}$	SE($\hat{\bar{y}}$)	Cost to Sample One Field	Relative Net Precision
1	1.12	0.15	31.67	0.24
2	1.01	0.10	33.33	0.30
3	0.96	0.08	35.00	0.34
4	0.91	0.07	36.67	0.35
5	0.91	0.06	38.33	0.40

The relative net precision is calculated as $1000/[(\text{cost})\ \text{CV}(\hat{\bar{y}})]$. For this example, since the cost to sample additional stems at a site is small compared with the time to traverse the field, the five-stem-per-site design is most efficient among those studied. ■

5.4.2 Choosing Subsampling Sizes

The goal in designing a survey is generally to get the most information possible for the least cost and inconvenience. In this section, we concentrate on designing a two-stage cluster survey when all psus have the same number, M, of ssus; designing cluster samples will be treated more generally in Chapters 6 and 7. One approach for equal-sized clusters, discussed in Cochran (1977), is to minimize the variance in (5.24) for a fixed cost. If $M_i = M$ and $m_i = m$ for all psus, then $V(\hat{\bar{y}}_{\text{unb}})$ may be rewritten (see Exercise 27) as:

$$V(\hat{\bar{y}}_{\text{unb}}) = \left(1 - \frac{n}{N}\right)\frac{\text{MSB}}{nM} + \left(1 - \frac{m}{M}\right)\frac{\text{MSW}}{nm}. \tag{5.34}$$

where MSB and MSW are the between and within mean squares, respectively, in Table 5.3, the population ANOVA table.

If MSW = 0 and hence $R_a^2 = 1$, for R_a^2 defined in (5.13), then each element within a psu equals the psu mean. In that case, you may as well take $m = 1$; examining more than one element per psu just costs extra time and money without increasing precision. For other values of R_a^2, the optimal allocation depends on the relative costs of sampling psus and ssus.

Consider the simple cost function

$$\text{total cost} = C = c_1 n + c_2 nm, \tag{5.35}$$

where c_1 is the cost per psu (not including the cost of measuring ssus) and c_2 is the cost of measuring each ssu. One can determine, using calculus, that the values

$$n_{\text{opt}} = \frac{C}{c_1 + c_2 m_{\text{opt}}}$$

and

$$m_{\text{opt}} = \sqrt{\frac{c_1 M(N-1)(1-R_a^2)}{c_2(NM-1)R_a^2}} \tag{5.36}$$

minimize the variance for fixed total cost C under this cost function (see Exercise 30); often, though, a number of different values will work about equally well, and graphing the projected variance of the estimator will give more information than merely computing one fixed solution. A graphical approach also allows you to perform what-if analyses on the designs: What if the costs or the cost function are slightly different? Or the value of R_a^2 is changed slightly? You can also explore different cost functions with this approach.

In (5.36), the value R_a^2 is from the population ANOVA table. In practice, we can estimate it from pilot survey data by $\hat{R}_a^2 = \widehat{\text{MSW}}/\hat{S}^2$. In large populations, the ratio $M(N-1)/(NM-1)$ will be close to 1, so we can use $\hat{m}_{\text{opt}} = \sqrt{c_1(1-\hat{R}_a^2)/(c_2\hat{R}_a^2)}$.

Example 5.11. Would subsampling have been more efficient for Example 5.2 than the one-stage cluster sample that was used? We do not know the population quantities, but have information from the sample that can be used for planning future studies. Recall that $\hat{S}^2 = 0.279$, and we estimated R_a^2 by 0.337. Figures 5.5 and 5.6 show the estimated variance that would be achieved for different subsample sizes for different values of c_1 and c_2, and for different values of R_a^2. ■

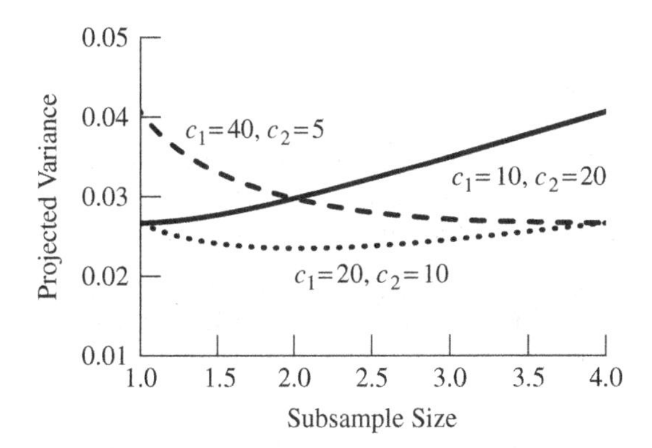

FIGURE 5.5
Estimated variance that would be obtained for the GPA example, for different values of c_1 and c_2 and different values of m. The sample estimate of 0.337 was used for R_a^2. The total cost used for this graph was $C = 300$. If it takes 40 minutes per suite and 5 minutes per person, then one-stage cluster sampling should be used; if it takes 10 minutes per suite and 20 minutes per person, then only one person should be sampled per suite; if it takes 20 minutes per suite and 10 minutes per person, the minimum is reached at $m \approx 2$, although the flatness of the curve indicates that any subsampling size would be acceptable.

For design purposes, we need only a rough estimate of R_a^2 or of MSW and MSB. The adjusted R^2 from the ANOVA table from sample data usually provides a good starting point, even though the table's mean square total underestimates S^2 when the number of psus in the sample is small (see Exercise 29).

Example 5.12. We obtain the following ANOVA table for the schools data in Example 5.7.

Source	df	SS	MS
Model	9	7018.48	779.83
Error	190	19538.40	102.83
Corrected Total	199	26556.88	

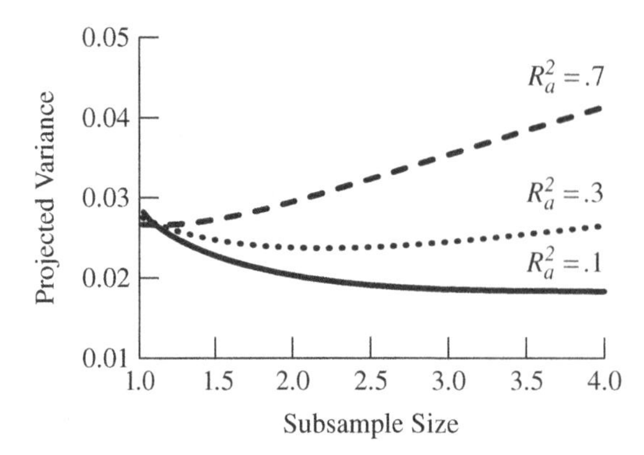

FIGURE 5.6
Estimated variance that would be obtained for the GPA example, for different values of R_a^2 and different values of m. The costs used in constructing this graph were $C = 300$, $c_1 = 20$, and $c_2 = 10$. The higher the value of R_a^2, the smaller the subsample size m should be.

If a future survey were planned to estimate average math scores, one might explore subsample sizes using R_a^2's around $1 - 102.83/(26556.88/199) = 0.23$. These data indicate a moderate degree of homogeneity within schools, and (5.36) suggests exploring subsampling sizes in the neighborhood of $m = \sqrt{3.3c_1/c_2}$. Of course, there will be many other considerations for setting the sample sizes, and these formulas just provide guidance. ■

Although we discussed only designs where all M_is are equal, we can use these methods with unequal M_is as well: just substitute $\bar{M}$ for M in the above work, and decide the average subsample size $\bar{m}$ to take. Then either take $\bar{m}$ observations in every cluster, or allocate observations so that

$$\frac{m_i}{M_i} = \text{constant}.$$

As long as the M_is do not vary too much, this should produce a reasonable design. If the M_is are widely variable, and the t_is are correlated with the M_is, a cluster sample with equal probabilities can be inefficient; an alternative design is presented in Chapter 6.

5.4.3 Choosing the Sample Size (Number of psus)

After the psu size is determined and the subsampling fraction set, we then look at the number of psus to sample, n. Like any survey design, design of a cluster sample is an iterative process: (1) Determine a desired precision, (2) choose the psu and subsample sizes, (3) conjecture the variance that will be achieved with that design, (4) set n to achieve the precision, and (5) iterate (adding stratification and auxiliary variables to use in ratio estimation) until the cost of the survey is within your budget.

If clusters are of equal size and we ignore the psu-level fpc, (5.34) implies that

$$V(\hat{\bar{y}}_{\text{unb}}) \leq \frac{1}{n}\left[\frac{\text{MSB}}{M} + \left(1 - \frac{m}{M}\right)\frac{\text{MSW}}{m}\right] = \frac{1}{n}v.$$

An approximate $100(1-\alpha)\%$ CI will be

$$\hat{\bar{y}}_{\text{unb}} \pm z_{\alpha/2}\sqrt{\frac{1}{n}v}.$$

Thus, to achieve a desired CI half-width e, set $n = z^2_{\alpha/2} v / e^2$. Of course, this approach presupposes that you have some knowledge of v, perhaps from a prior survey. In Section 7.4, we examine how to determine sample sizes for any situation in which you know the efficiency of the specified design relative to an SRS design.

5.5 Systematic Sampling

Systematic sampling, discussed briefly in Chapter 2, is really a special case of cluster sampling. Suppose we want to take a sample of size 3 from a population that has 12 elements:

$$1 \quad 2 \quad 3 \quad 4 \quad 5 \quad 6 \quad 7 \quad 8 \quad 9 \quad 10 \quad 11 \quad 12.$$

To take a systematic sample, choose a number randomly between 1 and 4. Draw that element and every fourth element thereafter. Thus, the population contains four psus (they are clusters even though the elements are not contiguous):

$$\{1, 5, 9\} \quad \{2, 6, 10\} \quad \{3, 7, 11\} \quad \{4, 8, 12\}.$$

Now we take an SRS of one psu.

In a population of NM elements, there are N possible choices for the systematic sample, each of size M. We observe only the mean of the one psu that comprises our systematic sample,

$$\bar{y}_i = \bar{y}_{i\mathcal{U}} = \hat{\bar{y}}_{\text{sys}}.$$

From the results in Section 5.2.1, $E[\hat{\bar{y}}_{\text{sys}}] = \bar{y}_{\mathcal{U}}$. For a simple systematic sample, we select $n = 1$ of the N psus, so by (5.5) and (5.12), the theoretical variance is

$$V(\hat{\bar{y}}_{\text{sys}}) = \left(1 - \frac{1}{N}\right)\frac{S_t^2}{M^2} = \left(1 - \frac{1}{N}\right)\frac{\text{MSB}}{M} \approx \frac{S^2}{M}[1 + (M - 1)\text{ICC}]. \tag{5.37}$$

In the notation for cluster sampling, M is the size of the systematic sample. Ignoring the fpc, we see that systematic sampling is more precise than an SRS of size M if the ICC is negative. Systematic sampling is more precise than simple random sampling when the variance within the possible systematic samples (psus) is *larger* than the overall population variance—then the psu means will be more similar. If there is little variation within the systematic samples relative to that in the population (that is, the ICC is large), then the elements in the sample all give similar information, and systematic sampling would be expected to have higher variance than an SRS.

Since $n = 1$, however, we cannot calculate $V(\hat{\bar{y}}_{\text{sys}})$ using (5.6); we need to know something about the structure of the population to estimate the variance. Let's look at three population structures.

1. *The sampling frame is in random order.* Systematic sampling is likely to produce a sample that behaves like an SRS. In many situations, the ordering of the population is unrelated to the characteristics of interest, as when the list of persons in the sampling frame is ordered by a randomly assigned identification number. There is no reason to believe that the persons in a systematic sample will be more or less similar than a random sample of persons: We expect that ICC ≈ 0. In this situation, simple random

and systematic sampling will give similar results. We can use SRS results and formulas to estimate $V(\hat{\bar{y}}_{\text{sys}})$.

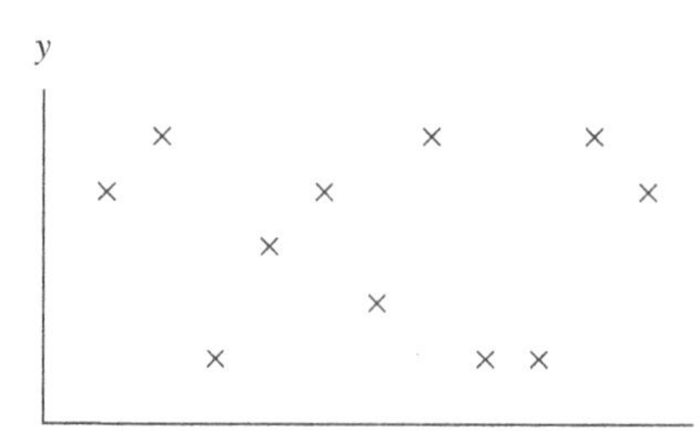

2. *The sampling frame is in increasing or decreasing order.* Systematic sampling is likely to be more precise than simple random sampling. Financial records may be listed with the largest amounts first and the smallest amounts last. In this case, $V(\hat{\bar{y}}_{\text{sys}})$ is less than the variance of the sample mean in an SRS of the same size since ICC < 0. A systematic sample forces the sample values to be spread out; it is possible that an SRS would consist of all low values or all high values. When the frame is in increasing or decreasing order, you may use the SRS formula for standard error, but it will likely be an overestimate and CIs constructed using the SRS standard error will be too wide.

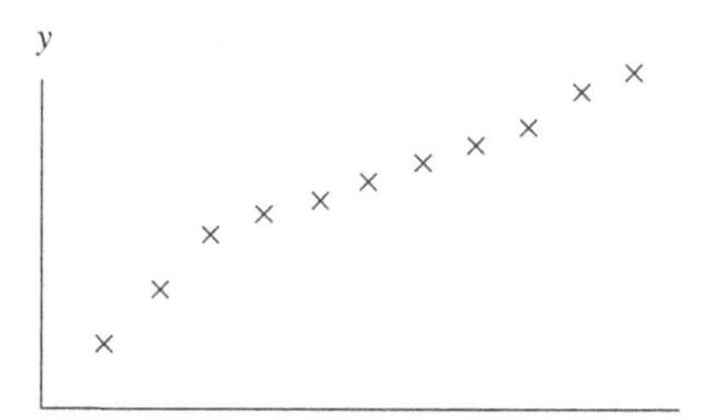

3. *The sampling frame has a periodic pattern.* If we sample at the same interval as the periodicity, systematic sampling will be less precise than simple random sampling. Systematic sampling is most dangerous when the population is in a cyclical or periodic order, and the sampling interval coincides with a multiple of the period.

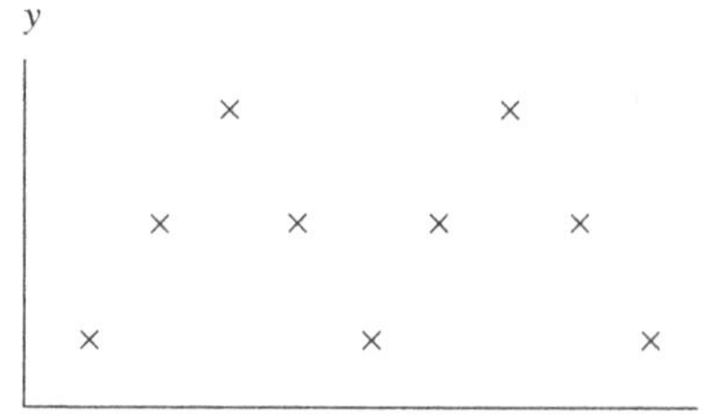

Suppose the population values (in sampling frame order) are

$$1 \quad 2 \quad 3 \quad 1 \quad 2 \quad 3 \quad 1 \quad 2 \quad 3 \quad 1 \quad 2 \quad 3$$

and the sampling interval is 3. Then all elements in the systematic sample will be the same; if we use the SRS formula to estimate the variance, we will have $\hat{V}(\hat{\bar{y}}_{\text{sys}}) = 0$. But the true value of $V(\hat{\bar{y}}_{\text{sys}})$ for this population is $2/3 = S^2$; this sample is no more precise than a single observation chosen randomly from the population.

Systematic sampling may be used when a researcher wants a representative sample of the population but does not have the resources to construct a sampling frame in advance. It is commonly used to select elements at the last stage of a cluster sample (see Exercise 28 of Chapter 7).

(a) Toxic waste is distributed randomly.

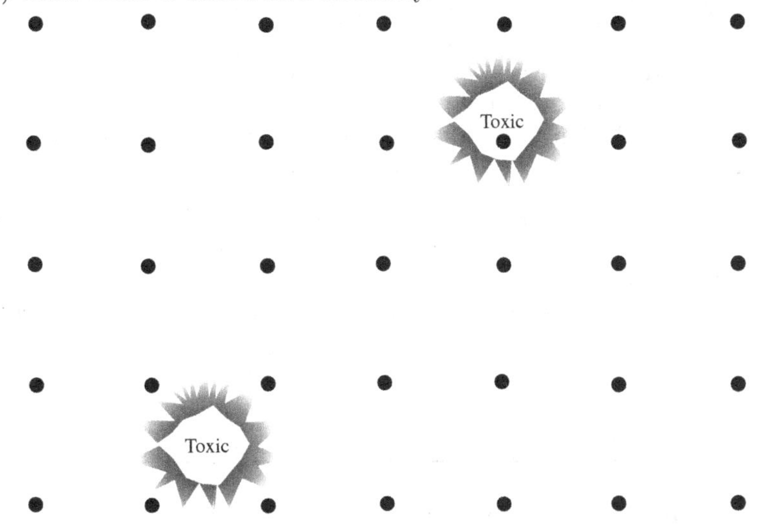

(b) Waste occurs in a similar pattern to the grid, so the systematic sample misses every deposit of toxic waste.

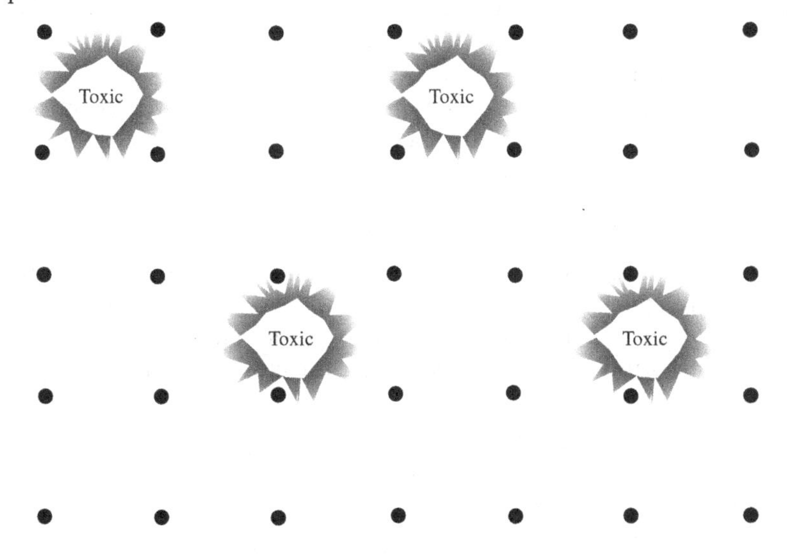

FIGURE 5.7
Using systematic samples to detect hazardous waste.

Example 5.13. *Sampling for hazardous waste sites.* Many dumps and landfills contain toxic materials. These materials may have been sealed in containers when deposited, but may now be suspected of leaking. But we no longer know where the materials were deposited—containers of hazardous waste may be randomly distributed throughout the landfill, or they may be concentrated in one area, or there may be none at all.

A common practice is to take a systematic sample of grid points and to take soil samples from each to look for evidence of contamination. Choose a point at random in the area, then construct a grid containing that point so that grid points are an equal distance apart. One such grid is shown in Figure 5.7(a). The advantages of taking a systematic sample rather than an SRS are that the systematic sample forces an even coverage of the region and is easier to implement in the field. If you are not worried about periodic patterns in the distribution of toxic materials, and you have little prior knowledge of where the toxic materials might be, a systematic sample is a good design.

With any grid in systematic sampling, you need to worry if the toxic materials are regularly placed so that the grid may miss all of them, as shown in Figure 5.7(b). If this is a concern, you would be better off taking a stratified sample. Lay out the grid, but select a point at random in each square at which to take the soil sample. ■

If periodicity is a concern in a population, one solution is to use **interpenetrating systematic samples** (Mahalanobis, 1946). Instead of taking one systematic sample, take several systematic samples from the population. Then you can use the formulas for cluster samples to estimate variances; each systematic sample acts as one psu. This approach is explored in Exercise 24.

5.6 Model-Based Theory for Cluster Sampling*

In most cluster samples, observations within the same cluster are more similar than observations selected randomly from the population as a whole. A sample of n clusters with m observations observed per cluster gives less information than an SRS of nm observations because the observations within a cluster are dependent. Any model for cluster sampling must include this dependence explicitly.

The one-way ANOVA model with fixed effects provides a theoretical framework for stratified sampling; an analogous model for cluster sampling is the one-way ANOVA model with random effects (Scott and Smith, 1969). Let's look at a simple version of this model:

$$\text{M1: } Y_{ij} = \mu + A_i + \varepsilon_{ij} \tag{5.38}$$

with A_i generated by a distribution with mean 0 and variance σ_A^2, ε_{ij} generated by a distribution with mean 0 and variance σ^2, and all A_is and ε_{ij}s independent.

A random effects model such as M1 allows observations in the same cluster to be positively correlated by specifying a probability distribution for the cluster means. With Model M1,

$$\text{Cov}_{\text{M1}}(Y_{ij}, Y_{kl}) = \begin{cases} \sigma^2 + \sigma_A^2 & \text{if } i = k \text{ and } j = l \\ \sigma_A^2 & \text{if } i = k \text{ and } j \neq l \\ 0 & \text{if } i \neq k \end{cases}.$$

The model-based intraclass correlation coefficient equals

$$\rho = \frac{\sigma_A^2}{\sigma_A^2 + \sigma^2}. \tag{5.39}$$

Note that ρ in (5.39) is always nonnegative, in contrast to the design-based ICC, which can take on negative values.

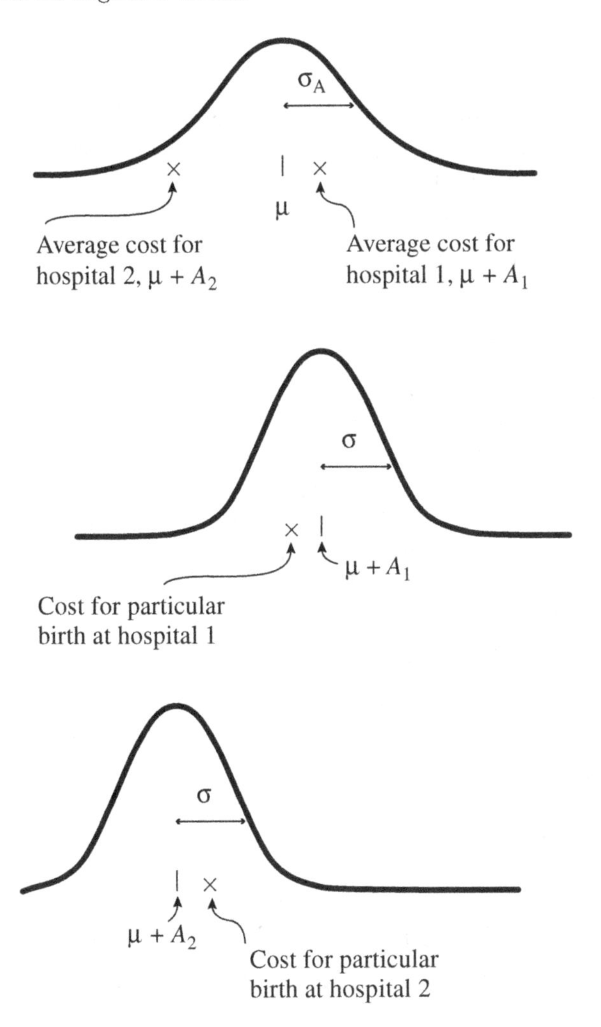

FIGURE 5.8
Illustration of random effects for hospitals and births.

Figure 5.8 illustrates Model M1, assuming that all random variables are normally distributed, for a two-stage cluster sample taken to estimate the total amount of hospital charges in a country for delivering babies. Hospitals are sampled at the first stage, and birth records from the selected hospitals are sampled at the second stage (twins and triplets count as one record). The average charge per birth varies from hospital to hospital—some hospitals may have higher personnel costs, and others may serve a higher-risk population or have more expensive equipment. That variation is reflected in model M1 by the random variables A_i: $\mu + A_i$ represents the average cost per birth in the ith hospital, and σ_A^2 is the population variance among the hospital means. In addition, costs vary from birth to birth within the hospitals; that variation is incorporated into the model by the term ε_{ij} with variance σ^2.

Costs for births in the same hospital tend to be more similar than costs for births selected randomly across the entire population of hospital births, because the cost for a birth in a given hospital incorporates the hospital-specific characteristics such as personnel costs. This similarity induces a positive correlation among observations in the same cluster.

5.6.1 Estimation Using Models

Now let's find properties of various estimators under Model M1. As in Sections 2.10 and 3.6, we want to predict the unsampled population members in the random variable representing the finite population total, which for cluster sampling is

$$T = \sum_{i=1}^{N} \sum_{j=1}^{M_i} Y_{ij}.$$

All of the estimators we consider have the form

$$\hat{T} = \sum_{i \in \mathcal{S}} \sum_{j \in \mathcal{S}_i} b_{ij} Y_{ij}, \tag{5.40}$$

where the b_{ij}s are a set of constants, so let's derive the model-based bias and variance for this general estimator and then substitute the values of b_{ij} for specific estimators later.

Inference in a model-based approach is conditional on the units selected to be in the sample; that is, the inference treats $\mathcal{S}$ and $\mathcal{S}_i$ as fixed, and treats Y_{ij} as a random variable. The bias of the general estimator $\hat{T}$ under model M1 is

$$E_{\text{M1}}[\hat{T} - T] = E_{\text{M1}} \left[\sum_{i \in \mathcal{S}} \sum_{j \in \mathcal{S}_i} b_{ij} Y_{ij} - \sum_{i=1}^{N} \sum_{j=1}^{M_i} Y_{ij} \right] = \mu \left(\sum_{i \in \mathcal{S}} \sum_{j \in \mathcal{S}_i} b_{ij} - M_0 \right). \tag{5.41}$$

Thus, $\hat{T}$ is model-M1-unbiased when $\sum_{i \in \mathcal{S}} \sum_{j \in \mathcal{S}_i} b_{ij} = M_0$. The model-based (for model M1) variance of $\hat{T} - T$ is (see Exercise 34):

$$V_{\text{M1}}[\hat{T} - T] = \sigma_A^2 \left[\sum_{i \in \mathcal{S}} \left(\sum_{j \in \mathcal{S}_i} b_{ij} - M_i \right)^2 + \sum_{i \notin \mathcal{S}} M_i^2 \right] + \sigma^2 \left[\sum_{i \in \mathcal{S}} \sum_{j \in \mathcal{S}_i} (b_{ij}^2 - 2b_{ij}) + M_0 \right]. \tag{5.42}$$

Properties of design-based estimators under Model M1. Now let's look at what happens with the design-based estimators we studied in Section 5.3 under Model M1. The random variable corresponding to the design-unbiased estimator $\hat{t}_{\text{unb}}$ in (5.21) is

$$\hat{T}_{\text{unb}} = \sum_{i \in \mathcal{S}} \sum_{j \in \mathcal{S}_i} \frac{NM_i}{nm_i} Y_{ij};$$

the coefficients b_{ij} are the sampling weights $(NM_i)/(nm_i)$. Using (5.41), the model-based bias of $\hat{T}_{\text{unb}}$ is

$$E_{\text{M1}}[\hat{T}_{\text{unb}} - T] = \mu \left(\sum_{i \in \mathcal{S}} \sum_{j \in \mathcal{S}_i} \frac{NM_i}{nm_i} - M_0 \right) = \mu \left(\frac{N}{n} \sum_{i \in \mathcal{S}} M_i - M_0 \right).$$

The bias depends on which sample is taken, and the estimator is model-unbiased under (5.38) only when the average of the M_is in the sample equals the average of the M_is in the population, such as will occur when all M_is are the same.

The ratio estimator corresponding to $M_0\hat{\bar{y}}_r$, for $\hat{\bar{y}}_r$ in (5.16), is

$$\hat{T}_r = \sum_{i\in\mathcal{S}}\sum_{j\in\mathcal{S}_i}\left(M_0\frac{M_i}{m_i\sum_{k\in\mathcal{S}}M_k}\right)Y_{ij}.$$

Since $\sum_{i\in\mathcal{S}}\sum_{j\in\mathcal{S}_i}(M_0M_i)/(m_i\sum_{k\in\mathcal{S}}M_k) = M_0$, $\hat{T}_r$ is model-unbiased under Model M1. If the sample consists of the population units with the largest values of M_i, the ratio estimator compensates by dividing by the sum of the M_i for the sample, which will also be large.

Best linear unbiased predictor under Model M1. If Model M1 really does describe the population and the M_is are unequal, one can find an estimator with smaller variance than both $\hat{T}_{\text{unb}}$ and $\hat{T}_r$.

In a model-based perspective, it is desired to predict the values of Y_{ij} for the unobserved population members. Exercises 35 and 36 show that among all model-unbiased predictors having the form of (5.40), $\hat{T}_{\text{BLUP}}$ in (5.43) has the smallest variance:

$$\begin{aligned}\hat{T}_{\text{BLUP}} &= \sum_{i\in\mathcal{S}}\left(\sum_{j\in\mathcal{S}_i}Y_{ij}+\sum_{j\notin\mathcal{S}_i}\hat{Y}_{ij}\right)+\sum_{i\notin\mathcal{S}}\sum_{j=1}^{M_i}\hat{Y}_{ij}\\ &= \sum_{i\in\mathcal{S}}\sum_{j\in\mathcal{S}_i}Y_{ij}+\sum_{i\in\mathcal{S}}\sum_{j\notin\mathcal{S}_i}\left[\rho\alpha_i\frac{1}{m_i}\sum_{l\in\mathcal{S}_i}Y_{il}+(1-\rho\alpha_i)\hat{\mu}\right]+\sum_{i\notin\mathcal{S}}M_i\hat{\mu},\end{aligned} \tag{5.43}$$

where $\alpha_i = m_i/[1+\rho(m_i-1)]$ and

$$\hat{\mu} = \left(\sum_{i\in\mathcal{S}}\alpha_i\sum_{j\in\mathcal{S}_i}\frac{Y_{ij}}{m_i}\right)\Bigg/\left(\sum_{i\in\mathcal{S}}\alpha_i\right). \tag{5.44}$$

Typically, ρ is unknown, and an estimator $\hat{\rho}$ is substituted into (5.43) and (5.44).

The unknown mean μ is estimated by a weighted average of the sample psu means, with α_i proportional to $1/V_{\text{M1}}(\sum_{j\in\mathcal{S}_i}Y_{ij}/m_i)$. Thus, the estimated mean $\hat{\mu}$ relies more heavily on the sample means from psus with larger sample sizes m_i, which have smaller variance under model M1. If psu i is not in the sample, then $\hat{Y}_{ij} = \hat{\mu}$. If psu i is in the sample but ssu j is not, $\hat{Y}_{ij}$ is predicted as a value somewhere between the sample mean of psu i and the overall mean $\hat{\mu}$.

If n/N is small, $\hat{T}_{\text{BLUP}} \approx M_0\hat{\mu}$. Under model M1, after all, every population unit has expected value μ, so if the number of psus in the sample is small relative to the number of psus in the population, it makes sense that the population mean would be approximately estimated by $\hat{\mu}$ and the total by the population size times $\hat{\mu}$.

If $M_i = M$ and $m_i = m$ for all i, then $\hat{\mu} = \sum_{i\in\mathcal{S}}\sum_{j\in\mathcal{S}_i}Y_{ij}/(nm)$, $\hat{T}_{\text{unb}} = \hat{T}_r = \hat{T}_{\text{BLUP}} = M_0\hat{\mu}$, and the variance in (5.42) simplifies to

$$V_{\text{M1}}[\hat{T}_{\text{unb}}-T] = M_0^2\left(1-\frac{n}{N}\right)\frac{\sigma_A^2}{n}+M_0^2\left(1-\frac{nm}{NM}\right)\frac{\sigma^2}{mn}. \tag{5.45}$$

Model assumptions. The assumptions for model-based inference are strong. For Model M1, we assume:

1. The model form is correct, that is, $Y_{ij} = \mu+A_i+\varepsilon_{ij}$ for every unit (i,j) in the population, whether sampled or not. One consequence is that the expected value of the total in psu i, $E\left[\sum_{j=1}^{M_i}Y_{ij}\right]$, is assumed to equal $M_i\mu$.

2. The proposed variance and correlation structures are correct, that is, $V(Y_{ij}) = \sigma_A^2 + \sigma^2$, $\text{Corr}(Y_{ij}, Y_{il}) = \rho$ for units $j \neq l$ in the same cluster, and $\text{Corr}(Y_{ij}, Y_{kl}) = 0$ for units in different clusters.

In a model-based approach, it does not matter which population units are in the sample because all population units are assumed to follow the model. For that reason, data from a nonprobability sample must be analyzed using models, as discussed in Chapter 15.

But what happens if the model is inappropriate for the population? We can (and should!) perform model diagnostics (see, for example, Demidenko, 2013; Loy et al., 2017) to examine how well the model fits the units in the sample, but unless additional information is available about the units not sampled, we cannot assess whether the model is appropriate for unobserved units. A model-based analyst must assume that the model fits the population units that are not observed. If psus and ssus are selected using simple random sampling, then it is reasonable to assume that a model that fits the observed units also fits the unobserved units since there is nothing "special" about the units that end up in the sample.

With a convenience sample, however, the psus in the sample may differ systematically from those not in the sample—for example, the sampled hospitals in Figure 5.8 might be larger or more urban than nonsampled hospitals—and hence predictions of the unobserved records in unsampled hospitals may be biased. The model-based variance, estimated from the data in the sample, does not capture that bias. Thus, if the model does not fit the unsampled units, the model-based variance can severely underestimate the mean squared error and give the impression that estimates are more accurate than they really are.

That said, Model M1 is a reasonable approximation for many real-life situations in which there is clustering, and it is widely used in practice for nonprobability samples of clusters. Available auxiliary information can be included in the model to improve prediction, as in Section 4.6. Of course, many other models have been proposed that assume units are dependent; see the For Further Reading section for references.

You can use various estimators in conjunction with a model-based approach. Thus, you can estimate the population total by $\hat{T}_{\text{unb}}$, $\hat{T}_r$, $\hat{T}_{\text{BLUP}}$, or another estimator. But the statistical properties of the estimator come from the assumed model. In practice, if estimating multiple quantities or statistics for domains, it may be desirable to use an estimator in which the constants b_{ij} are the same for every y variable, even if the variance is higher than that of $\hat{T}_{\text{BLUP}}$. The constants used in (5.43), which are given explicitly in Exercise 36, can differ for each response variable when the m_i vary from psu to psu. Thus, if you use $\hat{T}_{\text{BLUP}}$ to estimate the total hospital charges for California, the total hospital charges for Oregon, and the total hospital charges for California and Oregon together, the combined California–Oregon total might not equal the sum of the individual totals for the two states. If the constants b_{ij} are the same for every response variable, this problem will not occur.

Example 5.14. Let's fit Model M1 to the schools data from Example 5.7. Looking at Figure 5.3, it seems plausible that the within-school variances σ_i^2 are the same for all schools. Using statistical software for mixed models, we obtain estimated variance components $\hat{\sigma}_A^2 = 33.85$ and $\hat{\sigma}^2 = 102.83$. This results in $\hat{\rho} = 33.85/(33.85 + 102.83) = 0.25$, approximately the same as the value of R_a^2 in Example 5.8. Most software packages for mixed models estimate μ using (5.44). For the schools data, $m_i = 20$ students are sampled from each psu, so $\hat{\mu} = 34.66$ is the average of the individual school means in Table 5.7. The model-based standard error of $\hat{\mu}$ is 1.97. Note that $\hat{\mu}$ is slightly larger than $\hat{\bar{y}}_r = 33.12$ from Example 5.7. The model-based standard error, however, differs from the design-based standard error of 1.76; the model-based standard error estimates variability under M1, while the design-based standard error estimates variability under repeated sampling with the sampling design. ■

5.6.2 Design Using Models

Models are extremely useful for designing a cluster sample. Using a model for design does not mean you have to use a model for analysis of your survey data after it is collected; rather, the model provides a useful way of summarizing information you can use to make the survey design more efficient.

Suppose that Model M1 seems reasonable for your population and that all psu sizes in the population are equal. Then you would like to design the survey to minimize the variance in (5.45), subject to cost constraints. Using the cost function in (5.35), the model-based variance is minimized when

$$m = \sqrt{\frac{c_1 \sigma^2}{c_2 \sigma_A^2}}.$$

Suppose that the M_is are unequal and that Model M1 holds. We can use the variance in (5.42) to determine the optimal subsampling size m_i for each cluster. This approach was used by Royall (1976) for more general models than considered in this section. For $\hat{T}_r$, $b_{ij} = M_i / (m_i \sum_{k \in \mathcal{S}} M_k)$, and the variance is minimized when m_i is proportional to M_i (see Exercise 39).

5.7 Chapter Summary

Cluster sampling is commonly used in large surveys, but estimates obtained from cluster samples usually have greater variance than if we were able to measure the same number of observation units using an SRS. If it is much less expensive to sample clusters than individual elements, though, cluster sampling can provide more precision per dollar spent.

All of the formulas in this chapter for cluster sampling with equal probabilities are special cases of the general results for two-stage cluster sampling with unequal psu sizes, to be derived in Chapter 6. They can be applied to any two-stage cluster sample in which the psus are selected with equal probability. These formulas were given in (5.21), (5.27), (5.30), and (5.32) and are repeated here:

$$\hat{t}_{\text{unb}} = \frac{N}{n} \sum_{i \in \mathcal{S}} \hat{t}_i = \frac{N}{n} \sum_{i \in \mathcal{S}} M_i \bar{y}_i, \tag{5.46}$$

$$\hat{V}(\hat{t}_{\text{unb}}) = N^2 \left(1 - \frac{n}{N}\right) \frac{s_t^2}{n} + \frac{N}{n} \sum_{i \in \mathcal{S}} \left(1 - \frac{m_i}{M_i}\right) M_i^2 \frac{s_i^2}{m_i}, \tag{5.47}$$

$$\hat{\bar{y}}_r = \frac{\displaystyle\sum_{i \in \mathcal{S}} M_i \bar{y}_i}{\displaystyle\sum_{i \in \mathcal{S}} M_i}, \tag{5.48}$$

$$\hat{V}(\hat{\bar{y}}_r) = \frac{1}{\overline{M}^2} \left[\left(1 - \frac{n}{N}\right) \frac{s_r^2}{n} + \frac{1}{nN} \sum_{i \in \mathcal{S}} \left(1 - \frac{m_i}{M_i}\right) M_i^2 \frac{s_i^2}{m_i} \right], \tag{5.49}$$

with $\overline{M} = \sum_{i \in \mathcal{S}} M_i / n$,

$$s_t^2 = \frac{1}{n-1} \sum_{i \in \mathcal{S}} \left(\hat{t}_i - \frac{\hat{t}_{\text{unb}}}{N} \right)^2$$

and

$$s_r^2 = \frac{1}{n-1}\sum_{i\in\mathcal{S}} M_i^2(\bar{y}_i - \hat{\bar{y}}_r)^2.$$

In one-stage cluster sampling, the second term in (5.47) and (5.49) is zero since $m_i = M_i$. The variance estimators depend mostly on the variability between psus.

Point estimates of the population mean and total are usually calculated using weights. If an SRS of n of the N population psus is chosen, and an SRS of m_i of the M_i ssus in psu i is taken, then the sampling weight for observation j of psu i is:

$$w_{ij} = \frac{NM_i}{nm_i}.$$

Then

$$\hat{t}_{\text{unb}} = \sum_{i\in\mathcal{S}}\sum_{j\in\mathcal{S}_i} w_{ij}y_{ij}$$

and

$$\hat{\bar{y}}_r = \frac{\displaystyle\sum_{i\in\mathcal{S}}\sum_{j\in\mathcal{S}_i} w_{ij}y_{ij}}{\displaystyle\sum_{i\in\mathcal{S}}\sum_{j\in\mathcal{S}_i} w_{ij}}.$$

While weights can be used to find estimated means and totals, they do not provide sufficient information to estimate the variance in a cluster sample. You need to use the formulas in (5.47) and (5.49), or a method such as jackknife from Chapter 9, to calculate standard errors.

Key Terms

Cluster: A group of observation units that serve as a sampling unit.

Cluster sampling: A probability sampling design in which observations are grouped into clusters, and a probability sample of clusters is selected from the population of clusters.

Intraclass correlation coefficient (ICC): The Pearson correlation coefficient of all pairs of units within the same cluster.

One-stage cluster sampling: A cluster sampling design in which all ssus in selected psus are included in the sample.

Primary sampling unit (psu): The unit that is sampled from the population.

Secondary sampling unit (ssu): A subunit that is subsampled from a psu that has been selected to be in the sample.

Two-stage cluster sampling: A cluster sampling design in which the ssus in selected psus are subsampled.

For Further Reading

Stuart (1984) gives a great deal of intuition into cluster sampling with clear illustrations and examples. The books by Hansen et al. (1953b), Raj (1968), Cochran (1977), and Brewer (2002) thoroughly cover the theory of estimation in cluster samples. Skinner et al. (1989b) and Binder and Roberts (2003) delineate the issues involved in different approaches to inference in cluster samples.

Many authors have written about the problem of ignoring the clustering when analyzing data from a cluster sample. National Research Council (2004), Hurlbert (2009), Nelson (2014), and Dell-Kuster et al. (2018) are a few of the many books and articles that warn of this error and provide examples of research articles that have made it. Brogan (2015) illustrates how standard errors are underestimated when clustering is ignored.

Clustering may have an especially strong effect on standard errors for estimates of proportions, particularly when the population proportion is close to zero or one. Dean and Pagano (2015) review and evaluate alternative methods for constructing confidence intervals for proportions from cluster samples, including a modification of the Clopper-Pearson CI discussed in Exercise 34 of Chapter 2.

The classic paper by Mahalanobis (1946) gives insight into many issues in survey sampling. Among other concepts, Mahalanobis developed the technique of interpenetrating subsampling, in which the sample is drawn as two smaller, independent subsamples. We mentioned this technique briefly for estimating the variance of systematic samples. Ultimately, Mahalanobis' idea led to the replication methods (discussed in Sections 9.2 and 9.3) now commonly used for variance estimation in complex surveys.

There is a wealth of material on using models to analyze data from cluster samples. Scott and Smith (1969) derive estimators from a Bayesian perspective. Royall (1976) derives results for a general class of possible models, allowing unequal variances for different clusters, and Chapter 8 of Valliant et al. (2000) describes other correlation structures for clustered populations. Chambers and Clark (2012) describe maximum likelihood approaches to clustered data in which covariates are included to improve the predictions. Many excellent books tell how to use random effects models with survey and experimental data. These include Gelman and Hill (2007), Snijders and Bosker (2011), Goldstein (2011), and Demidenko (2013). Chapter 5 of Rao and Molina (2015) has a clear and concise description of the theory of mixed models, along with a description of methods for assessing model fit.

Much research has been done on using models for design: see Rao (1979b), Bellhouse (1984), Royall (1992), and Gonzalez and Eltinge (2010) for literature reviews.

5.8 Exercises

A. Introductory Exercises

1. A city council of a small city wants to know the proportion of eligible voters that oppose having an incinerator of Phoenix garbage opened just outside of the city limits. They randomly select 100 residential numbers from the city's telephone book that contains 3,000 such numbers. Each selected residence is then called and asked for (a) the total number of eligible voters and (b) the number of voters opposed to the incinerator. A total of 157 voters were surveyed; of these, 23 refused to answer the question. Of the remaining

134 voters, 112 opposed the incinerator, so the council estimates the proportion by

$$\hat{p} = 112/134 = 0.83582$$

with

$$\hat{V}(\hat{p}) = 0.83582(1 - 0.83582)/134 = 0.00102$$

Are these estimates valid? Why, or why not?

2. Senturia et al. (1994) described a survey taken to study how many children have access to guns in their households. Questionnaires were distributed to all parents who attended selected clinics in the Chicago area during a one-week period for well or sick child visits.
   (a) Suppose that the quantity of interest is the percentage of households containing children that own at least one gun. Describe why this is a cluster sample. What is the psu? The ssu? Is it a one-stage or two-stage cluster sample?
   (b) What is the sampled population for this study? Do you think this sampling procedure results in a representative sample of households with children? Why, or why not?

3. Kleppel et al. (2004) reported on a study of wetlands in upstate New York. Four wetlands were selected for the study: Two of the wetlands drain watersheds from small towns and the other two drain suburban watersheds. Quantities such as pH were measured at two to four randomly selected sites within each of the four wetlands.
   (a) Describe why this is a cluster sample. What are the psus? The ssus? How would you estimate the average pH in the suburban wetlands?
   (b) The authors used Student's two-sample t test to compare the average pH from the sites in the suburban wetlands with the average pH from the sites in the small town wetlands, treating all sites as independent. Is this analysis appropriate? Why, or why not?

4. Survey evidence is often introduced in court cases involving trademark violations and employment discrimination. There has been controversy, however, about whether nonprobability samples are acceptable as evidence in litigation. Jacoby and Handlin (1991) selected 26 from a list of 1,285 scholarly journals in the social and behavioral sciences. They examined all articles published during 1988 for the selected journals and recorded (1) the number of articles in the journal that described empirical research from a survey (they excluded articles in which the authors analyzed survey data which had been collected by someone else) and (2) the total number of articles for each journal which used probability sampling, nonprobability sampling, or for which the sampling method could not be determined. The data are in file `journal.csv`.
   (a) Explain why this is a cluster sample.
   (b) Estimate the proportion of articles in the 1,285 journals that use nonprobability sampling, and give the standard error of your estimate.
   (c) The authors concluded that, because "an overwhelming proportion of ... recognized scholarly and practitioner experts rely on non-probability sampling designs," courts "should have no problem admitting otherwise well-conducted non-probability surveys and according them due weight" (p. 175). Comment on this statement.

5. A language school owner takes an SRS of 10 of the 72 Introductory Spanish classes offered by the school. Each student in each of the sampled classes is given a vocabulary test and is also asked whether he or she is planning a trip to a Spanish-speaking country in the next year. The data are in file `spanish.csv`.

   (a) Estimate the total number of students planning a trip to a Spanish-speaking country in the next year, and give a 95% CI. Use the unbiased estimator in (5.14) and (5.19).

   (b) Estimate the mean vocabulary test score for Introductory Spanish students in the language school, and give a 95% CI. Use the ratio estimator in (5.16) and (5.20).

6. An inspector samples cans from a truckload of canned creamed corn to estimate the total number of worm fragments in the truckload. The truck has 580 cases; each case contains 24 cans. The inspector samples 12 cases at random, and subsamples 3 cans randomly from each selected case. The following table gives the number of worm fragments found in the sampled cans.

	Case											
	1	2	3	4	5	6	7	8	9	10	11	12
Can 1	1	4	0	3	4	0	5	3	7	3	4	0
Can 2	5	2	1	6	9	7	5	0	3	1	7	0
Can 3	7	4	2	6	8	3	1	2	5	4	9	0

   Using (5.23) and (5.27), estimate the total number of worm fragments, along with a 95% CI. Compare the estimated value of the variance from (5.27) with the approximation given in (5.29).

7. A market research firm took an SRS of 6 cities from the 45 cities in a region, and then took an SRS of m_i supermarkets from the M_i supermarkets in sampled city i. The following table gives the number of cases of the new candy Green Globules sold during March in each of the sampled supermarkets.

City	M_i	m_i	**Number of Cases Sold**
1	52	10	146, 180, 251, 152, 72, 181, 171, 361, 73, 186
2	19	4	99, 101, 52, 121
3	37	7	199, 179, 98, 63, 126, 87, 62
4	39	8	226, 129, 57, 46, 86, 43, 85, 165
5	8	2	12, 23
6	14	3	87, 43, 59

   Calculate the sampling weight of each sampled supermarket. Estimate the total number of cases sold in the region, and the average number sold per supermarket, along with standard errors.

8. A homeowner with a large library needs to estimate the purchase cost and replacement value of the book collection for insurance purposes. She has 44 shelves containing books and selects 12 shelves at random. To prepare for the second stage of sampling, she counts the number of books M_i, on the selected shelves. She then generates five random numbers between 1 and M_i for each selected shelf, to determine which specific books, numbered from left to right, to examine more closely. She looks up the replacement value for each sampled book. The data are given in the file `books.csv`.

(a) Draw side-by-side boxplots for the replacement costs of books on each shelf. Does it appear that the means are about the same? The variances?

(b) Estimate the total replacement cost for the library, and find the standard error of your estimate. What is the estimated coefficient of variation?

(c) Estimate the average replacement cost per book, along with the standard error. What is the estimated coefficient of variation?

9. Repeat Exercise 8 for the purchase cost for each book. Plot the data, and estimate the total and average amount she has spent on books, along with the standard errors.

10. Construct a sample ANOVA table for the replacement cost data in Exercise 8. What is your estimate for R_a^2? Do books on the same shelf tend to have more similar replacement costs? Suppose that $c_1 = 10$ and $c_2 = 4$. If all shelves had 30 books, how many books should be sampled per shelf?

B. Working with Survey Data

11. An accounting firm is interested in estimating the error rate in a compliance audit it is conducting. The population contains 828 claims, and the firm audits an SRS of 85 of those claims. In each of the 85 sampled claims, 215 fields are checked for errors. One claim has errors in 4 of the 215 fields, 1 claim has 3 errors, 4 claims have 2 errors, 22 claims have 1 error, and the remaining 57 claims have no errors. (Data courtesy of Fritz Scheuren.)

(a) Treating the claims as psus and the observations for each field as ssus, estimate the error rate, defined to be the average number of errors per field, along with the standard error for your estimate.

(b) Estimate (with standard error) the total number of errors in the 828 claims.

(c) Suppose that instead of taking a cluster sample, the firm had taken an SRS of $85 \times 215 = 18{,}275$ fields from the 178,020 fields in the population. If the estimated error rate from the SRS had been the same as in (a), what would the estimated variance $\hat{V}(\hat{p}_{\text{SRS}})$ be? How does this compare with the estimated variance from (a)?

12. Use the data in `coots.csv` to estimate the average egg length, along with its standard error. Be sure to plot the data appropriately.

13. A state program provides medical assistance to low-income households in the state. Each county determines whether households are eligible for assistance. Sometimes, however, households are certified to be eligible when they are actually not eligible. The certification error rate for a county is the number of persons who are erroneously certified to receive assistance divided by the total number of persons receiving assistance. Quality control audits are done by sampling household records; once a household record is selected and audited, it costs the same amount to evaluate one person in the household as to evaluate all persons in the household.

(a) Explain how to use cluster sampling to estimate the certification error rate for a county. Should one-stage or two-stage sampling be used?

(b) Suppose that a county certified 1,572 households to be eligible for medical assistance. In past years, the certification error rate per household has been about 10%. How many households should be included in your sample so that the margin of error for estimating the per-person certification error rate is less than 0.03? What assumptions did you make about the ICC to arrive at your sample size?

14. A researcher took an SRS of 4 high schools from a region with 29 high schools for a study on the prevalence of smoking among female high school students in the region. The results were as follows:

School	Number of Students	Female Students in School	Female Students Interviewed	Number of Smokers
1	1471	792	25	10
2	890	447	15	3
3	1021	511	20	6
4	1587	800	40	27

   (a) Estimate the percentage of female high school students in the region who smoke, along with a 95% CI.

   (b) Estimate the total number of female high school students in the region who smoke, along with a 95% CI.

   (c) The researcher now wants to study the prevalence of smoking and other risk behaviors among female high school students in a different region with 35 high schools. She intends to drive to n of the schools and then interview some or all of the female students in the selected schools. Assuming that MSB and MSW are similar in the two regions, use information from the study of 4 schools to estimate R_a^2 and design a cluster sample for the new study. Suppose it takes about 50 hours per school to contact school officials, obtain permission, obtain a list of female students, and travel back and forth. Although interviews themselves are only about 10 minutes, it takes about 30 minutes per interview obtained to allow for additional scheduling of no-shows, obtaining parental permission, and other administrative tasks. The investigator would like to spend 300 hours or less on the data collection.

15. Gnap (1995) conducted a survey to estimate the teacher workload in Maricopa County, Arizona, public school districts. Her target population was all first- through sixth-grade full-time public school teachers with at least one year of experience. At the time of the survey, Maricopa County had 46 school districts with 311 elementary schools and 15,086 teachers. Gnap stratified the schools by size of school district; the large stratum, consisting of schools in districts with more than 5,000 students, is considered in this exercise. The stratum contained 245 schools; 23 participated in the survey. All teachers in the selected schools were asked to fill out the questionnaire. Due to nonresponse, however, some questionnaires were not returned. (We shall examine possible effects of nonresponse in Exercise 13 of Chapter 8.) The data are in file `teachers.csv`, with psu information in `teachmi.csv`.

   (a) Why would a cluster sample be a better design than an SRS for this study? Consider issues such as cost, ease of collecting data, and confidentiality of respondent. What are some disadvantages of using a cluster sample?

   (b) Calculate the mean and standard deviation of *hrwork* for each school in the "large" stratum. Construct a graph of the means for each school and a separate graph of the standard deviations. Does there seem to be more variation within a school, or does more of the variability occur between different schools? How did you deal with missing values (coded as −9)?

   (c) Construct a scatterplot of the standard deviations versus the means for the schools, for the variable *hrwork*. Is there more variability in schools with higher workloads? Less? No apparent relation?

(d) Estimate the average of *hrwork* in the large stratum in Maricopa County, along with its standard error. Use *popteach* in `teachmi.csv` for the M_is.

16. The file `measles.csv` contains data consistent with that obtained in a survey of parents whose children had not been immunized for measles during a recent campaign to immunize all children between the ages of 11 and 15. During the campaign, 7,633 children from the 46 schools in the area were immunized; 9,962 children whose records showed no previous immunization were not immunized. In a follow-up survey to explore why the children had not been immunized during the campaign, Roberts et al. (1995) sent questionnaires to the parents of a cluster sample of the 9,962 children. Ten schools were randomly selected, then a sample of the M_i nonimmunized children from each school was selected, and the parents of those children were sent a questionnaire. Not all parents responded to the questionnaire; you will examine the effects of nonresponse in Exercise 14 of Chapter 8.

    (a) Estimate, separately for each school, the percentage of parents who returned a consent form (variable *returnf*). For this exercise, treat the "no answer" responses (value 9) as not returned.

    (b) Using the number of respondents in school i as m_i, construct the sampling weight for each observation.

    (c) Estimate the overall percentage of parents who received a consent form along with a 95% CI.

    (d) How do your estimate and interval in part (c) compare with the results you would have obtained if you had ignored the clustering and analyzed the data as an SRS? Find the ratio:

    $$\frac{\text{estimated variance from (c)}}{\text{estimated variance if the data were analyzed as an SRS}}.$$

    What is the effect of clustering?

17. Repeat Exercise 16, for estimating the percentage of children who had previously had measles.

18. Use the data in `schools.csv` (see Example 5.7) for this exercise.

    (a) Estimate the mean reading score for all students in the 75 schools, along with its SE, using the with-replacement variance estimator in (5.33).

    (b) Estimate the standard error of the mean reading score using the without-replacement formula in (5.32). How much difference does the fpc make?

19. For the data in Exercise 18, estimate the total number of students who would be expected to have *readlevel* $= 2$, along with a 95% CI. Use the with-replacement variance estimator.

20. (Model-based.) For the data in Exercise 18, calculate the model-based mean reading score $\hat{\mu}$ in (5.44) and its standard error under Model M1.

21. Refer to Example 5.10. Later in the potato growing season, it takes more time to inspect stems. Suppose that it takes two minutes to inspect each stem. Which psu size is most efficient?

22. Use the SRS from the Census of Agriculture data in the file `agsrs.csv` (discussed in Example 2.6) for this exercise.

(a) Find the sample ANOVA table of *acres92*, using *state* as the cluster variable. Estimate R_a^2 from the sample. Is there a clustering effect?

(b) Suppose that $c_1 = 15c_2$, where c_1 is the cost to sample a state, and c_2 is the cost to sample a county within a state. What should $\bar{m}$ be, if it is desired to sample a total of 300 counties? How many states would be sampled (that is, what is n)?

23. Using the value of n determined in Exercise 22, draw a self-weighting cluster sample of 300 counties from `agpop.csv`, using *state* as the cluster variable. Plot the data using side-by-side boxplots. Estimate the total number of acres devoted to farms in the United States, along with the standard error, using both the unbiased estimate and the ratio estimate. How do these values compare with each other and with values from the SRS and stratified sample from Examples 2.6 and 3.2?

24. The file `ozone.csv` contains hourly ozone readings from a site in Monterey County, California, for 2018 and 2019.

(a) Construct a histogram of the population values, excluding the missing values. Find the mean, standard deviation, and median of the population.

(b) Take a systematic sample with period 24. To do this, select a random integer k between 0 and 23, and let the sample consist of the ozone readings in the column corresponding to that hour. Construct a histogram of the sample values.

(c) Now suppose you treated your systematic sample as though it was an SRS. Find the sample mean, standard deviation, and median. Construct an interval estimate of the population mean using the procedure in Section 2.5. Does your interval contain the true value of the population mean from (a)?

(d) Take four independent systematic samples, each with period 96. Now use formulas from cluster sampling to estimate the population mean, and construct a 95% CI for the mean.

C. Working with Theory

25. The ICC is defined as the Pearson correlation coefficient for the $NM(M-1)$ pairs (y_{ij}, y_{ik}) for i between 1 and N and $j \neq k$:

$$\text{ICC} = \frac{\sum_{i=1}^{N}\sum_{j=1}^{M}\sum_{k \neq j}^{M}(y_{ij} - \bar{y}_{\mathcal{U}})(y_{ik} - \bar{y}_{\mathcal{U}})}{(NM-1)(M-1)\,S^2}. \tag{5.50}$$

Show that the above definition is equivalent to (5.10). HINT: First show that

$$\sum_{i=1}^{N}\sum_{j=1}^{M}\sum_{k \neq j}^{M}(y_{ij} - \bar{y}_{\mathcal{U}})(y_{ik} - \bar{y}_{\mathcal{U}}) + \sum_{i=1}^{N}\sum_{j=1}^{M}(y_{ij} - \bar{y}_{\mathcal{U}})^2 = M(\text{SSB}).$$

26. For the quantities in the population ANOVA table (Table 5.3), show that

$$\text{MSW} = \frac{NM-1}{NM}S^2(1 - \text{ICC})$$

and

$$\text{MSB} = \frac{NM-1}{M(N-1)}S^2[1 + (M-1)\text{ICC}].$$

27. Suppose in a two-stage cluster sample that all population cluster sizes are equal ($M_i = M$ for all i) and that all sample sizes for the clusters are equal ($m_i = m$ for all i).

    (a) Show (5.34).

    (b) Show that MSW $= S^2(1 - R_a^2)$ and that

    $$\text{MSB} = S^2\left[\frac{N(M-1)R_a^2}{N-1} + 1\right].$$

    (c) Using (a) and (b), express $V(\hat{\bar{y}})$ as a function of n, m, N, M, and R_a^2.

    (d) Show that if S^2 and the sample and population sizes are fixed, and if $(m-1)/m > n/N$, then $V(\hat{\bar{y}})$ is an increasing function of R_a^2.

28. Suppose in a two-stage cluster sample that all population cluster sizes are equal ($M_i = M$ for all i) and that all sample sizes for the clusters are equal ($m_i = m$ for all i).

    (a) Show that $\hat{t}_{\text{unb}} = \hat{t}_r$, and, hence, that $\hat{\bar{y}}_{\text{unb}} = \hat{\bar{y}}_r$.

    (b) Fill in the formulas for the sums of squares in the ANOVA table below, for the sample data.

Source	df	Sum of Squares	Mean Square
Between psus	$n-1$		msb
Within psus	$n(m-1)$		msw
Total	$nm-1$		msto

    (c) Show that $E[\text{msw}] =$ MSW and

    $$E[\text{msb}] = \frac{m}{M}\text{MSB} + \left(1 - \frac{m}{M}\right)\text{MSW},$$

    where MSB and MSW are the between and within mean squares, respectively, from the *population* ANOVA table given in Table 5.3.

    (d) Show that

    $$\widehat{\text{MSB}} = \frac{M}{m}\text{msb} - \left(\frac{M}{m} - 1\right)\text{msw}$$

    is an unbiased estimator of MSB.

    (e) Show, using (5.27) or (5.32), that

    $$\hat{V}(\hat{\bar{y}}_{\text{unb}}) = \left(1 - \frac{n}{N}\right)\frac{\text{msb}}{nm} + \frac{1}{N}\left(1 - \frac{m}{M}\right)\frac{\text{msw}}{m}.$$

29. For the situation in Exercise 28, let msto represent the mean square total from the sample ANOVA table.

    (a) Write msto as a function of msb and msw, and use the results of Exercise 28(c) to find $E[\text{msto}]$.

    (b) Show that $E[\text{msto}] \approx S^2$ if n and N are large.

    (c) Show that

    $$\hat{S}^2 = \frac{M(N-1)}{m(NM-1)}\text{msb} + \frac{(m-1)NM + M - m}{m(NM-1)}\text{msw}$$

    is an unbiased estimator of S^2.

30. (Requires calculus.) Show that if $M_i = M$ and $m_i = m$ for all i, and if the cost function is $C = c_1 n + c_2 nm$, then

$$m_{\text{opt}} = \sqrt{\frac{c_1 M(\text{MSW})}{c_2(\text{MSB} - \text{MSW})}} = \sqrt{\frac{c_1 M(N-1)(1-R_a^2)}{c_2(NM-1)R_a^2}}$$

minimizes the variance for fixed total cost C. HINT: Show the result with MSW and MSB first, then use Exercise 27(b).

31. *Using the with-replacement variance in two-stage cluster sampling.* The variance estimator in (5.27) requires calculating variances at both stages of sampling. Two alternative variance estimators are sometimes recommended for use in practice:

$$\hat{V}_{\text{psu}}\left(\hat{t}_{\text{unb}}\right) = N^2\left(1 - \frac{n}{N}\right)\frac{s_t^2}{n} \tag{5.51}$$

and, from (5.29),

$$\hat{V}_{\text{WR}}\left(\hat{t}_{\text{unb}}\right) = N^2\frac{s_t^2}{n}.$$

Using (5.28), show that

$$E\left[\hat{V}_{\text{psu}}\left(\hat{t}_{\text{unb}}\right)\right] = V\left(\hat{t}_{\text{unb}}\right) - \sum_{i=1}^{N} V\left(\hat{t}_i\right)$$

and

$$E\left[\hat{V}_{\text{WR}}\left(\hat{t}_{\text{unb}}\right)\right] = V\left(\hat{t}_{\text{unb}}\right) + NS_t^2,$$

where $V\left(\hat{t}_{\text{unb}}\right)$ is given in (5.24). Neither $\hat{V}_{\text{psu}}\left(\hat{t}_{\text{unb}}\right)$ nor $\hat{V}_{\text{WR}}\left(\hat{t}_{\text{unb}}\right)$ is exactly unbiased for estimating $V\left(\hat{t}_{\text{unb}}\right)$, but if n/N is small, the second term in each is negligible compared with the first.

32. (Requires trigonometry.) In Example 5.13, a systematic sampling scheme was proposed for detecting hazardous wastes in landfills. How far apart should sampling points be placed? Suppose that if there is leakage, it will spread to a circular region with radius R. Let D be the distance between adjacent sampling points in the same row or column.

    (a) Calculate the probability with which a contaminant will be detected. HINT: Consider three cases, with $R < D$, $D \le R \le \sqrt{2}D$, and $R > \sqrt{2}D$.

    (b) Propose a sampling design that gives a higher probability that a contaminant will be detected than the square grid, but does not increase the number of sampling points.

33. (Requires knowledge of random effects models.) Under the one-way random effects model M1 in (5.38), the intraclass correlation coefficient ρ may be estimated by

$$\hat{\rho} = \frac{\hat{\sigma}_A^2}{\hat{\sigma}_A^2 + \hat{\sigma}^2},$$

where $\hat{\sigma}_A^2$ and $\hat{\sigma}^2$ estimate the variance components σ_A^2 and σ^2. The methods of moments estimators for one-stage cluster sampling when all clusters are of the same size are $\hat{\sigma}^2 = \text{msw}$ and $\hat{\sigma}_A^2 = (\text{msb} - \text{msw})/M$, where msw and msb are the within and between mean squares from the sample ANOVA table.

(a) What is $\hat{\rho}$ in Example 5.4? How does it compare with $\widehat{\text{ICC}}$?

(b) Calculate $\hat{\rho}$ for Populations A and B in Example 5.3. Why do these differ from the ICC?

34. (Requires knowledge of random effects models.) Prove (5.42).

35. (Requires knowledge of mathematical statistics and linear models theory.) Assume that, in the expression of Model M1 in (5.38), the random variables A_i and ε_{ij} are normally distributed. Show that

$$E[T|\{Y_{ij}, i \in \mathcal{S}, j \in \mathcal{S}_i\}] = \sum_{i\in\mathcal{S}}\sum_{j\in\mathcal{S}_i} Y_{ij} + \sum_{i\in\mathcal{S}}\sum_{j\notin\mathcal{S}_i}\left[\rho\frac{\alpha_i}{m_i}\sum_{l\in\mathcal{S}_i} Y_{il} + (1-\rho\alpha_i)\mu\right] + \sum_{i\notin\mathcal{S}} M_i\mu$$

and that $\hat{\mu}$ in (5.44) is the best linear unbiased estimator of μ.

36. Consider the best linear unbiased predictor of T in (5.43).

(a) Show that $\hat{T}_{\text{BLUP}}$ can be written in the form of (5.40), with

$$\hat{T}_{\text{BLUP}} = \sum_{i\in\mathcal{S}}\sum_{j\in\mathcal{S}_i}\frac{\alpha_i}{m_i}\left[\rho M_i + \frac{M_0 - \rho\sum_{k\in\mathcal{S}}\alpha_k M_k}{\sum_{k\in\mathcal{S}}\alpha_k}\right] Y_{ij} = \sum_{i\in\mathcal{S}}\sum_{j\in\mathcal{S}_i} c_{ij}Y_{ij}.$$

(b) Show that $\sum_{i\in\mathcal{S}}\sum_{j\in\mathcal{S}_i} c_{ij} = M_0$, so that $E_{\text{M1}}\left[\hat{T}_{\text{BLUP}} - T\right] = 0$.

(c) Consider another unbiased estimator $\hat{T}_2 = \sum_{i\in\mathcal{S}}\sum_{j\in\mathcal{S}_i}(c_{ij} + a_{ij})Y_{ij}$. For $\hat{T}_2$ to be unbiased, the a_{ij} must satisfy $\sum_{i\in\mathcal{S}}\sum_{j\in\mathcal{S}_i} a_{ij} = 0$. Show that $V_{\text{M1}}(\hat{T}_2) \geq V_{\text{M1}}(\hat{T}_{\text{BLUP}})$, and hence that $\hat{T}_{\text{BLUP}}$ minimizes the variance in (5.42) among all unbiased predictors under Model M1 in (5.38).

37. *Model-based inference under an alternative model.* Model M1 implies that the expected total for y in a cluster is proportional to the size of the cluster. Under Model M2, each cluster total has expected value μ, regardless of the size M_i of the cluster:

$$\text{M2}: \; Y_{ij} = B_i + \varepsilon_{ij},$$

with $E[B_i] = \mu/M_i$, $V[M_iB_i] = \sigma_B^2$, $E[\varepsilon_{ij}] = 0$, $V[\varepsilon_{ij}] = \sigma^2$, and all B_i and ε_{ij} independent.

(a) Give an example where Model M2 might be appropriate.

(b) Find the bias $E_{\text{M2}}[\hat{T} - T]$ of the general linear estimator $\hat{T}$ in (5.40) under Model M2. Show that $E\hat{T}_{\text{unb}}$ is unbiased under Model M2.

(c) Find $V_{\text{M2}}[\hat{T} - T]$ under Model M2.

38. For the puppy homes discussed in Example 5.9, find the bias and variance of $\hat{T}_{\text{unb}}$, $\hat{T}_r$, and $\hat{T}_{\text{BLUP}}$, under Model M1. Calculate the MSE ($=$ bias^2 + variance) for each estimator. How do these compare with the MSEs calculated in Example 5.9 under the design-based approach?

39. (Requires calculus.) Suppose that the M_is are unequal and Model M1 holds. The budget allows you to take a total of L measurements on subunits. Show that the variance in (5.42) is minimized for $\hat{T}_r$ when m_i is proportional to M_i. HINT: Use Lagrange multipliers, with the constraint $\sum_{i\in\mathcal{S}} m_i = L$.

D. Projects and Activities

40. *Coverage probabilities of interval estimates when clustering is ignored* (requires computing). Alf and Lohr (2007) presented a computer program, *intervals*, that explores the effects of ignoring clustering on CIs. It generates a clustered population with a specified ICC in which each psu has size M, then draws K samples of psus from the population. Two interval estimates from each of the K samples: the first is calculated (incorrectly, when ICC $\neq 0$) assuming the individual elements are from an SRS, and the second is calculated using the correct standard error in (5.6). As in Example 2.10, one can estimate the coverage percentage by (number of samples whose interval estimate contains the true population mean) divided by the number of samples. The program is available as an R function (described in Lu and Lohr, 2022) and as a SAS macro (described in Lohr, 2022).

    (a) Run the program to generate $K = 100$ samples with $n = 10$ and $M = m = 5$, and the confidence interval for each sample, from a population with ICC$= 0$. What is the effect of ignoring clustering when ICC$= 0$?

    (b) Now see what happens with a population with ICC$= 1/2$. What percentage of the interval estimates, calculated ignoring the clustering, include the true mean? What about the intervals calculated using the formulae for cluster samples? Compare the average widths of the two interval estimates.

    (c) What do you think will happen if you try the intervals program with ICC$= 1$? What will happen to the widths of the correctly calculated confidence intervals? Do you expect the percentage of interval estimates that include the true mean to increase, or decrease, for the two methods? Test your predictions by running the program with ICC $= 1$.

    (d) Explore how the estimated coverage probability depends on the ICC and M. Run the program with all 12 combinations of $M \in \{2, 10, 25\}$ and ICC $\in \{0, .2, .7, 1\}$. Plot the coverage probability versus ICC, drawing a curve for each value of M.

41. Select your favorite newspaper columnist or blogger. Select an SRS of ten of the columnist's or blogger's columns that appeared in the past year. Use one-stage cluster sampling to estimate the proportion of total words taken up by "I," "me," and "myself." What is your psu? Your ssu?

42. *Online bookstore.* You may have noticed in Exercise 38 of Chapter 2 that it took quite a bit of time to locate the records chosen for the SRS. It may be faster to take an SRS of pages from the website, then look at some or all of the books listed on that page. Use the following procedure to take a cluster sample of books from the genre you studied in Chapter 2, recording the amount of time you spend selecting the sample and collecting the data. Take an SRS of 10 pages, then sample 5 books per page. For each sampled book, record the price, number of pages, and whether the book is paperback or hardback. Estimate the mean of each variable, and give a 95% CI. Do you think clustering decreased precision relative to an SRS? Compare the precision per unit time for the SRS and the cluster sample by calculating $1/[(\text{estimated variance}) \times \text{time}]$ for each method.

43. *Baseball data: One-stage sample.*

    (a) Use the population in the file `baseball.csv` to take a one-stage cluster sample with the teams as the psus. Your sample should have approximately 150 players altogether, as in the SRS from Exercise 37 of Chapter 2. Describe how you selected your sample.

(b) Draw side-by-side boxplots of *logsal* for the teams in your sample.

(c) Use your sample to estimate the mean of the variable *logsal*, and give a 95% CI.

(d) Estimate the proportion of players in the data set who are pitchers, and give a 95% CI.

(e) Compare your estimates with those from Exercise 37 of Chapter 2. Which estimates have smaller CIs? Why do you think that happened?

44. *Baseball data: Two-stage sample.*

(a) Use your SRS from Exercise 37 of Chapter 2 to estimate the population value of R_a^2. If we treat teams as the psus, and if all teams had the same size, what would the optimal subsampling size be (assume that $c_1 = c_2$).

(b) Use the population in the file `baseball.csv` to take a two-stage cluster sample with the teams as the psus, using the subsampling fraction from part (a). Your sample should have approximately 150 players altogether, as in the SRS from Exercise 37 of Chapter 2. Describe how you selected your sample.

(c) Draw side-by-side boxplots of *logsal* for the teams in your sample.

(d) Use your sample to estimate the mean of the variable *logsal*, and give a 95% CI.

(e) Estimate the proportion of players in the data set who are pitchers, and give a 95% CI.

(f) Compare your estimates with those from Exercise 37 of Chapter 2. Which estimates have smaller CIs? Why do you think that happened?

(g) Compare your estimates with those from Exercise 43.

45. *IPUMS exercises.*

(a) Generate a frequency table of the number of persons within each psu.

(b) Suppose that it costs $50 per interview to collect data using an SRS. If a cluster sample is taken, it costs $100 per psu chosen, plus $20 for each interview taken. Select an SRS of 10 psus. In each of the selected psus, take a subsample of persons with sample size proportional to the population size within that psu. Your total cost for the sample should be about the same as for the SRS you took in Chapter 2.

(c) Using the sample you selected, estimate the population mean of *inctot* and give the standard error of your estimate. Also, estimate the population total of *inctot* and give its standard error. How do these estimates compare with those from the SRS you took in Chapter 2?

6

Sampling with Unequal Probabilities

'Personally I never care for fiction or storybooks. What I like to read about are facts and statistics of any kind. If they are only facts about the raising of radishes, they interest me. Just now, for instance, before you came in'—he pointed to an encyclopædia on the shelves—'I was reading an article about "Mathematics." Perfectly pure mathematics.'

'My own knowledge of mathematics stops at "twelve times twelve," but I enjoyed that article immensely. I didn't understand a word of it; but facts, or what a man believes to be facts, are always delightful. That mathematical fellow believed in his facts. So do I. Get your facts first, and'—the voice dies away to an almost inaudible drone—'then you can distort 'em as much as you please.'

—Mark Twain, quoted in Rudyard Kipling, *From Sea to Sea*

Up to now, we have discussed sampling schemes in which the probabilities of choosing sampling units are equal. Equal probabilities give schemes that are often easy to design and explain. Such schemes are not, however, always possible or, if practicable, as efficient as schemes using unequal probabilities. We saw in Example 5.9 that a cluster sample with equal probabilities may result in a large variance for the design-unbiased estimator of the population mean and total.

Example 6.1. O'Brien et al. (1995) took a sample of nursing home residents in the Philadelphia area, with the objective of determining residents' preferences on life-sustaining treatments. Do they wish to have cardiopulmonary resuscitation (CPR) if the heart stops beating, or to be transferred to a hospital if a serious illness develops, or to be fed through an enteral tube if no longer able to eat? The target population was all residents of licensed nursing homes in the Philadelphia area. There were 294 such homes, with a total of 37,652 beds (before sampling, they knew only number of beds, not number of residents).

Because the survey was to be done in person, cluster sampling was essential for keeping survey costs manageable. Had the researchers chosen to use cluster sampling with equal probabilities of selection, they would have taken a simple random sample (SRS) of nursing homes, then another SRS of residents within each selected home.

In a cluster sample with equal probabilities, however, a nursing home with 20 beds is as likely to be chosen for the sample as a nursing home with 1,000 beds. The sample is then self-weighting only if the subsample size for each home is proportional to the number of beds in the home. Each bed sampled represents the same number of beds in the population if one-stage cluster sampling is used, or if 10% (or any other fixed percentage) of beds are sampled in each selected home.

Sampling homes with equal probabilities would result in a mathematically valid estimator, but it has three major shortcomings. First, you would expect that the total number of patients in a home who desire CPR (t_i) would be roughly proportional to the number of beds in the home (M_i), so estimators from Chapter 5 may have large variance. Second, a self-weighting equal-probability sample may be cumbersome to administer. It may require driving out to a nursing home just to interview one or two residents, and equalizing workloads of interviewers may be difficult. Third, the cost of the sample is unknown

DOI: 10.1201/9780429298899-6

in advance—a random sample of 40 homes may consist primarily of large nursing homes, which would lead to greater expense than anticipated.

Instead of taking a cluster sample of homes with equal probabilities, the investigators randomly drew a sample of 57 nursing homes with probabilities proportional to the number of beds. They then took an SRS of 30 beds (and their occupants) from a list of all beds within the nursing home. If the number of residents equals the number of beds, and if a home has the same number of beds when visited as are listed in the sampling frame, then the sampling design results in every resident having the same probability of being included in the sample. The cost is known before selecting the sample, the same number of interviews are taken at each home, and the estimator of a population total will likely have a smaller variance than estimators in Chapter 5.

Since this sample is self-weighting, you can easily obtain point estimates of desired quantities by usual methods. You can estimate the median age of the nursing home residents by finding the sample median of the residents in the sample, or the 70th percentile by finding the 70th percentile of the sample. If a sample is not self-weighting, point estimates are still easily calculated using weights. A warning, though: Always consider the cluster design when calculating the precision of your estimates. ■

In Chapter 3, we noted that sometimes stratified sampling is used to sample units with different probabilities. In a survey to estimate total business expenditures on advertising, we might want to stratify by company sales or income. The largest companies such as IBM would be in one stratum, medium-sized companies would be in a number of different strata, and very small companies such as Robin's Tailor Shop would be in yet other strata. An optimal allocation scheme would sample a very high fraction (perhaps 100%) in the stratum with the largest companies, and a small fraction of companies in the strata with the smallest companies; the variance from company to company will be much higher among IBM, General Motors, and Microsoft than among Robin's Tailor Shop, Pat's Shoe Repair, and Flowers by Leslie. The variance among the large companies is larger because the amounts of money involved are so much larger. Thus, the sampling variance is decreased by assigning unequal probabilities to sampling units in different strata.

To estimate the total spent on advertising using this stratified sample, we assign higher weights to companies with lower inclusion probabilities. As discussed in Section 3.3, the probability that a company in stratum h will be included in the sample is n_h/N_h; the sampling weight for that company is N_h/n_h. Each company sampled in stratum h represents N_h/n_h companies in the population, and $\hat{t}_{\text{str}} = \sum_{h=1}^{H} \sum_{j \in \mathcal{S}_h} (N_h/n_h) y_{hj}$.

We can also use unequal inclusion probabilities to decrease variances without (or in addition to) explicitly stratifying. When sampling with unequal probabilities, we deliberately vary the probabilities with which we will select different primary sampling units (psus) for the sample, and compensate by providing suitable weights in the estimation. The key is that we *know* the probabilities[1] with which we will select a given unit:

$$P(\text{unit } i \text{ selected on first draw}) = \psi_i \tag{6.1}$$

$$P(\text{unit } i \text{ in sample}) = \pi_i. \tag{6.2}$$

The deliberate selection of psus with known but unequal probabilities differs greatly from the selection bias discussed in Chapter 1. Many surveys with selection bias end up with unequal probabilities of selection (see Chapter 15), but these probabilities are unknown

[1] We consider two different probabilities in this chapter, because when sampling with unequal probabilities without replacement, as considered in Section 6.4, selecting a unit on the first draw can affect the selection probabilities for other units.

and unestimable, so the persons conducting the survey cannot compensate for the unequal probabilities in the weighting. If you take a survey of students by asking students who walk by the library to participate, you certainly are sampling with unequal probabilities—students who use the library frequently are more likely to be asked to participate in the survey, while other students never go by the library at all. But you have no idea how many students in the population are represented by a participant in your survey, and cannot correct for the unequal probabilities of selection in the estimation. In addition, some students in your target population never walk by the library, so they cannot be included in your sample.

When first presented with the idea of unequal-probability sampling, some people think of it as "unnatural" or "contrived." On the contrary, for many populations with clustering, unequal-probability sampling at the psu level produces a sample that mirrors the population better than an equal-probability sample. The examples in Section 6.5 show how an unequal-probability sample can be easier to administer as well as more efficient statistically.

We first consider with-replacement sampling in Sections 6.1 to 6.3, starting with the simple design of selecting one psu. Many large surveys are analyzed as though the sampling was done with replacement, even if a without-replacement sample was collected, because the estimators of the variance for with-replacement samples have simple form. In Section 6.4, we consider unequal-probability sampling without replacement. Notation used in this chapter is given in Tables 5.1 and 5.2 of Chapter 5.

6.1 Sampling One Primary Sampling Unit

As a special case, suppose we select just one ($n = 1$) of the N psus to be in the sample. The total for psu i is denoted by t_i, and we want to estimate the population total, $t = \sum_{i=1}^{N} t_i$. Sampling one psu will demonstrate the ideas of unequal-probability sampling without introducing the complications.

Let's start out by looking at what happens for a situation in which we know the whole population. A town has four supermarkets, ranging in size from 100 square meters (m^2) to 1,000 m^2. We want to estimate the total amount of sales in the four stores for last month by sampling just one of the stores. (Of course, this is just an illustration—if we really had only four supermarkets we would probably take a census.) You might expect that a larger store would have more sales than a smaller store, and that the variability in total sales among several 1,000-m^2 stores will be greater than the variability in total sales among several 100-m^2 stores.

Since we sample only one store, the probability that a store is selected on the first draw (ψ_i) is the same as the probability that the store is included in the sample (π_i). For this example, take

$$\pi_i = \psi_i = P(\text{Store } i \text{ selected})$$

proportional to the size of the store. Since Store A accounts for $1/16$ of the total floor area of the four stores, it is sampled with probability $1/16$. For illustrative purposes, we know the values of t_i for the whole population, shown in Table 6.1.

We could select a probability sample of size 1 with the probabilities given above by shuffling cards numbered 1 through 16 and choosing one card. If the card's number is 1, choose store A; if 2 or 3, choose B; if 4, 5, or 6, choose C; and if 7 through 16, choose D.

We compensate for the unequal probabilities of selection by also using ψ_i in the estimator. We have already seen such compensation for unequal probabilities in stratified

TABLE 6.1
Values of t_i for population of supermarkets.

Store	Size (m^2)	ψ_i	t_i (in Thousands)
A	100	$\frac{1}{16}$	11
B	200	$\frac{2}{16}$	20
C	300	$\frac{3}{16}$	24
D	1000	$\frac{10}{16}$	245
Total	1600	1	300

sampling: If we select 10% of the units in stratum 1 and 20% of the units in stratum 2, the sampling weight is 10 for each unit in stratum 1 and 5 for each unit in stratum 2. Here, we select store A with probability 1/16, so store A's sampling weight is 16. If the size of the store is roughly proportional to the total sales for that store, we would expect that store A also has about 1/16 of the total sales and that multiplying store A's sales by 16 would estimate the total sales for all four stores. As always, the sampling weight of unit i is the reciprocal of the probability of selection:

$$w_i = \frac{1}{P(\text{unit } i \text{ in sample})} = \frac{1}{\psi_i}.$$

Thus, our estimator of the population total from an unequal-probability sample of size 1 is

$$\hat{t}_\psi = \sum_{i \in \mathcal{S}} w_i t_i = \sum_{i \in \mathcal{S}} \frac{t_i}{\psi_i}.$$

Four samples of size 1 are possible from this simple population:

Sample	ψ_i	t_i	$\hat{t}_\psi$	$(\hat{t}_\psi - t)^2$
{A}	$\frac{1}{16}$	11	176	15,376
{B}	$\frac{2}{16}$	20	160	19,600
{C}	$\frac{3}{16}$	24	128	29,584
{D}	$\frac{10}{16}$	245	392	8,464

Using the definition of expected value in (2.3),

$$\begin{aligned} E[\hat{t}_\psi] &= \sum_{\text{samples } \mathcal{S}} P(\mathcal{S})\, \hat{t}_{\psi\mathcal{S}} \\ &= \frac{1}{16}(176) + \frac{2}{16}(160) + \frac{3}{16}(128) + \frac{10}{16}(392) = 300. \end{aligned}$$

Of course, $\hat{t}_\psi$ will always be unbiased because in general,

$$E[\hat{t}_\psi] = \sum_{i=1}^{N} \psi_i \frac{t_i}{\psi_i} = t. \tag{6.3}$$

The variance of $\hat{t}_\psi$ is

$$V[\hat{t}_\psi] = E[(\hat{t}_\psi - t)^2] = \sum_{\text{samples } \mathcal{S}} P(\mathcal{S}) \left(\hat{t}_{\psi\mathcal{S}} - t\right)^2 = \sum_{i=1}^{N} \psi_i \left(\frac{t_i}{\psi_i} - t\right)^2. \tag{6.4}$$

For this example,

$$V[\hat{t}_\psi] = \frac{1}{16}(15{,}376) + \frac{2}{16}(19{,}600) + \frac{3}{16}(29{,}584) + \frac{10}{16}(8{,}464) = 14{,}248.$$

Compare these results to those from an SRS of size 1, in which the probability of selecting each unit is $\psi_i = 1/4$, so $1/\psi_i = 4 = N$. Note that if all of the probabilities of selection are equal, as in simple random sampling, $1/\psi_i$ always equals N. For the SRS design:

Sample	ψ_i	t_i	$\hat{t}_\psi$	$(\hat{t}_\psi - t)^2$
{A}	$\frac{1}{4}$	11	44	65,536
{B}	$\frac{1}{4}$	20	80	48,400
{C}	$\frac{1}{4}$	24	96	41,616
{D}	$\frac{1}{4}$	245	980	462,400

As always, $\hat{t}_{\text{SRS}}$ is unbiased and thus has expectation 300, but for this example the SRS variance is much larger than the variance from the unequal-probability design:

$$V[\hat{t}_{\text{SRS}}] = \frac{1}{4}(65{,}536) + \frac{1}{4}(48{,}400) + \frac{1}{4}(41{,}616) + \frac{1}{4}(462{,}400) = 154{,}488.$$

The variance from the unequal-probability design, 14,248, is smaller because the design uses auxiliary information: We expect the store size to be related to the sales, and use that information in the sample design.

We believe that t_i is correlated to the size of the store, which is known. Since Store D accounts for 10/16 of the total floor area of supermarkets, it is reasonable to believe that Store D will account for about 10/16 of the total sales as well. Thus, if store D is chosen and is believed to account for about 10/16 of the total sales, we would have a good estimate of total sales by multiplying Store D's sales by 16/10.

What if Store D only accounts for 4/16 of the total sales? Then the unequal-probability estimator $\hat{t}_\psi$ will still be unbiased over repeated sampling, but it will have a large variance (see Exercise 3). The method still works mathematically, but is not as efficient as if t_i is roughly proportional to ψ_i.

Sampling only one psu is not as unusual as you might think. Many large complex surveys are so highly stratified that each stratum contains only a few psus. A large number of strata is used to increase the precision of the survey estimates. In such a survey, it may be perfectly reasonable to want to select only one psu from each stratum. But, with only one psu per stratum in the sample, we do not have an estimate of the variability between psus within a stratum. When large survey organizations sample only one psu per stratum, they often divide the psus into pseudo-psus for variance estimation.

6.2 One-Stage Sampling with Replacement

Now suppose $n > 1$, and we sample *with* replacement. Sampling with replacement means that the selection probabilities do not change after we have drawn the first unit. Let

$$\psi_i = P(\text{select unit } i \text{ on first draw}).$$

If we sample with replacement, then ψ_i is also the probability that unit i is selected on the second draw, or the third draw, or any other given draw.

The idea behind unequal-probability sampling is simple. Draw n psus with replacement. Then estimate the population total, using the estimator from the previous section, separately for each psu drawn. Some psus may be drawn more than once—the estimated population total, calculated using a given psu, is included as many times as the psu is drawn. Since the psus are drawn with replacement, we have n independent estimates of the population total. Estimate the population total t by averaging those n independent estimates of t. The estimated variance is the sample variance of the n independent estimates of t, divided by n.

6.2.1 Selecting Primary Sampling Units

Survey software packages will select samples of psus with unequal probabilities (see Lohr, 2022; Lu and Lohr, 2022, for how to do this in SAS and R software). Most algorithms for selecting with-replacement samples are based on either the cumulative size method or Lahiri's method, which are discussed below. All methods require that you have a measure of size for all psus in the population.

Cumulative size method. The cumulative-size method extends the method used in the previous section, in which random numbers are generated and psus corresponding to those numbers are included in the sample. For the supermarkets, we drew cards from a deck with cards numbered 1 through 16. If the card's number is 1, choose store A; if 2 or 3, choose B; if 4, 5, or 6, choose C; and if 7 through 16, choose D. To sample with replacement, put the card back after selecting a psu and draw again.

Example 6.2. Consider the population of introductory statistics classes at a college shown in Table 6.2. The college has 15 such classes (psus); class i has M_i students, for a total of 647 students in introductory statistics courses. We decide to sample 5 classes with replacement, with probability proportional to M_i, and then collect data from each student in the sampled classes. For this example, then, $\psi_i = M_i/647$.

To select the sample, generate five random integers with replacement between 1 and 647. Then the psus to be chosen for the sample are those whose range in the cumulative M_i includes the randomly generated numbers. The set of five random numbers {487, 369, 221, 326, 282} results in the sample of units {13, 9, 6, 8, 7}. The cumulative-size method allows the same unit to appear more than once: the five random numbers {553, 082, 245, 594, 150} leads to the sample {14, 3, 6, 14, 5}—psu 14 is included twice. ■

Of course, we can take an unequal-probability sample when ψ_i is not proportional to M_i: Simply form a cumulative ψ_i range instead, and sample uniform random numbers between 0 and 1. This variation of the method is discussed in Exercise 2.

Systematic sampling is often used to select psus, rather than generating random numbers with replacement. Systematic sampling usually gives a sample without replacement, but in large populations sampling without replacement and sampling with replacement are

TABLE 6.2
Population of introductory statistics classes.

Class Number	M_i	ψ_i	Cumulative M_i Range
1	44	0.068006	1 – 44
2	33	0.051005	45 – 77
3	26	0.040185	78 – 103
4	22	0.034003	104 – 125
5	76	0.117465	126 – 201
6	63	0.097372	202 – 264
7	20	0.030912	265 – 284
8	44	0.068006	285 – 328
9	54	0.083462	329 – 382
10	34	0.052550	383 – 416
11	46	0.071097	417 – 462
12	24	0.037094	463 – 486
13	46	0.071097	487 – 532
14	100	0.154560	533 – 632
15	15	0.023184	633 – 647
Total	647	1.000000	

very similar, as the probability that a unit will be selected twice is small. To sample psus systematically, list the population elements for the first psu in the sample, followed by the elements for the second psu, and so on. Then take a systematic sample of the elements. The psus to be included in the sample are those in which at least one element is in the systematic sample of elements. The larger the psu, the higher the probability it will be in the sample.

The statistics classes in Example 6.2 have a total of 647 students. To take a (roughly, because 647 is not a multiple of 5) systematic sample, choose a random number k between 1 and 129 and select the psu containing student k, the psu containing student $129 + k$, the psu containing student $2(129) + k$, and so on. Suppose the random number we select as a start value is 112. Then the systematic sample of elements results in the following psus being chosen:

Number in Systematic Sample	psu Chosen
112	4
241	6
370	9
499	13
628	14

Larger classes (psus) have a higher chance of being in the sample because it is more likely that a multiple of the random number chosen will be one of the numbered elements in a large psu. Systematic sampling does not give us a true random sample with replacement because it is impossible for classes with 129 or fewer students to occur in the sample more than once, and classes with more than 129 students are sampled with probability 1. In many populations, however, it is much easier to implement than methods that do give a random sample. If the psus are arranged geographically, taking a systematic sample may force the selected psus to be spread out over more of the region, and may result in estimates having smaller variance than if they were selected randomly with replacement (see Exercise 28 of Chapter 7).

TABLE 6.3
Lahiri's method, for Example 6.3.

First Random Number (psu i)	Second Random Number	M_i	Action
12	6	24	$6 < 24$; include psu 12 in sample
14	24	100	Include in sample
1	65	44	$65 > 44$; discard pair of numbers and try again
7	84	20	$84 > 20$; try again
10	49	34	Try again
14	47	100	Include
15	43	15	Try again
5	24	76	Include
11	87	46	Try again
1	36	44	Include

Lahiri's method. Lahiri's (1951) method is an example of a *rejective* method—you generate a pair of random numbers to select a psu and then to decide whether to accept or reject the psu for the sample. Let N = number of psus in population and $\max\{M_i\}$ = maximum psu size. You will show that Lahiri's method produces a with-replacement sample with the desired probabilities in Exercise 19.

1. Draw a random number between 1 and N. This indicates which psu you are considering.
2. Draw a random number between 1 and $\max\{M_i\}$. If this random number is less than or equal to M_i, then include psu i in the sample; otherwise go back to Step 1.
3. Repeat until desired sample size is obtained.

Example 6.3. Let's use Lahiri's method for the classes in Example 6.2. The largest class has $\max\{M_i\} = 100$ students, so we generate pairs of random integers, the first between 1 and 15 and the second between 1 and 100, until the sample has five psus (Table 6.3). The psus to be sampled are {12, 14, 14, 5, 1}. ■

6.2.2 Theory of Estimation

Because we are sampling with replacement, the sample may contain the same psu more than once. Let $\mathcal{W}$ denote the set of n units in the sample, including the repeats. For Example 6.3, $\mathcal{W} = \{12, 14, 14, 5, 1\}$; unit 14 is included twice in $\mathcal{W}$.

Estimating population totals. We saw in Section 6.1 that for a sample of size 1, $u_i = t_i/\psi_i$ is an unbiased estimator of the population total t. When we sample n psus with replacement, we have n independent estimators of t, so we average them:

$$\hat{t}_\psi = \frac{1}{n}\sum_{i\in\mathcal{W}}\frac{t_i}{\psi_i} = \frac{1}{n}\sum_{i\in\mathcal{W}} u_i = \bar{u}. \tag{6.5}$$

The estimator $\hat{t}_\psi$ in (6.5) is often referred to as the Hansen–Hurwitz (1943) estimator. We estimate $V(\hat{t}_\psi)$ by

$$\hat{V}(\hat{t}_\psi) = \frac{s_u^2}{n} = \frac{1}{n}\frac{1}{n-1}\sum_{i\in\mathcal{W}}(u_i - \bar{u})^2 = \frac{1}{n}\frac{1}{n-1}\sum_{i\in\mathcal{W}}\left(\frac{t_i}{\psi_i} - \hat{t}_\psi\right)^2. \tag{6.6}$$

Equation (6.6) is the estimated variance of the average $\bar{u}$ from a simple random sample with replacement. Where are the unequal probabilities in this formula? To prove that $\hat{t}_\psi$ and $\hat{V}(\hat{t}_\psi)$ are unbiased estimators of t and $V(\hat{t}_\psi)$, respectively, we need random variables to keep track of which psus occur multiple times in the sample. Define

$$Q_i = \text{number of times unit } i \text{ occurs in the sample;}$$

Q_i is a with-replacement analogue of the random variable Z_i used to indicate sample inclusion for without-replacement sampling in (2.33). Then $\hat{t}_\psi$ is the average of all t_i/ψ_i for units chosen to be in the sample, including each unit as many times as it appears in the sample:

$$\hat{t}_\psi = \frac{1}{n}\sum_{i\in\mathcal{W}}\frac{t_i}{\psi_i} = \frac{1}{n}\sum_{i=1}^{N} Q_i\frac{t_i}{\psi_i}. \tag{6.7}$$

If a unit appears k times in the sample, it is counted k times in the estimator. Note that $\sum_{i=1}^{N} Q_i = n$ and $E[Q_i] = n\psi_i$ (see Exercise 20), so $\hat{t}_\psi$ is an unbiased estimator of t.

To calculate the variance, note that the estimator in (6.7) is the average of n independent observations, each with variance $\sum_{i=1}^{N}\psi_i(t_i/\psi_i - t)^2$ [from (6.4)], so

$$V(\hat{t}_\psi) = \frac{1}{n}\sum_{i=1}^{N}\psi_i\left(\frac{t_i}{\psi_i} - t\right)^2. \tag{6.8}$$

To show that the variance estimator in (6.6) is unbiased for $V(\hat{t}_\psi)$, we write it in terms of the random variables Q_i:

$$\hat{V}(\hat{t}_\psi) = \frac{1}{n}\frac{1}{n-1}\sum_{i\in\mathcal{W}}\left(\frac{t_i}{\psi_i} - \hat{t}_\psi\right)^2 = \frac{1}{n}\frac{1}{n-1}\sum_{i=1}^{N} Q_i\left(\frac{t_i}{\psi_i} - \hat{t}_\psi\right)^2. \tag{6.9}$$

Equation (6.8) involves a weighted average of the N values of $(t_i/\psi_i - t)^2$, weighted by the unequal selection probabilities ψ_i. In taking the sample, we have already used the unequal probabilities—they appear in the random variables Q_i in (6.7). The ith psu appears Q_i times in the with-replacement sample. Because the n units are selected independently, $E[Q_i] = n\psi_i$, so including the squared deviation $\left(t_i/\psi_i - \hat{t}_\psi\right)^2$ a total of Q_i times in the variance estimator causes (6.9) to be an unbiased estimator of the variance in (6.8):

$$\begin{aligned}
E[\hat{V}(\hat{t}_\psi)] &= \frac{1}{n(n-1)}\sum_{i=1}^{N} E\left[Q_i\left(\frac{t_i}{\psi_i} - \hat{t}_\psi\right)^2\right] \\
&= \frac{1}{n(n-1)}E\left[\sum_{i=1}^{N} Q_i\left(\frac{t_i}{\psi_i} - t + t - \hat{t}_\psi\right)^2\right] \\
&= \frac{1}{n(n-1)}E\left[\sum_{i=1}^{N} Q_i\left(\frac{t_i}{\psi_i} - t\right)^2 + \sum_{i=1}^{N} Q_i(\hat{t}_\psi - t)^2 - 2\sum_{i=1}^{N} Q_i\left(\frac{t_i}{\psi_i} - t\right)(\hat{t}_\psi - t)\right] \\
&= \frac{1}{n(n-1)}E\left[\sum_{i=1}^{N} Q_i\left(\frac{t_i}{\psi_i} - t\right)^2 + n(\hat{t}_\psi - t)^2 - 2n(\hat{t}_\psi - t)^2\right] \\
&= \frac{1}{n(n-1)}\left[\sum_{i=1}^{N} n\psi_i\left(\frac{t_i}{\psi_i} - t\right)^2 - nV(\hat{t}_\psi)\right] \\
&= V(\hat{t}_\psi).
\end{aligned}$$

In line 4 of the argument, we use the facts that $\sum_{i=1}^{N} Q_i = n$ and $\sum_{i=1}^{N} Q_i t_i/\psi_i = n\hat{t}_\psi$. In Exercise 7, you will show that the variance estimator for simple random sampling with replacement is a special case of (6.9).

Estimating population means. We estimate $\bar{y}_\mathcal{U}$ by

$$\hat{\bar{y}}_\psi = \frac{\hat{t}_\psi}{\hat{M}_{0\psi}}, \tag{6.10}$$

where

$$\hat{M}_{0\psi} = \frac{1}{n}\sum_{i\in\mathcal{W}} \frac{M_i}{\psi_i} \tag{6.11}$$

estimates the total number of elements in the population. In (6.10), $\hat{\bar{y}}_\psi$ is a ratio; using results in Chapter 4, we calculate the residuals $t_i/\psi_i - \hat{\bar{y}}_\psi M_i/\psi_i$ to estimate the variance:

$$\hat{V}(\hat{\bar{y}}_\psi) = \frac{1}{(\hat{M}_{0\psi})^2}\frac{1}{n}\frac{1}{n-1}\sum_{i\in\mathcal{W}}\left(\frac{t_i}{\psi_i} - \frac{\hat{\bar{y}}_\psi M_i}{\psi_i}\right)^2. \tag{6.12}$$

Note that (6.12) has the same form as (6.6), but with the values $(t_i - \hat{\bar{y}}_\psi M_i)/\hat{M}_{0\psi}$ substituted for t_i.

Example 6.4. For the situation in Example 6.3, suppose we sample the psus selected by Lahiri's method, {12, 14, 14, 5, 1}. The response t_i is the total number of hours all students in class i spent studying statistics last week, with data in Table 6.4.

TABLE 6.4
Data for Example 6.4.

Class	ψ_i	t_i	t_i/ψ_i
12	$\frac{24}{647}$	75	2021.875
14	$\frac{100}{647}$	203	1313.410
14	$\frac{100}{647}$	203	1313.410
5	$\frac{76}{647}$	191	1626.013
1	$\frac{44}{647}$	168	2470.364

The numbers in the last column of Table 6.4 are the estimates of t that would be obtained if that psu were the only one selected in a sample of size 1. The population total is estimated by averaging the five values of t_i/ψ_i, using (6.5):

$$\hat{t}_\psi = \frac{2021.875 + 1313.410 + 1313.410 + 1626.013 + 2470.364}{5} = 1749.014.$$

The standard error of $\hat{t}_\psi$ is simply $s/\sqrt{n}$ [see Equation (6.6)], where s is the sample standard deviation of the five numbers in the rightmost column of the table:

$$\text{SE}(\hat{t}_\psi) = \frac{1}{\sqrt{5}}\sqrt{\frac{(2021.875 - 1749.014)^2 + \ldots + (2470.364 - 1749.014)^2}{4}} = 222.42.$$

Since $\psi_i = M_i/M_0$ for this sample, we have $\hat{M}_0 = M_0 = 647$. The average amount of time a student spent studying statistics is estimated as

$$\hat{\bar{y}}_\psi = \frac{1749.014}{647} = 2.70$$

hours with estimated variance, from (6.12),

$$\hat{V}(\hat{\bar{y}}_\psi) = \frac{1}{(647)^2}\frac{1}{5}\frac{1}{4}\sum_{i\in\mathcal{W}}\left(\frac{t_i}{\psi_i} - \frac{\hat{t}_\psi M_i}{M_0\psi_i}\right)^2 = \frac{49470.66}{(647)^2} = 0.118,$$

so the standard error of $\hat{\bar{y}}_\psi$ is 0.34 hours. ■

Warning: If N is small or some of the ψ_is are unusually large, it is possible that the sample will consist of one psu sampled n times. In that case, the estimated variance is zero; it is better to use sampling without replacement (see Section 6.4) if this may occur.

6.2.3 Designing the Selection Probabilities

We would like to choose the ψ_is so that the variance of $\hat{t}_\psi$ is as small as possible. Ideally, we would use $\psi_i = t_i/t$ (then $\hat{t}_\psi = t$ for all samples and $V[\hat{t}_\psi] = 0$), so if t_i is the annual income of the ith household, ψ_i would be the proportion of total income in the population that came from the ith household. But of course, the t_is are unknown until sampled. Even if the income were known before the survey was taken, we are often interested in more than one quantity; using income for designing the probabilities of selection may not work as well for estimating other quantities.

Because many totals in a psu are related to the number of elements in a psu, we often take ψ_i to be the proportion of elements in psu i or the relative size of psu i. Then, a large psu has a greater chance of being in the sample than a small psu. With M_i the number of elements in the ith psu and $M_0 = \sum_{i=1}^N M_i$ the number of elements in the population, we take $\psi_i = M_i/M_0$. With this choice of the probabilities ψ_i, we have **probability proportional to size** (pps) sampling. We used pps sampling in Example 6.2.

Then for one-stage pps sampling, $t_i/\psi_i = t_i M_0/M_i = M_0\bar{y}_i$, so $\hat{t}_\psi = \frac{1}{n}\sum_{i\in\mathcal{W}} M_0\bar{y}_i$ and $\hat{\bar{y}}_\psi = \frac{1}{n}\sum_{i\in\mathcal{W}}\bar{y}_i$. With $\psi_i = M_i/M_0$, $\hat{\bar{y}}_\psi$ is the average of the sampled psu means. Also, for $\psi_i = M_i/M_0$, (6.11) implies that $\hat{M}_{0\psi} = M_0$ for every possible sample, so from (6.12), $\hat{V}(\hat{\bar{y}}_\psi) = \frac{1}{n}\frac{1}{n-1}\sum_{i\in\mathcal{W}}(\bar{y}_i - \hat{\bar{y}}_\psi)^2$. Note that $\hat{V}(\hat{\bar{y}}_\psi)$ is of the form s^2/n, where s^2 is the sample variance of the psu means $\bar{y}_i$.

All of the work in pps sampling has been done in the sampling design itself. The pps estimates can be calculated simply by treating the $\bar{y}_i$s as individual observations, and finding their mean and sample variance. In practice, however, there are usually some deviations from a strict pps scheme, so you should use (6.5) and (6.6) for estimating the population total and its estimated variance.

What if you do not know the value of M_i for each psu in the population? In that case, you may know the value of a quantity that is related to M_i. If sampling fish, you may not know the number of fish in a haul but you may know the total weight of fish in a haul. You can then use $x_i =$ (total weight of fish in haul i) to set the selection probability for haul i as $\psi_i = x_i/t_x$. Since x_i/t_x is not exactly the same as M_i/M_0, you then must use (6.5) and (6.6) for estimating the population total of y and its standard error.

6.2.4 Weights in Unequal-Probability Sampling with Replacement

As in other types of sampling, we estimate the population total t using weights. In without-replacement sampling, we use the reciprocal of the inclusion probability ($= 1/E[Z_i]$) as the weight for a unit; $E[Z_i]$ is the expected number of times unit i appears in the sample (expected number of "hits"). In with-replacement sampling, we use the first-stage weight

$$w_i = \frac{1}{\text{expected number of hits}} = \frac{1}{E[Q_i]} = \frac{1}{n\psi_i}.$$

With this choice of weight, we have, for $\hat{t}_\psi$ in Equation (6.5),

$$\hat{t}_\psi = \sum_{i \in \mathcal{W}} w_i t_i.$$

In one-stage cluster sampling with replacement, we observe all of the M_i ssus every time psu i is selected, so we define the weight for ssu j of psu i as

$$w_{ij} = w_i = \frac{1}{n\psi_i}.$$

Then, in terms of the elements,

$$\hat{t}_\psi = \sum_{i \in \mathcal{W}} \sum_{j=1}^{M_i} w_{ij} y_{ij}$$

and

$$\hat{\bar{y}}_\psi = \frac{\sum_{i \in \mathcal{W}} \sum_{j=1}^{M_i} w_{ij} y_{ij}}{\sum_{i \in \mathcal{W}} \sum_{j=1}^{M_i} w_{ij}}.$$

Example 6.5. The weight for each psu in Example 6.4 is $w_i = 1/(n\psi_i)$. Thus, class 12 has weight 647/(24*5), each appearance of class 14 has weight 647/(100*5), class 5 has weight 647/(76*5), and class 1 has weight 647/(44*5). Since one-stage cluster sampling is used, these are also the weights for every student in the respective classes. The estimated population total is

$$\sum_{i \in \mathcal{W}} w_i t_i = \frac{647}{24(5)} 75 + \frac{647}{100(5)} 203 + \frac{647}{100(5)} 203 + \frac{647}{76(5)} 191 + \frac{647}{44(5)} 168 = 1749.014. \blacksquare$$

If the selection probabilities ψ_i are unequal, a one-stage cluster sample is not self-weighting; elements in large psus have smaller weights than elements in small psus. But we can obtain a self-weighting sample by selecting psus with probabilities proportional to size and then subsampling the same number of elements in each sampled psu. The next section shows how that works.

6.3 Two-Stage Sampling with Replacement

To select a two-stage cluster sample with replacement, first select a with-replacement sample of psus, where the ith psu is selected on a draw with known probability ψ_i. As in one-stage sampling with replacement, Q_i is the number of times psu i occurs in the sample. Then take a probability sample of m_i subunits in the ith psu. Simple random sampling without replacement or systematic sampling is often used to select the subsample, although any probability sampling method may be used.

The estimators for two-stage unequal-probability sampling with replacement are almost the same as those for one-stage sampling. The only difference is that in two-stage sampling, we must estimate t_i. If psu i is in the sample more than once, there are Q_i estimators of the total for psu i: $\hat{t}_{i1}, \hat{t}_{i2}, \ldots, \hat{t}_{iQ_i}$.

The subsampling procedure needs to meet two requirements:

1. Whenever psu i is selected to be in the sample, the same subsampling design is used to select secondary sampling units (ssus) from that psu. Different subsamples from the same psu, though, must be sampled independently. Thus, if you decide before sampling that you will take an SRS of size 5 from psu 42 if it is selected, every time psu 42 appears in the sample you must generate a different set of random numbers to select 5 of the ssus in psu 42. If you just take one subsample of size 5, and use it more than once for psu 42, you do not have independent subsamples and (6.14) will not be an unbiased estimator of the variance.

2. The jth subsample taken from psu i (for $j = 1, \ldots, Q_i$) is selected in such a way that $E[\hat{t}_{ij}] = t_i$. Because the same procedure is used each time psu i is selected for the sample, we can define $V[\hat{t}_{ij}] = V_i$ for all j.

The estimators from one-stage unequal sampling with replacement are modified slightly to allow for different subsamples in psus that are selected more than once:

$$\hat{t}_{\psi} = \frac{1}{n} \sum_{i=1}^{N} \sum_{j=1}^{Q_i} \frac{\hat{t}_{ij}}{\psi_i}. \tag{6.13}$$

$$\hat{V}(\hat{t}_{\psi}) = \frac{1}{n} \frac{1}{n-1} \sum_{i=1}^{N} \sum_{j=1}^{Q_i} \left(\frac{\hat{t}_{ij}}{\psi_i} - \hat{t}_{\psi} \right)^2. \tag{6.14}$$

In Exercise 20, you will show that (6.14) is an unbiased estimator of the variance $V(\hat{t}_{\psi})$, given in (6.46). Because sampling is with replacement, and hence it is possible to have more than one subsample from a given psu, the variance estimator captures both parts of the variance: the part due to the variability among psus, and the part that arises because t_i is estimated from a subsample rather than observed. When the psus are sampled with replacement, and when an independent subsample is chosen each time a psu is selected, the variance estimator can be calculated in the same way as if the psu totals were measured rather than estimated.

The estimator of the population mean $\bar{y}_{\mathcal{U}}$ has a form similar to (6.10):

$$\hat{\bar{y}}_{\psi} = \frac{\hat{t}_{\psi}}{\hat{M}_{0\psi}},$$

where

$$\hat{M}_{0\psi} = \frac{1}{n} \sum_{i \in \mathcal{W}} \frac{M_i}{\psi_i}$$

estimates the total number of elements in the population. The variance estimator again uses the ratio results in (6.12):

$$\hat{V}(\hat{\bar{y}}_\psi) = \frac{1}{(\hat{M}_{0\psi})^2} \frac{1}{n} \frac{1}{n-1} \sum_{i=1}^{N} \sum_{j=1}^{Q_i} \left(\frac{\hat{t}_{ij}}{\psi_i} - \frac{\hat{\bar{y}}_\psi M_i}{\psi_i} \right)^2 . \tag{6.15}$$

The weights for the observation units include a factor to reflect the subsampling within each psu. If an SRS of size m_i is taken in psu i, the weight for ssu j in psu i is

$$w_{ij} = \frac{1}{n\psi_i} \frac{M_i}{m_i}. \tag{6.16}$$

In a pps sample, in which the ith psu is selected with probability $\psi_i = M_i/M_0$, the weight for ssu j of psu i is $w_{ij} = \dfrac{M_0}{nM_i} \dfrac{M_i}{m_i} = \dfrac{M_0}{nm_i}$; a pps sample is self-weighting if all m_is are equal.

Summary: Steps for taking a two-stage unequal-probability sample with replacement

1. Determine the probabilities of selection ψ_i, the number n of psus to be sampled, and the subsampling procedure to be used within each psu. With any method of selecting the psus, we take a probability sample of ssus within the psus: often in two-stage cluster sampling, we take an SRS (without replacement) of elements within the chosen psus.

2. Select n psus with probabilities ψ_i and with replacement. The cumulative size method, Lahiri's method, or survey software may be used to select the psus for the sample.

3. Use the procedure determined in Step 1 to select subsamples from the psus chosen. If a psu occurs in the sample more than once, independent subsamples are used for each replicate.

4. Estimate the population total t from each psu in the sample as though it were the only one selected. The result is n estimates of the form $\hat{t}_{ij}/\psi_i$.

5. $\hat{t}_\psi$ is the average of the n estimates in Step 4. Alternatively, calculate $\hat{t}_\psi$ as $\sum_{i\in\mathcal{W}} \sum_{i\in\mathcal{S}_i} w_{ij} y_{ij}$ for w_{ij} in (6.16).

6. $\text{SE}(\hat{t}_\psi) = (1/\sqrt{n})$ (sample standard deviation of the n estimates in Step 4).

Example 6.6. Let's return to the situation in Example 6.4. Now suppose we subsample five students in each class rather than observing t_i. The estimation process is almost the same as in Example 6.4. The response y_{ij}, given in Table 6.5, is the total number of hours student j in class i spent studying statistics last week. Note that class 14 appears twice in the sample; each time it appears, a different subsample is collected.

Classes were selected with probability proportional to the number of students in the class, so $\psi_i = M_i/M_0$. Subsampling the same number of students in each class results in a self-weighting sample, with each student having weight

$$w_{ij} = \frac{M_0}{nM_i} \frac{M_i}{5} = \frac{647}{(5)(5)} = 25.88.$$

The population total is estimated as

$$\hat{t}_\psi = \sum_{i\in\mathcal{W}} \sum_{j\in\mathcal{S}_i} w_{ij} y_{ij} = 25.88(2 + 3 + 2.5 + \ldots + 3 + 2 + 5) = 1617.5,$$

TABLE 6.5
Calculations for Example 6.6.

Class	M_i	ψ_i	y_{ij}	$\bar{y}_i$	$\hat{t}_i$	$\hat{t}_i/\psi_i$
12	24	0.0371	2, 3, 2.5, 3, 1.5	2.4	57.6	1552.8
14	100	0.1546	2.5, 2, 3, 0, 0.5	1.6	160.0	1035.2
14	100	0.1546	3, 0.5, 1.5, 2, 3	2.0	200.0	1294.0
5	76	0.1175	1, 2.5, 3, 5, 2.5	2.8	212.8	1811.6
1	44	0.0680	4, 4.5, 3, 2, 5	3.7	162.8	2393.9
			average			1617.5
			std. dev.			521.628

or, equivalently, as $\hat{t}_\psi = \frac{1}{n}\sum_{i\in\mathcal{W}} \hat{t}_i/\psi_i = 1617.5$. The standard error is calculated using the standard deviation of the five values of $\hat{t}_i/\psi_i$ as in (6.14):

$$\text{SE}(\hat{t}_\psi) = \frac{521.628}{\sqrt{5}} = 233.28.$$

From this sample, the average amount of time a student spent studying statistics is

$$\hat{\bar{y}}_\psi = \frac{1617.5}{647} = 2.5$$

hours with standard error $233.28/647 = 0.36$ hour. A 95% CI for the average amount of time studying statistics is

$$2.5 \pm 2.78(0.36) = [1.50, 3.50];$$

because the sample has 5 psus, the critical value from a t distribution with 4 df is used to calculate the interval. ∎

Example 6.7. Let's see what happens if we use unequal-probability sampling on the puppy homes considered in Example 5.9. Take ψ_i proportional to the number of puppies in the home, so that Puppy Palace with 30 puppies is sampled with probability 3/4 and Dog's Life with 10 puppies is sampled with probability 1/4. As before, once a puppy home is chosen, take an SRS of two puppies in the home. Then if Puppy Palace is selected, $\hat{t}_\psi = \hat{t}_{\text{PP}}/(3/4) = (30)(4)/(3/4) = 160$. If Dog's Life is chosen, $\hat{t}_\psi = \hat{t}_{\text{DL}}/(1/4) = (10)(4)/(1/4) = 160$. Thus, either possible sample results in an estimated average of $\hat{\bar{y}}_\psi = 160/40 = 4$ legs per puppy, and the variance of the estimator is zero. ∎

Sampling with replacement has the advantage that it is easy to select the sample and to obtain estimates of the population total and its variance. If N is small, however, as occurs in many highly stratified complex surveys with few clusters in each stratum, sampling with replacement may be less efficient than sampling without replacement. The next section discusses advantages and challenges of sampling without replacement.

6.4 Unequal-Probability Sampling without Replacement

Generally, sampling with replacement is less efficient than sampling without replacement; with-replacement sampling is introduced first because of the ease in selecting and analyzing samples. Nevertheless, in large surveys with many small strata, the inefficiencies may wipe out the gains in convenience. Much research has been done on unequal-probability sampling without replacement; the theory is more complicated because the probability that a unit is selected is different for the first unit chosen than for the second, third, and subsequent units. When you understand the probabilistic arguments involved, however, you can find the properties of any sampling scheme.

Example 6.8. The supermarket example from Section 6.1 can be used to illustrate some of the features of unequal-probability sampling with replacement. Here is the population again:

Store	**Size** (m^2)	ψ_i	t_i **(in Thousands)**
A	100	$\frac{1}{16}$	11
B	200	$\frac{2}{16}$	20
C	300	$\frac{3}{16}$	24
D	1000	$\frac{10}{16}$	245
Total	1600	1	300

Let's select two psus without replacement and with unequal probabilities. As in Sections 6.1 to 6.3, let

$$\psi_i = P(\text{Select unit } i \text{ on first draw}).$$

Since we are sampling without replacement, though, the probability that unit j is selected on the second draw depends on which unit was selected on the first draw.

One way to select the units with unequal probabilities is to use ψ_i as the probability of selecting unit i on the first draw, and then adjust the probabilities of selecting the other stores on the second draw. Suppose we drew cards from a deck with cards numbered 1 through 16, choosing store A if the card drawn is 1, store B if the card is 2 or 3, C if the card is 4, 5, or 6, and D if the card is between 7 and 16. If store A was chosen on the first draw, then for selecting the second store we would remove Card 1, shuffle, and draw one of the remaining 15 cards. Thus,

$$P\text{ (store A chosen on first draw)} = \psi_A = \frac{1}{16}$$

and

$$P\text{ (B chosen on second draw} \mid \text{A chosen on first draw)} = \frac{\frac{2}{16}}{1-\frac{1}{16}} = \frac{\psi_B}{1-\psi_A}.$$

The denominator is the sum of the ψ_is for stores B, C, and D. In general,

$$\begin{aligned}
&P(\text{unit } i \text{ chosen first, unit } k \text{ chosen second})\\
&\qquad = P(\text{unit } i \text{ chosen first})\; P\text{ (unit } k \text{ chosen second} \mid \text{unit } i \text{ chosen first)}\\
&\qquad = \psi_i \frac{\psi_k}{1-\psi_i}.
\end{aligned}$$

TABLE 6.6
Joint inclusion probabilities (π_{ik}) for the possible samples of size 2 in Example 6.8. The margins give the inclusion probabilities π_i for the four stores.

		Store k				
		A	B	C	D	π_i
Store i	A	—	0.0173	0.0269	0.1458	0.1900
	B	0.0173	—	0.0556	0.2976	0.3705
	C	0.0269	0.0556	—	0.4567	0.5393
	D	0.1458	0.2976	0.4567	—	0.9002
	π_k	0.1900	0.3705	0.5393	0.9002	2.0000

Similarly,

$$P\text{ (unit } k \text{ chosen first, unit } i \text{ chosen second)} = \psi_k \frac{\psi_i}{1-\psi_k}.$$

Note that P(unit i chosen first, unit k chosen second) is not the same as P(unit k chosen first, unit i chosen second): The order of selection makes a difference! By adding the probabilities of the two choices, though, we can find the probability that a sample of size 2 consists of psus i and k:

$$\text{For } n=2, \;\; P\text{ (units } i \text{ and } k \text{ in sample)} = \pi_{ik} = \psi_i \frac{\psi_k}{1-\psi_i} + \psi_k \frac{\psi_i}{1-\psi_k}.$$

The probability that psu i is in the sample is then

$$\pi_i = \sum_{\mathcal{S}:\, i \in \mathcal{S}} P(\mathcal{S}).$$

Table 6.6 gives the π_is and π_{ik}s, rounded to four decimal places, for the supermarkets. ■

6.4.1 The Horvitz–Thompson Estimator for One-Stage Sampling

Assume we have a without-replacement sample of n psus, and we know the **inclusion probability**

$$\pi_i = P(\text{unit } i \text{ in sample})$$

and the **joint inclusion probability**

$$\pi_{ik} = P(\text{units } i \text{ and } k \text{ are both in the sample}).$$

The inclusion probability π_i can be calculated as the sum of the probabilities of all samples containing the ith unit and has the property that

$$\sum_{i=1}^{N} \pi_i = n. \tag{6.17}$$

For the π_{ik}s, as shown in Theorem 6.1 of Section 6.6,

$$\sum_{\substack{k=1 \\ k \neq i}}^{N} \pi_{ik} = (n-1)\pi_i. \tag{6.18}$$

Because the inclusion probabilities sum to n, we can think of π_i/n as the "average probability" that a unit will be selected on one of the draws. Recall that for one-stage sampling with replacement, $\hat{t}_\psi$ is the average of the values of t_i/ψ_i for psus in the sample. But when samples are drawn without replacement, the probabilities of selection depend on what was drawn before. Instead of dividing the total t_i from psu i by ψ_i, we divide by the *average* probability of selecting that unit in a draw, π_i/n. We then have the **Horvitz–Thompson (HT) estimator** of the population total (Horvitz and Thompson, 1952):

$$\hat{t}_{\text{HT}} = \sum_{i \in \mathcal{S}} \frac{t_i}{\pi_i} = \sum_{i=1}^{N} Z_i \frac{t_i}{\pi_i}, \tag{6.19}$$

where $Z_i = 1$ if psu i is in the sample, and 0 otherwise.

The Horvitz–Thompson estimator is an unbiased estimator of t because $P(Z_i = 1) = E[Z_i] = \pi_i$:

$$E[\hat{t}_{\text{HT}}] = \sum_{i=1}^{N} \pi_i \frac{t_i}{\pi_i} = t.$$

We shall show in Theorems 6.2 and 6.3 of Section 6.6 that the variance of the Horvitz–Thompson estimator in one-stage sampling is

$$V(\hat{t}_{\text{HT}}) = \sum_{i=1}^{N} \frac{1-\pi_i}{\pi_i} t_i^2 + \sum_{i=1}^{N} \sum_{k \neq i}^{N} \frac{\pi_{ik} - \pi_i \pi_k}{\pi_i \pi_k} t_i t_k \tag{6.20}$$

$$= \frac{1}{2} \sum_{i=1}^{N} \sum_{\substack{k=1 \\ k \neq i}}^{N} (\pi_i \pi_k - \pi_{ik}) \left(\frac{t_i}{\pi_i} - \frac{t_k}{\pi_k} \right)^2. \tag{6.21}$$

The expression in (6.21) is the Sen–Yates–Grundy (SYG) form of the variance (Sen, 1953; Yates and Grundy, 1953). You can see from (6.21) that the variance of the Horvitz–Thompson estimator is 0 if t_i is exactly proportional to π_i.

The expressions for the variance in (6.20) and (6.21) are algebraically identical (see Theorem 6.3 of Section 6.6). When the inclusion probabilities π_i or the joint inclusion probabilities π_{ik} are unequal, however, substituting sample quantities into (6.20) or (6.21) leads to different estimators of the variance.

The estimator of the variance starting from (6.20), suggested by Horvitz and Thompson (1952), is

$$\hat{V}_{\text{HT}}(\hat{t}_{\text{HT}}) = \sum_{i \in \mathcal{S}} (1-\pi_i) \frac{t_i^2}{\pi_i^2} + \sum_{i \in \mathcal{S}} \sum_{\substack{k \in \mathcal{S} \\ k \neq i}} \frac{\pi_{ik} - \pi_i \pi_k}{\pi_{ik}} \frac{t_i}{\pi_i} \frac{t_k}{\pi_k}. \tag{6.22}$$

The SYG estimator, working from (6.21), is

$$\hat{V}_{\text{SYG}}(\hat{t}_{\text{HT}}) = \frac{1}{2} \sum_{i \in \mathcal{S}} \sum_{\substack{k \in \mathcal{S} \\ k \neq i}} \frac{\pi_i \pi_k - \pi_{ik}}{\pi_{ik}} \left(\frac{t_i}{\pi_i} - \frac{t_k}{\pi_k} \right)^2. \tag{6.23}$$

Theorem 6.4 in Section 6.6 shows that (6.22) and (6.23) are both unbiased estimators of the variance in (6.21). The SYG form in (6.23) is generally more stable and is preferred for most applications.

TABLE 6.7
Variance estimates for all possible without-replacement samples of size 2 in Example 6.9.

Sample, $\mathcal{S}$	$P(\mathcal{S})$	$\hat{t}_{\text{HT}}$	$\hat{V}_{\text{HT}}(\hat{t}_{\text{HT}})$	$\hat{V}_{\text{SYG}}(\hat{t}_{\text{HT}})$
{A, B}	0.01726	111.87	−14,691.5	47.1
{A, C}	0.02692	102.39	−10,832.1	502.8
{A, D}	0.14583	330.06	4,659.3	7,939.8
{B, C}	0.05563	98.48	−9,705.1	232.7
{B, D}	0.29762	326.15	5,682.8	5,744.1
{C, D}	0.45673	316.67	6,782.8	3,259.8

Example 6.9. Let's look at the Horvitz–Thompson estimator for a sample of 2 supermarkets in Example 6.8 with joint inclusion probabilities given in Table 6.6. We use the draw-by-draw method to select the sample. To select the first psu, we generate a random integer from $\{1, \ldots, 16\}$: the random integer we generate is 12, which tells us that store D is selected on the first draw. We then remove the values $\{7, \ldots, 16\}$ corresponding to store D, and generate a second random integer from $\{1, \ldots, 6\}$; we generate 6, which tells us to select store C on the second draw.

Using the values of π_i in Table 6.6, the Horvitz–Thompson estimate of the total sales for sample {C, D} is

$$\hat{t}_{\text{HT}} = \sum_{i \in \mathcal{S}} \frac{t_i}{\pi_i} = \frac{245}{0.9002} + \frac{24}{0.5393} = 316.6639.$$

Since for this example we know the entire population, we can calculate the theoretical variance of $\hat{t}_{\text{HT}}$ using (6.21):

$$V(\hat{t}_{\text{HT}}) = \frac{1}{2} \sum_{i=1}^{N} \sum_{k \neq i}^{N} (\pi_i \pi_k - \pi_{ik}) \left(\frac{t_i}{\pi_i} - \frac{t_k}{\pi_k} \right)^2 = 4383.6.$$

[We obtain the same value, 4383.6, if we use the equivalent formulation in (6.20).]

We have two estimates of the variance from sample {C, D}. From (6.22),

$$\begin{aligned} \hat{V}_{\text{HT}}(\hat{t}_{\text{HT}}) &= \frac{(1-0.9002)(245)^2}{(0.9002)^2} + \frac{(1-0.5393)(24)^2}{(0.5393)^2} \\ &\quad + 2\,\frac{0.4567 - (0.9002)(0.5393)}{0.4567} \left(\frac{245}{0.9002} \right) \left(\frac{24}{0.5393} \right) \\ &= 6782.8. \end{aligned}$$

The SYG estimate, from (6.23), is

$$\hat{V}_{\text{SYG}}(\hat{t}_{\text{HT}}) = \frac{(0.9002)(0.5393) - 0.4567}{0.4567} \left(\frac{245}{0.9002} - \frac{24}{0.5393} \right)^2 = 3259.8.$$

Because all values in this population are known, we can examine the estimators for all possible samples selected according to the probabilities in Table 6.6. Results are given in Table 6.7. For three of the possible samples, $\hat{V}_{\text{HT}}(\hat{t}_{\text{HT}})$ is negative! This is true even though $\hat{V}_{\text{HT}}(\hat{t}_{\text{HT}})$ and $\hat{V}_{\text{SYG}}(\hat{t}_{\text{HT}})$ are unbiased estimators of $V_{\text{HT}}(\hat{t}_{\text{HT}})$; it is easy to check for this example that

$$\sum_{\text{possible samples } \mathcal{S}} P(\mathcal{S}) \hat{V}_{\text{HT}}(\hat{t}_{\text{HT},\mathcal{S}}) = \sum_{\text{possible samples } \mathcal{S}} P(\mathcal{S}) \hat{V}_{\text{SYG}}(\hat{t}_{\text{HT},\mathcal{S}}) = 4383.6. \quad \blacksquare$$

Example 6.9 demonstrates a problem that can arise in estimating the variance of $\hat{t}_{\text{HT}}$: The unbiased estimators in (6.22) or (6.23) can take on negative values in some unequal-probability designs! [See Exercise 30 for a situation in which (6.23) is negative.] In some designs, the estimates of the variance can be widely disparate for different samples—in other words, the variance estimator $\hat{V}$ has high variance itself. The stability can sometimes be improved by careful choice of the sampling design, but in general, the calculations are cumbersome.

In addition, the estimators in (6.22) and (6.23) can be difficult to use in practice because they require knowledge of the joint inclusion probabilities π_{ik} (see Särndal, 1996). Public-use data sets from large-scale surveys commonly include a variable of weights that can be used to calculate the Horvitz–Thompson estimator. But it is generally impractical to provide the joint inclusion probabilities π_{ik}—this would require an additional $n(n-1)/2$ values to be included in the data set, where n is often large. In addition, for many surveys it is challenging to calculate the joint inclusion probabilities π_{ik}.

Simplified variance estimation. An alternative suggested by Durbin (1953), which avoids some of the potential instability and computational complexity, is to pretend the units were selected with replacement and use the with-replacement variance estimator in (6.9) rather than (6.22) or (6.23). The with-replacement variance estimator, setting $\psi_i = \pi_i/n$, is

$$\hat{V}_{\text{WR}}(\hat{t}_{\text{HT}}) = \frac{1}{n}\frac{1}{n-1}\sum_{i\in\mathcal{S}}\left(\frac{t_i}{\psi_i} - \hat{t}_{\text{HT}}\right)^2 = \frac{n}{n-1}\sum_{i\in\mathcal{S}}\left(\frac{t_i}{\pi_i} - \frac{\hat{t}_{\text{HT}}}{n}\right)^2. \tag{6.24}$$

The variance estimator in (6.24) is always non-negative, so you can avoid the potential embarrassment of trying to explain a negative variance estimate. In addition, the with-replacement variance estimator does not require knowledge of the joint inclusion probabilities π_{ik}. If without-replacement sampling is more efficient than with-replacement sampling, the with-replacement variance estimator in (6.24) is expected to overestimate the variance and result in conservative confidence intervals, but the bias is expected to be small if the sampling fraction n/N is small. The commonly used computer-intensive methods described in Chapter 9 calculate the with-replacement variance.

In general, we recommend using the with-replacement variance estimator in (6.24). When the sampling fraction n/N is large, however, this can overestimate the variance. Some survey software packages will calculate the SYG variance estimate if the user provides the π_{ik}'s. Berger (2004), Brewer and Donadio (2003), and Thompson and Wu (2008) suggested alternatives for estimating $V(\hat{t}_{\text{HT}})$ when the joint inclusion probabilities are unknown (see Exercises 35 and 36).

Example 6.10. Let's select an unequal-probability sample without replacement of size 15 from the file `agpop.csv`. In Example 4.2, we used the variable *acres87* as auxiliary information in ratio estimation. We now use it in the sample design, selecting counties with probability proportional to *acres87*. The data for the sample, along with the joint inclusion probabilities, are in file `agpps.csv`.

The Horvitz–Thompson estimate of the total for *acres92* is

$$\hat{t}_{\text{HT}} = \sum_{i\in\mathcal{S}}\frac{t_i}{\pi_i} = 936{,}291{,}172,$$

where t_i is the value of *acres92* for county i in the sample. The three variance estimates are: $\hat{V}_{\text{HT}}(\hat{t}_{\text{HT}}) = 4.97\times10^{15}$, $\hat{V}_{\text{SYG}}(\hat{t}_{\text{HT}}) = 1.37\times10^{14}$, and $\hat{V}_{\text{WR}}(\hat{t}_{\text{HT}}) = 1.51\times10^{14}$. Because of the instability of $\hat{V}_{\text{HT}}(\hat{t}_{\text{HT}})$, we prefer to use either $\hat{V}_{\text{SYG}}(\hat{t}_{\text{HT}})$ or $\hat{V}_{\text{WR}}(\hat{t}_{\text{HT}})$ to estimate the variance. For this sample, $\hat{V}_{\text{WR}}(\hat{t}_{\text{HT}})$ is quite close to the SYG estimate because the sampling fraction n/N is small. The square root of the with-replacement variance estimate is $\text{SE}(\hat{t}_{\text{HT}}) = 12{,}293{,}010$.

Note the gain in efficiency from using unequal-probability sampling. From Example 2.7, an SRS of size 300 gave a standard error of 58,169,381 for the estimated total of *acres92*. The unequal-probability sample has a smaller standard error even though the sample size is only 15 because of the high correlation between *acres92* and *acres87*. Using the auxiliary information in the variable *acres87* in the design results in a large gain in efficiency. ■

6.4.2 Selecting the psus

For the supermarkets in Example 6.8, the draw-by-draw selection probabilities ψ_i are proportional to the store sizes. The inclusion probabilities π_is, however, are not proportional to the sizes of the stores—in fact, they cannot be proportional to the store sizes, because Store D accounts for more than half of the total floor area but cannot be sampled with a probability greater than one. The π_is that result from this draw-by-draw method due to Yates and Grundy (1953) may or may not be the desired probabilities of inclusion in the sample; you may need to adjust the ψ_is to obtain a pre-specified set of π_is. Such adjustments become difficult for large populations and for sample sizes larger than two.

Many methods have been proposed for selecting psus without replacement so that desired inclusion probabilities are attained. Systematic sampling can be used to draw a sample without replacement and is relatively simple to implement, but many of the π_{ik}'s for the population are zero. Brewer and Hanif (1983) presented more than 50 methods for selecting without-replacement unequal-probability samples. Most of these methods are for $n = 2$; three of the methods are described in Exercises 31, 33, and 34. Some methods are easier to compute, some are more suitable for specific applications, and some result in a more stable estimator of $V(\hat{t}_{\text{HT}})$. The sample in Example 6.10 was selected using a method developed by Hanurav (1967) and Vijayan (1968). Tillé (2006) gave general algorithms for selecting without-replacement unequal-probability samples, and most of these are implemented in SAS and R software (see Lohr, 2022; Lu and Lohr, 2022).

6.4.3 The Horvitz–Thompson Estimator for Two-Stage Sampling

The Horvitz–Thompson estimator for two-stage sampling is similar to the estimator for one-stage sampling in (6.19): We substitute an unbiased estimator $\hat{t}_i$ of the psu total for the unknown value of t_i, obtaining

$$\hat{t}_{\text{HT}} = \sum_{i \in \mathcal{S}} \frac{\hat{t}_i}{\pi_i} = \sum_{i=1}^{N} Z_i \frac{\hat{t}_i}{\pi_i}, \tag{6.25}$$

where $Z_i = 1$ if psu i is in the sample, and 0 otherwise.

The two-stage Horvitz–Thompson estimator is an unbiased estimator of t as long as $E[\hat{t}_i] = t_i$ for each psu i (see Theorem 6.2 in Section 6.6). We shall show in Section 6.6, using Equations (6.39) through (6.41), that the variance of the Horvitz–Thompson estimator for two-stage sampling is

$$V(\hat{t}_{\text{HT}}) = \sum_{i=1}^{N} \frac{1-\pi_i}{\pi_i} t_i^2 + \sum_{i=1}^{N} \sum_{k \neq i}^{N} \frac{\pi_{ik} - \pi_i \pi_k}{\pi_i \pi_k} t_i t_k + \sum_{i=1}^{N} \frac{V(\hat{t}_i)}{\pi_i} \tag{6.26}$$

$$= \frac{1}{2} \sum_{i=1}^{N} \sum_{\substack{k=1 \\ k \neq i}}^{N} (\pi_i \pi_k - \pi_{ik}) \left(\frac{t_i}{\pi_i} - \frac{t_k}{\pi_k} \right)^2 + \sum_{i=1}^{N} \frac{V(\hat{t}_i)}{\pi_i}. \tag{6.27}$$

The expression in (6.27) is again the SYG form. The first part of the variance is the same as for one-stage sampling [see (6.20) and (6.21)]. The last term is the additional variability due to estimating the t_is rather than measuring them exactly.

The Horvitz–Thompson estimator of the variance in two-stage cluster sampling is

$$\hat{V}_{\text{HT}}(\hat{t}_{\text{HT}}) = \sum_{i\in\mathcal{S}}(1-\pi_i)\frac{\hat{t}_i^2}{\pi_i^2} + \sum_{i\in\mathcal{S}}\sum_{\substack{k\in\mathcal{S}\\k\neq i}}\frac{\pi_{ik}-\pi_i\pi_k}{\pi_{ik}}\frac{\hat{t}_i}{\pi_i}\frac{\hat{t}_k}{\pi_k} + \sum_{i\in\mathcal{S}}\frac{\hat{V}(\hat{t}_i)}{\pi_i}, \tag{6.28}$$

and the SYG estimator is

$$\hat{V}_{\text{SYG}}(\hat{t}_{\text{HT}}) = \frac{1}{2}\sum_{i\in\mathcal{S}}\sum_{\substack{k\in\mathcal{S}\\k\neq i}}\frac{\pi_i\pi_k-\pi_{ik}}{\pi_{ik}}\left(\frac{\hat{t}_i}{\pi_i}-\frac{\hat{t}_k}{\pi_k}\right)^2 + \sum_{i\in\mathcal{S}}\frac{\hat{V}(\hat{t}_i)}{\pi_i}. \tag{6.29}$$

Theorem 6.4 in Section 6.6 shows that both are unbiased estimators of the variance in (6.27); however, just as in one-stage sampling, either can be negative in practice.

For most situations, we recommend using the with-replacement sampling variance estimator:

$$\hat{V}_{\text{WR}}(\hat{t}_{\text{HT}}) = \frac{1}{n}\frac{1}{n-1}\sum_{i\in\mathcal{S}}\left(\frac{n\hat{t}_i}{\pi_i}-\hat{t}_{\text{HT}}\right)^2 = \frac{n}{n-1}\sum_{i\in\mathcal{S}}\left(\frac{\hat{t}_i}{\pi_i}-\frac{\hat{t}_{\text{HT}}}{n}\right)^2. \tag{6.30}$$

The with-replacement variance estimator for two-stage sampling has exactly the same form as the estimator in (6.24) for one-stage sampling; the only difference is that we substitute the estimator $\hat{t}_i$ for the ith psu population total t_i. We saw in Section 6.3 that the with-replacement variance estimator captures the variability at both stages of sampling. This results in the tremendous practical advantage that the variance estimation method depends only on information at the first-stage level of the design. You do not have to use properties of the subsampling design at all for the variance estimation.

6.4.4 Weights in Unequal-Probability Samples

All without-replacement sampling designs discussed so far in the book can be considered as special cases of two-stage cluster sampling with (possibly) unequal probabilities. The formulas for unbiased estimation of totals in without-replacement sampling in Chapters 2, 3, 5, and 6 are special cases of (6.25) through (6.29). In Example 6.16, we will derive the formulas in Chapter 5 from the general Horvitz–Thompson results. You will show that the formulas for stratified sampling are a special case of Horvitz–Thompson estimation in Exercise 22.

As in earlier chapters, we can write the Horvitz–Thompson estimator using sampling weights. The first-stage sampling weight for psu i is

$$w_i = \frac{1}{\pi_i}.$$

Thus, the Horvitz–Thompson estimator for the population total is

$$\hat{t}_{\text{HT}} = \sum_{i\in\mathcal{S}} w_i\hat{t}_i.$$

For a without-replacement probability sample of ssus within psus, we define, using the notation of Särndal et al. (1992),

$$\pi_{j|i} = P(\, j\text{th ssu in } i\text{th psu included in sample} \mid i\text{th psu is in the sample}).$$

Then,

$$\hat{t}_i = \sum_{j \in \mathcal{S}_i} \frac{y_{ij}}{\pi_{j|i}}.$$

The overall probability that ssu j of psu i is included in the sample is $\pi_{j|i}\pi_i$. Thus, we can define the sampling weight for the (i, j)th ssu as

$$w_{ij} = \frac{1}{\pi_{j|i}\pi_i} \tag{6.31}$$

and the Horvitz–Thompson estimator of the population total as

$$\hat{t}_{\text{HT}} = \sum_{i \in \mathcal{S}} \sum_{j \in \mathcal{S}_i} w_{ij} y_{ij}. \tag{6.32}$$

The population mean is estimated by

$$\hat{\bar{y}}_{\text{HT}} = \frac{\displaystyle\sum_{i \in \mathcal{S}} \sum_{j \in \mathcal{S}_i} w_{ij} y_{ij}}{\displaystyle\sum_{i \in \mathcal{S}} \sum_{j \in \mathcal{S}_i} w_{ij}}. \tag{6.33}$$

The estimator $\hat{\bar{y}}_{\text{HT}}$ is a ratio, so, using the results from Chapter 4, we estimate its variance by forming the residuals from the estimated psu totals. Let

$$\hat{e}_i = \hat{t}_i - \hat{\bar{y}}_{\text{HT}} \hat{M}_i,$$

where $\hat{M}_i = \sum_{j \in \mathcal{S}_i} (1/\pi_{j|i})$ estimates the number of ssus in psu i. Note that $\hat{e}_i/\pi_i = \sum_{j \in \mathcal{S}_i} w_{ij}(y_{ij} - \hat{\bar{y}}_{\text{HT}})$ and $\sum_{i \in \mathcal{S}} \hat{e}_i/\pi_i = 0$. We then use the with-replacement variance in (6.30), with $\hat{e}_i/\hat{M}_0$ substituted for $\hat{t}_i$, to obtain:

$$\hat{V}_{\text{WR}}(\hat{\bar{y}}_{\text{HT}}) = \frac{n}{n-1} \sum_{i \in \mathcal{S}} \left(\frac{\hat{e}_i}{\hat{M}_0 \pi_i} \right)^2 = \frac{n}{n-1} \sum_{i \in \mathcal{S}} \left(\frac{\displaystyle\sum_{j \in \mathcal{S}_i} w_{ij}(y_{ij} - \hat{\bar{y}}_{\text{HT}})}{\displaystyle\sum_{k \in \mathcal{S}} \sum_{j \in \mathcal{S}_i} w_{kj}} \right)^2, \tag{6.34}$$

where $\hat{M}_0 = \sum_{i \in \mathcal{S}} \hat{M}_i = \sum_{i \in \mathcal{S}} \sum_{j \in \mathcal{S}_i} w_{ij}$ estimates M_0, the number of ssus in the population. Survey software will calculate these quantities for you.

Example 6.11. Let's take a two-stage unequal-probability sample without replacement from the population of statistics classes in Example 6.2. We want the psu inclusion probabilities to be proportional to the class sizes M_i given in Table 6.2. The data are in file `classpps.csv` and in Table 6.8.

The weight for each student in this sample is

$$w_{ij} = \frac{1}{\pi_i \pi_{j|i}} = \frac{1}{(5M_i/M_0)(4/M_i)} = \frac{M_0}{(5)(4)} = \frac{647}{20} = 32.35.$$

Since the same number of students ($m_i = 4$) is selected from each class and since the psu inclusion probabilities π_i are proportional to the class sizes M_i, the sample of students is self-weighting.

The estimated total number of hours spent studying statistics is

$$\hat{t}_{\text{HT}} = \sum_{i \in \mathcal{S}} \sum_{j \in \mathcal{S}_i} w_{ij} y_{ij} = 2232.15.$$

TABLE 6.8
Data from two-stage sample of introductory statistics classes.

Class	M_i	π_i	w_{ij}	y_{ij}	$w_{ij}y_{ij}$	$\hat{t}_i$	$\dfrac{\hat{t}_i}{\pi_i}$	$\left(\dfrac{\hat{t}_i}{\pi_i} - \dfrac{\hat{t}_{\text{HT}}}{5}\right)^2$	$\left(\dfrac{\hat{e}_i}{\hat{M}_0\pi_i}\right)^2$
4	22	0.17002	32.35	5	161.750	110.00	646.983	40,222.54	0.09609
4	22	0.17002	32.35	4.5	145.575				
4	22	0.17002	32.35	5.5	177.925				
4	22	0.17002	32.35	5	161.750				
10	34	0.26275	32.35	2	64.700	106.25	404.377	1,768.23	0.00423
10	34	0.26275	32.35	4	129.400				
10	34	0.26275	32.35	3	97.050				
10	34	0.26275	32.35	3.5	113.225				
1	44	0.34003	32.35	5	161.750	154.00	452.901	41.91	0.00010
1	44	0.34003	32.35	3	97.050				
1	44	0.34003	32.35	4	129.400				
1	44	0.34003	32.35	2	64.700				
9	54	0.41731	32.35	3.5	113.225	195.75	469.076	512.96	0.00123
9	54	0.41731	32.35	4	129.400				
9	54	0.41731	32.35	1	32.350				
9	54	0.41731	32.35	6	194.100				
14	100	0.77280	32.35	2	64.700	200.00	258.799	35,204.25	0.08410
14	100	0.77280	32.35	1.5	48.525				
14	100	0.77280	32.35	1.5	48.525				
14	100	0.77280	32.35	3	97.050				
Sum			647.00		2232.150		2232.150	77,749.90	0.18574

This can also be calculated by $\hat{t}_{\text{HT}} = \sum_{i\in\mathcal{S}} \hat{t}_i/\pi_i = 2232.15$. Using the with-replacement variance estimate in (6.30),

$$\hat{V}_{\text{WR}}(\hat{t}_{\text{HT}}) = \frac{n}{n-1}\sum_{i\in\mathcal{S}}\left(\frac{\hat{t}_i}{\pi_i} - \frac{\hat{t}_{\text{HT}}}{n}\right)^2 = \frac{5}{4}(77{,}749.9) = 97{,}187.4,$$

giving a standard error of $\sqrt{97{,}187.4} = 311.7$. For this example, since $n = 5$ is large relative to $N = 15$, this standard error is likely an overestimate; in Exercise 18 you will calculate the without-replacement variance estimates in (6.28) and (6.29).

We estimate the mean number of hours spent studying statistics by

$$\hat{\bar{y}}_{\text{HT}} = \frac{\sum_{i\in\mathcal{S}}\sum_{j\in\mathcal{S}_i} w_{ij}y_{ij}}{\sum_{i\in\mathcal{S}}\sum_{j\in\mathcal{S}_i} w_{ij}} = \frac{2232.15}{647} = 3.45.$$

Using (6.34),

$$\hat{V}_{\text{WR}}(\hat{\bar{y}}_{\text{HT}}) = \frac{n}{n-1}\sum_{i\in\mathcal{S}}\left(\frac{e_i}{\hat{M}_0\pi_i}\right)^2 = \frac{5}{4}(0.18574) = 0.23218,$$

so $\text{SE}(\hat{\bar{y}}_{\text{HT}}) = \sqrt{0.23218} = 0.482$. A 95% CI for the mean is [2.11, 4.79], calculated using a t distribution with 4 df. ■

6.5 Examples of Unequal-Probability Samples

Many sampling situations are well suited for unequal-probability samples. This section gives four examples of sampling designs in common use.

Example 6.12. *Large National Survey Conducted in Person.* In the early stages of the COVID-19 pandemic in 2020, one pressing question was how many people had been infected by the SARS-CoV-2 virus that caused the illness. Some persons infected by the virus were asymptomatic; other persons, with symptoms, were not counted in the official statistics if they were not tested or did not seek medical assistance. Thus, statistics about persons who had tested positive or received medical treatment would underestimate the number of persons who had been infected.

Pollán et al. (2020) reported results from a probability sample conducted to estimate the prevalence of infection in Spain. The virus had affected some areas of the country more than others, and virus transmission was thought to be related to population density, so the population was stratified by the 50 provinces and two autonomous cities (to capture regional differences and allow calculation of province-level estimates) and by size of municipality within the provinces (to capture differences by population density). The sample was disproportionately allocated to strata so that estimates for smaller provinces would achieve a pre-specified precision; each eligible person within a particular stratum, however, had approximately the same probability of being chosen for the sample.

Within each stratum, census tracts were sampled with probability proportional to population. Then 24 households were randomly selected from each sampled tract, and all residents of the sampled households were invited to participate in the survey. Altogether, 35,883 households, containing 102,562 persons, were selected for the sample; of these, 66,805 persons were eligible and agreed to participate. Participants were asked questions about history of symptoms that were compatible with COVID-19, their contact with suspected or confirmed cases, and risk factors. They also were given a rapid fingerprick test for antibodies to the virus, and asked to donate a blood sample for further laboratory analyses. Using the final survey weights, which adjusted the sampling weights for nonresponse, Pollán et al. (2020) estimated the national seroprevalence rate for the rapid antibody test to be 5%. They further estimated that about 1/3 of cases were asymptomatic. Seroprevalence varied in different regions, however, and was above 10% in central provinces but below 3% in most provinces along the coast.

This survey had three stages of clustering within each stratum. The psus were census tracts, the ssus were households, and the tsus (tertiary sampling units) were persons within households. Because Pollán et al. (2020) sampled the psus with probability proportional to size, and then sampled the same number of ssus in each psu (and thus approximately the same number of persons in each psu), the sample that was selected in each stratum—including both respondents and nonrespondents—was approximately self-weighting (note, though, that weights varied across strata because of the disproportional allocation). Moreover, the workload for interviewing and testing was approximately equal for each sampled census tract and known in advance.

Many large surveys that are conducted in person use pps sampling for the same reasons as this survey. A cluster sample is needed to control travel costs as well as limit the costs of establishing survey operations across the country. The pps sampling of psus gives a sample in which persons in a stratum have approximately equal probabilities of being selected for the sample, yet the survey workload is approximately the same for each psu. ■

Example 6.13. *Mitofsky–Waksberg Method for Random Digit Dialing.* In the United States, telephone numbers have the form

$$\begin{array}{ccccc} \text{area code} & + & \text{prefix} & + & \text{suffix} \\ \text{(3 digits)} & & \text{(3 digits)} & & \text{(4 digits).} \end{array}$$

An SRS of residential telephone numbers can be obtained by randomly generating 10-digit numbers and then discarding those that, when called, do not belong to a household. Random digit element sampling guarantees that all numbers, whether found in a telephone directory or not, have an equal chance of appearing in the sample. But much of the survey effort is wasted on calling numbers that do not belong to a household.

Mitofsky (1970) and Waksberg (1978) developed a cluster-sampling method for sampling residential telephone numbers that reduced the proportion of calling effort spent on nonresidential numbers. The following describes the "sampler's utopia" procedure in which everyone answers the phone (Brick and Tucker, 2007).

First, form the sampling frame of psus by appending each of the numbers 00 to 99 to each possible combination of area code and prefix in the region of interest. The resulting list of psus consists of the set of possible first eight digits for the 10-digit telephone numbers in the population. Each psu in the frame contains the numbers (abc)-def-gh00 to (abc)-def-gh99, and is called a 100-bank of numbers.

Select a psu at random from the list of all psus, and also select a number randomly between 00 and 99 to serve as the last two digits. Call that telephone number. If the selected number is residential, interview the household and choose its psu (the 100-bank of telephone numbers having the same first eight digits as the selected number) to be in the sample. Continue sampling in that psu until a total of k interviews are obtained (or all numbers in the psu have been called). If the original number selected in the psu is not residential, reject that psu. Repeat the psu selection and calling until the desired number of interviews is obtained.

Under ideal conditions, the Mitofsky–Waksberg procedure samples each psu with probability proportional to the size of its set of residential telephone numbers. If all psus in the sample have at least k residential telephone numbers, then the procedure gives each residential telephone number the same probability of being selected in the sample—the result is a self-weighting sample of residential telephone numbers.

To see this, let M_i be the number of residential telephone numbers in psu i, and let N be the total number of psus in the sampling frame. The probability that psu i is selected to be in the sample on the first draw is M_i/M_0, where $M_0 = \sum_{i=1}^{N} M_i$ (see Exercise 38), even though the values of M_i and M_0 are unknown. Then, if each psu in the population has either $M_i = 0$ or $M_i \geq k$,

$$\begin{aligned} P(\text{number selected}) &= P(\text{psu } i \text{ selected})\, P(\text{number selected} \mid \text{psu } i \text{ selected}) \\ &\propto \frac{M_i}{M_0} \frac{k}{M_i} = \frac{k}{M_0}. \end{aligned}$$

The sampling weight for each number in the sample is M_0/k; to estimate a population total, you would need to know M_0, the total number of residential telephone numbers in the population. To estimate an average or proportion, the typical goal of telephone surveys, you do not need to know M_0. You only need to know a relative weight w_{ij} for each response y_{ij} in the sample, and can estimate the population mean using (6.33). With a self-weighting sample, you can use relative weights of $w_{ij} = 1$.

In the 1970s and 1980s, the Mitofsky–Waksberg procedure increased calling efficiency because the psus of 100-banks for landline telephone numbers were clustered—some psus were unassigned, some tended to be assigned to commercial establishments, and some were largely residential. The procedure concentrated the calling effort in the residential psus. The method does not result in the same level of efficiency gains with cellular telephone numbers, however, and is less commonly used for telephone sampling where most persons have cellular telephones (McGeeney and Kennedy, 2017; Kennedy et al., 2018).

But there are many other situations in which one would like to sample psus with probability proportional to the number of units in a subpopulation of interest but that number is unknown. If the subpopulation is concentrated in a subset of the psus, the Mitofsky–Waksberg method can yield a larger sample size from that subpopulation without increasing costs. For example, retail workers who have had COVID-19 may be concentrated in certain retail establishments, students who have been crime victims may be concentrated in some colleges, or wells with high level of pollutants may be concentrated in some geographic regions. We'll discuss other methods for sampling members of subpopulations in Chapter 14. ■

Example 6.14. *3-P Sampling.* **P**robability **P**roportional to **P**rediction (3-P) sampling, described by Schreuder et al. (1968), has been used as a sampling scheme in forestry. Suppose an investigator wants to estimate the total volume of timber in an area. Several options are available: (1) Estimate the volume for each tree in the area. There may be thousands of trees, however, and this can be very time consuming. (2) Use a cluster sample in which plots of equal areas are selected, and measure the volume of every tree in the selected plots. (3) Use an unequal-probability sampling scheme in which points in the area are selected at random, and include the trees closest to the points in the sample. In this design, a tree is selected with probability proportional to the area of the region that is closer to that tree than to any other tree. (4) Estimate the volume of each tree by eye and then select trees with probability proportional to the estimated volume. When done in one pass, with trees selected as the volume is estimated, this is 3-P sampling—the prediction P stands for the predicted (estimated) volume used in determining the π_is.

The largest trees tend to produce the most timber and contribute most to the variability of the estimate of total volume. Thus, unequal-probability sampling can be expected to lead to less sampling effort. Theoretically, you could estimate the volume of each of the N trees in the forest by eye, obtaining a value x_i for tree i. Then, you could revisit trees randomly selected with probabilities proportional to x_i, and carefully measure the volume t_i. Such a procedure, however, requires you to make two trips through the forest and adds much work to the sampling process. In 3-P sampling, only one trip is made through the forest, and trees are selected for the sample at the same time the x_is are measured. The procedure is as follows:

1. Estimate or guess what the maximum value of x_i for the trees is likely to be. Define a value L that is larger than your estimated maximum value of x_i.

2. Proceed to a tree in the forest, and determine x_i for that tree. Generate a random number u_i in $[0, L]$. If $u_i \leq x_i$, then measure the volume y_i on that tree; otherwise, skip that tree and go on to the next tree.

3. Repeat Step 2 on every tree in the forest.

The unequal-probability sampling in this case essentially gives every board-foot of timber an equal chance of being selected for the sample. Note that the size of the unequal-probability sample is unknown until sampling is completed. The probability that tree i is included in the sample is $\pi_i = x_i/L$. The Horvitz–Thompson estimator is

$$\hat{t}_{\text{HT}} = \sum_{i \in \mathcal{S}} \frac{y_i}{\pi_i} = L \sum_{i \in \mathcal{S}} \frac{y_i}{x_i} = \sum_{i=1}^{N} Z_i \frac{y_i}{\pi_i},$$

where $Z_i = 1$ if tree i is in the sample, and 0 otherwise. The Z_is are independent random variables with $P(Z_i = 1) = \pi_i$ and $P(Z_i = 0) = 1 - \pi_i$, so 3-P sampling is a special case of a method known as **Poisson sampling**. The sample size is the random variable $\sum_{i=1}^{N} Z_i$ with expected value $\sum_{i=1}^{N} \pi_i$.

Because the sample size is variable rather than fixed, Poisson sampling provides a different method of unequal-probability sampling than those discussed in Sections 6.1 through 6.4. Exercise 27 examines some of the properties of Poisson sampling. ■

Example 6.15. *Dollar Unit Sampling.* An accountant auditing the accounts receivable amounts for a company often takes a sample to estimate the true total accounts receivable balance. The book value x_i is known for each account in the population; the audited value t_i will be known only for accounts in the sample. The auxiliary information x_i could be used in ratio or regression estimation to improve the precision from an SRS of accounts, as described in Chapter 4.

Or, instead of being used in the analysis, the book values could be used in the design of the sample. You could stratify the accounts by the value of x_i, or you could take an unequal-probability sample with inclusion probabilities proportional to x_i. (Or you could do both: First stratify, then sample with unequal probabilities within each stratum.) If you sample accounts with probabilities proportional to x_i, then each individual dollar in the book values has the same probability of being selected in the sample (hence the name *dollar unit sampling*). With each dollar equally likely to be included in the sample, an account with book value \$10,000 is ten times as likely to be in the sample as an account with book value \$1,000.

Consider a client with 87 accounts receivable, with a book balance of \$612,824. The auditor has decided that a sample of size 25 will be sufficient for estimating the error in accounts receivable and takes a random sample with replacement of the 612,824 dollars in the book value population. As individual dollars can only be audited as part of the whole account, each dollar selected serves as a "hook" to snag the whole account for audit. The cumulative-size method is used to select psus (accounts) for this example; often, in practice, auditors take a systematic sample of dollars and their accompanying psus. A systematic sample guarantees that accounts with book values greater than the sampling interval will be included in the sample. Table 6.9 shows the first few lines of the account selection; the full table is in file `auditselect.csv`. Here, accounts 3 and 13 are included once, and account 9 is included twice (but only needs to be audited once since this is a one-stage cluster sample). This is thus an example of one-stage pps sampling with replacement, as discussed in Section 6.2.

The selected accounts are audited, and the audit values are recorded in the file `auditresult.csv`. The overstatement in each sampled account is calculated as (book value − audit value). Table 6.10 gives part of a spreadsheet that may be used to estimate the total overstatement. Using the results from Section 6.2, the total overstatement is estimated from (6.5) to be \$4,334 with standard error $\$13{,}547/\sqrt{25} = \$2{,}709$ from (6.6). In many auditing situations, however, most of the audited values agree with the book values, so most of the differences are zeros. A CI based on a normal approximation does not perform well in this situation, so auditors typically use a CI based on the Poisson or multinomial distribution (see Neter et al., 1978) rather than a CI of the form $(\hat{t} \pm 1.96\,\text{SE})$.

Another way of looking at the unequal-probability estimate is to find the overstatement for each individual dollar in the sample. Account 24, for example, has a book value of \$7090 and an error of \$40. The error is prorated to every dollar in the book value, leading to an overstatement of \$0.00564 for each of the 7090 dollars. The average overstatement for the individual dollars in the sample is \$0.007071874, so the total overstatement for the population is estimated as $(0.007071874)(612{,}824) = 4{,}334$. ■

TABLE 6.9
Account selection for audit sample (partial listing).

Account	Book Value	Cumulative Book Value	Random Number	
1	2,459	2,459		
2	2,343	4,802		
3	6,842	11,644	11,016	
4	4,179	15,823		
5	750	16,573		
6	2,708	19,281		
7	3,073	22,354		
8	4,742	27,096		
9	16,350	43,446	31,056	38,500
10	5,424	48,870		
11	9,539	58,409		
12	3,108	61,517		
13	3,935	65,452	63,047	
14	900	66,352		

TABLE 6.10
Results of the audit on accounts in the sample.

Account (Audit Unit)	Book Value	ψ_i	Audit Value	Book − Audit Difference	Diff/ψ_i	Difference per Dollar
3	6,842	0.0111647	6,842	0	0	0.00000
9	16,350	0.0266798	16,350	0	0	0.00000
9	16,350	0.0266798	16,350	0	0	0.00000
13	3,935	0.0064211	3,935	0	0	0.00000
24	7,090	0.0115694	7,050	40	3,457	0.00564
29	5,533	0.0090287	5,533	0	0	0.00000
⋮	⋮	⋮	⋮	⋮	⋮	⋮
75	2,291	0.0037384	2,191	100	26,749	0.04365
79	4,667	0.0076156	4,667	0	0	0.00000
81	31,257	0.0510049	31,257	0	0	0.00000
		average			4,334	0.007071874
		std. dev.			13,547	0.02210527

6.6 Randomization Theory Results and Proofs*

In two-stage cluster sampling, we select the psus first and then select subunits within the sampled psus. One approach to calculate a theoretical variance for any estimator in multi-stage sampling is to condition on which psus are included in the sample. To do this, we need to use Properties 4 (successive conditioning) and 5 (calculating variances conditionally) of conditional expectation from Table A.2 in Appendix A.

In this section, we state and prove Theorem 6.2, the Horvitz–Thompson Theorem (Horvitz and Thompson, 1952), which gives the properties of the estimator in (6.25). In Theorem 6.4, we find unbiased estimators of the variance. We then show that the variance for cluster sampling with equal probabilities in (5.24) follows as a special case of these theorems. First, however, we prove (6.17) and (6.18) in Theorem 6.1.

Throughout this section, let

$$Z_i = \begin{cases} 1 & \text{if psu } i \text{ is in the sample} \\ 0 & \text{if psu } i \text{ is not in the sample} \end{cases} \tag{6.35}$$

denote the random variable specifying whether psu i is included in the sample or not. The probability that psu i is included in the sample is

$$\pi_i = P(Z_i = 1) = E(Z_i); \tag{6.36}$$

the probability that both psu i and psu k $(i \neq k)$ are included in the sample is

$$\pi_{ik} = P(Z_i = 1 \text{ and } Z_k = 1) = E(Z_i Z_k). \tag{6.37}$$

Theorem 6.1. For a without-replacement probability sample of n units, let Z_i, π_i, and π_{ik} be as defined in (6.35)–(6.37). Then

$$\sum_{i=1}^{N} \pi_i = n$$

and

$$\sum_{\substack{k=1 \\ k \neq i}}^{N} \pi_{ik} = (n-1)\pi_i.$$

Proof of Theorem 6.1. Since the sample size is n, $\sum_{i=1}^{N} Z_i = n$ for every possible sample. Also,

$$E\left[Z_i\right] = E\left[Z_i^2\right] = \pi_i$$

because $P(Z_i = 1) = \pi_i$. Consequently,

$$n = E\left[\sum_{i=1}^{N} Z_i\right] = \sum_{i=1}^{N} \pi_i.$$

In addition,

$$\sum_{\substack{k=1 \\ k \neq i}}^{N} \pi_{ik} = \sum_{\substack{k=1 \\ k \neq i}}^{N} E[Z_i Z_k] = E[Z_i(n - Z_i)] = \pi_i(n-1),$$

which completes the proof. ■

Theorem 6.2. (*Horvitz–Thompson*) Let Z_i, π_i, and π_{ik} be as defined in (6.35)–(6.37). Suppose that sampling is done at the second stage so that sampling in any psu is independent of the sampling in any other psu, and that $\hat{t}_i$ is independent of $(Z_1, \ldots, Z_N)$ with $E\left[\hat{t}_i\right] = E\left[\hat{t}_i \mid Z_1, \ldots, Z_N\right] = t_i$. Then

$$E\left[\sum_{i=1}^{N} Z_i \frac{\hat{t}_i}{\pi_i}\right] = \sum_{i=1}^{N} \pi_i \frac{t_i}{\pi_i} = t \tag{6.38}$$

and

$$V\left[\sum_{i=1}^{N} Z_i \frac{\hat{t}_i}{\pi_i}\right] = V_{\text{psu}} + V_{\text{ssu}}, \tag{6.39}$$

where

$$V_{\text{psu}} = V\left[\sum_{i=1}^{N} Z_i \frac{t_i}{\pi_i}\right] = \sum_{i=1}^{N}(1-\pi_i)\frac{t_i^2}{\pi_i} + \sum_{i=1}^{N}\sum_{\substack{k=1\\k\neq i}}^{N}(\pi_{ik}-\pi_i\pi_k)\frac{t_i}{\pi_i}\frac{t_k}{\pi_k} \tag{6.40}$$

and

$$V_{\text{ssu}} = \sum_{i=1}^{N}\frac{V(\hat{t}_i)}{\pi_i}. \tag{6.41}$$

Proof of Theorem 6.2. First note that

$$\text{Cov}\,(Z_i, Z_k) = \begin{cases} \pi_i(1-\pi_i) & \text{if } i = k \\ \pi_{ik} - \pi_i\pi_k & \text{if } i \neq k. \end{cases}$$

We use successive conditioning to show (6.38):

$$E\left[\sum_{i=1}^{N} Z_i\frac{\hat{t}_i}{\pi_i}\right] = E\left[E\left(\sum_{i=1}^{N} Z_i\frac{\hat{t}_i}{\pi_i}\,\middle|\, Z_1,\ldots,Z_N\right)\right] = E\left[\sum_{i=1}^{N} Z_i\frac{t_i}{\pi_i}\right] = \sum_{i=1}^{N}\pi_i\frac{t_i}{\pi_i} = t.$$

The first step above simply applies successive conditioning; in the second step, we use the independence of $\hat{t}_i$ and $(Z_1,\ldots,Z_N)$.

To find the variance, we use the expression for calculating the variance conditionally in Property 5 of Table A.2, and again use the independence of $\hat{t}_i$ and $(Z_1,\ldots,Z_N)$:

$$\begin{aligned}
V\left[\sum_{i=1}^{N} Z_i\frac{\hat{t}_i}{\pi_i}\right] &= V\left[E\left(\sum_{i=1}^{N} Z_i\frac{\hat{t}_i}{\pi_i}\,\middle|\, Z_1,\ldots,Z_N\right)\right] + E\left[V\left(\sum_{i=1}^{N} Z_i\frac{\hat{t}_i}{\pi_i}\,\middle|\, Z_1,\ldots,Z_N\right)\right] \\
&= V\left[\sum_{i=1}^{N} Z_i\frac{t_i}{\pi_i}\right] + E\left[\sum_{i=1}^{N} Z_i^2\frac{V(\hat{t}_i)}{\pi_i^2}\right] \\
&= \sum_{i=1}^{N}\sum_{k=1}^{N}\frac{t_i}{\pi_i}\frac{t_k}{\pi_k}\,\text{Cov}\,(Z_i,Z_k) + \sum_{i=1}^{N}\pi_i\frac{V(\hat{t}_i)}{\pi_i^2} \\
&= \sum_{i=1}^{N}\pi_i(1-\pi_i)\frac{t_i^2}{\pi_i^2} + \sum_{i=1}^{N}\sum_{\substack{k=1\\k\neq i}}^{N}(\pi_{ik}-\pi_i\pi_k)\frac{t_i}{\pi_i}\frac{t_k}{\pi_k} + \sum_{i=1}^{N}\frac{V(\hat{t}_i)}{\pi_i}. \quad \blacksquare
\end{aligned}$$

Equation (6.38) establishes that the Horvitz–Thompson estimator is unbiased, and (6.39) through (6.41) show that (6.26) is the variance of the Horvitz–Thompson estimator. In one-stage cluster sampling, $V(\hat{t}_i) = 0$ for $i \in \mathcal{S}$, so $V_{\text{ssu}} = 0$ and $V(\hat{t}_{\text{HT}}) = V_{\text{psu}}$ as given in (6.20).

We now show that the Horvitz–Thompson form of the variance in (6.20) and the SYG form in (6.21) are equivalent.

Theorem 6.3. Let V_{psu} be as defined in (6.40). Then

$$\begin{aligned}
V_{\text{psu}} &= \sum_{i=1}^{N}(1-\pi_i)\frac{t_i^2}{\pi_i} + \sum_{i=1}^{N}\sum_{\substack{k=1\\k\neq i}}^{N}(\pi_{ik}-\pi_i\pi_k)\frac{t_i}{\pi_i}\frac{t_k}{\pi_k} \\
&= \frac{1}{2}\sum_{i=1}^{N}\sum_{\substack{k=1\\k\neq i}}^{N}(\pi_i\pi_k-\pi_{ik})\left(\frac{t_i}{\pi_i}-\frac{t_k}{\pi_k}\right)^2.
\end{aligned}$$

Proof of Theorem 6.3. Starting with the SYG form in (6.21),

$$\frac{1}{2}\sum_{i=1}^{N}\sum_{\substack{k=1\\k\neq i}}^{N}(\pi_i\pi_k - \pi_{ik})\left(\frac{t_i}{\pi_i} - \frac{t_k}{\pi_k}\right)^2 = \frac{1}{2}\sum_{i=1}^{N}\sum_{\substack{k=1\\k\neq i}}^{N}(\pi_i\pi_k - \pi_{ik})\left(\frac{t_i^2}{\pi_i^2} + \frac{t_k^2}{\pi_k^2} - 2\frac{t_i}{\pi_i}\frac{t_k}{\pi_k}\right).$$

From results (6.17) and (6.18), proven in Theorem 6.1, and noting that $\pi_{ik} = \pi_{ki}$,

$$\frac{1}{2}\sum_{i=1}^{N}\sum_{\substack{k=1\\k\neq i}}^{N}\pi_i\pi_k\left(\frac{t_i^2}{\pi_i^2} + \frac{t_k^2}{\pi_k^2}\right) = \frac{1}{2}\sum_{i=1}^{N}\sum_{k=1}^{N}\pi_i\pi_k\left(\frac{t_i^2}{\pi_i^2} + \frac{t_k^2}{\pi_k^2}\right) - \sum_{i=1}^{N}\pi_i^2\frac{t_i^2}{\pi_i^2} = \sum_{i=1}^{N}(n-\pi_i)\frac{t_i^2}{\pi_i}$$

and

$$\sum_{i=1}^{N}\sum_{\substack{k=1\\k\neq i}}^{N}\pi_{ik}\frac{t_i^2}{\pi_i^2} = \sum_{i=1}^{N}\sum_{\substack{k=1\\k\neq i}}^{N}\pi_{ik}\frac{t_k^2}{\pi_k^2} = (n-1)\sum_{i=1}^{N}\frac{t_i^2}{\pi_i}.$$

Thus,

$$\frac{1}{2}\sum_{i=1}^{N}\sum_{\substack{k=1\\k\neq i}}^{N}(\pi_i\pi_k - \pi_{ik})\left(\frac{t_i}{\pi_i} - \frac{t_k}{\pi_k}\right)^2 = \sum_{i=1}^{N}[n - \pi_i - (n-1)]\frac{t_i^2}{\pi_i} - \sum_{i=1}^{N}\sum_{\substack{k=1\\k\neq i}}^{N}(\pi_i\pi_k - \pi_{ik})\frac{t_i}{\pi_i}\frac{t_k}{\pi_k},$$

which shows the equality of the two expressions for the variance. ■

Theorem 6.4 shows that (6.28) and (6.29) are unbiased estimators for the variance in (6.26) and (6.27); the one-stage variance estimators in (6.20) and (6.21) follow as a special case when $V(\hat{t}_i) = 0$.

Theorem 6.4. Suppose the conditions of Theorem 6.2 hold, and that $\hat{V}(\hat{t}_i)$ is an unbiased estimator of $V(\hat{t}_i)$ that is independent of Z_i. Then

$$E\left[\sum_{i=1}^{N} Z_i \frac{\hat{V}(\hat{t}_i)}{\pi_i^2}\right] = V_{\text{ssu}}, \tag{6.42}$$

$$\begin{aligned} E&\left[\sum_{i=1}^{N} Z_i(1-\pi_i)\frac{\hat{t}_i^2}{\pi_i^2} + \sum_{i=1}^{N}\sum_{\substack{k=1\\k\neq i}}^{N} Z_iZ_k\frac{\pi_{ik} - \pi_i\pi_k}{\pi_{ik}}\frac{\hat{t}_i}{\pi_i}\frac{\hat{t}_k}{\pi_k}\right] \\ &= E\left[\frac{1}{2}\sum_{i=1}^{N}\sum_{\substack{k=1\\k\neq i}}^{N} Z_iZ_k\frac{\pi_i\pi_k - \pi_{ik}}{\pi_{ik}}\left(\frac{\hat{t}_i}{\pi_i} - \frac{\hat{t}_k}{\pi_k}\right)^2\right] \\ &= V_{\text{psu}} + \sum_{i=1}^{N}(1-\pi_i)\frac{V(\hat{t}_i)}{\pi_i}, \end{aligned} \tag{6.43}$$

and

$$E\left[\hat{V}_{\text{HT}}(\hat{t}_{\text{HT}})\right] = E\left[\hat{V}_{\text{SYG}}(\hat{t}_{\text{HT}})\right] = V_{\text{psu}} + V_{\text{ssu}}. \tag{6.44}$$

Proof of Theorem 6.4. We prove (6.42) by using successive conditioning:

$$E\left[Z_i\frac{\hat{V}(\hat{t}_i)}{\pi_i^2}\right] = E\left[E\left(Z_i\frac{\hat{V}(\hat{t}_i)}{\pi_i^2}\;\middle|\; Z_i\right)\right] = E\left[Z_i\frac{V(\hat{t}_i)}{\pi_i^2}\right] = \frac{V(\hat{t}_i)}{\pi_i}.$$

Result (6.42) follows by summation.

To prove (6.43), note that because $\hat{t}_i$ and $(Z_1, \ldots, Z_N)$ are independent,

$$E\left[\hat{t}_i^2 \mid Z_1, \ldots, Z_N\right] = E\left[\hat{t}_i^2\right] = t_i^2 + V\left(\hat{t}_i\right).$$

Thus,

$$\begin{aligned}
E\left[\sum_{i=1}^N Z_i(1-\pi_i)\frac{\hat{t}_i^2}{\pi_i^2}\right] &= E\left[E\left(\sum_{i=1}^N Z_i(1-\pi_i)\frac{\hat{t}_i^2}{\pi_i^2}\;\middle|\; Z_1, \ldots, Z_N\right)\right] \\
&= E\left[\sum_{i=1}^N Z_i\frac{1-\pi_i}{\pi_i^2}\left\{t_i^2 + V(\hat{t}_i)\right\}\right] \\
&= \sum_{i=1}^N \frac{1-\pi_i}{\pi_i}[t_i^2 + V(\hat{t}_i)].
\end{aligned}$$

Because subsampling is done independently in different psus, $E[\hat{t}_i\,\hat{t}_k] = t_i\,t_k$ for $k \neq i$, so

$$\begin{aligned}
E\left[\sum_{i=1}^N\sum_{\substack{k=1\\k\neq i}}^N Z_iZ_k\frac{\pi_{ik}-\pi_i\pi_k}{\pi_{ik}}\frac{\hat{t}_i}{\pi_i}\frac{\hat{t}_k}{\pi_k}\right] &= E\left[E\left(\sum_{i=1}^N\sum_{\substack{k=1\\k\neq i}}^N Z_iZ_k\frac{\pi_{ik}-\pi_i\pi_k}{\pi_{ik}}\frac{\hat{t}_i}{\pi_i}\frac{\hat{t}_k}{\pi_k}\;\middle|\; Z_1, \ldots, Z_N\right)\right] \\
&= E\left[\sum_{i=1}^N\sum_{\substack{k=1\\k\neq i}}^N Z_iZ_k\frac{\pi_{ik}-\pi_i\pi_k}{\pi_{ik}}\frac{t_i}{\pi_i}\frac{t_k}{\pi_k}\right] \\
&= \sum_{i=1}^N\sum_{\substack{k=1\\k\neq i}}^N(\pi_{ik}-\pi_i\pi_k)\frac{t_i}{\pi_i}\frac{t_k}{\pi_k}.
\end{aligned}$$

Combining the two results, we see that

$$E\left[\sum_{i=1}^N Z_i(1-\pi_i)\frac{\hat{t}_i^2}{\pi_i^2} + \sum_{i=1}^N\sum_{\substack{k=1\\k\neq i}}^N Z_iZ_k\frac{\pi_{ik}-\pi_i\pi_k}{\pi_{ik}}\frac{\hat{t}_i}{\pi_i}\frac{\hat{t}_k}{\pi_k}\right] = V_{\text{psu}} + \sum_{i=1}^N\frac{1-\pi_i}{\pi_i}V(\hat{t}_i),$$

which proves the first part of (6.43). We show the second part of (6.43) similarly, using results from Theorem 6.1:

$$\begin{aligned}
&E\left[\frac{1}{2}\sum_{i=1}^N\sum_{\substack{k=1\\k\neq i}}^N Z_iZ_k\frac{\pi_i\pi_k-\pi_{ik}}{\pi_{ik}}\left(\frac{\hat{t}_i}{\pi_i}-\frac{\hat{t}_k}{\pi_k}\right)^2\right] \\
&\quad= E\left\{E\left[\sum_{i=1}^N\sum_{\substack{k=1\\k\neq i}}^N Z_iZ_k\frac{\pi_i\pi_k-\pi_{ik}}{\pi_{ik}}\left(\frac{\hat{t}_i^2}{\pi_i^2}-\frac{\hat{t}_i}{\pi_i}\frac{\hat{t}_k}{\pi_k}\right)\;\middle|\; Z_1, \ldots, Z_N\right]\right\} \\
&\quad= E\left[\sum_{i=1}^N\sum_{\substack{k=1\\k\neq i}}^N Z_iZ_k\frac{\pi_i\pi_k-\pi_{ik}}{\pi_{ik}}\left(\frac{t_i^2+V(\hat{t}_i)}{\pi_i^2}-\frac{t_i}{\pi_i}\frac{t_k}{\pi_k}\right)\right]
\end{aligned}$$

$$= \sum_{i=1}^{N}\sum_{\substack{k=1\\k\neq i}}^{N}(\pi_i\pi_k - \pi_{ik})\left(\frac{t_i^2 + V(\hat{t}_i)}{\pi_i^2} - \frac{t_i}{\pi_i}\frac{t_k}{\pi_k}\right)$$

$$= \sum_{i=1}^{N}[\pi_i(n-\pi_i) - (n-1)\pi_i]\left(\frac{t_i^2 + V(\hat{t}_i)}{\pi_i^2}\right) + \sum_{i=1}^{N}\sum_{\substack{k=1\\k\neq i}}^{N}(\pi_{ik} - \pi_i\pi_k)\frac{t_i}{\pi_i}\frac{t_k}{\pi_k}$$

$$= \sum_{i=1}^{N}\frac{1-\pi_i}{\pi_i}t_i^2 + \sum_{i=1}^{N}\sum_{\substack{k=1\\k\neq i}}^{N}(\pi_{ik} - \pi_i\pi_k)\frac{t_i}{\pi_i}\frac{t_k}{\pi_k} + \sum_{i=1}^{N}\frac{1-\pi_i}{\pi_i}V(\hat{t}_i)$$

$$= V_{\text{psu}} + \sum_{i=1}^{N}\frac{1-\pi_i}{\pi_i}V(\hat{t}_i).$$

Equation (6.44) follows because

$$E\left[\sum_{i\in\mathcal{S}}\frac{\hat{V}(\hat{t}_i)}{\pi_i}\right] = E\left[\sum_{i=1}^{N}Z_i\frac{\hat{V}(\hat{t}_i)}{\pi_i}\right] = \sum_{i=1}^{N}V(\hat{t}_i). \quad \blacksquare$$

Example 6.16. We now show that the results in Section 5.3 are special cases of Theorems 6.2 and 6.4. If psus are selected with equal probabilities,

$$P(Z_i = 1) = \pi_i = \frac{n}{N},$$

$$P(Z_i = 1 \text{ and } Z_k = 1) = \pi_{ik} = \frac{n}{N}\frac{n-1}{N-1},$$

and

$$\hat{t}_{\text{unb}} = \sum_{i\in\mathcal{S}}\frac{N}{n}\hat{t}_i = \sum_{i=1}^{N}Z_i\frac{N}{n}\hat{t}_i = \sum_{i=1}^{N}Z_i\frac{\hat{t}_i}{\pi_i},$$

so we can apply Theorem 6.2 with $\pi_i = n/N$. Then,

$$E[\hat{t}_{\text{unb}}] = \sum_{i=1}^{N}\frac{n}{N}\frac{N}{n}t_i = t,$$

and, from (6.40),

$$V_{\text{psu}}\left[\hat{t}_{\text{unb}}\right] = \sum_{i=1}^{N}\frac{1-\pi_i}{\pi_i}t_i^2 + \sum_{i=1}^{N}\sum_{k\neq i}^{N}\frac{\pi_{ik} - \pi_i\pi_k}{\pi_i\pi_k}t_it_k$$

$$= \sum_{i=1}^{N}\left(1-\frac{n}{N}\right)\left(\frac{N}{n}\right)t_i^2 + \sum_{i=1}^{N}\sum_{\substack{k=1\\k\neq i}}^{N}\left[\frac{n}{N}\frac{n-1}{N-1} - \left(\frac{n}{N}\right)^2\right]\left(\frac{N}{n}\right)^2 t_it_k$$

$$= \frac{N}{n}\left(1-\frac{n}{N}\right)\left[\sum_{i=1}^{N}t_i^2 - \frac{1}{N-1}\sum_{i=1}^{N}\sum_{\substack{k=1\\k\neq i}}^{N}t_it_k\right]$$

$$= \frac{N}{n(N-1)}\left(1 - \frac{n}{N}\right)\left[(N-1)\sum_{i=1}^{N} t_i^2 - \sum_{i=1}^{N}\sum_{k=1}^{N} t_i t_k + \sum_{i=1}^{N} t_i^2\right]$$

$$= \frac{N}{n(N-1)}\left(1 - \frac{n}{N}\right)\left[N\sum_{i=1}^{N} t_i^2 - t^2\right]$$

$$= N^2\left(1 - \frac{n}{N}\right)\frac{S_t^2}{n}.$$

By result (2.12) from SRS theory,

$$V(\hat{t}_i) = M_i^2\left(1 - \frac{m_i}{M_i}\right)\frac{S_i^2}{m_i},$$

so, using (6.41),

$$V_{\text{ssu}} = \sum_{i=1}^{N} \frac{N}{n} M_i^2\left(1 - \frac{m_i}{M_i}\right)\frac{S_i^2}{m_i}.$$

This completes the proof of (5.24). In the special case of an SRS, $t_i = y_i$ and S_t^2 is the variance among population elements, so V_{psu} reduces to the formula in (2.19).

For two-stage cluster sampling with equal probabilities, we defined

$$s_t^2 = \frac{1}{n-1}\sum_{i \in \mathcal{S}}\left(\hat{t}_i - \frac{\hat{t}_{\text{unb}}}{N}\right)^2$$

to be the sample variance among the estimated psu totals in (5.25). We now show that, when $\pi_i = n/N$ and $\pi_{ik} = n(n-1)/[N(N-1)]$,

$$\sum_{i=1}^{N} Z_i(1-\pi_i)\frac{\hat{t}_i^2}{\pi_i^2} + \sum_{i=1}^{N}\sum_{\substack{k=1\\k\neq i}}^{N} Z_i Z_k \frac{\pi_{ik} - \pi_i\pi_k}{\pi_{ik}}\frac{\hat{t}_i}{\pi_i}\frac{\hat{t}_k}{\pi_k} = N^2\left(1 - \frac{n}{N}\right)\frac{s_t^2}{n},$$

so that Theorem 6.4 can be applied. Substituting n/N for π_i and $n(n-1)/[N(N-1)]$ for π_{ik},

$$\sum_{i=1}^{N} Z_i(1-\pi_i)\frac{\hat{t}_i^2}{\pi_i^2} + \sum_{i=1}^{N}\sum_{\substack{k=1\\k\neq i}}^{N} Z_i Z_k \frac{\pi_{ik} - \pi_i\pi_k}{\pi_{ik}}\frac{\hat{t}_i}{\pi_i}\frac{\hat{t}_k}{\pi_k}$$

$$= \left(\frac{N}{n}\right)^2\left(1 - \frac{n}{N}\right)\sum_{i=1}^{N} Z_i\hat{t}_i^2 + \left(\frac{N}{n}\right)^2\left[1 - \frac{n(N-1)}{N(n-1)}\right]\sum_{i=1}^{N}\sum_{\substack{k=1\\k\neq i}}^{N} Z_i Z_k \hat{t}_i\hat{t}_k$$

$$= \left(\frac{N}{n}\right)^2\left(1 - \frac{n}{N}\right)\sum_{i=1}^{N} Z_i\hat{t}_i^2 - \left(\frac{N}{n}\right)^2\frac{1}{n-1}\left(1 - \frac{n}{N}\right)\sum_{i=1}^{N}\sum_{\substack{k=1\\k\neq i}}^{N} Z_i Z_k \hat{t}_i\hat{t}_k$$

$$= \left(\frac{N}{n}\right)^2\left(1 - \frac{n}{N}\right)\left(\sum_{i=1}^{N} Z_i\hat{t}_i^2 - \frac{1}{n-1}\sum_{i=1}^{N}\sum_{k=1}^{N} Z_i Z_k \hat{t}_i\hat{t}_k + \frac{1}{n-1}\sum_{i=1}^{N} Z_i\hat{t}_i^2\right)$$

$$= \left(\frac{N}{n}\right)^2\left(1 - \frac{n}{N}\right)\frac{1}{n-1}\left[n\sum_{i=1}^{N} Z_i\hat{t}_i^2 - \left(\sum_{k=1}^{N} Z_k\hat{t}_k\right)^2\right]$$

$$= \left(\frac{N}{n}\right)^2 \left(1 - \frac{n}{N}\right) \frac{1}{n-1} \left[n \sum_{i=1}^{N} Z_i \hat{t}_i^2 - n^2 \left(\frac{\hat{t}_{\text{unb}}}{N}\right)^2 \right]$$
$$= \left(\frac{N^2}{n}\right) \left(1 - \frac{n}{N}\right) \frac{1}{n-1} \sum_{i=1}^{N} Z_i \left(\hat{t}_i - \frac{\hat{t}_{\text{unb}}}{N}\right)^2$$
$$= N^2 \left(1 - \frac{n}{N}\right) \frac{s_t^2}{n}.$$

Thus, by (6.43),

$$E\left[N^2 \left(1 - \frac{n}{N}\right) \frac{s_t^2}{n}\right] = N^2 \left(1 - \frac{n}{N}\right) \frac{S_t^2}{n} + \frac{N}{n}\left(1 - \frac{n}{N}\right) \sum_{i=1}^{N} V(\hat{t}_i). \tag{6.45}$$

Note that the expected value of s_t^2 is larger than S_t^2: s_t^2 includes the variation from psu total to psu total, plus variation from not knowing the psu totals.

Because

$$\hat{V}(\hat{t}_i) = \left(1 - \frac{m_i}{M_i}\right) M_i^2 \frac{s_i^2}{m_i}$$

is an unbiased estimator of $V(\hat{t}_i)$, Theorem 6.4 implies that

$$E\left[\sum_{i=1}^{N} Z_i \left(\frac{N}{n}\right)^2 \hat{V}(\hat{t}_i)\right] = E\left[\sum_{i \in \mathcal{S}} \left(\frac{N}{n}\right)^2 \hat{V}(\hat{t}_i)\right] = V_{\text{ssu}}.$$

Using (6.45), then,

$$E\left[N^2 \left(1 - \frac{n}{N}\right) \frac{s_t^2}{n} + \frac{N}{n} \sum_{i \in \mathcal{S}} \hat{V}(\hat{t}_i)\right]$$
$$= N^2 \left(1 - \frac{n}{N}\right) \frac{S_t^2}{n} + \frac{N}{n}\left(1 - \frac{n}{N}\right) \sum_{i=1}^{N} V(\hat{t}_i) + \sum_{i=1}^{N} V(\hat{t}_i)$$
$$= N^2 \left(1 - \frac{n}{N}\right) \frac{S_t^2}{n} + \frac{N}{n} \sum_{i=1}^{N} V(\hat{t}_i),$$

so (5.27) is an unbiased estimator of (5.24). ■

The methods used in these proofs can be applied to any number of levels of clustering. You may want to sample schools, then classes within schools, then students within classes. Exercise 41 asks you to find an expression for the variance in three-stage cluster sampling. Rao (1979a) presented an alternative and elegant approach, relying on properties of nonnegative definite matrices, for deriving mean squared errors and variance estimators for linear estimators of population totals.

6.7 Model-Based Inference with Unequal-Probability Samples*

In general, data from a good sampling design should produce reasonable inferences from either a model-based or randomization approach. Let's see how the Horvitz–Thompson estimator performs for Model M1 from (5.38). The model is

$$\text{M1}: Y_{ij} = \mu + A_i + \varepsilon_{ij}$$

with the A_is generated by a distribution with mean 0 and variance σ_A^2, the ε_{ij}s generated by a distribution with mean 0 and variance σ^2, and all A_is and ε_{ij}s independent.

As we did for the estimators in Chapter 5, we can write the estimator as a linear combination of the random variables Y_{ij}. For a pps design, $\psi_i = M_i/M_0$ and $\pi_i = n\psi_i$, so

$$\hat{T}_P = \sum_{i \in \mathcal{S}} \frac{M_0}{nM_i} \hat{T}_i = \sum_{i \in \mathcal{S}} M_0 \frac{\bar{Y}_{\mathcal{S}_i}}{n} = \sum_{i \in \mathcal{S}} \sum_{j \in \mathcal{S}_i} \frac{M_0}{nm_i} Y_{ij}.$$

Note that $\sum_{i \in \mathcal{S}} \sum_{j \in \mathcal{S}_i} M_0/(nm_i) = M_0$, so by (5.41), $\hat{T}_P$ is unbiased under Model M1 in (5.38). In addition, from (5.42),

$$\begin{aligned} V_{\text{M1}}[\hat{T}_P - T] &= \sigma_A^2 \left[\sum_{i \in \mathcal{S}} \left(\sum_{j \in \mathcal{S}_i} \frac{M_0}{nm_i} - M_i \right)^2 + \sum_{i \notin \mathcal{S}} M_i^2 \right] \\ &\quad + \sigma^2 \left[\sum_{i \in \mathcal{S}} \sum_{j \in \mathcal{S}_i} \left\{ \left(\frac{M_0}{nm_i} \right)^2 - 2\frac{M_0}{nm_i} \right\} + M_0 \right] \\ &= \sigma_A^2 \left[\frac{M_0^2}{n} - 2\frac{M_0}{n} \sum_{i \in \mathcal{S}} M_i + \sum_{i=1}^{N} M_i^2 \right] + \sigma^2 \left[\sum_{i \in \mathcal{S}} \frac{M_0^2}{n^2 m_i} - M_0 \right]. \end{aligned}$$

The model-based variance for $\hat{T}_P$ has implications for design. Suppose a sample is desired that will minimize $V_{\text{M1}}[\hat{T}_P - T]$. The psu sizes M_i for the sample units appear only in the term $-2\sigma_A^2(M_0/n)\sum_{i \in \mathcal{S}} M_i$, so for fixed n the model-based variance is smallest when the n units with largest M_is are included in the sample. If, in addition, a constraint is placed on the number of subunits that can be examined, $\sum_{i \in \mathcal{S}}(1/m_i)$ is smallest when all m_is are equal.

Inference in the model-based approach does not depend on the sampling design. As long as model M1 holds for the population, $\hat{T}_P$ is model-unbiased with variance given above. In a model-based approach, an investigator with complete faith in the model can simply select the psus with the largest values of M_i to be the sample. In practice, however, this would not be done—no one has complete faith in a model, especially before data collection. Royall and Eberhardt (1975) suggested using balanced sampling, in which the sample is selected in such a way that inferences are robust to certain forms of model misspecification.

As described in Section 6.2, pps sampling can be thought of as a way of introducing randomness into the optimal design for model M1 and estimator $\hat{T}_P$. The self-weighting design of taking all m_is to be equal also minimizes the variance in the model-based approach. Thus, if model M1 is thought to describe the data, pps sampling and estimation should perform well in practice.

Basu's elephants. We conclude our discussion with a famous example from Basu (1971) that demonstrates that unequal-probability sampling and Horvitz–Thompson estimates can be as silly as any other statistical procedures when improperly applied. The story begins by describing the sampling problem faced by a fictional circus owner:

> The circus owner is planning to ship his 50 adult elephants and so he needs a rough estimate of the total weight of the elephants. As weighing an elephant is a cumbersome process, the owner wants to estimate the total weight by weighing just one elephant. Which elephant should he weigh? (Basu, 1971)

When all 50 elephants in the herd were weighed three years ago, it was found that the middle-sized elephant had weight equal to the average weight of the herd. The owner proposes weighing the middle-sized elephant and then multiplying that elephant's weight by 50 to estimate the total herd weight today. But the circus statistician is horrified by the proposed purposive sampling plan, and proposes a scheme where the probability of selection for the middle-sized elephant is 99/100, and the probability of selection for each of the other 49 elephants is 1/4900.

Under this scheme, however, if the middle-sized elephant is selected, the Horvitz-Thompson estimate of the total herd weight is (100/99) × (weight of middle-sized elephant). If one of the other elephants is selected, the total herd weight is estimated by 4900 times the weight of that elephant. These are both silly (although unbiased, when all possible samples are considered) estimates of the total weight for all 50 elephants, since the estimate is much too small if the middle-sized elephant is selected and much too large if one of the other elephants is selected. Basu (1971) concluded: "That is how the statistician lost his circus job (and perhaps became a teacher of statistics!)."

Should the circus statistician have been fired? A statistician desiring to use a model in analyzing survey data would say yes: The circus statistician is using the model $t_i \propto 99/100$ for the middle-sized elephant, and $t_i \propto 1/4900$ for all other elephants in the herd—certainly not a model that fits the data well. A randomization-inference statistician would also say yes: Even though models are not used explicitly in the Horvitz–Thompson theory, the estimator is most efficient (has the smallest variance) when the psu total t_i is proportional to the probability of selection. The silly design used by the circus statistician leads to a huge variance for the Horvitz–Thompson estimator. If that were not reason enough, the statistician proposes a sample of size 1—he can neither check the validity of the model in a model-based approach nor estimate the variance of the Horvitz–Thompson estimator!

Had the circus statistician used a ratio estimator in the design-based setting, he might have saved his job even though he used a poor design. Let y_i = weight of elephant i now, and x_i = weight of elephant i three years ago. The ratio estimator of the population total, t, is

$$\hat{t}_{yr} = \frac{\hat{t}_y}{\hat{t}_x} t_x.$$

Thus, if elephant i is selected,

$$\hat{t}_{yr} = \frac{y_i/\pi_i}{x_i/\pi_i} t_x = \frac{y_i}{x_i} t_x.$$

With the ratio estimator, the total weight of the elephants from three years ago is multiplied by the ratio of (weight now)/(weight 3 years ago) for the selected elephant.

6.8 Chapter Summary

Unequal-probability samples occur naturally in many situations, particularly in cluster sampling when the psus have unequal sizes. If the psu population totals t_i are highly correlated with the psu selection probabilities, then an unequal-probability sampling design can greatly increase efficiency.

We can draw an unequal-probability sample either with or without replacement. Selecting a with-replacement sample with unequal probabilities is simple; on each of the n draws, select one of the N psus with specified probability ψ_i, where $\sum_{i=1}^{N} \psi_i = 1$. Since any psu can be selected on each of the n draws, a psu can appear more than once in the sample.

Estimation is also simple in a with-replacement probability sample, and the estimators have the same form for either a one-stage or a multi-stage sample. The population total is estimated by $\hat{t}_\psi$ given in (6.5) and (6.13):

$$\hat{t}_\psi = \frac{1}{n}\sum_{i\in\mathcal{W}}\frac{\hat{t}_i}{\psi_i} = \sum_{i\in\mathcal{W}}\sum_{j\in\mathcal{S}_i} w_{ij}y_{ij},$$

where $\mathcal{W}$ denotes the set of psus selected for the sample (including psus as many times as they are selected). If an SRS of m_i of the M_i ssus is taken at stage 2, then $w_{ij} = [1/(n\psi_i)](M_i/m_i)$. An unbiased estimator of $V(\hat{t}_\psi)$ is given by

$$\hat{V}(\hat{t}_\psi) = \frac{1}{n}\frac{1}{n-1}\sum_{i\in\mathcal{W}}\left(\frac{\hat{t}_i}{\psi_i} - \hat{t}_\psi\right)^2,$$

which is simply the sample variance of the values of $\hat{t}_i/\psi_i$ divided by n. If a psu appears more than once in the sample, each time a different probability subsample of ssus is selected for estimating t_i. In with-replacement sampling, the population mean $\bar{y}_{\mathcal{U}}$ is estimated using (6.10) by $\hat{\bar{y}}_\psi = \hat{t}_\psi/\hat{M}_{0\psi}$, where $\hat{M}_{0\psi} = \sum_{i\in\mathcal{W}}\sum_{j\in\mathcal{S}_i} w_{ij}$. Equation (6.15) gives $\hat{V}(\hat{\bar{y}}_\psi)$ using ratio estimation methods.

Although the estimators of means and totals in without-replacement unequal-probability sampling have simple form, variance estimation and sample selection methods can be complicated. We recommend using survey software to select and analyze unequal-probability samples.

With $\pi_i = P(\text{psu } i \text{ is included in the sample})$, the Horvitz–Thompson estimator of the population total for a without-replacement sample is

$$\hat{t}_{\text{HT}} = \sum_{i\in\mathcal{S}}\frac{\hat{t}_i}{\pi_i},$$

where $\hat{t}_i$ is an unbiased estimator of the psu total t_i. The sampling weight for ssu j of psu i is

$$w_{ij} = \frac{1}{\pi_i}\frac{1}{P(\text{ssu } j \text{ of psu } i \text{ is in sample} \mid \text{psu } i \text{ is in sample})};$$

in terms of the weights,

$$\hat{t}_{\text{HT}} = \sum_{i\in\mathcal{S}}\sum_{j\in\mathcal{S}_i} w_{ij}y_{ij}$$

and

$$\bar{y}_{\text{HT}} = \frac{\displaystyle\sum_{i\in\mathcal{S}}\sum_{j\in\mathcal{S}_i} w_{ij}y_{ij}}{\displaystyle\sum_{i\in\mathcal{S}}\sum_{j\in\mathcal{S}_i} w_{ij}}.$$

The variance of the Horvitz–Thompson estimator is given in (6.20) and (6.21) for one-stage cluster sampling and in (6.26) and (6.27) for two-stage cluster sampling. The SYG unbiased estimator of $V(\hat{t}_{\text{HT}})$, given in (6.29), requires knowledge of the joint inclusion probabilities $\pi_{ik} = P(\text{psus } i \text{ and } j \text{ are included in the sample})$ and is often difficult to compute. In many situations, we recommend using the with-replacement variance estimators in (6.30) and (6.34), which do not require knowledge of the π_{ik}'s; if n/N is small, a with-replacement variance estimator performs well.

Key Terms

Horvitz–Thompson estimator: The Horvitz–Thompson estimator of a population total t is $\hat{t}_{\text{HT}} = \sum_{i\in\mathcal{S}} \hat{t}_i/\pi_i$. This is the most general form of the estimator of t in without-replacement samples with inclusion probabilities π_i.

Inclusion probability: π_i is the probability that psu i is included in the sample.

Joint inclusion probability: π_{ik} is the probability that psus i and k are both included in the sample.

Poisson sampling: A sampling process in which independent Bernoulli trials determine whether each unit in the population is to be included in the sample.

Probability proportional to size (pps) sampling: An unequal-probability sampling method in which the probability of sampling a unit is proportional to the number of elements in the unit.

Random digit dialing: A method used in telephone surveys in which a probability sample of telephone numbers is selected from a set of possible telephone numbers.

For Further Reading

Chapter 9 of Brewer (2002) discusses the Horvitz–Thompson estimator and methods for approximating its variance. Brewer and Hanif (1983) present more than 50 methods for drawing with- and without-replacement samples with unequal probabilities, and Tillé (2006) describes multiple algorithms for selecting unequal-probability samples. Traat et al. (2004) give a unifying perspective of unequal-probability sampling, treating with- and without-replacement samples as multivariate probability distributions on a set of N integers. Rao (2005) outlines the history of how practical problems have spurred development of survey methods, with an interesting section on the history of unequal-probability sampling.

Unequal-probability methods are common in natural resource sampling. Overton and Stehman (1995), Christman (2000), Buckland et al. (2001), and Lennert-Cody et al. (2018) review and give examples of unequal-probability sampling, with emphasis on examples in the biological sciences.

6.9 Exercises

A. Introductory Exercises

1. For each of the following situations, say what unit might be used as psu. Do you believe there would be a strong clustering effect? Would you sample psus with equal or unequal probabilities?

   (a) You want to estimate the percentage of patients of U.S. Air Force optometrists and ophthalmologists who wear contact lenses.

(b) Human taeniasis is acquired by ingesting larvae of the pork tapeworm in inadequately cooked pork. You have been asked to design a survey to estimate the percentage of inhabitants of a village who have taeniasis. A medical examination is required to diagnose the condition.

(c) You wish to estimate the total number of cows and heifers on all Ontario dairy farms; in addition, you would like to find estimates of the birth rate and stillbirth rate.

(d) You want to estimate the percentages of undergraduate students at U.S. universities who are registered to vote, and who are affiliated with each political party.

(e) A fisheries agency is interested in the distribution of carapace width of snow crabs. A trap hauled from a fishing boat has a limit of 30 crabs.

(f) You wish to conduct a customer satisfaction survey of persons who have taken guided bus tours of the Grand Canyon rim area. Tour groups range in size from 8 to 44 persons.

2. An investigator wants to take an unequal-probability sample of 10 of the 25 psus in the population listed below, and wishes to sample units with replacement.

psu	ψ_i	psu	ψ_i	psu	ψ_i
1	0.000110	10	0.022876	19	0.069492
2	0.018556	11	0.003721	20	0.036590
3	0.062998	12	0.024917	21	0.033853
4	0.078216	13	0.040654	22	0.016959
5	0.075245	14	0.014804	23	0.009066
6	0.073983	15	0.005577	24	0.021795
7	0.076580	16	0.070784	25	0.059186
8	0.038981	17	0.069635		
9	0.040772	18	0.034650		

(a) Adapt the cumulative-size method to draw a sample of size 10 with replacement with probabilities ψ_i. Instead of randomly selecting integers between 1 and $M_0 = \sum_{i=1}^{N} M_i$, select 10 random numbers between 0 and 1.

(b) Adapt Lahiri's method to draw a sample of size 10 with replacement with probabilities ψ_i.

3. For the supermarket example in Section 6.1, suppose that the ψ_is are the same, but that each store has $t_i = 75$. What is $E[\hat{t}_\psi]$? $V[\hat{t}_\psi]$?

4. For the supermarket example in Section 6.1, suppose that the ψ_is are 7/16 for store A, and 3/16 for each of stores B, C, and D. Show that $\hat{t}_\psi$ is unbiased, and find its variance. Do you think that the sampling scheme with these ψ_is is a good one?

5. Return to the supermarket example of Section 6.1. Now let's select two supermarkets with replacement. List the 16 possible samples (A,A), (A,B), etc., and find the probability with which each sample would be selected. Calculate $\hat{t}_\psi$ for each sample. What is $E[\hat{t}_\psi]$? $V[\hat{t}_\psi]$?

6. The file `azcounties.csv` gives data on population and housing unit counts for the counties in Arizona (excluding Maricopa County and Pima County, which are much larger than the other counties and would be placed in a separate stratum). For this exercise, suppose that the population (M_i) is known and you want to take a sample of counties to estimate the total number of housing units ($t = \sum_{i=1}^{13} t_i$). The file has the value of t_i for every county so you can calculate the population total and variance.

(a) Calculate the selection probabilities ψ_i for a sample of size 1 with probability proportional to population. Find $\hat{t}_\psi$ for each possible sample, and calculate the theoretical variance $V(\hat{t}_\psi)$.

(b) Repeat (a) for an equal probability sample of size 1. How do the variances compare? Why do you think one design is more efficient than the other?

(c) Now take a with-replacement sample of size 3. Find $\hat{t}_\psi$ and $\hat{V}(\hat{t}_\psi)$ for your sample.

7. For a simple random sample with replacement, with $\psi_i = 1/N$, show that (6.6) simplifies to

$$\hat{V}(\hat{t}_\psi) = \frac{N^2}{n}\frac{1}{n-1}\sum_{i \in \mathcal{W}}(t_i - \bar{t})^2,$$

where the sum is over all n units in the sample (including units as many times as they appear in the sample).

8. Let's return to the situation in Exercise 6 of Chapter 2, in which we took an SRS to estimate the average and total numbers of refereed publications of faculty and research associates. Now, consider a pps sample of faculty. The 27 academic units range in size from 2 to 92. We used Lahiri's method to choose 10 psus with probabilities proportional to size and with replacement, and took an SRS of four (or fewer, if $M_i < 4$) members from each psu. Note that academic unit 14 appears three times in the sample in Table 6.11; each time it appears, a different subsample was collected. Find the estimated total number of publications, along with its standard error.

TABLE 6.11
Data for Exercise 8.

Academic Unit	M_i	ψ_i	y_{ij}
14	65	0.0805452	3, 0, 0, 4
23	25	0.0309789	2, 1, 2, 0
9	48	0.0594796	0, 0, 1, 0
14	65	0.0805452	2, 0, 1, 0
16	2	0.0024783	2, 0
6	62	0.0768278	0, 2, 2, 5
14	65	0.0805452	1, 0, 0, 3
19	62	0.0768278	4, 1, 0, 0
21	61	0.0755886	2, 2, 3, 1
11	41	0.0508055	2, 5, 12, 3

B. Working with Survey Data

9. The file `statepps.csv` lists the number of counties, 2019 population, land area, and water area for the 50 states plus the District of Columbia.

(a) Use the cumulative-size method to draw a sample of size 10 with replacement, with probabilities proportional to land area. What is ψ_i for each state in your sample?

(b) Use the cumulative-size method to draw a sample of size 10 with replacement, with probabilities proportional to population. What is ψ_i for each state in your sample?

(c) How do the two samples differ? Which states tend to be in each sample?

10. Use your sample of states drawn with probability proportional to population, from Exercise 9, for this problem.

(a) Using the sample, estimate the total number of counties in the United States, and find the standard error of your estimate. How does your estimate compare with the true value of total number of counties (which you can calculate, since the file `statepps.csv` contains the data for the whole population)?

(b) Now suppose that your friend Tom finds the ten values of numbers of counties in your sample, but does not know that you selected these states with probabilities proportional to population. Tom then estimates the total number of counties using formulas for an SRS. What values for the estimated total and its standard error are calculated by Tom? How do these values differ from yours? Is Tom's estimator unbiased for the population total?

11. In Example 2.6, we took an SRS to estimate the total acreage devoted to farming in the United States in 1992. Now, use the sample of states drawn with probability proportional to land area in Exercise 9, and then subsample five counties randomly from each state using file `agpop.csv`. Estimate the total acreage devoted to farming in 1992, along with its standard error.

12. The file `statepop.csv` contains data from an unequal-probability sample of 100 counties from the 1994 *County and City Data Book* (U.S. Census Bureau, 1994), selected with probability proportional to population. The selection probabilities are given in variable *psii*. Sampling was done with replacement, so large counties occur multiple times in the sample: Los Angeles County, with the largest population in the United States, occurs four times. Let t_i represent the number of physicians in county i (variable *phys*).

    (a) Draw a scatterplot of t_i vs. ψ_i for the counties in the sample. Why would you expect pps sampling to work well for estimating the total number of physicians in the United States?

    (b) Calculate $u_i = t_i/\psi_i$ for each county in the sample. Calculate the mean $\bar{u}$ and standard deviation s_u for the 100 counties in the sample.

    (c) Using $\bar{u}$ and s_u from (b), calculate a 95% CI for the estimated total number of physicians. Does your CI include the value 532,638 that is given in the *County and City Data Book*?

    (d) Check your results by estimating the total number of physicians using survey software.

13. The file `statepop.csv`, used in Exercise 12, also contains information on total number of farms, number of veterans, and other items.

    (a) Plot the total number of farms versus the probabilities of selection ψ_i. Does your plot indicate that unequal-probability sampling will be helpful for this variable?

    (b) Estimate the total number of farms in the United States, along with a 95% CI.

14. Repeat Exercise 13 for estimating the total number of veterans.

15. The file `collshr.csv` contains a with-replacement pps sample of size 10 from the 347 colleges in the small, highly residential stratum (with *ccsizset* = 11) of the population of colleges discussed in Example 3.12. The size measure is the number of undergraduates at the college (variable *ugds*).

    (a) Why would sampling colleges with probability proportional to number of undergraduates be a good choice for estimating the total number of biology faculty members in the stratum?

(b) Estimate the total number of biology faculty members for the stratum, along with a 95% CI.

(c) Estimate the average number of biology faculty members per college in the stratum, along with a 95% CI.

16. For the sample in Exercise 15, estimate the total number of math instructors and the total number of psychology instructors for the stratum of small, highly residential colleges. Give a 95% CI for each estimate.

17. For the sample in Exercise 15, estimate the total number of female undergraduates (calculate this as *ugds* × *ugds_women*) for the stratum, along with a 95% CI. Plot the total number of female undergraduates versus the probabilities of selection ψ_i.

18. In Example 6.11, we calculated the with-replacement variance for $\hat{t}_{\text{HT}}$. In this example, 1/3 of the population psus are included in the sample, so the with-replacement variance is likely to overestimate the without-replacement variance. The joint inclusion probabilities for the psus are given in file `classppsjp.csv`.

(a) Calculate $\hat{V}_{\text{HT}}(\hat{t}_{\text{HT}})$ and $\hat{V}_{\text{SYG}}(\hat{t}_{\text{HT}})$ for this data set.

(b) Some survey software packages approximate the without-replacement variance in unequal-probability sampling using

$$\left(1 - \frac{n}{N}\right) \hat{V}_{\text{WR}}(\hat{t}_{\text{HT}}).$$

Calculate this approximation for the class data.

(c) How do these estimates compare, and how do they compare with the with-replacement variance calculated in Example 6.11?

C. Working with Theory
All exercises in this section require knowledge of probability

19. (a) Prove that Lahiri's method results in a probability proportional to size sample with replacement. HINT: Let J be an integer with $J \geq \max\{M_i\}$. Let $U_1, U_2, \ldots$ be discrete uniform $\{1, \ldots N\}$ random variables, let $V_1, V_2, \ldots$ be discrete uniform $\{1, \ldots, J\}$ random variables, and assume all U_i and V_j are independent. To select the first psu, we generate pairs (U_1, V_1), (U_2, V_2), ... until $V_j \leq M_{U_j}$.

(b) Suppose the population has N psus, with sizes $M_1, M_2, \ldots, M_N$. Let X represent the number of pairs of random numbers that must be generated to obtain a sample of size n. Find $E[X]$.

20. The random variables $Q_1, \ldots, Q_N$ in Section 6.3 have a joint multinomial distribution with probabilities $\psi_1, \psi_2, \ldots, \psi_N$. Use properties of the multinomial distribution to show that $\hat{t}_\psi$ in (6.13) is an unbiased estimator of t with variance given by

$$V(\hat{t}_\psi) = \frac{1}{n} \sum_{i=1}^{N} \psi_i \left(\frac{t_i}{\psi_i} - t\right)^2 + \frac{1}{n} \sum_{i=1}^{N} \frac{V_i}{\psi_i}. \tag{6.46}$$

Also show that (6.14) is an unbiased estimator of the variance in (6.46). HINT: Use properties of conditional expectation in Appendix A, and write

$$V(\hat{t}_\psi) = V(E[\hat{t}_\psi \mid Q_1, \ldots, Q_N]) + E(V[\hat{t}_\psi \mid Q_1, \ldots, Q_N]).$$

21. Show that (6.28) and (6.29) are equivalent when an SRS of psus is selected as in Chapter 5. Are they equivalent if psus are selected with unequal probabilities?

22. Show that the formulas for stratified random sampling in (3.1), (3.3), and (3.6) follow from the formulas for the Horvitz–Thompson estimator in Section 6.4.3. For a stratified random sample, we sample from every stratum in the population. Thus, if we treat strata as if they were psus, $\pi_i = 1$ for every stratum in the population.

23. Use the population in Exercise 2 of Chapter 4 for this exercise. Let ψ_i be proportional to x_i.

    (a) Using the draw-by-draw method illustrated in Example 6.8, calculate π_i for each unit and π_{ij} for each pair of units, for a without-replacement sample of size two.

    (b) What is $V(\hat{t}_{\text{HT}})$? How does it compare with the with-replacement variance using (6.46)?

24. *Covariance of estimated population totals in a cluster sample.* Suppose a one-stage cluster sample is taken from a population of N psus, with inclusion probabilities π_i. Let t_x and t_y be the population totals for response variables x and y, and let t_{ix} and t_{iy} be the totals of variables x and y in psu i.

    (a) Show that

$$\text{Cov}\,(\hat{t}_x, \hat{t}_y) = \sum_{i=1}^{N} \frac{1-\pi_i}{\pi_i} t_{ix} t_{iy} + \sum_{i=1}^{N} \sum_{\substack{k=1 \\ k \neq i}}^{N} (\pi_{ik} - \pi_i \pi_k) \frac{t_{ix}}{\pi_i} \frac{t_{ky}}{\pi_k}.$$

    (b) Suppose that an SRS of n of the N psus is selected, with $\pi_i = n/N$ and $\pi_{ik} = (n/N)[(n-1)/(N-1)]$. Show using part (a) that

$$\text{Cov}\,(\hat{t}_x, \hat{t}_y) = \frac{N^2}{(N-1)n} \left(1 - \frac{n}{N}\right) \left[\sum_{i=1}^{N} t_{ix} t_{iy} - \frac{t_x t_y}{N}\right].$$

25. *Comparing two domain means in a cluster sample.* In Exercise 26 of Chapter 4, you showed that in an SRS where $\bar{y}_1$ and $\bar{y}_2$ estimate respective population domain means $\bar{y}_{\mathcal{U}1}$ and $\bar{y}_{\mathcal{U}2}$, $V(\bar{y}_1 - \bar{y}_2) \approx V(\bar{y}_1) + V(\bar{y}_2)$ because $\text{Cov}\,(\bar{y}_1, \bar{y}_2) \approx 0$. Now let's explore what happens when a one-stage cluster sample is selected from a population of N psus. For simplicity, assume that each psu has M ssus and that an SRS of n psus is selected. Let $\hat{\bar{y}}_1$ and $\hat{\bar{y}}_2$ be the estimators of the domain means from the cluster sample. Let $x_{ij} = 1$ if ssu j of psu i is in domain 1 and $x_{ij} = 0$ if ssu j of psu i is in domain 2, and let $u_{ij} = x_{ij} y_{ij}$.

    (a) Find

$$\text{Cov}\left[\left(\hat{\bar{u}} - \frac{t_u}{t_x}\hat{\bar{x}}\right), \left\{\hat{\bar{y}} - \hat{\bar{u}} - \frac{t_y - t_u}{N - t_x}(1 - \hat{\bar{x}})\right\}\right].$$

    HINT: Use Exercise 24.

    (b) Show that the covariance in (a) is 0 if for each psu, all of the elements in that psu belong to the same domain—that is, either $t_{ix} = 0$ or $t_{ix} = M$ for each psu i. [If the covariance in (a) is 0, then (4.35) implies that $\text{Cov}\,(\hat{\bar{y}}_1, \hat{\bar{y}}_2) \approx 0$.]

    (c) Give an example in which the covariance in (a) is not 0.

26. *Indirect sampling.* Suppose you want to take a sample of students in a university but your sampling frame is a list of all classes offered by the university. A student may be in more than one class, so a probability sample of classes, which includes all students in those classes, may contain some students multiple times. Lavallée (2007) described a generalized weight share method for such situations, and this exercise is adapted from results in his book.

    Let $\mathcal{U}^A$ be the sampling frame population with N units. Let $Z_i = 1$ if unit i is in the sample $\mathcal{S}^A$ and 0 otherwise, with $\pi_i = P(Z_i = 1)$. The target population $\mathcal{U}^B$ has M elements. Each element in $\mathcal{U}^B$ is linked with one or more of the units in $\mathcal{U}^A$; let

$$\ell_{ik} = \begin{cases} 1 & \text{if element } k \text{ from } \mathcal{U}^B \text{ is linked to unit } i \text{ from } \mathcal{U}^A \\ 0 & \text{otherwise} \end{cases}$$

    and let $L_k = \sum_{i=1}^N \ell_{ik}$. We assume $L_k \geq 1$ for each k and that L_k is known. In our example, $\ell_{ik} = 1$ if student k is in class i and L_k is the number of classes taken by student k. Let y_k be a characteristic associated with element k of $\mathcal{U}^B$. We want to estimate $t_y = \sum_{k=1}^M y_k$.

    (a) Let $u_i = \sum_{k=1}^M (\ell_{ik} y_k / L_k)$ and let

$$\hat{t}_y = \sum_{i=1}^N \frac{Z_i}{\pi_i} u_i.$$

    Show that $\hat{t}_y$ is an unbiased estimator of t_y, with

$$V(\hat{t}_y) = \sum_{i=1}^N \frac{1-\pi_i}{\pi_i} u_i^2 + \sum_{i=1}^N \sum_{j \neq i}^N \frac{\pi_{ij} - \pi_i \pi_j}{\pi_i \pi_j} u_i u_j.$$

    (b) Let $\mathcal{S}^B$ be the set of distinct units sampled from $\mathcal{U}^B$ using this procedure. Show that $\hat{t}_y$ can be rewritten as

$$\hat{t}_y = \sum_{k \in \mathcal{S}^B} \frac{1}{L_k} \sum_{i=1}^N \frac{Z_i}{\pi_i} \ell_{ik}\, y_k.$$

    We can view $w_k^* = (1/L_k) \sum_{i=1}^N Z_i \ell_{ik} / \pi_i$ as a "weight" for y_k.

    (c) If $L_k = 1$ for all k, show that $\hat{t}_{\text{HT}}$ in (6.19) is a special case of $\hat{t}_y$. What is w_k^* in this case?

    (d) Suppose $\mathcal{U}^A = \{1, 2, 3\}$, $\mathcal{U}^B = \{1, 2\}$ and the values of ℓ_{ik} are given in the following table:

		Element k from $\mathcal{U}^B$	
ℓ_{ik}		1	2
Unit i	1	1	0
from $\mathcal{U}^A$	2	1	1
	3	0	1

    Suppose $y_1 = 4$ and $y_2 = 6$, so that $t_y = 10$. Find the value of $\hat{t}_y$ for each of the three possible SRSs of size 2 from $\mathcal{U}^A$. Using the sampling distribution of $\hat{t}_y$, show that $\hat{t}_y$ is unbiased but that $V(\hat{t}_y) > 0$. Even though each possible SRS from $\mathcal{U}^A$ contains both units from $\mathcal{U}^B$ (so in effect, a census is taken of $\mathcal{U}^B$), the variance of $\hat{t}_y$ is not zero.

(e) Data file `wtshare.csv` contains information from a hypothetical SRS of size $n = 100$ from a population of $N = 40{,}000$ adults. Each adult in the sample is asked about his or her children: how many children are between ages 0 and 5, whether those children attend preschool, and how many other adults in the population claim the child as part of their household. Estimate the total number of children in the population who attend preschool, along with an approximate 95% CI.

27. *Poisson sampling.* For the general case of Poisson sampling, introduced in Example 6.14, population unit i has inclusion probability π_i. Let $Z_1, Z_2, \ldots, Z_N$ be independent random variables with $P(Z_i = 1) = \pi_i$ and $P(Z_i = 0) = 1 - \pi_i$. The sample size from Poisson sampling is $\sum_{i=1}^N Z_i$ with expected value $\sum_{i=1}^N \pi_i$, and the estimated population total for variable y is $\hat{t}_{\text{HT}} = \sum_{i \in \mathcal{S}} y_i/\pi_i = \sum_{i=1}^N Z_i y_i/\pi_i$.

    (a) What is $V\left(\sum_{i=1}^N Z_i\right)$, the variance of the sample size, under Poisson sampling?

    (b) What is the probability that the procedure produces an empty sample, with $\sum_{i=1}^N Z_i = 0$?

    (c) Show that if $\pi_i = n/N$ for all units and N is large, $P(\sum_{i=1}^N Z_i = 0) \approx e^{-n}$.

    (d) Show that with Poisson sampling, $E[\hat{t}_{\text{HT}}] = t$.

    (e) Show that $V\left(\hat{t}_{\text{HT}}\right) = \sum_{i=1}^N \frac{y_i^2}{\pi_i}(1 - \pi_i)$.

    (f) Show that $\hat{V}(\hat{t}_{\text{HT}}) = \sum_{i \in \mathcal{S}} \frac{y_i^2}{\pi_i^2}(1 - \pi_i)$ is an unbiased estimator of the variance.

28. *Multi-purpose cluster samples.* Sometimes it is desired to estimate quantities at the psu level and at the ssu level. For example, consider a sample of colleges, where it is desired to conduct interviews with the college president and with a sample of the college's students. One might want to sample colleges with equal probabilities to calculate estimates about the presidents' views, and to sample with probability proportional to the number of students to calculate estimates about students' views. With both goals, one compromise would be to select colleges with probability proportional to the square root of the number of students.

    For the colleges population discussed in Example 3.12, consider the stratum consisting of the institutions with *ccsizset* = 11. Using (6.8), calculate the theoretical variance of the (i) estimated total number of colleges that are public institutions (variable *control*) and (ii) estimated total number of female undergraduate students (obtained by rounding *ugds* × *ugds_women* to the nearest integer) for an unequal probability sample of size 10 when:

    (a) Colleges are sampled with equal probabilities.

    (b) Colleges are sampled with probability proportional to *ugds*.

    (c) Colleges are sampled with probability proportional to the square root of *ugds*.

29. In simple random sampling, we know that a without-replacement sample of size n has smaller variance than a with-replacement sample of size n. The same result is not always true for unequal-probability sampling designs (Raj, 1968, p. 56). Consider a with-replacement design with selection probabilities ψ_i, and a corresponding without-replacement design with inclusion probabilities $\pi_i = n\psi_i$, assuming $n\psi_i < 1$ for $i = 1, \ldots, N$.

(a) Consider a population with $N = 4$ and $t_1 = -5$, $t_2 = 6$, $t_3 = 0$, and $t_4 = -1$. The joint inclusion probabilities for a without-replacement sample of size 2 are $\pi_{12} = 0.004$, $\pi_{13} = \pi_{23} = \pi_{24} = 0.123$, $\pi_{14} = 0.373$, and $\pi_{34} = 0.254$. Find the value of π_i for each unit. Show that for this design and population, $V(\hat{t}_\psi) < V(\hat{t}_{\text{HT}})$.

(b) Show that for $\pi_i = n\psi_i$ and $V(\hat{t}_\psi)$ in (6.8),

$$V(\hat{t}_\psi) = \frac{1}{2n}\sum_{i=1}^{N}\sum_{k=1}^{N}\pi_i\pi_k\left(\frac{t_i}{\pi_i} - \frac{t_k}{\pi_k}\right)^2.$$

(c) Using $V(\hat{t}_{\text{HT}})$ in (6.21), show that if

$$\pi_{ik} \geq \frac{n-1}{n}\pi_i\pi_k \quad \text{for all } i \text{ and } k,$$

then $V(\hat{t}_{\text{HT}}) \leq V(\hat{t}_\psi)$.

(d) Gabler (1984) showed that if

$$\sum_{i=1}^{N}\min_k\left(\frac{\pi_{ik}}{\pi_k}\right) \geq n - 1,$$

then $V(\hat{t}_{\text{HT}}) \leq V(\hat{t}_\psi)$. Show that if $\pi_{ik} \geq (n-1)\pi_i\pi_k/n$ for all i and k, then Gabler's condition is met.

(e) (Requires knowledge of linear algebra.) Show that if $V(\hat{t}_{\text{HT}}) \leq V(\hat{t}_\psi)$, then

$$\pi_{ik} \leq 2\frac{n-1}{n}\pi_i\pi_k \quad \text{for all } i \text{ and } k.$$

HINT: Use the results in Theorem 6.1 to simplify $V(\hat{t}_\psi) - V(\hat{t}_{\text{HT}})$ so that it may be written as $\sum_{i=1}^{N}\sum_{k=1}^{N} a_{ik}t_it_k$. Then $\mathbf{A}$, the matrix with elements a_{ik}, must be nonnegative definite and therefore all principal 2×2 submatrices must have determinant ≥ 0 (Narain, 1951).

30. Consider a without-replacement sample of size 2 from a population of size 4, with joint inclusion probabilities $\pi_{12} = \pi_{34} = 0.31$, $\pi_{13} = 0.20$, $\pi_{14} = 0.14$, $\pi_{23} = 0.03$, and $\pi_{24} = 0.01$.

    (a) Calculate the inclusion probabilities π_i for this design.

    (b) Suppose $t_1 = 2.5$, $t_2 = 2.0$, $t_3 = 1.1$, and $t_4 = 0.5$. Find $\hat{V}_{\text{HT}}(\hat{t}_{\text{HT}})$ and $\hat{V}_{\text{SYG}}(\hat{t}_{\text{HT}})$ for each possible sample.

31. *Brewer's (1963, 1975) procedure for without-replacement unequal-probability sampling.* For a sample of size $n = 2$, let π_i be the desired probability of inclusion for psu i, with the usual constraint that $\sum_{i=1}^{N}\pi_i = n$. Let $\psi_i = \pi_i/2$ and

$$a_i = \frac{\psi_i(1-\psi_i)}{1-\pi_i}.$$

Draw the first psu with probability $a_i/\sum_{k=1}^{N} a_k$ of selecting psu i. Supposing psu i is selected at the first draw, select the second psu from the remaining $N-1$ psus with probabilities $\psi_j/(1-\psi_i)$.

(a) Show that

$$\pi_{ij} = \frac{\psi_i \psi_j}{\sum_{k=1}^{N} a_k} \left(\frac{1}{1-\pi_i} + \frac{1}{1-\pi_j} \right).$$

(b) Show that P(psu i selected in sample) $= \pi_i$. HINT: First show that

$$2\sum_{k=1}^{N} a_k = 1 + \sum_{k=1}^{N} \frac{\psi_k}{1-\pi_k}.$$

(c) The SYG estimator of $V(\hat{t}_{\text{HT}})$ for one-stage sampling is given in (6.23). Show that $\pi_i\pi_j - \pi_{ij} \geq 0$ for Brewer's method, so that the SYG estimator of the variance is always nonnegative.

32. The following table gives population values for a small population of clusters:

psu, i	M_i	**Values,** y_{ij}	t_i
1	5	3, 5, 4, 6, 2	20
2	4	7, 4, 7, 7	25
3	8	7, 2, 9, 4, 5, 3, 2, 6	38
4	5	2, 5, 3, 6, 8	24
5	3	9, 7, 5	21

You wish to select two psus without replacement with probabilities of inclusion proportional to M_i. Using Brewer's method from Exercise 31, construct a table of π_{ij} for the possible samples. What is the variance of the one-stage Horvitz–Thompson estimator?

33. Rao (1963) discussed the following rejective method for selecting a pps sample without replacement: Select n psus with probabilities ψ_i and with replacement. If any psu appears more than once in the sample, reject the whole sample and select another n psus with replacement. Repeat until you obtain a sample of n psus with no duplicates.

Find π_{ij} and π_i for this procedure, for $n = 2$.

34. *The Rao–Hartley–Cochran (1962) method for selecting psus with unequal probabilities.* To take a sample of size n, divide the population into n random groups of psus, $\mathcal{U}_1, \mathcal{U}_2, \ldots, \mathcal{U}_n$. Then select one psu from each group with probability proportional to size. Let N_k be the number of psus in group k. If psu i is in group k, it is selected with probability $x_{ki} = M_i / \sum_{j \in \mathcal{U}_k} M_j$; the estimator is

$$\hat{t}_{\text{RHC}} = \sum_{k=1}^{n} \frac{t_i}{x_{ki}}.$$

Show that $\hat{t}_{\text{RHC}}$ is unbiased for t, and find its variance. HINT: Use two sets of indicator variables. Let $I_{ki} = 1$ if psu i is in group k and 0 otherwise, and let $Z_i = 1$ if psu i is selected to be in the sample.

35. *Approximating the joint inclusion probabilities.* The estimators of $V(\hat{t}_{\text{HT}})$ in (6.22) and (6.23) require knowledge of the joint inclusion probabilities π_{ik}. To use these formulas, the data file must contain an $n \times n$ matrix of the π_{ik}'s, which can dramatically increase its size; in addition, computing the variance estimator is complicated. If the joint inclusion

probabilities π_{ik} could be approximated as a function of the π_is, estimation would be simplified. Let $c_i = \pi_i(1-\pi_i)$. Hájek (1964) (see Berger, 2004, for extensions) suggested approximating π_{ik} by

$$\tilde{\pi}_{ik} = \pi_i \pi_k \left[1 - (1-\pi_i)(1-\pi_k) \Big/ \sum_{j=1}^{N} c_j \right].$$

(a) Does the set of $\tilde{\pi}_{ik}$'s satisfy condition (6.18)?

(b) What is $\tilde{\pi}_{ik}$ if an SRS is taken? Show that if N is large, $\tilde{\pi}_{ik}$ is close to π_{ik}.

(c) Show that if $\tilde{\pi}_{ik}$ is substituted for π_{ik} in (6.21), the expression for the variance can be written as

$$V_{\text{Haj}}(\hat{t}_{\text{HT}}) = \sum_{i=1}^{N} c_i e_i^2,$$

where $e_i = t_i/\pi_i - A$ and

$$A = \left(\sum_{j=1}^{N} c_j \frac{t_j}{\pi_j} \right) \Big/ \left(\sum_{j=1}^{N} c_j \right).$$

HINT: Write (6.21) as

$$\frac{1}{2} \sum_{i=1}^{N} \sum_{k=1}^{N} (\pi_i \pi_k - \tilde{\pi}_{ik}) \left(\frac{t_i}{\pi_i} - \frac{t_k}{\pi_k} \right)^2.$$

(d) We can estimate $V_{\text{Haj}}(\hat{t}_{\text{HT}})$ by

$$\hat{V}_{\text{Haj}}(\hat{t}_{\text{HT}}) = \sum_{i \in \mathcal{S}} \tilde{c}_i \hat{e}_i^2,$$

where $\tilde{c}_i = (1-\pi_i)n/(n-1)$, $\hat{e}_i = t_i/\pi_i - \hat{A}$, and $\hat{A} = \sum_{j \in \mathcal{S}} \tilde{c}_j \frac{t_j}{\pi_j} \Big/ \sum_{j \in \mathcal{S}} \tilde{c}_j$. Show that if an SRS of size n is taken, then $\hat{V}_{\text{Haj}}(\hat{t}_{\text{HT}}) = N^2(1-n/N)s_t^2/n$.

36. This exercise is based on results in Brewer and Donadio (2003).

(a) Show, using the results in Theorem 6.1, that the variance in (6.21) can be rewritten as:

$$V(\hat{t}_{\text{HT}}) = \sum_{i=1}^{N} \pi_i \left(\frac{t_i}{\pi_i} - \frac{t}{n} \right)^2 - \sum_{i=1}^{N} \pi_i^2 \left(\frac{t_i}{\pi_i} - \frac{t}{n} \right)^2 + \sum_{i=1}^{N} \sum_{\substack{k=1 \\ k \neq i}}^{N} (\pi_{ik} - \pi_i \pi_k) \left(\frac{t_i}{\pi_i} - \frac{t}{n} \right) \left(\frac{t_k}{\pi_k} - \frac{t}{n} \right). \quad (6.47)$$

HINT: Write $t_i/\pi_i - t_k/\pi_k = t_i/\pi_i - t/n + t/n - t_k/\pi_k$.

(b) The first term in (6.47) is the variance that would result if a with-replacement sample with selection probabilities $\psi_i = \pi_i/n$ were taken. Brewer and Donadio (2003) suggested that the second term may be viewed as a finite population correction for unequal-probability sampling, so that the first two terms in (6.47) approximate $V(\hat{t}_{\text{HT}})$ without depending on the joint inclusion probabilities π_{ik}. Calculate the three terms in (6.47) for an SRS of size n.

(c) Suppose that there exist constants c_i such that $\pi_{ik} \approx \pi_i \pi_k (c_i + c_k)/2$. Show that with this substitution, the third term in (6.47) can be approximated by

$$\sum_{i=1}^{N} \pi_i^2 (1 - c_i) \left(\frac{t_i}{\pi_i} - \frac{t}{n} \right)^2$$

so that

$$V(\hat{t}_{\text{HT}}) \approx \sum_{i=1}^{N} \pi_i (1 - c_i \pi_i) \left(\frac{t_i}{\pi_i} - \frac{t}{n} \right)^2. \tag{6.48}$$

Two choices suggested for c_i are $c_i = (n-1)/(n - \pi_i)$ or, following Hartley and Rao (1962),

$$c_i = \frac{n-1}{\left(1 - 2\pi_i + \frac{1}{n} \sum_{k=1}^{N} \pi_k^2 \right)}.$$

Calculate the variance approximation in (6.48) for an SRS with each of these choices of c_i.

37. (Requires calculus.) Suppose that for (6.46) in Exercise 20, the variance of the estimator of the total in psu i is $V(\hat{t}_i) = M_i^2 S_i^2 / m_i$. What values of m_i minimize the variance subject to the constraint that $E[\sum_{i \in \mathcal{S}} m_i] \leq C$?

38. Consider the Mitofsky–Waksberg method, discussed in Example 6.13. Show that the probability that psu i is selected as the first psu in the sample is

$$P(\text{select psu } i) = \frac{M_i}{M_0}.$$

HINT: See Exercise 19 and argue that the Mitofsky–Waksberg method for selecting psus is a special case of Lahiri's method.

39. One drawback of the Mitofsky–Waksberg method as described in Example 6.13 is that the sequential sampling procedure of selecting numbers in the psu until one has a total of k residential numbers can be cumbersome to implement. Suppose in the second stage you dial an additional $(k-1)$ numbers whether they are residential or not, and let x be the number of residential lines among the $(k-1)$. What are the relative weights for the residential telephone numbers?

40. The Mitofsky–Waksberg method, described in Example 6.13, gives a self-weighting sample of telephone numbers under ideal circumstances. Does it give a self-weighting sample of adults? Why or why not? If not, what relative weights should be used?

41. Suppose a three-stage cluster sample is taken from a population with N psus, M_i ssus in the ith psu, and L_{ij} tsus in the jth ssu of the ith psu. To draw the sample, n psus are randomly selected, then m_i ssus from the selected psus, then l_{ij} tsus from the selected ssus.

(a) Show that the sample weights are

$$w_{ijk} = \frac{N}{n} \frac{M_i}{m_i} \frac{L_{ij}}{l_{ij}}.$$

(b) Let $\hat{t} = \sum_{i \in \mathcal{S}} \sum_{j \in \mathcal{S}_i} \sum_{k \in \mathcal{S}_{ij}} w_{ijk} y_{ijk}$. Show that $E[\hat{t}] = t = \sum_{i=1}^{N} \sum_{j=1}^{M_i} \sum_{k=1}^{L_{ij}} y_{ijk}$.

(c) Using the properties of conditional expectation in Section A.4, find an expression for $V(\hat{t})$.

42. (Model-based.) Suppose the entire population is observed in the sample, so that $n = N$ and $m_i = M_i$. Examine the three estimators $\hat{T}_{\text{unb}}$, $\hat{T}_{\text{ratio}}$ (from Section 5.6) and $\hat{T}_P$ (from Section 6.7). If the entire population is observed, which of these estimators equal $T = \sum_{i=1}^{N} \sum_{j=1}^{M_i} Y_{ij}$?

D. Projects and Activities

43. *Rectangles.* Use the population of rectangles in Exercise 35 of Chapter 2 for the exercise. The file `rectlength.csv` contains information on the vertical length of each of the 100 rectangles in the population.

(a) Select a sample of 10 rectangles with replacement from the 100 rectangles, with probability proportional to the length of the rectangle.

(b) For your sample, plot t_i, the area of the rectangle, vs. the selection probability ψ_i. What is the correlation between t_i and ψ_i?

(c) Estimate the total area of all 100 rectangles, and find a 95% confidence interval for the total area. Compare your answers with the estimate and confidence interval from the SRS in Exercise 35 of Chapter 2. Did unequal-probability sampling result in a smaller variance estimate?

44. Repeat Exercise 43(a), using a without-replacement sample of 10 rectangles selected with probability proportional to the length of the rectangle.

(a) What are the inclusion probabilities π_i for the rectangles in your sample?

(b) Estimate the total area of all 100 rectangles using the Horvitz–Thompson estimator $\hat{t}_{\text{HT}}$.

(c) Find the with-replacement variance estimate for $\hat{t}_{\text{HT}}$.

(d) (Requires knowledge of the joint inclusion probabilities.) Find the SYG variance estimate for $\hat{t}_{\text{HT}}$. How does this compare with the estimate in (b)?

45. The sum of the sampling weights ($\hat{N} = \sum_{i \in \mathcal{W}} w_i$) for the sample of classes in Example 6.5 is an unbiased estimator of N, but does not exactly equal N. Calculate $\hat{N}$ for the sample. Investigate the sampling distribution of $\hat{N}$ by drawing 5,000 independent samples from the population (each using a different random number seed) and calculating the sum of the sampling weights for each sample. Draw a histogram of the 5,000 values of $\hat{N}$ and calculate the mean, standard deviation, median, and quartiles. What could be done to obtain a pps sample where the sampling distribution of $\hat{N}$ has smaller variance?

46. Historians wanting to use data from United States censuses collected in the pre-computer age faced the daunting task of poring over reels of handwritten records on microfilm, arranged in geographical order. The Public Use Microdata Samples (PUMS) were constructed by taking samples of the records and typing those records into the computer. Ruggles (1995, p. 44) described the PUMS construction for the 1940 Census:

> The population schedules of the 1940 census are preserved on 4,576 microfilm reels. Each census page contains information on forty individuals. Two lines on each page were designated as "sample lines" by the Census Bureau: the

individuals falling on those lines—5 percent of the population—were asked a set of supplemental questions that appear at the bottom of the census page.

Two of every five census pages were systematically selected for examination. On each selected census page, one of the two designated sample lines was then randomly selected. Data-entry personnel then counted the size of the sample unit containing the targeted sample line. Units size six or smaller were included in the sample in inverse proportion to their size. Thus, every one-person unit was included in the sample, every second two-person unit, every third three-person unit, and so on. Units with seven or more persons were included with a probability of 1-in-7: every seventh household of size seven or more was selected for the sample.

(a) Explain why this is a cluster sample. What are the psus? The ssus?

(b) What effect do you think the clustering will have on estimates of race? Age? Occupation?

(c) Construct a table for the inclusion probabilities for persons in one-person units, two-person units, and so on.

(d) What happens if you estimate the mean age of the population by the average age of all persons in the sample? What estimator should you use?

(e) Do you think that taking a systematic sample was a good idea for this sample? Why or why not?

(f) Does this method provide a representative sample of households? Why or why not?

(g) What type of sample is taken of the individuals with supplementary information? Explain.

47. Ruggles (1995, p. 45) also described the 1950 PUMS:

> The 1950 census schedules are contained on 6,278 microfilm reels. Each census page contains information on thirty individuals. Every fifth line on the census page was designated as a sample line, and additional questions for the sample-line individuals appear at the bottom of the form. For the last sample-line individual on each page, there was a block of additional supplemental questions. Thus, 20 percent of individuals were asked a basic set of supplemental questions, and 3.33 percent of individuals were asked a full set of supplemental questions.
>
> One-in-eleven pages within enumeration districts was selected randomly. On each selected census page, the sixth sample-line individual (the one with the full set of questions) was selected for inclusion in the sample. Any other members of the sample unit containing the selected individual were also included.

Answer the same questions from Exercise 46 for the 1950 PUMS.

48. *Estimating the size of an audience.* In Exercise 40 of Chapter 2, you estimated the size of an audience by taking an SRS. Explain how this is a special case of cluster sampling. Obtain a seating chart for an auditorium in which the rows have different numbers of seats. Using the seating chart, select an unequal-probability sample of 10 or 20 rows, with probabilities proportional to the numbers of seats in each row. Why might you expect the unequal-probability sample to have a smaller variance for the estimated audience size than an SRS of the same number of rows?

Estimate the audience size for this auditorium using your unequal-probability sample; count the number of people in each selected row. Give a 95% CI for the total number of people in the auditorium.

49. *Create your own stock market index fund.* The data file `sp500.csv` contains a listing of the companies in the S&P 500® Index, along with the market capitalization of each company, as of September 2020. The market capitalization of a company is the market value of its outstanding shares, calculated as (price per share) × (number of shares outstanding).

    There are several ways you could own a self-weighting sample of dollars represented by all the companies in this index. You could take an SRS of the individual dollars in the stock market, buying shares in each company for which you have at least one dollar in your SRS. This can be cumbersome, however, and would mean buying shares of a large (and random) number of companies.

    An easier way is to take a sample of companies with probability proportional to market capitalization. Suppose you have $300,000 to invest. Select a sample of 40 companies with probability proportional to market capitalization. Create a file of the companies in your sample; for each company, state how much money you will invest in that company so that you have a self-weighting sample of dollars in the index.

50. *Baseball data.*

    (a) Use the population in the file `baseball.csv` to take a two-stage cluster sample (without replacement) with the teams as the psus, with probabilities proportional to the total number of runs scored for the teams. Your sample should have approximately 150 players altogether, as in the SRS from Exercise 37 of Chapter 2. Describe how you selected your sample.

    (b) Construct the sampling weights for your sample.

    (c) Let $\hat{t}_i$ be the estimated total of the variable *logsal* for team i in your sample, and let π_i be the inclusion probability for team i. Plot $\hat{t}_i$ vs. π_i.

    (d) Use your sample to estimate the mean of the variable *logsal*, and give a 95% CI.

    (e) Estimate the proportion of players in the data set who are pitchers, and give a 95% CI.

    (f) Do you think that unequal-probability sampling resulted in more efficiency for your estimators? Why, or why not?

51. *IPUMS exercises.* Exercise 42 of Chapter 2 described the IPUMS data.

    (a) Select an unequal-probability sample of 10 psus, with probability proportional to number of persons. Take a subsample of 20 persons in each of the selected psus.

    (b) Using the sample you selected, estimate the population mean and total of *inctot* and give the standard errors of your estimates.

7

Complex Surveys

> There is no more effective medicine to apply to feverish public sentiment than figures. To be sure, they must be properly prepared, must cover the case, not confine themselves to a quarter of it, and they must be gathered for their own sake, not for the sake of a theory. Such preparation we get in a national census.
>
> —Ida Tarbell, *The Ways of Woman*

Most large surveys involve several of the ideas we have discussed. A probability sample may be stratified with several stages of clustering and rely on ratio and regression estimation to calibrate to known population characteristics. The formulas for estimating standard errors can become complicated, especially if there are several stages of clustering.

This chapter tells how to put all these concepts together—how to use sampling weights to estimate any quantity that could be calculated from the full population (Sections 7.2 and 7.3), how to calculate design effects to summarize the effects of the survey design on estimates' precision (Section 7.4), and how to produce graphical displays of survey data (Section 7.6). These concepts are illustrated using data from the U.S. National Health and Nutrition Examination Survey (NHANES), whose design is described in Section 7.5.

7.1 Assembling Design Components

We have seen most of the components of a complex survey: random sampling, ratio estimation, stratification, and clustering. Now, let's see how to assemble them into one sampling design and a unified estimation procedure. Although in practice weights (Section 7.2) are often used to find point estimates and computer-intensive methods (Chapter 9) are used to calculate variances of the estimates, understanding the basic principles of how the components work together is important. Here are the concepts you already know, in a modular form ready for assembly.

7.1.1 Building Blocks for Surveys

1. *Cluster sampling with replacement.* Select a sample of n clusters with replacement; primary sampling unit (psu) i is selected with probability ψ_i on a draw. Estimate the total for psu i using an unbiased estimator $\hat{t}_i$. Then treat the n values (the sample is with replacement, so some of the values in the set may be from the same psus) of $u_i = \hat{t}_i/\psi_i$ as observations. Estimate the population total by $\bar{u}$, and estimate the variance of the estimated total by s_u^2/n.

2. *Cluster sampling without replacement.* Select a sample of n psus without replacement; π_i is the probability that psu i is included in the sample. Estimate the total for psu i using an unbiased estimator $\hat{t}_i$, and calculate an unbiased estimator of the variance

DOI: 10.1201/9780429298899-7

of $\hat{t}_i$, $\hat{V}(\hat{t}_i)$. Then estimate the population total with the Horvitz-Thompson estimator[1] from (6.19):

$$\hat{t}_{\text{HT}} = \sum_{i \in \mathcal{S}} \frac{\hat{t}_i}{\pi_i}.$$

Use a formula from Chapters 5 or 6 or a method from Chapter 9 to estimate the variance. We often estimate the variance assuming that psus were selected with replacement, as discussed in Section 6.4.

3. *Stratification.* Let $\hat{t}_1, \ldots, \hat{t}_H$ be unbiased estimators of the stratum totals $t_1, \ldots, t_H$, and let $\hat{V}(\hat{t}_1), \ldots, \hat{V}(\hat{t}_H)$ be unbiased estimators of the variances. Then estimate the population total by

$$\hat{t} = \sum_{h=1}^{H} \hat{t}_h$$

and its variance by

$$\hat{V}(\hat{t}) = \sum_{h=1}^{H} \hat{V}(\hat{t}_h).$$

Stratification usually forms the coarsest classification: Strata may be, for example, areas of the country, different area codes, or types of habitat. Clusters (sometimes several stages of clusters) are sampled from each stratum in the design, and additional stratification may occur within clusters. Many surveys have a **stratified multistage survey design**, in which a stratified sample is taken of psus, and subsamples of secondary sampling units (ssus) are selected within each selected psu. With several stages of clustering and stratification, it helps to draw a diagram or construct a table of the survey design, as illustrated in the following example.

Example 7.1. Malaria has long been a serious health problem in the Gambia. Malaria morbidity can be reduced by using bed nets that are impregnated with insecticide, but this is only effective if the bed nets are in widespread use. In 1991, a nationwide survey was designed to estimate the prevalence of bed net use in rural areas (D'Alessandro et al., 1994).

The sampling frame consisted of all rural villages of fewer than 3,000 people in the Gambia. The villages were stratified by three geographic regions (eastern, central, and western) and by whether the village had a public health clinic (PHC) or not. In each region five districts were chosen with probability proportional to the district population as estimated in the 1983 national census. In each district four villages were chosen, again with probability proportional to census population: two PHC villages and two non-PHC villages. Finally, six compounds were chosen more or less randomly from each village, and a researcher recorded the number of beds and nets, along with other information, for each compound.

In summary, the sample design is the following:

Stage	Sampling Unit	Stratification
1	District	Region
2	Village	PHC/non-PHC
3	Compound	

[1]Recall that the Horvitz–Thompson estimator encompasses the other without-replacement, unbiased estimators of the total as special cases, as discussed in Section 6.4.4.

To calculate estimates or standard errors using formulas from the previous chapters, start at Stage 3 and work up. The following are steps you would follow to estimate the total number of bed nets (without using ratio estimation):

1. Record the total number of nets for each compound.

2. Estimate the total number of nets for each village by (number of compounds in the village) $\times$ (average number of nets per compound). Find the estimated variance of the total number of nets, for each village.

3. Estimate the total number of nets for the PHC villages in each district. Villages were sampled from the district with probabilities proportional to population, so formulas from Chapter 6 need to be used to estimate the total and the variance of the estimated total. Repeat for the non-PHC villages in each district.

4. Add the estimates from the two strata (PHC and non-PHC) to estimate the number of nets in each district; sum the estimated variances from the two strata to estimate the variance for the district.

5. At this point you have the estimated total number of nets and its estimated variance, for each district. Now use two-stage cluster sampling formulas to estimate the total number of nets for each region.

6. Finally, add the estimated totals for each region to estimate the total number of bed nets in the Gambia. Add the region variances as called for in stratified sampling.

Sounds a little complicated, doesn't it? And we have not even included ratio estimation, which would almost certainly be incorporated here because we know approximate population numbers for the numbers of beds at each stage.

Fortunately, we do not always have to go to this much work to analyzed data from a complex survey. As we shall see later in this chapter and in Chapter 9, we can use sampling weights and computer-intensive methods to avoid much of this effort. Using a with-replacement variance estimator allows us to estimate a variance using only the weighting, stratification, and psu information. ■

7.1.2 Ratio Estimation in Complex Surveys

Ratio estimation is part of the analysis, not the design, and may be used at almost any level of the survey. We discussed ratio estimation with stratified random sampling in Section 4.5. The principles are the same for any probability sampling design used within the strata in a stratified multistage sample.

Suppose that the population total t_x is known for an auxiliary variable x, and that $\hat{t}_y$ and $\hat{t}_x$ are unbiased estimators for t_y and t_x, respectively, from the sample. The **combined ratio estimator** of the population total for variable y is

$$\hat{t}_{yrc} = \hat{B} t_x,$$

where

$$\hat{B} = \frac{\hat{t}_y}{\hat{t}_x};$$

in Section 9.1 we show that the mean squared error (MSE) of $\hat{t}_{yrc}$ can be estimated by

$$\hat{V}(\hat{t}_{yrc}) = \left(\frac{t_x}{\hat{t}_x}\right)^2 \left[\hat{V}(\hat{t}_y) + \hat{B}^2 \hat{V}(\hat{t}_x) - 2\hat{B}\widehat{\text{Cov}}\,(\hat{t}_y, \hat{t}_x)\right].$$

The **separate ratio estimator** applies ratio estimation within each stratum first, then combines the strata:

$$\hat{t}_{yrs} = \sum_{h=1}^{H} \hat{t}_{yhr} = \sum_{h=1}^{H} t_{xh} \frac{\hat{t}_{yh}}{\hat{t}_{xh}},$$

with

$$\hat{V}(\hat{t}_{yrs}) = \sum_{h=1}^{H} \hat{V}(\hat{t}_{yhr}).$$

As we saw in Section 5.2.3, we often use ratio estimation for estimating means, letting the auxiliary variable x_i equal 1 for all observation units. Then $\hat{t}_x$ estimates the population size, and the ratio $\hat{B} = \hat{t}_y/\hat{t}_x$ divides the estimated population total by the estimated population size.

Other ratios are often of interest as well. One quantity of interest for the survey in Example 7.1 was the proportion of beds that have nets. In this case, x refers to beds and y refers to nets. Then, t_x is the total number of beds in the population and t_y is the total number of bed nets in the population. We estimate the proportion of beds that have nets by $\hat{B} = \hat{t}_y/\hat{t}_x$. Alternatively, the ratio can be estimated separately for each region if it is desired to compare the bed net coverage for the regions.

7.1.3 Simplicity in Survey Design

All of these design and estimation components have been shown to increase efficiency in survey after survey. Sometimes, though, a complex sampling design may be used simply because it is there or has been used in the past, not because it has been demonstrated to be more efficient.

Make sure you know from pretests or previous research that a complex design really is more efficient and practical. A simpler design giving the same amount of information per dollar spent is almost always to be preferred to a more complicated design: It is often easier to administer and analyze, and data from the survey are less likely to be analyzed incorrectly by subsequent analysts. A design should be efficient for estimating *all* quantities of primary interest—an optimal allocation in stratified sampling for estimating the total amount U.S. businesses spend on health care benefits may be inefficient for estimating the percentage of businesses that declare bankruptcy in a year.

7.2 Sampling Weights

7.2.1 Constructing Sampling Weights

In most large sample surveys, weights are used to calculate point estimates. We have already seen how sampling weights are used in stratified sampling and in cluster sampling. In without-replacement sampling, the sampling weight for an observation unit is always the reciprocal of the probability that the observation unit is included in the sample.

Recall that for stratified random sampling,

$$\hat{t}_{\text{str}} = \sum_{h=1}^{H} \sum_{j \in \mathcal{S}_h} w_{hj} y_{hj},$$

where the sampling weight $w_{hj} = (N_h/n_h)$ can be thought of as the number of observations in the population represented by the sample observation y_{hj}. The probability of selecting the jth unit in the hth stratum to be in the sample is $\pi_{hj} = n_h/N_h$, so the sampling weight is simply the inverse of the probability of selection: $w_{hj} = 1/\pi_{hj}$.

The sum of the sampling weights in stratified random sampling equals the population size N; each sampled unit "represents" a certain number of units in the population, so the whole sample "represents" the whole population. The stratified sampling estimator of $\bar{y}_{\mathcal{U}}$ is

$$\bar{y}_{\text{str}} = \frac{\sum_{h=1}^{H} \sum_{j \in \mathcal{S}_h} w_{hj} y_{hj}}{\sum_{h=1}^{H} \sum_{j \in \mathcal{S}_h} w_{hj}}.$$

The same forms of the estimators were used in cluster sampling in Section 5.3, and in the general form of weighted estimators in Section 6.4.4. In cluster sampling with equal probabilities, for example,

$$w_{ij} = \frac{NM_i}{nm_i} = \frac{1}{\text{probability that the } j\text{th ssu in the } i\text{th psu is in the sample}}.$$

Again,

$$\hat{t} = \sum_{i \in \mathcal{S}} \sum_{j \in \mathcal{S}_i} w_{ij} y_{ij}$$

and the estimator of the population mean is

$$\frac{\hat{t}}{\sum_{i \in \mathcal{S}} \sum_{j \in \mathcal{S}_i} w_{ij}} = \frac{\sum_{i \in \mathcal{S}} \sum_{j \in \mathcal{S}_i} w_{ij} y_{ij}}{\sum_{i \in \mathcal{S}} \sum_{j \in \mathcal{S}_i} w_{ij}}.$$

For cluster sampling with unequal probabilities, when π_i is the probability that the ith psu is in the sample, and $\pi_{j|i}$ is the probability that the jth ssu is in the sample given that the ith psu is in the sample, the sampling weights are $w_{ij} = 1/(\pi_i \pi_{j|i})$.

For three-stage cluster sampling, the principle extends: Let w_p be the weight for the psu, $w_{s|p}$ be the weight for the ssu, and $w_{t|s,p}$ be the weight associated with the tsu (tertiary sampling unit). Then the overall sampling weight for an observation unit is

$$w = w_p \times w_{s|p} \times w_{t|s,p}.$$

All the information needed to construct point estimates is contained in the sampling weights; when computing point estimates, the sometimes cumbersome probabilities with which psus, ssus, and tsus are selected appear only through the weights. But the sampling weights are not sufficient for finding standard errors of the estimates. Variances of estimates depend on the probabilities that any pair of observation units is selected to be in the sample, requiring more knowledge of the sampling design than given by weights alone.

Very large weights are often truncated or smoothed, so that no single observation has an overpowering contribution to the overall estimate. While this biases the estimators, it can reduce the MSE (Elliott and Little, 2000). Truncation is often used when weights are used to adjust for nonresponse, as described in Chapter 8.

Since we consider stratified multistage designs in the remainder of this book, from now on we will adopt a unified notation for estimators of population totals. We consider y_i to be a measurement on observation unit i, and w_i to be the sampling weight of observation unit i. Thus, for a stratified random sample, y_i is an observation unit within a particular stratum, and $w_i = N_h/n_h$, where unit i is in stratum h. This allows us to write the general estimator of the population total as

$$\hat{t}_y = \sum_{i \in \mathcal{S}} w_i y_i, \tag{7.1}$$

where all measurements are at the observation unit level. The general estimator of the population mean is

$$\hat{\bar{y}} = \frac{\hat{t}_y}{\sum\limits_{i \in \mathcal{S}} w_i} = \frac{\sum\limits_{i \in \mathcal{S}} w_i y_i}{\sum\limits_{i \in \mathcal{S}} w_i}; \tag{7.2}$$

$\sum_{i \in \mathcal{S}} w_i$ estimates the number of observation units in the population.

Example 7.2. The Gambia bed net survey in Example 7.1 was designed so that within each region each compound would have approximately the same probability of being included in the sample; probabilities varied only because different districts had different numbers of persons in PHC villages and because number of compounds might not always be exactly proportional to village population. For the central region PHC villages, for example, the probability that a given compound would be included in the survey was

$$\begin{aligned} P(\text{compound selected}) &= P(\text{district selected}) \times P(\text{village selected} \mid \text{district selected}) \\ &\quad \times P(\text{compound selected} \mid \text{district and village selected}) \\ &\propto \frac{D1}{R} \times \frac{V}{D2} \times \frac{1}{C}, \end{aligned}$$

where

$$\begin{aligned} C &= \text{number of compounds in the village} \\ V &= \text{number of people in the village} \\ D1 &= \text{number of people in the district} \\ D2 &= \text{number of people in the district in PHC villages} \\ R &= \text{number of people in PHC villages in all central districts.} \end{aligned}$$

Since the number of compounds in a village will be roughly proportional to the number of people in a village, V/C should be approximately the same for all compounds. The value of R is also the same for all compounds within a region. The weights for each region, the reciprocals of the inclusion probabilities, differ largely because of the variability in $D1/D2$. As R varies from stratum to stratum, though, compounds in more populous strata have higher weights than those in less populous strata. ■

Example 7.3. The American Community Survey (ACS) housing unit sample in the 50 states and District of Columbia consists of more than 3.5 million housing unit addresses each year. Here is a broad overview of the design used to select housing units for the sample; see U.S. Census Bureau (2014) for more details, including descriptions of how newly constructed housing units are included in the sampling frame and how the sample design guarantees that a housing unit address is sampled at most once during a five-year period.

Addresses for the sample are selected via stratified sampling. Within each county or county-equivalent, census blocks are assigned to one of 16 sampling strata that are defined by a measure of size (based on the number of housing units in the county and in geographic entities for which estimates are desired, such as municipalities and school districts) and by the anticipated response rate in the block. A stratified sample is drawn independently from each county, according to the 16 sampling rates assigned to the sampling strata (some counties contain only some of the sampling strata).

Sampling fractions are higher in sampling strata containing blocks with small measure of size, to ensure that areas with smaller population sizes have a large enough sample to give estimates with the desired level of accuracy. Selection probabilities are also higher in strata with lower anticipated response rates. The base weight for addresses selected for the sample equals 1 divided by the probability of selection.

Because of the large sample size, the basic ACS survey design is not clustered even though nonresponse follow-up is done in person (see Example 8.3). Estimates are desired for every area of the country, so personnel must be available to interview nonrespondents across the country. For this survey, clustering would increase variances without compensating decreases in data collection costs. ■

7.2.2 Self-Weighting and Non-Self-Weighting Samples

Each observation unit in a self-weighting sample has the same sampling weight. Self-weighting samples can, in the absence of nonsampling errors, be viewed as a miniature version of the population because each observed unit represents the same number of unobserved units in the population. Standard statistical methods may then be applied to the sample to obtain point estimates. A histogram of the sample values displays the approximate frequencies of occurrence in the population; the sample mean, median, and other sample statistics estimate the corresponding population quantities.

Most large self-weighting samples used in practice are not simple random samples (SRSs), however. Stratification is used to reduce variances and obtain separate estimates for domains of interest; clustering, often with unequal probabilities, is used to reduce costs. You need to use statistical software that is specifically designed for survey data to obtain valid statistics from complex survey data. If you instead use statistical software that is intended for data fulfilling the usual statistical assumptions that observations are independent and identically distributed, the standard errors, hypothesis test results, and confidence intervals (CIs) produced by the software will be wrong (Brogan, 2015). If the sample is not self-weighting, estimates of means and percentiles that are calculated without using the weights will also be biased. When you read a paper or book in which the authors analyze data from a complex survey, see whether they accounted for the data structure in the analysis, or whether they simply ran the raw data through a non-survey statistical package procedure and reported the results. If the latter, their results must be viewed with suspicion; it is possible that they report statistical significance only because they fail to account for the survey design in the standard errors.

Many surveys, of course, purposely sample observation units with different probabilities. The disproportional sampling probabilities often occur in the stratification: a higher sampling fraction is used for a stratum of large business establishments than for a stratum of small business establishments.

7.3 Estimating Distribution Functions and Quantiles

So far, we have concentrated on estimating population means, totals, and ratios. Historically, sampling theory was developed primarily to find these basic statistics and to answer questions such as "What percentage of adult males are unemployed?" or "What is the total amount of money spent on health care in the United States?" or "What is the ratio of the numbers of exotic to native birds in an area?"

But one may also want to estimate the median income in Canada, find the 95th percentile of test scores, or construct a histogram to show the distribution of fish lengths. An insurance company may set reimbursements for a medical procedure using the 75th percentile of charges for the procedure. We can estimate any of these quantities with sampling weights. The sampling weights allow us to estimate what the full population looks like, and thus to estimate any quantity that could be calculated from the full population—this property is what makes a probability sample representative of the population. The weights do not, however, give sufficient information for calculating standard errors of the statistics—you need to know the details of the survey design for that.

Let's look at how we would calculate population quantities from the population distribution functions (if we knew them), and then see how we estimate those functions from the sample. You can think of estimating a population characteristic as following three steps. First, express the population characteristic as a function of the population distribution function. Second, estimate the population distribution from the sample. And third, apply the same function to the sample distribution function to obtain the sample estimate of the population characteristic.

Population cumulative distribution and probability mass functions. Suppose the values of y_i for the entire population of N units are known. Then any quantity of interest may be calculated from the **probability mass function**,

$$f(y) = \frac{\text{number of population units whose value is } y}{N}$$

or the **cumulative distribution function** (cdf)

$$F(y) = \frac{\text{number of population units with value } \leq y}{N} = \sum_{x \leq y} f(x).$$

In probability theory, these are the probability mass function and cdf for the random variable Y, where Y is the value obtained from a random sample of size one from the population. Then $f(y) = P\{Y = y\}$ and $F(y) = P\{Y \leq y\}$. Of course, $\sum f(y) = F(\infty) = 1$.

Any population quantity can be calculated from the probability mass function or cdf. The population mean is

$$\bar{y}_{\mathcal{U}} = \sum_{\substack{\text{values of } y \\ \text{in population}}} y\, f(y).$$

The population variance, too, can be written using the probability mass function:

$$\begin{aligned} S^2 &= \frac{1}{N-1}\sum_{i=1}^{N}(y_i - \bar{y}_{\mathcal{U}})^2 \\ &= \frac{N}{N-1}\sum_{y} f(y)\left[y - \sum_{x} x f(x)\right]^2 \\ &= \frac{N}{N-1}\left[\sum_{y} y^2 f(y) - \left(\sum_{x} x f(x)\right)^2\right]. \end{aligned}$$

Example 7.4. Consider the artificial population of 1,000 men and 1,000 women in file `htpop.csv`. Each person's height is measured to the nearest centimeter (cm). The frequency table in file `htcdf.csv` gives the probability mass function and cdf for the 2,000 persons in the population. The population mean is $\bar{y}_{\mathcal{U}} = \sum_y y f(y) = 168.6$ and the population variance is $S^2 = 124.5$. Figure 7.1 shows the graphs of $f(y)$ and $F(y)$. The population members have 65 different values of the height, y, and the cdf $F(y)$ jumps at each value. ■

Empirical probability mass and cumulative distribution functions. Sampling weights allow us to estimate probability mass functions and cdfs from the sample data, and thus to estimate any statistic that could be calculated from the population. Define the **empirical probability mass function** to be the sum of the weights for all observations taking on the value y, divided by the sum of all the weights:

$$\hat{f}(y) = \frac{\displaystyle\sum_{i\in\mathcal{S}:\, y_i = y} w_i}{\displaystyle\sum_{i\in\mathcal{S}} w_i}. \tag{7.3}$$

The **empirical cumulative distribution function** (empirical cdf) $\hat{F}(y)$ is the sum of all weights for observations with values $\leq y$, divided by the sum of all weights:

$$\hat{F}(y) = \sum_{x \leq y} \hat{f}(x). \tag{7.4}$$

The functions $\hat{f}(y)$ and $\hat{F}(y)$ estimate the population functions f and F. The weight w_i is the number of population units represented by unit i, so the sum of the weights for observations having $y_i = y$ estimates the total number of units in the population that have value y. For a self-weighting sample, $\hat{f}(y)$ reduces to the relative frequency of y in the sample.

Example 7.5. The data file `htsrs.csv` contains an SRS of size 200 from the population in Example 7.4. Each person in the sample represents $w_i = 10$ persons in the population. We can view $\hat{F}(y)$ as the cdf of a "pseudo-population" constructed by making w_i copies of observation y_i (see Exercise 7). The first three values of y in `htsrs.csv` are 159, 174, and 186, so the pseudo-population has the value 159 repeated ten times, then the value 174 repeated ten times, then the value 186 repeated ten times, and so on. The 2,000 pseudo-observations are not the true population, of course, but they represent an estimate of the population. In practice, we almost never physically construct the pseudo-population—for most samples, weights are not integers and the population size is large—but it is helpful to think of $\hat{F}(y)$ as estimating the population cdf $F(y)$.

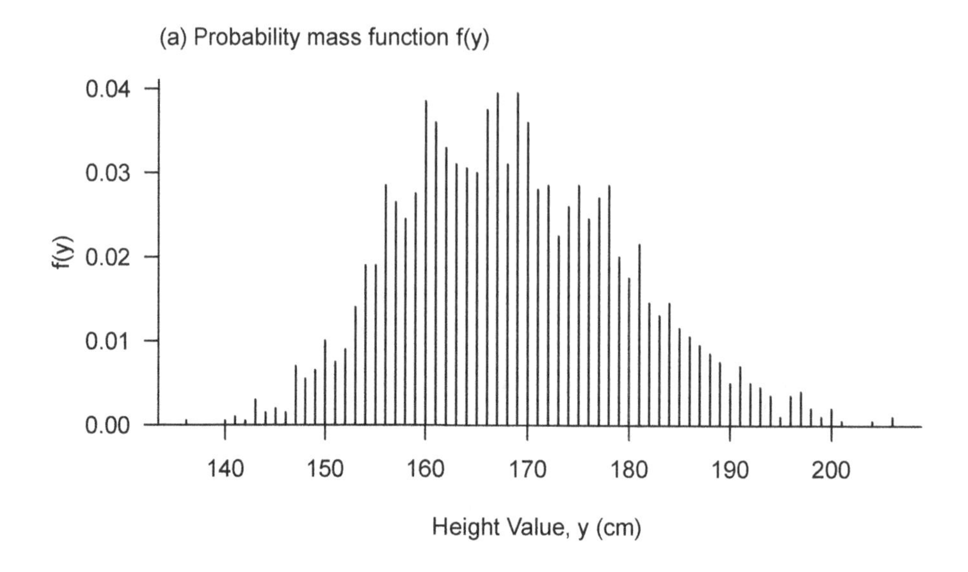

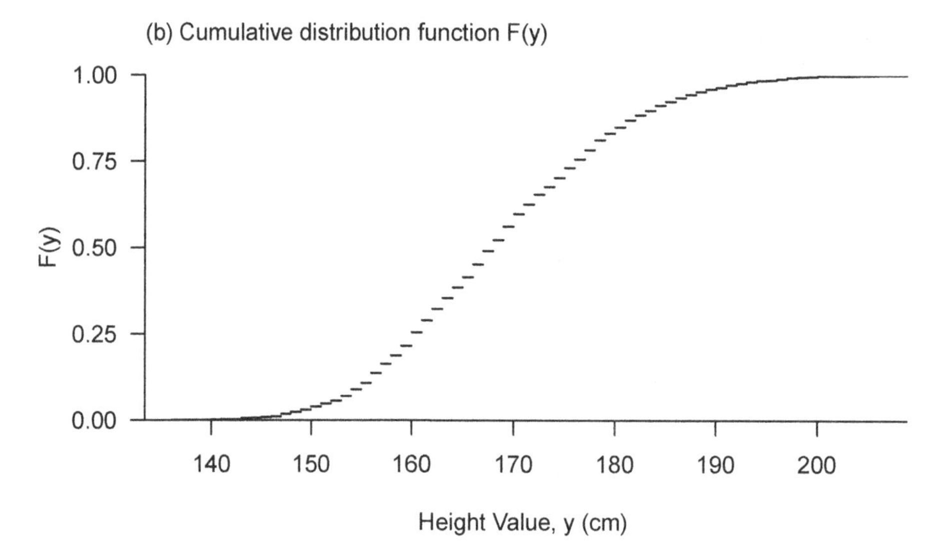

FIGURE 7.1
The functions $f(y)$ and $F(y)$ for the population of heights.

Because the SRS is self-weighting, a histogram of the sample data should resemble $f(y)$ from the population; Figure 7.2 shows that it does.

But suppose a disproportional stratified random sample of 160 women and 40 men (file `htstrat.csv`) is taken instead of a self-weighting sample. In the stratified sample, each woman has weight $1000/160 = 6.25$ and each man has weight $1000/40 = 25$. A histogram of the raw data (without weights) will distort the population distribution, as illustrated in Figure 7.3. The sample mean and median are too low because men are underrepresented in the sample.

Each woman in the stratified sample in Example 7.4 has sampling weight 6.25; each man has sampling weight 25. The empirical probability mass function from the stratified

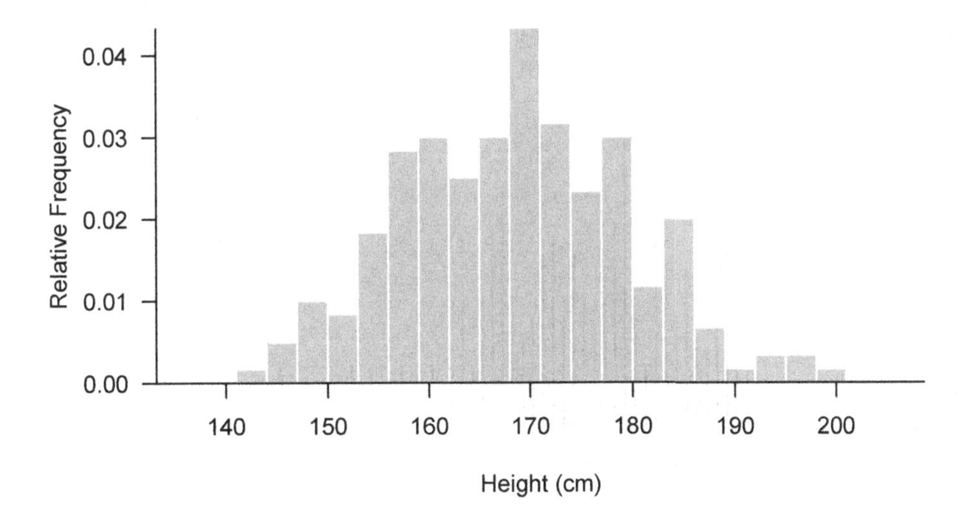

FIGURE 7.2
A histogram of raw data from an SRS of size 200. The general shape is similar to that of $f(y)$ for the population because the sample is self-weighting.

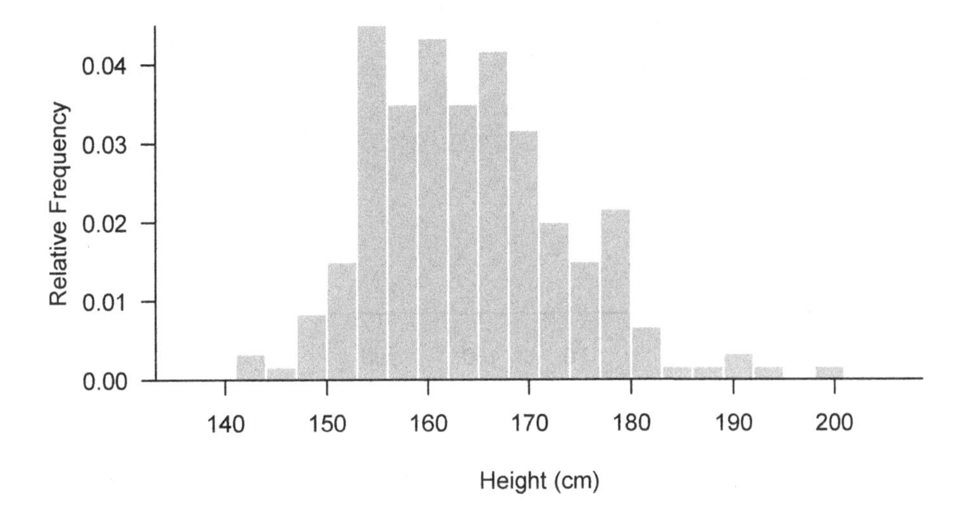

FIGURE 7.3
A histogram of raw data (not using weights) from a disproportionally allocated stratified sample of 160 women and 40 men. Tall persons are underrepresented in the sample, and this histogram presents a misleading picture of the population distribution.

sample is in Figure 7.4. The weights correct the underrepresentation of taller people found in the histogram in Figure 7.3. You can think of the empirical probability mass function and empirical cdf as describing the distribution of the pseudo-population that would be constructed if we copied each sampled woman's value 6.25 times and each man's value 25 times. The weight w_i is used as though there were actually w_i persons in the population exactly like person i.

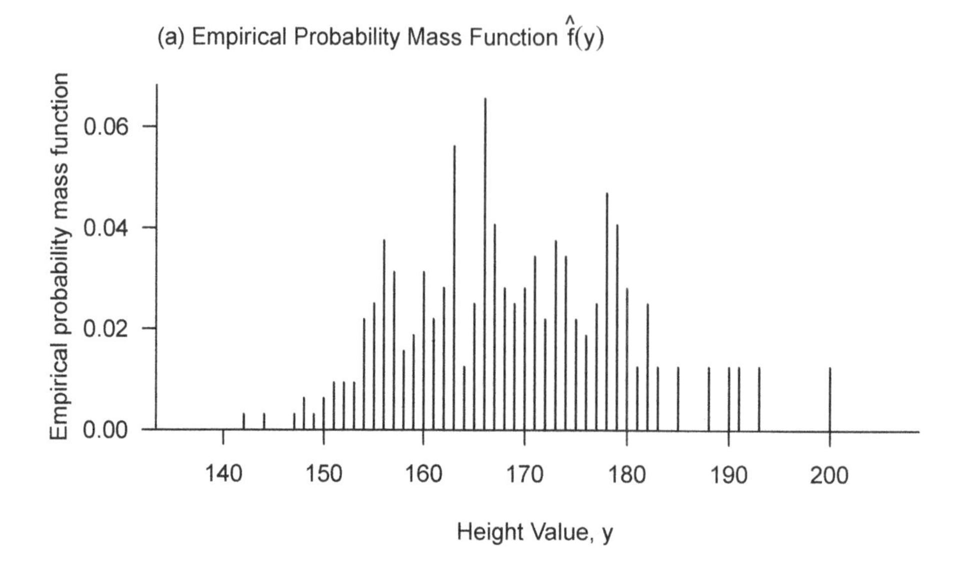

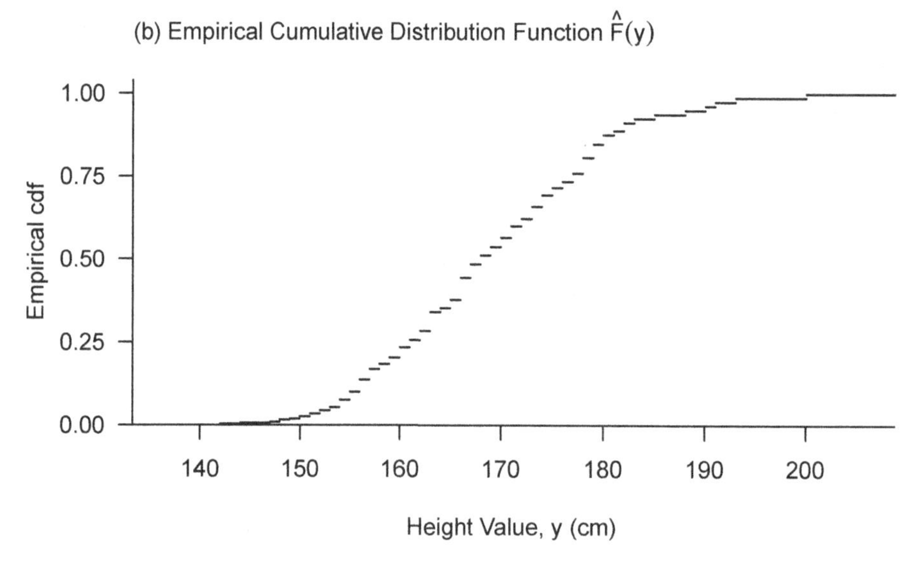

FIGURE 7.4
The estimates $\hat{f}(y)$ and $\hat{F}(y)$ for the stratified sample of 160 women and 40 men.

The scarcity of men in the sample, however, demands a price: The right tail of $\hat{f}(y)$ has a few spikes of size 0.0125=25/2000, each spike coming from one man in the stratified sample, rather than a number of values tapering off. ■

The empirical probability mass function $\hat{f}(y)$ can be used to find estimates of population characteristics. First, express the population characteristic in terms of $f(y)$. For example, the population mean and variance can be written as $\bar{y}_{\mathcal{U}} = \sum y f(y)$ and

$$S^2 = \frac{N}{N-1}\left[\sum_y f(y)\left\{y - \sum_x x f(x)\right\}^2\right] = \frac{N}{N-1}\left[\sum_y y^2 f(y) - \left\{\sum_y y f(y)\right\}^2\right].$$

Then, substitute $\hat{f}(y)$ for every appearance of $f(y)$ and $\hat{N} = \sum_{i\in\mathcal{S}} w_i$ for every appearance of N (which may be unknown) to obtain an estimate of the population characteristic. Using this method, then,

$$\hat{\bar{y}} = \sum_y y\hat{f}(y) = \frac{\sum_{i\in\mathcal{S}} w_i y_i}{\sum_{i\in\mathcal{S}} w_i} \tag{7.5}$$

and

$$\hat{S}^2 = \frac{\hat{N}}{\hat{N}-1}\left[\sum_y y^2\hat{f}(y) - \left(\sum_y y\hat{f}(y)\right)^2\right]. \tag{7.6}$$

Estimating population quantiles. The same method is used to estimate population quantiles. First, define a population quantile in terms of the population cdf F, and then estimate it by substituting the empirical cdf $\hat{F}$ for F.

The finite population is a data set with N units. The qth quantile, θ_q, is calculated by ordering the N units from smallest to largest, then setting θ_q equal to the value of y in the Nqth position in the ordered list. When Nq is not an integer, set θ_q equal to the value of y in the position of the smallest integer $\geq Nq$. Thus about $100q\%$ of the observations in the population are less than θ_q (which is also referred to as the $(100q)$th percentile), and about $100(1-q)\%$ of the observations are greater than θ_q. For a finite population, $F(y)$ is the proportion of the N population units that have value less than or equal to y, so θ_q is the smallest value of y having $F(y) \geq q$.

Substituting the empirical cdf $\hat{F}$ for F, then, we can estimate θ_q by

$$\hat{\theta}_q = \text{smallest value of } y \text{ satisfying } \hat{F}(y) \geq q. \tag{7.7}$$

Example 7.6. Consider again the stratified sample of height data from Example 7.5 in file `htstrat.csv`. We use the sampling weights to estimate the quantiles of the data. Using all 200 observations, we note that $\hat{F}(167) = \sum_{y_i\leq 167} w_i / \sum_{i\in\mathcal{S}} w_i = 0.4844$ and $\hat{F}(168) = \sum_{y_i\leq 168} w_i / \sum_{i\in\mathcal{S}} w_i = 0.5125$. Thus, the estimated median is $\hat{\theta}_{0.5} = 168$. We find the 25th and 75th percentiles similarly. For the 25th percentile, $\hat{F}(160) = 0.2344$, $\hat{F}(161) = 0.2563$, and $\hat{\theta}_{0.25} = 161$. For the 75th percentile, $\hat{F}(176) = 0.7344$, $\hat{F}(177) = 0.7594$, and $\hat{\theta}_{0.75} = 177$.

Table 7.1 shows the population statistics and sample statistics for the population, SRS, and stratified random sample discussed in Examples 7.4 and 7.5. Two sets of estimates are given for the stratified sample: the first incorrectly ignores the weights, and the second incorporates the sampling weights through the function $\hat{f}(y)$. The statistics calculated using weights are much closer to the population quantities. ■

Stratification is the only complex design feature in Example 7.6, but the method is the same for any survey design. The sampling weights allow you to estimate almost anything through the empirical cdf. If desired, you can smooth or interpolate the empirical cdf before estimating quantiles and other statistics. Survey software packages often use interpolation to calculate quantiles (see Exercise 19).

Although the weights may be used to find point estimates through the empirical cdf, calculating standard errors is much more complicated and requires knowledge of the sampling design. We'll discuss how to calculate standard errors of quantiles in Section 9.5.2.

TABLE 7.1
Estimates from samples in Example 7.6.

Quantity	Population	SRS	Stratified, No Weights	Stratified, with Weights
Mean	168.6	168.9	164.6	169.0
Variance, S^2	124.5	122.6	93.4	116.8
25th percentile	160	160	157	161
Median	168	169	163	168
75th percentile	176	176	170	177
90th percentile	184	184	178	182

7.4 Design Effects

Cornfield (1951) suggested measuring the efficiency of a sampling plan by the ratio of the variance that would be obtained from an SRS of n observation units to the variance obtained from the complex sampling plan with n observation units. Kish (1965) named the reciprocal of Cornfield's ratio the **design effect** (abbreviated **deff**) of a sampling plan and estimator and used it to summarize the effect of the sampling design on the variance of the estimator:

$$\text{deff(plan,statistic)} = \frac{V(\text{estimator from sampling plan})}{V(\text{estimator from an SRS with same number of observation units})}. \tag{7.8}$$

For estimating a mean from a sample with n observation units,

$$\text{deff}(\text{plan},\hat{\bar{y}}) = \frac{V(\hat{\bar{y}})}{\left(1 - \frac{n}{N}\right)\frac{S^2}{n}}. \tag{7.9}$$

The design effect provides a measure of the precision gained or lost by using the more complicated design instead of an SRS. Of course, different quantities in the same survey may have different design effects. The design effect for the percentage of persons who are unemployed may differ from the design effect for the percentage of persons with college degrees.

The SRS variance is usually easier to obtain than $V(\hat{\bar{y}})$. If estimating a proportion, the SRS variance is approximately $p(1-p)/n$; if estimating another type of mean, the SRS variance is approximately S^2/n. So if the design effect is approximately known, the variance of the estimator from the complex sample approximately equals (deff $\times$ SRS variance). The variance of an estimated proportion $\hat{p}$ (ignoring the fpc) is approximately

$$\hat{V}(\hat{p}) = \text{deff} \times \frac{\hat{p}(1-\hat{p})}{n}$$

and an approximate 95% CI for p is:

$$\hat{p} \pm 1.96\sqrt{\text{deff}}\sqrt{\frac{\hat{p}(1-\hat{p})}{n}}.$$

Design effects for stratified and cluster samples. We have already seen design effects for several sampling plans. In Section 3.4, the design effect for stratified sampling with proportional allocation was shown to be approximately

$$\frac{V_{\text{prop}}}{V_{\text{SRS}}} \approx \frac{\sum_{h=1}^{H} N_h S_h^2}{N S^2} \approx \frac{\sum_{h=1}^{H} N_h S_h^2}{\sum_{h=1}^{H} N_h \left[S_h^2 + (\bar{y}_{\mathcal{U}h} - \bar{y}_{\mathcal{U}})^2\right]}. \tag{7.10}$$

Unless all of the stratum means are equal, the design effect for a proportionally allocated stratified sample will usually be less than 1—stratification will give more precision per observation unit than an SRS.

We also looked extensively at design effects in cluster sampling, particularly in Section 5.2.2. From (5.12), the design effect for single-stage cluster sampling when all psus have M ssus is approximately

$$1 + (M - 1)\ \text{ICC}.$$

The intraclass correlation coefficient (ICC) is usually positive in cluster sampling, so the design effect is usually larger than 1; cluster samples usually give less precision per observation unit than an SRS.

In surveys with both stratification and clustering, we cannot say before calculating variances for our sample whether the design effect for a given quantity will be less than 1 or greater than 1. Stratification tends to increase precision and clustering tends to decrease it, so the overall design effect depends on whether more precision is lost by clustering than gained by stratification.

Example 7.7. For the bed net survey discussed in Example 7.1, the design effect for the proportion of beds with nets was calculated to be 5.89. This means that about six times as many observations are needed with the complex sampling design used in the survey to obtain the same precision that would have been achieved with an SRS. The high design effect in this survey is due to the clustering: Villages tend to be homogeneous in bed net use. If you ignored the clustering and analyzed the sample as though it were an SRS, the estimated standard errors would be much too low, and you would think you had much more precision than really existed. ■

Design effects and sample sizes. Design effects are extremely useful for estimating the sample size needed for a survey. That is the purpose for which they were introduced by Cornfield (1951), who used them to estimate the sample size that would be needed if the sampling unit in a survey estimating the prevalence of tuberculosis was a census tract or block rather than an individual. The maximum allowable error was specified to be 20% of the true prevalence, or $0.2 \times p$. If the prevalence of tuberculosis was 1%, the sample size for an SRS would need to be

$$n = \frac{1.96^2 p(1-p)}{(0.2p)^2} = 9508.$$

Cornfield recommended increasing the sample size for an SRS to 20,000, to give more precision in separate estimates for subpopulations. He estimated the design effect for sampling census tracts rather than individuals to be 7.4 and concluded that if census tracts, which averaged 4,600 individuals, were used as a sampling unit, a sample size of 148,000 adults, rather than 20,000 adults, would be needed.

If you have estimates of design effects for variables from a similar survey, you only need to be able to estimate the sample size you would take using an SRS. Then multiply that sample size by the anticipated deff for a key survey variable (or the largest deff for a set of key variables) to obtain the number of observation units you need to observe with the complex design.

Conversely, to assess the amount of independent information from a complex sample, compute the **effective sample size**. The effective sample size from a survey that measures n observation units and has deff d is $n_{\text{eff}} = n/d$. A sample of size n_{eff} from an SRS would be expected to give same precision as the n observation units taken in the complex sample.

Example 7.8. When planning the seroprevalence study described in Example 6.12, Pollán et al. (2020) determined the sample size needed within each province by positing a design effect of 2 to account for the within-tract and within-household correlations. They wanted to be able to estimate a seroprevalence thought to be about 5% with margin of error $e = 2.5\%$. Using (2.30), the effective sample size needed to attain that precision is $n_{\text{eff}} = (1.96)^2(0.05)(0.95)/(0.025)^2 = 292$. The sample size needed to achieve the desired precision within a province with the cluster sample, then, is $2n_{\text{eff}} = 584$, which the investigators rounded up to 600. Anticipating that about 2/3 of sampled persons would respond, the investigators specified a minimum sample size of 900 persons for each province. ■

Alternative definitions used for design effect. The deff in Equation (7.9) is the ratio of the variance of a statistic under the complex sampling design to the variance of that statistic under a without-replacement SRS with the same number of observation units. Note, however, that some authors use different definitions. Some authors omit the fpc in (7.9), and use deff2 $= V(\hat{\bar{y}}_{\text{plan}})/(S^2/n)$, dividing by the with-replacement variance instead of the without-replacement variance. Still others (see, for example, Kish, 1995) use the term "design effect" to refer to the square root of deff2. All of these definitions provide information about the variance of an estimator from a specific sampling design relative to the variance that would be obtained from an SRS of the same size, but you need to pay attention to which definition a survey uses.

7.5 The National Health and Nutrition Examination Survey

Let's look at how all of these ideas are used in a large national survey. The goal of the National Health and Nutrition Examination Survey (NHANES), administered by the U.S. National Center for Health Statistics, is to collect information about the health and nutrition of a nationally representative sample of U.S. residents. Data are collected through in-person interviews and a standardized physical examination. The target population is the noninstitutionalized civilian population of the 50 states plus the District of Columbia. This section describes the main features of the 2015–2018 NHANES stratified multistage cluster design; see Chen et al. (2020) for a full description with additional details.

Because the NHANES conducts a physical examination of survey participants, it has much less measurement error than health surveys that ask people to report their own health information. Heights and weights that are measured by trained personnel using uniform procedures will, in general, be more accurate than self-reported heights and weights. Many persons may be unaware that they have diabetes, high blood pressure, or hepatitis; the examination includes tests to diagnose these and other conditions. Indeed, one of the incentives for persons to participate in the survey is that participants receive a complete report of the medical examination results.

The physical examination also makes the survey extraordinarily expensive to conduct, and budgetary considerations limit the sample size to about 5,000 persons each year. A sample of size 5,000 would be ample if estimates were desired only for the population as a whole (not for subpopulations) and if the data were collected as an SRS or stratified random sample with no clustering. But separate estimates are desired for 87 demographic domains defined by age, race, ethnicity, gender, and income. To obtain sufficient sample size in each domain of interest, the survey oversamples Hispanic, Black, and Asian persons; persons with low income; and persons aged 0–11 years or 80 years and over. Even with the oversampling, data from several years of the survey must be combined to obtain a large enough sample to produce reliable estimates for some of the domains.

In addition, the physical examination necessitates in-person data collection, which is only feasible if a cluster sampling design is used. The use of clustering decreases the effective sample size of the survey, so that 5,000 persons selected with the cluster design do not provide as much information as if one could take an SRS of 5,000 persons.

Sample design for the NHANES. A psu in the NHANES is a county, a group of adjacent counties, or a group of census tracts within a county. For the 2015–2018 NHANES design, four psus are so large that they are put in a certainty stratum. One of those psus is sampled each year so that over a four-year period, the sample includes all of the certainty psus.

All other psus are grouped into 14 major strata based on (a) a state health index calculated from information related to overall health in the state including death rate, infant mortality rate, percentage of adults with high blood pressure, percentage of adults who are overweight or obese, percentage of adults with poor nutrition, and percentage of adults who smoke and (b) the percentage of the psu's population living in rural areas. The stratification information comes from information on state-level mortality rates and from the Behavioral Risk Factor Surveillance System, a large health survey conducted by telephone in each state. Each year, 15 psus are sampled: one psu from the certainty stratum plus one psu from each of the 14 major strata, so that 60 psus are sampled over a four-year period.

The psus in the non-certainty strata are selected with probability proportional to size, where the measure of size gives psus with large numbers of persons in domains of interest higher probabilities of selection. To accomplish this, the measure of size is calculated as a weighted average of five population counts for the psu: (a) number of non-Hispanic, non-Black Asian persons, (b) number of non-Hispanic Black persons, (c) number of Hispanic persons, (d) number of persons with low income not in one of the first three categories, and (e) all other persons. Category (a) has the highest contribution toward the measure of size, and category (e) has the lowest. Thus a psu of 50,000 persons with 20% in each category (a)–(e) will have a higher selection probability than a psu with all 50,000 persons in category (e).

Stage 2 of sampling selects ssus (subareas composed of census blocks or groups of census blocks) from the sampled psus. The same number of ssus (24) is sampled in each non-certainty psu because the psus are selected with probability proportional to size. This results in approximately equal workloads for interviewing and conducting medical examinations for each non-certainty psu. The number of ssus in certainty psus depends on the size of the psu. The ssus are also sampled with probability proportional to size.

Stage 3 of sampling selects households from the sampled ssus. Stage 4 selects persons from the sampled households to participate in the interview and medical examination. Persons in different age-race-gender-income domains have different probabilities of selection so that the sample size targets for the domains will be met.

The design is set up so that each one-year, two-year, and four-year period of data collection gives a nationally representative sample. Four years of data are needed to obtain reliable estimates in some of the smaller demographic domains, but, to facilitate more timely estimates, the data are weighted and released in two-year cycles.

Weighting the NHANES data. Weights are calculated in three steps for each two-year period. Step 1 calculates the base weight for each person as the inverse of the probability of selection, Step 2 adjusts the weights for nonresponse, and Step 3 calibrates the weights to known population totals.

Step 1. The base weight of a person in a particular sampling domain equals 1 divided by the probability that a person within that domain is selected for the sample, accounting for the selection probabilities at all four stages of sampling. Persons in oversampled domains have lower base weights because they have higher probabilities of being selected for the sample. For example, the base weight of a 45-year-old non-low-income white woman is about three times as high as the base weight of a 45-year-old Hispanic woman. Because Hispanic persons are oversampled, a Hispanic woman selected for the sample represents fewer persons in the population than a woman in a demographic group that is not oversampled.

Step 2. Some of the households selected for the sample at Stage 3 cannot be reached or refuse to participate in the survey, and some of the persons selected in Stage 4 refuse to be interviewed or do not participate in the medical examination. The interviewer gathers demographic information on the nonrespondents, and that demographic information is used to adjust the weights in an attempt to counteract the nonresponse. This is an example of weighting class adjustments for nonresponse, which will be discussed in more detail in Section 8.5.

Persons who participate in the sample are placed in weighting classes using information that is available for both respondents and nonrespondents such as characteristics of the census tract where the person resides. Then, weights for the respondents in each weighting class are increased so that they represent the nonrespondents in that class as well as themselves. After the nonresponse adjustment for the interview, the weight for a person in a particular weighting class is

$$\text{base weight} \times \frac{\text{sum of base weights for all sampled persons in class}}{\text{sum of base weights for respondents in class}}.$$

The goal of Step 2 is to redistribute the weights of persons who are not interviewed to interviewed persons who are similar to them with respect to the demographic variables.

Step 3. Because the NHANES is a sample, estimates of demographic characteristics calculated using the weights from Step 2 (such as the percentage of persons who are female or the percentage who are between ages 40 and 49) differ from demographic estimates calculated from the ACS. In the last step of weighting, the Step-2 weights for survey respondents are calibrated to estimated population totals for demographic groups from the ACS. Even though the ACS is a survey, its large sample size means that ACS estimates have very small sampling error for large subpopulations and can be used as population totals for calibration (see Exercise 41 of Chapter 11). This step reduces the variance of estimates through ratio estimation and may also reduce undercoverage bias.

File `nhanes.csv` contains selected variables from the 2015–2016 NHANES. The file has two sets of weights: one for the set of persons who have data from the interview, and a second set of weights for the smaller set of persons who participate in the medical examination (the medical examination weights are constructed by applying Steps 2 and 3 to the final interview weights). Both sets of weights are highly variable because of the oversampling and nonresponse adjustments. The interview weights range from 3,294 to 233,756, with median weight of 20,160. The medical examination weights range from 3,419 to 242,387, with median 21,034. Estimates of population quantities *must* be calculated using weights. If the weights are ignored, estimates will be biased.

Example 7.9. Body mass index (BMI) is calculated from the height and weight measurements in the NHANES medical examination as weight/height^2. Let's look at the mean value and quantiles of BMI for adults age 20 and over, as well as the percentage of adults who have BMI greater than 30 kg/m^2 (often used as a criterion for obesity). One could calculate these estimates using the formulas in Chapters 3 to 6, but it is much easier to use software designed for survey estimates. Table 7.2 gives the estimates calculated from SAS software, along with their standard errors and 95% CIs. The quantiles are calculated using interpolation (see Exercise 19).

TABLE 7.2
Statistics for adults age 20 and over from NHANES.

BMI Statistic	Estimate	Standard Error	95% Confidence Interval
Mean	29.39	0.253	[28.85, 29.93]
5th percentile	20.30	0.180	[19.92, 20.68]
25th percentile	24.35	0.214	[23.90, 24.81]
Median	28.23	0.317	[27.56, 28.91]
75th percentile	33.07	0.308	[32.41, 33.72]
95th percentile	42.64	0.354	[41.89, 43.40]
Percent with BMI > 30	39.22	1.586	[35.84, 42.60]

The estimates in Table 7.2 use almost every feature discussed so far in this book. The psus are stratified and selected with unequal probabilities. The oversampling, nonresponse adjustments, and calibration cause the observations to have unequal weights. In addition, the estimates are computed for the domain of adults age 20 and over, so domain estimation is used as well.

Although the survey has multiple stages of cluster sampling, only the psu-level information is needed to calculate the with-replacement variance within each stratum, as described in Equation (6.34) of Section 6.4.4. The confidence intervals are calculated using critical values from a t distribution with 15 degrees of freedom—the number of psus (30) minus the number of strata (15). We'll discuss how to calculate the standard errors and CIs for estimated quantiles in Section 9.5.2.

We can also assess the variance penalty for unequal weighting and clustering by examining design effects. These are most easily computed for proportions, where the SRS variance is estimated as $\hat{p}(1-\hat{p})/n$. For the proportion of adults with BMI > 30, the estimated variance that would have been obtained under an SRS of 5,406 adults (the number of adults in the sample with medical examination data for BMI) is $0.392(1-0.392)/5406 = 0.0000441$. The variance of the proportion, estimated using the NHANES survey design and weights, is 0.000251. The estimated design effect for the proportion of adults with BMI > 30, then, is $0.000251/0.0000441 = 5.7$. For this response, the variance is nearly six times as high as it would have been had an SRS been taken. ■

7.6 Graphing Data from a Complex Survey

Simple plots reveal much information about data from a small SRS or representative systematic sample. Histograms or smoothed density estimates display the shape of the data; scatterplots and scatterplot matrices show relationships between variables.

As we saw in Figure 7.3, however, graphical displays commonly used for SRSs can mislead when applied to raw, unweighted, data from non-self-weighting samples. They display the characteristics of the individuals in the sample, but can distort the picture of the population when the weights are unequal. Incorporating the sampling weights into the graphics, however, gives a picture of the shape of the population and displays relationships between variables. This section provides some examples of how to do that.

The univariate plots are illustrated using the stratified sample of heights from Example 7.5 because the simplicity of that example allows showing the calculations. Scatterplots are illustrated using data from the NHANES. Lu and Lohr (2022) and Lohr (2022) provide code for drawing graphs similar to those in this section with R and SAS software.

7.6.1 Univariate Plots

Histograms. One of the simplest plots for displaying the shape of data is the histogram. To construct a relative frequency histogram for an SRS of size n, divide the range of the data into k bins with each bin having width b. Then the height of the histogram in the jth bin is

$$\text{height}(j) = \frac{\text{relative frequency for bin } j}{b} = \frac{1}{bn} \sum_{i \in \mathcal{S}} u_i(j),$$

where $u_i(j) = 1$ if observation i is in bin j and 0 otherwise. If a sample is self-weighting, as with an SRS, an unweighted histogram of the sample data will estimate the population probability mass function.

We saw in Figure 7.3, though, that if a sample is not self-weighting a histogram of the raw data may underrepresent some parts of the population in the display. We can use the sampling weights to construct a histogram that estimates the population histogram. As before, divide the range of the data into k bins with each bin having width b. Now use the sampling weights w_i to find the height of the histogram in bin j:

$$\text{height}(j) = \frac{1}{b \sum_{i \in \mathcal{S}} w_i} \sum_{i \in \mathcal{S}} w_i u_i(j). \tag{7.11}$$

Dividing by the quantity $b \sum_{i \in \mathcal{S}} w_i$ ensures that the total area under the histogram equals 1.

Example 7.10. To construct a histogram of the height data from the stratified sample in `htstrat.csv` (Example 7.4), first decide on a bin width, b. We decide to use $b = 3$ as in Figure 7.3. This choice gives 20 histogram bins. The cutpoints for the histogram bins are at 141, 144, 147, 150, ..., 198, and 201. The first histogram bar includes persons in the sample whose heights are in the interval (141, 144]; the sample contains one woman with height 142 and one woman with height 144. Each woman in the sample has sampling weight 6.25, so the height of the first histogram bar is

$$\frac{2(6.25)}{b \sum_{i \in \mathcal{S}} w_i} = \frac{12.5}{(3)(2000)} = 0.00208.$$

The tallest histogram bar includes the persons in the sample who have heights in the interval (165, 168]; the sample contains 19 women and 6 men with heights in this range, so the height for the bar corresponding to (165, 168] is

$$\frac{19(6.25) + 6(25)}{b \sum_{i \in \mathcal{S}} w_i} = \frac{268.75}{(3)(2000)} = 0.04479.$$

The heights for the other histogram bars are computed similarly.

The histogram for the stratified sample, incorporating the weights, is in Figure 7.5. The histogram with the weights shows higher relative frequencies for heights over 165 than does the histogram without weights in Figure 7.3. The histogram without weights in Figure 7.3 displays the frequencies of persons in different height categories in the sample but, because the weights are unequal, it does not estimate the shape of the population. Figure 7.5, which incorporates the weights, estimates the shape of the population distribution. ■

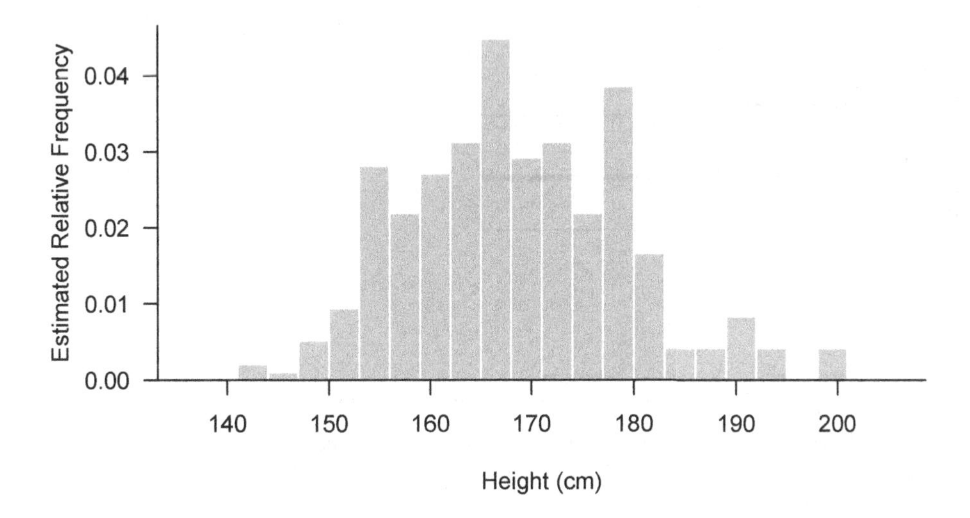

FIGURE 7.5
Histogram of height data from stratified sample, incorporating the sampling weights.

Boxplots. Side-by-side boxplots, sometimes called box-and-whisker plots, are a useful way to display the distribution of a population or to compare domain distributions visually. The box in a boxplot has lines at the 25th, 50th, and 75th quantiles, and whiskers that extend to the extremes of the data (or, alternatively, to a multiple of the interquartile range). If the sample is not self-weighting, the weights should be used to calculate the quantiles in the display.

Example 7.11. The last column of Table 7.1 gave the estimated quantiles for the stratified sample in `htstrat.csv`. Side-by-side boxplots of the data, using these estimated quantiles and extending the whiskers to the range of the data, are shown in Figure 7.6. ■

Density estimates. Smoothed density estimates (Scott, 2015) are useful for displaying the estimated shape of a population. The idea is to create a smooth version of a histogram. Instead of having bars in a histogram, one could create a plot by connecting the heights at the midpoints of the histogram bins. Such a plot would not be particularly smooth, however, and could be improved by using each possible value of y as the midpoint of a histogram bin of width b, finding the height for that bin, and then drawing a line through those values. In essence, the histogram bars slide continuously along the horizontal axis; as points enter and leave the bar, the height corresponding to the midpoint changes.

A symmetric density function K, called a kernel function, is used to allow more flexibility in the smoothing method. Bellhouse and Stafford (1999) and Buskirk and Lohr (2005)

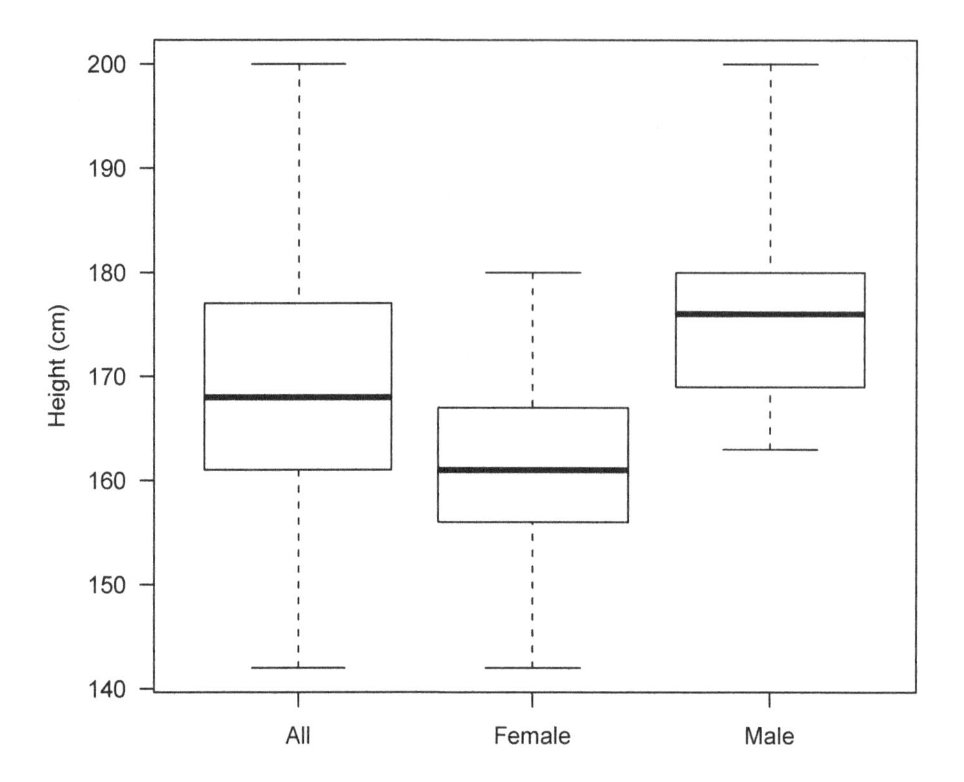

FIGURE 7.6
Boxplots of height data from stratified sample, incorporating the sampling weights. The first box uses data from the entire sample, the second box uses data from the women, and the third box uses data from the men.

adapted kernel density estimation to survey data by incorporating the weights, with

$$\hat{f}(y; b) = \frac{1}{b \sum_{i \in \mathcal{S}} w_i} \sum_{i \in \mathcal{S}} w_i K \left[\frac{y - y_i}{b} \right].$$

Commonly used kernel functions include the normal kernel function $K_N(t) = \exp(-t^2/2)/\sqrt{2\pi}$ and the quadratic kernel function $K_Q(t) = \frac{3}{4}(1 - t^2)$ for $|t| < 1$. The sliding histogram described above corresponds to a box kernel with $K_B(t) = 1$ for $|t| \leq 1/2$ and $K_B(t) = 0$ for $|t| > 1/2$; in that case, $\hat{f}(y; b)$ corresponds to the histogram height given in (7.11) for a point y in the middle of a bin of width b.

Example 7.12. Figure 7.7 shows a smoothed density estimate for the stratified sample of heights from Examples 7.4 and 7.10, using a quadratic kernel function with $b = 7$. As with the histogram in Figure 7.5, using the sampling weights increases the estimated density in the right tail despite the paucity of data in that region. ■

The choice of b, called the bandwidth, determines the amount of smoothing to be used. Small values of b use little smoothing since the sliding window is small. A large value of b provides much smoothing since each point in the plot represents the weighted average of many points from the data. One potential problem in some surveys is that respondents may round their answers. For example, some respondents may round their height to 165 or 170

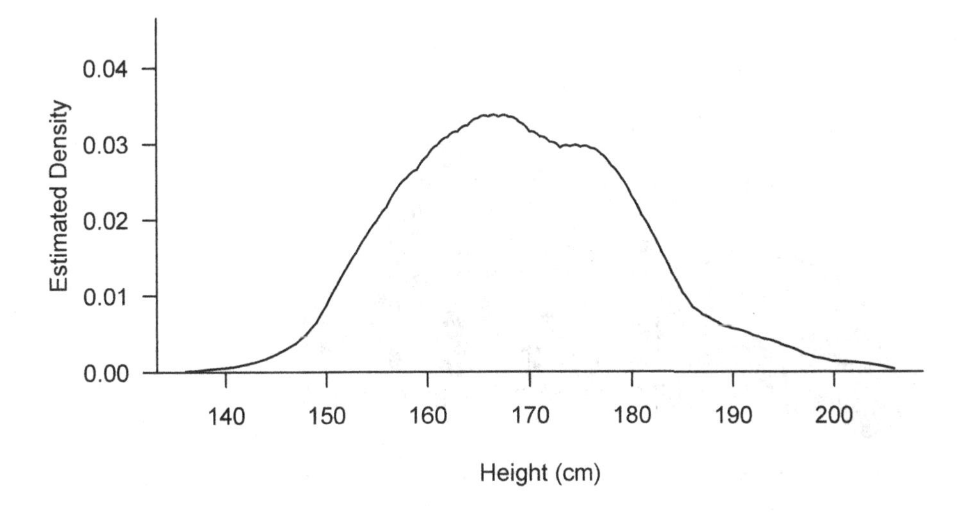

FIGURE 7.7
Estimated density function for the stratified sample of heights.

cm, causing spikes at those values. You may want to choose a larger bandwidth b to increase the amount of smoothing, or you may want to adopt a model for the effect of rounding by the respondent.

7.6.2 Bivariate Plots

You may also be interested in bivariate relationships among variables. We typically explore such relationships visually through a scatterplot. With complex survey data, unequal weights should be considered for interpreting bivariate relationships.

Since they involve two variables, scatterplots are more complicated than univariate displays. Many government surveys have large amounts of data. The ACS, for example, collects data from millions of housing units each year. A scatterplot of two continuous variables from the ACS will have so many data points that the graph may be solid black and useless for visual inspection of the data. In addition, if both variables take on integer values, for example if x = number of persons in household and y = number of rooms in dwelling unit, many observations will share the same x and y values.

The challenging part for scatterplots is how to incorporate the weights. In a histogram, only the horizontal axis uses the data values so the weights can be incorporated in the relative frequencies displayed on the vertical axis. But in a scatterplot, the horizontal axis displays information about the x variable and the vertical axis displays information about the y variable, so the weight information must be incorporated by some other means. It is generally a good idea to construct several types of scatterplots for survey data since some will work better with a particular data set than others.

This section displays a variety of graphs displaying the relationship between body mass index and age in the 2015–2016 NHANES data. These graphs display the relationship for persons of all ages, in contrast to the estimates in Example 7.9 that were limited to adults age 20 and over. Age is topcoded at 80 to protect confidentiality of the respondents; any person with age greater than 80 is assigned age value 80. Figure 7.8 shows a plot of the raw data without weights; as you can see, the data set is so large that it is difficult to see

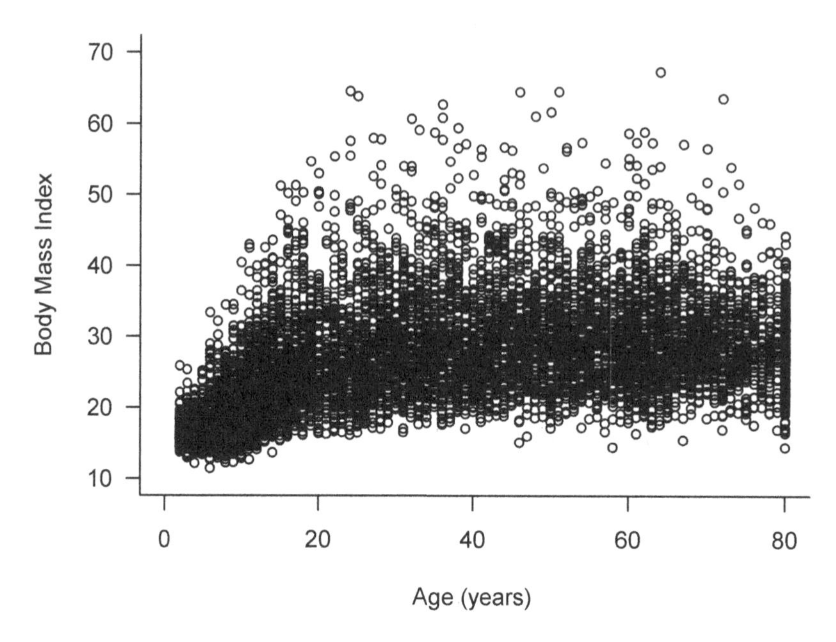

FIGURE 7.8
Plot of raw data from NHANES. There are so many data points that it is difficult to see patterns in the plot; in addition, no weighting information is used. This plot is not recommended for complex survey data with unequal weights.

the structure of the bivariate relationship from the graph. In addition, this graph shows the relationship between the variables for the sample; the relationship between the variables in the population may be different.

Weighted bubble plot. The plot in Figure 7.8 does not include information about the weights. The NHANES is designed to oversample persons in certain demographic domains, and persons in the oversampled domains have smaller sampling weights. To get a better view of the data, we should incorporate the unequal weights. One way of doing that is to use a circle as a plotting symbol, and, for each observation, make the area of the circle proportional to the weight of the observation. This plot for the NHANES data is shown in Figure 7.9.

The data are easier to see on this plot than in Figure 7.8 because the symbols are smaller; however, there are still so many data points that some features may be obscured. Observations with small weights have very small circles and are nearly invisible. In data sets containing many distinct values for x or y, a weighted bubble plot will still have such high data density in areas that the plot will appear to be a solid mass.

Binned bubble plot. If there are too many distinct (x, y) pairs to display each of them individually on the plot, you can aggregate the points in bins first and then plot bubbles with area proportional to the sum of the weights in each bin.

This idea is similar to creating bins for a histogram. In a histogram, the y values are grouped into a bin, and the sum of the weights is found for the y values falling into each bin. To extend this idea to a scatterplot, divide the region into rectangles. Find the sum of the weights for the (x, y) values falling in each rectangle. Then, plot a circle (or other shape) with area proportional to the sum of the weights at the midpoint of the rectangle. Figure 7.10 shows the NHANES data in bins formed by rounding the x and y values to

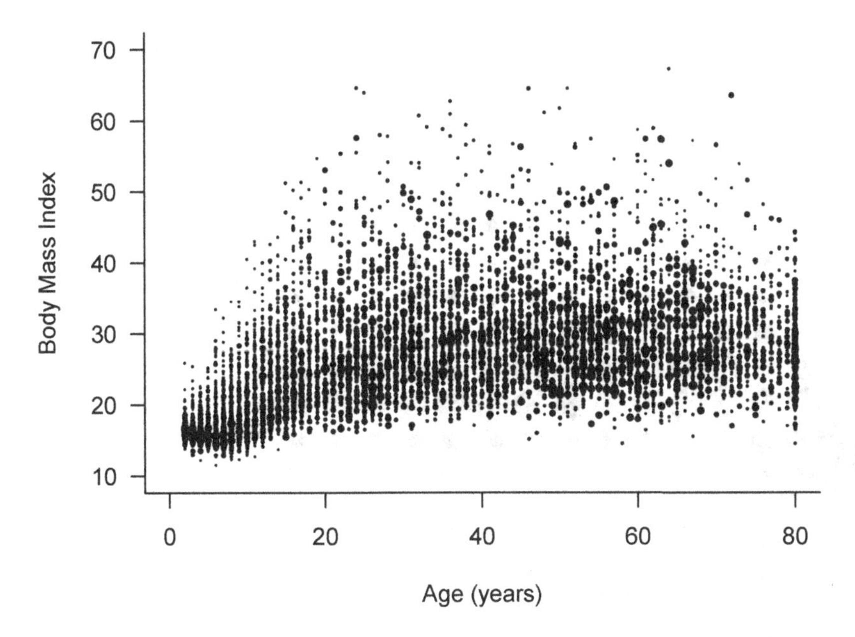

FIGURE 7.9
Weighted bubble plot of NHANES data. The circle size for each point is proportional to the sum of the weights at that point.

the closest multiple of 5. This type of plot is especially useful if the data set contains many observations having the same value (x, y), since the plot displays the multiplicity of the data.

Shaded plot. Similarly to the binned bubble plot in Figure 7.10, group the data into bins formed by the values of x and y. Find the sum of the weights for the (x, y) values falling in each bin, then use shading to indicate the sum of weights. This type of plot is sometimes called a heat map.

For the plot in Figure 7.11, we form bins by rounding the x and y values to the nearest integer, creating a grid of x and y values. The shading value is proportional to the sum of the weights for observations in the square, with darker colors representing larger numbers of estimated population observations.

Plot a subsample of points. Instead of plotting all the data, we can plot a subset of the data. Since the sampling weight of an observation can be interpreted as the number of population units represented by that unit, a plot of a subsample selected with probabilities proportional to the weights can be interpreted much the same way as a regular scatterplot (see Exercise 24). Figure 7.12 shows a scatterplot of a Poisson sample of about 500 points selected with probability proportional to the weight variable.

This plot can be repeated with different subsamples, and each plot will be different. Each plot, however, has less information than the full data set since it is based on a subsample of the data. Unusual observations such as outliers might not appear on a particular plot.

Side-by-side boxplots. Instead of using shading, we can group the x variable into bins and draw a boxplot at the midpoint of each x bin, calculating the quantiles of each bin as in Example 7.6. Figure 7.13 shows side-by-side boxplots of the NHANES data, where the age (x) variable bins have width 5.

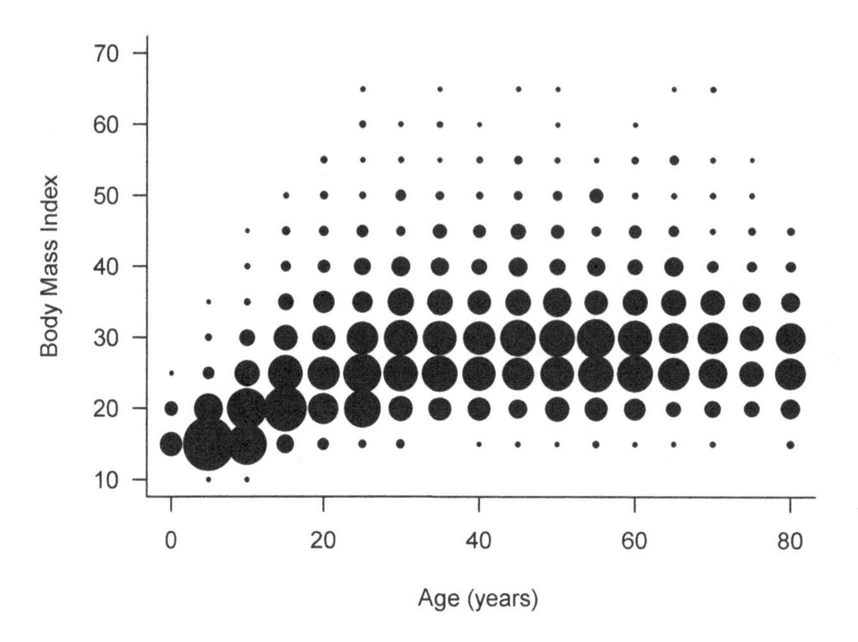

FIGURE 7.10
Binned bubble plot of NHANES data. The area of each circle is proportional to the sum of the weights of the set of observations in the corresponding bin.

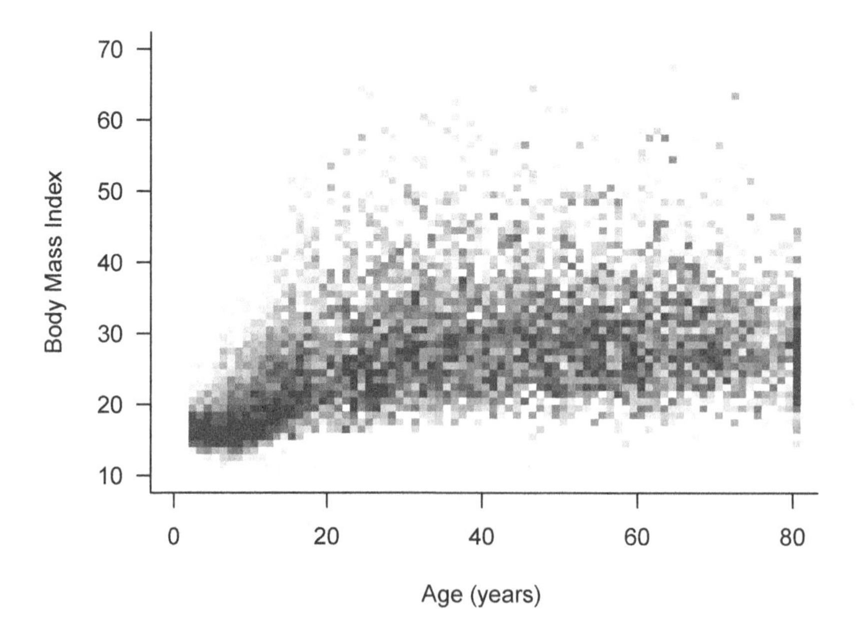

FIGURE 7.11
Shaded plot of NHANES data. The shading is proportional to the sum of the weights for the rectangle.

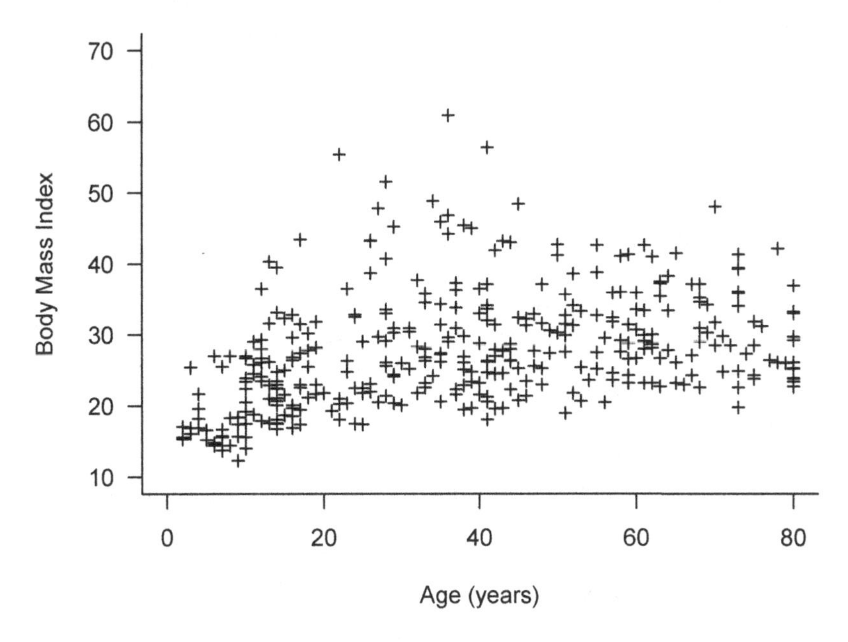

FIGURE 7.12
Plot of subsample of NHANES data, selected with probability proportional to the weight variable.

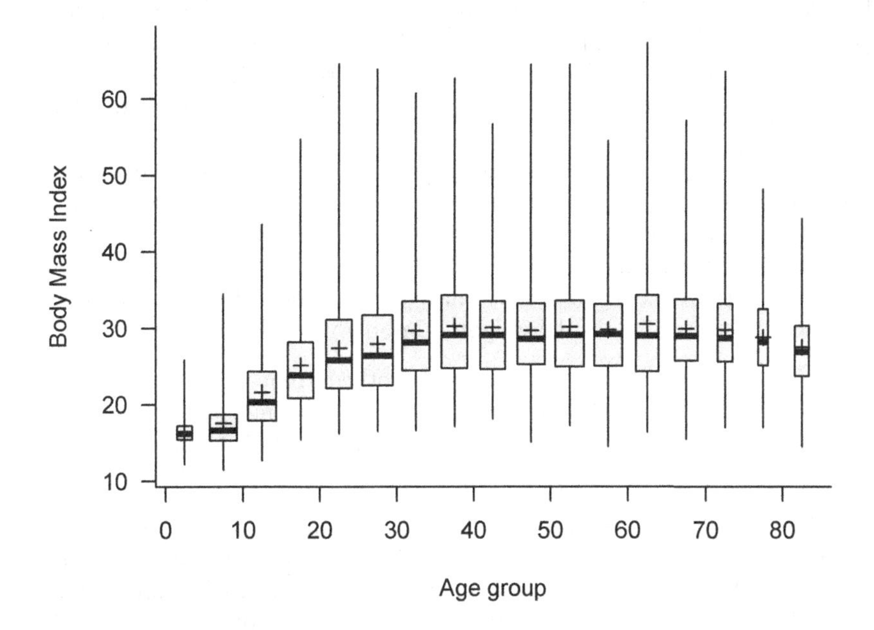

FIGURE 7.13
Side-by-side boxplots of NHANES data. The width of each box is proportional to the sum of the weights of the set of observations used for the box. The + in each box denotes the mean.

Smoothed trend line for mean. As with the smoothed density estimates in Figure 7.7, we can fit a smoothed trend line to the data. This can be thought of as a smooth line

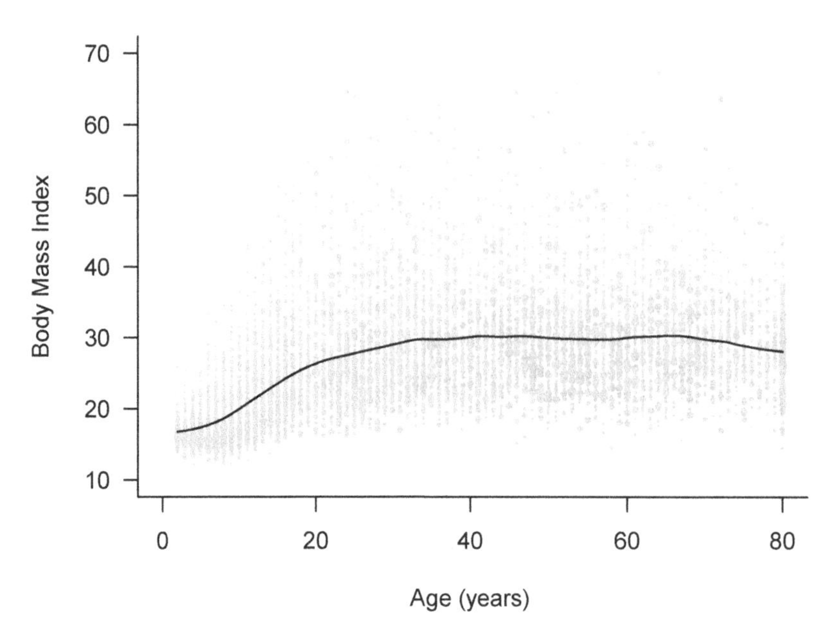

FIGURE 7.14
Weighted bubble plot of NHANES data, with trend line for mean.

that approximately follows the line of means of the boxplots in Figure 7.13. Many methods can be used to calculate a smoothed trend line. One simple method, similar to the kernel density estimation used in Figure 7.7, calculates a weighted average of the y_i values that fall in each data window. Other methods fit a straight line or polynomial in each window (see Exercise 34 of Chapter 11).

The trend line in Figure 7.14 was computed using smoothing splines (Zhang et al., 2015). Essentially, this method divides the horizontal axis into segments and fits a polynomial regression model to each piece (using the weights, as detailed in Section 11.2), while requiring the curves to be smooth where the segments join.

The trend line can be displayed by itself, or as an overlay on one of the other plots. Figure 7.14 superimposes the trend line on the weighted bubble plot from Figure 7.9.

Smoothed trend line for quantiles. We can also fit smoothed trend lines for different quantiles. Figure 7.15 shows lines estimating the quantiles corresponding to $q = 0.95$, 0.75, 0.5 (median), 0.25, and 0.05. The middle three lines approximately follow the horizontal lines of the boxes (showing the 25th, 50th, and 75th percentiles) in Figure 7.13.

Summary. All of the plots displayed in this section (except for the unweighted scatterplot in Figure 7.8, when used with a non-self-weighting sample) provide graphical representations of the estimated relationship between x and y in the population. A variety of different plots are given because some will work better with a particular data set than others. For a small data set, or a data set with relatively few distinct (x, y) pairs, you may want to use a weighted bubble plot. For a larger data set, or one with many distinct (x, y) pairs, a shaded plot or side-by-side boxplots may work better. Smoothed trend lines summarize the shape of the relationship for any data set.

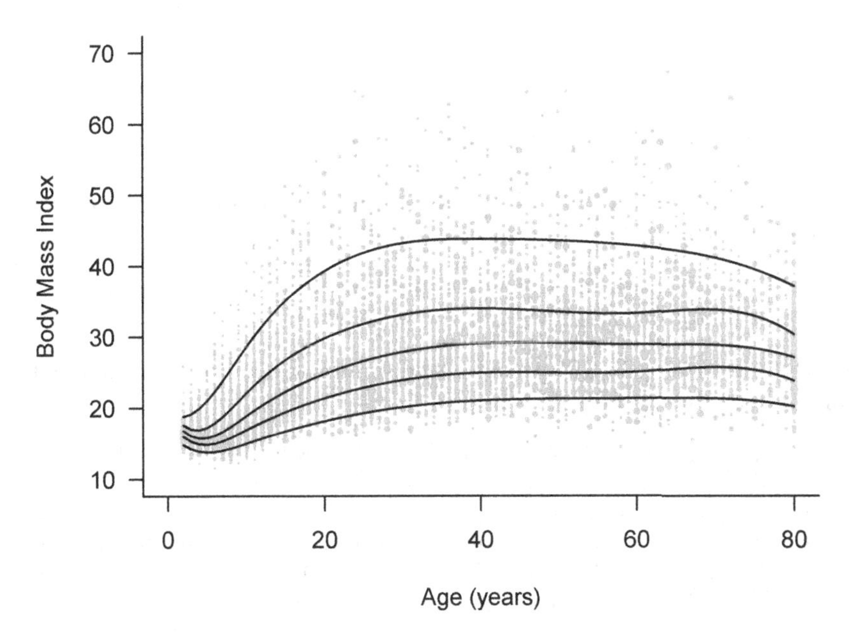

FIGURE 7.15
Weighted bubble plot of NHANES data, with trend lines for (from top to bottom) 95th, 75th, 50th, 25th, and 5th percentiles.

7.7 Chapter Summary

Many large surveys have a stratified multistage sampling design, in which the psus are selected by stratified sampling and then subsampled. Estimators of population quantities from a stratified multistage sample are calculated using the principles from Chapters 2 to 6. In most instances, only the stratification and information from the first stage of clustering are used to calculate standard errors of estimates.

Any population quantity can be estimated from the sample using the weights. The empirical cdf and the empirical probability mass function estimate the cdf and probability mass function of the population by incorporating the weights w_i. Since w_i can be thought of as the number of observations units in the population represented by observation unit i in the sample, the empirical cdf can be thought of as the cdf of a pseudo-population in which observation i in the sample is replicated w_i times.

Although the survey weights can be used to find a point estimate of any population quantity, the weights are not sufficient information to calculate standard errors of statistics. Standard errors depend on the stratification and clustering in the survey design. The design effect, which is the ratio of the variance of a statistic calculated using the complex survey design to the variance that would have been obtained from an SRS with the same number of observation units, is useful for assessing the effect of design features on the variance. The design effect is often used to determine the sample size needed for a complex survey.

Graphs that are commonly used for displaying data from an SRS can be adapted for complex survey data by incorporating the survey weights. Histograms, boxplot, and scatterplots that use the survey weights display features of the data that might not be apparent from the summary statistics.

Key Terms

Cumulative distribution function: $F(y)$ = (number of observation units in the population whose value is less than or equal to $y)/N$ describes the distribution of the finite population. It can be thought of as $P(Y \leq y)$, where Y is the random variable describing the distribution of a sample of size one taken from the population.

Design effect: Ratio of the variance of an estimator from the sampling plan to the variance of the corresponding estimator from an SRS with the same number of observation units.

Effective sample size: The size of an SRS that has the same variance for the estimated mean as the complex sampling design.

Empirical probability mass function: An estimator of the probability mass function using sampling weights: $\hat{f}(y) = \sum_{i \in \mathcal{S}, y_i = y} w_i / \sum_{i \in \mathcal{S}} w_i$.

Probability mass function: $f(y)$ = (number of observation units in the population whose value is $y)/N$. It can be thought of as $P(Y = y)$, where Y is the random variable describing the distribution of a sample of size one taken from the population.

Stratified multistage sample: A sampling design in which primary sampling units are grouped into strata and a probability sample is taken of the psus in each stratum. Secondary sampling units are then subsampled within each selected psus. In some cases, the selected ssus are also clusters and are themselves subsampled.

For Further Reading

The volumes edited by Pfeffermann and Rao (2009a,b) give a wealth of information on topics for designing and computing estimates from complex surveys. Heeringa et al. (2017) discuss practical issues involved when analyzing data from surveys. Haziza and Beaumont (2017) present an overview of weighting procedures, including nonresponse weighting. Valliant et al. (2018) provide multiple examples, along with computer code, for how to construct weights for survey samples. Beaumont (2008) and Santacatterina and Bottai (2018) discuss methods for smoothing and trimming weights, and their effects on survey estimates.

For a comprehensive guide to the theory of survey sampling, see Fuller (2009). This book, written as a sequence of theorems and proofs, requires a strong background in measure-theoretic probability theory and asymptotic theory of statistics.

The idea of using design effects for sample size estimation was introduced by Cornfield (1951); the paper gives an interesting example of sampling in practice. Kish (1992, 1995), Gabler et al. (1999, 2006), and Park and Lee (2004) further explain the ideas underlying design effects.

There are many alternative ways of defining population and sample quantiles, and some of these, for data from an SRS, are discussed by Hyndman and Fan (1996) and Parzen (2004). Francisco and Fuller (1991), Shao (1994), and Wang (2021) discuss estimating quantiles from complex surveys.

Graphs for complex survey data are extensions of those that have been developed for SRSs. Many of the graphs for survey data discussed in this chapter are presented in Korn and Graubard (1998b, 1999) and Lohr (2012).

Scott (2015), Kloke and McKean (2014), and Eilers and Marx (2021) present methods for constructing histograms, density estimates, and trend lines for data assumed to be independent and identically distributed. Numerous articles have extended smoothing techniques to complex survey data. Bellhouse and Stafford (1999) and Buskirk and Lohr (2005) examine properties of smoothed density estimates for survey data; Korn and Graubard (1998b), Bellhouse and Stafford (2001), and Harms and Duchesne (2010) calculate trend lines using kernel regression methods. Opsomer and Miller (2005) provide methods for determining the amount of smoothing to be used. For use of smoothing splines with survey data, see Goga (2005), McConville and Breidt (2013), and Zhang et al. (2015).

The theory of probability sampling shares many features with the theory of design of randomized experiments. Both areas rely on randomization for inference. Stratified sampling is analogous to blocking, cluster sampling shares features of nested and split-plot designs, and calibration in surveys has similar purpose as covariate adjustments in designed experiments. Parallels between sample surveys and designed experiments are discussed by Yates (1981) and Fienberg and Tanur (2018).

7.8 Exercises

A. Introductory Exercises

1. You are asked to design a survey to estimate the total number of cars without permits that park in handicapped parking places on your campus. What variables (if any) would you consider for stratification? For clustering? What information do you need to aid in the design of the survey? Describe a survey design that you think would work well for this situation.

2. Repeat Exercise 1 for a survey to estimate the total number of books in a library that need rebinding.

3. Repeat Exercise 1 for a survey to estimate the percentage of persons in your city who speak more than one language.

4. Repeat Exercise 1 for a survey to estimate the distribution of number of eggs laid by Canada geese in a region.

5. Repeat Exercise 1 for a survey to estimate the percentage of NFL players who had CTE, as described in Exercise 21 of Chapter 1.

6. The organization "Women tired of waiting in line" wants to estimate statistics about restroom usage. Design a survey to estimate the average amount of time spent by women in a restroom and the average time spent by men in a restroom at your university.

7. Use the data in file `integerwt.csv` for this exercise. The strata are constructed with $N_1 = 200$, $N_2 = 800$, $N_3 = 400$, $N_4 = 600$.

   (a) Take a stratified random sample with $n_1 = 50$, $n_2 = 50$, $n_3 = 20$ and $n_4 = 25$. Calculate the sampling weight w_i for each observation in your sample (the sample sizes were selected so that each weight is an integer).

   (b) Using the weights, estimate $\bar{y}_{\mathcal{U}}$, S^2, and the 25th, 50th, and 75th percentiles of the population.

(c) Now create a "pseudo-population" by constructing a data set in which the data value y_i is replicated w_i times. Your pseudo-population should have $N = 2000$ observations. Estimate the same quantities in (b) using the pseudo-population and usual formulas for an SRS. How do the estimates compare with the estimates from (b)?

B. Working with Survey Data

8. Using the data in `nybight.csv` (see Exercise 22 of Chapter 3), find the empirical probability mass function of number of species caught per trawl in 1974. Be sure to use the sampling weights.

9. Using the data in `teachers.csv` (see Exercise 15 of Chapter 5), use the sampling weights to find the empirical probability mass function of the number of hours worked. What is the design effect?

10. Using the data in `measles.csv` (see Exercise 16 of Chapter 5), what is the design effect for percentage of parents who received a consent form? For the percentage of children who had previously had measles?

11. For the data in `schools.csv` (see Example 5.7), estimate the design effect for the proportion of students with *mathlevel* = 2.

12. Using the data in file `statepop.csv` (see Exercise 12 of Chapter 6), draw a histogram, using the weights, of the number of veterans. How does this compare with a histogram that does not use the weights?

13. Using the data in file `statepop.csv` (see Exercise 12 of Chapter 6), draw one of the scatterplots, using the weights, of the number of veterans versus the number of physicians.

14. Use the data in file `nhanes.csv` for this exercise. The sagittal abdominal diameter (variable *bmdavsad*), measures the distance from the small of the back to the upper abdomen.

    (a) Draw a histogram of *bmdavsad*, using the weights. Do the data appear to be normally distributed?

    (b) Estimate the mean value of *bmdavsad* for the domain of adults age 20 and over, along with a 95% CI.

    (c) Find the minimum, 25th, 50th, and 75th percentiles, and maximum of *bmdavsad*. Calculate the same quantities separately for each gender (variable *riagendr*). Use these to construct side-by-side boxplots of the data as in Figure 7.6.

    (d) Construct a weighted bubble plot with smoothed trend line for y = *bmdavsad* and x = *bmxbmi*. Does there appear to be a linear relationship? What other features do you see in the data?

15. Answer the questions in Exercise 14, for y = waist circumference (variable *bmxwaist*) and x = upper arm circumference (variable *bmxarmc*).

C. Working with Theory

16. *Effects of disproportional allocation.* Suppose that a stratified sample of size n is to be taken from H strata, and that the population sizes are sufficiently large that the fpcs can be ignored. Consider a disproportionally allocated sample of size n where observation j in stratum h has weight $w_{hj} = N_h/n_h$ ($\sum_{h=1}^{H} n_h = n$); the stratified estimator is $\bar{y}_{str,d}$ from (3.11). Let $\bar{y}_{str,p}$ be the stratified estimator of the population mean with proportional allocation, that is, the sample size in stratum h is nN_h/N. Show that if $S_1^2 = S_2^2 = \ldots = S_H^2$, then

$$\frac{V(\bar{y}_{str,d})}{V(\bar{y}_{str,p})} = \left(\sum_{h=1}^{H} \frac{N_h}{N} w_h\right)\left(\frac{n}{N}\right) = \left(\sum_{h=1}^{H} \frac{N_h}{N} w_h\right)\left(\sum_{h=1}^{H} \frac{N_h}{N w_h}\right).$$

17. *Design effect from unequal weighting* (requires mathematical statistics). Suppose that $Y_1, \ldots, Y_n$ are independent random variables with mean μ and variance σ^2 (model M). Let $\bar{Y} = \sum_{i=1}^{n} Y_i/n$ be the unweighted mean and let $\bar{Y}_w = \sum_{i=1}^{n} w_i Y_i / \sum_{i=1}^{n} w_i$, where $w_1, \ldots, w_n$ are non-negative weights, denote the weighted mean.

   (a) Show that under model M,

$$V_M\left(\frac{\sum_{i=1}^{n} w_i Y_i}{\sum_{i=1}^{n} w_i}\right) = \frac{\sum_{i=1}^{n} w_i^2 \sigma^2}{(\sum_{i=1}^{n} w_i)^2}.$$

   (b) Show that the ratio of $V(\bar{Y}_w)$ to $V(\bar{Y})$ is

$$\frac{V_M(\bar{Y}_w)}{V_M(\bar{Y})} = n \frac{\sum_{i=1}^{n} w_i^2}{\left(\sum_{i=1}^{n} w_i\right)^2} \approx 1 + \text{CV}_w^2, \tag{7.12}$$

   where CV_w is the coefficient of variation of the weights.

   The variance ratio in (7.12) is sometimes called the *weighting design effect* (Kish, 1992; Gabler et al., 1999). It measures the increase in variance that results from using a weighted rather than unweighted sample mean to estimate the population mean.

18. *Trimmed means.* Some statisticians recommend using trimmed means to estimate a population mean $\bar{y}_{\mathcal{U}}$ if there are outliers. The procedure used to find an α-trimmed mean in an SRS of size n is to remove the largest $n\alpha$ observations and the smallest $n\alpha$ observations, and then calculate the mean of the $n(1-2\alpha)$ observations that remain.

   Show that the α-trimmed mean for a finite population $\mathcal{U}$ of N observation units is

$$\bar{y}_{\mathcal{U}\alpha} = \left(\sum_{q_1 \le y \le q_2} y f(y)\right) \Big/ \left(\sum_{q_1 \le y \le q_2} f(y)\right),$$

   where q_1 and q_2 are the α and $(1-\alpha)$ quantiles, respectively, of the population. Now propose an estimator of the population α-trimmed mean for data from a complex survey using $\hat{F}(y)$ and $\hat{f}(y)$.

19. *Interpolated quantiles.* For a continuous and strictly increasing cdf F, such as the cdf of a normal distribution, the qth quantile is the unique value θ_q that satisfies $F(\theta_q) = q$. The empirical cdf $\hat{F}$, however, is neither continuous nor strictly increasing because it

has jumps at the values of y in the sample. Consequently, for many values of q there is no sample value y satisfying $\hat{F}(y) = q$.

The empirical cdf can be approximated, though, by an interpolated function that is continuous and strictly increasing. Let $y_{(1)} < y_{(2)} < \ldots < y_{(K)}$ be the distinct values of y in the sample (these are the values at which $\hat{F}$ jumps, ordered from smallest to largest). Define the interpolated empirical cdf as

$$\tilde{F}(y) = \begin{cases} \hat{F}(y_{(1)}) & \text{if } y < y_{(1)} \\ \hat{F}(y_{(k)}) + \dfrac{y - y_{(k)}}{y_{(k+1)} - y_{(k)}} \left[\hat{F}(y_{(k+1)}) - \hat{F}(y_{(k)})\right] & \text{if } y_{(k)} \leq y < y_{(k+1)} \\ 1 & \text{if } y \geq y_{(K)} \end{cases}. \tag{7.13}$$

Then $\tilde{F}$ is continuous and strictly increasing for y between $y_{(1)}$ and $y_{(K)}$ and the interpolated qth quantile, $\tilde{\theta}_q$, for $q \geq \hat{F}(y_{(1)})$, is defined as the value of y satisfying $\tilde{F}(y) = q$.

(a) Show, by verifying that $\tilde{F}(\tilde{\theta}_q) = q$, that

$$\tilde{\theta}_q = \begin{cases} y_{(1)} & \text{if } q < \hat{F}(y_{(1)}) \\ y_{(k)} + \dfrac{q - \hat{F}(y_{(k)})}{\hat{F}(y_{(k+1)}) - \hat{F}(y_{(k)})}[y_{(k+1)} - y_{(k)}] & \text{if } \hat{F}(y_{(k)}) \leq q < \hat{F}(y_{(k+1)}) \\ y_{(K)} & \text{if } q = 1 \end{cases} \tag{7.14}$$

is the interpolated qth quantile for $q \geq \hat{F}(y_{(1)})$.

(b) Draw a graph of the interpolated cdf $\tilde{F}$ for the stratified sample of height data in `htstrat.csv`.

(c) Calculate the interpolated quantiles for `htstrat.csv` for $q \in \{0.25, 0.5, 0.75, 0.9\}$. How do these compare with the values in Table 7.1?

(d) For what types of finite populations will the interpolation make a difference?

20. *Probability–probability plots.* A probability–probability plot compares the empirical cdf from a sample with a specified theoretical cdf G such as the cdf of a normal distribution with specified mean and variance. If the proposed cdf G describes the data well (including the specification of the mean and variance), the points in a probability–probability plot of $\hat{F}(y)$ versus $G(y)$ will lie approximately on a straight line with intercept 0 and slope 1.

    Construct a probability–probability plot for the height data in `htstrat.csv`, used in Example 7.4. Use a normal distribution for G, with the mean and standard deviation estimated from the sample. Draw in the line with intercept 0 and slope 1. Is G a reasonable distribution to use to summarize the data?

21. *Quantile–quantile plots.* Quantile–quantile plots are often used to assess how well a theoretical probability distribution fits a data set. To construct a quantile–quantile plot from an SRS of size n, order the sample values so that $y_{(1)} \leq y_{(2)} \leq \cdots \leq y_{(n)}$. Then, to compare with a continuous theoretical cdf G, calculate $x_{(i)} = G^{-1}[(i - 0.375)/(n + 0.25)]$ and draw a scatterplot of the n pairs $(x_{(i)}, y_{(i)})$. If G is a good fit for the data, the points in the quantile–quantile plot will approximate a straight line.

To use a quantile–quantile plot with survey data, let $w_{(1)}, \ldots, w_{(n)}$ be the weight values corresponding to the ordered sample $y_{(1)} \leq y_{(2)} \leq \ldots \leq y_{(n)}$. Let

$$v_{(i)} = \frac{\left(\sum_{j=1}^{i} w_{(j)}\right)\left(1 - \frac{0.375}{i}\right)}{\left(\sum_{j \in \mathcal{S}} w_{(j)}\right)\left(1 + \frac{0.25}{n}\right)}.$$

Then plot $y_{(i)}$ versus $G^{-1}(v_{(i)})$ and assess whether the values appear to be approximately on a straight line.

(a) Show that the plot of $y_{(i)}$ versus $G^{-1}(v_{(i)})$ gives the SRS quantile–quantile plot when the sample is self-weighting.

(b) Construct a quantile–quantile plot for the height data in `htstrat.csv`, used in Example 7.4. Use a standard normal cdf for G. Do you think the normal distribution describes these data well?

22. Show that in a stratified sample, $\sum y\hat{f}(y)$ produces the estimator in (3.2).

23. What is $\hat{S}^2$ in (7.6) for an SRS? How does it compare with the sample variance s^2?

24. Consider a probability sample $\mathcal{S}$ of n observation units from a population $\mathcal{U}$ of N observation units. The weights are $w_i = 1/\pi_i$, where π_i is the probability that unit i is in the sample. Now let $\mathcal{S}_2$ be a subsample of $\mathcal{S}$ of size n_2, with units selected with probability proportional to w_i. Show that $\mathcal{S}_2$ is a self-weighting sample from $\mathcal{U}$.

25. In a two-stage cluster sample of rural and urban areas in Nepal, Rothenberg et al. (1985) found that the design effect for common contagious diseases was much higher than for rare contagious diseases. In the urban areas measles, with an estimated incidence of 123.9 cases per 1,000 children per year, had a design effect of 7.8; diphtheria, with an estimated incidence of 2.1 cases per 1,000 children per year, had a design effect of 1.9.

    Explain why one would expect this disparity in the design effects. (HINT: Suppose a sample of 1,000 children is taken, in 50 clusters of 20 children each. Also suppose that the disease is as aggregated as possible, so if the estimated incidence were 40 per 1,000, all children in two clusters would have the disease, and no children in the remaining 38 clusters would have the disease. Now calculate the deffs for incidences varying from 1 per 1,000 to 200 per 1,000.)

26. In the 1990s, the British Crime Survey (BCS) had a stratified, multistage design in which inner-city areas were sampled at about twice the rate of non-inner-city areas (Aye Maung, 1995). Households were selected using probability sampling, but only one adult (selected at random) was interviewed in each responding household. Set the relative sampling weight for an inner-city household to be 1.

    (a) Consider the BCS as a sample of households. What is the relative sampling weight for a non-inner-city household?

    (b) Consider the BCS as a sample of adults. Find the relative sampling weight for an adult in an inner-city household with k adults, for $k = 1, \ldots, 5$.

    (c) Repeat part (b) for adults in non-inner-city households.

27. *Clopper-Pearson Confidence Intervals for Complex Survey Data.* Korn and Graubard (1998a) suggested adapting the Clopper-Pearson interval from Exercise 34 of Chapter 2 by substituting the effective sample size from the survey for n in (2.39). Show that this gives approximately the correct coverage probability.

28. *Implicit Stratification.* In stratified random sampling, as described in Chapter 3, the population is divided among well-defined strata and then an SRS is taken from each stratum. The variance of an estimated mean from a stratified random sample is estimated using the formula in (3.6).

    With implicit stratification, a list of the population is sorted, and a systematic sample is drawn from the sorted list (Lynn, 2019). This forces the sample to have representation across the spectrum of the sorting variable or variables. Thus, if a list of university students is sorted by date of birth, a systematic sample is expected to have an age distribution that is closer to that of the population than if an SRS were selected. If y is related to age, then the variance of the estimated mean $\bar{y}$ is likely to be smaller than the SRS variance in (2.12).

    Recall that a systematic sample is, structurally, a cluster sample in which one cluster is selected by choosing a random starting place. The sample variance of the psu totals, required for the standard error formula in (5.6), cannot be calculated for a sample with only one psu.

    One method that has been proposed for estimating the variance of an implicitly stratified sample calls for acting as though consecutive pairs in the systematic sample are from the same stratum. In other words, treat the sample as if it were a stratified random sample of size n from $H = n/2$ strata, where the first two units in the sample are considered to be in stratum 1, then next two are in stratum 2, and so on.

    (a) Give a formula for the estimated variance of $\bar{y}$ under this method.

    (b) Sort the population in `college.csv`, described in Example 3.12, by variable *ugds*. Now draw 200 independent systematic samples, each of size 100. Let y_i be the average net price of attendance (variable *npt4*). For each sample, calculate $\bar{y}$, the estimated variance of $\bar{y}$ using the SRS formula in (2.12), and the estimated variance of $\bar{y}$ using the formula in part (a).

    (c) Calculate the average of the 200 variances from part (b), separately for each method.

    (d) Because 200 independent samples were taken, the sample variance of the 200 values of $\bar{y}$ provides an unbiased estimate of $V(\bar{y})$. Calculate this. Which average variance, from part (c), is closer?

D. Projects and Activities

29. *College scorecard data.* Exercise 15 of Chapter 6 described a probability-proportional-to-size sample of the small, highly residential stratum discussed in Example 3.12. Obtain a probability sample of the entire population by taking a pps sample in each stratum using the Neyman allocation in Table 3.8.

    Use your sample to estimate the total number of female undergraduate students in the population of all colleges (calculate the number of female undergraduates as *ugds* × *ugds_women*). Draw a histogram of the estimated distribution of number of female students for the population. How does that compare with an unweighted histogram?

30. *Trucks.* The U.S. Vehicle Inventory and Use Survey (VIUS) was described in Exercise 49 of Chapter 3.

    (a) Draw a histogram, using the weights, of the number of miles driven (variable *miles_annl*) for the five truck class strata.

    (b) Draw side-by-side boxplots, using the weights, of miles per gallon (*MPG*) for each class of gross vehicle weight (*vius_gvw*).

    (c) Draw two of the scatterplots that incorporate weights, described in Section 7.6.2, for y variable *miles_annl* and x variable model year (*adm_modelyear*). How do these differ from scatterplots that do not use the weights?

31. *IPUMS exercises.*

    (a) Use the file `ipums.csv` to select a two-stage stratified cluster sample from the population. Select two psus from each stratum, with probability proportional to size. Then take a simple random sample of persons from each selected psu; use the same subsampling size within each psu. Your final sample should have about 1200 persons.

    (b) Construct the column of sampling weights for your sample.

    (c) Draw a histogram of the variable *inctot* for your sample, using the weights.

    (d) Construct side-by-side boxplots of the variable *inctot* for each level of marital status (variable *marstat*).

    (e) Draw two of the scatterplots that incorporate weights, described in Section 7.6.2, for y variable *inctot* and x variable *age*. How do these differ from scatterplots that do not use the weights?

    (f) Using the sample you selected, estimate the population mean of *inctot* and give the standard error of your estimate. Also estimate the population total of *inctot* and give its standard error.

    (g) Compare your results with those from an SRS with the same number of persons. Find the design effect of your response (the ratio of your variance from the unequal-probability sample to the variance from the SRS).

32. *Baseball data.* Use the two-stage sample from Exercise 44 of Chapter 5 for this exercise.

    (a) Draw a histogram of the variable *salary* for your sample, using the weights.

    (b) Construct side-by-side boxplots of the variable *salary* for each position.

    (c) Draw two of the scatterplots that incorporate weights, described in Section 7.6.2, for y variable *salary* and x variable number of games played (g). How do these differ from scatterplots that do not use the weights?

    (d) Draw two of the scatterplots that incorporate weights, described in Section 7.6.2, for y variable *salary* and x variable number of home runs (*hr*). What do you see in these graphs?

    (e) Draw quantile-quantile plots (see Exercise 21) for the variables *salary* and log(*salary*). Does either variable appear to follow, approximately, a normal distribution?

33. Obtain a research article based on a survey employing a complex survey design, and write a short critique. Your critique should include:

    (a) a brief summary of the design and analysis

(b) a discussion of the effectiveness of the design and the appropriateness of the analysis

(c) your recommendations for future studies of this type.

34. Many governmental statistical organizations and other collectors of survey data have websites providing information on the designs of surveys that they conduct. Look up a website describing a complex survey. Write a summary of the survey purpose, design, and methods used for analysis.

35. *Activity for course project.* Find a survey data set that has been collected by a federal government or large survey organization. Many of these are available online, and contain information about stratification and clustering that you can use to calculate standard errors of survey estimates. Some examples in the United States include the National Crime Victimization Survey, the National Health Interview Survey, the Current Population Survey, the Commercial Buildings Energy Consumption Survey, and the General Social Survey.

    Read the documentation for the survey. What is the survey design? What stratification and clustering variables are used? (Some surveys do not release stratification and clustering information to protect the confidentiality of data respondents, so make sure your survey provides that information.)

    Select variables that you are interested in to analyze. If possible, find at least one variable that is not categorical. Draw a histogram, using the final weight variable, for that variable. Use the weights to estimate the summary statistics of the mean and 25th, 50th, and 75th percentiles. We'll return to this data set in subsequent chapters so that you will have an opportunity to study multivariate relationships among your variables of interest.

8

Nonresponse

Miss Schuster-Slatt said she thought English husbands were lovely, and that she was preparing a questionnaire to be circulated to the young men of the United Kingdom, with a view to finding out their matrimonial preferences.

"But English people won't fill up questionnaires," said Harriet.

"Won't fill up questionnaires?" cried Miss Schuster-Slatt, taken aback.

"No," said Harriet, "they won't. As a nation we are not questionnaire-conscious."

—Dorothy Sayers, *Gaudy Night*

The best way to deal with nonresponse is to prevent it. After nonresponse has occurred, it is sometimes possible to construct models to predict the missing data, but predicting the missing observations is never as good as observing them in the first place. Nonrespondents often differ in critical ways from respondents; if the nonresponse rate is not negligible, inference based upon the respondents alone may be seriously flawed.

We discuss two types of nonresponse in this chapter: **unit nonresponse**, in which the entire observation unit is missing, and **item nonresponse**, in which some measurements are present for the observation unit but at least one item is missing. In a survey of persons, unit nonresponse means that the person provides no information for the survey; item nonresponse means that the person does not respond to a particular item on the questionnaire. Unit nonresponse can arise for a variety of reasons: The interviewer may not be able to contact the household; someone may be ill and cannot respond to the survey; a person who is contacted may refuse to participate in the survey. Item nonresponse often occurs because someone declines to answer a question (a person may skip the question about income, for example) or does not finish the survey.

In agriculture or wildlife surveys, the term *missing data* is generally used instead of *nonresponse,* but the concepts and remedies are similar. In a survey of breeding ducks, for example, some birds will not be found by the researchers; they are, in a sense, nonrespondents. The nest may be raided by predators before the investigator can determine how many eggs were laid; this is comparable to item nonresponse. Lesser and Kalsbeek (1999) and Gentle et al. (2006) discussed nonresponse and other nonsampling errors in environmental surveys.

In this chapter, we discuss four major approaches for dealing with nonresponse:

1. Prevent it. Design the survey so that nonresponse is low. This is by far the best method (Section 8.2).

2. Take a representative subsample of the nonrespondents; use that subsample to make inferences about the other nonrespondents (Section 8.3).

3. Use a statistical model to predict values for the nonrespondents (Section 8.4). Weighting class methods (Section 8.5) and calibration (Section 8.6) use models to adjust the weights of respondents so that they also represent the population share of nonrespondents thought to be similar to them. Imputation models (Section 8.7) fill in missing values of y variables resulting from item (and sometimes unit) nonresponse.

DOI: 10.1201/9780429298899-8

4. Ignore the nonresponse (not recommended, but unfortunately common in practice). Section 8.1 shows what can happen when you ignore the nonresponse.

8.1 Effects of Ignoring Nonresponse

Example 8.1. Thomsen and Siring (1983) reported results from a 1969 survey on voting behavior carried out by the Central Bureau of Statistics in Norway. In this survey, three calls were followed by a mail survey. The final response rate was 90.1%, which is often considered to be a high response rate. Did the nonrespondents differ from the respondents?

In the Norwegian voting register, it was possible to find out whether a person voted in the election. The percentage of persons who voted could then be compared for respondents and nonrespondents; Table 8.1 shows the results. The selected sample is all persons selected to be in the sample, including data from the Norwegian voting register for both respondents and nonrespondents.

TABLE 8.1
Percentage of persons who voted, from Thomsen and Siring (1983).

	Age Group					
	All	20–24	25–29	30–49	50–69	70–79
Nonrespondents	71	59	56	72	78	74
Selected sample	88	81	84	90	91	84

The difference in voting rate between the nonrespondents and the selected sample was largest in the younger age groups. Among the nonrespondents, the voting rate varied with the type of nonresponse. The overall voting rate for the persons who refused to participate in the survey was 81%, the voting rate for the not-at-homes was 65% and the voting rate for the mentally and physically ill was 55%, implying that absence or illness were the primary causes of nonresponse bias. ■

It has been demonstrated repeatedly that nonresponse can have large effects on the results of a survey—in Example 8.1, the voting rate in Norway was overestimated even with a response rate exceeding 90%. Lorant et al. (2007), comparing characteristics of participants in the compulsory Belgian census (response rate 96.5%) and a stratified multistage household survey (response rate 61.4%) that asked some of the same questions, found that the survey participants were more likely to be homeowners, highly educated, of Belgian nationality, and in good health. The survey nonresponse also affected relationships among variables: Among the census participants in poor health, persons with low education and those not working were less likely to respond to the survey than persons with higher education and employed persons. Because of the nonresponse, estimates from the survey understated the association between lower socio-economic status and poor health.

Moreover, increasing the sample size without targeting nonresponse might not reduce nonresponse bias; a larger sample size might merely provide more observations from the types of persons most likely to respond to the survey. Increasing the sample size may actually worsen the nonresponse bias, as the larger sample size may divert resources that could have been used to reduce or remedy the nonresponse, or it may result in less care in the data collection. Recall that the *Literary Digest* poll (Example 1.1) had nearly 2.4 million respondents but, because Roosevelt supporters were less likely to participate in the poll, the percentage of poll respondents supporting Roosevelt severely underestimated the vote for Roosevelt.

Many persons analyzing data from surveys (like the editors of the *Literary Digest* in 1936) ignore nonresponse and report results based on complete records only. They often estimate a population mean by the raw sample mean. An analysis of complete records—with no adjustments for nonresponse—has the underlying assumptions that the nonrespondents are similar to the respondents, and that units with missing items are similar to units that have responses for every question. Much evidence indicates that this assumption does not hold true in practice (Särndal and Lundström, 2005).

Results reported from an analysis of complete records alone, without considering the nonresponse, should be taken as representative of the population of persons who would respond to the survey. This is rarely the same as the target population.

Nonresponse increases the variance of estimators (because the sample size is smaller than anticipated) but the main concern is potential bias. Think of the population as being divided into two somewhat artificial strata of respondents and nonrespondents. The population respondents are the units that would respond if they were chosen to be in the sample; the number of population respondents, N_R, is unknown. Similarly, the N_M (M for missing) population nonrespondents are the units that would not respond. We then have the following population quantities:

Stratum	**Size**	**Total**	**Mean**	**Variance**
Respondents	N_R	t_R	$\bar{y}_{R\mathcal{U}}$	S_R^2
Nonrespondents	N_M	t_M	$\bar{y}_{M\mathcal{U}}$	S_M^2
Entire population	N	t	$\bar{y}_{\mathcal{U}}$	S^2

The population as a whole has total t, mean $\bar{y}_{\mathcal{U}}$, and variance $S^2 = \sum_{i=1}^{N}(y_i - \bar{y}_{\mathcal{U}})^2/(N-1)$. A probability sample from the population will likely contain some respondents and some nonrespondents. But we do not observe y_i for any of the units in the nonrespondent stratum. If the population mean in the nonrespondent stratum differs from that in the respondent stratum, the sample mean from the respondents will be a biased estimator of the population mean.[1]

Let $\bar{y}_R$ be an approximately unbiased estimator of the mean in the respondent stratum, using only the respondents. Because

$$\bar{y}_{\mathcal{U}} = \frac{N_R}{N}\bar{y}_{R\mathcal{U}} + \frac{N_M}{N}\bar{y}_{M\mathcal{U}},$$

the bias is approximately

$$E[\bar{y}_R] - \bar{y}_{\mathcal{U}} \approx \frac{N_M}{N}\left(\bar{y}_{R\mathcal{U}} - \bar{y}_{M\mathcal{U}}\right).$$

The bias is small if either (1) the mean for the nonrespondents is close to the mean for the respondents, or (2) N_M/N is small—there is little nonresponse. But we can never be assured of (1), as we do not observe y for the nonrespondents. Eliminating nonresponse is the only sure way to control nonresponse bias.

[1] A population variance that estimated from respondents alone is often biased as well. In income surveys, for example, the rich and the poor may be more likely to be nonrespondents on the income questions. In that case, S_R^2, from the respondent stratum, is smaller than S^2. The point estimator of the mean may be biased, and the estimator of the variance S^2 may be biased, too.

8.2 Designing Surveys to Reduce Nonresponse

A common feature of poor surveys is a lack of time spent on design and nonresponse follow-up. Many persons new to surveys (and some, unfortunately, not new) simply jump in and start collecting data without considering potential problems in the data collection process; they send questionnaires to everyone in the target population and analyze those that are returned. It is not surprising that such surveys have poor response rates.

A researcher who knows the target population well will be able to anticipate some of the reasons for nonresponse and prevent some of it. Most investigators, however, do not know as much about reasons for nonresponse as they think they do. They need to discover why the nonresponse occurs and resolve as many of the problems as possible before commencing the survey.

Designed experiments and quality improvement methods can help increase response rates and reduce nonresponse bias. You do not know why surveys similar to yours have a low response rate? Design an experiment to find out. You think errors are introduced in the data recording and processing? Use a nested design to find the sources of errors. And, of course, you can rely on previous researchers' experiments to help you minimize nonsampling errors.

Example 8.2. Biemer et al. (2018) designed an experiment to investigate effects of survey recruitment strategies on response rates and costs. Four data collection protocols were studied, with the goals of having (1) a high response rate, (2) a high percentage of respondents who choose the lower-cost internet response option, and (3) low nonresponse bias:

- Online only: All survey invitations tell how to respond to the survey online over the internet. Persons are not given the option of a paper questionnaire.
- Online/Choice: The first survey invitation tells how to respond to the survey online. If there is no response to the first invitation, subsequent invitations give the option of responding either online or by returning a paper questionnaire.
- Choice: All survey invitations give the option of responding either online or by returning a paper questionnaire.
- Choice Plus: As with Choice, all survey invitations give the option of responding either online or by returning a paper questionnaire. However, an extra \$10 is promised to persons who respond online instead of on paper.

In addition, two levels of incentives were studied, promising either \$10 (low incentive) or \$20 (high incentive) for completing the survey. Each of the 9,650 addresses in the sample was randomly assigned to one of the eight treatment groups formed by the combination of the four data collection protocols with the two incentive amounts. Each treatment group had about 1,206 addresses, which gave a margin of error of approximately 3 percentage points for the estimated response rate. Table 8.2 gives the response rate, and the percentage of respondents who responded using the internet option, for each of the eight treatment groups.

The response rate was higher with the \$20 incentive than the \$10 incentive for each of the four protocols. But was the higher response rate bringing in previously underrepresented persons from the sample, or was it just getting more observations from the set of persons most likely to respond to the survey?

Biemer et al. (2018) could not investigate nonresponse bias for all variables measured in the survey. But 12 variables, including race and ethnicity, household size, household income, and internet access, were measured both in this study and in the ACS. The ACS

TABLE 8.2
Results for experiment in Example 8.2, from Biemer et al. (2018).

Survey Protocol	Incentive	Response Rate (%)	% Internet Respondents
Online only	$10	32.0	100
Online only	$20	38.9	100
Online/Choice	$10	38.6	64
Online/Choice	$20	42.3	71
Choice	$10	38.8	28
Choice	$20	43.4	27
Choice Plus	$10	42.8	64
Choice Plus	$20	45.0	64

was considered to be the gold standard and thus, a protocol whose distributions of these 12 variables more closely matched the distributions from the ACS was deemed to be more "representative."

The "Choice Plus" protocol with $20 incentive had the highest response rate, a relatively high percentage of respondents choosing the internet, and the highest average degree of similarity to the ACS for the 12 variables studied. Of course, the response rate for this option was still less than 50 percent, and methods such as those in Sections 8.3 to 8.7 would be needed to attempt to reduce potential nonresponse bias after the survey data were collected. ■

Nonresponse can have many different causes; as a result, no single method can be recommended for reducing it in every survey. Figure 8.1 displays some of the sources of nonresponse that can be attributed to survey and questionnaire design, methods of data collection, interviewers, and respondent characteristics (see also Platek, 1977; Hidiroglou et al., 1993; Groves et al., 2011; Dillman et al., 2014).

- *Survey content.* A survey on drug use or financial matters may have a large number of refusals. Sometimes the response rate can be increased for surveys involving sensitive subjects by careful ordering of the questions, by using a randomized response technique (see Section 16.3.3), or by using a self-administered questionnaire on the computer to protect the respondents' privacy.

- *Questionnaire design.* We saw in Chapter 1 that question wording has a large effect on the responses received; it can also affect whether a person responds to the survey or to a particular item on the questionnaire. Beatty and Herrmann (2002) reviewed research on applying cognitive research to questionnaire design. A well-designed questionnaire form may increase data accuracy and reduce item nonresponse (Dillman, 2008).

- *Survey introduction.* The survey introduction, sometimes sent in an advance letter telling a household or business that they have been selected to participate in the survey, provides the first contact between the interviewer and potential respondent. A good introduction, giving the recipient motivation to respond, can increase response rates dramatically. The potential respondent should be told who is conducting the survey (a survey sponsored by a respected organization may achieve a higher response rate), for what purpose the data will be used (unscrupulous persons often pretend to be taking a survey when they are really trying to attract customers or converts), and assured confidentiality.

- *Sampling frame.* Some sampling frames have more detailed or accurate information than others. A sampling frame of addresses for a household survey may also contain telephone

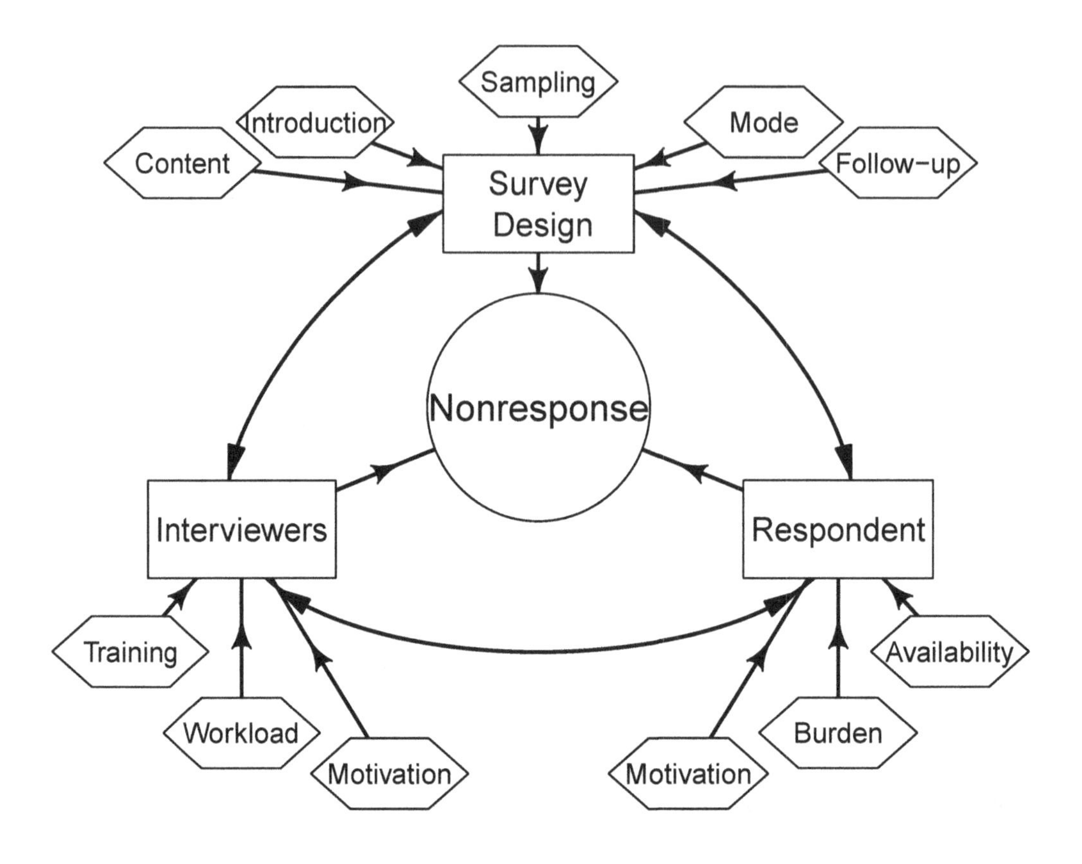

FIGURE 8.1
Some factors affecting nonresponse.

numbers for some of the addresses; these, if accurate, could be used for following up with nonrespondents. A sampling frame for a business survey may contain contact information for persons within the businesses who typically fill out the surveys. Having detailed, accurate information about units in the frame may make it easier to contact and obtain responses from sampled units.

- *Sampling design.* Surveys of businesses or farms are often stratified by a number of variables related to responses of interest, including size. Sampling the strata containing large businesses at a higher rate than strata containing small businesses may reduce the variance of estimates of quantities such as total payroll (see Example 3.8). If large businesses are more likely to respond to the initial survey invitation (for example, if they have designated employees who fill out surveys), the disproportional allocation may also increase the response rate. The sample has fewer initial nonrespondents than a proportionally allocated sample would have, and consequently each can be targeted with more nonresponse follow-up efforts.

  For surveys that are repeated annually, choice of a cross-sectional design (in which a new sample is drawn each year) or a longitudinal or panel design (in which at least some of the units in the year-1 sample are kept for the year-2 sample) can affect response rates. Once a household is recruited to be in a panel survey such as the U.S. Current Population Survey, which measures characteristics related to employment, it stays in the survey for subsequent interviews. This reduces the need to recruit an entirely new set of households for every data collection, and allows more of the data collection effort to be spent on the initial recruitment and on reducing nonresponse.

- *Survey mode.* Surveys of all modes have reported steep decreases in response rates since 2000 (National Research Council, 2013; Williams and Brick, 2018). But household surveys conducted in person (sometimes called face-to-face surveys) have in general had higher response rates than surveys conducted by telephone, mail, e-mail, or internet. In a high-quality in-person survey, interviewers are sent to the households selected in a probability sample of areas or addresses. This, not surprisingly, is much more expensive than conducting a survey by another mode but often yields a higher response rate.

  Sometimes using multiple modes can combine the higher response rate of an in-person survey with the lower cost of a survey conducted by another mode. The ACS employs three modes of data collection: internet, mail, and in-person (U.S. Census Bureau, 2014). First, an invitation is mailed to addresses selected for the sample, containing instructions for how to complete the survey online. A paper questionnaire is mailed about three weeks later to those who have not already responded over the internet. Later, a subsample of households that have not responded to earlier attempts is selected, and each household in the subsample is visited by an interviewer who has been trained to be "polite but persistent." Using the less-expensive mail and internet modes for the initial attempts lowers costs for the survey, while using in-person interviews for the nonresponse follow-up yields a higher response rate and reduces nonresponse bias.

  The survey mode can also affect item nonresponse. A person filling out a paper survey can skip questions more easily than a person who is prompted by an interviewer. An internet survey can require participants to answer a question before moving to the next screen, although participants can still break off the survey or provide untruthful answers.

- *Follow-up.* Almost all high-quality household surveys follow up with households or persons who do not respond to the initial survey request. In some surveys, more than 30 attempts are made to reach a household or person selected to be in the sample. If the initial survey is by mail, a reminder may increase the response rate. Not everyone responds to follow-ups, though; some persons may refuse to respond to the survey no matter how often they are contacted. You need to decide how many follow-up calls to make before the marginal returns do not justify the expense.

- *Interviewers.* Interviewers, in surveys that employ them, contact persons in the sample and attempt to persuade them to participate in the survey. Some interviewers are better at this than others. Experience, workload, training, attitudes and motivation, speech patterns, persistence, and personal characteristics can all affect the response rate achieved by an interviewer (West and Blom, 2017). Similar considerations apply in non-household surveys. Some field investigators in a bird survey, for example, may be better at spotting and identifying birds than others. Quality improvement methods (see Chapter 16) can be applied to increase the response rate and accuracy for interviewers.

- *Respondent burden.* Persons who respond to a survey are doing you an immense favor, and the survey should be as nonintrusive as possible. Respondent burden is high when the respondent perceives the task of participating in the survey to be difficult, stressful, or time-consuming (Bradburn, 1978; Yan et al., 2020). This can occur when the questionnaire is lengthy, the survey involves sensitive or disturbing topics, questions are perceived to be difficult (for example, questions requiring respondents to look up detailed financial records), or the survey is administered multiple times (a panel survey such as the Current Population Survey may be perceived as more burdensome because each household is interviewed multiple times).

A shorter questionnaire, requiring less detail, may reduce the burden to the respondent, but may also provide less information. A split-questionnaire design, in which different groups of the survey respondents are given different subsets of the questionnaire, may reduce respondent burden. With a split-questionnaire design, each individual receives a shortened questionnaire yet every question is administered to at least a subsample of respondents (Raghunathan and Grizzle, 1995). Pretesting the survey questionnaire and introduction can reduce burden, as can exploring alternative data sources for information. It may be possible, for example, to obtain property tax information from public records instead of asking the survey respondent.

- *Respondent motivation.* Much research has been done on how to motivate sampled persons to respond to the survey (see Section 16.2). Incentives, financial or otherwise, may increase the response rate (Singer and Ye, 2013; Mercer et al., 2015). Disincentives may work as well: Physicians who refused to be assessed by peers after selection in a stratified sample from the College of Physicians and Surgeons of Ontario registry had their medical licenses suspended. Not surprisingly, nonresponse was low (McAuley et al., 1990).

- *Respondent availability and access.* Some calling periods, times of day, or seasons of the year may yield higher response rates than others. The survey protocols for many in-person and telephone surveys specify attempting to reach the sampled household or person at different times of day and different days of the week.

  You should also try to remove other barriers to survey participation, for example by making the survey accessible for persons with disabilities and available in multiple languages.

You should try to obtain at least some information about nonrespondents that can be used later to adjust for nonresponse bias, and include surrogate items that can be used for item nonresponse. True, there is no complete compensation for not having the data, but partial information is better than none. Information about the race, gender, or age of a nonrespondent may be used to weight the data, as will be discussed in Section 8.5. Questions about income may well lead to refusals, but questions about cars, employment, or education may be answered and can be used to predict income. If the pretests of the survey indicate a nonresponse problem that you do not know how to prevent, try to design the survey so that at least some information is collected for each observation unit.

The quality of survey data is largely determined at the design stage. Fisher's (1938) words about experiments apply equally well to the design of sample surveys: "To call in the statistician after the experiment is done may be no more than asking him to perform a postmortem examination: he may be able to say what the experiment died of." Any survey budget needs to allocate sufficient resources for survey design and for nonresponse follow-up. Do not scrimp on the survey design; every hour spent on design may save weeks of remorse later.

8.3 Two-Phase Sampling

Nonresponse occurs even with the most carefully designed surveys. Fortunately, statistical methods exist that can reduce nonresponse bias, and these are discussed in the remainder of the chapter. This section discusses using **two-phase sampling** (also called **double sampling**) to obtain a representative subsample of the nonrespondents. If everyone in the subsample responds, then the procedure essentially eliminates the nonresponse bias.

We shall look at the theory of two-phase sampling for general survey designs in Chapter 12; here, we illustrate how it can be used for nonresponse. In two-phase sampling for nonresponse, proposed by Hansen and Hurwitz (1946), you can think of the population as divided into two strata: respondents and initial nonrespondents. The respondents are the population members who, if selected for the sample, would respond to initial attempts to gain their cooperation for the survey. The initial nonrespondents are the population members who would not respond to these initial attempts.

In the simplest form of two-phase sampling, randomly select an SRS of n units in the population. Of these, n_R respond and n_M (M stands for "missing") do not respond. The sample sizes n_R and n_M are random variables; they will change if a different SRS is selected. Then make a second call on a random subsample of $100\nu\%$ of the n_M nonrespondents in the sample, where the subsampling fraction ν does not depend on the data collected.

Suppose that through some superhuman effort all of the targeted nonrespondents are reached and provide data. Let $\bar{y}_R$ be the sample average of the original respondents, and $\bar{y}_M$ be the average of the subsampled nonrespondents. The two-phase sampling estimators of the population mean and total are

$$\hat{\bar{y}} = \frac{n_R}{n}\bar{y}_R + \frac{n_M}{n}\bar{y}_M \tag{8.1}$$

and

$$\hat{t} = N\hat{\bar{y}} = \frac{N}{n}\sum_{i\in\mathcal{S}_R} y_i + \frac{N}{n}\frac{1}{\nu}\sum_{i\in\mathcal{S}_M} y_i, \tag{8.2}$$

where $\mathcal{S}_R$ is the set of sampled units in the respondent stratum and $\mathcal{S}_M$ is the set of subsampled units in the nonrespondent stratum. Note that $\hat{t}$ is a weighted sum of the observed units; the weights are N/n for the respondents and $N/(n\nu)$ for the subsampled nonrespondents. Because only a subsample was taken in the nonrespondent stratum, each subsampled unit in that stratum represents more units in the population than does a unit in the respondent stratum.

The expected value and variance of these estimators are given in Section 12.2. If everyone in the subsample responds, the estimators in (8.1) and (8.2) are unbiased because the final sample is a full-response probability sample of the population with known probabilities of selection. Two-phase sampling not only removes the nonresponse bias but also accounts for the first-phase nonresponse in the estimated variance.

Example 8.3. The ACS uses two-phase sampling to reduce nonresponse bias. The initial attempts to reach a household selected for the sample are by mail, and households are asked to respond by internet or mail. A subsample of addresses from which no response has been obtained is selected for in-person interviewing. The subsampling rate is higher in census tracts where low levels of self-response to the survey are predicted. In a tract where the subsampling rate is 50%, a systematic sample of every other address in the sorted list of nonresponding addresses is selected for an in-person visit. Each address in the subsample for that tract is assigned a sampling weight of the initial weight multiplied by two (U.S. Census Bureau, 2014). ■

8.4 Response Propensities and Mechanisms for Nonresponse

Most surveys have some residual nonresponse even after careful design, follow-up of nonrespondents, and two-phase sampling. All methods for fixing up the remaining nonresponse are necessarily model-based. If we are to make any inferences about the nonrespondents, we must assume that we can predict their data from the respondents' data or other information.

Dividing population members into two fixed strata of would-be respondents and would-be nonrespondents, as in Section 8.1, is fine for thinking about potential nonresponse bias and for two-phase methods. To adjust for nonresponse that remains after all other measures have been taken, we need a more elaborate setup. We also need some form of additional knowledge about the nonrespondents, usually through auxiliary variables measured in the survey.

8.4.1 Auxiliary Information for Treating Nonresponse

Suppose that y_i is a characteristic of interest for unit i (measured only for respondents), and that $\mathbf{x}_i$ is a vector of auxiliary variables for unit i. To be able to use auxiliary variables $\mathbf{x}_i$ for nonresponse modeling, $\mathbf{x}_i$ must be measured for each respondent and *something* must be known about the $\mathbf{x}$ variables for units outside the set of sample respondents. This extra knowledge may come in various forms:

A) The auxiliary variables are known for every unit in the sampling frame. A list of registered voters, used as a sampling frame, may contain each voter's age, gender, date of registering to vote, and history of participating in previous elections.

B) The auxiliary variables are known for every unit in the selected sample (for both respondents and nonrespondents). Of course, information known for all units in the population (as in item A) is also known for all units in the sample. In other situations, variables may be known for every unit in the sample but not necessarily for everyone in the population. For example, the sampler knows the location of each address in a randomly selected sample of addresses, and thus can look up census characteristics about the neighborhood of each respondent and nonrespondent. In an in-person survey, the interviewer visiting the address may observe characteristics of the housing unit such as estimated number of rooms, and this information can be recorded both for respondents and nonrespondents.

C) Population means or totals of the auxiliary variables are known from an external data source. The sampler may not know the age or race for every person selected for the sample, but may know the total number of population persons in each age or race group from a population registry or census.

8.4.2 Methods to Adjust for Nonresponse

The three main methods for dealing with nonresponse in probability samples all attempt to reduce nonresponse bias through use of the auxiliary variables $\mathbf{x}_i$. These can be used singly or in combination.

- Treat nonresponse as a second phase of data collection (following the first phase of selecting the probability sample), and use information from the $\mathbf{x}$ variables to estimate the probability that each sampled unit i responds. This method is discussed in the remainder of this section and in Section 8.5.

- Use poststratification (or, more generally, calibration) to adjust the weights of the sample so that estimates of population totals for the $\mathbf{x}$ variables agree with known population totals for those variables from an external data source. This method will be discussed in Section 8.6.
- Use information from the $\mathbf{x}$ variables to predict the value of y_i for each nonrespondent. Imputation of missing data will be discussed in Section 8.7.

8.4.3 Response Propensities

When thinking of nonresponse as a second phase of data collection, two sets of random variables determine whether population unit i is a respondent: the first set describes selection for the probability sample, and the second set describes whether units in the sample respond. In Chapters 2 to 7, we have used the random variable Z_i to indicate selection for the sample, with $P(Z_i = 1) = \pi_i$. Now suppose that once a unit is in the sample, there is a random choice about whether to respond (Section 15.2 will discuss some ways of thinking about the decision to participate as a random choice). Define the random variable

$$R_i = \begin{cases} 1 & \text{if unit } i \text{ responds} \\ 0 & \text{if unit } i \text{ does not respond.} \end{cases} \tag{8.3}$$

After data collection, the sampler knows which units in the sample have responded. A value for y_i is recorded if R_i takes on the value 1 for unit i; otherwise, y_i is missing. The probability that sampled unit i will respond,

$$\phi_i = P(R_i = 1), \tag{8.4}$$

is unknown but assumed to be greater than zero; ϕ_i is called the **response propensity** or **propensity score** for unit i (Rosenbaum and Rubin, 1983).

Under this setup, if Z_i and R_i are independent then the probability that unit i will be a respondent is

$$P(\text{unit } i \text{ selected in sample and responds}) = E[Z_i R_i] = \pi_i \phi_i.$$

If we knew the true response propensity ϕ_i, we could estimate the population total $t_y = \sum_{i=1}^N y_i$ using an estimator of the same form as the Horvitz–Thompson estimator:

$$\hat{t}_\phi = \sum_{i=1}^{N} Z_i R_i \frac{y_i}{\pi_i \phi_i}. \tag{8.5}$$

8.4.4 Types of Missing Data

In practice the probabilities of responding to the survey, ϕ_i are unknown (unlike the probabilities π_i of being selected for the sample, which are known), and the best we can do is to try to estimate them from data. Whether we can come up with a good estimate of ϕ_i depends on the relationship between responding to the survey and the auxiliary information available. We consider three types of missing data, using the Little and Rubin (2019) classification.

Missing completely at random. If ϕ_i, the probability of responding, does not depend on $\mathbf{x}_i$, y_i, or the survey design, the missing data are **missing completely at random** (MCAR). Such a situation occurs if, for example, someone at the laboratory drops a test tube containing the blood sample of one of the survey participants—there is no reason to think that the dropping of the test tube had anything to do with the white blood cell count.[2] If data are MCAR, the respondents are representative of the selected sample.

Consider a survey of student satisfaction taken at a university. The university's database, which is used as the sampling frame, contains each student's race, gender, age, courses taken, major, grade point average, high school background, and other information—these $\mathbf{x}$ variables are known for every student in the sample, respondent or nonrespondent. The missing data from nonrespondents in this survey would be MCAR if the probability of being a respondent is completely unrelated to the student's race, gender, age, and every other variable in $\mathbf{x}$, *and* if the probability of being a respondent is unrelated to the y variables measured in the survey about satisfaction with the university. Nonrespondents are essentially selected at random from the sampled students.

If the response probabilities ϕ_i are all equal and the events $\{R_i = 1\}$ are independent of each other and of the sample selection process, then the data are MCAR. If an SRS of size n is taken, then under this mechanism the respondents form a simple random subsample of variable size n_R and the sample mean of the respondents, $\bar{y}_R$, is approximately unbiased for the population mean. The MCAR mechanism is implicitly adopted when nonresponse is ignored.

Missing at random given covariates. If ϕ_i depends on $\mathbf{x}_i$ but not on y_i, the data are **missing at random given covariates** (MAR).[3] In mathematical terms, the response mechanism for MAR data satisfies

$$P(R_i = 1|y_i, \mathbf{x}_i) = P(R_i = 1|\mathbf{x}_i)$$

for all sampled units—after accounting for $\mathbf{x}_i$, the variable y_i has no additional information about the probability that unit i responds to the survey. If we know the values of $\mathbf{x}_i$ for all sample units, or we know the population totals for the $\mathbf{x}$ variables from the sampling frame or an external source, we can construct a statistical model to estimate the response propensities.

Persons in the student satisfaction survey would be MAR if the probability of responding to the survey is related to race, gender, and college major—all known quantities—but for the set of students in a particular race/gender/major group, satisfied students are just as likely to respond to the survey as dissatisfied students. In the MAR situation, engineering majors might be more likely to return the survey than liberal arts majors, and, as a group, engineering majors might also be more likely to be satisfied with their education. But engineering majors who are satisfied with their education are just as likely to return the survey as engineering majors who are dissatisfied with their education, and liberal arts majors who are satisfied with their education are just as likely to return the survey as liberal arts majors who are dissatisfied with their education. Thus, the respondent engineering majors can be thought of as a random subsample of the engineering majors in the full sample, and the respondent liberal arts majors can be thought of as a random subsample

[2] Even here, though, the suspicious mind can create a scenario in which the nonresponse might be related to quantities of interest. Perhaps laboratory workers are less likely to drop test tubes that they believe contain infectious material.

[3] Little and Rubin (2019), and many of the statistical articles about missing data that adopted their terminology from the first edition published in 1987, referred to this type of nonresponse simply as "missing at random," which is why the acronym you will see in the literature is "MAR" and not "MARGC." Bhaskaran and Smeeth (2014) explained the term "missing at random" and gave examples of the differences between "missing completely at random" and "missing at random."

Nonresponse researchers sometimes use a related term, **ignorable nonresponse**: Ignorable means that a model can explain the nonresponse mechanism and that the nonresponse can be ignored after the model accounts for it, not that the nonresponse can be completely ignored and complete-data methods used.

of the liberal arts majors in the full sample. The views of the nonrespondent engineering majors are represented by the engineering majors who responded.

Not missing at random. If the probability of response depends on the value of y (unobserved for nonrespondents), and cannot be completely explained by values that are observed, then the nonresponse is **not missing at random** (NMAR). Nonrespondents for the student satisfaction survey are NMAR if a satisfied student is more likely to respond to the survey than a dissatisfied student who has the same values for race, gender, college major, and other variables in $\mathbf{x}$. Models can help reduce nonresponse bias in this situation, because the probability of responding may depend in part on the known $\mathbf{x}$ variables, but in general cannot completely remove it unless the modeler happens to guess the true mechanism generating the nonresponse.

The probabilities of responding, ϕ_i, are useful for thinking about the type of nonresponse. Unfortunately, since these probabilities are unknown, we do not know for sure which type of nonresponse is present. We can sometimes distinguish between MCAR and MAR by fitting a model attempting to predict the observed probabilities of response for subgroups from known covariates; if the coefficients in a logistic regression model predicting nonresponse are significantly different from 0, the missing data are likely not MCAR. Distinguishing between MAR and NMAR is more difficult. In practice, we expect most nonresponse in surveys to be of the NMAR type. It is unreasonable to expect that we can construct a perfect model from the $\mathbf{x}$ information that will completely explain the nonresponse mechanism. But we can try to reduce the bias due to nonresponse. In the next section, we discuss a method that is commonly used to estimate the ϕ_is.

8.5 Adjusting Weights for Nonresponse

In previous chapters we have seen how weights can be used in calculating estimates for various sampling designs (see Sections 2.4, 3.3, 5.3, and 7.2). The sampling weights are the reciprocals of the inclusion probabilities, with $w_i = 1/P(Z_i = 1)$, so that an estimator of the population total is $\sum_{i \in \mathcal{S}} w_i y_i$. For stratification, the weights are $w_i = N_h/n_h$ if unit i is in stratum h; for sampling units with unequal probabilities, $w_i = 1/\pi_i$.

Weights can also be used to adjust for nonresponse, using the random variables R_i defined in (8.3). If we knew the values of $\phi_i = P(R_i = 1)$ for each unit, and if we knew that $\phi_i > 0$ for all units in the sample, then we could estimate the population total using $\hat{t}_\phi$ in (8.5), which assigns a weight of $1/(\pi_i \phi_i)$ to each respondent. But the values of ϕ_i are unknown, and must be estimated from the data.

Weighting adjustment methods for nonresponse assume that the response propensities ϕ_i can be estimated from auxiliary information $\mathbf{x}$; they assume nonrespondents' data are MAR. With an estimate $\hat{\phi}_i$ of ϕ_i for each respondent, the population total may be estimated by

$$\hat{t}_{\hat{\phi}} = \sum_{i=1}^{N} Z_i R_i \frac{y_i}{\pi_i \hat{\phi}_i} = \sum_{i \in \mathcal{R}} \frac{y_i}{\pi_i \hat{\phi}_i}, \tag{8.6}$$

where $\mathcal{R}$ is the set of respondents in the sample.

For the methods discussed in the remainder of this section, suppose that the values of $\mathbf{x}$ are known for every unit in the selected sample, both respondents and nonrespondents—that is, the auxiliary information is of type (A) or (B) according to the classification in

Section 8.4.1. Then there is no nonresponse for the auxiliary variables, and they can be used to estimate the probability that unit i in the sample responds to the survey. The simplest method for doing that is through forming weighting classes.

8.5.1 Weighting Class Adjustments

The sampling weight w_i can be interpreted as the number of units in the population represented by unit i of the sample. Taken together, the sample represents all N members of the population, and the sum of the sampling weights for everyone in the sample, $\sum_{i \in \mathcal{S}} w_i$, is approximately equal to N. With nonresponse, however, the nonrespondents provide no data for the population units that are supposed to be represented by them. Weighting class adjustments shift the nonrespondents' weights to respondents. At the end of the process, the modified weight $\tilde{w}_i$ estimates how many population units are represented by respondent i, and the sum of the modified weights over all respondents approximately equals the population size N.

Variables known for all units in the selected sample are used to form weighting-adjustment classes that consist of units with similar values of the $\mathbf{x}$ variables, and it is hoped that respondents and nonrespondents in the same weighting-adjustment class are also similar with respect to the y variables of interest. Weights of respondents in the weighting-adjustment class are increased so that the respondents represent the nonrespondents' share of the population in that class as well as their own. If the data are MCAR within weighting classes—that is, the respondents in a weighting class are essentially a random subsample of the sampled units in that weighting class—then estimates calculated using the adjusted weights will be approximately unbiased.

Steps for calculating weighting class adjustments for nonresponse

Step 1. Divide the units in the selected sample $\mathcal{S}$ among C weighting classes so that each unit in $\mathcal{S}$, whether respondent or nonrespondent, is in exactly one of the classes. The weighting classes are formed using information that is known for every unit in $\mathcal{S}$. The auxiliary variables are $x_{i1}, x_{i2}, \ldots, x_{iC}$, where $x_{ic} = 1$ if unit i is in class c and 0 otherwise.

Step 2. For each class c (for $c = 1, \ldots, C$), redistribute the sampling weights of the nonrespondents in class c to the respondents in class c.

Each unit in $\mathcal{S}$ has a sampling weight because it was selected for the sample. The sampling weight of unit i is $w_i = 1/\pi_i$, where π_i is the probability unit i was selected to be in the sample.

Set the nonresponse-adjusted weight for respondent i in class c equal to:

$$\begin{aligned}\tilde{w}_i &= w_i \frac{\text{sum of weights for selected sample in class } c}{\text{sum of weights for respondents in class } c} \\ &= w_i \left(1 + \frac{\text{sum of weights for nonrespondents in class } c}{\text{sum of weights for respondents in class } c}\right). \end{aligned} \tag{8.7}$$

Set the nonresponse-adjusted weight for each nonrespondent as $\tilde{w}_i = 0$.

After Step 2, the weights for the nonrespondents in class c have been transferred to the respondents in class c, so that respondent i in class c represents its original share of the population (w_i) plus a fraction (w_i/[sum of weights for respondents in class c]) of the class c nonrespondents' share of the population. Each nonrespondent's weight in class c has been spread among the respondents in that class, proportionally to each respondent's sampling weight w_i.

TABLE 8.3
Illustration of weighting class adjustment factors.

	Age					
	15–24	25–34	35–44	45–64	65+	Total
Sample size	202	220	180	195	203	1,000
Respondents	124	187	162	187	203	863
Sum of weights for sample	30,322	33,013	27,046	29,272	30,451	150,104
Sum of weights for respondents	18,693	28,143	24,371	28,138	30,451	129,796
$\hat{\phi}_c$	0.6165	0.8525	0.9011	0.9613	1.0000	
Weight factor	1.6221	1.1730	1.1098	1.0403	1.0000	

Step 3. Check that the weight adjustments have been done correctly by verifying that

$$\sum_{i \in \mathcal{R}} \tilde{w}_i = \sum_{i \in \mathcal{S}} w_i$$

and

$$\sum_{i \in \mathcal{R}} x_{ic} \tilde{w}_i = \sum_{i \in \mathcal{S}} x_{ic} w_i \text{ for each weighting class } c.$$

The nonrespondents' weights have been shifted to respondents, so the sum of the new weights $\tilde{w}_i$ for the respondents should approximately equal the population size N.

This step is especially important if you are using a software package to calculate the weight adjustments; it provides a way to check for possible mistakes in the weighting process, which, in many applications, is much more complicated than the simple procedure outlined here.

Estimating population means and totals after weighting class adjustments. Weighting class methods rely on the assumption that the probability of response depends solely on the class membership, so that $P(R_i = 1) = \phi_c$ for every unit in weighting class c. The values of ϕ_c are assumed to differ across classes, however. The response propensity ϕ_c in class c is estimated using the sampling weights:

$$\hat{\phi}_c = \frac{\text{sum of weights for respondents in class } c}{\text{sum of weights for selected sample in class } c}. \tag{8.8}$$

The nonresponse-adjusted weight for a respondent in weighting class c is $\tilde{w}_i = 1/(\pi_i \hat{\phi}_c)$. The denominator $\pi_i \hat{\phi}_c$ estimates the probability that unit i is selected for the sample and responds, so $\tilde{w}_i$ can be thought of as the number of population units represented by respondent unit i—if the assumption of $P(R_i = 1) = \phi_c$ is true.

Example 8.4. Table 8.3 shows information about the selected sample and respondents for a fictitious sample of persons. Person i in the selected sample has sampling weight $w_i = 1/\pi_i$. Since the age is known for every member of the selected sample, weighting classes can be formed by dividing the selected sample among different age classes.

We estimate the response propensity for each class c by (8.8). Then the sampling weight for each respondent in class c is multiplied by $1/\hat{\phi}_c$, the weight factor in Table 8.3. The weight of each respondent with age between 15 and 24, for example, is multiplied by 1.6221. Since there is no nonresponse in the 65+ age group, their weights are unchanged.

After multiplying each respondent's weight by the appropriate weight factor, check that the sum of the weights $\tilde{w}_i$ for the respondents in an age group equals the sum of the sampling weights w_i for the selected sample in that age group. For age group 15–24, for example, each respondent's weight is multiplied by (30,322/18,693) = 1.6221015, so the sum of the nonresponse-adjusted weights $\tilde{w}_i$ for the respondents in that age group is 1.6221015 × 18,693 = 30,322. ■

After the nonresponse-adjusted weights are calculated, they are used to estimate population totals and means in the same way as sampling weights, summing over the set of respondents $\mathcal{R}$:

$$\hat{t}_{\hat{\phi}} = \sum_{i \in \mathcal{R}} \tilde{w}_i y_i \tag{8.9}$$

and

$$\hat{\bar{y}}_{\hat{\phi}} = \frac{\hat{t}_{\hat{\phi}}}{\sum_{i \in \mathcal{R}} \tilde{w}_i} = \frac{\sum_{i \in \mathcal{R}} \tilde{w}_i y_i}{\sum_{i \in \mathcal{R}} \tilde{w}_i}. \tag{8.10}$$

Other statistics, such as the median, are calculated similarly, substituting the nonresponse-adjusted weight $\tilde{w}_i$ for the sampling weight in the formula used to calculate the statistic.

In an SRS, if n_c is the number of sample units in class c, n_{cR} is the number of respondents in class c, and $\bar{y}_{cR}$ is the average for the respondents in class c, then $\hat{\phi}_c = n_{cR}/n_c$ and

$$\hat{t}_{\hat{\phi}} = \sum_{i \in \mathcal{R}} \sum_{c} \frac{N}{n} \frac{n_c}{n_{cR}} x_{ic} y_i = N \sum_{c} \frac{n_c}{n} \bar{y}_{cR}.$$

Weighting class adjustments also affect the variances of estimates. Exercise 22 looks at the variance of $\hat{\bar{y}}_{\hat{\phi}}$ for an SRS; in general, we use methods described in Chapter 9 to estimate variances after nonresponse weight adjustments. If the weighting adjustments have removed the nonresponse bias, then the estimated variance of $\hat{\bar{y}}_{\hat{\phi}}$ also estimates the mean squared error (see Section 2.9) of $\hat{\bar{y}}_{\hat{\phi}}$. If, however, nonresponse bias still remains after the weighting, then the mean squared error contains an additional term for the squared bias and is larger than the variance.

Constructing weighting classes. We want the weight adjustments to reduce nonresponse bias as much as possible for estimates of key population quantities. How should the weighting classes be formed in order to accomplish this?

The mathematical theory (see Exercise 20) says that weighting class adjustments approximately remove nonresponse bias for estimating a population total t_y (for example, in a health survey, t_y might be the total number of persons in the population who smoke cigarettes) if at least one of the following situations occurs:

A. Within each weighting class c, every sampled unit has the same propensity to respond to the survey, that is, $\phi_i = \phi_j$ for any pair of sampled units in class c. The response propensities are constant within each weighting class, and thus the variance of response propensities within class c equals 0.

If every unit in the weighting class has the same propensity to respond, then the respondents are essentially a randomly sampled subset of the units from that weighting class that were selected to be in the sample. In this case, the weighting adjustments have removed nonresponse bias for *any* variable y that is measured in the survey.

OR

B. Within each weighting class c, all population members have the same value of y, that is, $y_i = y_j$ for each pair of sampled units in class c. Another way of looking at this is to say the variance of y in class c, including both respondents and nonrespondents, equals 0. In the context of our example, this would mean that in class c, either everyone (respondents and nonrespondents) smokes cigarettes, or no one smokes cigarettes.

OR

C. Within each weighting class, the response propensity is uncorrelated with the characteristic y being measured. Once you know that someone is in class c, knowing the person's smoking status gives no additional information about the probability the person will respond to the survey.

It is unlikely that any of conditions A, B, or C is met exactly. But they provide guidance for forming weighting classes, because the nonresponse bias will be reduced if the conditions are even approximately true.

Consider condition B. In practice, it is unlikely that you can construct weighting classes such that $y_i = y_j$ for every pair of units in class c. But you can try to construct classes such that the population variance of y within a weighting class is lower than S_y^2, the variance of y for everyone in the population. This means forming weighting classes similarly to forming strata, so that the units (respondents and nonrespondents) within each weighting class are as homogeneous as possible. When the nonrespondents in a weighting class are similar to the respondents, the weighting reduces the nonresponse bias. The homogeneity of y within weighting classes may also reduce the variance of the estimator $\hat{\bar{y}}_{\hat{\phi}}$ in (8.10).

Similarly, condition A will be approached if the weighting classes are formed so that the response propensities in each class are similar (that is, the variance of the ϕ_is in each class c is low). Again, this can be thought of as being similar to stratification, where the goal is to form weighting classes such that response propensities in each weighting class are as homogeneous as possible.

Thus, to reduce nonresponse bias, it is best to use auxiliary variables that are associated with the response propensities and that are also associated with many of the key y variables measured in the survey. At the same time, increasing the variability of the weights can increase the variance of estimated means and totals (see Exercise 17 of Chapter 7), so avoid using $\mathbf{x}$ variables that are unrelated to response propensities and key y variables—these can increase the variance without any corresponding decrease in bias.

Ideally, weighting classes should be formed so that both of the following conditions are met:

1. Response propensities vary from class to class but are relatively homogeneous within a class. If this condition is met, the weighting class adjustments reduce bias for *all* variables measured in the survey.

2. For each key outcome (y) variable, the values of y_i are similar for nonrespondents and respondents within a weighting class. If this condition is met, the weighting classes reduce the bias (and may also reduce the variance) of the estimated population mean for this y variable.

Of course the challenge in practice is that we do not know the true response propensities for sampled units, nor do we know the values of the outcome variables for nonrespondents. In addition, a typical survey has many outcome variables, and a set of weighting classes that reduces the within-class variances, or the correlation with the response propensities, for one outcome variable may have little effect on another outcome variable.

For these reasons, most survey researchers emphasize condition (1) when forming weighting classes, since meeting this condition will reduce the nonresponse bias for all y variables.

Although the true response propensities are unknown, $\hat{\phi}_c$ in (8.8) is an approximately unbiased estimator of the average response propensity for that group (see Exercise 23). Then, condition (1) is approached when the weighting adjustment classes have different response rates. For example, in a survey to estimate the total number of persons who smoke, perhaps men have lower response rates than women, and younger persons have lower response rates than older persons. Then it would be reasonable to form weighting classes based on age and gender, particularly if these variables are also thought to be related to the key variable of smoking status.

For large surveys, constructing weighting classes and calculating weighting class adjustments can require multiple steps.

Example 8.5. The ACS still has unit nonresponse after the two-phase sampling described in Example 8.3. Auxiliary information that can be used to form weighting classes is limited to information that is known for *all* households in the sample. This information includes the census tract where the household is located (a census tract is a subdivision of a city or county that typically has a population between 1,200 and 8,000 people) and building type (single-housing-unit structure or multi-housing-unit structure). The ACS weighting adjustments are performed in several steps (U.S. Census Bureau, 2014).

In the first step, housing units in the sample are divided among weighting classes formed by the cross-classification of census tract and building type. For weighting class c, let W_{Rc} be the sum of the weights for the respondents, and W_{Mc} be the sum of the weights for the nonrespondents. Then the new weight for a respondent in class c will be the sampling weight multiplied by the weighting adjustment factor $(W_{Mc} + W_{Rc})/W_{Rc}$. Thus the weights that would be assigned to nonrespondents are reallocated among respondents with similar (we hope) characteristics.

A problem occurs if the number of respondents in a weighting class is too small. Then the weighting class may contain many more nonrespondents than respondents, and the weighting adjustment factor for this class will be very large (or infinite, if the class has no respondents) relative to the adjustment factors from other classes. To improve the stability of estimates, and reduce the weight variation, the Census Bureau collapses weighting classes across adjoining census tracts until each final weighting class meets a predetermined threshold for minimum number of respondents. ■

8.5.2 Regression Models for Response Propensities

The weighting class adjustment model in Section 8.5.1 is commonly adopted to estimate the response propensities ϕ, but other models can also be used. Consider the general model

$$\phi_i = f(\mathbf{x}_i, \boldsymbol{\beta}), \tag{8.11}$$

where $\mathbf{x}_i$ is a vector of auxiliary variables known for all respondents and nonrespondents in the sample, $\boldsymbol{\beta}$ is a vector of parameters that must be estimated from the data, and f represents a function. Examples of functions f include:

- Linear regression, with

$$f(\mathbf{x}_i, \boldsymbol{\beta}) = \beta_0 + \beta_1 x_{i1} + \beta_2 x_{i2} + \ldots + \beta_p x_{ip}.$$

  We'll see in Exercise 22 of Chapter 11 that weighting class adjustment is a special case of a linear regression model. For weighting class adjustments, the estimated response propensities are always between 0 and 1, but other linear regression models can lead to estimated response propensities that are negative—an undesirable feature for an estimated probability to have.

- Logistic regression, with

$$f(\mathbf{x}_i, \boldsymbol{\beta}) = \frac{\exp(\beta_0 + \beta_1 x_{i1} + \beta_2 x_{i2} + \ldots + \beta_p x_{ip})}{1 + \exp(\beta_0 + \beta_1 x_{i1} + \beta_2 x_{i2} + \ldots + \beta_p x_{ip})}.$$

  With logistic regression, and related techniques such as probit regression, the estimated response propensities are always between 0 and 1.

- Other types of regression models. Regression trees use a set of successive yes-no questions about the $\mathbf{x}$ variables to partition the selected sample into smaller groups. The estimated response propensity for each group is the proportion of units in that group that responded to the survey (Phipps and Toth, 2012). These, and related models based on data-mining techniques (Lohr et al., 2015; McConville and Toth, 2019), can account for complicated interactions among the $\mathbf{x}$ variables without having to specify them explicitly in a model. Da Silva and Opsomer (2009) used a nonparametric regression model to predict response propensities.

Each of these types of models gives an estimated response propensity $\hat{\phi}_i$ for each respondent. You can then use these directly to form weight $\tilde{w}_i = w_i/\hat{\phi}_i$ respondent i. Alternatively, if some of the $\hat{\phi}_i$s are negative or if the estimated response propensities are highly variable (causing final estimates to have higher variance), you can use the $\hat{\phi}_i$s to divide the selected sample into weighting classes (Little, 1986). For example, you might form five weighting classes based on the predicted values $\hat{\phi}_i$ obtained using a logistic regression model, with the 20% of units having the lowest values of $\hat{\phi}_i$ forming one weighting class, the 20% of units with the next lowest values of $\hat{\phi}_i$ forming the next class, and so on. Then follow the steps for calculating weighting class adjustments described in Section 8.5.1. With this approach, the predicted response propensities from the regression model are used only to form the classes; the final weight for a respondent in class c is calculated by (8.7). This approach allows the $\hat{\phi}_i$s to be calculated from multiple auxiliary variables, but then smooths out the final weights.

The key issue for any model is that the auxiliary information needs to be rich enough to allow accurate estimation of the true response propensities. Kim and Kim (2007) showed that $\hat{t}_{\hat{\phi}}$ in (8.6) is approximately unbiased for estimating the population total, for *any* characteristic y, if the model in (8.11) is specified correctly and if all of the random variables R_i are independent.

Thus, if the model specifying the nonresponse mechanism gives accurate estimates of the true response propensities, adjusting the weights using estimates of the ϕ_is from that model will reduce nonresponse bias for *all* outcome variables measured in the survey. The same set of nonresponse-adjusted weights is used for all analyses of the survey data.

8.6 Poststratification

Poststratification was introduced in Section 4.4; it is a form a ratio adjustment. After the sample is collected, units are grouped into H different poststrata, often based on demographic variables such as race or gender. The population is known to have N_h units in poststratum h. To use poststratification with nonresponse, we modify the respondents' weights so that the sum of the final weights in each poststratum agrees with the known population count for that poststratum.

Poststratification is a special case of a more general type of adjustment called **calibration**, which we'll discuss in Section 11.6. Calibration can be used when the auxiliary information, in the framework of Section 8.4.1, is of type (A) or (C). For a set of variables $\mathbf{x}$ measured in the survey, the vector of population totals $\mathbf{t_x}$ is known from the frame or, more commonly, an external source such as a population census. Weights from the survey are modified so that estimates of $\mathbf{t_x}$ from the survey, using the modified weights, equal the known population totals. In poststratification, the population totals that are known from an external source are the poststratum sizes N_h.

8.6.1 Poststratification Using Weights

In a general survey design, the sum of the weights in a subgroup estimates the population count for that subgroup. Poststratification uses the ratio estimator within each subgroup to adjust by the true population count.

As in Section 4.4, let $x_{ih} = 1$ if unit i is in poststratum h, and 0 otherwise. Then let

$$w_i^* = w_i \sum_{h=1}^{H} x_{ih} \frac{N_h}{\sum_{j \in \mathcal{R}} w_j x_{jh}}, \tag{8.12}$$

where $\mathcal{R}$ is the set of respondents in the sample. The weights w_i in (8.12) may be the sampling weights, or they may be the nonresponse-adjusted weights $\tilde{w}_i$ calculated in Section 8.5. For unit i in poststratum h, w_i is multiplied by N_h divided by the sum of the weights w_i for the respondents in poststratum h. The poststratified weights w_i^* satisfy

$$\sum_{i \in \mathcal{R}} w_i^* x_{ih} = N_h. \tag{8.13}$$

The poststratified estimators of the population total and mean for variable y are

$$\hat{t}_{\text{post}} = \sum_{i \in \mathcal{R}} w_i^* y_i \quad \text{and} \quad \hat{\bar{y}}_{\text{post}} = \hat{t}_{\text{post}} \Big/ \sum_{i \in \mathcal{R}} w_i^*. \tag{8.14}$$

The poststratified weights w_i^* depend on the particular sample selected and on which units in the sample respond to the survey. For poststratification, the auxiliary information x_{ih} only needs to be known for the respondents to the survey; for weighting class adjustments, the auxiliary information needs to be known for all units in the selected sample, including the nonrespondents.

Poststratification adjusts for undercoverage as well as nonresponse if the population count N_h includes individuals not in the sampling frame for the survey. As shown in Chapter 4, poststratification can also reduce the variance of estimated population quantities. A variance estimator for poststratification will be given in Exercise 21 of Chapter 9.

Example 8.6. The nonresponse adjustments for the ACS described in Example 8.5 result in weight $\tilde{w}_i$ for housing unit i. The survey is then poststratified to independent population control totals. This step adjusts for possible undercoverage and overcoverage of the survey (U.S. Census Bureau, 2014).

The Census Bureau produces independent estimates of the total number of housing units for subcounty areas in the United States based on the most recent decennial census, new construction, and housing units lost since the census. Letting N_h represent the number of housing units in poststratum h from this independent estimate, the weight $\tilde{w}_i$ for a housing unit in poststratum h is multiplied by

$$\frac{N_h}{\text{sum of weights for all housing units in poststratum } h}.$$

With weighting class adjustments, the weighting factor in (8.7) is always at least one because the sum of weights for respondents in a weighting class is always less than or equal to the sum of weights for the selected sample. With poststratification, because weights are adjusted so that they sum to a known population total, the weighting factor can be any positive number, although weighting factors of two or less are desirable. ■

Poststratification in an SRS. Poststratification is similar to weighting class adjustment, except that known population totals are used to adjust the weights instead of estimates from the sample. Let's look at the difference between them for an SRS of n units. The population has N_h units in group h; of these, n_h were selected for the sample and n_{hR} responded. The sample mean of y for the respondents in group h is $\bar{y}_{hR}$. The poststratified estimator for $\bar{y}_{\mathcal{U}}$ is

$$\bar{y}_{\text{post}} = \sum_{h=1}^{H} \frac{N_h}{N} \bar{y}_{hR};$$

the weighting-class estimator for $\bar{y}_{\mathcal{U}}$ is

$$\bar{y}_{\hat{\phi}} = \sum_{h=1}^{H} \frac{n_h}{n} \bar{y}_{hR}.$$

The two estimators are similar in form. The only difference is that with poststratification, respondent i in poststratum h has weight $w_i^* = N_h/n_{hR}$. With weighting class adjustments using the same H groups, respondent i in weighting class h has weight $\tilde{w}_i = (N/n)(n_h/n_{hR})$; the weight $\tilde{w}_i$ calibrates to the estimated size of group h, Nn_h/n. Thus, although $\bar{y}_{\text{post}}$ and $\bar{y}_{\hat{\phi}}$ have similar form, the control totals N_h for poststratification are fixed constants. The estimated totals Nn_h/n used for weighting class adjustments are random variables that would take on a different value if a different sample were chosen.

8.6.2 Raking Adjustments

Raking is a poststratification method that may be used when poststrata are formed using more than one variable, but only the marginal population totals are known. Raking was first used with the 1940 U.S. census to make sure that the complete census and samples taken from it gave consistent results (Deming and Stephan, 1940).

Consider the following table of sums of weights from a sample; each entry in the table is the sum of the sampling weights for persons in the sample falling in that classification (for example, the sum of the sampling weights for Black females is 300).

	Black	White	Asian	Native American	Other	Sum of Weights
Female	300	1200	60	30	30	1620
Male	150	1080	90	30	30	1380
Sum of Weights	450	2280	150	60	60	3000

Now suppose we know the true population counts for the marginal totals: We know that the population has 1510 women and 1490 men, 600 Blacks, 2120 Whites, 150 Asians, 100

Native Americans, and 30 persons in the "Other" category. The population counts for each cell in the table, however, are unknown; we do not know the number of Black females in this population. Raking allows us to adjust the weights so that the sums of weights in the margins equal the population counts. We do this by poststratifying to each dimension in turn.

First, adjust the rows. Multiply each entry by (true row population)/(estimated row population). Multiplying the cells in the female row by 1510/1620 and the cells in the male row by 1490/1380 results in the following table:

	Black	**White**	**Asian**	**Native American**	**Other**	**Sum of Weights**
Female	279.63	1118.52	55.93	27.96	27.96	1510
Male	161.96	1166.09	97.17	32.39	32.39	1490
Total	441.59	2284.61	153.1	60.35	60.35	3000

The row totals are fine now, but the column totals do not yet equal the population totals. Repeat the same procedure with the columns in the new table. The entries in the first column are each multiplied by 600/441.59. The following table results:

	Black	**White**	**Asian**	**Native American**	**Other**	**Sum of Weights**
Female	379.94	1037.93	54.51	46.33	13.90	1532.61
Male	220.06	1082.07	94.70	53.67	16.10	1466.60
Total	600.00	2120.00	150.00	100.00	30.00	3000

But this has thrown the row totals off again. Repeat the procedure until both row and column totals equal the population counts. The procedure converges as long as all cell counts are positive. In this example, the final table of adjusted counts is:

	Black	**White**	**Asian**	**Native American**	**Other**	**Sum of Weights**
Female	375.59	1021.47	53.72	45.56	13.67	1510
Male	224.41	1098.53	96.28	54.44	16.33	1490
Total	600.00	2120.00	150.00	100.00	30.00	3000

The entries in the last table may be better estimates of the cell populations (that is, with smaller variance) than the original weighted estimates, simply because they use more information about the population. The weighting adjustment factor for each white male in the sample is 1098.53/1080; the weight of each white male is increased a little to adjust for nonresponse and undercoverage. Likewise, the weights of white females are decreased because they are overrepresented in the sample.

The assumptions for raking are the same as for poststratification, with the additional assumption that the response propensities depend only on the row and column and not on the particular cell. Raking works best when all cells are nonzero, since the algorithm may not converge if some of the cell estimates are zero.

Example 8.7. The ACS has one more major weighting step: constructing weights for each person in responding households. Each person in an occupied housing unit is initially assigned the weight for the household obtained after the poststratification described in Example 8.6. But, because of sampling variability, undercoverage, and nonresponse, estimates of the total number of persons in different demographic categories differ from the Census

Bureau's counts for those categories derived from information in the most recent decennial census along with data on births, deaths, and migration.

The Census Bureau rakes the initial person weights on three dimensions in turn: (1) population count for the subcounty area containing the person, (2) position in household, with four categories: person 1 in two-person relationship, person 2 in two-person relationship, head of household for other type of household, and other person, and (3) age, race, sex, and Hispanic origin. The Census Bureau raking procedure cycles through these three dimensions until the weighted estimates of population counts from the survey agree with the independent population counts, and the estimated number of persons using the person weights in each of the household categories agree with the corresponding estimates formed from the housing unit weights. This ensures a consistency between the housing unit weights and the person weights so that, for example, the estimated number of householders (tabulated using the person weight) agrees with the survey-estimated number of households and occupied housing units (tabulated using the housing unit weight). ■

Sources of poststratification totals. Poststratification and, more generally, calibration, require knowledge of the poststratum sizes or calibration totals. These generally come from an independent data source in which the population totals are assumed to be known without error. The ACS is calibrated to independent population counts estimated from birth, death, and migration information. The Canadian Survey of Employment, Payrolls and Hours, discussed in Example 3.8, is calibrated so that estimates match total monthly payroll employment statistics from the Canada Revenue Agency (Statistics Canada, 2020). If the sampling frame coincides with the target population—there is no undercoverage or overcoverage—then population quantities calculated from the frame can be used as poststratification totals.

The population totals for poststratification often come from an external source, so it is important to make sure the external source defines and measures the poststratification variables the same way as the survey. Estimates of characteristics such as race or ethnicity may differ when questions are asked differently.

The main difference between weighting class adjustments and poststratification is that the poststratification totals are assumed to be known exactly (or with high accuracy, as in Section 7.5). The sums of the sampling weights used in weighting class adjustments, however, may be random variables that depend on the particular sample that was selected from the population.

8.6.3 Steps for Constructing Final Survey Weights

We have now seen two methods for adjusting weights to attempt to reduce bias from nonresponse and undercoverage. The weighting class adjustments of Section 8.5 modify the weights using information known for the selected sample. Calibration methods, which include poststratification and raking, adjust the weights so that estimates of the population totals for auxiliary variables $\mathbf{x}$ agree with known population totals from an independent data source.

Either method can be used by itself. If you have a lot of information about the sampled units, weighting class adjustments alone may remove much of the nonresponse bias. If you have almost no information about the nonrespondents in the sample but have information about population totals from another source, then you might skip weighting class adjustments and use poststratification alone to try to reduce the nonresponse bias. A random-digit-dialing telephone survey, for example, may have little information about the nonrespondents but you can poststratify to population counts using information collected from the respondents about age, race, gender, education, and other variables. If poststratification is used by itself, a poststratified estimator is approximately unbiased under the same conditions that give unbiasedness for weighting class adjustments—that is, within each poststratum h, (a) each unit has the same probability of responding, (b) the value of y_i is the same for every unit, or (c) the response y_i is uncorrelated with the response propensity ϕ_i (see Exercise 21).

If, however, you have both types of information, it is often better to first adjust for nonresponse using weighting class adjustments and then calibrate to known population totals. With two steps of weighting, there is a better chance that at least one of them will reduce nonresponse bias (Haziza and Lesage, 2016). If weighting adjustments are done first, the calibration weight adjustments should be relatively small, and calibration can be done much the same way as in the complete-response case described in Sections 4.4 and 11.6.

With both types of information, here are the steps for constructing weights for a survey with nonresponse. Section 7.5 illustrated these steps for the National Health and Nutrition Examination Survey (NHANES).

1. Determine the sampling weights for units in the selected sample. Let $w_i = 1/\pi_i$.

2. Adjust the weights for nonresponse by obtaining an estimate $\hat{\phi}_i$ of the response propensity for each respondent. Then set $\tilde{w}_i = w_i/\hat{\phi}_i$. If a weighting class adjustment adjustment is used, then

$$\tilde{w}_i = w_i \frac{\text{sum of weights for selected sample units in class } c}{\text{sum of weights for respondents in class } c}.$$

   Check that the weights sum to the estimated population size for each weighting class.

3. Calibrate the nonresponse-adjusted weights $\tilde{w}_i$ to a set of known population totals. If poststratification is used as the calibration method, check that the final weights w_i^* satisfy (8.13) for the poststrata. If raking is used, check that the sum of the final weights for each raking characteristic equals the population total for that characteristic.

8.6.4 Advantages and Disadvantages of Weighting Adjustments

Weighting class adjustments and poststratification can both help reduce bias from unit nonresponse. These methods adjust the survey weights so that the sum of the weights for respondents in a group agree with the sum of the weights for sampled units in that group (for weighting class adjustments) or agree with estimates of the total number of population members in that group from an external source (for poststratification).

If the population subgroups in the weighting are associated with survey outcome y, then the weighting adjustments also reduce the variance of the estimated mean or total for that outcome, as shown in Section 4.4. Chapter 9 will tell how to estimate variances when nonresponse-adjusted weights are used.

The end result of the procedures is one set of final weights to be used for all analyses. This ensures that estimates are consistent with each other—for example, the sum of the estimated number of persons aged 18–54 who smoke and the estimated number of persons aged 55 and over who smoke equals the estimated number of persons aged 18 and over who smoke. Some of the other methods that have been proposed for missing data—in particular, models that address bias for each outcome variable separately—do not have this property.

The models for weighting adjustments for nonresponse are strong: In each weighting class or poststratum, it is assumed that the response propensity is uncorrelated with the variables of interest y. These models never exactly describe the true state of affairs, and you should always consider their plausibility and implications. Weights may improve many of the estimates, but they rarely eliminate all nonresponse bias. If weighting adjustments are made (and remember, making no adjustments is itself a model about the nature of the nonresponse), practitioners should always state the assumed response model and give evidence to justify it.

8.7 Imputation

Missing items may occur in surveys for several reasons. An interviewer may fail to ask a question; a respondent may refuse to answer the question or cannot provide the information; a clerk entering the data may skip the value. Sometimes, items with responses are changed to missing when the data set is edited or cleaned—a data editor may not be able to resolve the discrepancies for a 3-year-old who voted in the last election, and may set both age and voting status to missing.

Imputation is commonly used to assign values to the missing items. A replacement value, often from another unit in the survey that is similar to the item nonrespondent on other variables, is imputed (filled in) for the missing value. When imputation is used, an additional variable should be created for each imputed variable in the data set that indicates which records have imputed values.

Imputation procedures are used not only to reduce the nonresponse bias but to produce a "clean," rectangular data set—one without holes for the missing values. We may want to look at tables for subgroups of the population, and imputation allows us to do that without considering the item nonresponse separately each time we construct a table.

This section briefly describes some simple imputation methods; see the For Further Reading section for references and guides on how to do imputation in practice.

Example 8.8. Missing values in variables used to construct weighting classes or poststrata must be imputed before those weighting adjustment methods can be used to adjust for nonresponse. The ACS person-level weighting described in Example 8.7 requires values for the age, race, ethnicity, and sex of each respondent. The ACS assigns values when appropriate (for example, a person who has given birth within the past 12 months would be imputed to be female) and supplies values for other missing items from persons in the sample who have similar characteristics through a hot-deck imputation procedure (see Section 8.7.3).

For surveys such as the ACS, if imputation is to be done, the agency collecting the data has more information to guide it in filling in the missing values than does an independent analyst, because the agency has information about respondents that is not released to the public. The goal is to have the final edited and imputed data be as consistent and complete as possible, and ready for tabulation. ■

We use the small data set in Table 8.4 to illustrate some of the imputation methods. This artificial data set is only used for illustration; in practice, a much larger data set is needed for imputation. A "1" means the respondent answered yes to the question, a "0" means the respondent answered no, and a "?" means the respondent provided no answer.

8.7.1 Deductive Imputation

Some values may be imputed in the data editing, using logical relations among the variables. Person 9 is missing the response for whether she was a victim of violent crime. But she had responded that she was not a victim of any crime, so the violent crime response should be changed to 0.

TABLE 8.4
Small data set used to illustrate imputation methods.

Person	Age	Gender	Years of Education	Crime Victim?	Violent Crime Victim?
1	47	M	16	0	0
2	45	F	?	1	1
3	19	M	11	0	0
4	21	F	?	1	1
5	24	M	12	1	1
6	41	F	?	0	0
7	36	M	20	1	?
8	50	M	12	0	0
9	53	F	13	0	?
10	17	M	10	?	?
11	53	F	12	0	0
12	21	F	12	0	0
13	18	F	11	1	?
14	34	M	16	1	0
15	44	M	14	0	0
16	45	M	11	0	0
17	54	F	14	0	0
18	55	F	10	0	0
19	29	F	12	?	0
20	32	F	10	0	0

Deductive imputation may sometimes be used in longitudinal surveys. If a woman has two children in year 1 and two children in year 3, but is missing the value for year 2, the logical value to impute would be two.

8.7.2 Cell Mean Imputation

Respondents are divided into classes (cells) based on known variables, as in weighting class adjustments. Then the average of the values for the responding units in cell c, $\bar{y}_{Rc}$, is substituted for each missing value. Cell mean imputation assumes that missing items are missing completely at random within the cells.

Example 8.9. The four cells for the data in Table 8.4 are constructed using the variables age and gender. (In practice, of course, you would want to have many more individuals in each cell.)

	Age $\leq$ 34	Age $\geq$ 35
Male	Persons 3, 5, 10, 14	Persons 1, 7, 8, 15, 16
Female	Persons 4, 12, 13, 19, 20	Persons 2, 6, 9, 11, 17, 18

Persons 2 and 6, missing the value for years of education, would be assigned the mean value for the four women aged 35 or older who responded to the question: 12.25. The mean for each cell after imputation is the same as the mean of the respondents. Note, however, that the imputed value of 12.25 is not one of the possible responses to the question about education because it is not an integer. ■

Cell mean imputation gives the same point estimates for means, totals, and proportions as would weighting class adjustments performed with the cells. Cell mean imputation methods fail to reflect the variability of the nonrespondents, however—all missing observations in a class are given the same imputed value. The distribution of y will be distorted because of a "spike" at the value of the sample mean of the respondents. As a consequence, the estimated variance in the subclass will be too small.

To avoid the spike, a stochastic cell mean imputation could be used. If the response variable were approximately normally distributed, the missing values could be imputed with a randomly generated value from a normal distribution with mean $\bar{y}_{cR}$ and standard deviation s_{cR}.

Cell mean imputation, stochastic or otherwise, distorts relationships among variables, because the imputation is done separately for each missing item. Other imputation methods are preferred if regression coefficients or correlations are to be computed from the imputed data set.

8.7.3 Hot-Deck Imputation

In *hot-deck imputation*, as in cell mean imputation and weighting adjustment methods, the sample units are divided into classes. The value of one of the responding units in the class is substituted for each missing response. Often, the values for a set of related missing items are taken from the same donor, to preserve some of the multivariate relationships. The name is from the days when computer programs and data sets were punched on cards—the deck of cards containing the data set being analyzed was warmed by the card reader, so the term *hot deck* was used to refer to imputations made using the same data set. Haziza (2009) and Andridge and Little (2010) reviewed methods for hot-deck imputation with large surveys.

How is the donor unit to be chosen? Several methods are possible.

Sequential hot-deck imputation. Some hot-deck imputation procedures impute the value in the same subgroup that was last read by the computer. This is partly a carryover from the days when computers used punch cards (imputation could be done in one pass), and partly a belief that if the data are arranged in geographical or another meaningful order, adjacent units in the same subgroup will tend to be more similar than randomly chosen units in the subgroup. One problem with using the value on the previous "card" is that often the nonrespondents also tend to occur in clusters, so one person may be a donor multiple times, in a way that the sampler cannot control. One of the other hot-deck imputation methods is usually used today for most surveys.

In our example, person 19 is missing the response for crime victimization. Person 13 had the last response recorded in her subclass, so the value 1 is imputed.

Nearest-neighbor hot-deck imputation. Define a distance measure between observations, and impute the value of a respondent who is "closest" to the person with the missing item, where closeness is defined using the distance function.

If age and gender are used for the distance function, so that the person of closest age with the same gender is selected to be the donor, the victimization responses of person 3 will be imputed for person 10.

Random hot-deck imputation. A donor is randomly chosen from the persons in the cell with information on all the missing items. To preserve multivariate relationships, usually values from the same donor are used for all missing items of a person.

Example 8.10. In the data in Table 8.4, person 10 is missing both variables for victimization. Persons 3, 5, and 14 in his cell have responses for both crime questions, so one of the three is chosen randomly as the donor. In this case, person 14 is chosen, and his values are imputed for both missing variables. ■

8.7.4 Regression Imputation and Chained Equations

Regression imputation predicts the missing value using a regression of the item of interest on variables observed for all cases. A variation is *stochastic regression imputation*, in which the missing value is replaced by the predicted value from the regression model plus a randomly generated error term.

We only have 18 complete observations for the response crime victimization (not really enough for fitting a model to our data set), but a logistic regression with explanatory variable age gives the following model for predicted probability of victimization, $\hat{p}$:

$$\ln \frac{\hat{p}}{1-\hat{p}} = 2.5643 - 0.0896 \times \text{ age}.$$

The predicted probability of being a crime victim for a person aged 17 is 0.74; because that is greater than a predetermined cutoff of 0.5, the value 1 is imputed for Person 10.

Multivariate imputation by chained equations (MICE). The MICE method relies on a sequence of regression models that predict the missing values for each variable in turn. For the data in Table 8.4, the imputation would be carried out as follows. First, fill in the missing data with an initial set of imputed values; one choice might be to impute cell means for each variable. Then, using the set of records that have an observed (non-imputed) value for education, fit a regression model to predict education from the variables age, gender, crime, and violent crime. Replace the previously imputed values for education with predictions from the model. Now, using the records that have observed data for crime, fit a model predicting crime from the other four variables, and use predictions from that model to replace the previously imputed values for crime. Continue the procedure, cycling through the variables with missing data until the models stabilize. Each model is fit to the set of observations that have observed values for the response variable, using the observed values or current imputations for the explanatory variables.

Azur et al. (2011) and van Buuren (2018) describe how to impute data using MICE. It is often used together with multiple imputation (Section 8.7.6). MICE preserves multivariate relationships that are included in the imputation model, but relationships not in that model may be distorted.

8.7.5 Imputation from Another Data Source

Sometimes an item is unanswered on your survey, but the information may be available from another data source (possibly with errors). Imputed values can come from a previous survey or other information, such as from historical data or administrative records. This is sometimes called *cold-deck imputation*, from the days when the card deck containing the data used for imputation was not currently running through the computer and hence was "cold."

Example 8.11. Golbeck et al. (2019) described the imputation procedures used in the 2017 Annual Survey of the Mathematical Sciences, which reports on faculty composition and degrees awarded by departments of mathematical sciences in U.S. colleges and universities. They argued that year-to-year changes within the same department tend to be small. Thus, when a department selected for the sample does not respond, the most recent prior year's response is imputed when available. Weighting methods are used to adjust for nonresponse remaining after the imputation. ■

Some surveys substitute a convenient member of a population for a designated member who does not respond. For example, if no one is at home in the designated household, a field representative might try next door. In a wildlife survey, the investigator might substitute an area next to a road for a less accessible area. In each case, the sampled but nonparticipating units most likely differ on a number of characteristics from the substituted units that respond so that the substitution results in selection bias. The substituted household may be more likely to have a member who does not work outside of the house than the originally selected household. The area by the road may have fewer frogs than the area that is harder to reach.

8.7.6 Multiple Imputation

In multiple imputation, introduced by Rubin (1987), each missing value is imputed m (≥ 2) different times. These create m different "data" sets with no missing values. Each of the m data sets is analyzed as if no imputation had been done; the results from the different imputations give a measure of the additional variance due to the imputation.

Example 8.12. The U.S. Survey of Consumer Finances asks respondents to provide detailed information about their income, financial assets, and liabilities (Bricker et al., 2017). But there is item nonresponse to questions that respondents deem sensitive, or to which they may not know the answer (for example, the value of a privately-held business may be known only after it is sold). Other questions have partially missing information—a respondent may report a range of values for municipal bond fund holding instead of a single dollar amount.

Values are imputed for missing items using a version of MICE. An iterative process imputes each outcome variable in turn, using a model that estimates the probability distribution of that variable from the other variables measured in the survey. A value is randomly sampled from that distribution to impute the missing item, and the iterations continue until the imputations are stable. The process is carried out five times, each using different random samplings from the probability distributions, to create five imputed data sets. Because imputed values are randomly selected from the predicted probability distributions, the multiple imputation allows users to estimate uncertainty associated with the missing data (Kennickell, 2017; U.S. Federal Reserve Board, 2017). ■

8.7.7 Advantages and Disadvantages of Imputation

Imputation creates a "clean," rectangular data set with no missing values. Analyses of different subsets of the data will produce consistent results. If the data are missing at random given the covariates used in the imputation procedure, imputation can substantially reduce the bias due to item nonresponse. Some researchers also use imputation for unit nonresponse, by imputing all variables in missing units.

Imputation must be performed when variables used in weighting class adjustments or poststratification have missing values, so that every unit can be assigned to a weighting class or poststratum.

One danger of using imputation is that future data analysts will not distinguish between the original and the imputed values. The imputer should create a "flag" variable that records which observations are imputed and perform analyses to check the sensitivity of imputations to the model adopted. The imputed values may be good predictions, but they are not real data.

If you analyze the data in a standard survey software program, treating the imputed values as though they were observed in the survey, estimated variances will be too small. This is partly because of the artificial increase in the sample size and partly because the imputed values are predictions but are analyzed as though they were really obtained in the data collection. Multiple imputation, or methods described by Shao (2003) and Haziza and Rao (2006), can be used to estimate variances when missing data have been imputed.

Imputation, like all methods for treating missing data, depends on the type of missing data and the model used for the imputation. Imputers implicitly assume that the missing data are MAR, so that predictions can be made from the data that are available, and that the model used for imputation gives good predictions for all analyses that will be performed with the imputed data. If those assumptions are not met, then results calculated from imputed data can be misleading.

8.8 Response Rates and Nonresponse Bias Assessments

8.8.1 Calculating and Reporting Response Rates

Every survey should be accompanied by a detailed breakdown of the response rate and the reasons for nonresponse. A response rate has the general form

$$\text{RR} = 100 \times \frac{\text{number of units that provide data for the survey}}{\text{number of sampled units that are eligible for the survey}}$$

but there are many ways of calculating the numerator and denominator. The numerator depends on how one defines "provide data." Is someone who answered the first question on the survey but ignored all the other questions a respondent? Similarly, for the denominator, how is eligibility for the survey determined?

The American Association of Public Opinion Research (2016b) definitions classify each unit in the selected sample from a telephone, in-person, mail, or internet survey into one of four response status categories:

Respondents: Units that provide data for the survey. The survey conductors need to define what counts as "providing data" before starting data collection. For example, in a telephone survey, the survey conductors might define an interview in which the respondent answers all of the key questions and at least 80% of the other questions to be "complete," an interview that is not complete but obtains answers to at least 50% of the questions to be "partial," and an interview in which the respondent answers fewer than 50% of the questions to be a "break-off" that will be counted as nonresponse.

Nonrespondents: Units eligible for the survey that are not complete or partial respondents. It is also important to document the type of nonresponse for each unit in the sample that does not provide data; this can help with developing strategies for reducing nonresponse on future surveys. Households or persons selected for a household survey might be nonrespondents because (1) they could not be contacted, (2) they were contacted but refused to participate or broke off the interview, or (3) there was some other impediment to responding such as a disability or language barrier.

Not eligible for survey: Units in the sampling frame population but not the target population (see Figure 1.1). The survey conductors should set eligibility criteria before selecting the sample. For example, the target population for one survey may be persons age 65 and older; for another, it may be households with children under age 6; for another, it may be adult residents of California. In a telephone survey of households, a telephone number in the sample that is non-working, belongs to a business, or belongs to a household not in the target population is ineligible. A sampled address that is not a housing unit or is vacant is ineligible for an in-person or mail survey.

Unknown eligibility: If no contact is made with a sampled unit, it may difficult to tell whether the unit is eligible for the survey. A number sampled in a random-digit-dialing household telephone survey has unknown eligibility if the number is always busy, no one answers, or the automated message is unclear about whether the number belongs to a household or a business. A cell phone number in the sample may have a California area code, but someone must answer the telephone for you to ascertain that the number's owner actually lives in California. In a mail survey, you will not know if a sampled address contains children under age 6 if no one returns the questionnaire.

Unweighted response rates. The American Association of Public Opinion Research (2016b) lists six standard response rate definitions:

$$\text{RR1} = 100 \frac{\sum_{i \in \mathcal{S}} C_i}{\sum_{i \in \mathcal{S}} [C_i + P_i + (NR)_i + U_i]}, \qquad \text{RR2} = 100 \frac{\sum_{i \in \mathcal{S}} (C_i + P_i)}{\sum_{i \in \mathcal{S}} [C_i + P_i + (NR)_i + U_i]},$$

$$\text{RR3} = 100 \frac{\sum_{i \in \mathcal{S}} C_i}{\sum_{i \in \mathcal{S}} [C_i + P_i + (NR)_i + e * U_i]}, \qquad \text{RR4} = 100 \frac{\sum_{i \in \mathcal{S}} (C_i + P_i)}{\sum_{i \in \mathcal{S}} [C_i + P_i + (NR)_i + e * U_i]},$$

$$\text{RR5} = 100 \frac{\sum_{i \in \mathcal{S}} C_i}{\sum_{i \in \mathcal{S}} [C_i + P_i + (NR)_i]}, \qquad \text{RR6} = 100 \frac{\sum_{i \in \mathcal{S}} (C_i + P_i)}{\sum_{i \in \mathcal{S}} [C_i + P_i + (NR)_i]},$$

where $C_i = 1$ if unit i gave a completed interview and 0 otherwise, $P_i = 1$ if unit i gave a partial interview, $(NR)_i = 1$ if unit i was a nonrespondent, and $U_i = 1$ if unit i had unknown eligibility. Response rates RR1–RR6 report the percentage of sampled units that respond, irrespective of the sample design. RR1, for example, divides the number of completed interviews by the total number of units in the sample that are eligible or of unknown eligibility.

In RR3 and RR4, the number of units of unknown eligibility is multiplied by a factor e that estimates the proportion of those units thought to be eligible, and the method used to estimate e should be reported along with the response rate definition. The American Association of Public Opinion Research (2016b) states: "In estimating e, one must be guided by the best available scientific information on what share eligible cases make up among the unknown cases and one must not select a proportion in order to boost the response rate." Note that RR1 $\leq$ RR3 $\leq$ RR5, and RR2 $\leq$ RR4 $\leq$ RR6.

It may be desired to report more than one response rate, particularly if it is desired to compare response rates with those from other surveys. RR5 and RR6, however, are only appropriate when it is reasonable to assume that $e = 0$, that is, all cases of unknown eligibility are ineligible for the survey.

The reported response rate depends on which definition is used, so it is important to choose the definition that is most appropriate for the survey and to document all steps of the calculation.

Some surveys report a **cooperation rate**, which is the number of respondents divided by the number of eligible sample units that are contacted. This can provide useful information, but should be reported as a supplement to the response rate, not as a substitute. A cooperation rate is always higher than the corresponding response rate, because it excludes units that could not be contacted from the denominator.

Weighted response rates. The unweighted response rates RR1–RR6 describe the percentage of the units in the sample that respond, and thus measure the success of the survey efforts in obtaining responses from sampled units. In some situations, a weighted response rate may be preferred. In a two-phase design, for example, follow-up efforts are targeted on a subsample of the initial nonrespondents, and a response rate reported for two-phase sampling should account for the subsampling. A weighted response rate was used to estimate $\hat{\phi}_c$ in Section 8.5.1.

A weighted response rate has general form (sum of weights for respondents) divided by (sum of weights for eligible units in sample), and it estimates the percentage of the population that would respond to the survey. The weighted response rate corresponding to RR1, for example, is

$$\text{RR1}_w = 100 \frac{\sum_{i \in \mathcal{S}} w_i C_i}{\sum_{i \in \mathcal{S}} w_i [C_i + P_i + (NR)_i + U_i]},$$

where w_i is the sampling weight for unit i. The numerator estimates the number of population units who would complete the survey if contacted, and the denominator estimates the number of eligible units and units of unknown eligibility in the population. Weighted response rates RR2_w to RR6_w can be calculated similarly.

Let's look at the weighted response rate for the situation in Section 8.3 in which an SRS of size n is selected, and then an SRS of size νn_M is taken of the n_M nonrespondents. For simplicity, assume that all n units in the first-phase SRS are eligible for the survey and all respondents give completed interviews. The sampling weight for each initial respondent is $w_i = N/n$ and the sampling weight for each unit subsampled in phase II is $w_i = N/(n\nu)$. The weight for an eligible unit that did not respond when initially contacted, and was not selected for the phase II subsample, is set equal to 0. Then, if $n_M^{(2)}$ of the νn_M units in the phase II subsample respond, the weighted RR1 response rate is

$$\text{RR1}_w = 100 \frac{\dfrac{N}{n} n_R + \dfrac{N}{n\nu} n_M^{(2)}}{\dfrac{N}{n} n_R + \dfrac{N}{n} n_M} = 100 \frac{n_R + n_M^{(2)}/\nu}{n_R + n_M}.$$

If all units in the phase II sample respond (that is, $n_M^{(2)} = \nu n_M$), the weighted response rate is 100%. The unweighted response rate for this situation is $100(n_R + n_M^{(2)})/n\%$, which is not appropriate here since it does not consider the effects of the second phase of sampling which, when $n_M^{(2)} = \nu n_M$, results in a full-response probability sample.

If in doubt about which response rate to use, report both unweighted and weighted response rates. The unweighted response rate gives the percentage of the sampled units that respond; the weighted response rate estimates the percentage of the population that would respond under the survey protocol.

8.8.2 What Is an Acceptable Response Rate?

Often an investigator will say, "I expect to get a 60% response rate in my survey. Is that acceptable, and will the survey give me valid results?" As we have seen in this chapter, the answer to that question depends on the nature of the nonresponse: If the nonrespondents are MCAR, then we can largely ignore the nonresponse and use the respondents as a representative sample of the population. If the nonrespondents tend to differ from the respondents, then the biases in the results from using only the respondents may make the entire survey worthless.

Many references give advice on cut-offs for acceptability of response rates. Babbie (2007, p. 262), for example, wrote: "A review of the published social research literature suggests that a response rate of at least 50 percent is considered adequate for analysis and reporting. A response of 60 percent is good; a response rate of 70 percent is very good." I believe that giving such absolute guidelines for acceptable response rates is dangerous and has led many survey investigators to unfounded complacency about nonresponse; many examples exist of severely biased estimates from surveys with a 70% response rate. The ACS needs corrections for nonresponse bias even with high response rates.

Increasing the response rate from 20% to 30% does not necessarily decrease nonresponse bias. Exercise 18 provides an example where increasing the response rate worsened the bias. Sometimes efforts to increase the response rate merely yield more of the easy-to-recruit people, who may differ from the harder-to-get members of the selected sample. Suppose, for example, that the survey has too few high-income households among the respondents. Offering a monetary incentive, say $50, for completing the survey may spur response among lower-income households but lead to increased bias for estimates of quantities related to income.

8.8.3 Nonresponse Bias Assessments

Although it is important to report response rates for a survey, these do not tell how much nonresponse bias there is. The amount of nonresponse bias for estimating t_y depends on how strongly the y variable is correlated with the propensity to participate in the survey (see Exercise 19), and on how well the methods discussed in this chapter deal with that correlation. Because the correlation between y and ϕ depends on the particular y variable studied, the amount of nonresponse bias differs for different y variables. If knowing the value of y_i gives no information about unit i's likelihood of responding to the survey, then a survey with a 5% response rate will have low bias for estimating $\bar{y}_{\mathcal{U}}$. On the other hand, a survey with a 95% response rate can have a large bias if there are still large differences between the respondents and nonrespondents after the weighting adjustments (Groves, 2006; Groves and Peytcheva, 2008).

A nonresponse bias assessment can give some information about how much bias remains after the weighting adjustments. Now, in practice, we cannot know the nonresponse bias remaining for a variable y that is measured only in the survey, because we do not know the value of y for the nonrespondents. But we sometimes have external information on variables that are related to y, or we can examine information in the survey data collection process. Montaquila and Olson (2012) listed three types of analyses that may give information about potential nonresponse bias.

Compare survey results with statistics from another data source. Sometimes a variable of interest from the survey is measured in another survey that has a higher response rate. Comparing statistics for the same (or similar) characteristics across multiple data sources can provide information about systematic differences among the sources. Be careful, though, that your survey uses the same definitions and measurements as the comparison data source. Your survey may have different estimates of unemployment than the ACS because you define unemployment differently, or because your survey covers a different time period.

Example 8.13. Kohut et al. (2012) compared results from three surveys taken in 2011 and 2012. The first telephone survey of 1,507 adults, taken over a five-day period, achieved a response rate of 9%. The second telephone survey, of 2,226 adults, was conducted over a two-month period; by using a longer time period, offering monetary incentives for participation, and employing interviewers with proven records for persuading reluctant sample members to participate, it achieved a response rate of 22%. Both telephone surveys selected landline and cellular telephone numbers using random digit dialing. The Current Population Survey (CPS) was used for comparison; its estimates were considered to be the gold standard because of the survey's high response rate and data quality.

Table 8.5 shows estimated percentages of adults with various characteristics for the three surveys. All percentages were calculated after the surveys were weighted to match the demographic composition of the population. Kohut et al. (2012) reported a margin of error of 2.9 percentage points for the standard survey and 2.7 percentage points for the high-effort survey. The margin of error for estimates from the CPS is less than 1 percentage point.

TABLE 8.5
Estimated percentages of adults having specific characteristics from three surveys.

Characteristic	Standard Survey (%)	High-Effort Survey (%)	CPS (%)
Lived at current address 5 or more years	56	59	59
Internet user	80	80	74
Education of college degree of higher	39	33	28
Received unemployment compensation in 2011	11	13	11
Received Social Security payments in 2011	32	33	27
Received food stamps or nutrition assistance in 2011	17	18	10
Registered to vote (among citizens)	78	79	75
Contacted a public official in past year to express opinion	31	29	10
Volunteered for an organization in past year	55	56	27
Talked with neighbors in past week	58	58	41

Questions about residence and receiving unemployment compensation had similar estimated percentages for the three surveys. But the estimated percentage of persons who had volunteered in the past year was more than twice as high in both telephone surveys than in the CPS, indicating that these surveys have substantial nonresponse bias for this question (see Exercise 17). ■

This method assesses potential nonresponse bias after the nonresponse weighting adjustments have been performed, but it should be applied only to variables that were not used in the weighting. In Example 8.13, all estimates were poststratified or raked to the population's demographic characteristics. Thus the weighted estimates for the percentage of the population that is female or in a particular age group will equal the population percentage for that characteristic—agreement is expected for the weighting variables and tells you nothing about possible nonresponse bias for other variables.

You may be able to obtain comparison data on the nonrespondents (or a random subsample of them) from an external data source. Beebe et al. (2011) linked persons in a sample of residents of Olmstead County, Minnesota to a database of medical records from health care providers in the county. They were able to compare number of medical office visits for respondents and nonrespondents, and found that respondents were more likely to have at least 3 office visits.

Study internal variation in the data collection. The survey itself often has information about the nonrespondents. You can study potential nonresponse bias by comparing response rates across population subgroups (Peytcheva and Groves, 2009). In some surveys, interviewers may be able to collect information about nonrespondents (for example, interviewers in an in-person survey can record information about the dwelling unit) that can be compared with respondents' information. In a longitudinal survey, some people drop out between the first and second interviews; you can compare the first-interview responses for the dropouts with those of persons who participated in the second interview.

You can also compare respondents who agreed to participate in the survey right away with respondents who were more reluctant to participate. If nonrespondents are similar to reluctant respondents, then a difference between early and later respondents can signal nonresponse bias that is not corrected by the weighting.

Example 8.14. Cantor et al. (2015) explored effects of nonresponse in a survey about sexual assault that was administered at 27 universities. All undergraduate, graduate, and professional school students age 18 and over were e-mailed an invitation to take the survey. Students who did not respond to the initial contact were sent up to two reminder e-mails. Overall, about 19% of students who were contacted responded to the survey.

The researchers compared responses of students who participated in the survey before a reminder was sent with those who participated later. They found that the later respondents had statistically significantly lower rates for four types of sexual victimization than the early respondents. Of course, the nonrespondents' experiences are unknown, but if nonrespondents are more similar to the late respondents then the survey estimates, even after weighting, may still have some nonresponse bias. ■

Contrast survey estimates from alternative weighting and imputation models. There are often several choices for nonresponse adjustment models. Weighting classes, for example, can be formed with different variables or with different numbers of classes. Estimates from weighting class adjustments can be compared with those from other response propensity models. If two equally defensible models give very different estimates for key outcome variables, then the survey conductors will want to explore reasons for the differences.

Example 8.15. Cohn (2018) reported polling estimates from different weighting schemes for the 2018 U.S. Congressional elections. In Illinois Congressional District 6, a probability sample of 36,455 likely voters was taken from voter registration files; 512 of these responded to the survey. With such a low response rate (1.4%), a nonresponse bias assessment is essential. The predictions for this district changed substantially with different weighting schemes. With weights that were calibrated to turnout model projections for the composition of the likely electorate by age, political party, gender, region, race, and education, the Republican candidate was predicted to win by one percentage point. Weighting the respondents to census data instead of voter records predicted the Republican candidate would lose by six percentage points. Both of these were reasonable weighting schemes to use, and the difference indicates that the estimate is sensitive to the weighting model.

Of course, even if estimates from different weighting schemes agreed there still could be nonresponse bias. There may be systematic differences between people who are willing to take a telephone poll and those who are unwilling to take a poll that cannot be explained by the available auxiliary information. ■

All weighting schemes are based on the auxiliary information that is available, and that may not be rich enough to provide accurate estimates of response propensities or to detect nonresponse bias that remains after weighting adjustments. In Example 8.13, none of the available auxiliary variables explained the higher representation of persons who regularly volunteer among the respondents. But when different (and equally reasonable) weighting models lead to substantially different estimates for key variables, there may be increased concern about nonresponse bias for those variables.

8.9 Chapter Summary

Nonresponse and undercoverage present serious challenges for survey inference. The main concern is that failure to obtain information from some units in the selected sample (nonresponse), or failure to include parts of the population in the sampling frame (undercoverage), can result in biased estimates of population quantities.

The survey design should include features to minimize nonresponse and nonresponse bias. Designed experiments can give insight into methods for increasing response rates. If possible, the survey frame should contain some information on everyone in the selected sample so that respondents and nonrespondents can be compared on those variables, and so that the auxiliary information can be used to adjust for nonresponse.

Weighting methods attempt to reduce nonresponse bias by adjusting the weights of respondents. In weighting class methods, the weights of respondents in a grouping class are increased to compensate for the nonrespondents in that class. In poststratification, the weights of respondents in a poststratum are increased so that they sum to an independent count of the population in that poststratum. The nonresponse mechanism can also be modeled explicitly.

Imputation methods create a "complete" data set by filling in values for data that are missing because of item or unit nonresponse. Many methods are available for performing imputations. Hot-deck methods impute values from a sample respondent for the missing data items of another sample member. Other methods rely on statistical models that predict the missing values from relationships in the observed data.

All surveys should report response rates and assess potential nonresponse bias that remains after weighting adjustments and imputation. If imputation is used, the imputed values should be flagged so that data analysts know which values were measured for the sampled unit and which values were imputed.

Key Terms

Calibration: A method used to adjust for nonresponse and undercoverage in which the weighted sums of certain variables are forced to equal their known population totals.

Imputation: Methods used to "fill in" values for missing items so that the data set appears complete.

Item nonresponse: Nonresponse that occurs when a sampled unit has responses to some but not all of the items in the survey instrument.

Missing at random given covariates (MAR): A missing data mechanism in which the probability of being missing can be fully explained by variables for which you have full information.

Missing completely at random (MCAR): A missing data mechanism in which the probability of being missing is unrelated to all variables measured in the survey.

Nonresponse bias: Bias that occurs because nonrespondents differ from survey respondents.

Not missing at random (NMAR): A missing data mechanism in which the probability of being missing depends on y, the outcome variable of interest, even after accounting for all auxiliary variables $\mathbf{x}$.

Poststratification: A method used for nonresponse adjustment in which respondents are divided into subgroups, and their weights are adjusted so that sample estimates for the subgroups agree with population totals known from an independent data source.

Raking: A weighting adjustment method in which weights are iteratively adjusted to match row and column population totals.

Respondent: A unit in the selected sample that provides data for the survey.

Response propensity: The probability that a unit will respond to the survey.

Selected sample: The set of population units selected to be in the sample; this includes respondents and nonrespondents.

Two-phase sampling: A method of sampling in which, after an initial probability sample is selected, a probability subsample is selected using inclusion probabilities that may depend on results from the initial sample.

Undercoverage: Failure of the sampling frame to include all of the population of interest.

Unit nonresponse: A failure to obtain any information from the observation unit.

Weighting class adjustment: A method of adjusting weights of respondents so they also represent the population share of nonrespondents in the same weighting class.

For Further Reading

This chapter introduced you to the problems caused by nonresponse, and methods used to prevent and adjust for it. The references listed in this section provide detailed guidance for preventing nonresponse and for carrying out weighting adjustments and imputation.

Of course, the best way to deal with nonresponse is to prevent it through careful survey design and experimentation. Kreuter (2013) reviews methods that can be used to prevent nonresponse. Chapter 6 of Groves et al. (2011) discusses reasons for nonresponse and survey design features that can reduce it. Groves and Couper (2002) present a framework for survey design to reduce nonresponse, in the context of surveys of low-income populations, and Bethlehem et al. (2011) provide guidance for preventing and dealing with nonresponse in household surveys. The references on designing experiments and quality improvement listed in Chapter 16 will tell you how to collect data for improving the quality of survey data.

Little and Rubin (2019) and Kim and Shao (2013) are standard references on missing data (for all types of data collections, not just surveys). Groves et al. (2002) and the National Research Council (2013) discuss issues of nonresponse in social science surveys.

For more information on weighting adjustments for nonresponse, see Bethlehem (2002), Brick (2013), Haziza and Lesage (2016), Haziza and Beaumont (2017), and Valliant et al. (2018). Eltinge and Yansaneh (1997) discuss methods for choosing the number of weighting classes to use, and Mohadjer and Choudhry (2002) go through practical considerations for constructing weighting adjustment classes. Han and Valliant (2021) review research and further develop the theory for raking methods. Lavallée and Beaumont (2015) explain, intuitively, why weights work. Lohr (2019a) provides a nontechnical description of weighting methods, illustrated by the steps used to calculate weights for the National Crime Victimization Survey.

Stuart (2010) reviews a number of different methods for propensity score estimation, and Kim and Riddles (2012) present theoretical results underlying propensity score estimation for survey data. Deville and Särndal (1992), Särndal and Lundström (2005), Särndal (2007), Kott (2016), and Wu and Thompson (2020) discuss methods for adjusting for nonresponse under the unifying umbrella of calibration; Devaud and Tillé (2019) give an excellent summary of the theory behind calibration.

References for imputation include Rässler et al. (2008), Haziza (2009), Berg et al. (2016), Murray (2018), and Chen and Haziza (2019). Reiter et al. (2006) emphasize that the imputation model should incorporate features of the design. Azur et al. (2011) and van Buuren (2018) describe how to impute data using MICE. Yang and Kim (2016) review methods for fractional imputation, a variation of hot-deck imputation in which several donors each contribute a fraction of their data values to the recipient. Kim and Shao (2013) give a rigorous treatment of the theory behind response propensity modeling and imputation. See He et al. (2010) and Carpenter and Kenward (2012) for guides on implementing multiple imputation.

The American Association of Public Opinion Research (2016b) lists standard definitions and provides spreadsheets for calculating response rates. Groves (2006), Halbesleben and Whitman (2013), Lineback and Thompson (2010), and Montaquila and Olson (2012) provide comprehensive guides on performing nonresponse bias assessments.

This chapter focused on approaches to nonresponse within a design-based survey framework. An alternative approach, beyond the scope of this book, integrates nonresponse into a model-based framework for survey inference (see Sections 2.10, 3.6, 4.6, 5.6, and 6.7). In a completely model-based approach, components that account for the proposed nonresponse mechanism are added to a model for the complete data. Such an approach can be adapted to include assumptions about possible NMAR mechanisms, and has the advantage that the assumptions about nonresponse must be stated explicitly in the model and sometimes can be evaluated within the context of the model. Little and Rubin (2019) discuss likelihood-based methods for missing data in general, and Chambers et al. (2012) present methods for estimating survey quantities using maximum likelihood, without or with nonresponse. Other references are given in the "For Further Reading" section of Chapter 15.

8.10 Exercises

A. Introductory Exercises

1. The poll discussed in Example 8.15 used a stratified random sampling design, where the frame of registered voters was stratified by region, age, race, gender, political party, and 2014 turnout. Selection probabilities were higher in strata where the anticipated response propensities were lower.

(a) Any poll taken, regardless of sampling frame or mode, would have nonresponse. What are relative advantages of sampling from a list of registered voters, when compared with a random-digit-dialing survey or an online survey, for dealing with nonresponse and undercoverage bias? The disadvantages?

(b) After the weighting, the estimated percentages of women and persons in different age groups agreed with those from the model projections. Does this agreement indicate that the weighting removes nonresponse bias for estimated percentage of voters preferring the Republican candidate? Explain.

(c) For an election poll, the true population value eventually becomes known. Two months after this poll was taken, the Republican candidate lost the election by about seven percentage points. How might this information, and information from the other polls taken during 2018, be used to modify the weighting methods for future polls?

2. Investigators selected an SRS of 200 high school seniors from a population of 2000 for a survey of television-viewing habits, with an overall response rate of 75%. By checking school records, they were able to find the grade point average (GPA) for the nonrespondents, and classify the sample accordingly:

GPA	**Sample Size**	**Number of Respondents**	**Hours of Television** $\bar{y}$	s_y
3.00–4.00	75	66	32	15
2.00–2.99	72	58	41	19
Below 2.00	53	26	54	25
Total	200	150		

(a) What is the estimate for the average number of hours of television watched per week if only respondents are analyzed? What is the standard error of the estimate?

(b) Perform a χ^2 (chi-square) test for the null hypothesis that the three GPA groups have the same response rates. What do you conclude? What do your results say about the type of missing data: Do you think they are MCAR? MAR? NMAR?

(c) Perform a one-way analysis of variance to test the null hypothesis that the three GPA groups have the same mean level of television viewing. What do you conclude? Does your ANOVA indicate that GPA would be a good variable for constructing weighting cells? Why, or why not?

(d) Use the GPA classification to adjust the weights of the respondents in the sample. What is the weighting-class estimate of the average viewing time?

(e) The population counts are 700 students with GPA between 3 and 4; 800 students with GPA between 2 and 3; and 500 students with GPA less than 2. Use these population counts to construct a poststratified estimate of the mean viewing time.

(f) What other methods might you use to adjust for the nonresponse?

(g) What other variables might be collected that could be used in nonresponse models?

3. The following description and assessment of nonresponse is from a study of Hamilton, Ontario, homeowners' attitudes on composting toilets:

> The survey was carried out by means of a self-administered mail questionnaire. Twelve hundred questionnaires were sent to a randomly selected sample of house-dwellers. Follow-up thank you notes were sent a week later. In total, 329 questionnaires were returned, representing a response rate of 27%. This was deemed satisfactory since many mail surveyors consider a 15 to 20% response rate to be a good return (Wynia et al., 1993, p. 362).

Do you agree that the response rate of 27% is satisfactory? Suppose the investigators came to you for statistical advice on analyzing these data and designing a follow-up survey. What would you tell them?

4. Kosmin and Lachman (1993) had a question on religious affiliation included in 56 consecutive weekly household surveys; the subject of household surveys varied from week to week from cable TV use, to preference for consumer items, to political issues. After four callbacks, the unit nonresponse rate was 50%; an additional 2.3% refused to answer the religion question. The authors wrote:

> Nationally, the sheer number of interviews and careful research design resulted in a high level of precision ... Standard error estimates for our overall national sample show that we can be 95% confident that the figures we have obtained have an error margin, plus or minus, of less than 0.2%. This means, for example, that we are more than 95% certain that the figure for Catholics is in the range of 25.0% to 26.4% for the U.S. population (p. 286).

   (a) Critique the preceding statement.

   (b) If you anticipated item nonresponse, do you think it would be better to insert the question of interest in different surveys each week, as was done here, or to use the same set of additional questions in each survey? Explain your answer. How would you design an experiment to test your conjecture?

5. Mohadjer and Choudhry (2002) used a fictional survey of low-income families to illustrate weight adjustments. Each of the $n = 820$ families selected in a self-weighting probability sample had sampling weight 50, but not all families responded to the survey. However, auxiliary information was known about the gender and race of the person considered to be the "head" of each family, for each family in the sample. That information is summarized in Table 8.6.

TABLE 8.6
Sample size and number of respondents in weighting classes for Exercise 5.

Family Head's Gender	Family Head's Race	Sample Size	Number of Respondents	Mean Income of Respondents ($)
Male	White	160	77	1,712
Male	Nonwhite	51	35	1,509
Female	White	251	209	982
Female	Nonwhite	358	327	911

   (a) What is the response rate for each group in Table 8.6? What is the overall response rate?

   (b) Given the information in the table, do you think these two variables (family head's gender and race) are good choices for forming nonresponse-adjustment weighting classes? Explain.

(c) Calculate the nonresponse-adjusted weight for a family that responded to the survey in each of the four weighting classes. Make sure you check that the sum of the respondents' nonresponse-adjusted weights equals the sum of the sampling weights for the selected sample.

(d) Calculate the average income using the sampling weights with the respondents, and contrast that with the estimated mean income calculated with the nonresponse-adjusted weights.

6. *Continuation of Exercise 5.* Now suppose that information about employment and education status is known from an external source for the population. According to that information, given in Table 8.7, the population actually has 46,000 families, a value larger than the sum of the sampling weights in Exercise 5.

   Starting with the nonresponse-adjusted weights you developed in Exercise 5, find poststratified weights using the 4 poststrata in Table 8.7. The nonresponse-adjusted weights differ for the weighting classes used in Exercise 5, and Table 8.8 gives the number of respondents in each (gender × race × education × employment) category. Give the final weight and the estimated number of families in the population for each of these 16 categories.

TABLE 8.7
Population counts by employment and education status, for Exercise 6.

	No High School Diploma	High School Diploma	Total
Unemployed	10,596	6,966	17,562
Employed	6,313	22,125	28,438
Total	16,909	29,091	46,000

TABLE 8.8
Number of respondents in each gender × race × education × employment category, for Exercise 6. Table entries are number of respondents in each category.

Family Head's	**No High School**		**High School**	
Race, Gender	**Unemployed**	**Employed**	**Unemployed**	**Employed**
White male	16	11	12	38
Nonwhite male	9	6	5	15
White female	45	30	33	101
Nonwhite female	71	47	51	158

B. Working with Survey Data

7. Guo et al. (2016) performed a randomized experiment to explore effects of survey design features on response rates in a mail survey. Table 8.9 gives results for four of the survey variants studied, with 1,000 households randomly assigned to each. The control group was offered no incentive, the lottery group was entered in a lottery with a chance to win $100 or $1000, the cash group was sent $2 in the invitation letter, and the lottery/cash group was offered both the lottery and cash incentives. The household adult with the most recent birthday was asked to fill out the survey.

   (a) Perform a χ^2 test for the null hypothesis that the response rate is the same in all four groups.

(b) Perform a χ^2 test for the null hypothesis that the proportion of respondents who are male is the same for all four groups.

(c) Construct a one-way analysis of variance table to test the null hypothesis that respondents in all four groups have the same mean age.

(d) Which combination of incentives would you recommend, and why?

TABLE 8.9
Results of randomized experiment described in Exercise 7.

	Control	Lottery	Cash	Lottery/Cash
Number of respondents	168	198	205	281
Male	57	79	86	116
Female	111	119	119	165
Average age of respondents	53.4	51.0	50.4	51.9
Standard deviation of age	16.0	16.5	16.4	15.4

8. Leitch et al. (2018) took a stratified random sample (stratified by location and practice size) from the 988 general medical practices in the New Zealand Primary Health Organisation Database. Table 8.10 lists the outcomes.

TABLE 8.10
Results from medical practice recruitment, for Exercise 8.

Location	Size	N_h	n_h	Refused	Ineligible	Respondents
Urban	Small	255	12	3	3	6
Rural	Small	75	12	5	2	5
Urban	Medium	263	12	2	2	8
Rural	Medium	66	12	2	2	8
Urban	Large	271	12	5	0	7
Rural	Large	58	12	0	1	11

(a) Calculate the unweighted response rate RR1 for this survey.

(b) Calculate the weighted response rate RR1_w.

(c) What do the two response rates measure?

9. Calculate the response rates RR1 through RR6 for the U.S. National Intimate Partner and Sexual Violence Survey discussed in Exercise 18 of Chapter 1. Treat the 1,542 persons who were interviewed but did not answer the questions on sexual violence as "partial completes," and assume that the ratio of eligible to ineligible households is the same in the households of unknown eligibility as in the households of known eligibility. Which response rate is most appropriate for a report on statistics about sexual violence?

10. The issue of nonresponse in the Winter Break Closure Survey (in file `winter.csv`) was briefly mentioned in Exercise 23 of Chapter 3. What model is adopted for nonresponse when the formulas from stratified sampling are used to estimate the proportion of university employees who would answer "yes" to the question "Would you want to have Winter Break Closure again?" Do you think this is a reasonable model? How else might you model the effects of nonresponse in this survey? What additional information could be collected to adjust for unit nonresponse?

11. Select an SRS of 300 colleges from the college scorecard data (`college.csv`), discussed in Example 3.12. Now calculate raked weights for your sample so that the estimated totals for variables *control* and *ccsizset* agree with the marginal population totals for those variables (you may want to collapse categories that have few sample members). What are the raked weights for colleges in each combination of the raking variables? Estimate the average and 75th percentile of the out-of-state tuition fee, with the sampling weights and then with the raked weights. How much do the estimates, and their standard errors, change with the raking?

12. The ACLS survey in Example 3.4 had nonresponse. Calculate the response rate in each stratum for the survey. What model was adopted for the nonresponse in Example 3.4? Is there evidence that the nonresponse rate varies among the strata, or that it is related to the percentage female membership?

13. Gnap (1995) conducted a survey on teacher workload which was used in Exercise 15 of Chapter 5.

    (a) The original survey was intended as a one-stage cluster sample. What was the overall response rate?

    (b) Would you expect nonresponse bias in this study? If so, in which direction would you expect the bias to be? Which teachers do you think would be less likely to respond to the survey?

    (c) Gnap also collected data on a random subsample of the nonrespondents in the "large" stratum, in file `teachnr.csv`. How do the respondents and nonrespondents differ?

    (d) Is there evidence of nonresponse bias, when you compare the subsample of nonrespondents to the respondents in the original survey?

14. Not all of the parents surveyed in the study discussed in Exercise 16 of Chapter 5 returned the questionnaire. In the original sampling design, 50 questionnaires were mailed to parents of children in each school, for a total planned sample size of 500. We know that of the 9,962 children who were not immunized during the campaign, the consent form had not been returned for 6,698 of the children, the consent form had been returned but immunization refused for 2,061 of the children, and 1,203 children whose parents had consented were absent on immunization day.

    (a) Calculate the response rate for each cluster. What is the correlation of the response rate and the percentage of respondents in the school who returned the consent form? Of the response rate and the percentage of respondents in each school who refused consent?

    (b) Overall, about 67% (6698/9962) of the parents in the target population did not return the consent form. Using the data from the respondents, calculate a 95% CI for the proportion of parents in the sample who did not return the consent form. Calculate two additional interval estimates for this quantity: one assuming that the missing values are all 0's, and one assuming that the missing values are all 1's. What is the relation between your estimates and the population quantity?

    (c) Repeat part (b), examining the percentage of parents that returned the form but refused to have their children immunized.

    (d) Do you think nonresponse bias is a problem for this survey?

15. Use the data in file `agpop.csv` for this exercise. Let y_i be the value of *acres92* for unit i and x_i be the value of *acres87* for unit i. Draw an SRS of size 400. Now generate missing data from your sample by generating a standard uniform random variable U_i for each observation and deleting the observation if $16U_i \geq \ln(x_i)$.

    (a) If you ignore the missing data, do you expect the mean of y to be too large or too small?

    (b) Compute the mean from the data set with missing values. Does a 95% CI, computed ignoring the nonresponse, contain the true mean from the population?

    (c) Now poststratify the sample by region, using the stratum population sizes in Example 3.2 to adjust the sampling weight in each region. Does this appear to reduce the bias?

    (d) Try a different poststratification. This time, form 4 groups based on the value of x_i in the population, and find the number of population counties in each group. How do the results of this poststratification compare with the poststratification by region?

16. Repeat Exercise 15, using weighting class methods instead of poststratification. With weighting class methods, you adjust the weights using counts from the selected sample rather than the population.

17. Example 8.13 presented statistics from three surveys.

    (a) Using the margin of error for each survey to provide a (conservative) estimate of the standard error, for which of the quantities reported in Table 8.5 are the results from the standard survey significantly different from those of the CPS? The high-effort survey?

    (b) Which questions have the largest differences among the three survey estimates?

    (c) Why do you think the differences are small for some characteristics but large for others?

18. The probability sample for the 2009 Swedish Survey of Living Conditions was selected from adults age 16 and over in the Sweden Total Population Register, which contains nearly everyone in the country. The survey was conducted by telephone, and in the initial fieldwork period up to 30 attempts were made to reach persons selected for the sample. Then, after a break, an additional three weeks was spent following up with persons who had not yet responded, using the same survey procedures as in the initial fieldwork period.

    Lundquist and Särndal (2013) compared estimates from the survey respondents with estimates from the full sample (using values from the population register). Data file `swedishlcs.csv` shows the relative bias for three statistics—percentage of persons receiving sickness insurance *benefits*, percentage of persons who are *employed*, and average *income* (including both employment and retirement income)—at different points of the survey. The relative bias for each statistic at a time point is calculated as

    $$\frac{\text{estimated mean from respondents at that point} - \text{estimated mean from full sample}}{\text{estimated mean from full sample}}.$$

    (a) For each of the variables *benefits*, *employed*, and *income*, graph the relative bias as a function of the number of call attempts.

    (b) When is the bias smallest for each response? Do you think the three-week follow-up period helped reduce the nonresponse bias? Why, or why not?

C. Working with Theory

19. Let $Z_i = 1$ if unit i is included in the sample and 0 otherwise, with $P(Z_i = 1) = \pi_i$. Let $R_i = 1$ if unit i responds to the survey and 0 otherwise, with $P(R_i = 1) = \phi_i$ and $\bar{\phi}_{\mathcal{U}} = \sum_{i=1}^{N} \phi_i/N$. Assume R_i is independent of Z_i for each $i = 1, \ldots, N$. Let $\hat{\bar{y}}_R$ estimate the population mean $\bar{y}_{\mathcal{U}} = \sum_{i=1}^{N} y_i/N$ using only the respondents:

$$\hat{\bar{y}}_R = \frac{\sum_{i=1}^{N} Z_i R_i w_i y_i}{\sum_{i=1}^{N} Z_i R_i w_i},$$

where $w_i = 1/\pi_i$. Show that the bias of $\hat{\bar{y}}_R$ is approximately

$$E[\hat{\bar{y}}_R] - \bar{y}_{\mathcal{U}} \approx \frac{1}{N}\sum_{i=1}^{N} \frac{\phi_i y_i}{\bar{\phi}_{\mathcal{U}}} \approx \frac{\text{Cov}\,(\phi, y)}{\bar{\phi}_{\mathcal{U}}}, \tag{8.15}$$

where $\text{Cov}\,(\phi, y) = \sum_{i=1}^{N}(\phi_i - \bar{\phi}_{\mathcal{U}})(y_i - \bar{y}_{\mathcal{U}})/(N-1)$. As a consequence, the nonresponse bias is approximately zero if either (1) $\phi_i = \bar{\phi}_{\mathcal{U}}$ for all i, that is, the response propensity is the same for all units, or (2) the propensity to respond is uncorrelated with the response y_i (Tremblay, 1986).

20. Let Z_i and R_i be as defined in Exercise 19. Divide the sample into C weighting classes and define $x_{ic} = 1$ if unit i is in class c and 0 otherwise. Let $\bar{\phi}_c = \sum_{i=1}^{N} \phi_i x_{ic}/\sum_{i=1}^{N} x_{ic}$,

$$\hat{\phi}_c = \frac{\sum_{i=1}^{N} Z_i R_i w_i x_{ic}}{\sum_{i=1}^{N} Z_i w_i x_{ic}},$$

and

$$\hat{t}_{\hat{\phi}} = \sum_{i=1}^{N} Z_i R_i w_i y_i \sum_{c=1}^{C} \frac{x_{ic}}{\hat{\phi}_c}.$$

Show that if the weighting classes are sufficiently large,

$$\text{Bias}\,(\hat{t}_{\hat{\phi}}) \approx \sum_{c=1}^{C}\sum_{i=1}^{N} x_{ic}\phi_i(y_i - \bar{y}_{c\mathcal{U}})/\bar{\phi}_c,$$

where $\bar{y}_{c\mathcal{U}} = \sum_{i=1}^{N} y_i x_{ic}/\sum_{i=1}^{N} x_{ic}$. Thus, the weighting class adjustments for nonresponse in Section 8.5 produce an approximately unbiased estimator if (a) $\phi_i = \bar{\phi}_c$ for all units in class c, (b) $y_i = \bar{y}_{c\mathcal{U}}$ for all units in class c, or (c) within each class c, the propensity to respond is uncorrelated with y_i.

21. *Bias of poststratified estimator.* Let Z_i and R_i be as defined in Exercise 19. Divide the sample into H poststrata. Let N_h be the number of population units in poststratum h, obtained from an independent source such as a population register or census. Show that if the poststrata are sufficiently large,

$$\text{Bias}\,(\hat{t}_{\text{post}}) \approx \sum_{h=1}^{H} \frac{N_h}{N}\text{Cov}_h(\phi, y)/\bar{\phi}_h,$$

where $\bar{\phi}_h = \sum_{i=1}^N x_{hi}\phi_i/N_h$ and $\text{Cov}_h(\phi, y)$ is the population covariance of the y_is and ϕ_is for population units in poststratum h.

22. *Effect of weighting class adjustment on variances.* Suppose that an SRS of size n is taken. Let $Z_i = 1$ if unit i is included in the sample and 0 otherwise, with $P(Z_i = 1) = n/N$. Two weighting classes are used to adjust for nonresponse; define $x_i = 1$ if unit i is in class 1 and 0 if unit i is in class 2. Let $R_i = 1$ if unit i responds to the survey and 0 otherwise. Assume that the R_is are independent Bernoulli random variables with $P(R_i = 1) = x_i\phi_1 + (1 - x_i)\phi_2$, and that R_i is independent of $Z_1, \ldots, Z_N$. The sample sizes in the two classes are $n_1 = \sum_{i=1}^N Z_i x_i$ and $n_2 = \sum_{i=1}^N Z_i(1 - x_i)$; note that n_1 and n_2 are random variables. Similarly, the number of respondents in the two classes are $n_{1R} = \sum_{i=1}^N Z_i R_i x_i$ and $n_{2R} = \sum_{i=1}^N Z_i R_i(1 - x_i)$. Assume the number of respondents in each group is sufficiently large so that $E[n_c/n_{cR}] \approx 1/\phi_c$ for $c = 1, 2$. With these assumptions, the weighting class adjusted estimator of the mean,

$$\hat{\bar{y}}_{\hat{\phi}} = \frac{n_1}{n}\frac{1}{n_{1R}}\sum_{i=1}^N Z_i R_i x_i y_i + \frac{n_2}{n}\frac{1}{n_{2R}}\sum_{i=1}^N Z_i R_i(1 - x_i)y_i$$

is approximately unbiased for the population mean $\bar{y}_{\mathcal{U}}$ (see Exercise 20). Find the approximate variance of $\hat{\bar{y}}_{\hat{\phi}}$. HINT: Use Property 5 of Table A.2.

23. Equation (8.3) defined the response indicators for the sampled units. Extend that notion to the population by letting $R_i = 1$ if unit i would respond to the survey if selected for the sample, with $\phi_i = P(R_i = 1)$. Let $x_{ic} = 1$ if unit i is in weighting class c and 0 otherwise, for $i = 1, \ldots, N$. Show that, if Z_i and R_i are independent, the weighted response rate in class c, from (8.8), is an approximately unbiased estimator of $\sum_{i=1}^N (x_{ic}\phi_i)/\sum_{i=1}^N x_{ic}$.

24. *The Hartley (1946) and Politz–Simmons (1949) method.* Suppose that all calls are made during Monday through Friday evenings. Each respondent is asked whether he or she was at home at the time of the interview, on each of the four preceding weeknights. Respondent i replies that she was home k_i of the four nights. It is then assumed that the probability of response is proportional to the number of nights at home during interviewing hours, so the probability of response is estimated by $\hat{\phi}_i = (k_i + 1)/5$. Let

$$\hat{\bar{y}}_{\text{HPS}} = \frac{\sum_{i \in \mathcal{S}} w_i y_i/\hat{\phi}_i}{\sum_{i \in \mathcal{S}} w_i/\hat{\phi}_i}.$$

   (a) Under what circumstances would you expect the method to reduce bias due to nonresponse? What assumptions must be made for the estimator to be approximately unbiased?

   (b) What are some potential drawbacks of the method for use in practice? How does it adjust for persons who were not at home during any of the five nights, or who refused to participate in the survey?

25. *R-indicators.* Section 8.8.3 discussed studying internal variation in the data collection to assess potential nonresponse bias. Schouten et al. (2009) noted that the response rate may be unrelated to nonresponse bias, and proposed using internal variation in response propensities as a supplemental measure of survey quality. The R-indicator (R stands for

"representativeness") for the population values is defined as:

$$R(\boldsymbol{\phi}) = 1 - 2\sqrt{\frac{1}{N-1}\sum_{i=1}^{N}\left(\phi_i - \frac{1}{N}\sum_{j=1}^{N}\phi_j\right)^2}. \tag{8.16}$$

(a) Show that $0 \leq R(\boldsymbol{\phi}) \leq 1$, and that $R(\boldsymbol{\phi}) = 1$ if and only if all response propensities are equal.

(b) Schouten et al. (2009) argued that when $R(\boldsymbol{\phi}) = 1$, there is no bias from nonresponse. Why is this true?

(c) In practice, the true response propensities are unknown, and the R-indicator is estimated by

$$R(\hat{\boldsymbol{\phi}}) = 1 - 2\sqrt{\frac{1}{N-1}\sum_{i\in\mathcal{S}} w_i\left(\hat{\phi}_i - \frac{1}{N}\sum_{j\in\mathcal{S}} w_j\hat{\phi}_j\right)^2},$$

where w_i is the sampling weight for unit i and $\hat{\phi}_i$ is an estimate of the response propensity for unit i based on auxiliary information known for all sample units. Calculate $R(\hat{\boldsymbol{\phi}})$ for the response propensities in Table 8.3, using $N = 150{,}104$.

(d) There is no nonresponse bias when $R(\boldsymbol{\phi}) = 1$. But there can be nonresponse bias when the response propensities are estimated from the data using a statistical model. Give an example where $R(\hat{\boldsymbol{\phi}}) = 1$, yet $\sum_{i\in\mathcal{S}} w_i y_i/\hat{\phi}_i$ is biased for estimating $t = \sum_{i=1}^{N} y_i$.

D. Projects and Activities

26. *Trucks.* Use the data from the Vehicle Inventory and Use Survey (VIUS), described in Exercise 49 of Chapter 3, for this exercise.

    The variable giving the miles per gallon for each vehicle (*MPG*) has missing data. Develop a set of weights for the set of trucks that have values for *MPG* (the "respondents") that sum to the total number of trucks in each stratum.

27. Find the methodology report of a recent poll. How do they describe the sources of error in the survey? Do they give the nonresponse rate, or reference a document that details the treatment of nonresponse? How do they adjust for nonresponse in their estimates?

28. Find an example of a survey in a popular newspaper or magazine. Is the nonresponse rate given? If so, how was it calculated? How do you think the nonresponse might have affected the conclusions of the survey? Give suggestions for how the journalist could discuss nonresponse problems in the article.

29. Find an example of a survey in a scholarly journal. How did the authors calculate the nonresponse rate? How did the survey deal with nonresponse? How do you think the nonresponse might have affected the conclusions of the study? Do you think the authors adequately account for potential nonresponse biases? What suggestions do you have for future studies?

30. The U.S. Survey of Doctorate Recipients (National Science Foundation, 2019) provides information about the characteristics and employment of persons who hold doctorate degrees in science, engineering, and health fields. Read the technical notes that accompany one of the data sets. What is the response rate, and what is done to adjust for nonresponse? Do you think that nonresponse bias is a problem for this survey?

31. How did the survey you critiqued in Exercise 33 of Chapter 7 deal with nonresponse? In your opinion, did the investigators adequately address the problems of nonresponse? What suggestions do you have for improvement?

32. Answer the questions in Exercise 31 for the survey you examined in Exercise 34 of Chapter 7.

33. *Activity for course project.* Return to the data you chose in Exercise 35 of Chapter 7. What kinds of nonresponse occur in your data set? How does the survey define nonresponse rate, and what are the nonresponse rates for the survey? What methods are used to try to adjust for the nonresponse?

9

Variance Estimation in Complex Surveys

And infinite in earth's dominions
Arts, climates, wonders, and opinions.

—Phyllis McGinley, In Praise of Diversity

Population means and totals are easily estimated using weights. Estimating variances is more intricate: In Chapter 7, we noted that in a complex survey with several levels of stratification and clustering, variances for estimated means and totals are calculated at each level and then combined as the survey design is ascended. Poststratification and nonresponse adjustments also affect the variance.

In previous chapters, we have presented and derived variance formulas for a variety of sampling plans. Some of the variance formulas, such as those for simple random samples (SRSs), are relatively simple. Other formulas, such as $\hat{V}(\hat{t})$ from a two-stage cluster sample without replacement, are more complicated. All work for estimating variances of estimated totals. But we often want to estimate other quantities from survey data for which we have presented no variance formula. For example, in Chapter 4, we derived an approximate variance for a ratio of two means when an SRS is taken. What if you want to estimate a ratio, but the survey is not an SRS? How would you estimate the variance?

This chapter describes several methods for estimating variances of estimated totals and other statistics from complex surveys. Section 9.1 describes the commonly used linearization method for calculating variances of estimators. Sections 9.2 and 9.3 present random group and resampling methods for calculating variances of linear and nonlinear statistics. Section 9.4 describes the calculation of generalized variance functions, and Section 9.5 describes constructing confidence intervals (CIs).

9.1 Linearization (Taylor Series) Methods

Most of the variance formulas in Chapters 2 through 6 were for estimators of means and totals. Those formulas can be used to find variances for any linear combination of estimated means and totals. Let y_{ij} be the response of unit i to item j. Suppose $\hat{t}_1, \ldots, \hat{t}_k$ are unbiased estimators of the k population totals $t_1, \ldots, t_k$, with $\hat{t}_j = \sum_{i \in \mathcal{S}} w_i y_{ij}$. Then, for any constants $a_1, \ldots, a_k$, we can define a new variable

$$q_i = \sum_{j=1}^{k} a_j y_{ij}$$

so that

$$\hat{t}_q = \sum_{i \in \mathcal{S}} w_i q_i = \sum_{j=1}^{k} a_j \hat{t}_j$$

DOI: 10.1201/9780429298899-9

and

$$V\left(\sum_{j=1}^{k} a_j \hat{t}_j\right) = V(\hat{t}_q) = \sum_{j=1}^{k} a_j^2 V\left(\hat{t}_j\right) + 2\sum_{j=1}^{k-1}\sum_{l=j+1}^{k} a_j a_l \operatorname{Cov}\left(\hat{t}_j, \hat{t}_l\right). \tag{9.1}$$

Thus, if t_1 is the total number of dollars robbery victims reported stolen, t_2 is the number of days of work they missed because of the crime, and t_3 is their total medical expenses, one measure of financial consequences of robbery (assuming \$150 per day of work lost) might be $\hat{t}_1 + 150\,\hat{t}_2 + \hat{t}_3$. By (9.1), the variance is

$$\begin{aligned} V(\hat{t}_1 + 150\,\hat{t}_2 + \hat{t}_3) &= V(\hat{t}_q) \\ &= V(\hat{t}_1) + 150^2 V(\hat{t}_2) + V(\hat{t}_3) \\ &\quad + 300 \operatorname{Cov}(\hat{t}_1, \hat{t}_2) + 2 \operatorname{Cov}(\hat{t}_1, \hat{t}_3) + 300 \operatorname{Cov}(\hat{t}_2, \hat{t}_3), \end{aligned}$$

where $q_i = y_{i1} + 150y_{i2} + y_{i3}$ is the financial loss from robbery for person i.

Suppose, though, that we are interested in the proportion of total loss accounted for by the stolen property, t_1/t_q. This is not a linear statistic, as t_1/t_q cannot be expressed in the form $a_1t_1 + a_2t_q$ for constants a_1 and a_2. But Taylor's theorem from calculus allows us to **linearize** a smooth nonlinear function $h(t_1, t_2, \ldots, t_k)$ of the population totals; Taylor's theorem gives the constants $a_0, a_1, \ldots, a_k$ so that

$$h(t_1, \ldots, t_k) \approx a_0 + \sum_{j=1}^{k} a_j t_j.$$

Then $V[h(\hat{t}_1, \ldots, \hat{t}_k)]$ may be approximated by $V(\sum_{j=1}^{k} a_j \hat{t}_j)$, which we know how to calculate using (9.1).

Taylor series approximations have long been used in statistics to calculate approximate variances. Woodruff (1971) illustrated their use in complex surveys. Binder (1983) gave a rigorous treatment of Taylor series methods for complex surveys and told how to use linearization when the parameter of interest θ solves $h(\theta, t_1, \ldots, t_k) = 0$, but θ is an implicit function of $t_1, \ldots, t_k$.

Example 9.1. The quantity $\theta = p(1-p)$, where p is a population proportion, may be estimated by $\hat{\theta} = \hat{p}(1-\hat{p})$. Assume that $\hat{p}$ is an unbiased estimator of p and that $V(\hat{p})$ is known. Let $h(x) = x(1-x)$, so $\theta = h(p)$ and $\hat{\theta} = h(\hat{p})$. Now h is a nonlinear function of x, but the function can be approximated at any nearby point a by the tangent line to the function; the slope of the tangent line is given by the derivative, as illustrated in Figure 9.1.

The first-order version of Taylor's theorem states that if the second derivative of h is continuous, then

$$h(x) = h(a) + h'(a)(x-a) + \int_a^x (x-t)h''(t)dt;$$

under conditions commonly satisfied in statistics, the last term is small relative to the first two and we use the approximation

$$\begin{aligned} h(\hat{p}) &\approx h(p) + h'(p)(\hat{p} - p) \\ &= p(1-p) + (1-2p)(\hat{p}-p). \end{aligned}$$

Then,

$$V[h(\hat{p})] \approx (1-2p)^2 V(\hat{p} - p),$$

and $V(\hat{p})$ is known, so the approximate variance of $h(\hat{p})$ can be estimated by

$$\hat{V}[h(\hat{p})] = (1-2\hat{p})^2 \hat{V}(\hat{p}). \quad \blacksquare$$

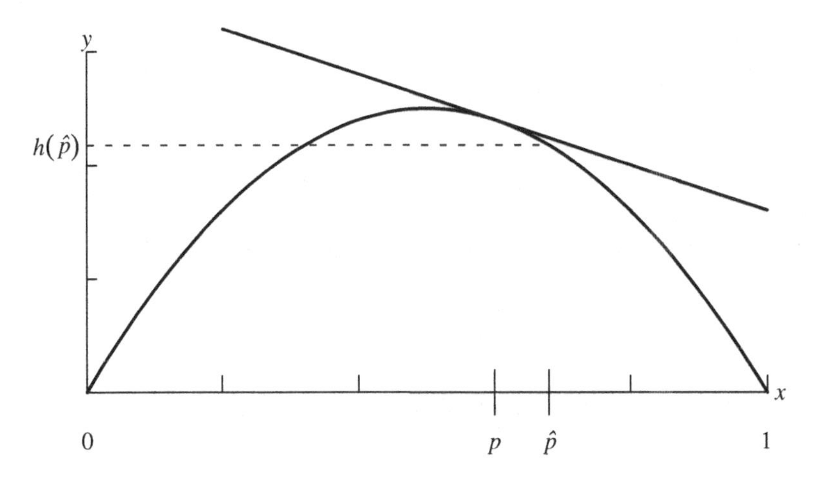

FIGURE 9.1
The function $h(x) = x(1-x)$, along with the tangent to the function at point p. If $\hat{p}$ is close to p, then $h(\hat{p})$ will be close to the tangent line. The slope of the tangent line is $h'(p) = 1 - 2p$.

Steps for constructing a linearization estimator of the variance

1. Express the quantity of interest as a twice differentiable function of means or totals of variables measured or computed in the sample. In general, $\theta = h(t_1, t_2, \ldots, t_k)$ or $\theta = h(\bar{y}_{1\mathcal{U}}, \ldots, \bar{y}_{k\mathcal{U}})$. In Example 9.1, $\theta = h(\bar{y}_{\mathcal{U}}) = h(p) = p(1-p)$ and $\hat{\theta} = h(\hat{p})$.

2. Find the partial derivative of h with respect to each argument. The partial derivatives, evaluated at the population quantities, form the linearizing constants a_j.

3. Apply Taylor's theorem to linearize the estimate:
$$h(\hat{t}_1, \hat{t}_2, \ldots, \hat{t}_k) \approx h(t_1, t_2, \ldots, t_k) + \sum_{j=1}^{k} a_j(\hat{t}_j - t_j),$$
where
$$a_j = \left.\frac{\partial h(c_1, c_2, \ldots, c_k)}{\partial c_j}\right|_{t_1, t_2, \ldots, t_k}.$$

4. Define the new variable q by
$$q_i = \sum_{j=1}^{k} a_j y_{ij}.$$
Now find the estimated variance of $\hat{t}_q = \sum_{i \in \mathcal{S}} w_i q_i$, substituting estimators for unknown population quantities. This will generally approximate the variance of $\hat{\theta} = h(\hat{t}_1, \ldots, \hat{t}_k)$.

Example 9.2. *Ratio estimation.* We used linearization methods to approximate the variance of the ratio and regression estimators in Chapter 4. In Chapter 4, we used an SRS, estimator $\hat{B} = \bar{y}/\bar{x} = \hat{t}_y/\hat{t}_x$, and the approximation
$$\hat{B} - B = \frac{\bar{y} - B\bar{x}}{\bar{x}} \approx \frac{\bar{y} - B\bar{x}}{\bar{x}_{\mathcal{U}}} = \sum_{i \in \mathcal{S}} \frac{y_i - Bx_i}{n\bar{x}_{\mathcal{U}}}.$$

In (4.12), we estimated the variance by

$$\hat{V}(\hat{B}) = \left(1 - \frac{n}{N}\right) \frac{s_e^2}{n\bar{x}^2},$$

where s_e^2 is the sample variance of the residuals $e_i = y_i - \hat{B}x_i$. Essentially, we used Taylor's theorem to obtain this estimator.

The following steps give the variance of the ratio estimator for general probability sampling designs.

1. Express B as a function of the population totals. Let $h(c, d) = d/c$, so

$$B = h(t_x, t_y) = \frac{t_y}{t_x} \quad \text{and} \quad \hat{B} = h(\hat{t}_x, \hat{t}_y) = \frac{\hat{t}_y}{\hat{t}_x}.$$

Assume that the estimators $\hat{t}_x$ and $\hat{t}_y$ are unbiased.

2. The partial derivatives are

$$\frac{\partial h(c,d)}{\partial c} = -\frac{d}{c^2} \quad \text{and} \quad \frac{\partial h(c,d)}{\partial d} = \frac{1}{c};$$

evaluated at $c = t_x$ and $d = t_y$, these are $-t_y/t_x^2$ and $1/t_x$.

3. By Taylor's theorem,

$$\begin{aligned}\hat{B} &= h(\hat{t}_x, \hat{t}_y) \\ &\approx h(t_x, t_y) + \left.\frac{\partial h(c,d)}{\partial c}\right|_{t_x,t_y} (\hat{t}_x - t_x) + \left.\frac{\partial h(c,d)}{\partial d}\right|_{t_x,t_y} (\hat{t}_y - t_y).\end{aligned}$$

Using the partial derivatives from Step 2,

$$\hat{B} - B \approx -\frac{t_y}{t_x^2}(\hat{t}_x - t_x) + \frac{1}{t_x}(\hat{t}_y - t_y).$$

4. The approximate mean squared error (MSE) of $\hat{B}$ is

$$E\left[(\hat{B} - B)^2\right] \approx E\left[\left\{-\frac{t_y}{t_x^2}(\hat{t}_x - t_x) + \frac{1}{t_x}(\hat{t}_y - t_y)\right\}^2\right] \tag{9.2}$$

$$= \frac{1}{t_x^2}\left[B^2 V(\hat{t}_x) + V(\hat{t}_y) - 2B\,\text{Cov}(\hat{t}_x, \hat{t}_y)\right]. \tag{9.3}$$

Substitute estimators of the unknown quantities into (9.2) to define

$$q_i = \frac{1}{\hat{t}_x}\left[y_i - \hat{B}x_i\right] = \frac{1}{\hat{t}_x} e_i, \tag{9.4}$$

and find $\hat{V}(\hat{B}) = \hat{V}(\hat{t}_q) = \hat{V}(\hat{t}_e)/\hat{t}_x^2$ using the survey design. For an SRS, this results in the variance estimator in (4.12). ■

The method in Example 9.2 requires substituting estimators for quantities such as B. Note that alternative variance estimators can be derived from (9.2). In particular, if t_x is known, it can be used in place of an estimator $\hat{t}_x$ in the denominator of q_i, giving $\hat{V}_2(\hat{B}) = \hat{V}(\hat{t}_e)/t_x^2$. The estimators $\hat{V}(\hat{B})$ and $\hat{V}_2(\hat{B})$ are asymptotically equivalent since we expect $t_x/\hat{t}_x \approx 1$ for large sample sizes. For small samples, $\hat{V}(\hat{B})$ works slightly better than $\hat{V}_2(\hat{B})$ in many situations (see Exercise 21 of Chapter 4). An alternative procedure for deriving linearization variance estimators that results in a unique estimator is discussed in Exercise 29.

Advantages. If the partial derivatives are known, linearization almost always gives a variance estimate for a statistic and can be applied in general sampling designs. Linearization methods have been used for a long time in statistics, and the theory is well developed. Survey software will calculate linearization variance estimates for many nonlinear functions of interest such as ratios and regression coefficients.

Disadvantages. The linearization method works only for statistics that can be expressed as differentiable functions of population means or totals. A different variance estimation method must be used for statistics such as quantiles that cannot be expressed in that form. Calculations can be messy, and the method is sometimes difficult to apply for complex functions involving weights. You must either find analytical expressions for the partial derivatives of h or calculate the partial derivatives numerically. A separate variance formula is needed for each statistic that is estimated, and that can require much special programming; a different method is needed for each statistic. The accuracy of the linearization approximation depends on the sample size—the variance estimator can be biased downwards in small samples.

9.2 Random Group Methods

Resampling and replication variance estimation methods allow you to calculate variance estimates without having to find partial derivatives or compute formulas. They specify drawing multiple replicate subsamples from the full sample and using the variability among those subsamples to estimate the variances of statistics computed from the full sample. In essence, replication methods substitute computation for the theoretical work of calculating a separate variance formula for each statistic—all the theoretical work has already been done in the general theorems that justify the methods.

In this section, we look at a simple resampling method to see how replication methods work, then move on in Section 9.3 to methods that are more commonly used in practice.

9.2.1 Replicating the Survey Design

Suppose the basic survey design is replicated independently R times. *Independently* here means that each of the R sets of random variables used to select the sample is independent of the other sets—after each sample is drawn, the sampled units are replaced in the population so they are available for later samples. Then the R replicate samples produce R independent estimates of the quantity of interest; the variability among those estimates can be used to estimate the variance of $\hat{\theta}$. Mahalanobis (1939, 1946) described early uses of the method, which he called "replicated networks of sample units" and "interpenetrating sampling."

Let

$$\begin{aligned}\theta &= \text{ parameter of interest}\\ \hat{\theta}_r &= \text{ estimate of } \theta \text{ calculated from } r\text{th replicate}\\ \tilde{\theta} &= \frac{1}{R}\sum_{r=1}^{R}\hat{\theta}_r.\end{aligned}$$

The values $\hat{\theta}_1, \ldots, \hat{\theta}_R$ are independent because the samples are selected independently, and they are identically distributed because the same sampling design is used to select each sample. If $\hat{\theta}_r$ is an unbiased estimator of θ, so is $\tilde{\theta}$, and

$$\hat{V}_1(\tilde{\theta}) = \frac{1}{R}\frac{1}{R-1}\sum_{r=1}^{R}\left(\hat{\theta}_r - \tilde{\theta}\right)^2 \tag{9.5}$$

is an unbiased estimator of $V(\tilde{\theta})$. Note that $\hat{V}_1(\tilde{\theta})$ is the sample variance of the R independent estimators of θ divided by R—the usual estimator of the variance of a sample mean.

Example 9.3. Suppose we want to estimate the ratio of nonresident tuition to resident tuition for the population of 500 public colleges and universities in file `college.csv` (see Example 3.12). In a typical implementation of the random group method, independent samples are chosen using the same design from the population, and $\hat{\theta}$ is found for each sample.

Let's take five SRSs, each with $n = 10$, from the population. Each of the five SRSs is drawn without replacement, but each is drawn from the full population so a college could appear in more than one replicate sample. Indeed, the data in file `collegerg.csv` contains the University of South Carolina at Beaufort twice—once in replicate 2 and again in replicate 3.

TABLE 9.1
Summary statistics for five replicate SRSs of colleges in Example 9.3.

Replicate Number, r	Average In-state Tuition, $\bar{x}_r$	Average Out-of-state Tuition, $\bar{y}_r$	$\hat{\theta}_r$
1	8913.3	21614.7	2.4250
2	9542.0	21497.5	2.2529
3	10210.6	21323.4	2.0884
4	9004.7	18469.0	2.0510
5	9467.1	22844.0	2.4130

Table 9.1 shows the value of $\bar{x}_r$ (average of in-state tuitions for colleges in replicate r), $\bar{y}_r$ (average of out-of-state tuitions for colleges in replicate r), and $\hat{\theta}_r = \bar{y}_r/\bar{x}_r$ for each replicate sample, r. The sample average of the five independent estimates of θ is $\tilde{\theta} = 2.246$. The sample standard deviation is 0.175, so the standard error of $\tilde{\theta}$ is $0.175/\sqrt{5} = 0.0784$ and a 95% CI for the ratio is

$$2.247 \pm 2.78(0.0784) = [2.03, 2.46],$$

where 2.78 is the t critical value with 4 degrees of freedom (df). ■

Note that the variance calculation in this example relies only on the simple formula for the variance of a sample mean. You do not need to derive a mathematical formula giving the variance of a ratio, a correlation coefficient, a median, or another statistic—Equation 9.5 gives the variance estimate for almost any statistic that you would want to calculate.

There is a tradeoff between the number of replicates R and the number of observations per replicate n. For a statistic such as a ratio, whose sampling distribution is skewed, it is desirable to have a large value of n so the $\hat{\theta}_r$'s are approximately unbiased. On the other hand, if nR, the total number of observations, is fixed by the budget, a large n means R is small and the t critical value has few df. In Example 9.3, the small number of replicates causes the CI to be wider than it would be if more replicate samples were taken.

9.2.2 Dividing the Sample into Random Groups

In practice, subsamples are usually not drawn independently, but the complete sample is selected according to the survey design. The sample is then divided among R groups so that each group forms a miniature version of the survey, mirroring the sample design. The groups are then treated as though they are independent replicates of the basic survey design. This method was first described by Hansen et al. (1953a, p. 440).

If the sample is an SRS of size n, the groups are formed by randomly apportioning the n observations into R groups, each of size n/R. These pseudo-random groups are not quite independent replicates because an observation unit can only appear in one of the groups; if the population size is large relative to the sample size, however, the groups can be treated as though they are independent replicates.

In a cluster sample, the psus are randomly divided among the R groups. The psu takes all its observation units with it to the random group, so each random group is still a cluster sample. In a stratified multistage sample, a random group contains a sample of psus from each stratum. Note that if k psus are sampled in the smallest stratum, at most k random groups can be formed.

If θ is a nonlinear quantity, $\tilde{\theta}$ will not, in general, be the same as $\hat{\theta}$, the estimator calculated directly from the complete sample. For example, in ratio estimation, $\tilde{\theta} = (1/R)\sum_{r=1}^{R} \hat{\bar{y}}_r/\hat{\bar{x}}_r$, while $\hat{\theta} = \hat{\bar{y}}/\hat{\bar{x}}$. Usually, $\hat{\theta}$ is a more stable estimator than $\tilde{\theta}$. Sometimes $\hat{V}_1(\tilde{\theta})$ is used to estimate $V(\hat{\theta})$, although it is an overestimate. Another estimator of the variance is slightly larger but is often used:

$$\hat{V}_2(\hat{\theta}) = \frac{1}{R}\frac{1}{R-1}\sum_{r=1}^{R}(\hat{\theta}_r - \hat{\theta})^2. \tag{9.6}$$

Example 9.4. The 1987 Survey of Youth in Custody (Beck et al., 1988; U.S. Department of Justice, 1989) sampled juveniles and young adults in long-term, state-operated juvenile institutions. Residents of facilities were interviewed about family background, previous criminal history, and drug and alcohol use.

The 206 facilities in the sampling frame were divided into 16 strata by number of residents. Each of the 11 facilities with 360 or more residents formed its own stratum (strata 6–16); the facility was sampled with certainty and its residents were subsampled. The remaining 195 facilities were divided among strata 1–5. In those strata, facilities were sampled with probability proportional to number of residents and then residents were subsampled with predetermined sampling fractions. Thus, in strata 1–5, facilities serve as the psus; in strata 6–16, residents are the psus.

For variance estimation using the random group method, it was desired that each random group be a miniature of the sampling design. In strata 6–16, random group numbers were assigned as follows: The first resident selected from the facility was assigned a number between 1 and 7. Let's say the first resident was assigned number 6. Then the second resident in that facility would be assigned number 7, the third resident 1, the fourth resident 2, and so on. In strata 1–5, all residents in a facility (psu) were assigned to the same random group. Thus for the seven facilities sampled in stratum 2, all residents in facility 33 were assigned random group number 1, all residents in facility 9 were assigned random group number 2, and so on. Only seven random groups could be formed because strata 2 through 5 each have seven psus.

Each random group had the same basic design as the original sample. Random group 1, for example, formed a stratified sample in which a (roughly) random sample of residents was taken from the facilities in strata 6–16, and an unequal-probability sample of facilities was taken from each of strata 1–5. Selected variables from the survey are in file `syc.csv`.

The following table shows estimates of the mean age of residents for each random group; each estimate was calculated using

$$\hat{\theta}_r = \frac{\sum w_i y_i}{\sum w_i},$$

where w_i is the final weight for resident i, and the summations are over observations in random group r.

Random Group	1	2	3	4	5	6	7
$\hat{\theta}_r$	16.55	16.66	16.83	16.06	16.32	17.03	17.27

The seven estimates of θ are treated as independent observations, so $\tilde{\theta} = \sum_{r=1}^{7} \hat{\theta}_r / 7 =$ 16.67 and

$$\hat{V}_1(\tilde{\theta}) = \frac{1}{7}\left\{\frac{1}{6}\sum_{r=1}^{7}(\hat{\theta}_r - \tilde{\theta})^2\right\} = \frac{0.1704}{7} = 0.024.$$

Using the entire data set, we calculate $\hat{\theta} = 16.64$ with

$$\hat{V}_2(\hat{\theta}) = \frac{1}{7}\left\{\frac{1}{6}\sum_{r=1}^{7}(\hat{\theta}_r - \hat{\theta})^2\right\} = \frac{0.1716}{7} = 0.025.$$

We can use either $\hat{V}_1$ or $\hat{V}_2$ to calculate CIs; using $\hat{V}_1(\hat{\theta})$, a 95% CI for mean age is

$$16.64 \pm 2.45\sqrt{0.025} = [16.3, 17.0]$$

(2.45 is the t critical value with 6 df, where the df is calculated as the number of random groups minus 1). ■

Advantages. No special software is necessary to calculate the variance estimate, and it is easy to calculate and explain. The method is well-suited to multiparameter or nonparametric problems. It can be used to estimate variances for quantiles and nonsmooth functions as well as variances of smooth functions of the population totals.

Disadvantages. The number of random groups is often small—this gives imprecise estimates of the variances (see Exercise 22). If $\hat{\theta}$ is a nonlinear statistic, $\tilde{\theta}$ can have large bias if the number of observations in each group is small. Generally, one would like at least ten random groups to obtain a more stable estimate of the variance with more df. Setting up the random groups can be difficult in complicated designs, as each random group must have the same design structure as the complete survey.

The survey design may limit the number of random groups that can be constructed; if two psus are selected in each stratum, then only two random groups can be formed. Because of this limitation on the number of groups that can be formed, most stratified multistage surveys use one of the methods in Section 9.3 for replication variance estimation. These methods extend the ideas underlying the random group method.

9.3 Resampling and Replication Methods

Random group methods are easy to compute and explain but are unstable if a complex sample can only be split into a small number of groups. Resampling methods treat the sample as if it were itself a population; we take different samples from this new "population" and use the subsamples to estimate the variance. All of the methods in this section calculate variance estimates for a sample in which psus are sampled with replacement. If psus are sampled without replacement, these methods may still be used but are expected to overestimate the variance and result in conservative CIs, as discussed in Section 6.4.3.

9.3.1 Balanced Repeated Replication (BRR)

Some surveys are stratified to the point that only two psus are selected from each stratum. This gives the highest degree of stratification possible while still allowing calculation of variance estimates in each stratum.

BRR in a stratified random sample. We illustrate BRR for a problem we already know how to solve—calculating the variance for $\bar{y}_{\text{str}}$ from a stratified random sample. More complex statistics from stratified multistage samples are discussed later in this section.

Suppose an SRS of two observation units is chosen from each of seven strata. We arbitrarily label one of the sampled units in stratum h as y_{h1}, and the other as y_{h2}. The sampled values are given in Table 9.2.

TABLE 9.2
A small stratified random sample used to illustrate BRR.

Stratum	N_h/N	y_{h1}	y_{h2}	$\bar{y}_h$	$y_{h1} - y_{h2}$
1	0.30	2,000	1,792	1,896	208
2	0.10	4,525	4,735	4,630	−210
3	0.05	9,550	14,060	11,805	−4,510
4	0.10	800	1,250	1,025	−450
5	0.20	9,300	7,264	8,282	2,036
6	0.05	13,286	12,840	13,063	446
7	0.20	2,106	2,070	2,088	36

The estimated population mean is

$$\bar{y}_{\text{str}} = \sum_{h=1}^{H} \frac{N_h}{N} \bar{y}_h = 4451.7.$$

Ignoring the finite population corrections (fpcs) in (3.6) gives the variance estimator

$$\hat{V}_{\text{str}}(\bar{y}_{\text{str}}) = \sum_{h=1}^{H} \left(\frac{N_h}{N}\right)^2 \frac{s_h^2}{n_h};$$

when $n_h = 2$, as here, $s_h^2 = (y_{h1} - y_{h2})^2/2$, so

$$\hat{V}_{\text{str}}(\bar{y}_{\text{str}}) = \sum_{h=1}^{H} \left(\frac{N_h}{N}\right)^2 \frac{(y_{h1} - y_{h2})^2}{4}.$$

Here, $\hat{V}_{\text{str}}(\bar{y}_{\text{str}}) = 55{,}892.75$. This may overestimate the variance if sampling is without replacement.

To use the random group method, we would randomly select one of the observations in each stratum for group 1 and assign the other to group 2. The groups in this situation are half-samples. For example, group 1 might consist of $\{y_{11}, y_{22}, y_{32}, y_{42}, y_{51}, y_{62}, y_{71}\}$ and group 2 of the other seven observations. Then,

$$\hat{\theta}_1 = (0.3)(2000) + (0.1)(4735) + \cdots + (0.2)(2106) = 4824.7$$

and

$$\hat{\theta}_2 = (0.3)(1792) + (0.1)(4525) + \cdots + (0.2)(2070) = 4078.7.$$

The random group estimate of the variance—in this case, 139,129—has only 1 df for a two-psu-per-stratum design and is unstable in practice. If a different assignment of observations to groups had been made—had, for example, group 1 consisted of y_{h1} for strata 2, 3, and 5 and y_{h2} for strata 1, 4, 6, and 7—then $\hat{\theta}_1 = 4508.6$, $\hat{\theta}_2 = 4394.8$, and the random group estimate of the variance would have been 3238.

McCarthy (1966, 1969) noted that altogether 2^H possible half-samples could be formed, and suggested using a balanced sample of the 2^H possible half-samples to estimate the variance. **Balanced repeated replication** uses the variability among R replicate half-samples that are selected in a balanced way to estimate the variance of $\hat{\theta}$.

To define balance, let's introduce the following notation. Half-sample r can be defined by a vector $\boldsymbol{\alpha}_r = (\alpha_{r1}, \ldots, \alpha_{rH})$: Let

$$y_h(\boldsymbol{\alpha}_r) = \begin{cases} y_{h1} & \text{if } \alpha_{rh} = 1 \\ y_{h2} & \text{if } \alpha_{rh} = -1. \end{cases}$$

Equivalently,

$$y_h(\boldsymbol{\alpha}_r) = \frac{\alpha_{rh} + 1}{2} y_{h1} - \frac{\alpha_{rh} - 1}{2} y_{h2}.$$

If group 1 contains observations $\{y_{11}, y_{22}, y_{32}, y_{42}, y_{51}, y_{62}, y_{71}\}$ as above, then $\boldsymbol{\alpha}_1 = (1, -1, -1, -1, 1, -1, 1)$. Similarly, $\boldsymbol{\alpha}_2 = (-1, 1, 1, 1, -1, 1, -1)$. The set of R replicate half-samples is **balanced** if

$$\sum_{r=1}^{R} \alpha_{rh}\alpha_{rl} = 0 \quad \text{for all } l \neq h.$$

For replicate r, calculate $\hat{\theta}(\boldsymbol{\alpha}_r)$ the same way as $\hat{\theta}$ but using only the observations in the half-sample selected by $\boldsymbol{\alpha}_r$. For estimating the mean of a stratified random sample, $\hat{\theta}(\boldsymbol{\alpha}_r) = \sum_{h=1}^{H} (N_h/N) y_h(\boldsymbol{\alpha}_r)$. Define the BRR variance estimator to be

$$\hat{V}_{\text{BRR}}(\hat{\theta}) = \frac{1}{R} \sum_{r=1}^{R} \left[\hat{\theta}(\boldsymbol{\alpha}_r) - \hat{\theta}\right]^2.$$

If the set of half-samples is balanced, then for stratified random sampling $\hat{V}_{\text{BRR}}(\bar{y}_{\text{str}}) = \hat{V}_{\text{str}}(\bar{y}_{\text{str}})$. (The proof of this is left as Exercise 23.) If, in addition, $\sum_{r=1}^{R} \alpha_{rh} = 0$ for $h = 1, \ldots, H$, then $\frac{1}{R} \sum_{r=1}^{R} \bar{y}_{\text{str}}(\boldsymbol{\alpha}_r) = \bar{y}_{\text{str}}$.

Example 9.5. For the data in Table 9.2, the set of $\boldsymbol{\alpha}$'s in the following table meets the balancing condition $\sum_{r=1}^{8} \alpha_{rh}\alpha_{rl} = 0$ for all $l \neq h$. The 8×7 matrix of -1's and 1's has orthogonal columns; in fact, it is the design matrix (excluding the column of ones) for a fractional factorial design (Box et al., 1978), called a Hadamard matrix. See Wolter (2007) for more detail on constructing these matrices.

		Stratum (h)						
		1	2	3	4	5	6	7
	$\boldsymbol{\alpha}_1$	−1	−1	−1	1	1	1	−1
	$\boldsymbol{\alpha}_2$	1	−1	−1	−1	−1	1	1
	$\boldsymbol{\alpha}_3$	−1	1	−1	−1	1	−1	1
Half-sample	$\boldsymbol{\alpha}_4$	1	1	−1	1	−1	−1	−1
(r)	$\boldsymbol{\alpha}_5$	−1	−1	1	1	−1	−1	1
	$\boldsymbol{\alpha}_6$	1	−1	1	−1	1	−1	−1
	$\boldsymbol{\alpha}_7$	−1	1	1	−1	−1	1	−1
	$\boldsymbol{\alpha}_8$	1	1	1	1	1	1	1

The estimate from each half-sample, $\hat{\theta}_r = \bar{y}_{\text{str}}(\boldsymbol{\alpha}_r)$, is calculated from the data in Table 9.2.

Half-sample	$\hat{\theta}(\boldsymbol{\alpha}_r)$	$\left[\hat{\theta}(\boldsymbol{\alpha}_r) - \hat{\theta}\right]^2$
1	4732.4	78,792.5
2	4439.8	141.6
3	4741.3	83,868.2
4	4344.3	11,534.8
5	4084.6	134,762.4
6	4592.0	19,684.1
7	4123.7	107,584.0
8	4555.5	10,774.4
average	4451.7	55,892.8

The average of $[\hat{\theta}(\boldsymbol{\alpha}_r) - \hat{\theta}]^2$ for the eight replicate half-samples is 55,892.75, which is the same as $\hat{V}_{\text{str}}(\bar{y}_{\text{str}})$ for sampling with replacement. ■

Note that we can calculate the BRR variance estimate by creating a new variable of weights for each replicate half-sample. The sampling weight for observation i in stratum h is $w_{hi} = N_h/n_h$, and

$$\bar{y}_{\text{str}} = \frac{\sum_{h=1}^{H}\sum_{i=1}^{2} w_{hi}y_{hi}}{\sum_{h=1}^{H}\sum_{i=1}^{2} w_{hi}}.$$

Define

$$w_{hi}(\boldsymbol{\alpha}_r) = \begin{cases} 2w_{hi} & \text{if observation } i \text{ of stratum } h \text{ is in} \\ & \text{the half-sample selected by } \boldsymbol{\alpha}_r \\ 0 & \text{otherwise.} \end{cases}$$

Then

$$\bar{y}_{\text{str}}(\boldsymbol{\alpha}_r) = \frac{\sum_{h=1}^{H}\sum_{i=1}^{2} w_{hi}(\boldsymbol{\alpha}_r)y_{hi}}{\sum_{h=1}^{H}\sum_{i=1}^{2} w_{hi}(\boldsymbol{\alpha}_r)}.$$

Similarly, for any statistic $\hat{\theta}$ calculated using the weights w_{hi}, $\hat{\theta}(\boldsymbol{\alpha}_r)$ is calculated exactly the same way, but using the new weights $w_{hi}(\boldsymbol{\alpha}_r)$. Using the new weight variables instead of selecting the subset of observations simplifies calculations for surveys with many response variables—the same column $w(\boldsymbol{\alpha}_r)$ can be used to find the rth half-sample estimate for

TABLE 9.3
Data structure, showing first 11 observations, after sorting.

Observation Number	Stratum Number	psu Number	ssu Number	Weight, w_i	Response Variable 1	Response Variable 2	Response Variable 3
1	1	1	1	w_1	y_1	x_1	u_1
2	1	1	2	w_2	y_2	x_2	u_2
3	1	1	3	w_3	y_3	x_3	u_3
4	1	1	4	w_4	y_4	x_4	u_4
5	1	2	1	w_5	y_5	x_5	u_5
6	1	2	2	w_6	y_6	x_6	u_6
7	1	2	3	w_7	y_7	x_7	u_7
8	1	2	4	w_8	y_8	x_8	u_8
9	1	2	5	w_9	y_9	x_9	u_9
10	2	1	1	w_{10}	y_{10}	x_{10}	u_{10}
11	2	1	2	w_{11}	y_{11}	x_{11}	u_{11}

all quantities of interest. The modified weights also make it easy to extend the method to stratified multistage samples.

BRR in a stratified multistage survey. When $\bar{y}_{\mathcal{U}}$ is the only quantity of interest in a stratified random sample, BRR is simply a fancy method of calculating the variance in (3.6) and adds little extra to the procedure in Chapter 3. BRR's value in a complex survey comes from its ability to estimate the variance of a general population quantity θ, where θ may be a ratio of two variables, a correlation coefficient, a quantile, or another quantity of interest.

Suppose the population has H strata, and two psus are selected from stratum h with unequal probabilities and with replacement. (In replication methods, we like sampling with replacement because the subsampling design does not affect the variance estimator, as we saw in Section 6.3.) The same method may be used when sampling is done without replacement in each stratum, but the estimated variance is expected to be larger than the without-replacement variance. The data file for a complex survey with two psus per stratum often resembles that shown in Table 9.3, after sorting by stratum and psu.

The vector $\boldsymbol{\alpha}_r$ defines the half-sample r: If $\alpha_{rh} = 1$, then all observation units in psu 1 of stratum h are in half-sample r. If $\alpha_{rh} = -1$, then all observation units in psu 2 of stratum h are in half-sample r. The vectors $\boldsymbol{\alpha}_r$ are selected in a balanced way, exactly as in stratified random sampling. For half-sample r, create a new column of weights $w(\boldsymbol{\alpha}_r)$:

$$w_i(\boldsymbol{\alpha}_r) = \begin{cases} 2w_i & \text{if observation unit } i \text{ is in half-sample } r \\ 0 & \text{otherwise.} \end{cases} \tag{9.7}$$

For the data structure in Table 9.3 with $\alpha_{r1} = -1$ and $\alpha_{r2} = 1$, the column $w(\boldsymbol{\alpha}_r)$ will be

$$(0, 0, 0, 0, 2w_5, 2w_6, 2w_7, 2w_8, 2w_9, 2w_{10}, 2w_{11}, \ldots).$$

Now use the column $w(\boldsymbol{\alpha}_r)$ instead of w to estimate quantities for half-sample r. The estimate of the population total of y for the full sample is $\sum_{i\in\mathcal{S}} w_i y_i$; the estimate of the population total of y for half-sample r is $\sum_{i\in\mathcal{S}} w_i(\boldsymbol{\alpha}_r) y_i$. If $\theta = t_y/t_x$, then $\hat{\theta} = \sum_{i\in\mathcal{S}} w_i y_i / \sum_{i\in\mathcal{S}} w_i x_i$, and $\hat{\theta}(\boldsymbol{\alpha}_r) = \sum_{i\in\mathcal{S}} w_i(\boldsymbol{\alpha}_r) y_i / \sum_{i\in\mathcal{S}} w_i(\boldsymbol{\alpha}_r) x_i$. We saw in Section 7.3 that the empirical cumulative distribution function is calculated using the weights:

$$\hat{F}(y) = \frac{\text{sum of } w_i \text{ for all observations with } y_i \le y}{\text{sum of } w_i \text{ for all observations}}.$$

Then the empirical distribution using half-sample r is

$$\hat{F}_r(y) = \frac{\text{sum of } w_i(\boldsymbol{\alpha}_r) \text{ for all observations with } y_i \le y}{\text{sum of } w_i(\boldsymbol{\alpha}_r) \text{ for all observations}}.$$

If θ is the population median, then $\hat{\theta}$ may be defined as the smallest value of y for which $\hat{F}(y) \ge 1/2$, and $\hat{\theta}(\alpha_r)$ is the smallest value of y for which $\hat{F}_r(y) \ge 1/2$.

For a population quantity θ, define

$$\hat{V}_{\text{BRR}}(\hat{\theta}) = \frac{1}{R}\sum_{r=1}^{R}\left[\hat{\theta}(\alpha_r) - \hat{\theta}\right]^2. \tag{9.8}$$

BRR can also be used to estimate covariances of statistics: If θ and η are two quantities of interest, then

$$\widehat{\text{Cov}}_{\text{BRR}}(\hat{\theta}, \hat{\eta}) = \frac{1}{R}\sum_{r=1}^{R}\left[\hat{\theta}\left(\boldsymbol{\alpha}_r\right) - \hat{\theta}\right]\left[\hat{\eta}\left(\boldsymbol{\alpha}_r\right) - \hat{\eta}\right].$$

Other BRR variance estimators, variations of (9.8), are described in Exercise 26.

While the exact equivalence of $\hat{V}_{\text{BRR}}(\bar{y}_{\text{str}}(\boldsymbol{\alpha}))$ and $\hat{V}_{\text{str}}(\bar{y}_{\text{str}})$ does not extend to nonlinear statistics, Rao and Wu (1985) showed that if θ is a smooth function of the population totals, the variance estimator from BRR is asymptotically equivalent to that from linearization. BRR also provides a consistent estimator of the variance for quantiles when a stratified random sample is taken (Shao and Wu, 1992).

When a replication method such as BRR is used, data analysts can calculate variances from data files without needing to know the stratification and clustering information. The public-use data set can consist of the response (y) variables, final full-sample weights, and the columns of replicate weights. The statistic $\hat{\theta}$ is calculated by using the final weights w_i with the data vector of y_is. Then the columns of replicate weights are used to perform the variance estimation: We calculate $\hat{\theta}(\boldsymbol{\alpha}_r)$, for $r = 1, \ldots, R$, by performing the same calculations used to find $\hat{\theta}$, with weights $w_i(\boldsymbol{\alpha}_r)$ substituted for the original weights w_i. Equation (9.8) is then applied to estimate the variance of $\hat{\theta}$. Weighting adjustments for nonresponse, such as those discussed in Section 8.5, can be incorporated into the replicate weights so that the BRR estimate of variance includes the effects of the nonresponse and calibration adjustments.

Example 9.6. Let's use BRR to estimate variances from the NHANES data discussed in Section 7.5. The public-use data set includes variables for pseudo-strata and pseudo-psus that can be used for variance estimation. (The original strata and psu variables are not released to the public to preserve the confidentiality of the respondents' data.) Each pseudo-stratum has two pseudo-psus, so BRR weights can be constructed using the methods in this section.

The BRR replicate weights can be used to calculate standard errors for almost any statistic. For example, to estimate the interquartile range $\theta_{.75} - \theta_{.25}$, the final weight variable can be used to calculate the estimate $\hat{\theta}_{.75} - \hat{\theta}_{.25}$, and the replicate weights can then be used to estimate the variance of $\hat{\theta}_{.75} - \hat{\theta}_{.25}$.

Since our replicate weights are based on the final weight variable included in the NHANES data, however, they do not incorporate effects of nonresponse adjustments on the variance. Many data sets that are made available to the public have replicate weights that account for the nonresponse adjustments, and those are preferred if available.

The data set has 15 pseudo-strata, so we use a 16×16 Hadamard matrix (16 is the first multiple of 4 after 15 for which a Hadamard matrix exists). The replicate weights for a few of the observations are given in Table 9.4. Each entry in the replicate weight columns is either 0 or $2w_i$. Note that the pattern of replicate weights is the same for all observations that belong to the same psu. Both observations in psu 1 of stratum 125 have the same pattern. The observation in psu 1 of stratum 128 has the opposite pattern from the observation in psu 2 of stratum 128.

TABLE 9.4
NHANES data with final and replicate weights (rounded to nearest integer).

Stratum	psu	Weight	Repl. Wt. 1	Repl. Wt. 2	Repl. Wt. 3	···	Repl. Wt. 15	Repl. Wt. 16
125	1	135630	271259	0	0	···	271259	0
125	1	25282	50565	0	0	···	50565	0
131	1	12576	25152	0	25152	···	25152	0
131	1	102079	204157	0	204157	···	204157	0
126	2	18235	36469	36469	36469	···	0	0
128	1	10879	21757	21757	0	···	21757	21757
120	1	9861	19721	19721	0	···	0	0
124	2	46173	92347	92347	0	···	92347	92347
119	1	10963	21927	0	21927	···	21927	0
128	2	39353	0	0	78707	···	0	0

We estimate the mean and the median first using the original weight vector and then using each of the 16 vectors of replicate weights. Using the vector of final weights for the full sample, we estimate the mean body mass index (variable *bmxbmi*) for adults age 20 and over as $\hat{\bar{y}} = 29.389$ and the median as $\hat{m} = 28.235$. Table 9.5 shows the values calculated using the replicate weights.

TABLE 9.5
Estimates of mean and median BMI using the 16 NHANES replicate weights.

Replicate	1	2	3	4	5	6	7	8
Mean	29.669	29.239	28.981	29.315	29.910	29.332	29.020	29.146
Median	28.566	27.849	27.631	28.196	28.737	28.279	27.803	27.898

Replicate	9	10	11	12	13	14	15	16
Mean	29.490	29.131	29.338	29.342	29.613	29.613	29.336	29.776
Median	28.362	27.845	28.096	28.183	28.630	28.456	28.310	28.656

Using (9.8) we estimate $\hat{V}(\hat{\bar{y}}) = 0.0673$ and $\hat{V}(\hat{m}) = 0.1098$. In practice, survey software packages will perform the calculations in Table 9.5 for commonly requested statistics, but the table shows how the calculations may be done for statistics not implemented in the software packages. ■

Advantages. BRR gives a variance estimator that is asymptotically equivalent to that from linearization methods for smooth functions of population totals. It can also be used for estimating variances of quantiles. The data analyst only needs the columns of replicate weights, and does not need the original sampling design information, to calculate variances. It requires relatively few computations (and relatively few columns of replicate weights) when compared with jackknife and bootstrap.

Disadvantages. As defined above, BRR can only be used in situations in which there are two psus per stratum. In practice, though, it may be extended to other sampling designs by using more complicated balancing schemes (Valliant et al., 2018, Chapter 15). BRR, like the jackknife and bootstrap, estimates the with-replacement variance, and may overestimate the without-replacement variance.

The BRR method as described in this section can be unstable for domain estimates when some psus contain no members from the domain. Because psus are either completely in or completely out of each half-sample, it is possible that some replicates will contain no domain members. Fay's variant of BRR (see Exercise 24) provides a way to avoid this problem.

9.3.2 Jackknife

The **jackknife** method, like BRR, extends the random group method by allowing the replicate groups to overlap. The jackknife was introduced by Quenouille (1956) as a method of reducing bias; Tukey (1958) proposed using it to estimate variances and calculate CIs. In this section, we describe the *delete-one jackknife*; Shao and Tu (1995) discuss other forms of the jackknife and give theoretical results.

Jackknife in an SRS. For an SRS, let $\hat{\theta}_{(j)}$ be the estimator of the same form as $\hat{\theta}$, but not using observation j. Thus if $\hat{\theta} = \bar{y}$, then $\hat{\theta}_{(j)} = \bar{y}_{(j)} = \sum_{i \neq j} y_i/(n-1)$. For an SRS, define the delete-one jackknife estimator (so called because we delete one observation in each replicate) as

$$\hat{V}_{\text{JK}}(\hat{\theta}) = \frac{n-1}{n} \sum_{j=1}^{n} \left(\hat{\theta}_{(j)} - \hat{\theta}\right)^2. \tag{9.9}$$

Why the multiplier $(n-1)/n$? Let's look at $\hat{V}_{\text{JK}}(\hat{\theta})$ when $\hat{\theta} = \bar{y}$. When $\hat{\theta} = \bar{y}$,

$$\bar{y}_{(j)} = \frac{1}{n-1} \sum_{i \neq j} y_i = \frac{1}{n-1} \left(\sum_{i=1}^{n} y_i - y_j \right) = \bar{y} - \frac{1}{n-1}(y_j - \bar{y}).$$

Then,

$$\sum_{j=1}^{n} \left(\bar{y}_{(j)} - \bar{y}\right)^2 = \frac{1}{(n-1)^2} \sum_{j=1}^{n} \left(y_j - \bar{y}\right)^2 = \frac{1}{n-1} s_y^2,$$

so $\hat{V}_{\text{JK}}(\bar{y}) = s_y^2/n$, the with-replacement estimator of the variance of $\bar{y}$.

Example 9.7. Let's use the jackknife to estimate the ratio of out-of-state tuition (y) to in-state tuition (x) for the first replicate group of colleges in Example 9.3. Here, $\hat{\theta} = \bar{y}/\bar{x}$, $\hat{\theta}_{(j)} = \hat{B}_{(j)} = \bar{y}_{(j)}/\bar{x}_{(j)}$, and

$$\hat{V}_{\text{JK}}(\hat{B}) = \frac{n-1}{n} \sum_{j \in \mathcal{S}} \left(\hat{B}_{(j)} - \hat{B}\right)^2.$$

For each jackknife group in Table 9.6, omit one observation. Thus, $\bar{x}_{(1)}$ is the average of all x's except for x_1: $\bar{x}_{(1)} = (1/9)\sum_{i=2}^{9} x_i$. Here, $\hat{B} = 2.425$, $\sum(\hat{B}_{(j)} - \hat{B})^2 = 0.0595$, and $\hat{V}_{\text{JK}}(\hat{B}) = (0.9)(0.0595) = .05358$. ■

TABLE 9.6
Jackknife calculations for Example 9.7.

j	x	y	$\bar{x}_{(j)}$	$\bar{y}_{(j)}$	$\hat{B}_{(j)}$
1	9912	23640	8802.3	21389.7	2.4300
2	7140	14810	9110.3	22370.8	2.4555
3	9808	26648	8813.9	21055.4	2.3889
4	8987	35170	8905.1	20108.6	2.2581
5	7930	8674	9022.6	23052.6	2.5550
6	7200	17550	9103.7	22066.3	2.4239
7	8929	21692	8911.6	21606.1	2.4245
8	11976	22488	8573.0	21517.7	2.5099
9	8935	27199	8910.9	20994.2	2.3560
10	8316	18276	8979.7	21985.7	2.4484

Jackknife in a stratified multistage sample. How can we extend this to a cluster sample? You might think that you could just delete one observation unit at a time, but that will not work—deleting one observation unit at a time destroys the cluster structure and gives an estimate of the variance that is only correct if the intraclass correlation coefficient is zero. In any resampling method and in the random group method, keep observation units within a psu together while constructing the replicates—this preserves the dependence among observation units within the same psu. For a cluster sample, then, we would apply the jackknife variance estimator in (9.9) by letting n be the number of psus, and letting $\hat{\theta}_{(j)}$ be the estimate of θ that we would obtain by deleting all the observations in psu j.

In a stratified multistage cluster sample, the jackknife is applied separately in each stratum at the first stage of sampling, with one psu deleted at a time. Suppose there are H strata, and n_h psus are chosen for the sample from stratum h. Assume these psus are chosen with replacement.

To apply the jackknife, delete one psu at a time. Let $\hat{\theta}_{(hj)}$ be the estimator of the same form as $\hat{\theta}$ when psu j of stratum h is omitted. To calculate $\hat{\theta}_{(hj)}$, define a new weight variable: Let

$$w_{i(hj)} = \begin{cases} w_i & \text{if observation unit } i \text{ is not in stratum } h \\ 0 & \text{if observation unit } i \text{ is in psu } j \text{ of stratum } h \\ \dfrac{n_h}{n_h - 1} w_i & \text{if observation unit } i \text{ is in stratum } h \text{ but not in psu } j. \end{cases} \tag{9.10}$$

Then use the weights $w_{i(hj)}$ to calculate $\hat{\theta}_{(hj)}$, and

$$\hat{V}_{\text{JK}}(\hat{\theta}) = \sum_{h=1}^{H} \frac{n_h - 1}{n_h} \sum_{j=1}^{n_h} \left(\hat{\theta}_{(hj)} - \hat{\theta} \right)^2. \tag{9.11}$$

Example 9.8. Here we use the jackknife to calculate the variance of the mean egg volume from Example 5.8. We calculated $\hat{\theta} = \bar{y}_r = 4375.947/1757 = 2.49$. In that example, since we did not know the number of clutches in the population, we calculated the with-replacement variance.

The jackknife can also be used to calculate the variance. Here, there is one stratum with 184 clusters, so $h = 1$ for all observations. The first step is to calculate the 184 vectors of replicate weights. For $\hat{\theta}_{(1,1)}$, delete the first psu. Thus the new weights for the observations in the first psu are 0; the weights in all remaining psus are the previous weights times $n_h/(n_h - 1) = 184/183$. Using the weights from Example 5.8, the new jackknife weight columns are shown in Table 9.7.

TABLE 9.7
Jackknife weights for Example 5.8. The values w_i are the relative weights; $w_{i(k)}$ is observation i's jackknife weight for the replication omitting psu k.

Clutch	Size	w_i	$w_{i(1)}$	$w_{i(2)}$	...	$w_{i(184)}$
1	13	6.5	0	6.535519	...	6.535519
1	13	6.5	0	6.535519	...	6.535519
2	13	6.5	6.535519	0	...	6.535519
2	13	6.5	6.535519	0	...	6.535519
3	6	3	3.016393	3.016393	...	3.016393
3	6	3	3.016393	3.016393	...	3.016393
⋮	⋮	⋮	⋮	⋮		⋮
184	12	6	6.032787	6.032787	...	0
184	12	6	6.032787	6.032787	...	0
Sum		1758	1754.535519	1754.535519	...	1755.540984

Note that the sums of the jackknife weights vary from column to column because the original sample is not self-weighting. We calculated $\hat{\theta}$ as $(\sum w_i y_i)/\sum w_i$; to find $\hat{\theta}_{(hj)}$, we follow the same procedure but use $w_{i(hj)}$ in place of w_i. Thus, $\hat{\theta}_{(1,1)} = 4352.205466/1754.535519 = 2.480546$ and the values of $\hat{\theta}$ for the other 183 replicates are calculated similarly. Using (9.11), then, we calculate $\hat{V}_{\text{JK}}(\hat{\theta}) = 0.00373$. This results in a standard error of 0.061, the same as calculated in Example 5.8. In general, the jackknife standard error will not be exactly the same as the linearization standard error, but is expected to be similar. ■

Advantages. The jackknife is an all-purpose method. The same procedure is used to estimate the variance for every statistic for which jackknife can be used. The jackknife works in stratified multistage samples in which BRR does not apply because more than two psus are sampled in each stratum. The jackknife provides a consistent estimator of the variance when θ is a smooth function of population totals (Krewski and Rao, 1981). Replication methods such as the jackknife can be used to account for some of the effects of imputation on the variance estimates (Rao and Shao, 1992; Yung and Rao, 2000).

Disadvantages. For some sampling designs, the delete-one jackknife may require a large amount of computation. The jackknife performs poorly for estimating the variances of some statistics that are not smooth functions of population totals. For example, the delete-one jackknife does not give a consistent estimator of the variance of quantiles in an SRS.

9.3.3 Bootstrap

As with the jackknife, theoretical results for the **bootstrap** were first developed for areas of statistics other than survey sampling. We first describe the bootstrap for an SRS with replacement, as described by Davison and Hinkley (1997), and then show how the method can be extended for stratified multistage sampling designs.

Bootstrap for SRS. Suppose $\mathcal{S}$ is an SRS with replacement of size n. A large sample is expected to reproduce properties of the whole population. We then treat the sample $\mathcal{S}$ as if it were a population, and take resamples from $\mathcal{S}$. If the sample really is similar to the population—if the empirical probability mass function of the sample is similar to the probability mass function of the population—then samples generated from the empirical probability mass function should behave like samples taken from the population.

Example 9.9. Let's use the bootstrap to estimate the variance of the median height, θ, for the population from Example 7.4, using the sample in the file `htsrs.csv`. The population median height is $\theta = 168$; the sample median from `htsrs.csv` is $\hat{\theta} = 169$. The population probability mass function in Figure 7.1 and the histogram from the SRS in Figure 7.2 are similar in shape (mostly because the sample size for the SRS is large), so we would expect that taking an SRS of size n with replacement from $\mathcal{S}$ would be like taking an SRS with replacement from the population. A resample from $\mathcal{S}$, though, will not be exactly the same as $\mathcal{S}$ because the sample is with replacement—some observations in $\mathcal{S}$ may occur twice or more in the resample, while other observations in $\mathcal{S}$ may not occur at all.

We take an SRS of size 200 with replacement from $\mathcal{S}$ to form the first resample. The first resample from $\mathcal{S}$ has an empirical probability mass function similar to but not identical to that of $\mathcal{S}$; the resample median is $\hat{\theta}_1^* = 170$. Repeating the process, the second resample from $\mathcal{S}$ has median $\hat{\theta}_2^* = 169$. We take a total of $R = 2000$ resamples from $\mathcal{S}$ and calculate the sample median from each sample, obtaining $\hat{\theta}_1^*$, $\hat{\theta}_2^*, \ldots, \hat{\theta}_R^*$. We obtain the following frequency table for the 2000 resample medians:

Median of Resample	165	166	166.5	167	167.5	168	168.5	169	169.5	170	170.5	171	171.5	172
Frequency	1	5	2	40	15	268	87	739	111	491	44	188	5	4

The sample mean of these 2000 values is 169.3 and the sample variance of these 2000 values is 0.9148; this is the bootstrap estimate of the variance of the sample median. An approximate 95% CI may be constructed using the bootstrap variance as $169.3 \pm 1.96\sqrt{0.9148} = [167.4, 171.2]$. Alternatively, the bootstrap distribution may be used to calculate a CI directly. The bootstrap distribution estimates the sampling distribution of $\hat{\theta}$, so a 95% percentile CI may be calculated by finding the 2.5th percentile and the 97.5th percentile of the bootstrap distribution. For this example, a 95% percentile CI for the median is [167.5, 171.0]. Manly and Navarro Alberto (2020) described other methods for finding CIs using the bootstrap. ■

Bootstrap for stratified multistage samples. To apply bootstrap to a complex survey, we take bootstrap resamples of the psus within each stratum. As with BRR and the jackknife, observations within a psu are always kept together in the bootstrap iterations.

Here are steps for using the rescaling bootstrap of Rao and Wu (1988) for a stratified multistage sample. Let n_h be the number of psus sampled from stratum h. Let R be the number of bootstrap replicates to be created. Typically, $R = 500$ or 1,000, although some statisticians use smaller values of R.

1. For bootstrap replicate r ($r = 1, \ldots, R$), select an SRS of $n_h - 1$ psus with replacement from the n_h sample psus in stratum h. Do this independently for each stratum. Let $m_{hj}(r)$ be the number of times psu j of stratum h is selected in replicate r.

2. Create the replicate weight vector for replicate r as

$$w_i(r) = w_i \times \frac{n_h}{n_h - 1} m_{hj}(r), \text{ for observation } i \text{ in psu } j \text{ of stratum } h.$$

   The result is R vectors of replicate weights.

3. Use the vectors of replicate weights to estimate $V(\hat{\theta})$. Let $\hat{\theta}_r^*$ be the estimator of θ, calculated the same way as $\hat{\theta}$ but using weights $w_i(r)$ instead of the original weights w_i.

Then,

$$\hat{V}_B(\hat{\theta}) = \frac{1}{R-1}\sum_{i=1}^{R}\left(\hat{\theta}_r^* - \hat{\theta}\right)^2. \tag{9.12}$$

Example 9.10. We use the bootstrap to estimate variances from the data in file `htstrat.csv`, discussed in Example 7.6. The bootstrap weights are constructed by taking 1000 stratified random samples with replacement from the data set; we select 159 women and 39 men with replacement in each resample. The average height is estimated by $\bar{y}_{\text{str}} = 169.02$ with bootstrap standard error 0.737; the standard error calculated using the stratified sampling formula in (3.6), ignoring the fpc, is 0.739. ■

Advantages. The bootstrap will work for smooth functions of population means and for some nonsmooth functions such as quantiles in general sampling designs. It can be used to estimate the distribution of a statistic $\hat{\theta}$ in addition to its variance. The bootstrap is well suited for finding CIs directly: To calculate a 95% CI, one can take the 2.5th and 97.5th percentiles from $\hat{\theta}_1^*$, $\hat{\theta}_2^*, \ldots, \hat{\theta}_R^*$, or can use a bootstrap-t or other method described in Mashreghi et al. (2016).

Disadvantages. In some settings, the bootstrap may require more computations than BRR or jackknife, since R is typically a large number. In other surveys, however, for example if a stratified random sample is taken, the bootstrap may require fewer computations than the jackknife. The bootstrap variance estimate differs when a different set of bootstrap samples is taken.

9.3.4 Creating and Using Replicate Weights

This section has discussed three methods for creating replicate weights for estimating variances. All of them, like the linearization variance method, produce consistent variance estimators for statistics that are smooth functions of means or totals, in the sense that $\hat{V}(\hat{\theta})/V(\hat{\theta})$ is, with high probability, close to 1 when the number of psus is large. The BRR and bootstrap methods also produce consistent variance estimators for some nonsmooth statistics such as quantiles.

Steps for constructing replicate weights for a survey

1. Choose the method for replicate variance estimation. Since all methods are asymptotically equivalent to linearization for variance estimates of smooth functions of means or totals, the choice is often based on computational considerations or convenience.

   BRR works well, in general, for two-psu-per-stratum designs and can be extended for other designs. Fay's variant (see Exercise 24) may be used if there is concern that domain members may be concentrated in a subset of the psus.

   The jackknife is an all-purpose method and can be used for any probability sampling design. The mathematical theory for the jackknife method (see, for example, Shao and Tu, 1995) shows that it works for estimating variances whenever the linearization method will work. It has two drawbacks, however. The number of replicate weights can be large for some designs (a stratified random sample of 20,000 observations leads to 20,000 replicate weights using the method described in this section), and the method does not produce valid variance estimates for statistics such as quantiles. There are, however, modifications of the jackknife method that can deal with both of these drawbacks (see Exercise 30, Shao and Wu, 1989, and Fuller, 2009).

The bootstrap can also be used for any probability sampling design. It produces consistent variance estimators for smooth functions of population means and for quantiles. It also allows the data analyst to view the estimated sampling distribution for statistics calculated from the data. The main disadvantage is that for some designs, the bootstrap requires a large number of replicates in order to produce a stable estimate of the variance.

2. Construct the columns of sampling weights (w_i) and final nonresponse-adjusted weights (w_i^*) as in Section 8.6.3. The sampling weight for unit i is the reciprocal of the probability the unit was selected to be in the sample. The final weight for unit i incorporates the weighting class adjustments and poststratification, raking, or calibration.

3. Construct the replicate sampling weight vectors $\mathbf{w}_1, \mathbf{w}_2, \ldots, \mathbf{w}_R$ using the method chosen in Step 1. For BRR, the ith element of $\mathbf{w}_1$ is $w_i(\boldsymbol{\alpha}_1)$; for jackknife, the ith element of $\mathbf{w}_1$ is $w_{i(11)}$; for bootstrap, $\mathbf{w}_1$ is the vector of weights created from the first bootstrap resample. This step applies the replicate weight method to the sampling weights from the full sample. Always keep the members of a psu together when constructing replicate weights.

4. For each replicate weight vector $\mathbf{w}_r$, for $r = 1, \ldots, R$, apply the same nonresponse adjustments that were used to arrive at final nonresponse-adjusted weights w_i^* in Step 2. This results in a set of final replicate weights $\mathbf{w}_1^*, \ldots \mathbf{w}_R^*$.

All of the methods produce variance estimates of the form

$$\hat{V}(\hat{\theta}) = \sum_{r=1}^{R} c_r \left(\hat{\theta}_r - \hat{\theta}\right)^2, \tag{9.13}$$

where the coefficients c_r depend on the method used. For BRR, from (9.8), $c_r = 1/R$, the reciprocal of the number of replicates; for bootstrap, from (9.12), $c_r = 1/(R-1)$; for jackknife, $c_r = (n_h - 1)/n_h$, where n_h is the number of psus for the stratum considered in replicate r. The replicate variance estimator in (9.13) has the same form for any statistic $\hat{\theta}$, whether that statistic is an estimated mean, population total, ratio, regression or correlation coefficient, or quantile (for BRR and bootstrap).

Step 4 of the replicate weight construction allows variance estimates calculated using the replicate weights to capture the effects of the nonresponse adjustments. For example, suppose that $x_i = 1$ if person i is between 18 and 24 years old and 0 otherwise, and the survey is poststratified to independent population counts for the age groups, where the population total for the number of persons between 18 and 24 is $t_x = 20{,}000$. The final poststratified weights w_i^* will result in $\hat{t}_x = \sum_{i \in \mathcal{S}} w_i^* x_i = 20{,}000$. But each vector $\mathbf{w}_r^*$ of replicate weights is also poststratified to the independent population counts for the age groups, so that the estimate of t_x from replicate r also equals 20,000. Then, with $\theta = t_x$, we have $\hat{\theta}_r = \hat{\theta} = 20{,}000$ for each replicate r, and the estimated variance in (9.13) equals 0, which is the correct variance for characteristics that are poststrata. Variables that are associated with the poststratification variables will also have reduced variance, and the replicate variance estimation captures this variance reduction as well.

Using replicate weights for analysis. Once the replicate weights have been created, anyone can use them to calculate standard errors for almost any statistic desired. With replicate weight methods, the data analyst does not need to know stratification and clustering information for the original survey, or information about the nonresponse adjustments. The salient information from all of these features (for variance-calculation purposes) is contained in the replicate weights. When nonresponse adjustments are performed on the replicate weights, standard errors calculated with them automatically incorporate the effects of the nonresponse adjustments.

To estimate statistics and their variances using replicate weights, you need: (1) the final weight variable, w_i^* (incorporating the nonresponse adjustments, if appropriate), (2) the R replicate weight variables that have been created (these should also incorporate the nonresponse adjustments, if those have been used for the final weights), and (3) the coefficients c_r in (9.13). You can then use a software package to estimate the variances with the replicate weights (see Lohr, 2022; Lu and Lohr, 2022, for how to do this with SAS and R software, respectively), or you can use Equation (9.13) directly.

Some variants of BRR, jackknife, and bootstrap (see, for example, Exercises 24, 26, and 30) use coefficients c_r that differ from the coefficients presented in this section. The methodology report for a survey will usually describe the method used to calculate the replicate weights, and give the coefficients explicitly. For example, persons analyzing data from the American Community Survey (ACS) are instructed to use coefficients $c_r = 4/80$ for the 80 replicate weight columns that are provided on the public-use data files (U.S. Census Bureau, 2020f). These coefficients are then used with survey analysis software, or directly in (9.13), to calculate variances.

9.4 Generalized Variance Functions

In many large government surveys such as the U.S. Current Population Survey or the Canadian Labour Force Survey, hundreds or thousands of estimates are calculated and published each year. Data users calculate additional estimates from public-use data sets or from online tools that produce user-requested statistical tables. Each estimate needs to be accompanied by a standard error, but, as we have seen, variance estimation can be complicated. Sometimes published tables or public-use data files do not provide enough information to allow standard errors to be calculated. And sometimes linearization or replication-based variance estimates are themselves highly variable; this occurs, for example, when a statistic is calculated for a domain that is found in only a few psus.

Generalized variance functions (GVFs) are provided in a number of surveys to calculate standard errors. A GVF is a statistical model that relates the variance of a statistic to the statistic's expected value and to other information such as the size of the domain to which the statistic applies. The model coefficients are estimated using a set of statistics and their variances that have been calculated using linearization or replication, and the estimated coefficients are then used to predict the variance for other statistics. GVFs have been used for the Current Population Survey since 1947 (U.S. Census Bureau, 2019b).

Suppose t_i is the total number of observation units belonging to a class, say the total number of persons in the United States who were employed during a particular time period. Let $p_i = t_i/N$, the proportion of persons in the population belonging to that class. If d_i is the design effect (deff) in the survey for estimating p_i (see Section 7.4), then

$$V(\hat{p}_i) \approx d_i \frac{p_i(1-p_i)}{n} = \frac{b_i}{N} p_i(1-p_i), \tag{9.14}$$

where $b_i = d_i \times (N/n)$. Similarly,

$$V(\hat{t}_i) \approx d_i N^2 \frac{p_i(1-p_i)}{n} = a_i t_i^2 + b_i t_i,$$

where $a_i = -d_i/n$. If estimating a proportion in a domain, say the proportion of persons in the 20–24 age group who were employed, the denominator in (9.14) is changed to the estimated population size of the domain (see Section 4.3).

If the deffs are similar for different estimates so that $a_i \approx a$ and $b_i \approx b$, then constants a and b can be estimated using the following steps:

1. Using replication or some other method, estimate variances for the estimated population totals of k variables of special interest, $\hat{t}_1, \hat{t}_2, \ldots, \hat{t}_k$. Let v_i be the relative variance for $\hat{t}_i$, $v_i = \hat{V}(\hat{t}_i)/\hat{t}_i^2$, for $i = 1, 2, \ldots, k$.

2. Postulate a model relating v_i to $\hat{t}_i$. Many surveys adopt a linear regression model with response variable v_i and explanatory variable $1/\hat{t}_i$:

$$v_i = \alpha + \frac{\beta}{\hat{t}_i}. \tag{9.15}$$

3. Use regression techniques to estimate α and β by a and b. Valliant (1987) suggested using weighted least squares to estimate the parameters, giving higher weight to items with small v_i.

4. Use the estimated regression equation to predict the relative variance of an estimated total $\hat{t}_{\text{new}}$: $\hat{v}_{\text{new}} = a + b/\hat{t}_{\text{new}}$. Since $\hat{v}_{\text{new}}$ is the predicted value of the relative variance $\hat{V}(\hat{t}_{\text{new}})/\hat{t}^2_{\text{new}}$, the GVF estimate of $V(\hat{t}_{\text{new}})$ is

$$\hat{V}(\hat{t}_{\text{new}}) = a\,\hat{t}^2_{\text{new}} + b\,\hat{t}_{\text{new}}. \tag{9.16}$$

   The basic GVF model can also be used to estimate the variance of an estimated proportion $\hat{p}$, with

$$\hat{V}(\hat{p}) = \frac{b}{\hat{D}}\,\hat{p}(1-\hat{p}), \tag{9.17}$$

   where $\hat{D}$ is the estimated number of units in the denominator of the proportion (see Exercise 32).

Valliant (1987) found that if deffs for the k estimated totals are similar, the GVF variances are often more stable than the direct estimates of variance, as they smooth out some of the fluctuations from item to item. If the variance of a statistic of interest does not follow the model postulated in Step 2, however, the GVF estimate of the variance may be inaccurate, and you can only know that it is inaccurate by calculating the variance directly.

Example 9.11. Couzens et al. (2015) described GVFs for estimating the variances of totals, rates, and proportions calculated from the U.S. National Crime Victimization Survey. The model in (9.15) assumes that the deff is constant for all quantities. To allow more flexibility for modeling possibly nonconstant deffs (see Krenzke, 1995), they added a third term to the model in (9.15): $v_i = \alpha + \beta/\hat{t}_i + \gamma/\sqrt{\hat{t}_i}$. Under this model, a GVF estimate for the variance of an estimated total is

$$\hat{V}(\hat{t}) = a\,\hat{t}^2 + b\,\hat{t} + c\,\hat{t}^{3/2}.$$

The estimated total number of robberies in 2015, calculated using the final survey weights, was 578,578. Using the published GVF model coefficients for 2015, $a = -0.000482$, $b = 4840$, and $c = 7.618$, the GVF standard error for number of robberies is

$$\text{SE}_{\text{GVF}}(\hat{t}) = \sqrt{-0.000482(578578)^2 + 4840(578578) + 7.618(578578)^{3/2}} = 77{,}405.$$

For comparison, the standard error computed from the public-use data file with BRR is $\text{SE}_{\text{BRR}}(\hat{t}) = 70{,}612$. ■

Advantages. The GVF may be used when insufficient information is provided in the public-use data files to allow direct calculation of standard errors, or to smooth out estimated variances. The data collector can calculate the GVF, and often has more information for estimating variances than is released to the public. A GVF saves a great deal of time and speeds production of annual reports. It is also useful for designing similar surveys in the future.

Disadvantages. The model relating v_i to $\hat{t}_i$ may not be appropriate for the quantity you are interested in, resulting in an unreliable estimate of the variance. You must be careful about using GVFs for estimates not included when calculating the regression parameters. If a subpopulation has an unusually high degree of clustering (and hence higher deffs than the subpopulations considered when developing the model), the GVF estimate of the variance for that subpopulation may be too small.

9.5 Confidence Intervals

9.5.1 Confidence Intervals for Smooth Functions of Population Totals

Theoretical results exist for most of the variance estimation methods discussed in this chapter, stating that under certain assumptions $(\hat{\theta} - \theta)/\sqrt{\hat{V}(\hat{\theta})}$ asymptotically follows a standard normal distribution. These results and conditions are given in Binder (1983) for linearization estimates, in Krewski and Rao (1981) and Rao and Wu (1985) for jackknife and BRR, and in Rao and Wu (1988) and Sitter (1992) for bootstrap. Consequently, when the assumptions are met, an approximate 95% CI for θ may be constructed as

$$\hat{\theta} \pm 1.96\sqrt{\hat{V}(\hat{\theta})}.$$

Alternatively, a t percentile may be substituted for 1.96, with df = (number of groups − 1) for the random group method, and df = (number of psus − number of strata) for the other methods. Rust and Rao (1996) and Valliant and Rust (2010) gave guidelines for appropriate dfs when domains are analyzed. The bootstrap method may also be used to calculate CIs directly.

Roughly speaking, the assumptions for constructing a CI using variances estimated via linearization, jackknife, BRR, or bootstrap are as follows:

1. The quantity of interest θ can be expressed as a smooth function of the population totals; more precisely, $\theta = h(t_1, t_2, \ldots, t_k)$, where the second-order partial derivatives of h are continuous.

2. The sample sizes are large: Either the number of psus sampled in each stratum is large, or the survey contains a large number of strata. (See Rao and Wu, 1985, for the precise technical conditions needed.) Also, to construct a CI using the normal distribution, the sample sizes must be large enough so that the sampling distribution of $\hat{\theta}$ is approximately normal.

A number of simulation studies, summarized by Wolter (2007) and Mashreghi et al. (2016), indicate that these CIs behave well in practice, in the sense that if repeated samples are taken from a population, approximately 95% of the CIs include the population characteristic. Sometimes a transformation may be used so that the sampling distribution of a statistic is closer to a normal distribution: if estimating total income, for example, a log transformation may be used because the distribution of income is extremely skewed.

9.5.2 Confidence Intervals for Population Quantiles

The theoretical results described above for BRR, jackknife, bootstrap and linearization methods do not apply to population quantiles, however, because they are not smooth functions of population totals. Special methods have been developed to construct CIs for quantiles.

Let q be between 0 and 1. Then define the quantile θ_q as $\theta_q = F^{-1}(q)$, where $F^{-1}(q)$ is defined to be the smallest value y satisfying $F(y) \geq q$. Similarly, define $\hat{\theta}_q = \hat{F}^{-1}(q)$. (Alternatively, as in Exercise 19 of Chapter 7, interpolation can be used to define quantiles.) Now F^{-1} and $\hat{F}^{-1}$ are *not* smooth functions, but we assume the population and sample are large enough that they can be well approximated by continuous functions.

Replication variance estimates for quantiles. Some of the methods already discussed work well for constructing CIs for quantiles. The random group method works well if the number of random groups, R, is sufficiently large. Let $\hat{\theta}_q(r)$ be the estimated quantile from random group r. Then, a CI for θ_q is

$$\hat{\theta}_q \pm t\sqrt{\frac{1}{R(R-1)}\sum_{r=1}^{R}\left[\hat{\theta}_q(r) - \hat{\theta}_q\right]^2},$$

where t is the appropriate percentile from a t distribution with $R-1$ degrees of freedom. Similarly, studies by McCarthy (1993), Kovar et al. (1988), Sitter (1992), Rao et al. (1992), Shao and Chen (1998), and Conti and Marella (2015) indicated that in certain designs an approximate 95% CI can be formed using

$$\hat{\theta}_q \pm 1.96\sqrt{\hat{V}(\hat{\theta}_q)},$$

where the variance estimate is calculated using BRR or bootstrap. This method does not work for the delete-one jackknife, but Section 4.2.3 of Fuller (2009) describes modifications that allow the jackknife to be used to calculate confidence intervals for quantiles.

Woodruff's method. An alternative interval can be constructed based on a method introduced by Woodruff (1952). For any y, $\hat{F}(y)$ is a function of population totals: $\hat{F}(y) = \sum_{i\in\mathcal{S}} w_i u_i / \sum_{i\in\mathcal{S}} w_i$, where $u_i = 1$ if $y_i \leq y$ and $u_i = 0$ if $y_i > y$. Thus, one of the methods in this chapter can be used to estimate $V[\hat{F}(y)]$ for any value y, and an approximate 95% CI for $F(y)$ is given by

$$\hat{F}(y) \pm 1.96\sqrt{\hat{V}[\hat{F}(y)]}.$$

Now let's use the CI for $q = F(\theta_q)$ to obtain an approximate CI for θ_q. Since we have a 95% CI,

$$\begin{aligned}
0.95 &\approx P\left\{\hat{F}(\theta_q) - 1.96\sqrt{\hat{V}[\hat{F}(\hat{\theta}_q)]} \leq q \leq \hat{F}(\theta_q) + 1.96\sqrt{\hat{V}[\hat{F}(\hat{\theta}_q)]}\right\} \\
&= P\left\{q - 1.96\sqrt{\hat{V}[\hat{F}(\hat{\theta}_q)]} \leq \hat{F}(\theta_q) \leq q + 1.96\sqrt{\hat{V}[\hat{F}(\hat{\theta}_q)]}\right\} \\
&= P\left(\hat{F}^{-1}\left\{q - 1.96\sqrt{\hat{V}[\hat{F}(\hat{\theta}_q)]}\right\} \leq \theta_q \leq \hat{F}^{-1}\left\{q + 1.96\sqrt{\hat{V}[\hat{F}(\hat{\theta}_q)]}\right\}\right).
\end{aligned}$$

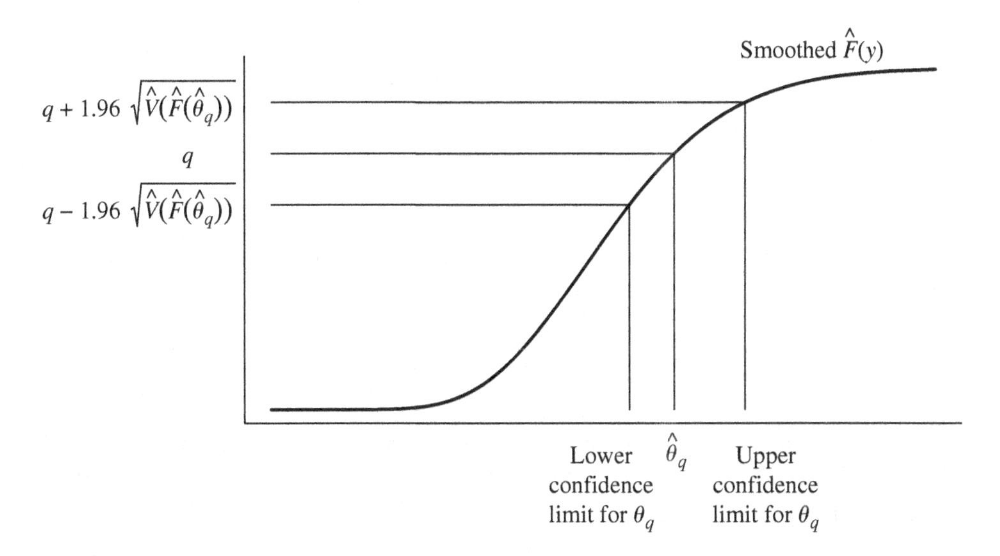

FIGURE 9.2
Woodruff's confidence interval for the quantile θ_q if the empirical distribution function is continuous. Since $F(y)$ is a proportion, we can calculate a confidence interval for any value of y, shown on the vertical axis. The corresponding points on the horizontal axis form a confidence interval for θ_q.

Thus, an approximate 95% CI for the quantile θ_q is

$$\left[\hat{F}^{-1}\left\{q - 1.96\sqrt{\hat{V}[\hat{F}(\hat{\theta}_q)]}\right\}, \hat{F}^{-1}\left\{q + 1.96\sqrt{\hat{V}[\hat{F}(\hat{\theta}_q)]}\right\}\right]. \tag{9.18}$$

The derivation of the CI in (9.18) is illustrated in Figure 9.2. An appropriate t critical value may be substituted for 1.96 if desired.

We need several technical assumptions to use the Woodruff CI. These assumptions are stated by Rao and Wu (1987) and Francisco and Fuller (1991), who studied a similar CI. Essentially, the problem is that both F and $\hat{F}$ are step functions; they have jumps at the values of y in the population and sample. The technical conditions basically say that the jumps in F and in $\hat{F}$ should be small (so that they can be approximated by a continuous function whose inverse exists), and that the sampling distribution of $\hat{F}(y)$ is approximately normal. Sitter and Wu (2001) showed that the Woodruff method gives CIs with approximately correct coverage probabilities even when q is large or small.

Example 9.12. Let's use Woodruff's method to construct a 95% CI for the median height in the file `htstrat.csv`, discussed in Example 7.6. The following values were obtained for the empirical distribution function:

y	165	166	167	168	169	170	171
$\hat{F}(y)$	0.3781	0.4438	0.4844	0.5125	0.5375	0.5656	0.6000

Because the values of height are rounded to the nearest integer, we use the interpolated quantiles discussed in Exercise 19 of Chapter 7 to estimate the median and CI. The interpolated median is

$$\hat{\theta}_{0.5} = 167 + \frac{0.5 - 0.4844}{0.5125 - 0.4844}(168 - 167) = 167.6.$$

Note that

$$\hat{F}(\hat{\theta}_q) = \frac{\sum_{h=1}^{2} \sum_{i \in \mathcal{S}_h} w_{hi} u_{hi}}{\sum_{h=1}^{2} \sum_{i \in \mathcal{S}_h} w_{hi}} = \frac{\sum_{h=1}^{2} \sum_{i \in \mathcal{S}_h} w_{hi} u_{hi}}{2000}$$

where $u_{hi} = 1$ if $y_{hi} \leq \hat{\theta}_{0.5}$ and 0 otherwise, so, using the variance for the combined ratio estimator in Section 4.5,

$$\hat{V}[\hat{F}(\hat{\theta}_q)] = \frac{1}{(2000)^2} \left[(1000)^2 \sum \left(1 - \frac{160}{1000}\right) \frac{s_{e1}^2}{160} + (1000)^2 \sum \left(1 - \frac{40}{1000}\right) \frac{s_{e2}^2}{40} \right],$$

where s_{eh}^2 is the sample variance of the values $e_{hi} = u_{hi} - 0.5$ for stratum h. Using the values $s_{e1}^2 = 0.1789$ and $s_{e2}^2 = 0.1641$ results in $\hat{V}[\hat{F}(\hat{\theta}_{0.5})] = 0.00121941$. Thus, for this sample, $1.96\sqrt{\hat{V}[\hat{F}(\hat{\theta}_{0.5})]} = 0.0684$.

The lower confidence bound for the median is then $\hat{F}^{-1}(0.5 - 0.0684)$, and the upper confidence bound for the median is $\hat{F}^{-1}(0.5 + 0.0684)$. We again use linear interpolation to obtain

$$\hat{F}^{-1}(0.4316) = 165 + \frac{0.4316 - 0.3781}{0.4438 - 0.3781}(166 - 165) = 165.8$$

and

$$\hat{F}^{-1}(0.5684) = 170 + \frac{0.5684 - 0.5656}{0.6 - 0.5656}(171 - 170) = 170.1$$

Thus, an approximate 95% CI for the median is [165.8, 170.1]. ■

Some software packages use a slightly different method for calculating Woodruff CIs for quantiles; these CIs are asymptotically equivalent to the CIs derived in this section if the underlying population distribution function is sufficiently smooth. Lohr (2022) and Lu and Lohr (2022) describe the CIs calculated by SAS and R software.

9.6 Chapter Summary

This chapter has briefly introduced you to some basic types of variance estimation methods that are used in practice: linearization, random groups, replication, and generalized variance functions.

Linearization methods have been thoroughly researched and widely used to find variance estimates in complex surveys. The main drawback of linearization, though, is that the derivatives need to be calculated for each statistic of interest, and this complicates the programs for estimating variances. If the statistic you are interested in is not handled in the software, you must write your own code.

The random group method is an intuitively appealing method for estimating variances. It is easy to explain and to compute, and can be used for almost any statistic of interest. Its main drawback is that we generally need enough random groups to have a stable estimate of the variance, and the number of random groups we can form is limited by the number of psus sampled in a stratum.

Resampling methods for stratified multistage surveys avoid the need to calculate partial derivatives by computing estimates for subsamples of the complete sample. They must be constructed carefully, however, so that the correlation of observations in the same cluster is preserved in the resampling. Resampling methods require more computing time than linearization but less programming time: The same method is used on all statistics. They have been shown to be equivalent to linearization for large samples when the characteristic of interest is a smooth function of population totals. When the replicate weights are constructed following the same procedure for nonresponse adjustments as in the final weights, they also capture the effects of nonresponse adjustments on variances.

The BRR method, like the random group method, can be used with almost any statistic but is usually employed in two-psu-per-stratum designs, or for designs that can be reformulated into two psu per strata. The jackknife and bootstrap can also be used for most estimators likely to be used in surveys (exception: the delete-one jackknife does not work well for estimating the variance of quantiles), and may be used in stratified multistage samples in which more than two psus are selected in each sample.

Generalized variance functions fit a model predicting the variance of a quantity from other characteristics. They are easy to use, but may give incorrect inferences for a statistic that does not follow the model used to develop the GVF.

Key Terms

Balanced repeated replication (BRR): A resampling method for variance estimation used when there are two psus sampled per stratum.

Bootstrap: A resampling method for variance estimation in which samples of psus with replacement are taken within each stratum.

Generalized variance function: A formula for variance estimation constructed by fitting a regression model to variance estimates that have been computed using replication or linearization methods.

Jackknife: A resampling method for variance estimation in which each psu is deleted in turn.

Linearization: A method for estimating the variance of a statistic that is a differentiable function of estimated population totals by using a Taylor series expansion.

For Further Reading

Bruch et al. (2011) provide a concise overview of methods for variance estimation. Wolter (2007) gives theoretical results and practical advice for the different methods of variance estimation. Valliant et al. (2018) provide step-by-step instructions for how to create and use replicate weights.

For more information on the mathematical theory underlying methods for variance estimation, see the books by Shao and Tu (1995) and Fuller (2009). Thousands of technical papers have been written about estimating variances in surveys. These include the articles by Binder (1983, 1996), presenting a general theory for using the linearization method of estimating variances, even when the quantities of interest are defined implicitly; Demnati and Rao (2004), deriving linearization variance estimators using the weights; Krewski and Rao (1981) and Rao and Wu (1985, 1988), showing the asymptotic equivalence of linearization, jackknife, BRR and bootstrap; and Wang and Opsomer (2011), discussing variance estimation for nondifferentiable functions such as quantiles. For reviews of the literature on variance estimation methods, see Rao (1988), Rao (1997), Rust and Rao (1996), Shao (2003), and Mashreghi et al. (2016).

We noted in Section 8.7 that if a data set has imputed values and then is analyzed as if those imputed values were real, the resulting variance estimate is too low. Replication methods such as bootstrap can be used to account for some of the effects of imputation on the variance estimates. Rao and Shao (1992), Shao and Sitter (1996), Shao and Steel (1999), Yung and Rao (2000), Cai and Rao (2019), Chen et al. (2019), Haziza and Vallée (2020), and the multiple imputation references given in Chapter 8 describe methods that can be used to estimate variances after imputation.

The replication variance estimation methods presented in this chapter all provide estimates of the with-replacement variance as discussed in Section 6.4. If sampling fractions within strata are small, the with-replacement and without-replacement variance estimates should be similar. If sampling fractions are large, however, it may be desirable to use a jackknife or bootstrap estimator that gives a closer approximation to the without-replacement variance. Gross (1980) proposed a without-replacement version of the bootstrap by creating a pseudopopulation of N/n copies of a without-replacement SRS and then drawing R SRSs without replacement from the pseudopopulation. Ranalli and Mecatti (2012) reviewed subsequent developments and proposed a unified framework for with-replacement and without-replacement bootstrap methods; Conti et al. (2020) derived a framework for variance estimation based on taking replicate samples from a pseudo-population constructed from the sample. Campbell (1980) proposed a generalized jackknife to estimate the variance from without-replacement samples; see Berger and Skinner (2005) and Berger (2007) for additional results and extensions.

9.7 Exercises

A. Introductory Exercises

1. Which of the variance estimation methods in this chapter would be suitable for estimating the proportion of beds that have bednets for the Gambia bednet survey in Example 7.1? Explain why each method is or is not appropriate.

2. Use the jackknife to estimate $V(\bar{y})$ for the data in `srs30.csv`, and verify that $\hat{V}_{\text{JK}}(\bar{y}) = s^2/30$ for these data. What are the jackknife weights for jackknife replicate j?

3. Use Woodruff's method to construct a 95% CI for the median of the data in file `srs30.csv`.

4. Estimate the 25th percentile, median, and 75th percentile for the variable *acres92* in file `agstrat.csv`, used in Example 3.2. Give a 95% CI for each parameter.

5. Construct jackknife weights for your stratified sample from Exercise 9 of Chapter 3, and use those weights to calculate 95% CIs for the average area of objects in the bin and the total number of gray objects in the bin.

B. Working with Survey Data

6. Take five replicate stratified samples of size 10 from the population of 500 public college and universities in `college.csv`, described in Example 9.3. Select each sample using proportional allocation with two strata, where stratum 1 consists of the large institutions (with *ccsizset* ≥ 15) and stratum 2 consists of the smaller institutions (with *ccsizset* $<= 14$). Use the sampling weights to estimate the ratio of out-of-state to in-state tuition for each replicate sample, and then use the five estimates to compute a point estimate and 95% CI for the ratio of out-of-state to in-state tuition.

7. Use the data in file `syc.csv` (see Example 9.4) with weight variable *finalwt* to estimate each of the following population quantities. Calculate two variance estimates for each statistic: the first using the random group method, and the second using linearization (or Woodruff's method, for quantiles). Calculate a 95% CI with each variance estimate. How do the CIs calculated using the two methods differ?

   (a) Average age at first arrest (variable *agefirst*)
   (b) Median age at first arrest
   (c) Proportion of youths who are held for a violent offense (variable *crimtype*)
   (d) Proportion of youths who lived with both parents when growing up (variable *livewith*)
   (e) Proportion of youths who are male (variable *gender*)
   (f) Proportion of youths who have used illegal drugs (variable *everdrug*).

8. Example 9.4 used the random group method to estimate the variance of the estimated average age for the data in `syc.csv`. Create jackknife weights for the survey, and use them to estimate the variance for average age. The psus are facilities in strata 1–5 and persons in strata 6–16; if you delete one psu at a time, you will have 861 replicate weights. (See Exercise 30 for an approach that reduces the number of replicate weights.)

9. Calculate the jackknife estimate of the variance for the regression estimate of the population mean age of trees in a stand for the data in Exercise 3 of Chapter 4. How does the jackknife variance compare with the variance calculated using linearization methods?

10. Use the jackknife to estimate the variances of your estimates in parts (b) and (c) of Exercise 16 of Chapter 5.

11. Use the jackknife to estimate the variance of the ratio estimator used in Example 4.2. How does it compare with the linearization estimator?

12. Use Woodruff's method to construct a 95% CI for the median weekday greens fee for nine holes, using the SRS in file `golfsrs.csv`.

13. Construct jackknife weights for the sample of 60 books in `mysteries.csv` and use these to answer the questions in Exercise 18 of Chapter 4.

14. Use the data in `nhanes.csv` along with the BRR method to estimate the variance of the ratio of average sagittal abdominal diameter (variable *bmdavsad*) to body mass index (variable *bmxbmi*) for adults age 20 and over (see Exercise 14 of Chapter 7).

15. Use the data in `nhanes.csv` along with the BRR method to estimate the variance of the estimated median value of waist circumference for adults age 20 and over (see Exercise 15 of Chapter 7).

C. Working with Theory
Exercises in this section require probability theory and calculus

16. As in Example 9.1, let $h(p) = p(1-p)$.

    (a) Find the remainder term in the Taylor expansion, $\int_a^x (x-t)h''(t)dt$, and use it to find an exact expression for $h(\hat{p})$.

    (b) Is the remainder term likely to be smaller than the other terms? Explain.

    (c) Find an exact expression for $V[h(\hat{p})]$ for a simple random sample with replacement. How does it compare with the approximation in Example 9.1? HINT: Use moments of the Binomial distribution to find $E(\hat{p}^4)$ and be prepared to do a lot of algebra.

17. The straight-line regression slope for the population is

$$B_1 = \frac{\sum_{i=1}^{N}(x_i - \bar{x}_U)(y_i - \bar{y}_U)}{\sum_{i=1}^{N}(x_i - \bar{x}_U)^2}.$$

    (a) Express B_1 as a function of population totals $t_1 = \sum_{i=1}^N x_i y_i$, $t_2 = \sum_{i=1}^N x_i$, $t_3 = \sum_{i=1}^N y_i$, $t_4 = \sum_{i=1}^N x_i^2$, and $t_5 = \sum_{i=1}^N 1 = N$, so that $B_1 = h(t_1, t_2, t_3, t_4, t_5)$.

    (b) Let $\hat{B}_1 = h(\hat{t}_1, \hat{t}_2, \hat{t}_3, \hat{t}_4, \hat{t}_5)$, and suppose that $E[\hat{t}_i] = t_i$ for $i = 1, 2, 3, 4, 5$. Use the linearization method to find an approximation to the variance of $\hat{B}_1$. Express your answer in terms of $V(\hat{t}_i)$ and Cov $(\hat{t}_i, \hat{t}_j)$.

    (c) What is the linearization approximation to the variance for an SRS of size n?

    (d) Find a linearized variate q_i so that $\hat{V}(\hat{B}_1) = \hat{V}(\hat{t}_q)$.

18. The variance of a population is

$$S^2 = \frac{1}{N-1}\sum_{i=1}^{N}(y_i - \bar{y}_U)^2.$$

    (a) Express S^2 as a function h of population totals $t_1 = \sum_{i=1}^N y_i^2$, $t_2 = \sum_{i=1}^N y_i$, and $t_3 = \sum_{i=1}^N (1)$.

    (b) Find an estimator $\hat{S}^2$ by substituting estimators for t_1, t_2, and t_3.

    (c) Find the linearization variance estimator of $\hat{S}^2$.

19. The correlation coefficient for the population is

$$R = \frac{\sum_{i=1}^{N}(x_i - \bar{x}_U)(y_i - \bar{y}_U)}{\sqrt{\sum_{i=1}^{N}(x_i - \bar{x}_U)^2 \sum_{i=1}^{N}(y_i - \bar{y}_U)^2}}.$$

    (a) Express R as a function of population totals $t_1 = \sum_{i=1}^N x_i$, $t_2 = \sum_{i=1}^N y_i$, $t_3 = \sum_{i=1}^N x_i^2$, $t_4 = \sum_{i=1}^N x_i y_i$, and $t_5 = \sum_{i=1}^N y_i^2$, so that $R = h(t_1, t_2, t_3, t_4, t_5)$.

(b) Let $r = h(\hat{t}_1, \ldots, \hat{t}_5)$, and suppose that $E[\hat{t}_i] = t_i$ for $i = 1, \ldots, 5$. Use the linearization method to find an approximation to the variance of r. Express your answer in terms of $V(\hat{t}_i)$ and Cov $(\hat{t}_i, \hat{t}_j)$.

(c) What is the linearization approximation to the variance for an SRS of size n?

20. *Ratio estimation with Poisson sampling.* Exercise 27 of Chapter 6 explored properties of the Horvitz-Thompson estimator for Poisson sampling. The variance of the estimator is large, however, because of the random sample size. An alternative is to use ratio estimation. Let x_i be an auxiliary variable for which t_x is known, and, assuming that the sample $\mathcal{S}$ is not empty, define

$$\hat{t}_{yr} = t_x \frac{\sum_{i \in \mathcal{S}} y_i/\pi_i}{\sum_{i \in \mathcal{S}} x_i/\pi_i} = t_x \frac{\hat{t}_{y\text{HT}}}{\hat{t}_{x\text{HT}}}.$$

(a) Use the results in Example 9.2 and Exercise 27 of Chapter 6 to show that

$$V(\hat{t}_{yr}) \approx \sum_{i=1}^{N} \frac{(y_i - x_i t_y/t_x)^2}{\pi_i}(1 - \pi_i).$$

(b) Show that

$$\hat{V}(\hat{t}_{yr}) = \frac{t_x^2}{\hat{t}_x^2} \sum_{i \in \mathcal{S}} \frac{e_i^2}{\pi_i^2}(1 - \pi_i),$$

where $e_i = y_i - x_i \hat{t}_{y\text{HT}}/\hat{t}_{x\text{HT}}$, is an approximately unbiased estimator of $V(\hat{t}_{yr})$.

(c) Find the estimator $\hat{t}_{yr}$ and its variance (using the result in part a) when $x_i = 1$ for $i = 1, \ldots, N$. Then $t_x = N$ and $\hat{t}_{x\text{HT}} = \sum_{i \in \mathcal{S}} 1/\pi_i$ is an unbiased estimator of N.

(d) Find the estimator $\hat{t}_{yr}$ and its variance (using the result in part a) when $x_i = \pi_i$. Then $t_x = \sum_{i=1}^{N} \pi_i$ is the expected sample size, and $\hat{t}_{x\text{HT}} = \sum_{i \in \mathcal{S}} \pi_i/\pi_i$ is the actual sample size.

(e) Which of the three estimators ($\hat{t}_{y\text{HT}}$, ratio estimator in part c, ratio estimator in part d) would you expect to have the smallest variance if (i) $y_i > 0$ and $\pi_i = n/N$ for all units? (ii) $y_i = k\pi_i$ for some constant k?

21. *Variance estimation with poststratification.* Suppose we poststratify the sample into L poststrata, with population counts $N_1, N_2, \ldots, N_L$. Then the poststratified estimator for the population total is

$$\hat{t}_{\text{post}} = \sum_{l=1}^{L} \frac{N_l}{\hat{N}_l} \hat{t}_l = h(\hat{t}_1, \ldots, \hat{t}_L, \hat{N}_1, \ldots, \hat{N}_L),$$

where

$$\hat{t}_l = \sum_{i \in \mathcal{S}} w_i x_{il} y_i, \quad \hat{N}_l = \sum_{i \in \mathcal{S}} w_i x_{il},$$

and $x_{il} = 1$ if unit i is in poststratum l and 0 otherwise. Show, using linearization, that

$$V(\hat{t}_{\text{post}}) \approx V\left\{\sum_{l=1}^{L}\left(\hat{t}_l - \frac{t_l}{N_l}\hat{N}_l\right)\right\}.$$

We can thus estimate $V(\hat{t}_{\text{post}})$ by $\hat{V}(\hat{t}_{\text{post}}) = \hat{V}\left(\sum_{i \in \mathcal{S}} w_i q_i\right)$, where

$$q_i = \sum_{l=1}^{L} x_{il}\left(y_i - \hat{t}_l/\hat{N}_l\right).$$

22. Consider the random group estimator of the variance from Section 9.2.2. The parameter of interest is $\theta = \bar{y}_{\mathcal{U}}$. A simple random sample with replacement of size n is taken from the population. The sample is divided into R random groups, each of size m. Let $\hat{\theta}_r$ be the sample mean of the m observations in random group r, let $\hat{\theta} = \bar{y} = \sum_{r=1}^{R} \hat{\theta}_r / R$, and let $\hat{V}_2(\hat{\theta})$ be the variance estimator defined in (9.6). Show that

$$\text{CV}[\hat{V}_2(\hat{\theta})] = \left[\frac{\kappa}{m} + 3\frac{m-1}{m} - \frac{R-3}{R-1}\right]^{1/2} \frac{1}{\sqrt{R}}$$

where $\kappa = \sum_{i=1}^{N} (y_i - \bar{y}_{\mathcal{U}})^4 / [(N-1)S^4]$.

23. Suppose a stratified random sample is taken with two observations per stratum. Show that if $\sum_{r=1}^{R} \alpha_{rh}\alpha_{rl} = 0$ for $l \neq h$, then

$$\hat{V}_{\text{BRR}}(\bar{y}_{\text{str}}) = \hat{V}_{\text{str}}(\bar{y}_{\text{str}}).$$

HINT: First note that

$$\bar{y}_{\text{str}}(\alpha_i) - \bar{y}_{\text{str}} = \sum_{h=1}^{H} \frac{N_h}{N} \alpha_{ih} \frac{y_{h1} - y_{h2}}{2}.$$

Then express $\hat{V}_{\text{BRR}}(\bar{y}_{\text{str}})$ directly using y_{h1} and y_{h2}.

24. *Fay's variation of BRR.* Variances computed by BRR using (9.7) may be unstable for domain statistics when the domain members are concentrated in a subset of the sampled psus. With BRR, a psu is either completely in or completely out of a half-sample, and it is possible for a half-sample to contain few or no members from the domain. In some cases, a statistic $\hat{\theta}$ might be well-defined for the full sample but not be defined for one or more of the replicates. Fay (see Dippo et al., 1984; Fay, 1984, 1989; Judkins, 1990; Rao and Shao, 1999) proposed a gentler version of BRR that defines replicate weights as

$$w_i(\boldsymbol{\alpha}_r, \varepsilon) = \begin{cases} (1 + \alpha_{rh} - \alpha_{rh}\varepsilon)\, w_i & \text{if observation unit } i \text{ is in psu 1} \\ (1 - \alpha_{rh} + \alpha_{rh}\varepsilon)\, w_i & \text{if observation unit } i \text{ is in psu 2,} \end{cases} \tag{9.19}$$

where ε is a number satisfying $0 \leq \varepsilon < 1$. Then $\hat{\theta}(\alpha_r, \varepsilon)$ is the estimate of θ of the same form as $\hat{\theta}$, but using weights $w_i(\boldsymbol{\alpha}_r, \varepsilon)$ instead of weights w_i. A domain found in only one psu of a stratum will thus be in every replicate under this modification. The modified BRR variance estimate is

$$\hat{V}_{\text{BRR},\varepsilon}(\hat{\theta}) = \frac{1}{R(1-\varepsilon)^2} \sum_{r=1}^{R} \left[\hat{\theta}(\alpha_r, \varepsilon) - \hat{\theta}\right]^2. \tag{9.20}$$

(a) Show that (9.7) is a special case of (9.19) when $\varepsilon = 0$.

(b) Show that when the set of half-samples is balanced, that is, $\sum_{r=1}^{R} \alpha_{rh}\alpha_{rl} = 0$ for all $l \neq h$, and when a stratified random sample is taken with two observations per stratum, $\hat{V}_{\text{BRR},\varepsilon}(\bar{y}_{\text{str}}) = \hat{V}_{\text{str}}(\bar{y}_{\text{str}})$.

25. Using Fay's variant of BRR in Exercise 24 with $\varepsilon = 1/2$, calculate replicate weights for the data in `nhanes.csv`. Calculate the standard errors of the estimates in Example 7.9 of Chapter 7 using these replicate weights, and compare these to standard errors obtained using linearization.

26. Other BRR estimators of the variance are

$$\frac{1}{4R}\sum_{r=1}^{R}[\hat{\theta}(\alpha_r) - \hat{\theta}(-\alpha_r)]^2$$

and

$$\frac{1}{2R}\sum_{r=1}^{R}[\{\hat{\theta}(\alpha_r) - \hat{\theta}\}^2 + \{\hat{\theta}(-\alpha_r) - \hat{\theta}\}^2].$$

For a stratified random sample with two observations per stratum, show that if $\sum_{r=1}^{R} \alpha_{rh}\alpha_{rl} = 0$ for $l \neq h$, then each of these variance estimators is equivalent to $\hat{V}_{\text{str}}(\bar{y}_{\text{str}})$.

27. Suppose the parameter of interest is $\theta = h(t)$, where $h(t) = at^2 + bt + c$ and t is the population total. Let $\hat{\theta} = h(\hat{t})$. Show, in a stratified random sample with two observations per stratum, that if $\sum_{r=1}^{R} \alpha_{rh}\alpha_{rl} = 0$ for $l \neq h$, then

$$\frac{1}{4R}\sum_{r=1}^{R}\left[\hat{\theta}(\alpha_r) - \hat{\theta}(-\alpha_r)\right]^2 = \hat{V}_L(\hat{\theta}),$$

the linearization estimator of the variance (see Rao and Wu, 1985).

28. The linearization method in Section 9.1 is the one historically used to find variances. Binder (1996) proposed proceeding directly to the estimate of the variance by evaluating the partial derivatives at the sample estimates rather than at the population quantities. What is Binder's estimate for the variance of the ratio estimator? Does it differ from that in Section 9.1?

29. *An alternative approach to linearization variance estimators.* Demnati and Rao (2004) derived a unified theory for linearization variance estimation using weights. Let θ be the population quantity of interest, and define the estimator $\hat{\theta}$ to be a function of the vector of sampling weights and the population values:

$$\hat{\theta} = g(\mathbf{w}, \mathbf{y}_1, \mathbf{y}_2, \ldots, \mathbf{y}_k),$$

where $\mathbf{w} = (w_1, \ldots, w_N)^T$ with w_i the sampling weight of unit i ($w_i = 0$ if i is not in the sample), and $\mathbf{y}_j$ is the vector of population values for the jth response variable. Then a linearization variance estimator can be found by taking the partial derivatives of the function with respect to the *weights*. Let

$$z_i = \frac{\partial g(\mathbf{w}, \mathbf{y}_1, \mathbf{y}_2, \ldots, \mathbf{y}_k)}{\partial w_i}$$

evaluated at the sampling weights w_i. Then we can estimate $V(\hat{\theta})$ by

$$\hat{V}(\hat{\theta}) = \hat{V}(\hat{t}_z) = \hat{V}\left(\sum_{i\in\mathcal{S}} w_i z_i\right).$$

For example, considering the ratio estimator of the population total,

$$\hat{\theta} = g(\mathbf{w}, \mathbf{x}, \mathbf{y}) = \frac{\hat{t}_y}{\hat{t}_x}t_x = \frac{\sum_{k\in\mathcal{S}} w_k y_k}{\sum_{k\in\mathcal{S}} w_k x_k}t_x.$$

The partial derivative of $\hat{\theta} = g(\mathbf{w}, \mathbf{x}, \mathbf{y})$ with respect to w_i is

$$z_i = \frac{\partial g(\mathbf{w}, \mathbf{x}, \mathbf{y})}{\partial w_i} = \frac{y_i}{\sum_{k \in \mathcal{S}} w_k x_k} t_x - \frac{x_i \sum_{k \in \mathcal{S}} w_k y_k}{\left(\sum_{k \in \mathcal{S}} w_k x_k\right)^2} t_x = (y_i - \hat{B} x_i) \frac{t_x}{\hat{t}_x}.$$

For an SRS, finding the estimated variance of $\hat{t}_z$ gives (4.13).

Consider the poststratified estimator in Exercise 21.

(a) Write the estimator as $\hat{t}_{\text{post}} = g(\mathbf{w}, \mathbf{y}, \mathbf{x}_1, \ldots, \mathbf{x}_L)$, where $x_{li} = 1$ if observation i is in poststratum l and 0 otherwise.

(b) Find an estimator of $V(\hat{t}_{\text{post}})$ using the Demnati–Rao (2004) approach.

30. *Grouped jackknife.* For some data sets, the delete-one jackknife described in Section 9.3.2 can result in large sets of replicate weights. For example, a stratified random sample with $n = 10{,}000$ would have 10,000 columns of replicate weights. This number is unwieldy to store and also can slow computations. You can reduce the number of jackknife replicates by grouping observations and creating pseudo-psus; in a stratified random sample, this should result in an approximately unbiased variance estimate because the intraclass correlation in these pseudo-psus is expected to be close to zero (see Kott, 2001; Valliant et al., 2008, for group formation and weighting in stratified multistage samples).

    To construct grouped jackknife weights for a stratified random sample, randomly assign the observations in stratum h to one of g_h groups. Then treat these groups as psus, using the method in Section 9.3.2 to form the jackknife weights as in (9.10) by deleting one group at a time, and calculate

$$\hat{V}_{\text{GJK}}(\hat{\theta}) = \sum_{h=1}^{H} \frac{g_h - 1}{g_h} \sum_{j=1}^{g_h} (\hat{\theta}_{(hg)} - \hat{\theta})^2.$$

    Construct grouped jackknife weights for strata 6–16 for the Survey of Youths in Custody `syc.csv`, using 7 groups in each stratum. Estimate the variance for the average age, and compare your answer to that from the full jackknife in Exercise 8.

31. Consider the one-stage cluster design studied in Section 5.2.2, in which each psu has size M and an SRS of n psus is selected from the N psus in the population. Assume that N is large and n/N is small.

    (a) For a binary (taking on values of 0 and 1) response with $p = \bar{y}_{\mathcal{U}}$, show that, for the GVF model in (9.15),

$$V(\hat{t}) \approx (NM)^2 \frac{p(1-p)}{nM} \left[1 + (M-1)\text{ICC}\right].$$

    (b) Show that the relative variance $v = V(\hat{t})/t^2$ can be written as $v \approx \alpha + \beta/t$, and give α and β. Consequently, if the intraclass correlation coefficient is similar for responses in the survey, the GVF method should work well.

32. Let b be an estimator for β in the model for GVFs in (9.15). Let $B = t_y/t_x$ and $\hat{B} = \hat{t}_y/\hat{t}_x$. Suppose that $\hat{B}$ and $\hat{t}_x$ are independent.

(a) Using the model in (9.15) and the result in (9.2), show that we can estimate $V(\hat{B})$ by

$$\hat{V}(\hat{B}) = \hat{B}^2 \left[\frac{b}{\hat{t}_y} - \frac{b}{\hat{t}_x} \right].$$

(b) Now let B be a proportion for a subpopulation, where t_x is the size of the subpopulation and t_y is the number of units in that subpopulation having a certain characteristic. Show that $\hat{V}(\hat{B}) = b\hat{B}(1 - \hat{B})/\hat{t}_x$ and that $\hat{V}(\hat{B}) = \hat{V}(1 - \hat{B})$.

D. Projects and Activities

33. *Index fund.* In Exercise 49 of Chapter 6, you selected a sample of size 40 from the S&P 500 companies with probability proportional to market capitalization. Construct jackknife weights for this sample.

34. *Trucks.* Use the data from the Vehicle Inventory and Use Survey (VIUS), described in Exercise 49 of Chapter 3, for this problem. The survey design is stratified random sampling, with a sample size of 136,113 trucks.

    (a) Which of the variance estimation methods in this chapter can be used to estimate the variance of the estimated ratio of miles driven in 2002 (*miles_annl*) to lifetime miles driven (*miles_life*)? What are the advantages and drawbacks of each method?

    (b) Which of the variance estimation methods can be used to estimate the variance of the estimated median number of miles driven in 2002? What are the advantages and drawbacks of each method?

    (c) Use the bootstrap with 500 replications to estimate the variances of the estimates in (a) and (b).

35. *Baseball data.* Construct jackknife weights for your dataset from Exercise 50 of Chapter 3. Use these weights to estimate the variance of the estimated mean of the variable *logsal*, and of the ratio (total number of home runs)/(number of runs scored).

36. *IPUMS exercises.* Construct the jackknife weights for your dataset from Exercise 45 of Chapter 5. Use these weights to estimate the variances of the estimated population mean and total of *inctot.*

37. Find a survey that releases replicate weights for variance estimation. What method was used to construct the replicate weights? How are nonresponse adjustments incorporated into the replicate weights?

38. *Activity for course project.* Describe the method used for variance estimation for the survey you looked at in Exercise 35 of Chapter 7. If the survey releases replicate weight variables, what method was used to construct them? Do the replicate weights incorporate nonresponse adjustments? If so, how? Now do either (a) or (b) for your survey:

    (a) If the survey releases replicate weight variables, use them to estimate the variance of the estimated means you found in Exercise 35 of Chapter 7. If the replicate weights are formed by BRR or bootstrap, also estimate the variances of the estimated quantiles.

    (b) If the survey releases stratification and clustering information, use these to construct replicate weights using one of the resampling methods described in this chapter. Use the replicate weights to estimate the variance of the estimated means you found in Exercise 35 of Chapter 7.

10

Categorical Data Analysis in Complex Surveys

But Statistics must be made otherwise than to prove a preconceived idea.

—Florence Nightingale, Annotation in *Physique Sociale* by A. Quetelet

Up to now, we have mostly been looking at how to estimate summary quantities such as means, totals, and percentages in different sampling designs. Totals and percentages are important for many surveys to provide a description of the population: for instance, the percentage of the population having high cholesterol or the total number of unemployed persons in the United States. Often, though, researchers are interested in multivariate questions: Is age associated with criminal victimization, or can we predict unemployment status from demographic variables? Such questions are typically answered in statistics using techniques in categorical data analysis or regression. The techniques you learned in an introductory statistics course, though, assumed that observations were all independent and identically distributed from some population distribution. These assumptions are no longer met in data from complex surveys; in this and the following chapter, we examine the effects of the complex sampling design on commonly used statistical analyses.

Since much information from sample surveys is collected in the form of percentages, categorical data methods are extensively used in the analysis. In fact, many of the data sets used to illustrate the chi-square test in introductory statistics textbooks originate in complex surveys. Our greatest concern is with the effects of clustering on hypothesis tests and models for categorical data since clustering usually decreases precision. We begin by reviewing various chi-square tests when a simple random sample (SRS) is taken from a large population.

10.1 Chi-Square Tests with Multinomial Sampling

Example 10.1. Each couple in an SRS of 500 married couples from a large population is asked whether (1) the household owns at least one personal computer and (2) the household subscribes to cable television. Table 10.1 displays the cross-classification.

TABLE 10.1
Cross-classified table from Example 10.1.

Observed Count		Computer? Yes	Computer? No	
Cable?	Yes	119	188	307
	No	88	105	193
		207	293	500

DOI: 10.1201/9780429298899-10

Are households with a computer more likely to subscribe to cable? A chi-square test for independence is often used for such questions. Under the null hypothesis that owning a computer and subscribing to cable are independent, the expected counts for each cell in the contingency table are the following:

Expected Count		**Computer?** Yes	No	
Cable?	Yes	127.1	179.9	307
	No	79.9	113.1	193
		207	293	500

Pearson's chi-square test statistic is

$$X^2 = \sum_{\text{all cells}} \frac{(\text{observed count} - \text{expected count})^2}{\text{expected count}} = 2.281.$$

The **likelihood ratio chi-square test statistic** is

$$G^2 = 2 \sum_{\text{all cells}} (\text{observed count}) \ln \left(\frac{\text{observed count}}{\text{expected count}} \right) = 2.275.$$

The two test statistics are asymptotically equivalent; for large samples, each approximately follows a chi-square (χ^2) distribution with 1 degree of freedom (df) under the null hypothesis. The p-value for each statistic is 0.13, giving no reason to doubt the null hypothesis that owning a computer and subscribing to cable television are independent.

If owning a computer and subscribing to cable are independent events, the odds that a cable subscriber will own a computer should equal the odds that a non-cable-subscriber will own a computer. We estimate the odds of owning a computer if the household subscribes to cable as 119/188 and estimate the odds of owning a computer if the household does not subscribe to cable as 88/105. The **odds ratio** is therefore estimated as

$$\frac{119/118}{88/105} = 0.755.$$

If the null hypothesis of independence is true, we expect the odds ratio to be close to one. Equivalently, we expect the logarithm of the odds ratio to be close to zero. The log odds is -0.28 with asymptotic standard error

$$\sqrt{\frac{1}{119} + \frac{1}{88} + \frac{1}{188} + \frac{1}{105}} = 0.186;$$

an approximate 95% confidence interval (CI) for the log odds ratio is $-0.28 \pm 1.96(0.186) = [-0.646, 0.084]$. This CI includes 0, and confirms the result of the hypothesis test that there is no evidence against independence. ■

Chi-square tests are commonly used in three situations: testing independence of factors, testing homogeneity of proportions, and testing goodness of fit. Each assumes a form of random sampling. These tests are discussed in more detail in Agresti (2013) and Simonoff (2006).

10.1.1 Testing Independence of Factors

Each of n independent observations is cross-classified by two factors: row factor R with r levels and column factor C with c levels. Each observation has probability p_{ij} of falling into row category i and column category j, giving the following table of true probabilities. Here, $p_{i+} = \sum_{j=1}^{c} p_{ij}$ is the probability that a randomly selected unit will fall in row category i, and $p_{+j} = \sum_{i=1}^{r} p_{ij}$ is the probability that a randomly selected unit will fall in column category j.

		C				
		1	2	$\cdots$	c	
	1	p_{11}	p_{12}	$\cdots$	p_{1c}	p_{1+}
	2	p_{21}	p_{22}	$\cdots$	p_{2c}	p_{2+}
R	$\vdots$	$\vdots$	$\vdots$		$\vdots$	$\vdots$
	r	p_{r1}	p_{r2}	$\cdots$	p_{rc}	p_{r+}
		p_{+1}	p_{+2}	$\cdots$	p_{+c}	1

The observed count in cell (i,j) from the sample is x_{ij}. If all units in the sample are independent, the x_{ij}s are from a multinomial distribution with rc categories; this sampling scheme is known as **multinomial sampling**. In surveys, the assumptions for multinomial sampling are met in an SRS with replacement; they are approximately met in an SRS without replacement when the sample size is small compared with the population size. The latter situation occurred in Example 10.1: Independent multinomial sampling means we have a sample of 500 (approximately) independent households, and we observe to which of the four categories each household belongs.

The null hypothesis of independence is

$$H_0 : p_{ij} = p_{i+}p_{+j} \quad \text{for} \quad i = 1, \ldots, r \quad \text{and} \quad j = 1, \ldots, c. \tag{10.1}$$

Let $m_{ij} = np_{ij}$ represent the expected counts. If H_0 is true, $m_{ij} = np_{i+}p_{+j}$, and m_{ij} can be estimated by

$$\hat{m}_{ij} = n\hat{p}_{i+}\hat{p}_{+j} = n\frac{x_{i+}}{n}\frac{x_{+j}}{n},$$

where $\hat{p}_{ij} = x_{ij}/n$, $\hat{p}_{+j} = \sum_{i=1}^{r} \hat{p}_{ij}$, and $\hat{p}_{i+} = \sum_{j=1}^{c} \hat{p}_{ij}$. Pearson's chi-square test statistic is

$$X^2 = \sum_{i=1}^{r}\sum_{j=1}^{c} \frac{(x_{ij} - \hat{m}_{ij})^2}{\hat{m}_{ij}} = n\sum_{i=1}^{r}\sum_{j=1}^{c} \frac{(\hat{p}_{ij} - \hat{p}_{i+}\hat{p}_{+j})^2}{\hat{p}_{i+}\hat{p}_{+j}}. \tag{10.2}$$

The likelihood ratio test statistic is

$$G^2 = 2\sum_{i=1}^{r}\sum_{j=1}^{c} x_{ij} \ln\left(\frac{x_{ij}}{\hat{m}_{ij}}\right) = 2n\sum_{i=1}^{r}\sum_{j=1}^{c} \hat{p}_{ij} \ln\left(\frac{\hat{p}_{ij}}{\hat{p}_{i+}\hat{p}_{+j}}\right). \tag{10.3}$$

If multinomial sampling is used with a sufficiently large sample size, X^2 and G^2 are approximately distributed as a χ^2 random variable with $(r-1)(c-1)$ df under the null hypothesis. How large is "sufficiently large" depends on the number of cells and expected probabilities; Fienberg (1979) argued that p-values will be approximately correct if (a) the expected count in each cell is greater than 1 and (b) $n \geq 5 \times$ (number of cells).

An equivalent statement to (10.1) is that all odds ratios equal 1:

$$H_0 : \frac{p_{11}p_{ij}}{p_{1j}p_{i1}} = 1 \text{ for all } i \geq 2 \text{ and } j \geq 2.$$

We can estimate any odds ratio $(p_{ij}p_{kl})/(p_{il}p_{kj})$ by substituting in estimated proportions: $(\hat{p}_{ij}\hat{p}_{kl})/(\hat{p}_{il}\hat{p}_{kj})$. If the sample is sufficiently large, the *logarithm* of the estimated odds ratio is approximately normally distributed with estimated variance (see Exercise 21)

$$\hat{V}\left[\ln\left(\frac{\hat{p}_{ij}\hat{p}_{kl}}{\hat{p}_{il}\hat{p}_{kj}}\right)\right] = \frac{1}{x_{ij}} + \frac{1}{x_{kl}} + \frac{1}{x_{il}} + \frac{1}{x_{kj}}.$$

10.1.2 Testing Homogeneity of Proportions

The Pearson and likelihood ratio test statistics in (10.2) and (10.3) may also be used when independent random samples from r populations are each classified into c categories. Multinomial sampling is done within each population, so the sampling scheme is called **product-multinomial sampling**. Product-multinomial sampling is equivalent to stratified random sampling when the sampling fraction for each stratum is small or when sampling is with replacement.

The difference between product-multinomial sampling and multinomial sampling is that the row totals p_{i+} and x_{i+} are fixed quantities in product-multinomial sampling—x_{i+} is the predetermined sample size for stratum i. The null hypothesis that the proportion of observations falling in class j is the same for all strata is

$$H_0 : \frac{p_{1j}}{p_{1+}} = \frac{p_{2j}}{p_{2+}} = \cdots = \frac{p_{rj}}{p_{r+}} = p_{+j} \quad \text{for all } j = 1, \ldots, c. \tag{10.4}$$

If the null hypothesis in (10.4) is true, again $m_{ij} = np_{i+}p_{+j}$ and the expected counts under H_0 are $\hat{m}_{ij} = np_{i+}\hat{p}_{+j}$, exactly as in the test for independence.

Example 10.2. The sample sizes used in Exercise 18 of Chapter 3, the stratified sample of nursing students and tutors, were the sample sizes for the respondents. Let's use a chi-square test for homogeneity of proportions to test the null hypothesis that the response rate is the same for each stratum. The four strata form the rows in the following contingency table.

	Nonrespondent	**Respondent**	**Total**
General student	46	222	268
General tutor	41	109	150
Psychiatric student	17	40	57
Psychiatric tutor	8	26	34
	112	397	509

The two chi-square test statistics are $X^2 = 8.218$, with p-value 0.042 and $G^2 = 8.165$, with p-value 0.043, providing evidence that the response rates differ among the four groups. ■

10.1.3 Testing Goodness of Fit

In the classical goodness of fit test, multinomial sampling is again assumed, with independent observations classified into k categories. The null hypothesis is

$$H_0 : p_i = p_i^{(0)} \text{ for } i = 1, \ldots, k,$$

where $p_i^{(0)}$ is prespecified or is a function of parameters θ to be estimated from the data.

Example 10.3. Webb (1955) examined the safety records for 17,952 Air Force pilots for an 8-year period around World War II and constructed the following frequency table.

Number of Accidents	0	1	2	3	4	5	6	7
Number of Pilots	12,475	4,117	1,016	269	53	14	2	2

If accidents occur randomly—if no pilots are more or less "accident-prone" than others—a Poisson distribution should fit the data well. We estimate the mean of the Poisson distribution by the mean number of accidents per pilot in the sample, 0.40597. The observed and expected probabilities under the null hypothesis that the data follow a Poisson distribution are given in the following table. The expected probabilities are computed using the Poisson probabilities $e^{-\lambda}\lambda^x/x!$ with $\lambda = 0.40597$.

Number of Accidents	**Observed Proportion,** $\hat{p}_i$	**Expected Probability Under** H_0, $\hat{p}_i^{(0)}$
0	0.6949	0.6663
1	0.2293	0.2705
2	0.0566	0.0549
3	0.0150	0.0074
4	0.0030	0.0008
5+	0.0012	0.0001

The two chi-square test statistics are

$$X^2 = \sum_{\text{all cells}} \frac{(\text{observed count} - \text{expected count})^2}{\text{expected count}} = n \sum_{i=1}^{k} \frac{\left(\hat{p}_i - \hat{p}_i^{(0)}\right)^2}{\hat{p}_i^{(0)}} \tag{10.5}$$

and

$$G^2 = 2n \sum_{i=1}^{k} \hat{p}_i \ln\left(\frac{\hat{p}_i}{\hat{p}_i^{(0)}}\right). \tag{10.6}$$

For the pilots, $X^2 = 756$ and $G^2 = 400$. If the null hypothesis is true, both statistics follow a χ^2 distribution with 4 df (2 df are spent on n and $\hat{\lambda}$). Both p-values are less than 0.0001, providing evidence that a Poisson model does not fit the data. More pilots have no accidents, or more than two accidents, than would be expected under the Poisson model. There is thus evidence that some pilots are more accident-prone than would occur under the Poisson model. ■

All of the chi-square test statistics in (10.2), (10.3), (10.5), and (10.6) grow with n. If the null hypothesis is not exactly true in the population—if households with cable are even infinitesimally more likely to own a personal computer than households without cable—we can almost guarantee rejection of the null hypothesis by taking a large enough random sample. This property of the hypothesis test means that it will be sensitive to artificially inflating the sample size by ignoring clustering.

10.2 Effects of Survey Design on Chi-Square Tests

The survey design can affect both the estimated cell probabilities and the tests of association or goodness of fit. In complex survey designs, we no longer have the random sampling that gives both X^2 and G^2 an approximate χ^2 distribution. Thus, if we ignore the survey design and use the chi-square tests described in Section 10.1, the significance levels and p-values will be wrong. Clustering, especially, can have a strong effect on the p-values of chi-square tests. In a cluster sample with a positive intraclass correlation coefficient (ICC), the true p-value will often be much larger than the p-value reported by a statistical package using the assumption of independent multinomial sampling. Let's see what can happen to hypothesis tests if the survey design is ignored in a cluster sample.

Example 10.4. Suppose that both husband and wife are asked about the household's cable and computer status for the survey discussed in Example 10.1, and both give the same answer. While the assumptions of multinomial sampling were met for the SRS of couples, they are not met for the cluster sample of persons—far from being independent units, the husband and wife from the same household agree completely in their answers. The ICC for the cluster sample is 1.

What happens if we ignore the clustering? The contingency table for the observed frequencies is as follows:

Observed Count		**Computer?** Yes	No	
Cable?	Yes	238	376	614
	No	176	210	386
		414	586	1000

The estimated proportions and odds ratio are identical to those in Example 10.1: $\hat{p}_{11} = 238/1000 = 119/500$ and the odds ratio is

$$\frac{238/376}{176/210} = 0.755.$$

But $X^2 = 4.562$ and $G^2 = 4.550$ are twice the values of the test statistics in Example 10.1. If you ignored the clustering and compared these statistics to a χ^2 distribution with 1 df, you would report a "p-value" of 0.033 and conclude that the data provided evidence that having a computer and subscribing to cable are not independent. If playing this game, you could lower the "p-value" even more by interviewing both children in each household as well, thus multiplying the original test statistics by 4.

Can you attain an arbitrarily low p-value by observing more ssus per psu? Absolutely not. The statistics X^2 and G^2 have a null χ_1^2 distribution *when multinomial sampling is used*. When a cluster sample is taken instead, and when the intraclass correlation coefficient is positive, X^2 and G^2 do *not* follow a χ_1^2 distribution under the null hypothesis. For the 1,000 husbands and wives, $X^2/2$ and $G^2/2$ follow a χ_1^2 distribution under H_0—this gives the same p-value found in Example 10.1. ■

10.2.1 Contingency Tables for Data from Complex Surveys

The observed counts x_{ij} do not necessarily reflect the relative frequencies of the categories in the population unless the sample is self-weighting. Suppose an SRS of elementary school classrooms in Denver is taken, and each of ten randomly selected students in each classroom is evaluated for self-concept (high or low) and clinical depression (present or not). Students are selected for the sample with unequal probabilities—students in small classes are more likely to be in the sample than students from large classes. A table of observed counts from the sample, ignoring the inclusion probabilities, would not give an accurate picture of the association between self-concept and depression in the population if the degree of association differs with class size. Even if the association between self-concept and depression is the same for different class sizes, the estimates of numbers of depressed students using the margins of the contingency table may be wrong.

Remember, though, that sampling weights can be used to estimate any population quantity. Here, they can be used to estimate the cell proportions. Estimate p_{ij} by

$$\hat{p}_{ij} = \frac{\sum_{k \in \mathcal{S}} w_k y_{kij}}{\sum_{k \in \mathcal{S}} w_k}, \tag{10.7}$$

where

$$y_{kij} = \begin{cases} 1 & \text{if observation unit } k \text{ is in cell } (i,j) \\ 0 & \text{otherwise} \end{cases}$$

and w_k is the weight for observation unit k. Thus,

$$\hat{p}_{ij} = \frac{\text{sum of weights for observation units in cell } (i,j)}{\text{sum of weights for all observation units in sample}}.$$

If the sample is self-weighting, $\hat{p}_{ij}$ will be the proportion of observation units falling in cell (i,j). Using the estimates $\hat{p}_{ij}$, construct the table

		C				
		1	2	$\cdots$	c	
	1	$\hat{p}_{11}$	$\hat{p}_{12}$	$\cdots$	$\hat{p}_{1c}$	$\hat{p}_{1+}$
	2	$\hat{p}_{21}$	$\hat{p}_{22}$	$\cdots$	$\hat{p}_{2c}$	$\hat{p}_{2+}$
R	$\vdots$	$\vdots$	$\vdots$		$\vdots$	$\vdots$
	r	$\hat{p}_{r1}$	$\hat{p}_{r2}$	$\cdots$	$\hat{p}_{rc}$	$\hat{p}_{r+}$
		$\hat{p}_{+1}$	$\hat{p}_{+2}$	$\cdots$	$\hat{p}_{+c}$	1

to examine associations, and estimate odds ratios by $(\hat{p}_{ij}\hat{p}_{kl})/(\hat{p}_{il}\hat{p}_{kj})$. A CI for p_{ij} may be constructed by using any method of variance estimation from Chapter 9.

Do not throw the observed counts away, however. If the odds ratios calculated using the $\hat{p}_{ij}$ differ appreciably from the odds ratios calculated using the observed counts x_{ij}, you should explore why they differ. Perhaps the odds ratio for depression and self-concept differs for larger classes or depends on socioeconomic factors related to class size. If that is the case, you should include these other factors in a model for the data or perhaps test the association separately for large and small classes.

10.2.2 Effects on Hypothesis Tests and Confidence Intervals

We can estimate contingency table proportions and odds ratios using weights. The weights, however, are not sufficient for constructing hypothesis tests and CIs—these depend on the clustering and (sometimes) stratification of the survey design.

Let's look at the effect of stratification first. If the strata in a stratified random sample are the row categories, the stratification poses no problem—we essentially have product-multinomial sampling as described in Section 10.1 and can test for homogeneity of proportions the usual way.

Often, though, we want to study association between factors that are not stratification variables. In general, stratification (with proportional allocation) increases precision of the estimates. For an SRS, (10.2) gives

$$X^2 = n \sum_{i=1}^{r} \sum_{j=1}^{c} \frac{(\hat{p}_{ij} - \hat{p}_{i+}\hat{p}_{+j})^2}{\hat{p}_{i+}\hat{p}_{+j}}.$$

A stratified sample with n observation units provides the same precision for estimating p_{ij} as an SRS with n/d_{ij} observation units, where d_{ij} is the design effect (deff) for estimating p_{ij}. With proportional allocation, the deffs are expected to be less than 1. Consequently, if we use the SRS test statistics in (10.2) or (10.3) with the $\hat{p}_{ij}$ from the stratified sample, X^2 and G^2 will be smaller than they should be to follow a null $\chi^2_{(r-1)(c-1)}$ distribution; "p-values" calculated ignoring the stratification will be too large and H_0 will not be rejected as often as it should be. Thus, while a statistics program for non-survey data may give you a p-value of 0.04, the actual p-value may be 0.02. Ignoring the stratification results in a conservative test. Similarly, a CI constructed for a log odds ratio is generally too large if the stratification is ignored. Your estimates are really more precise than the SRS CI indicates.

Clustering usually has the opposite effect. Design effects for $\hat{p}_{ij}$ with a cluster sample are usually greater than 1—a cluster sample with n observation units gives the same precision as an SRS with fewer than n observations. If the clustering is ignored, X^2 and G^2 are expected to be larger than if the equivalently sized SRS were taken, and "p-values" calculated ignoring the clustering are likely to be too small. An analysis ignoring the survey design may give you a p-value of 0.04, while the actual p-value may be 0.25. If you ignore the clustering, you may well declare an association to be statistically significant when it is really just due to random variation in the data. CIs for log odds ratios will be narrower than they should be—the estimates are not as precise as the CIs from an SRS-based analysis would lead you to believe.

Ignoring clustering in chi-square tests is often more dangerous than ignoring stratification. An SRS-based chi-square test using data from a proportionally allocated stratified sample will still indicate strong associations; it just may not uncover all weaker associations. Ignoring clustering, however, will lead to declaring associations statistically significant that really are not. Ignoring the clustering in goodness-of-fit tests may lead to adopting an unnecessarily complicated model to describe the data.

An investigator ignorant of sampling theory may analyze a stratified sample correctly by using the strata as one of the classification variables. But the investigator may not even record the clustering, and too often simply runs the observed counts through a program that calculates a chi-square test for data assumed independent and reports the p-value output from that program. Consider an investigator wanting to replicate the study discussed in Example 5.1 on how male and female professors are evaluated by college students. The investigator selects a stratified sample of male and female professors at the college and asks each student in those professors' classes to evaluate the professor's teaching. More than 2,000 student responses are obtained, and the investigator cross-classifies those responses by professor gender and student rating (high or low). Comparing Pearson's X^2 statistic on the observed counts to a χ^2_1 distribution, the investigator declares a statistically significant association between professor gender and student rating. This reported p-value is almost certainly incorrect because it does not account for the clustering of students within a class. If student evaluations reflect teaching quality, students of a "good" professor would be expected to give higher ratings than students of a "bad" professor. The ICC for students is positive, and the effective sample size is less than 2,000. The p-value reported by the investigator is then much too small, and the investigator may be wrong in concluding faculty men and women receive significantly different student evaluations.

The same problem occurs outside of sample surveys, particularly in medical or biological applications, where it is often referred to as pseudoreplication (Hurlbert, 1984, 2009). Clusters may correspond to pairs of eyes, to patients in the same hospital, repeated measurements of blood pressure on the same person, plants taken from the same small field, or sets of neurons taken from the same mouse. A study of 160 neurons taken from a random sample of 16 mice has a sample size of 16 independent units, not 160. And if the mice themselves are from a cluster sample, the effective sample size may be smaller than 16.

Is the clustering problem serious in surveys taken in practice? A number of studies have found that it can be. Holt et al. (1980) found that the actual significance levels for tests nominally conducted at the $\alpha = 0.05$ level ranged from 0.05 to 0.50. Fay (1985) referenced a number of studies demonstrating that the SRS-based test statistics "may give extremely erroneous results when applied to data arising from a complex sample design." The simulation study in Thomas et al. (1996) calculated actual significance levels attained for X^2 and G^2 when the nominal significance level was set at $\alpha = 0.05$—they found actual significance levels of about 0.30 to 0.40.

10.3 Corrections to Chi-Square Tests

In this section, we outline some of the basic approaches for testing independence with data from a complex survey. The theory for goodness-of-fit tests and tests for homogeneity of proportions is similar. In complex surveys, though, unlike in multinomial and product multinomial sampling, the tests for independence and homogeneity of proportions are not necessarily the same. Holt et al. (1980) noted that often (but not always) clustering has less effect on tests for independence than on tests for goodness of fit or homogeneity of proportions.

Recall from (10.1) that the null hypothesis of independence is

$$H_0 : p_{ij} = p_{i+}p_{+j} \quad \text{for} \quad i = 1, \ldots, r \quad \text{and} \quad j = 1, \ldots, c.$$

For a 2×2 table, $p_{ij} = p_{i+}p_{+j}$ for all i and j is equivalent to $p_{11}p_{22} - p_{12}p_{21} = 0$, so the null hypothesis reduces to a single equation. In general, the null hypothesis can be expressed as $(r-1)(c-1)$ distinct equations, which leads to $(r-1)(c-1)$ df for the χ^2 tests used for multinomial sampling. Let

$$\theta_{ij} = p_{ij} - p_{i+}p_{+j}.$$

Then the null hypothesis of independence is

$$H_0 : \theta_{11} = 0, \; \theta_{12} = 0, \; \ldots, \; \theta_{r-1,c-1} = 0.$$

10.3.1 Wald Tests

The Wald (1943) test was the first to be used for testing independence in complex surveys (Koch et al., 1975). For the 2×2 table, the null hypothesis involves one quantity,

$$\theta = \theta_{11} = p_{11} - p_{1+}p_{+1} = p_{11}p_{22} - p_{12}p_{21},$$

and θ is estimated by

$$\hat{\theta} = \hat{p}_{11}\hat{p}_{22} - \hat{p}_{12}\hat{p}_{21}.$$

The quantity θ is a smooth function of population totals, so we estimate $V(\hat{\theta})$ using one of the methods in Chapter 9. If the sample sizes are sufficiently large and $H_0 : \theta = 0$ is true, then $\hat{\theta}/\sqrt{\hat{V}(\hat{\theta})}$ approximately follows a standard normal distribution. Equivalently, under H_0, the **Wald statistic**

$$X_W^2 = \frac{\hat{\theta}^2}{\hat{V}(\hat{\theta})} \tag{10.8}$$

approximately follows a χ^2 distribution with 1 df. In practice, we often compare X_W^2 to an F distribution with 1 and κ df, where κ is the df associated with the variance estimator. If the random group method is used to estimate the variance, then κ equals (number of groups) − 1; if another method is used, κ equals (number of psus) − (number of strata).

Example 10.5. Let's look at the association between variables "Was anyone in your family ever incarcerated?" (variable *famtime*) and "Have you ever been put on probation or sent to a correctional institution for a violent offense?" (variable *everviol*) using data from the Survey of Youth in Custody discussed in Example 9.4. A total of $n = 2588$ persons in the survey had responses for both items. The following table gives the sum of the weights for each category.

		Ever Violent?		
		No	Yes	Total
Family Member	No	4,761	7,154	11,915
Incarcerated?	Yes	4,838	7,946	12,784
	Total	9,599	15,100	24,699

This results in the following table of estimated proportions:

		Ever Violent?		
		No	Yes	Total
Family Member	No	0.1928	0.2896	0.4824
Incarcerated?	Yes	0.1959	0.3217	0.5176
	Total	0.3886	0.6114	1.0000

Thus,

$$\hat{\theta} = \hat{p}_{11}\hat{p}_{22} - \hat{p}_{12}\hat{p}_{21} = \hat{p}_{11} - \hat{p}_{1+}\hat{p}_{+1} = 0.0053.$$

We can write $\theta = h(p_{11}, p_{12}, p_{21}, p_{22})$ and $\hat{\theta} = h(\hat{p}_{11}, \hat{p}_{12}, \hat{p}_{21}, \hat{p}_{12})$ for $h(a, b, c, d) = ad - bc$, so we can use linearization (see Exercise 20) or a resampling method to estimate $V(\hat{\theta})$. Using linearization to estimate the variance, we obtain $X_W^2 = (0.0053)^2/\hat{V}(\hat{\theta}) = 0.995$ with p-value $= 0.32$.

This test gives no evidence of an association between the two factors, when we look at the population as a whole. But of course the hypothesis test does not say anything about possible associations among the two variables in subpopulations—it could occur, for example, that violence and incarceration of a family member are positively associated among older youth, and negatively associated among younger youth—we would need to look at the subpopulations separately or fit a loglinear model. ■

For larger tables, let $\theta_{ij} = p_{ij} - p_{i+}p_{+j}$ and let $\boldsymbol{\theta} = [\theta_{11}\, \theta_{12}\, \ldots\, \theta_{r-1,c-1}]^T$ be the $(r-1)(c-1)$-vector of θ_{ij}s (the superscript T denotes the transpose of the vector), so that the null hypothesis is

$$H_0 : \boldsymbol{\theta} = \mathbf{0}.$$

The Wald statistic is then

$$X_W^2 = \hat{\boldsymbol{\theta}}^T \hat{V}(\hat{\boldsymbol{\theta}})^{-1}\hat{\boldsymbol{\theta}},$$

where $\hat{V}(\hat{\boldsymbol{\theta}})$ is the estimated covariance matrix of $\hat{\boldsymbol{\theta}}$. In very large samples, X_W^2 approximately follows a $\chi^2_{(r-1)(c-1)}$ distribution under H_0. But "large" in a complex survey refers to a large number of psus, not necessarily to a large number of observation units. In a 4×4 contingency table, $\hat{V}(\hat{\boldsymbol{\theta}})$ is a 9×9 matrix, and requires calculation of 45 different variances and covariances. If a cluster sample has only 50 psus, the estimated covariance matrix will be unstable. We do not recommend the Wald test for large contingency tables because it often performs poorly in that setting. The method in the next section works much better for large tables.

10.3.2 Rao–Scott Tests

The test statistics X^2 and G^2 do not follow a $\chi^2_{(r-1)(c-1)}$ distribution in a complex survey under the null hypothesis of independence. But both statistics have a skewed distribution that resembles the general shape of a χ^2 distribution. Rao–Scott tests (Rao and Scott, 1981, 1984) multiply X^2 or G^2 by a factor so that the adjusted test statistic approximately follows a χ^2 distribution.

First-order design correction. We can obtain a first-order correction by matching the mean of the test statistic to the mean of the $\chi^2_{(r-1)(c-1)}$ distribution. The mean of a $\chi^2_{(r-1)(c-1)}$ distribution is $(r-1)(c-1)$; we can calculate $E[X^2]$ or $E[G^2]$ under the complex sampling design when H_0 is true and compare the test statistic

$$X_F^2 = \frac{(r-1)(c-1)X^2}{E[X^2]}$$

or

$$G_F^2 = \frac{(r-1)(c-1)G^2}{E[G^2]}$$

to a $\chi^2_{(r-1)(c-1)}$ distribution. Bedrick (1983) and Rao and Scott (1984) showed that under H_0,

$$E[X^2] \approx E[G^2] \approx \sum_{i=1}^{r}\sum_{j=1}^{c}(1-p_{ij})d_{ij} - \sum_{i=1}^{r}(1-p_{i+})d_i^R - \sum_{j=1}^{c}(1-p_{+j})d_j^C, \qquad (10.9)$$

where d_{ij} is the deff for estimating p_{ij}, d_i^R is the deff for estimating p_{i+}, and d_j^C is the deff for estimating p_{+j}. In practice, if the estimator of the cell variances has κ df, it works slightly better to compare $X_F^2/(r-1)(c-1)$ or $G_F^2/(r-1)(c-1)$ to an F distribution with $(r-1)(c-1)$ and $(r-1)(c-1)\kappa$ df.

The first-order correction can often be used with published tables because variance estimates are needed only for the proportions in the contingency table—you need not estimate the full covariance matrix of the $\hat{p}_{ij}$ as is required for the Wald test. But the first-order correction only adjusts the test statistic so that its mean under H_0 is $(r-1)(c-1)$; p-values of interest come from the tail of the reference distribution, and it does not necessarily follow that the tail of the distribution of X_F^2 matches the tail of the $\chi^2_{(r-1)(c-1)}$ distribution. Rao and Scott (1981) showed that X_F^2 and G_F^2 have a null χ^2 distribution if and only if all the deffs for the variances and covariances of the $\hat{p}_{ij}$ are equal. Otherwise, the variance of X_F^2 is larger than the variance of a $\chi^2_{(r-1)(c-1)}$ distribution, and p-values from X_F^2 are often a bit smaller than they should be (but closer to the actual p-values than if no correction was done at all).

Second-order design correction. Rao and Scott (1981, 1984) also proposed a second-order correction—matching the mean and variance of the test statistic to the mean and variance of a χ^2 distribution, as done for analysis of variance (ANOVA) model tests by Satterthwaite (1946). Satterthwaite compared a test statistic T with skewed distribution to a χ^2 reference distribution by choosing a constant k and df ν so that $E[kT] = \nu$ and $V[kT] = 2\nu$ (ν and 2ν are the mean and variance of a χ^2 distribution with ν df). Here, letting $m = (r-1)(c-1)$, we know that $E[kX_F^2] = km$ and

$$V[kX_F^2] = V\left[\frac{kmX^2}{E(X^2)}\right] = \frac{k^2m^2\,V[X^2]}{[E(X^2)]^2},$$

so matching the moments gives

$$\nu = 2\frac{[E(X^2)]^2}{V[X^2]} \quad \text{and} \quad k = \frac{\nu}{m}.$$

Then,

$$X_S^2 = \frac{\nu X_F^2}{(r-1)(c-1)} \tag{10.10}$$

is compared to a χ^2 distribution with ν df. The statistic G_S^2 is formed similarly. Again, if the estimator of the variances of the $\hat{p}_{ij}$ has κ df, it works slightly better to compare X_S^2/ν or G_S^2/ν to an F distribution with ν and $\nu\kappa$ df.

Estimating $V[X^2]$ requires the complete covariance matrix of the $\hat{p}_{ij}$s, and is best done with software. If the deffs are all similar, the first- and second-order corrections will behave similarly. When the deffs vary appreciably, however, p-values using X_F^2 may be too small, and X_S^2 is preferred. Exercise 25 derives the theoretical results for the first- and second-order Rao–Scott tests.

Example 10.6. In the Survey of Youth in Custody, let's look at the relationship between age and whether the youth was known to be sent to the institution for a violent offense (variable *currviol* was defined to be 1 if variable *crimtype* = 1 and 0 otherwise). Using the final weights, we estimate the proportion of the population falling in each cell:

Estimated Proportions		**Age Class**			
		≤ 15	16 or 17	≥ 18	Total
Violent offense?	No	0.1698	0.2616	0.1275	0.5589
	Yes	0.1107	0.1851	0.1453	0.4411
	Total	0.2805	0.4467	0.2728	1.0000

First, let's look at what happens if we ignore the clustering and pretend that the test statistic in (10.2) follows a χ^2 distribution with 2 df. With $n = 2621$ youths in the table, Pearson's X^2 statistic is

$$X^2 = n\sum_{i=1}^{2}\sum_{j=1}^{3}\frac{(\hat{p}_{ij} - \hat{p}_{i+}\hat{p}_{+j})^2}{\hat{p}_{i+}\hat{p}_{+j}} = 33.99.$$

Comparing this unadjusted statistic to a χ_2^2 distribution yields an incorrect "p-value" of 4.2×10^{-8}.

Now let's look at the Rao–Scott tests. The following design effects were estimated using the stratification and clustering information in the survey:

Design Effects		**Age Class**			
		≤ 15	16 or 17	≥ 18	All Ages
Violent	No	14.9	4.0	3.5	6.8
Offense?	Yes	4.7	6.5	3.8	6.8
	Total	14.5	7.5	6.6	

Several of the deffs are very large, as might be expected because some facilities had mostly violent or mostly nonviolent offenders. We would expect the clustering to have a substantial effect on the hypothesis test.

Using (10.9), we estimate $E[X^2]$ by 4.89 and use $X_F^2 = 2X^2/4.89 = 13.9$. Comparing 13.9 to a χ_2^2 distribution (or comparing 13.9/2 to an $F_{2,1690}$ distribution) gives an approximate p-value of 0.001. This p-value may still be a bit too small, though, because of the wide disparity in the deffs.

The second-order Rao–Scott correction in (10.10), calculated using statistical software, has $\nu = 1.75$ and $X_S^2 = 10.86$. Comparing X_S^2 to a χ^2 distribution with 1.75 df results in a second-order corrected p-value of 0.003. For this example, because the deffs are highly variable, the second-order correction provides a better approximation to the distribution of the test statistic. ■

10.3.3 Model-Based Methods for Chi-Square Tests

The methods in Sections 10.3.1 and 10.3.2 use the covariance estimates of the proportions to adjust the χ^2 tests. A model-based approach may also be used. We describe a model due to Cohen (1976) for a cluster sample with two observation units per cluster. Extensions and other models that have been used for cluster sampling are described by Altham (1976), Brier (1980), Morel and Neerchal (2012), and Alonso-Revenga et al. (2017).

Example 10.7. Cohen (1976) presented an example exploring the relationship between gender and diagnosis with schizophrenia. The data consisted of 71 hospitalized pairs of siblings. Many mental illnesses tend to run in families, so we might expect that if one sibling is diagnosed with schizophrenia, the other sibling is more likely to be diagnosed with schizophrenia. Thus, any analysis that ignores the dependence among siblings is likely to give p-values that are much too small. If we just categorize the 142 patients by gender and diagnosis and ignore the correlation between siblings, we get the following table. Here, S means the patient was diagnosed with schizophrenia, and N means the patient was not diagnosed with schizophrenia.

	S	**N**	
Male	43	15	58
Female	32	52	85
	75	67	142

If analyzed using the assumption of multinomial sampling, $X^2 = 17.89$ and $G^2 = 18.46$. Such an analysis, however, assumes that all the observations are independent, so the "p-value" of 0.00002 is incorrect.

We know the clustering structure for the 71 clusters, though. You can see in Table 10.2 that most of the pairs fall in the diagonal blocks: If one sibling has schizophrenia, the other is more likely to have it. In 52 of the sibling pairs, either both siblings are diagnosed as having schizophrenia, or both siblings are diagnosed as not having schizophrenia.

TABLE 10.2
Cluster information for the 71 pairs of siblings.

		Younger Sibling				
		SM	SF	NM	NF	Total
	SM	13	5	1	3	22
Older	SF	4	6	1	1	12
Sibling	NM	1	1	2	4	8
	NF	3	8	3	15	29
	Total	21	20	7	23	71

Let q_{ij} be the probability that a pair falls in the (i, j) cell in the classification of the pairs. Thus, q_{11} is the probability that both siblings are schizophrenic and male, q_{12} is the probability that the younger sibling is a schizophrenic female and the older sibling is a schizophrenic male, etc. Then model the q_{ij}s by

$$q_{ij} = \begin{cases} aq_i + (1-a)q_i^2 & \text{if } i = j \\ (1-a)q_iq_j & \text{if } i \neq j \end{cases} \tag{10.11}$$

where a is a clustering effect and q_i is the probability that an individual is in class i ($i =$ SM, SF, NM, NF). If $a = 0$, members of a pair are independent, and we can just do the regular chi-square test using the individuals—the usual Pearson's X^2, calculated ignoring the clustering, would be compared to a $\chi^2_{(r-1)(c-1)}$ distribution. If $a = 1$, the two siblings are perfectly correlated so we essentially have only one piece of information from each pair—$X^2/2$ would be compared to a $\chi^2_{(r-1)(c-1)}$ distribution. For a between 0 and 1, if the model holds, $X^2/(1+a)$ approximately follows a $\chi^2_{(r-1)(c-1)}$ distribution under the null hypothesis.

The model may be fit by maximum likelihood (see Cohen, 1976, for details). Then, $\hat{a} = 0.3006$, and the estimated probabilities for the four cells are the following:

	S	**N**	**Total**
Male	0.2923	0.1112	0.4035
Female	0.2330	0.3636	0.5966
Total	0.5253	0.4748	1.0000

We can check the model by using a goodness-of-fit test for the clustered data in Table 10.2. This model does not exhibit significant lack of fit, while the model assuming independence does. For testing whether gender and schizophrenia are independent in the 2×2 table, $X^2/1.3006 = 13.76$, which we compare to a χ^2_1 distribution. The resulting p-value is 0.0002, about 10 times as large as the p-value from the analysis that pretended siblings were independent. ■

10.4 Loglinear Models

If there are more than two classification variables, we are often interested in seeing if there are more complex relationships in the data. Loglinear models are commonly used to study these relationships.

10.4.1 Loglinear Models with Multinomial Sampling

In a two-way table, if the row variable and the column variable are independent, then $p_{ij} = p_{i+}p_{+j}$ or, equivalently, $\ln(p_{ij}) = \ln(p_{i+}) + \ln(p_{+j})$. If the factors are independent, we would expect the model

$$\ln(p_{ij}) = \mu + \alpha_i + \beta_j, \tag{10.12}$$

where

$$\sum_{i=1}^{r} \alpha_i = 0 \quad \text{and} \quad \sum_{j=1}^{c} \beta_j = 0,$$

to describe the observed cell probabilities. This is called a **loglinear model** because the logarithms of the cell probabilities follow a linear model.

The observed cell probabilities $\hat{p}_{ij}$, $\hat{p}_{i+}$, and $\hat{p}_{+j}$ are calculated using the survey weights. For the data in Example 10.1, these are calculated by dividing each number in Table 10.1 by 500, and are shown in Table 10.3.

TABLE 10.3
Observed cell probabilities for data in Table 10.1.

		Computer?		
		Yes	No	
Cable?	Yes	0.238	0.376	0.614
	No	0.176	0.210	0.386
		0.414	0.586	1.000

The parameter estimates for the model in (10.12), calculated using statistical software with the observed probabilities in Table 10.3, are $\hat{\mu} = -1.428$, $\hat{\alpha}_1 = 0.232$ (corresponding to *cable* = "yes"), $\hat{\alpha}_2 = -0.232$ (*cable* = "no"), $\hat{\beta}_1 = -0.174$ (*computer* = "yes"), and $\hat{\beta}_2 = 0.174$ (*computer* = "no"). The fitted values of $\tilde{p}_{ij}$ under the model of independence are then

$$\tilde{p}_{ij} = \exp(\hat{\mu} + \hat{\alpha}_i + \hat{\beta}_j),$$

and are shown in Table 10.4.

TABLE 10.4
Estimated cell probabilities under independence model.

		Computer?		
		Yes	No	
Cable?	Yes	0.254	0.360	0.614
	No	0.160	0.226	0.386
		0.414	0.586	1.000

We would like to see how well the model in (10.12), where the two factors are independent, fits the data. We can do that in two ways:

1. Test the goodness of fit of the model using either X^2 in (10.5) or G^2 in (10.6): For a two-way contingency table, these statistics are equivalent to the statistics for testing independence. For the computer/cable example, the likelihood ratio statistic for goodness of fit is $G^2 = 2.28$. In multinomial sampling, X^2 and G^2 approximately follow a $\chi^2_{(r-1)(c-1)}$ distribution if the model is correct.

2. A full, or saturated, model for the data can be written as

$$\ln(p_{ij}) = \mu + \alpha_i + \beta_j + (\alpha\beta)_{ij}$$

with $\sum_{i=1}^{r}(\alpha\beta)_{ij} = \sum_{j=1}^{c}(\alpha\beta)_{ij} = 0$. The last term is analogous to the interaction term in a two-way ANOVA model. The saturated model will give a perfect fit to the observed cell probabilities because it has rc parameters. The null hypothesis of independent factors is equivalent to

$$H_0 : (\alpha\beta)_{ij} = 0 \quad \text{for} \quad i = 1, \ldots, r-1 \quad \text{and} \quad j = 1, \ldots, c-1.$$

Example 10.8. The following estimates (for the "no" levels of the variables) are obtained from statistical software for the saturated model with the computer/cable example:

Effect	**Estimate**	**Std. Error**	**Chi-Square**	***p*-value**
cable	−0.2211	0.0465	22.59	<0.0001
computer	0.1585	0.0465	11.61	0.0007
cable*computer	−0.0702	0.0465	2.28	0.1313

The predicted cell probabilities under the saturated model are those calculated from the cell counts in Table 10.3. The values in the column "Chi-Square" are the Wald chi-square statistics for testing whether each parameter is zero (these are the squared values of $\hat{\theta}_i/\text{SE}(\hat{\theta}_i)$ for each parameter). Note that $2.28 = G^2$, the value of the likelihood ratio test statistic for independence. Thus the p-value, under multinomial sampling, for testing whether the interaction term is zero is 0.1313—again, for this example, this is exactly the same as the p-value from the test for independence. The predicted cell probabilities under the saturated model are those calculated from the cell counts in Table 10.3. ■

10.4.2 Loglinear Models in a Complex Survey

What happens in a complex survey? We obtain point estimates of the model parameters like we always do, by using weights. Thus, we estimate p_{ij} by (10.7), and calculate pseudo-maximum likelihood estimates of the loglinear model parameters incorporating the weights. We can estimate standard errors for the model coefficients through linearization or a replication method, and carry out hypothesis tests using extensions of the Wald or Rao–Scott tests to loglinear models (Lumley and Scott, 2014, 2017).

Example 10.9. Let's look at a three-dimensional table from the Survey of Youth in Custody, to examine relationships among the variables used in Example 10.5 (*everviol*, was the offender ever put on probation or sent to a correctional institution for a violent offense; and *famtime*, was anyone in the family ever incarcerated), along with age group. The cell probabilities are p_{ijk}. The estimated probabilities $\hat{p}_{ijk}$, estimated using the survey weights, are in the following table:

		Family Member Incarcerated?				
		No		Yes		
		Ever Violent?		**Ever Violent?**		
		No	Yes	No	Yes	Total
Age	≤ 15	0.0588	0.0698	0.0659	0.0856	0.2801
Class	16–17	0.0904	0.1237	0.0944	0.1375	0.4461
	≥ 18	0.0435	0.0962	0.0355	0.0986	0.2738
	Total	0.1928	0.2896	0.1959	0.3217	1.0000

The saturated model for the three-way table is

$$\ln(p_{ijk}) = \mu + \alpha_i + \beta_j + \gamma_k + (\alpha\beta)_{ij} + (\alpha\gamma)_{ik} + (\beta\gamma)_{jk} + (\alpha\beta\gamma)_{ijk}.$$

Table 10.5 shows the parameter estimates, calculated using R (see Lu and Lohr, 2022), for the 11 terms in the saturated model, along with standard errors and p-values for the Wald tests that the individual parameters equal 0. There are two parameters for each term involving the three-category factor *ageclass*. The p-values can be used to test individual terms; alternatively, Rao–Scott corrections can be used to perform hypothesis tests comparing sets of nested models. ■

TABLE 10.5
Estimates for saturated loglinear model, Example 10.9.

Parameter	Estimate	Standard Error	p-Value
ageclass	−0.1149	0.1159	0.322
	0.3441	0.0726	<0.001
everviol	−0.2446	0.0453	<0.001
famtime	0.0242	0.0327	0.459
ageclass*everviol	0.1366	0.0582	0.019
	0.0724	0.0322	0.024
ageclass*famtime	0.0555	0.0368	0.132
	0.0128	0.0284	0.652
everviol*famtime	−0.0317	0.0230	0.168
ageclass*everviol*famtime	0.0089	0.0289	0.759
	0.0161	0.0278	0.562

10.5 Chapter Summary

Since many surveys collect categorical data, we often want to perform chi-square tests to explore association among variables. We can estimate probabilities in contingency tables using the sampling weights. Pearson and likelihood-ratio chi-square tests for association must be modified to account for the stratification and clustering in the survey design. Wald tests use the design-based variance so that the Wald test statistic approximately follows a chi-square distribution. The Rao–Scott test modifies the usual Pearson or likelihood-ratio test statistics by the average design effect to obtain corrected p-values.

Key Terms

Loglinear model: A model used for associations in categorical data.

Rao–Scott correction: A modification to a chi-square test statistic to account for the complex survey design.

Wald test: A Wald test statistic for the null hypothesis $H_0 : \boldsymbol{\theta} = \boldsymbol{\theta}_0$ has the form $X_W^2 = (\hat{\boldsymbol{\theta}} - \boldsymbol{\theta}_0)^T[\hat{V}(\hat{\boldsymbol{\theta}})]^{-1}(\hat{\boldsymbol{\theta}} - \boldsymbol{\theta}_0)$, where $\hat{V}(\hat{\boldsymbol{\theta}})$ accounts for the complex survey design.

For Further Reading

Agresti (2013) and Simonoff (2006) are general references for the analysis of categorical data in non-survey situations. Bilder and Loughin (2014) have a thorough treatment of analyzing categorical data from complex surveys.

A number of methods have been proposed to account for the survey design when testing for goodness of fit, homogeneity of populations, and independence of variables. Some of these methods are described in Rao and Thomas (1988, 1989); Thomas et al. (1996); Rao and Thomas (2003). Scott (2007) reviews Rao–Scott corrections for chi-square tests and outlines other areas of application. Lumley and Scott (2014, 2017) present methods for fitting and testing loglinear models in survey data.

10.6 Exercises

A. Introductory Exercises

1. Find an example or exercise in an introductory statistics textbook that performs a chi-square test on data from a survey. What design do you think was used for the survey? Is a chi-square test for multinomial sampling appropriate for the data? Why, or why not?

2. Read a research article in which a categorical data analysis is performed on survey data. Describe the sampling design and the method of analysis. Did the authors account for the design in their data analysis? Should they have analyzed the data differently?

3. Schei and Bakketeig (1989) took an SRS of 150 women between 20 and 49 years of age from Trondheim, Norway. Their goal was to investigate the relationship between sexual and physical abuse by a spouse and certain gynecological symptoms in the women. Of the 150 women selected to be in the sample, 15 had moved, 1 had died, 3 were excluded because they were not eligible for the study, and 13 refused to participate.

   Of the 118 women who participated in the study, 20 reported some type of sexual or physical abuse from their spouse: eight reported being hit, two being kicked or bitten, seven being beaten up, and three being threatened or cut with a knife. Seventeen of the women in the study reported a gynecological symptom of irregular bleeding or pelvic pain. The numbers of women falling into the four categories of gynecological symptom and abuse by spouse are given in the following contingency table:

		Abuse		
		No	Yes	
Gynecological	No	89	12	101
Symptom Present?	Yes	9	8	17
		98	20	118

   (a) If abuse and presence of gynecological symptoms are not associated, what are the expected probabilities in each of the four cells?

   (b) Perform a χ^2 test of independence for the variables abuse and presence of gynecological symptoms.

(c) What is the response rate for this study? Which definition of response rate did you use? Do you think that the nonresponse might affect the conclusions of the study? Explain.

4. Samuels (1996) collected data to examine how well students do in follow-up courses if the prerequisite course is taught by a part-time or full-time instructor. The following table gives results for students in Math I and Math II.

Instructor for Math I	Instructor for Math II	Grade in Math II A, B, C	Grade in Math II D, F, Withdraw	Total
Full Time	Full Time	797	461	1258
Full Time	Part Time	311	181	492
Part Time	Full Time	570	480	1050
Part Time	Part Time	909	449	1358
Total		2587	1571	4158

(a) The null hypothesis here is that the proportion of students receiving an A, B, or C is the same for each of the four combinations of instructor type. Is this a test of independence, homogeneity, or goodness of fit?

(b) Perform a hypothesis test for the null hypothesis in (a), assuming students are independent.

(c) Do you think the assumption that students are independent is valid? Explain.

B. Working with Survey Data

5. For the data in `schools.csv` (see Example 5.7), construct an estimate of the cross-classified contingency table of *mathlevel* by *readlevel*, and carry out a chi-square test of independence.

6. Use the data set `mysteries.csv` (see Exercise 13 of Chapter 3) with weight *p1weight*, for this exercise. Define variable *fdet* as 1 if the book has at least one female detective, and 0 otherwise.

(a) Estimate the population contingency table cross-classifying *fdet* with *authorgender*. (Table entries should sum to 655, the total number of novels in the population.)

(b) Perform a Rao–Scott chi-square test to evaluate the association between the two variables in (a). Do not use an fpc. Give the test statistic and p-value. Is there an association between the author gender and the presence of at least one female detective in the book?

7. Answer the questions in Exercise 6 for variables *authorgender* and *genre*.

8. Answer the questions in Exercise 6 for variables *authorgender* and *historical*.

9. In Example 10.5, we used linearization to estimate $V(\hat{\theta})$. We can alternatively use the random group method to estimate the variance of $\hat{\theta}$. Calculate $\hat{p}_{11}\hat{p}_{22} - \hat{p}_{12}\hat{p}_{21}$ for each of the seven random groups, as discussed in Example 9.4, and find the variance of the seven nearly independent estimates of θ. Form the Wald statistic based on your estimated variance. Since the estimate of the variance from the random groups method has only 6 df, the test statistic should be compared to an $F_{1,6}$ distribution rather than to a χ^2_1 distribution. Compare your p-value with that in Example 10.5.

10. *Public health articles.* Use the data in `healthjournals.csv` (see Exercise 12 in Chapter 3) for this exercise.

    (a) Estimate the population contingency for the cross-classification of variables *RandomSel* and *ConfInt*. Calculate a p-value for a test of independence using the Rao–Scott method.

    (b) Create a binary variable *authors_cat* that takes on value 1 if $Numauthors \geq 6$ and 0 if $NumAuthors \leq 5$. What is the odds ratio, along with a 95% CI, for the association between *authors_cat* and *RandomSel*?

11. Use the file `winter.csv` for this exercise (see Exercise 23 of Chapter 3).

    (a) Test the null hypothesis that *class* is not associated with *breakaga*. In the context of Section 10.1, what type of sampling was done?

    (b) Now construct a 2×2 contingency table for the variables *breakaga* and *work*. Use the sampling weights to estimate the probabilities p_{ij} for each cell.

    (c) Calculate the odds ratio using the $\hat{p}_{ij}$ from (b). How does this compare with an odds ratio calculated using the observed counts (and ignoring the sampling weights)?

    (d) Estimate $\theta = p_{11}p_{22} - p_{21}p_{12}$ using the $\hat{p}_{ij}$ you calculated in (b).

    (e) Test the null hypothesis $H_0 : \theta = 0$.

    (f) How did the stratification affect the hypothesis test?

12. Use the file `teachers.csv` for this exercise (see Exercise 15 of Chapter 5).

    (a) Construct a new variable *zassist*, which takes on the value 1 if a teacher's aide spends any time assisting the teacher, and 0 otherwise. Construct another new variable *zprep*, which takes on values Low, Medium, and High based on the amount of time the teacher spends in school on preparation.

    (b) Construct a 2×3 contingency table for the variables *zassist* and *zprep*. Use the sampling weights to estimate the probabilities p_{ij} for each cell.

    (c) Using the Rao–Scott method, test the null hypothesis that *zassist* is not associated with *zprep*.

13. The following data, given in Rao and Thomas (1989, p. 107), are from the Canada Health Survey. They relate smoking status (current smoker, occasional smoker, never smoked) to fitness level for 2,505 persons. Smokers who had quit were not included in the analysis. The estimated proportions in the table below were calculated with the sample weights. The design effects are in brackets. We would like to test whether smoking status and fitness level are independent.

Smoking Status	**Fitness level** Recommended	Minimum acceptable	Unacceptable	
Current	0.220 [3.50]	0.150 [4.59]	0.170 [1.50]	0.540 [1.44]
Occasional	0.023 [3.45]	0.010 [1.07]	0.011 [1.09]	0.044 [2.32]
Never	0.203 [3.49]	0.099 [2.07]	0.114 [1.51]	0.416 [2.44]
Total	0.446 [4.69]	0.259 [5.96]	0.295 [1.71]	1.000

(a) What is the value of X^2 if you assume the 2,505 observations were collected in a multinomial sample? Of G^2? What is the p-value for each statistic under multinomial sampling, and why are these p-values incorrect?

(b) Using (10.9), find the approximate expected value of X^2 and G^2.

(c) Calculate the corrected statistics X^2_F and G^2_F for these data, and find p-values for the hypothesis tests. Does the clustering in the Canada Health Survey make a difference in the p-value you obtain?

14. The following data, from Rao and Thomas (1988), were collected in the Canadian Class Structure Survey, a stratified multistage sample taken to study employment and social structure. Canada was divided into 35 strata by region and population size; two psus were sampled in 34 of the strata, and one psu was sampled in the 35th stratum. Variances were estimated using balanced repeated replication. Estimated design effects are in brackets behind the estimated proportion for each cell.

	Males	Females	Total
Decision-making managers	0.103 [1.20]	0.038 [1.31]	0.141 [1.09]
Advisor-managers	0.018 [0.74]	0.016 [1.95]	0.034 [1.95]
Supervisors	0.075 [1.81]	0.043 [0.92]	0.118 [1.30]
Semi-autonomous workers	0.105 [0.71]	0.085 [1.85]	0.190 [1.44]
Workers	0.239 [1.42]	0.278 [1.15]	0.516 [1.86]
Total	0.540 [1.29]	0.460 [1.29]	

(a) What is the value of X^2 if you assume the 1463 persons were surveyed in a simple random sample? Of G^2? What is the p-value for each statistic under multinomial sampling, and why are these p-values incorrect?

(b) Using (10.9), find the approximate expected value of X^2 and G^2.

(c) How many df are associated with the BRR variance estimates?

(d) Calculate the first-order corrected statistics X^2_F and G^2_F for these data, and find approximate p-values for the hypothesis tests. Does the clustering in the survey make a difference in the p-value you obtain?

(e) The second-order Rao–Scott correction gave test statistic $X^2_S = 38.4$, with 3.07 df. How does the p-value obtained using the X^2_S compare with the p-value from X^2_F?

15. Using the data in `syc.csv`, define the variable *currprop* as 1 if *crimtype* = 2 and 0 otherwise. Perform a Rao–Scott test of whether *currprop* is associated with age group, for the groups given in Example 10.6. Also give the table of estimated probabilities for the cross-classification.

16. Using the data in **`nhanes.csv`**, categorize the respondents into three groups: (1) body mass index (*bmxbmi*) < 25; (2) $25 \leq$ *bmxbmi* < 30, and (3) *bmxbmi* ≥ 30. Also create three age groups: 20–39, 40–59, and 60 and over. Using the weights in variable *wtmec2yr*, create a table of the estimated probabilities and design effects for the cross-classification of these two categorical variables, and perform a Rao–Scott test of association.

17. For the data in **`nhanes.csv`** form two categories for total cholesterol: high (200 mg/dL or higher) and low (less than 200 mg/dL). Form the age groups as in Exercise 16, and, using the weights in variable *wtmec2yr*, create a table of the estimated probabilities and design effects for the cross-classification of these two categorical variables. Perform a Rao–Scott test of association.

18. Fit a saturated loglinear model to the data in `nhanes.csv` using the categories of age, cholesterol, and body mass index in Exercises 16 and 17. Which factors appear to be associated?

C. Working with Theory

19. Some researchers have used the following method to perform tests of association in two-way tables. Instead of using the original observation weights w_k, they define

$$w_k^* = nw_k \Big/ \left(\sum_{i \in \mathcal{S}} w_i \right),$$

where n is the number of observation units in the sample. The sum of the new weights w_k^*, then, is n. The "observed" count for cell (i,j) is

$$x_{ij} = \text{sum of the } w_k^*\text{'s for observations in cell } (i,j)$$

and the "expected" count for cell (i,j) is $\hat{m}_{ij} = (x_{i+}x_{+j})/n$. The test statistic

$$\sum_{i=1}^{r} \sum_{j=1}^{c} \frac{(x_{ij} - \hat{m}_{ij})^2}{\hat{m}_{ij}}$$

is compared to a $\chi^2_{(r-1)(c-1)}$ distribution.

Does this test give correct p-values for data from a complex survey? Why, or why not? HINT: Try it out on the data in Examples 10.1 and 10.4.

20. *Linearization variance for Wald statistic* (requires calculus). Consider X_W^2 in (10.8).

    (a) Use the linearization method of Section 9.1 to approximate $V(\hat{\theta})$ in terms of $V(\hat{p}_{ij})$ and $\text{Cov}(\hat{p}_{ij}, \hat{p}_{kl})$. Show that if we let $y_{ijk} = 1$ if observation k is in cell (i,j) and 0 otherwise, then $\hat{V}(\hat{\theta}) = \hat{V}(\hat{\bar{q}})$, where $q_k = \hat{p}_{22}y_{11k} + \hat{p}_{11}y_{22k} - \hat{p}_{12}y_{21k} - \hat{p}_{21}y_{12k}$.

    (b) What is the Wald statistic, using the linearization estimate of $V(\hat{\theta})$ in (a), when multinomial sampling is used? Is this the same as Pearson's X^2 statistic? HINT: Under multinomial sampling, $V(\hat{p}_{ij}) = p_{ij}(1 - p_{ij})/n$ and $\text{Cov}(\hat{p}_{ij}, \hat{p}_{kl}) = -p_{ij}p_{kl}/n$.

21. *Estimating the log odds ratio in a complex survey* (requires calculus). Let

$$\theta = \ln\left(\frac{p_{11}p_{22}}{p_{12}p_{21}}\right) \text{ and } \hat{\theta} = \ln\left(\frac{\hat{p}_{11}\hat{p}_{22}}{\hat{p}_{12}\hat{p}_{21}}\right).$$

    (a) Use the linearization method of Section 9.1 to approximate $V(\hat{\theta})$ in terms of $V(\hat{p}_{ij})$ and $\text{Cov}(\hat{p}_{ij}, \hat{p}_{kl})$.

    (b) Using (a), show that $\hat{V}(\hat{\theta}) = \dfrac{1}{x_{11}} + \dfrac{1}{x_{12}} + \dfrac{1}{x_{21}} + \dfrac{1}{x_{22}}$ under multinomial sampling.

22. In Section 10.3.1, we used a Wald test for $H_0 : \theta = 0$, where $\theta = \theta_{11} = p_{11}p_{22} - p_{12}p_{21}$. An equivalent null hypothesis is $H_0 : \eta = 0$, where $\eta = \ln\left[(p_{11}p_{22})/(p_{12}p_{21})\right]$. Using the result of Exercise 21, derive the Wald test statistic for $H_0 : \eta = 0$.

23. *Bonferroni tests.* The Wald test often performs poorly for testing independence in an $R \times C$ table when R or C is large. Thomas (1989) suggested using the Bonferroni inequality in that situation. The null hypothesis of independence,

$$H_0 : \theta_{11} = 0,\ \theta_{12} = 0,\ \ldots,\ \theta_{r-1,c-1} = 0,$$

has $m = (r-1)(c-1)$ components:

$$\begin{aligned} H_0(1):&\quad \theta_{11} = 0 \\ H_0(2):&\quad \theta_{12} = 0 \\ &\quad \vdots \\ H_0(m):&\quad \theta_{(r-1)(c-1)} = 0, \end{aligned}$$

where $\theta_{ij} = p_{ij} - p_{i+}p_{+j}$. Using the Bonferroni inequality, we can test each component $H_0(k)$ separately with significance level α/m. The null hypothesis H_0 will be rejected at level α if any of the $H_0(k)$ is rejected at level α/m—that is, if for any i or j, $\hat{\theta}_{ij}^2/\hat{V}(\hat{\theta}_{ij}) > F_{1,\kappa,\alpha/m}$, where the estimator of the variance has κ df.

Carry out the Bonferroni test for the data in Example 10.6. Do you reach the same conclusion as in the example?

24. Show that for multinomial sampling, $X_F^2 = X^2$. HINT: what is $E[X^2]$ in (10.9) for a multinomial sample?

25. *Deriving the first- and second-order corrections to Pearson's* X^2 (requires mathematical statistics and theory of linear models; see Rao and Scott, 1981).

    (a) Suppose the random vector $\mathbf{Y}$ is normally distributed with mean $\mathbf{0}$ and covariance matrix $\boldsymbol{\Sigma}$. Then, if $\mathbf{C}$ is symmetric and positive definite, show that $\mathbf{Y}^T\mathbf{C}\mathbf{Y}$ has the same distribution as $\sum \lambda_i W_i$, where the W_is are independent χ_1^2 random variables and the λ_is are the eigenvalues of $\mathbf{C}\boldsymbol{\Sigma}$.

    (b) Let $\hat{\boldsymbol{\theta}} = (\hat{\theta}_{11}, \ldots, \hat{\theta}_{1,(c-1)}, \ldots, \hat{\theta}_{(r-1),1}, \ldots, \hat{\theta}_{(r-1),(c-1)})^T$, where $\hat{\theta}_{ij} = \hat{p}_{ij} - \hat{p}_{i+}\hat{p}_{+j}$. Let $\mathbf{A}$ be the covariance matrix of $\hat{\boldsymbol{\theta}}$ if a multinomial sample of size n is taken and the null hypothesis is true. Using (a), argue that $\hat{\boldsymbol{\theta}}^T\mathbf{A}^{-1}\hat{\boldsymbol{\theta}}$ asymptotically has the same distribution as $\sum \lambda_i W_i$, where the W_i are independent χ_1^2 random variables, and the λ_is are the eigenvalues of $\mathbf{A}^{-1}V(\hat{\boldsymbol{\theta}})$.

    (c) What are $E[\hat{\boldsymbol{\theta}}^T\mathbf{A}^{-1}\hat{\boldsymbol{\theta}}]$ and $V[\hat{\boldsymbol{\theta}}^T\mathbf{A}^{-1}\hat{\boldsymbol{\theta}}]$ in terms of the λ_is?

    (d) Find $E[\hat{\boldsymbol{\theta}}^T\mathbf{A}^{-1}\hat{\boldsymbol{\theta}}]$ and $V[\hat{\boldsymbol{\theta}}^T\mathbf{A}^{-1}\hat{\boldsymbol{\theta}}]$ for a 2×2 table. You may want to use your answer in Exercise 20.

26. We know the clustering structure for the data in Example 10.7. Use results from Chapter 5 (assume one-stage cluster sampling) to estimate the proportion for each cell and margin in the 2×2 table, and find the variance for each estimated proportion. Now use estimated design effects to perform a hypothesis test of independence using X_F^2. How do the results compare to the model-based test?

D. Projects and Activities

27. *Trucks.* Use the VIUS data described in Exercise 49 of Chapter 3. Define the variable *heavy* to be 1 if the gross vehicle weight is higher than 10,000 pounds and 0 otherwise, and define the variable *autotran* to be 1 if the vehicle has automatic transmission and 0 otherwise. Using the sample weights, construct a 2×2 table of estimated probabilities for the cross-classification of *heavy* and *autotran*. What is the design effect for each estimated proportion? Carry out a Rao–Scott test for independence. How do the results compare with a Wald test for independence?

28. *Baseball data.* For the sample you selected in Exercise 44 of Chapter 5, define the variable *pitcher* to be 1 if the player is a pitcher and 0 otherwise, and the variable *million* to be 1 if the salary is greater than $1 million and 0 otherwise.

    (a) Test whether the variables *pitcher* and *million* are associated, using the first-order Rao–Scott test.

    (b) Using the sampling weights, estimate the log odds ratio.

29. *IPUMS exercises.* Use your sample from Exercise 31 of Chapter 7 for this problem.

    (a) Create a categorical variable from *inctot* with two categories: low income and high income. Use the median income as the dividing point for the categories in the new variable *catinc*.

    (b) Conduct hypothesis tests to explore whether *catinc* is associated with (i) *race* or (ii) *sex*. What method did you use to account for the complex sampling design?

30. *Activity for course project.* Return to the survey you explored in Exercise 35 of Chapter 7. Now consider two categorical responses in the survey. Construct a two-way table of estimated proportions, using the weight variable. Conduct a hypothesis test to explore whether these variables are associated. What method did you use to account for the complex sampling design?

11

Regression with Complex Survey Data

Now he knew that he knew nothing fundamental and, like a lone monk stricken with a conviction of sin, he mourned, "If I only knew more! ... Yes, and if I could only remember statistics!"

—Sinclair Lewis, *It Can't Happen Here*

Example 11.1. How is fruit and vegetable consumption related to diagnosed or undiagnosed diabetes? What lifestyle factors are associated with higher levels of blood pressure? How does the mean value of total cholesterol vary by age?

In most of this book, we have emphasized estimating population means and totals—for example, how many adults have total cholesterol exceeding 240 mg/dL? Questions on the relation between variables, however, are often answered in statistics by using some form of a regression analysis. A response variable (for example, total cholesterol) is related to a number of explanatory variables (for example, age, fruit and vegetable consumption, and physical activity). We would like to be able to use the resulting regression equation to identify relationships among variables for the population and, sometimes, to predict the value of the response variable for units not in the sample.

You know how to fit regression models if the "usual assumptions," reviewed in Section 11.1, are met. These assumptions are often not met for data from complex surveys, however. To answer the questions above, for example, you might want to use data from the National Health and Nutrition Examination Survey (NHANES), discussed in Section 7.5. But, like most large-scale surveys, the NHANES is not a simple random sample (SRS). It is highly clustered and has unequal weights. Regression analyses that ignore the sampling design may give misleading impressions about relationships among variables. ■

As we found for analysis of contingency tables in the previous chapter, unequal probabilities of selection and the clustering and stratification of the sample complicate a statistical analysis. In NHANES, for example, the unequal weights may affect estimated regression coefficients. In addition, standard errors for the regression coefficients need to account for the stratification and clustering in the survey design; standard errors that are calculated under the assumption that observations are independent will be incorrect.

In this chapter, we explore how to do regression in complex sample surveys, accounting for unequal selection probabilities, stratification, and clustering in the survey design. We review the traditional model-based approach to regression analysis, as taught in introductory statistics courses, in Section 11.1. In Section 11.2, we discuss a design-based approach to linear regression, and present methods for calculating standard errors of regression coefficients; Section 11.5 applies these ideas to logistic regression. Section 11.3 uses regression methods to compare domain means. Section 11.4 contrasts design-based and model-based approaches to regression.

In Sections 11.1 to 11.5, our primary interest is in exploring the relation among different variables, and thus in estimating the regression coefficients. Section 11.6 revisits the use of regression estimation (see Chapter 4) for improving the precision of estimated totals, through calibration.

DOI: 10.1201/9780429298899-11

11.1 Model-Based Regression in Simple Random Samples

As usually exposited in areas of statistics other than sampling, regression inference is based on a model that is assumed to describe the relationship between the explanatory variable, x, and the response variable, y. The straight-line model commonly used for a single explanatory variable is

$$Y_i \mid x_i = \beta_0 + \beta_1 x_i + \varepsilon_i, \tag{11.1}$$

where Y_i is a random variable for the response, x_i is an explanatory variable, and β_0 and β_1 are unknown parameters. The Y_is are random variables; the data collected in the sample of size n are one realization of those n random variables, $\{y_i, i \in \mathcal{S}\}$. The ε_is, the deviations of the response variable about the line described by the model, are assumed to satisfy conditions (A1) through (A3):

(A1) $E_M[\varepsilon_i] = 0$ for all i. In other words, $E_M[Y_i|x_i] = \beta_0 + \beta_1 x_i$. The subscript M refers to taking the expected value under the model. This assumption is a way of saying that the model in (11.1) is appropriate for the population.

(A2) $V_M[\varepsilon_i] = \sigma^2$ for all i. The variance about the regression line is the same for all values of x.

(A3) $\text{Cov}_M[\varepsilon_i, \varepsilon_j] = 0$ for $i \neq j$. Observations are uncorrelated.

Often, (A4) is also assumed: It implies (A1) through (A3), and adds the additional assumption of normally distributed ε_is.

(A4) Conditionally on the x_is, the ε_is are independent and identically distributed from a normal distribution with mean 0 and variance σ^2.

The **ordinary least squares** (OLS) **estimators** of the parameters are the values $\hat{\beta}_0$ and $\hat{\beta}_1$ that minimize the residual sum of squares $\sum[y_i - (\beta_0 + \beta_1 x_i)]^2$. Estimators of the slope β_1 and intercept β_0 are obtained by solving the **normal equations**. For the model in (11.1), the normal equations are:

$$\begin{aligned} \beta_0\, n + \beta_1 \sum x_i &= \sum y_i \\ \beta_0 \sum x_i + \beta_1 \sum x_i^2 &= \sum x_i y_i. \end{aligned}$$

Solving the normal equations gives the parameter estimators

$$\begin{aligned} \hat{\beta}_1 &= \frac{\sum x_i y_i - \frac{1}{n}\left(\sum x_i\right)\left(\sum y_i\right)}{\sum x_i^2 - \frac{1}{n}\left(\sum x_i\right)^2} \\ \hat{\beta}_0 &= \frac{1}{n}\sum y_i - \hat{\beta}_1 \frac{1}{n}\sum x_i. \end{aligned} \tag{11.2}$$

Both $\hat{\beta}_1$ and $\hat{\beta}_0$ are linear in y, as we can write each in the form $\sum a_i y_i$ for known constants a_i. Although not usually taught in this form, it is equivalent to (11.2) to write

$$\hat{\beta}_1 = \sum_{i \in \mathcal{S}} \left[\frac{x_i - \frac{1}{n}\sum x_j}{\sum x_j^2 - \frac{1}{n}\left(\sum x_j\right)^2} \right] y_i$$

and

$$\hat{\beta}_0 = \sum_{i \in \mathcal{S}} \frac{1}{n} \left[1 - \frac{x_i \sum x_j - \frac{1}{n}\left(\sum x_j\right)^2}{\sum x_j^2 - \frac{1}{n}\left(\sum x_j\right)^2} \right] y_i.$$

If assumptions (A1) to (A3) are satisfied, then $\hat{\beta}_0$ and $\hat{\beta}_1$ are the **best linear unbiased estimators**—among all linear estimators that are unbiased under model (11.1), $\hat{\beta}_0$ and $\hat{\beta}_1$ have the smallest variance. If assumption (A4) is met, we can use the t distribution to construct confidence intervals (CIs) and hypothesis tests for the slope and intercept of the "true" regression line. Under assumption (A4),

$$\frac{\hat{\beta}_1 - \beta_1}{\sqrt{\hat{V}_M(\hat{\beta}_1)}}$$

follows a t distribution with $n-2$ degrees of freedom (df). For model (11.1), a model-unbiased estimator of the variance is

$$\hat{V}_M(\hat{\beta}_1) = \frac{\sum_{i \in \mathcal{S}} (y_i - \hat{\beta}_0 - \hat{\beta}_1 x_i)^2/(n-2)}{\sum_{i \in \mathcal{S}} (x_i - \bar{x})^2} \tag{11.3}$$

The coefficient of determination R^2 in model-based straight-line regression is the proportion of variability about the mean that is explained by the regression line:

$$R^2 = 1 - \frac{\sum_{i \in \mathcal{S}} \left(y_i - \hat{\beta}_0 - \hat{\beta}_1 x_i\right)^2}{\sum_{i \in \mathcal{S}} \left(y_i - \bar{y}\right)^2}. \tag{11.4}$$

Example 11.2. We illustrate straight-line regression models with one of the most famous data sets in the history of statistics, used by Student (1908) when developing properties of the t distribution. The anthropometric data reported by Macdonell (1901) contained the length of the left middle finger (cm) and height (inches) for 3,000 criminals. Many of the early investigations in statistical theory involved the collection of anthropometric data; see Albrizio (2007) and Langkjær-Bain (2019) for discussions of the role and legacy of anthropometry in the historical development of statistical methods. The data set for the population of the 3,000 men is in file `anthrop.csv`.

An SRS of 200 men (file `anthsrs.csv`) was taken from the 3,000 observations. Table 11.1 gives the output from fitting a straight line model with y = height and x = (finger length) using statistical software.

TABLE 11.1
Output for model-based regression in Example 11.2.

Parameter	Estimate	Standard Error	t Value	Pr > $\lvert t \rvert$	95% Confidence Interval	
Intercept	30.31625	2.56681	11.81	<0.0001	25.25445	35.37805
finger	3.04525	0.22172	13.73	<0.0001	2.60802	3.48248

The sample data are plotted along with the OLS regression line in Figure 11.1. The model appears to be a good fit to the data ($R^2 = 0.49$), and, using the model-based analysis, a

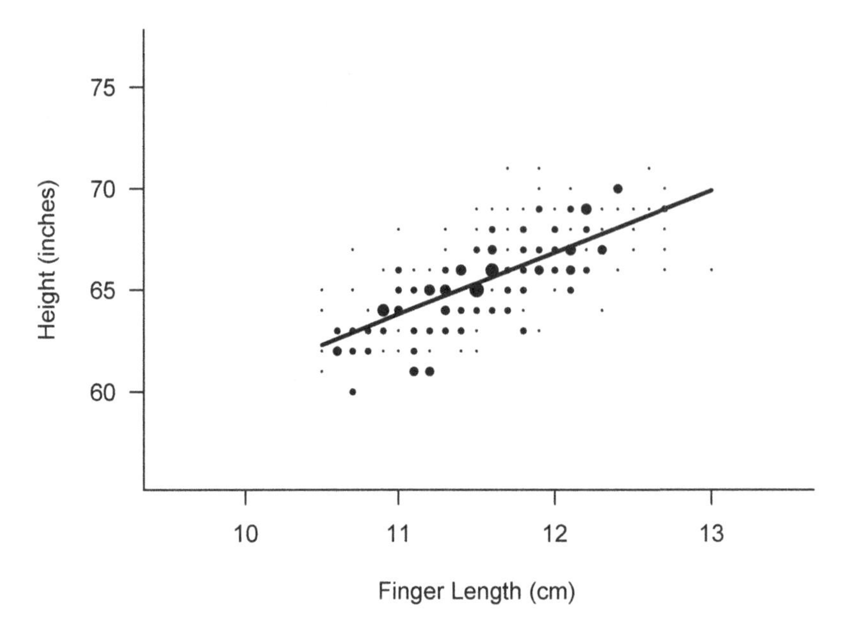

FIGURE 11.1
A plot of height vs. finger length for an SRS of 200 observations. The area of each circle is proportional to the number of observations at that value of (x, y). The OLS regression line, drawn in, has equation $y = 30.32 + 3.05x$.

95% CI for the slope of the line is

$$3.0453 \pm 1.972(0.2217) = [2.61, 3.48].$$

Under a model-based analysis, our sample can be thought of as resulting from the process of generating y values from the model in (11.1) for the set of x values in the sample. If we repeated that process over and over again, each time generating a new set of y values for the values of x in the sample $\mathcal{S}$, and if we constructed a CI for the slope for each sample, we would expect 95% of the resulting CIs to include the true value of β_1. ■

Remarks on applying regression to data from a sample

1. No assumptions whatsoever are needed to calculate the estimates $\hat{\beta}_0$ and $\hat{\beta}_1$ from the data; these are simply formulas. The assumptions in (A1) to (A4) are needed to make *inferences* about the "true" but unknown parameters β_0 and β_1 and about predicted values of the response variable. So the assumptions are used only when we construct a CI for β_0 or β_1 or for a predicted value, or when we want to say, for example, that $\hat{\beta}_1$ is the best linear unbiased estimator of β_1.

   The same holds true for other statistics we calculate. If we take a convenience sample of 100 persons, we can always calculate the average of those persons' incomes. But we cannot assess the accuracy of that statistic unless we make model assumptions about the population and sample. With a probability sample, however, we can use the sample design itself to make inferences and do not need to make assumptions about the model.

2. If the assumptions are not at least approximately satisfied, model-based inferences about parameters and predicted values will likely be wrong. For example, if observations are positively correlated rather than independent, the variance estimate from (11.3) is likely

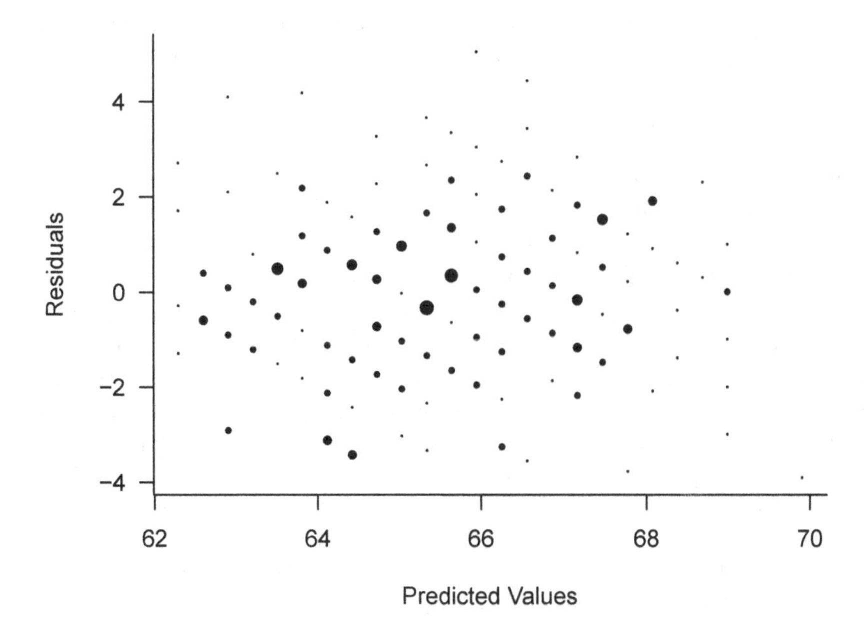

FIGURE 11.2
A plot of residuals for the model-based analysis of the SRS in Figure 11.1.

to be smaller than it should be. Consequently, regression coefficients are likely to be deemed statistically significant more often than they should be.

3. We can partially check the assumptions of the model by plotting the residuals and using various diagnostic statistics as described in the regression books listed in the For Further Reading section. One commonly used plot is that of residuals versus predicted values, used to check (A1) and (A2). For the data in Example 11.2, this plot is shown in Figure 11.2, and gives no indication that the data in the sample violate assumptions (A1) or (A2). This does not mean that the assumptions are true, just that we see nothing in the plot to indicate that they do not hold. Some of the assumptions, particularly independence, are quite difficult to check in practice. If a nonprobability sample is taken, we may have no way of knowing whether unsampled observations follow the model.

4. Regression techniques can be used to compare domain means. Let y be total cholesterol, and let x take on the value 1 if the person is female and 0 if the person is not female. Then the regression slope estimates the difference in mean cholesterol for females and non-females, and the test statistic for $H_0 : \beta_1 = 0$ is the pooled t-test statistic for the null hypothesis that the mean cholesterol is the same for both genders. Thus comparison of means for subpopulations, or domains, can be treated as a special case of regression analysis, as will be seen in Section 11.3.

11.2 Regression with Complex Survey Data

Many investigators performing regression analyses on complex survey data simply run the data through standard software for the model in (11.1) and report the parameter estimates and standard errors given by the software. One may debate whether to take a model-based or design-based approach (and we shall, in Section 11.4), but the data structure needs to be taken into account in either approach.

What can happen in complex surveys?

1. Observations may have different inclusion probabilities, π_i. If π_i is related to the response variable y_i, then an analysis that does not account for the different probabilities of selection may lead to biases in the estimated regression parameters. This problem is discussed in detail by Nathan (2005), who gives a bibliography of related literature.

   For example, suppose that an unequal-probability sample of 200 men is taken from the population described in Example 11.2 and that the inclusion probabilities are higher for the shorter men. (For illustration purposes, I used the y_is to set the inclusion probabilities, with π_i proportional to 24 for $y < 65$, 12 for $y = 65$, 2 for $y = 66$ or 67, and 1 for $y > 67$, with data in file `anthuneq.csv`.) Figure 11.3 shows a scatterplot of the data from this sample, along with the ordinary least squares regression line described in Section 11.1. The OLS regression equation is $y = 43.41 + 1.79x$, compared with the equation $y = 30.32 + 3.05x$ for the SRS in Example 11.2. Ignoring the inclusion probabilities in this example leads to a very different estimate of the regression line and distorts the relationship in the population. The scatterplot and line in Figure 11.3 describe the sample but not the population.

   Nonresponse can distort the relationship for much the same reason. If nonrespondents to a health survey are more likely to have high cholesterol, then a model predicting cholesterol from explanatory variables may not fit the nonrespondents.

2. Even if the sample is self-weighting, the standard errors given by non-survey regression software will likely be wrong if the survey design involves stratification or clustering. Usually, with clustering, the design effect (deff) for regression coefficients will be greater than 1.

11.2.1 Point Estimation

Traditionally, design-based sampling theory has been concerned with estimating quantities from a finite population such as $t_y = \sum_{i=1}^{N} y_i$ or $\bar{y}_{\mathcal{U}} = t_y/N$. In that descriptive spirit, then, the finite population quantities of interest for regression are the OLS coefficients for the population, B_0 and B_1, that minimize

$$\sum_{i=1}^{N}(y_i - B_0 - B_1 x_i)^2$$

over the entire finite population. It would be nice if the equation $y = B_0 + B_1 x$ summarizes useful information about the population (otherwise, why are you interested in B_0 and B_1?), but no assumptions are necessary to say that these are the quantities of interest.

As in Section 11.1, the normal equations are

$$\begin{aligned} B_0 N + B_1 \sum_{i=1}^{N} x_i &= \sum_{i=1}^{N} y_i \\ B_0 \sum_{i=1}^{N} x_i + B_1 \sum_{i=1}^{N} x_i^2 &= \sum_{i=1}^{N} x_i y_i, \end{aligned}$$

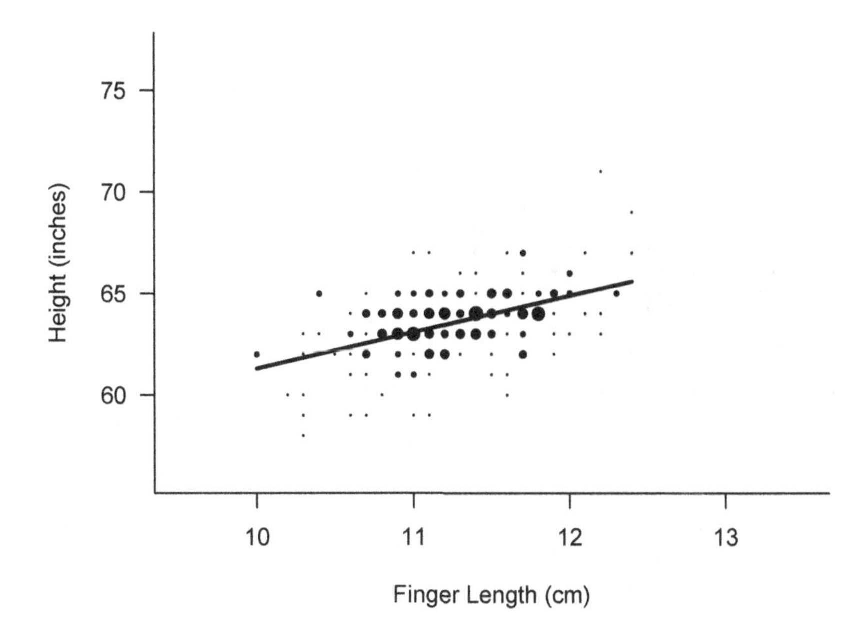

FIGURE 11.3
A scatterplot of y vs. x for the unequal-probability sample in `anthuneq.csv`. The circle area is proportional to the number of observations at that data point—not to the sum of weights at the point. The OLS line, ignoring the sampling weights, is $y = 43.41 + 1.79x$. The smaller slope of this line, when compared to the slope 3.05 for the SRS in Figure 11.1, reflects the undersampling of tall men. The OLS regression estimators are biased for the population quantities because they do not incorporate the unequal sampling weights.

and B_0 and B_1 can be expressed as functions of the population totals:

$$B_1 = \frac{\sum_{i=1}^{N} x_i y_i - \frac{1}{N}\left(\sum_{i=1}^{N} x_i\right)\left(\sum_{i=1}^{N} y_i\right)}{\sum_{i=1}^{N} x_i^2 - \frac{1}{N}\left(\sum_{i=1}^{N} x_i\right)^2} = \frac{t_{xy} - \frac{t_x t_y}{N}}{t_{x^2} - \frac{(t_x)^2}{N}} \tag{11.5}$$

$$B_0 = \frac{1}{N}\sum_{i=1}^{N} y_i - B_1 \frac{1}{N}\sum_{i=1}^{N} x_i = \frac{t_y - B_1 t_x}{N}. \tag{11.6}$$

We know the values for the entire population for the sample drawn in Example 11.2. These population values are plotted in Figure 11.4, along with the population least squares line $y = 30.179 + 3.056x$.

As both B_0 and B_1 are functions of population totals, we can use methods derived in earlier chapters to estimate each total separately and then substitute the estimators into (11.5) and (11.6). We estimate each population total in (11.5) and (11.6) using weights, with $\hat{N} = \sum_{i \in \mathcal{S}} w_i$, $\hat{t}_y = \sum_{i \in \mathcal{S}} w_i y_i$, $\hat{t}_x = \sum_{i \in \mathcal{S}} w_i x_i$, $\hat{t}_{xy} = \sum_{i \in \mathcal{S}} w_i x_i y_i$, and $\hat{t}_{x^2} = \sum_{i \in \mathcal{S}} w_i x_i^2$.

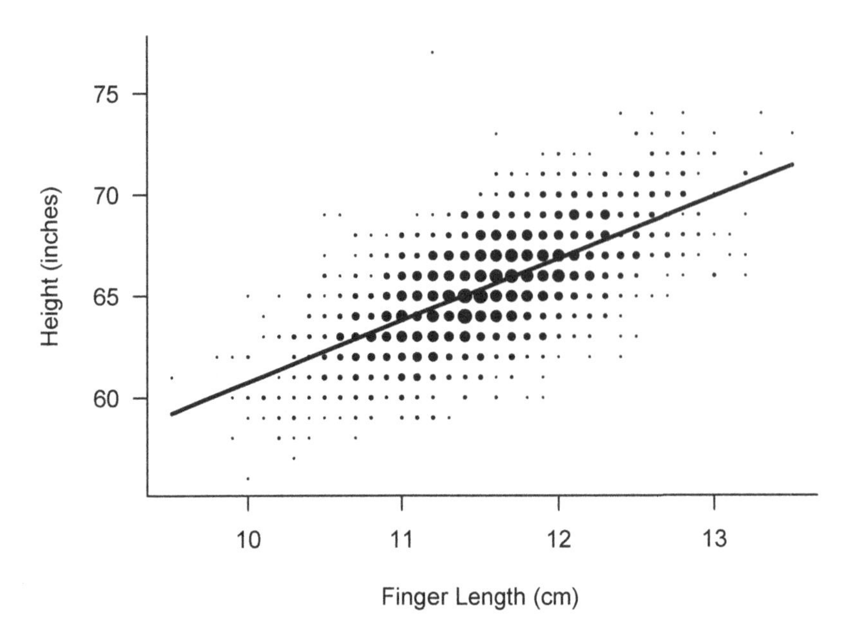

FIGURE 11.4
A plot of the population of the 3,000 observations in `anthrop.csv`. The area of each circle is proportional to the number of population observations at those coordinates. The population OLS regression line is $y = 30.18 + 3.06x$.

Then,

$$\hat{B}_1 = \frac{\sum_{i\in\mathcal{S}} w_i x_i y_i - \frac{1}{\sum_{i\in\mathcal{S}} w_i}\left(\sum_{i\in\mathcal{S}} w_i x_i\right)\left(\sum_{i\in\mathcal{S}} w_i y_i\right)}{\sum_{i\in\mathcal{S}} w_i x_i^2 - \frac{1}{\sum_{i\in\mathcal{S}} w_i}\left(\sum_{i\in\mathcal{S}} w_i x_i\right)^2} \tag{11.7}$$

and

$$\hat{B}_0 = \frac{\sum_{i\in\mathcal{S}} w_i y_i - \hat{B}_1 \sum_{i\in\mathcal{S}} w_i x_i}{\sum_{i\in\mathcal{S}} w_i}. \tag{11.8}$$

Computation and software. Although (11.7) and (11.8) are correct expressions for the estimators, they are subject to roundoff error and are not as good for computation as other algorithms that have been developed. In practice, use statistical software designed for estimating regression parameters in complex surveys, such as the packages described on pages xvi–xviii of the Preface. Lu and Lohr (2022) and Lohr (2022) give the code used to produce the estimates and graphs in this chapter using R and SAS software.

Before you use a software package to perform a regression analysis on sample survey data, investigate how it deals with missing data. Most packages exclude an observation from the analysis if it is missing the weight, the y variable, or any of the explanatory variables. If your survey has a large amount of item nonresponse on different variables, it is possible

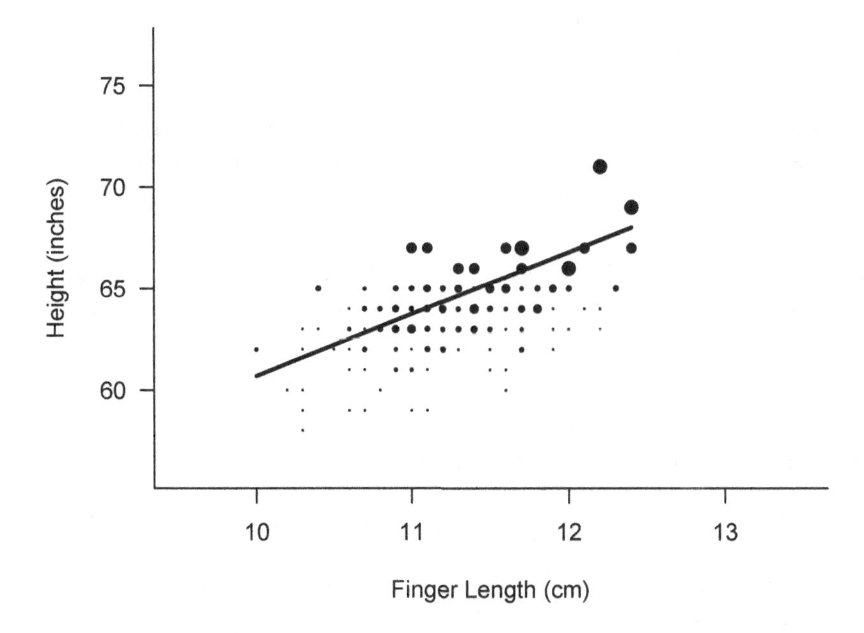

FIGURE 11.5
A scatterplot from the unequal-probability sample in `anthuneq.csv`. The area of each circle is proportional to the sum of the weights for observations with that value of x and y. Taller men have larger weights, so the regression line calculated using weights differs from the OLS line shown in Figure 11.3.

that you may end up performing your regression analysis using only 20 of the observations in your sample.

Example 11.3. Let's estimate the finite population quantities B_0 and B_1 for the unequal-probability sample in `anthuneq.csv` (plotted without weights in Figure 11.3). The point estimates, using the weights, are $\hat{B}_0 = 30.19$ and $\hat{B}_1 = 3.05$. If we ignored the weights and simply ran the observed data through a regression program for data assumed to meet model-based assumptions (A1)–(A3) from Section 11.1, we would get very different estimates: $\hat{\beta}_0 = 43.41$ and $\hat{\beta}_1 = 1.79$, which are the values in Figure 11.3.

The weighted bubble plot in Figure 11.5 shows why the weights, which are related to y, make a difference. Taller men had lower inclusion probabilities and thus not as many of them appeared in the unequal-probability sample. However, the taller men that were selected have higher sampling weights; a 69-inch man in the sample represents 24 times as many population units as a 60-inch man in the sample. When the weights are incorporated, estimates of the parameters are computed as though there were actually w_i data points with values (x_i, y_i). ■

11.2.2 Standard Errors

Let's now examine the effect of the complex sampling design on standard errors. As $\hat{B}_0$ and $\hat{B}_1$ are functions of estimated population totals, methods from Chapter 9 may be used to calculate variance estimates.

For any method of estimating the variance, under certain regularity conditions an approximate $100(1-\alpha)\%$ CI for B_1 is given by

$$\hat{B}_1 \pm t_{\alpha/2}\sqrt{\hat{V}(\hat{B}_1)},$$

where $t_{\alpha/2}$ is the upper $\alpha/2$ point of a t distribution with df associated with the variance estimate. For linearization, jackknife, bootstrap, or balanced repeated replication in a stratified multistage sample, we generally use (number of sampled psus) – (number of strata) as the df.

Standard errors using linearization. The linearization variance estimator for the slope may be used because B_1 is a smooth function of five population totals: From (11.5), $B_1 = h(t_{xy}, t_x, t_y, t_{x^2}, N)$, where

$$h(a,b,c,d,e) = \frac{a - bc/e}{d - b^2/e} = \frac{ea - bc}{ed - b^2}.$$

Using linearization, then, as you showed in Exercise 17 from Chapter 9,

$$\begin{aligned} V(\hat{B}_1) &\approx V\left[\frac{\partial h}{\partial a}(\hat{t}_{xy} - t_{xy}) + \frac{\partial h}{\partial b}(\hat{t}_x - t_x) + \frac{\partial h}{\partial c}(\hat{t}_y - t_y) + \frac{\partial h}{\partial d}(\hat{t}_{x^2} - t_{x^2}) + \frac{\partial h}{\partial e}(\hat{N} - N)\right] \\ &= V\left[\left\{t_{x^2} - \frac{(t_x)^2}{N}\right\}^{-1} \sum_{i\in\mathcal{S}} w_i(y_i - B_0 - B_1 x_i)(x_i - \bar{x}_{\mathcal{U}})\right]. \end{aligned}$$

Define

$$q_i = (y_i - \hat{B}_0 - \hat{B}_1 x_i)(x_i - \hat{\bar{x}}),$$

where $\hat{\bar{x}} = \hat{t}_x/\hat{N}$. Then, we may use

$$\hat{V}_L(\hat{B}_1) = \frac{\hat{V}\left(\sum_{i\in\mathcal{S}} w_i q_i\right)}{\left[\sum_{i\in\mathcal{S}} w_i x_i^2 - \frac{\left(\sum_{i\in\mathcal{S}} w_i x_i\right)^2}{\sum_{i\in\mathcal{S}} w_i}\right]^2} \tag{11.9}$$

to estimate the variance of $\hat{B}_1$.

Design- and model-based variance estimators in an SRS. The design-based variance estimator in (11.9) differs from the model-based variance estimator in (11.3), even if an SRS is taken. In an SRS of size n, ignoring the finite population correction (fpc),

$$\hat{V}\left(\sum_{i\in\mathcal{S}} w_i q_i\right) = \hat{V}(\hat{t}_q) = N^2 \frac{s_q^2}{n}$$

with

$$s_q^2 = \frac{1}{n-1}\sum_{i\in\mathcal{S}}(q_i - \bar{q}_{\mathcal{S}})^2 = \frac{1}{n-1}\sum_{i\in\mathcal{S}}(x_i - \bar{x}_{\mathcal{S}})^2(y_i - \hat{B}_0 - \hat{B}_1 x_i)^2.$$

Thus, for an SRS with replacement, (11.9) gives

$$\hat{V}_L(\hat{B}_1) = \frac{n}{n-1}\frac{\sum_{i\in\mathcal{S}}(x_i - \bar{x}_{\mathcal{S}})^2(y_i - \hat{B}_0 - \hat{B}_1 x_i)^2}{\left[\sum_{i\in\mathcal{S}}(x_i - \bar{x}_{\mathcal{S}})^2\right]^2}$$

but from (11.3),

$$\hat{V}_M(\hat{\beta}_1) = \frac{\sum_{i \in \mathcal{S}} (y_i - \hat{\beta}_0 - \hat{\beta}_1 x_i)^2}{(n-2) \sum_{i \in \mathcal{S}} (x_i - \bar{x})^2}.$$

Why the difference? The design-based estimator of the variance $\hat{V}_L$ comes from the inclusion probabilities of the design, while $\hat{V}_M$ comes from the average squared deviation over all possible realizations of the model. CIs constructed from the two variance estimates have different interpretations. With the design-based CI, $\hat{B}_1 \pm t_{\alpha/2}\sqrt{\hat{V}_L(\hat{B}_1)}$, the confidence level is $\sum u(\mathcal{S})P(\mathcal{S})$, where the sum is over all possible samples $\mathcal{S}$ that can be selected using the sampling design, $P(\mathcal{S})$ is the probability that sample $\mathcal{S}$ is selected, and $u(\mathcal{S}) = 1$ if the CI constructed from sample $\mathcal{S}$ contains the population characteristic B_1 and $u(\mathcal{S}) = 0$ otherwise. In an SRS, the design-based confidence level is the proportion of possible samples that result in a CI that includes B_1, from the set of all SRSs of size n from the finite population of fixed values $\{(x_1, y_1), (x_2, y_2), \ldots, (x_N, y_N)\}$.

For the model-based CI $\hat{\beta}_1 \pm t_{\alpha/2}\sqrt{\hat{V}_M(\hat{\beta}_1)}$, the confidence level is the expected proportion of CIs that will include β_1, from the set of (x, y) pairs that could be generated from the model in (A1) to (A4). Thus the model-based estimator is fit under the assumption that (A1) to (A4) hold for the infinite population mechanism that generates the values of y for the values of x in the sample. The SRS design of the sample makes assumption (A3) of uncorrelated observations reasonable. If a straight line model describes the relation between x and y, then (A1) is also plausible. A violation of assumption (A2) (equal variances), however, can have a large effect on inferences. The linearization design-based estimator of the variance is more robust to assumption (A2), as explored in Exercise 24.

Example 11.4. For the SRS in Example 11.2, the model-based and design-based estimates of the variance are quite similar, as the model assumptions appear to be met for the sample and population. For these data, $\hat{B}_1 = \hat{\beta}_1$ because $w_i = 3000/200$ for all i; $\hat{V}_L(\hat{B}_1) = 0.048$ and $\hat{V}_M(\hat{\beta}_1) = (0.2217)^2 = 0.049$. In other situations, however, the estimates of the variance can be quite different; usually, if there is a difference, the linearization estimate of the variance is larger than the model-based estimate of the variance because the linearization estimate in (11.9) is valid whether the model is "correct" or not. ■

Model assumptions (A2) and (A3) are usually not met for designs with stratification and clustering. If the sample is not self-weighting, model assumption (A1) may be violated as well, as the sampled units may have a different regression relationship than the non-sampled units.

Example 11.5. To estimate standard errors of the regression coefficients for the unequal-probability sample in Example 11.3 (file `anthuneq.csv`), define the new variable

$$q_i = (y_i - \hat{B}_0 - \hat{B}_1 x_i)(x_i - \hat{\bar{x}}) = (y_i - 30.1859 - 3.0541 x_i)(x_i - 11.51359).$$

(Note that $\hat{\bar{x}} = 11.51359$ is the estimate of $\bar{x}_{\mathcal{U}}$ calculated using the weights; the sample average of the 200 x_is in the sample is 11.2475, which is quite a bit smaller.) Then $\hat{V}(\sum_{i \in \mathcal{S}} w_i q_i) = 264{,}031$ and, using (11.9), $\hat{V}_L(\hat{B}_1) = 0.35$. If the weights are ignored, then the OLS analysis gives $\hat{\beta}_1 = 1.79$ and $\hat{V}_M(\hat{\beta}_1) = 0.05$. The estimated variance is much smaller using the model, but $\hat{\beta}_1$ is biased as an estimator of B_1. Since an unequal-probability sample was taken, $\hat{B}_1$ and $\hat{V}_L(\hat{B}_1)$ should be used, giving a 95% CI of $[1.89, 4.22]$ for B_1.

In practice, of course, we use survey software to compute the estimates. Table 11.2 shows the parameter estimates calculated using SAS software (Lohr, 2022). ■

TABLE 11.2
Output for straight-line regression in Example 11.5.

Parameter	Estimate	Standard Error	t Value	Pr > $\|t\|$	95% Confidence Interval	
Intercept	30.1858583	6.64323949	4.54	<0.0001	17.0856787	43.2860379
finger	3.0540995	0.58962334	5.18	<0.0001	1.8913879	4.2168111

Standard errors using jackknife. Suppose we have a stratified multistage sample, with weights w_i and H strata. A total of n_h psus are sampled in stratum h. Recall (see Section 9.3.2) that for jackknife iteration j in stratum h, we omit all observation units in psu j and recalculate the estimate using the remaining units. Define

$$w_{i(hj)} = \begin{cases} w_i & \text{if observation unit } i \text{ is not in stratum } h \\ 0 & \text{if observation unit } i \text{ is in psu } j \text{ of stratum } h \\ \dfrac{n_h}{n_h - 1} w_i & \text{if observation unit } i \text{ is in stratum } h \text{ but not in psu } j. \end{cases}$$

The jackknife estimator of the with-replacement variance of $\hat{B}_1$ is

$$\hat{V}_{\text{JK}}(\hat{B}_1) = \sum_{h=1}^{H} \frac{n_h - 1}{n_h} \sum_{j=1}^{n_h} (\hat{B}_{1(hj)} - \hat{B}_1)^2, \tag{11.10}$$

where $\hat{B}_1$ is defined in (11.7) and $\hat{B}_{1(hj)}$ is of the same form but with $w_{i(hj)}$ substituted for every occurrence of w_i in (11.7).

Example 11.6. For our two probability samples of size 200, in `anthsrs.csv` and `anthuneq.csv`,

$$\hat{V}_{\text{JK}}(\hat{B}_1) = \frac{199}{200} \sum_{j=1}^{200} \left(\hat{B}_{1(j)} - \hat{B}_1\right)^2,$$

where $\hat{B}_{1(j)}$ is the estimated slope when observation j is deleted and the other observations reweighted accordingly. The SRS and the unequal-probability sample have different weights and replicate weight values. For the SRS, the original weights are $w_i = 3000/200$; consequently, $w_{i(j)} = 200w_i/199 = 3000/199$ for $i \neq j$. Thus for the SRS, $\hat{B}_{1(j)}$ is the OLS estimate of the slope when observation j is omitted. For the SRS, we calculate $\hat{V}_{\text{JK}}(\hat{B}_1) = 0.050$.

For the unequal-probability sample, the original weights are $w_i = 1/\pi_i$ and the jackknife weights are $w_{i(j)} = 200w_i/199$ for $i \neq j$. The new weights $w_{i(j)}$ are used to calculate $\hat{B}_{1(j)}$ for each jackknife iteration, giving $\hat{V}_{\text{JK}}(\hat{B}_1) = 0.461$. The jackknife estimated variance is usually very close to the linearization variance. For this sample, the difference between the jackknife and linearization variances is larger than in a typical data set because the observations with large weights are highly influential for determining the fitted regression line (see Exercise 32). ■

11.2.3 Multiple Regression

Now let's give results for multiple regression in general. We rely heavily on matrix results found in the linear models and regression books listed in the For Further Reading section.

Suppose we wish to find a relation between y_i and a p-dimensional vector of explanatory variables $\mathbf{x}_i$, where $\mathbf{x}_i = [x_{i1}, x_{i2}, \ldots, x_{ip}]^T$ and the superscript T denotes the transpose of the vector. The finite population characteristic to be estimated is the p-dimensional vector of population parameters, $\mathbf{B}$, in the model $y = \mathbf{x}^T\mathbf{B}$. Define

$$\mathbf{y}_{\mathcal{U}} = \begin{bmatrix} y_1 \\ y_2 \\ \vdots \\ y_N \end{bmatrix} \text{ and } \mathbf{X}_{\mathcal{U}} = \begin{bmatrix} \mathbf{x}_1^T \\ \mathbf{x}_2^T \\ \vdots \\ \mathbf{x}_N^T \end{bmatrix}.$$

The normal equations for the entire population are

$$\mathbf{X}_{\mathcal{U}}^T\mathbf{X}_{\mathcal{U}}\mathbf{B} = \mathbf{X}_{\mathcal{U}}^T\mathbf{y}_{\mathcal{U}},$$

and, assuming that $(\mathbf{X}_{\mathcal{U}}^T\mathbf{X}_{\mathcal{U}})^{-1}$ exists, the finite population quantities of interest are

$$\mathbf{B} = (\mathbf{X}_{\mathcal{U}}^T\mathbf{X}_{\mathcal{U}})^{-1}\mathbf{X}_{\mathcal{U}}^T\mathbf{y}_{\mathcal{U}}, \tag{11.11}$$

the least squares estimates for the entire population.

Both $\mathbf{X}_{\mathcal{U}}^T\mathbf{X}_{\mathcal{U}}$ and $\mathbf{X}_{\mathcal{U}}^T\mathbf{y}_{\mathcal{U}}$ are matrices of population totals: $\mathbf{X}_{\mathcal{U}}^T\mathbf{X}_{\mathcal{U}} = \sum_{i=1}^N \mathbf{x}_i\mathbf{x}_i^T$ and $\mathbf{X}_{\mathcal{U}}^T\mathbf{y}_{\mathcal{U}} = \sum_{i=1}^N \mathbf{x}_i y_i$. The (j,k)th element of the $p \times p$ matrix $\mathbf{X}_{\mathcal{U}}^T\mathbf{X}_{\mathcal{U}}$ is $\sum_{i=1}^N x_{ij}x_{ik}$, and the kth element of the p-vector $\mathbf{X}_{\mathcal{U}}^T\mathbf{y}_{\mathcal{U}}$ is $\sum_{i=1}^N x_{ik}y_i$.

Thus, we can estimate the matrices $\mathbf{X}_{\mathcal{U}}^T\mathbf{X}_{\mathcal{U}}$ and $\mathbf{X}_{\mathcal{U}}^T\mathbf{y}_{\mathcal{U}}$ using weights. We estimate $\mathbf{X}_{\mathcal{U}}^T\mathbf{X}_{\mathcal{U}} = \sum_{i=1}^N \mathbf{x}_i\mathbf{x}_i^T$ by $\sum_{i\in\mathcal{S}} w_i\mathbf{x}_i\mathbf{x}_i^T$, and we estimate $\mathbf{X}_{\mathcal{U}}^T\mathbf{y}_{\mathcal{U}} = \sum_{i=1}^N \mathbf{x}_i y_i$ by $\sum_{i\in\mathcal{S}} w_i\mathbf{x}_i y_i$. Then, analogously to (11.7) and (11.8), define the estimator of $\mathbf{B}$ to be

$$\hat{\mathbf{B}} = \left(\sum_{i\in\mathcal{S}} w_i\mathbf{x}_i\mathbf{x}_i^T\right)^{-1} \sum_{i\in\mathcal{S}} w_i\mathbf{x}_i y_i. \tag{11.12}$$

Let

$$\mathbf{q}_i = \mathbf{x}_i(y_i - \mathbf{x}_i^T\hat{\mathbf{B}}).$$

Then, using linearization (see Exercise 27),

$$\hat{V}_L\left(\hat{\mathbf{B}}\right) = \left(\sum_{i\in\mathcal{S}} w_i\mathbf{x}_i\mathbf{x}_i^T\right)^{-1} \hat{V}\left(\sum_{i\in\mathcal{S}} w_i\mathbf{q}_i\right)\left(\sum_{i\in\mathcal{S}} w_i\mathbf{x}_i\mathbf{x}_i^T\right)^{-1}. \tag{11.13}$$

Confidence intervals for individual parameters may be constructed as

$$\hat{B}_k \pm t\sqrt{\hat{V}_L(\hat{B}_k)},$$

where t is the appropriate percentile from the t distribution.

R-squared for regression with complex survey data. In model-based regression, R^2 is the proportion of variability in the y variable that is explained by the regression model. When performing regression analyses with complex survey data, we estimate the statistics that would be obtained if we could fit the regression model to the entire population. Thus, $\hat{\mathbf{B}}$, the vector of estimated regression coefficients from the sample, estimates $\mathbf{B}$, the vector of coefficients that would be obtained by fitting the regression model to every observation in the population. We do the same thing for R^2—first, we define $R^2_{\mathcal{U}}$ to be the population value, and then we estimate it from the sample.

If we fit a regression model using OLS to the entire population, we would have, from (11.4), $R^2_{\mathcal{U}} = 1 - \text{SSW}_{\mathcal{U}}/\text{SSTO}_{\mathcal{U}}$, where $\text{SSTO}_{\mathcal{U}} = \sum_{i=1}^N (y_i - \bar{y}_{\mathcal{U}})^2$ is the corrected total sum of squares and $\text{SSW}_{\mathcal{U}} = \sum_{i=1}^N (y_i - \hat{y}_{i\mathcal{U}})^2$ is the residual sum of squares from an analysis of variance (ANOVA) table for the population. Here, $\bar{y}_{\mathcal{U}}$ is the population mean of y, and $\hat{y}_{i\mathcal{U}} = \mathbf{x}_i^T \mathbf{B}$ is the predicted value of y for the ith population member using the model with the population regression coefficients $\mathbf{B}$ in (11.11).

We can estimate $\text{SSTO}_{\mathcal{U}}$ and $\text{SSW}_{\mathcal{U}}$ using the weights as $\widehat{\text{SSTO}} = \sum_{i \in \mathcal{S}} w_i (y_i - \hat{\bar{y}})^2$ and $\widehat{\text{SSW}} = \sum_{i \in \mathcal{S}} w_i (y_i - \hat{y}_i)^2$, where $\hat{\bar{y}}$ is the Horvitz-Thompson estimate of $\bar{y}_{\mathcal{U}}$ and $\hat{y}_i = \mathbf{x}_i^T \hat{\mathbf{B}}$ is the predicted value of y for observation i using the estimated regression coefficients in (11.12). Then $R^2 = 1 - \widehat{\text{SSW}}/\widehat{\text{SSTO}}$ is an estimate of $R^2_{\mathcal{U}}$; it estimates the proportion of variability in the y variable, considering all the population members, that is explained by the population regression model.

Example 11.7. Return to the NHANES data that we plotted in Section 7.6.2. Figure 7.14 displayed a trend line for body mass index (BMI) plotted against age. We can, alternatively, fit a polynomial regression model to the sample. It appears from the data plots in Section 7.6.2 that a quadratic model might be reasonable. For this model, y_i = BMI (variable *bmxbmi*) for person i and $\mathbf{x}_i = [1, x_i, x_i^2]^T$, where x_i = age for person i.

The regression coefficients and fit statistics in Table 11.3 were estimated using SAS software, with standard errors calculated using linearization. The CIs were calculated using a t distribution with 15 df, setting the df equal to the number of psus (30) minus the number of strata (15). The value of R^2 for this model is 0.2834; we estimate that the quadratic regression model explains about 28% of the variability in BMI in the population.

TABLE 11.3
Output for quadratic model in Example 11.7.

Parameter	Estimate	Standard Error	t Value	Pr > $\|t\|$	95% Confidence Interval	
Intercept	15.2981488	0.22938064	66.69	<0.0001	14.8092355	15.7870621
x	0.6031084	0.01883971	32.01	<0.0001	0.5629525	0.6432643
x^2	−0.0057488	0.00023109	−24.88	<0.0001	−0.0062413	−0.0052563

Both the linear and the quadratic terms are statistically significant in this model. (In fact, with large data sets, it is common to have many of the predictors be statistically significant because the sample size is large.) The predicted value of y_i from the regression model is

$$\hat{y}_i = 15.298 + 0.603 x_i - 0.0057 x_i^2.$$

Although the regression model explains some of the variability in BMI, it is not a perfect fit to the data, as you will see when examining the residual plots in Exercise 30.

In design-based inference, however, no assumptions are made about the model and how well it describes the population. The values $\hat{B}_0$, $\hat{B}_1$, and $\hat{B}_2$ estimate the population quantities B_0, B_1, and B_2, which are the values that would minimize the sum of squares $\sum_{i=1}^{N} (y_i - B_0 - B_1 x_i - B_2 x_i^2)^2$ if the entire population were measured. Thus, the design-based estimates and standard errors are correct for inference about B_0, B_1, and B_2 even if the model does not describe the data well. We shall return to this issue in Section 11.4. ■

11.2.4 Regression Using Weights versus Weighted Least Squares

Many regression textbooks discuss regression estimation using weighted least squares (WLS) as a remedy for unequal variances. If the model generating the data is

$$Y_i = \mathbf{x}_i^T \beta + \varepsilon_i$$

with ε_i independent and normally distributed with mean 0 and variance σ_i^2, then ε_i/σ_i follows a normal distribution with mean 0 and variance 1. The WLS estimator is

$$\hat{\beta}_{\text{WLS}} = (\mathbf{X}^T\mathbf{\Sigma}^{-1}\mathbf{X})^{-1}\mathbf{X}^T\mathbf{\Sigma}^{-1}\mathbf{y},$$

with $\mathbf{\Sigma} = \text{diag}(\sigma_1^2, \sigma_2^2, \ldots, \sigma_n^2)$, the $n \times n$ matrix with diagonal entries $\sigma_1^2, \sigma_2^2, \ldots, \sigma_n^2$. The WLS estimator minimizes $\sum(y_i - x_i^T\beta)^2/\sigma_i^2$, and gives observations with smaller variance more weight in determining the regression equation. If the model holds, then, under WLS theory,

$$V_M(\hat{\beta}_{\text{WLS}}) = (\mathbf{X}^T\mathbf{\Sigma}^{-1}\mathbf{X})^{-1}.$$

We are *not* using WLS in this sense, even though our point estimator is the same: $\hat{\mathbf{B}}$ is the value that minimizes $\sum w_i(y_i - x_i^T\mathbf{B})^2$. Our weights come from the sampling design, not from an assumed covariance structure. Our estimated variance of the coefficients is not $(\mathbf{X}^T\hat{\mathbf{\Sigma}}^{-1}\mathbf{X})^{-1}$, the estimated variance under WLS theory, but is

$$\left(\sum_{i\in\mathcal{S}} w_i\mathbf{x}_i\mathbf{x}_i^T\right)^{-1} \hat{V}\left[\sum_{i\in\mathcal{S}} w_i\mathbf{x}_i(y_i - \mathbf{x}_i^T\hat{\mathbf{B}})\right]\left(\sum_{i\in\mathcal{S}} w_i\mathbf{x}_i\mathbf{x}_i^T\right)^{-1}.$$

One may, of course, combine the WLS approach as taught in regression courses with the finite population approach by defining the population quantities of interest to be

$$\mathbf{B} = (\mathbf{X}_\mathcal{U}^T\mathbf{\Sigma}_\mathcal{U}^{-1}\mathbf{X}_\mathcal{U})^{-1}\mathbf{X}_\mathcal{U}^T\mathbf{\Sigma}_\mathcal{U}^{-1}\mathbf{y}_\mathcal{U},$$

thus generalizing the regression model. This is essentially what is done in ratio estimation, using $\mathbf{\Sigma}_\mathcal{U} = \text{diag}(x_1, x_2, \ldots, x_N)$, as will be shown in Example 11.13.

11.3 Using Regression to Compare Domain Means

We often want to compare subgroups in a population. In Exercise 25 of Chapter 6, you showed that the method used to compare domain means in an SRS (namely, to compare the statistic $(\hat{\bar{y}}_1 - \hat{\bar{y}}_2)/\sqrt{\hat{V}(\hat{\bar{y}}_1) + \hat{V}(\hat{\bar{y}}_2)}$ to a t distribution) can give incorrect inferences with data from a complex survey. If clusters contain units from both domains, then $\hat{\bar{y}}_1$ and $\hat{\bar{y}}_2$ are correlated so that $V(\hat{\bar{y}}_1 - \hat{\bar{y}}_2) \neq V(\hat{\bar{y}}_1) + V(\hat{\bar{y}}_2)$.

But have no fear—we can use regression to compare domain means and to fit one-way and factorial analysis of variance (ANOVA) models. To compare the means for two domains which together comprise the entire population, define a new variable x with $x_i = 1$ if observation i is in domain 1 and $x_i = 0$ if observation i is in domain 2. Then the population slope B_1 in a straight-line regression model is $B_1 = \bar{y}_{1\mathcal{U}} - \bar{y}_{2\mathcal{U}}$ and $\hat{B}_1 = \hat{\bar{y}}_1 - \hat{\bar{y}}_2$ (see Exercise 21). Consequently, $\hat{V}(\hat{B}_1) = \hat{V}(\hat{\bar{y}}_1 - \hat{\bar{y}}_2)$, and the 95% confidence interval (CI) for B_1 is the 95% CI for the difference in domain means, $\bar{y}_{1\mathcal{U}} - \bar{y}_{2\mathcal{U}}$.

Example 11.8. Let's compare the mean value of body mass index for males and females (including persons of all ages) using the data in **nhanes**.csv (see Section 7.6.2 and Example 11.7). Create a variable x with $x_i = 1$ if person i is female and $x_i = 0$ if person i is male. Then fit the model $y = B_0 + B_1x$.

The parameter estimates are given in Table 11.4. The slope $\hat{B}_1$ is the difference in domain means, $\hat{\bar{y}}_{\text{female}} - \hat{\bar{y}}_{\text{male}} = 27.617 - 26.928 = 0.690$. A 95% CI for $\bar{y}_{\mathcal{U},\text{female}} - \bar{y}_{\mathcal{U},\text{male}}$ is [0.150, 1.229]. This interval does not contain 0, showing that the means are significantly different at the 0.05 level. ■

TABLE 11.4
Estimates for comparing domain means for Example 11.8.

Parameter	Estimate	Standard Error	t Value	Pr > $\|t\|$	95% Confidence Interval	
Intercept	26.9276308	0.20702536	130.07	<0.0001	26.4863667	27.3688949
x	0.6896696	0.25314568	2.72	0.0157	0.1501024	1.2292369

Comparing k domain means is similar. Let $x_{ij} = 1$ if observation i is in domain j and 0 otherwise, for $j = 1, \ldots, (k-1)$. The kth domain mean is estimated by $\hat{B}_0$, and $\hat{B}_j$ estimates the difference between the mean of domain j and the mean of domain k.

Example 11.9. Let's compare the mean value of body mass index for adults (age 20 and over) in five race and ethnicity groups that can be defined using variable *ridreth3* in the NHANES data: Hispanic, Asian non-Hispanic, Black non-Hispanic, White non-Hispanic, and Other. Figure 11.6 contains a side-by-side boxplot showing the quantiles and mean for each domain.

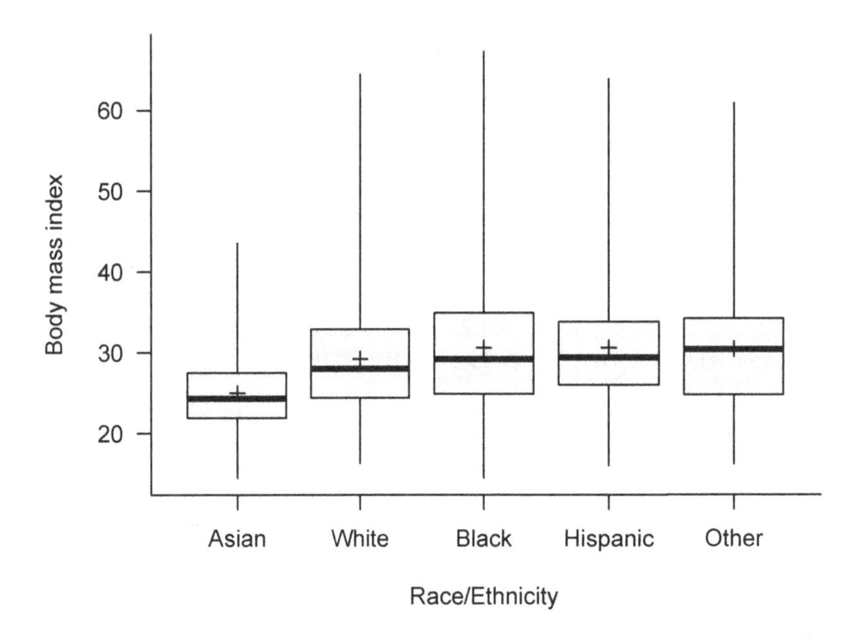

FIGURE 11.6
Boxplots of body mass index for adults in five race-ethnicity groups, incorporating the sampling weights in NHANES. The "+" indicates the estimated mean.

Table 11.5 shows the regression coefficients. For this analysis, "Asian" is the reference category, and the intercept estimates the mean BMI for Asian American adults. Each other regression coefficient represents the difference between the mean from that group and the mean of the reference category. Thus the estimated mean BMI for Hispanic adults is $24.97 + 5.63 = 30.60$ kg/m^2.

The F statistic for testing the null hypothesis that the means of all five groups are equal, calculated using the Wald method described in Section 10.3.1, is 131.57. This is compared to an F distribution with 4 and 15 df, giving a p-value that is less than 0.0001—strong evidence against the null hypothesis. You can also use the coefficients of the regression

TABLE 11.5
Regression coefficients for domain analysis in Example 11.9.

Parameter	Estimate	Standard Error	t Value	Pr > $\|t\|$	95% Confidence Interval	
Intercept	24.9706788	0.14116854	176.89	<0.0001	24.6697852	25.2715724
Black	5.6205312	0.33808373	16.62	<0.0001	4.8999227	6.3411396
Hispanic	5.6265680	0.33683525	16.70	<0.0001	4.9086207	6.3445154
Other	5.4714137	0.53378751	10.25	<0.0001	4.3336726	6.6091549
White	4.2598721	0.23723360	17.96	<0.0001	3.7542207	4.7655235

analysis to compare all pairs of group means, and adjust the p-values for multiple testing by using a multiple comparison method such as Bonferroni. ■

11.4 Interpreting Regression Coefficients from Survey Data

11.4.1 Purposes of Regression Analyses

In most areas of statistics, a regression analysis has one of three purposes:

1. It describes the relationship between two or more variables. Of interest may be the relationship between age and body mass index or the relationship between education level, income, and the probability of being a victim of violent crime. The interest is simply in a summary statistic that describes the association between the explanatory and response variables.

2. It predicts the value of y for a future observation. If we know the values for a number of demographic and health variables for an expectant mother, can we predict the birth weight of the infant, or the probability of the infant's survival?

3. It tells how to control future values of y by changing the values of the explanatory variables. For this purpose, we would like the regression equation to describe a cause-and-effect relationship between x and y.

Survey data can be used for the first and second purposes, but they generally cannot be used to establish causal relationships among variables. As Pearl (2009) explained, regression models deal with static relationships. A fitted regression model $\hat{y} = 2x$ says that the estimated mean value of y for units with $x = 2$ is twice as large as the estimated mean value of y for units with $x = 1$. To be able to conclude that changing x from 1 to 2 for a particular unit will cause y to double, you need additional information or assumptions. If, for example, the regression equation came from an experiment in which units were randomly assigned to different values of x, then, in some situations, you could conclude that changing x causes y to change. Otherwise, the adage "correlation does not imply causation" applies.

Sample surveys provide observational, not experimental, data. We observe a subset of possible explanatory variables, and these do not necessarily include the variables that are the root causes of changes in y. In a health survey intended to study the relationship between nutrition, exercise, and cancer incidence, survey participants may be asked about their diet and exercise habits (or the researcher may observe them) and be followed up later to see whether they have contracted cancer. Suppose that a regression analysis later indicates a significant negative association between vitamin E intake and cancer incidence, after adjusting for other variables such as age. The analysis only establishes association, not causation; you cannot conclude that cancer incidence will decrease if you start feeding people vitamin E. Although vitamin E could be the cause of the decreased cancer incidence, the cause could also be one of the unmeasured variables that is associated with both vitamin E intake and cancer incidence. To conclude that vitamin E affects cancer incidence, you need to perform an experiment: randomly assign study participants to vitamin E and no-vitamin-E groups, and observe the cancer incidence at a later time.

The purpose of a regression analysis often differs from that of an analysis to estimate population means and totals. When estimating the total number of unemployed persons from a survey, we are interested in the finite population quantity t_y; we want to estimate how many persons in the population were unemployed during a specific period of time.

But in a regression analysis, are you interested in B_0 and B_1, the summary statistics for the finite population? Or are you interested in uncovering a "universal truth"—to be able to say, for example, that not only do you find a positive association between dietary fat intake and systolic blood pressure for the population studied, but that you would expect a similar association in other populations? Cochran noted this point for comparison of domain means: "It is seldom of scientific interest to ask whether [the finite population domain means are equal], because these means would not be exactly equal in a finite population, except by rare chance. Instead, we test the null hypothesis that the two domains were drawn from *infinite* populations having the same mean" (1977, p. 39). Cochran's comments apply equally well to linear regression in general.

11.4.2 Model-Based and Design-Based Inference

Two basic approaches have been advocated for inference in surveys. Section 4.6 described model-based and design-based inference for use in regression estimation, where the interest is in using information from auxiliary variables to improve the estimation of population totals. Now let's look at a pure model-based approach and a pure design-based approach from the perspective of estimating regression coefficients.

1. *Model-based.* A stochastic model describes the relation between y_i and x_i that holds for every observation in the population. One possible model is

$$Y_i|\mathbf{x}_i = \mathbf{x}_i^T\boldsymbol{\beta} + \varepsilon_i, \tag{11.14}$$

where the ε_is are independent and normally distributed with mean zero and constant variance σ^2. If the observations in the population really follow the model, then the sample design should have no effect as long as the inclusion probabilities depend on y only through the x's. Since the model is assumed to hold for every data point in the sample, the OLS estimators

$$\hat{\boldsymbol{\beta}}_{\text{OLS}} = \left(\sum_{i\in\mathcal{S}} \mathbf{x}_i\mathbf{x}_i^T\right)^{-1} \sum_{i\in\mathcal{S}} \mathbf{x}_i Y_i \tag{11.15}$$

are unbiased for the parameters $\boldsymbol{\beta}$ that are assumed to generate the data.

Much is attractive about the model-based approach for regression: It links with sociological theories of the investigator, is consistent with other areas of statistics, and provides a mechanism for accounting for nonresponse. The model-based approach provides a framework for comparing theories about structural relationships. In addition, model-based estimates can be used with relatively small samples and with nonprobability samples. The standard errors of the model-based parameter estimates are usually smaller than those of the corresponding design-based estimates.

The properties of the OLS estimators $\hat{\boldsymbol{\beta}}_{\text{OLS}}$ hold only if the model assumptions are met, however. Thus, if the assumption that $E_M[\varepsilon_i] = 0$ is violated, then $\hat{\boldsymbol{\beta}}_{\text{OLS}}$ may be a biased estimator of $\boldsymbol{\beta}$. If the error terms ε_i are correlated, then the variance estimator in (11.3) is likely to underestimate the variance of $\hat{\beta}_1$.

A data analyst taking a model-based approach must examine the model assumptions carefully and do everything possible to check the adequacy of the model for the sample data. This includes plotting the data and residuals, performing diagnostic tests, and investigating possible violations of the model assumptions. But these activities may not catch all violations of the model assumptions—in particular, the assumption that the ε_is are independent is difficult to check for observational data. More importantly, there is no guarantee that the model applies to population units that are not in the sample; a model-based adherent must assume—often with no evidence—that it does.

2. *Design-based.* The design-based approach was presented in Section 11.2. The quantities of interest are the finite population characteristics $\mathbf{B}$ defined in (11.11), which are the OLS estimators of $\boldsymbol{\beta}$ from the data set that consists of the entire population. No assumptions are made about how well a model such as (11.14) fits the observations of the population, or about the distribution of error terms ε_i. Inferences are based on repeated sampling from the finite population, and the probability structure used for inference is that defined by the random variables indicating inclusion in the sample.

   Thus, under design-based theory, the regression coefficients $\hat{\mathbf{B}}$ given in (11.12) are approximately unbiased estimators of the finite population characteristics $\mathbf{B}$, and the linearization variance estimator in (11.13) estimates $E\left[(\hat{\mathbf{B}} - \mathbf{B})(\hat{\mathbf{B}} - \mathbf{B})^T\right]$, where the expectation is under repeated sampling from the population using the same sampling design. Even if the model in (11.14) provides a horrible fit to the population data, the inference under design-based theory will still be valid.

   Of course, design-based analysts usually decide to fit a particular model because they believe it a plausible candidate for describing the population (and may also believe it applies to units outside of the particular finite population studied). But they use the sampling weights to estimate the parameters and the sample design to estimate variances of the regression coefficients. All inference is based on the survey design.

The distinction between the approaches is important for the survey analyst because most regression programs use either a design-based or a model-based approach. The REG or GLM procedures in SAS software or the R function *lm* assume a model-based approach to regression, as exposited in Section 11.1. Survey software estimates the finite population parameters using the approach in Section 11.2. Thus it is important for you to know which approach you wish to take. Blindly running your data through software, without understanding what the software is estimating, can lead you to misinterpret the results.

With these issues in mind, let's look at the effects of survey weighting, stratification, and clustering on estimated regression coefficients and inference.

11.4.3 Survey Weights and Regression

Most statisticians agree that it is a good thing if a regression model describes the true state of nature. Thus, if it were known that a model would describe the relationship between y and $\mathbf{x}$ for every possible data point, then that model should be adopted. In the physical sciences, many models such as force = mass × acceleration can be theoretically derived. As long as you stay away from near-light velocity, any observation for which force, mass, and acceleration are accurately measured should be fit by the model. The design for how observations are sampled, then, should make little difference for finding the point estimates of regression coefficients, as every possible subset of the data is described by the model.

Unfortunately, theoretically derived models known to hold for all observations seldom exist for survey situations. An economist may conjecture a relationship between number of children, income, and amount spent on food, but there is no guarantee that this model will be appropriate for every subgroup in the population. Other variables may be related to the amount spent on food (such as educational level or amount of time away from home) but not measured in the survey. For many regression analyses, the x variables that are measured explain only a small part of the variability in the response variable.

Let's look at this in the context of the population anthropometry data plotted in Figure 11.4. The stochastic model in (11.1), $Y_i|x_i = \beta_0 + \beta_1 x_i + \varepsilon_i$, explains 43% (the value of $R^2_{\mathcal{U}}$) of the variability in the y variable, height, for the population of size 3,000. But it does not explain *all* of the variability. We might be able to obtain better predictions of y if the data set contained more explanatory variables (for example, genetic data, nutrition as a child, other anthropometric measurements), but even the best statistical models do not describe all of the factors associated with or influencing y. George Box expressed this idea in the famous aphorism: "All models are wrong but some are useful" (Box, 1979, p. 202).

Because the population regression model is not a perfect fit (in Box's words, the model is "wrong"), the sample selection method can cause the OLS estimates in (11.15) to be biased for estimating $\mathbf{B}$. We saw this bias in Figure 11.3 for the sample in `anthuneq.csv`, where tall men had lower inclusion probabilities than short men. The estimated OLS regression line was much flatter than the regression line fit to the entire population in Figure 11.4. For this example, the sampling weights contain information about the population relationship between y and x.

In the design-based approach, the estimates $\hat{\mathbf{B}}$, using the weights, are approximately unbiased for the finite population parameters $\mathbf{B}$, with approximate variance $V(\hat{\mathbf{B}})$ given in Exercise 27. This is true for sufficiently large samples even if the proposed population model in (11.14) is a poor description of the data. In Example 11.7, for instance, the model explains only about a quarter of the variability in BMI. A straight line model would describe even less of the variability, and would clearly not be a good choice for the data. Yet for either model (if nonresponse is not distorting the relationship among variables) CIs constructed as $\hat{B}_j \pm t\sqrt{\hat{V}(\hat{B}_j)}$ have approximately correct coverage probabilities.

A *model-assisted approach*, advocated by Särndal et al. (1992), combines features of a pure design-based and pure model-based approach. A model such as that in (11.14) is assumed to generate the finite population, but all inference is based on the survey design—that means that the survey weights are used to calculate the regression coefficients $\hat{\mathbf{B}}$ in (11.12) and standard errors are calculated using the survey design. If the model you posit really does describe the mechanism generating the population data, then the finite population quantity $\mathbf{B}$ should be close to the theoretical parameter $\boldsymbol{\beta}$. Then $\hat{\mathbf{B}}$ estimates the finite population quantity $\mathbf{B}$, which in turn estimates the theoretical regression parameters $\boldsymbol{\beta}$. Exercises 30 to 33 describe residual analyses and diagnostics that can be used to investigate possible inadequacies of the model for describing the finite population.

11.4.4 Survey Design and Standard Errors

Stratification can affect regression inference in two ways. First, unequal weights from a disproportional allocation may affect the regression coefficients $\hat{\mathbf{B}}$, as discussed in Section 11.4.3. The stratification also affects the standard errors of the coefficients.

Clustering in the survey design does not affect the point estimates $\hat{\mathbf{B}}$, which are calculated using the survey weights, but it does affect the standard errors of the regression coefficients. Ignoring the clustering in the inference is equivalent to assuming that observations in the same cluster are uncorrelated, which is not generally true. That said, the deffs of clustering for regression coefficients are often smaller than deffs for estimates of means or totals in the same survey. This is because the $\mathbf{x}$ variables in a regression model may explain some of the between-cluster variability in y.

Consider an education survey in which schools are the psus. You would expect the mean score from a standardized mathematics test to have a deff greater than 1 because students in the same school share environmental factors such as having the same teachers, living in the same neighborhood, having similar pre-school opportunities, and so on. The intraclass correlation coefficient (ICC) for math scores is expected to be positive (Hedges and Hedberg, 2007). Now suppose you fit a regression model predicting the math score from a variable related to the socio-economic status of the student's family. The mean of the variable varies from school to school, and it may well be associated with part of the school-to-school variability in math scores. The residuals from the model, therefore, are expected to exhibit less of a clustering effect than the test scores, and the estimated regression slope from the model is expected to have a smaller deff than the estimated mean test score (Lohr, 2014).

Because no regression model is "perfect," however, in most cluster samples the residuals from the model will still exhibit some clustering effects. With a design-based approach to inference, the clustering among the residuals is automatically incorporated into $\hat{V}(\hat{\mathbf{B}})$ in (11.13).

In a model-based approach, the dependence among observations in the same cluster must be explicitly included in the model. The next section shows one model that is commonly used for this.

11.4.5 Mixed Models for Cluster Samples

Chapter 5 discussed using a random effects model as a superpopulation model for cluster sampling. We can use this approach for regression analyses as well, by allowing different clusters to have their own regression equations, but relating the different regression equations for the clusters through a model.

Example 11.10. Consider the two-stage sample of students discussed in Example 5.7, where an SRS of students is taken from each school selected at the first stage of sampling. Let Y_{ij} be the math score of student j at school i in the sample, and let x_{ij} be the student's score from an assessment three years ago.

We expect a clustering effect in these data—measuring all variables that might be associated with mathematics scores is impossible, and the unmeasured characteristics of the schools, teachers, and neighborhoods induce a positive correlation among the scores within a school. For example, the math teachers in one school might be superb at inspiring students to learn mathematics, but that excellence would not be recorded in the survey. The students from that school might then all perform better than average on the test, so their scores are more similar, even after adjusting for known covariates, than scores of a random sample of students from the population. When unmeasured characteristics such as these are considered across all schools, the result is a positive intraclass correlation coefficient.

Thus a model $Y_{ij} = \beta_0 + x_{ij}\beta_1 + \varepsilon_{ij}$, where the ε_{ij}s are independent random variables having mean 0 and variance σ^2, is likely to be inappropriate for these data because it does not account for the positive intraclass correlation. If this inappropriate model is adopted, the p-values calculated for parameter estimates will be far too small. In addition, the model assumes that the relationship between current and previous score is the same for all schools.

A model that incorporates cluster effects and allows schools to have different slopes and intercepts is:

$$Y_{ij} = \beta_{0i} + (x_{ij} - \bar{x}_i)\beta_{1i} + \varepsilon_{ij}.$$

Here, the ε_{ij}s are assumed to be independent $N(0, \sigma^2)$ random variables; the mean of x_{ij} for school i, $\bar{x}_i$, is subtracted from each x_{ij} so that β_{0i} can be interpreted as the average test score in school i. School i has its own straight-line regression model with intercept β_{0i} and slope β_{1i}. But the slopes and intercepts from different schools are also related through a model. A simple model for the slopes and intercepts allows them to essentially be randomly distributed about a mean:

$$\beta_{0i} = \beta_0 + \delta_{0i}; \quad \beta_{1i} = \beta_1 + \delta_{1i},$$

with δ_{0i} and δ_{1i} following a bivariate normal distribution with $E_M[\delta_{0i}] = E_M[\delta_{1i}] = 0$, $V_M[\delta_{0i}] = \tau_{00}, V_M[\delta_{1i}] = \tau_{11}$, and $\text{Cov}\,(\delta_{0i}, \delta_{1i}) = \tau_{01}$. Under this situation, the model may be written as

$$Y_{ij} = \beta_0 + (x_{ij} - \bar{x}_i)\beta_1 + \delta_{0i} + (x_{ij} - \bar{x}_i)\delta_{1i} + \varepsilon_{ij}. \tag{11.16}$$

The parameter β_0 represents the mean score for schools; β_1 represents the mean slope for schools. The random effects δ_{0i} and δ_{1i} represent the difference in the intercept and slope between school i and the average values for intercept and slope for all schools; they measure the school effect. Finally, ε_{ij} refers to additional deviation from the mean for the individual student, after accounting for the previous test score and school effect. ■

The model in (11.16) is an example of a **mixed linear model**; it has both fixed (β_0 and β_1) and random (δ_{0i}, δ_{1i}, and ε_{ij}) coefficients. In econometrics, (11.16) is often referred to as a **random-coefficient** regression model; in the social sciences, it is called a **multilevel** or **hierarchical** linear model. Demidenko (2013) describes the theory of mixed models.

The mixed model in (11.16) is a superpopulation model and is assumed to hold for all schools and students in the population. The model can be fit to data from any sample—it is not limited to probability samples—as long as the model describes all schools in the population of interest.

The previous discussion described a completely model-based approach to fitting mixed models. One can also take a design-based approach when fitting mixed models to probability samples by using the survey weights to obtain estimates of the model parameters and the survey design to estimate standard errors. Such an approach also allows examination of models where the psus in the survey design may differ from the clusters in the model. For example, a researcher may be interested in the within-school intraclass correlation, so that it is desired to fit a hierarchical model with schools as the clusters, but the psus in the survey might be cities or counties. Pfeffermann et al. (1998), Rabe-Hesketh and Skrondal (2006), Rao et al. (2013), Yi et al. (2016), and Dumitrescu et al. (2021) discuss the use of mixed models with survey data.

11.5 Logistic Regression

In linear regression, the response variable is usually considered to be approximately continuous—for example, body mass index, income, or leaf area. In surveys, however, many variables of interest are dichotomous, with y_i taking values of 1 (yes) or 0 (no). Logistic regression is often used to predict probabilities of having response 1 for dichotomous variables.

First, let's review logistic regression from a model-based viewpoint. Let $\mathbf{x}$ be a vector of explanatory variables and $\boldsymbol{\beta}$ be the vector of unknown parameters. Then the standard logistic regression model takes the form

$$p(\mathbf{x}) = \frac{\exp(\mathbf{x}^T\boldsymbol{\beta})}{1+\exp(\mathbf{x}^T\boldsymbol{\beta})}, \tag{11.17}$$

where $p(\mathbf{x})$ represents the probability that a unit with covariates $\mathbf{x}$ will have a response of 1. Alternatively, the model may be expressed in logit scale, where $\text{logit}(p) = \ln[p/(1-p)]$:

$$\text{logit}[p(\mathbf{x})] = \mathbf{x}^T\boldsymbol{\beta}. \tag{11.18}$$

Example 11.11. For the data in Example 10.1, let $y_i = 1$ if household i has a computer and $y_i = 0$ if household i does not have a computer. Let $x_i = 1$ if household i subscribes to cable and $x_i = 0$ if household i does not subscribe to cable. The fitted logistic regression model is

$$\widehat{\text{logit}}\,[p_i] = -0.177 - 0.281x_i.$$

Note that the slope, -0.281, is the log odds ratio from Example 10.1. It is easy to transform back to predicted probabilities: When $x = 1$, then $\ln[\hat{p}/(1-\hat{p})] = -0.4573184$ so that

$$\hat{p}(1) = \frac{\exp(-0.4573184)}{1+\exp(-0.4573184)} = 0.388 = \frac{119}{307}. \blacksquare$$

Much of the discussion in this chapter on linear regression also applies to logistic regression—a complex sample design will affect standard errors of the logistic regression coefficients, just as it affects standard errors of the linear regression coefficients. Logistic regression with one dichotomous independent variable is essentially equivalent to finding the odds ratio in a 2×2 contingency table, so the discussion in Chapter 10 about how the sampling design affects standard goodness-of-fit tests also applies to significance tests of logistic regression coefficients.

Just as for linear regression, the population quantities of interest are defined to be the coefficients $\mathbf{B}$ that would be obtained from fitting the logistic regression model to the entire population. Logistic regression models are usually fit using maximum likelihood, and the likelihood function (assuming independence) for the N units in the population is

$$\mathcal{L}(\boldsymbol{\beta}) = \prod_{i=1}^{N} p_i^{y_i}(1-p_i)^{1-y_i}, \tag{11.19}$$

where $p_i = \exp(\mathbf{x}_i^T\boldsymbol{\beta})/[1+\exp(\mathbf{x}_i^T\boldsymbol{\beta})]$ represents the probability that a unit with covariates $\mathbf{x}_i$ has a response of 1. The finite population parameter $\mathbf{B}$ is then defined to be the maximum likelihood estimate of $\boldsymbol{\beta}$ using (11.19). The parameter $\mathbf{B}$ is the solution to the system of equations

$$\sum_{i=1}^{N} x_{ij}\left[y_i - \frac{\exp(\mathbf{x}_i^T\mathbf{B})}{1+\exp(\mathbf{x}_i^T\mathbf{B})}\right] = 0 \quad \text{for } j = 1,\ldots,p \tag{11.20}$$

if all elements in the population could be observed.

Now that $\mathbf{B}$ is defined, calculate $\hat{\mathbf{B}}$ by substituting estimators for the population totals in (11.20). A design-based estimator of $\mathbf{B}$ is given by the solution $\hat{\mathbf{B}}$ to

$$\sum_{i\in\mathcal{S}} w_i x_{ij}\left[y_i - \frac{\exp(\mathbf{x}_i^T\hat{\mathbf{B}})}{1+\exp(\mathbf{x}_i^T\hat{\mathbf{B}})}\right] = 0 \quad \text{for } j = 1,\ldots,p, \tag{11.21}$$

where $\mathcal{S}$ denotes the units included in the sample. The ith observation in the sample represents w_i observations in the population.

The coefficients $\hat{\mathbf{B}}$ are defined implicitly in (11.21), so a linearization variance estimator may be obtained using methods in Binder (1983). Any of the resampling methods in Chapter 9 can be used to estimate the variance of logistic regression coefficients.

Example 11.12. Let's fit a logistic regression model with the NHANES data to predict the event that body mass index in an adult exceeds 30, using covariates waist circumference (variable *bmxwaist*) and gender (*female* = 1 if female and 0 if male).

Table 11.6 gives the regression coefficients calculated by SAS software. It shows a Wald test (see Section 10.3.1) for each coefficient, where the t value equals $\hat{B}_j/\sqrt{\hat{V}(\hat{B}_j)}$ and is compared to a t distribution with 15 df. We can also carry out a Rao–Scott likelihood ratio test for the null hypothesis that the coefficients of *bmxwaist* and *female* are both 0. Using the second-order correction, the F statistic is 1836.11, and is compared to an F distribution with 1.80 and 27.04 df, giving a p-value less than 0.0001.

TABLE 11.6
Regression coefficients for Example 11.12.

Parameter	Estimate	Standard Error	t Value	Pr > $\|t\|$	95% Confidence Interval	
Intercept	−29.9557	1.1921	−25.13	<0.0001	−32.4967	−27.4148
bmxwaist	0.2809	0.0115	24.43	<0.0001	0.2564	0.3054
female	1.5786	0.1666	9.48	<0.0001	1.2236	1.9337

The coefficients in Table 11.6 can be used to estimate the predicted probability that a person has BMI > 30. For males, where the value of covariate *female* is 0, we estimate that $\ln(\hat{p}) = -29.9557 + 0.2809\, bmxwaist$ and thus that

$$\hat{p}(\text{male}) = \exp(-29.9557 + 0.2809\, bmxwaist)/[1 + \exp(-29.9557 + 0.2809\, bmxwaist)].$$

Similarly, we have intercept $-29.9557 + 1.5786 = -28.3771$ for females, and

$$\hat{p}(\text{female}) = \exp(-28.3771 + 0.2809\, bmxwaist)/[1 + \exp(-28.3771 + 0.2809\, bmxwaist)].$$

The top graph in Figure 11.7 shows the predicted probabilities that BMI > 30 from the model for each value of the covariates. The boxplots in the bottom panel show the distribution, calculated using the weights, of the covariate *bmxwaist* for each gender and BMI response category. The mean and quantiles of waist circumference are higher for males than for females, and much higher for persons with BMI > 30 than for persons with BMI <= 30, and these differences can be seen in the predicted probabilities in the top panel.

The model with coefficients in Table 11.6 assumes that the coefficient of *bmxwaist* is the same for each gender; see Exercise 18 for a model that allows this coefficient to differ for the two genders. ■

11.6 Calibration to Population Totals

In Chapter 4, we introduced ratio and regression estimation in the setting of SRSs, with estimators

$$\hat{t}_{yr} = \frac{\hat{t}_y}{\hat{t}_x} t_x$$

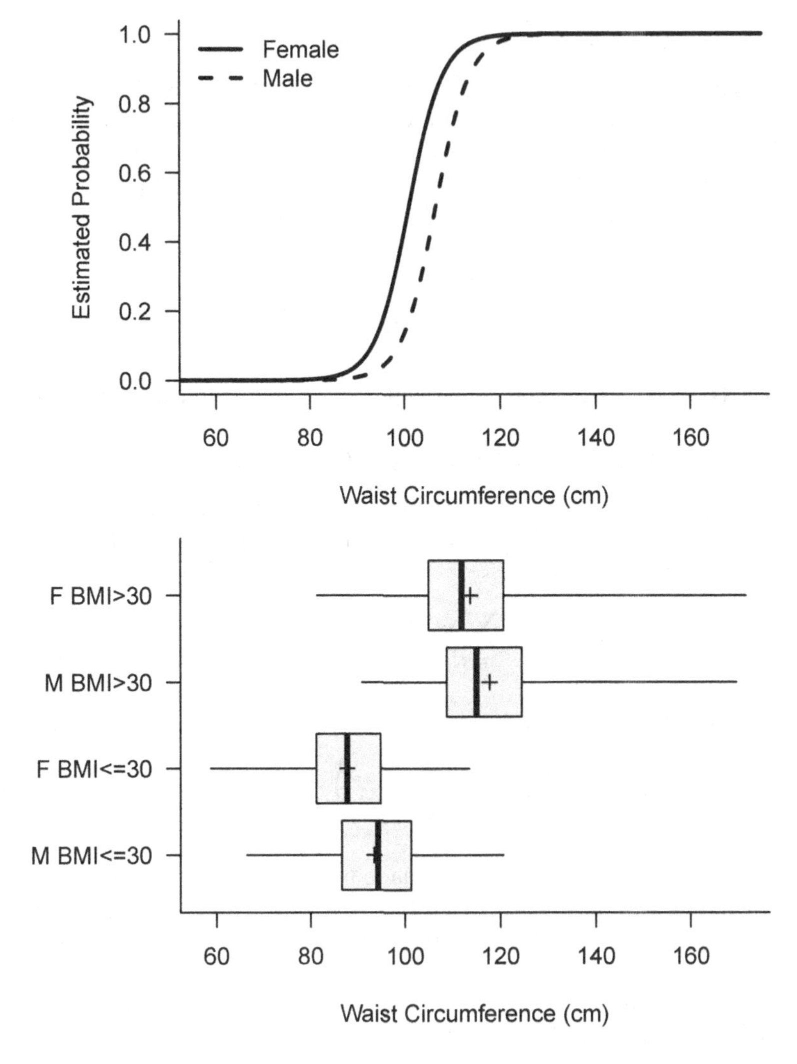

FIGURE 11.7
Predicted probability that BMI > 30, from the model in Table 11.6, and boxplots showing the distribution of waist circumference for each level of the response variable and each gender (F or M).

and

$$\hat{t}_{y\text{reg}} = \hat{t}_y + \hat{B}_1(t_x - \hat{t}_x).$$

Now let's extend these estimators to complex survey samples. We want to reduce the mean squared error of the estimator $\hat{t}_y = \sum_{i \in \mathcal{S}} w_i y_i$ by including auxiliary information through the working model

$$Y_i | \mathbf{x}_i = \mathbf{x}_i^T \boldsymbol{\beta} + \varepsilon_i, \tag{11.22}$$

with $\mathbf{x}_i^T = (x_{i1}, x_{i2}, \ldots, x_{ip})$ and $V_M(\varepsilon_i) = \sigma_i^2$ for σ_i^2 known. We assume that the vector of true population totals $\mathbf{t}_\mathbf{x}$ is known and thus can be used to adjust the estimator $\hat{t}_y$. Using a working model in (11.22), but relying on the sampling design for inference, is an example of the model-assisted approach described in Section 11.4.3.

Define

$$\mathbf{B} = (\mathbf{X}_{\mathcal{U}}^T \mathbf{\Sigma}_{\mathcal{U}}^{-1} \mathbf{X}_{\mathcal{U}})^{-1} \mathbf{X}_{\mathcal{U}}^T \mathbf{\Sigma}_{\mathcal{U}}^{-1} \mathbf{y}_{\mathcal{U}},$$

where $\mathbf{\Sigma}_{\mathcal{U}}$ is a diagonal matrix with ith diagonal element σ_i^2. The finite population parameter $\mathbf{B}$ is the weighted least squares estimate of $\boldsymbol{\beta}$ for observations in the population, using the model in (11.22). Thus the form of $\mathbf{B}$ is inspired by (11.22), but we then treat $\mathbf{B}$ as a finite population quantity to be estimated using information in the sample. Note that $\mathbf{X}_{\mathcal{U}}^T \mathbf{\Sigma}_{\mathcal{U}}^{-1} \mathbf{X}_{\mathcal{U}} = \sum_{i=1}^N \mathbf{x}_i \mathbf{x}_i^T / \sigma_i^2$ and $\mathbf{X}_{\mathcal{U}}^T \mathbf{\Sigma}_{\mathcal{U}}^{-1} \mathbf{y}_{\mathcal{U}} = \sum_{i=1}^N \mathbf{x}_i y_i / \sigma_i^2$. Thus, $\mathbf{B}$ may be estimated by

$$\hat{\mathbf{B}} = \left(\sum_{i \in \mathcal{S}} w_i \frac{1}{\sigma_i^2} \mathbf{x}_i \mathbf{x}_i^T \right)^{-1} \sum_{i \in \mathcal{S}} w_i \frac{1}{\sigma_i^2} \mathbf{x}_i y_i. \tag{11.23}$$

Generalized regression (GREG) estimator. The generalized regression estimator of the population total is

$$\hat{t}_{y\text{GREG}} = \hat{t}_y + (\mathbf{t}_{\mathbf{x}} - \hat{\mathbf{t}}_{\mathbf{x}})^T \hat{\mathbf{B}}, \tag{11.24}$$

where $\hat{\mathbf{B}}$ is given in (11.23). The term $(\mathbf{t}_{\mathbf{x}} - \hat{\mathbf{t}}_{\mathbf{x}})^T \hat{\mathbf{B}}$ in (11.24) is a regression adjustment to the Horvitz–Thompson estimator, $\hat{t}_y = \sum_{i \in \mathcal{S}} w_i y_i$. Note that $\hat{t}_{y\text{GREG}}$ is a weighted sum of the y_i values in the sample: we can write

$$\hat{t}_{y\text{GREG}} = \sum_{i \in \mathcal{S}} w_i g_i \, y_i, \tag{11.25}$$

where

$$g_i = 1 + (\mathbf{t}_{\mathbf{x}} - \hat{\mathbf{t}}_{\mathbf{x}})^T \left(\sum_{j \in \mathcal{S}} w_j \frac{1}{\sigma_j^2} \mathbf{x}_j \mathbf{x}_j^T \right)^{-1} \frac{1}{\sigma_i^2} \mathbf{x}_i. \tag{11.26}$$

The values g_i are the adjustments to the weights made by using the regression estimator. For large samples with no nonresponse, we expect $\hat{\mathbf{t}}_{\mathbf{x}}$ to be close to $\mathbf{t}_{\mathbf{x}}$ so that g_i will be close to 1 for many observations. When there is nonresponse, the weighting adjustments g_i may be larger.

For any choice of the constants σ_i^2, the GREG estimator calibrates the sample to the population total of each x variable used in the regression. To see this, look at the GREG estimator of $\mathbf{t}_{\mathbf{x}}$. From (11.25) and (11.26),

$$\begin{aligned}
\hat{\mathbf{t}}_{\mathbf{x}\text{GREG}} &= \sum_{i \in \mathcal{S}} w_i g_i \, \mathbf{x}_i \\
&= \hat{\mathbf{t}}_{\mathbf{x}} + \sum_{i \in \mathcal{S}} w_i \left[(\mathbf{t}_{\mathbf{x}} - \hat{\mathbf{t}}_{\mathbf{x}})^T \left(\sum_{j \in \mathcal{S}} w_j \frac{1}{\sigma_j^2} \mathbf{x}_j \mathbf{x}_j^T \right)^{-1} \frac{1}{\sigma_i^2} \mathbf{x}_i \right] \mathbf{x}_i \\
&= \hat{\mathbf{t}}_{\mathbf{x}} + \sum_{i \in \mathcal{S}} w_i \frac{1}{\sigma_i^2} \mathbf{x}_i \left[\mathbf{x}_i^T \left(\sum_{j \in \mathcal{S}} w_j \frac{1}{\sigma_j^2} \mathbf{x}_j \mathbf{x}_j^T \right)^{-1} (\mathbf{t}_{\mathbf{x}} - \hat{\mathbf{t}}_{\mathbf{x}}) \right] \\
&= \hat{\mathbf{t}}_{\mathbf{x}} + (\mathbf{t}_{\mathbf{x}} - \hat{\mathbf{t}}_{\mathbf{x}}) \\
&= \mathbf{t}_{\mathbf{x}}.
\end{aligned}$$

Variance of the GREG estimator. Using linearization,

$$V(\hat{t}_{y\text{GREG}}) = V[\hat{t}_y + (\mathbf{t}_x - \hat{\mathbf{t}}_x)^T \hat{\mathbf{B}}] \approx V(\hat{t}_y - \hat{\mathbf{t}}_x^T \mathbf{B}).$$

Let $e_i = y_i - \mathbf{x}_i^T\hat{\mathbf{B}}$ be the ith residual. Then the variance may be estimated by

$$\hat{V}_1(\hat{t}_{y\text{GREG}}) = \hat{V}\left(\sum_{i\in\mathcal{S}} w_i e_i\right). \tag{11.27}$$

An alternative estimator of the variance (see Exercise 28) is

$$\hat{V}_2(\hat{t}_{y\text{GREG}}) = \hat{V}\left(\sum_{i\in\mathcal{S}} w_i g_i e_i\right). \tag{11.28}$$

If the model is a good one, we expect the variability in the residuals to be smaller than the variability in the original observations, so that the GREG estimator will be more efficient than $\hat{t}_y$. In an SRS, for example,

$$\hat{V}(\hat{t}_y) = \frac{N^2}{n}\left(1 - \frac{n}{N}\right)\frac{1}{n-1}\sum_{i\in\mathcal{S}}(y_i - \bar{y})^2$$

but

$$\hat{V}_1(\hat{t}_{y\text{GREG}}) = \frac{N^2}{n}\left(1 - \frac{n}{N}\right)\frac{\sum_{i\in\mathcal{S}} e_i^2}{n-1};$$

if the residuals tend to be smaller than the deviations of y_i about the mean, then the estimated variance is smaller for the GREG estimator.

Example 11.13. Ratio estimation. For ratio estimation, we adopt the working model

$$y_i = \beta x_i + \varepsilon_i, \quad V_M(\varepsilon_i) = \sigma^2 x_i.$$

The population quantity B is the weighted least squares estimate of β using the whole population. Then, using (11.23),

$$\hat{B} = \left(\sum_{i\in\mathcal{S}} w_i \frac{x_i^2}{x_i}\right)^{-1}\sum_{i\in\mathcal{S}}\frac{w_i x_i y_i}{x_i} = \frac{\sum_{i\in\mathcal{S}} w_i y_i}{\sum_{i\in\mathcal{S}} w_i x_i} = \frac{\hat{t}_y}{\hat{t}_x}.$$

The GREG estimator of the population total is

$$\hat{t}_{y\text{GREG}} = \hat{t}_y + (t_x - \hat{t}_x)\frac{\hat{t}_y}{\hat{t}_x} = t_x\frac{\hat{t}_y}{\hat{t}_x},$$

which is the usual ratio estimator. ■

Example 11.14. Poststratification. We discussed poststratification in Sections 4.4 and 8.6 as a method of calibrating estimates to population totals of subgroups and as a method of adjusting for nonresponse. Suppose we know the population counts N_c for C poststrata, $c = 1, \ldots, C$. Define the variables $x_{ic} = 1$ if observation unit i is in poststratum c and 0 otherwise, and let $\mathbf{x}_i = [x_{i1}, \ldots, x_{iC}]^T$. Consider the working model

$$Y_i = \beta_1 x_{i1} + \beta_2 x_{i2} + \ldots + \beta_C x_{iC} + \varepsilon_i$$

with $V_M(\varepsilon_i) = \sigma^2$. Then,

$$\sigma^2\mathbf{X}_{\mathcal{U}}^T\boldsymbol{\Sigma}_{\mathcal{U}}^{-1}\mathbf{X}_{\mathcal{U}} = \mathbf{X}_{\mathcal{U}}^T\mathbf{X}_{\mathcal{U}} = \text{ diag }(N_1, \ldots, N_C)$$

and

$$\sigma^2 \sum_{i \in \mathcal{S}} w_i \frac{1}{\sigma^2} \mathbf{x}_i \mathbf{x}_i^T = \text{diag}\,(\hat{N}_1, \ldots, \hat{N}_C).$$

As a result, $\hat{B}_c = \hat{t}_{yc}/\hat{N}_c$, where $\hat{t}_{yc} = \sum_{i \in \mathcal{S}} w_i x_{ic} y_i$ is the estimated population total in poststratum c and $\hat{N}_c = \sum_{i \in \mathcal{S}} w_i x_{ic}$ is the estimated population count in poststratum c. The generalized regression estimator is

$$\hat{t}_{y\text{GREG}} = \hat{t}_y + \sum_{c=1}^{C} (N_c - \hat{N}_c) \frac{\hat{t}_{yc}}{\hat{N}_c} = \sum_{c=1}^{C} \frac{N_c}{\hat{N}_c} \hat{t}_{yc}. \quad \blacksquare$$

The x variables are often chosen so as to reduce the variance (and bias, if there is nonresponse) for key y variables. The generalized regression estimator is a linear estimator in y, as seen in (11.25): $\hat{t}_{y\text{GREG}} = \sum_{i \in \mathcal{S}} w_i g_i \, y_i$. The weight adjustments g_i in (11.26) depend on the xs but they do not depend on values of the response variable y. To use generalized regression estimation, form a new column in the data with values $w_i^* = w_i g_i$. Then use the vector of w_i^*s as the weight vector for every statistic calculated from the sample. The effects of the calibration on variances may be incorporated using linearization, as in (11.27) and (11.28), or through use of replicate variance estimation methods, as described in Section 9.3.4.

11.7 Chapter Summary

In regression methods with complex survey data, the population characteristic of interest is $\mathbf{B}$, the vector of least squares or logistic regression coefficients that would be estimated if we knew the entire population. Since $\mathbf{B}$ is a function of population totals, it is estimated by $\hat{\mathbf{B}}$ using the sampling weights. Ideally, the finite population values $\mathbf{B}$ reflect an underlying relationship between y and $\mathbf{x}$, but inferences about $\mathbf{B}$, using the survey design, are valid whether the regression model is a good one or not.

The generalized regression estimator provides a method for using auxiliary information to reduce the mean squared error of estimators. It can also be used to reduce bias due to nonresponse.

Key Terms

Calibration: A procedure in which weights are adjusted so that estimated population totals of auxiliary variables agree with the known population totals of those variables.

Generalized Regression (GREG) Estimator: An estimator of a population total that uses auxiliary information through a regression model.

Model-Assisted Estimation: An approach to inference in which a population model motivates the form of estimators, but all inference is based on the survey design.

Ordinary Least Squares: A form of regression in which the coefficients minimize the sum of squares of the residuals, without weights.

For Further Reading

Many books have been written about model-based linear and logistic regression in non-survey contexts. See, for example, Kutner et al. (2005), Weisberg (2014), Christensen (2020), and Zimmerman (2020) for theoretical results and practical considerations for linear regression and Hosmer et al. (2013) for logistic regression.

For general references about regression with survey data, see Lehtonen and Pahkinen (2004) and Heeringa et al. (2017), who discuss practical issues for performing regression analyses with survey data. Lumley and Scott (2017) review regression models for survey data. The model-assisted approach to inference adopted in this chapter is discussed in more detail by Särndal et al. (1992). In this approach, a model is used to specify the parameters of interest, but all inference is based on the survey design.

Kish and Frankel (1974) wrote one of the first papers showing that the sample design affects estimates of regression parameters. Other references on the theory of linear regression with survey data include Korn and Graubard (1999) and Fuller (2002). Binder (1983), Chambless and Boyle (1985), and Roberts et al. (1987) derived the design-based theory for estimating logistic regression parameters. The books edited by Skinner et al. (1989a) and Chambers and Skinner (2003) contain several chapters on modeling data from complex surveys.

If you want to learn more about inferential philosophies in sample surveys, start with the entertaining debate by Brewer and Mellor (1973) between "Harry," a design-based survey statistician, and "Fred," who promotes a model-based approach. For additional insight on the contrasts among the approaches, see Brewer (2002, 2013). Pfeffermann and Holmes (1985), DuMouchel and Duncan (1983), Rubin (1985), Smith (1988), Little (1991), Kott (1991), and Pfeffermann (1993, 1996) discuss the role of sampling weights in model-based analyses. Korn and Graubard (1995) provide an example where weighted and unweighted regression coefficients exhibit large differences, and discuss the role of weights and model misspecification. Binder and Roberts (2003) discuss inference for regression models, and Rubin-Bleuer and Kratina (2005) provide a rigorous mathematical framework for inference under both model-based and design-based approaches. Zanutto (2006) discusses domain means in complex surveys, and compares linear regression estimates with those obtained by applying propensity score methods to survey data.

Sometimes, when the main interest is in estimating a regression relationship, alternative estimators can be considered that reduce the variance of the regression parameters by (1) including survey design variables among the model covariates or (2) constructing modified weights that are conditioned on the information in the explanatory variables. Pfeffermann and Sverchkov (1999) and Pfeffermann (2011) describe these options.

Särndal (2007) and Wu and Lu (2016) give overviews of generalized regression estimation and calibration in survey sampling. Montanari (1987) and Rao (1994) present alternative methods of using regression for estimating population totals. Beaumont and Alavi (2004) derive a robust generalized regression estimator, and Breidt and Opsomer (2000) and Montanari and Ranalli (2005) use nonparametric regression methods for estimating population totals. Silva and Skinner (1997) and McConville et al. (2017) discuss methods for selecting the model to use in regression estimation. Gelman (2007) adopts a hierarchical Bayesian approach to weight adjustment and Valliant (2009) discusses a model-based approach to using auxiliary information when estimating population totals.

11.8 Exercises

A. Introductory Exercises

1. Read a research article in which regression or logistic regression is used on data from a complex survey. Write a critique of the article. What is the purpose and design of the survey? What is the goal of the analysis? How do the authors use information from the survey design in the analysis? Do you think that the data analysis is done well? If so, why? If not, how could it have been improved? Are the conclusions drawn in the article justified?

2. An investigator wants to study the relationship between a child's age and number of siblings, and the dollar amount of the child's Christmas list presented to Santa Claus. She also wants to estimate the total number of children that visit Santa Claus, and the total dollar amount of all childrens' requests. It would be very difficult to construct a sampling frame of children who will visit Santa Claus between December 1 and December 24, but the investigator has a list of shopping malls and stores in which Santa will appear in the city, as well as the times that Santa will be at each location. The Santa sites are divided into four categories: 23 department stores, 19 discount stores, 15 toy stores, and 5 shopping malls. The investigator wants you to help design the sample of children.

   (a) What questions would you ask the investigator to clarify the problem?

   (b) Assuming any answers you like to the questions you asked, suggest a design for the survey.

   (c) How will your survey design affect the regression analysis of the data? How do you propose to analyze the data? Are there other explanatory variables that you would suggest to the investigator?

3. For the data in file `spanish.csv` (see Exercise 5 of Chapter 5), let domain 1 consist of students who are planning a trip to a Spanish-speaking country in the next year and domain 2 consist of the students who are not planning such a trip. We are interested in whether the mean vocabulary score (y) differs in the two domains. The population domain mean in domain 1 is $\bar{y}_{\mathcal{U}1}$ and the population domain mean in domain 2 is $\bar{y}_{\mathcal{U}2}$. Using regression, estimate $\bar{y}_{\mathcal{U}1} - \bar{y}_{\mathcal{U}2}$ and give a 95% CI. Is there evidence that the domain means differ?

B. Working with Survey Data

4. Use the data in `anthrop.csv` for this problem.

   (a) Construct a population from the 3,000 observations in `anthrop.csv` in which the 1,000 individuals with the highest value of y have been removed. Now take an SRS of size 200 from the remaining 2,000 individuals, and plot the data along with the OLS regression line. How does this line compare to the population regression line in Figure 11.4?

   (b) Repeat (a), but now take the SRS from the 2,000 individuals with the lowest values of x.

   (c) Is there a difference in the regression equations in (a) and (b)? Explain, and relate your findings to the model in (11.1).

5. For the data in `asafellow.csv`, described in Example 3.7 and Exercise 14 of Chapter 3, calculate the variable *timelag* as the number of years between the year of terminal degree (*degreeyr*) and the year of award (*awardyr*).

   (a) Calculate the mean of *timelag* for male Fellows. Do the same for female Fellows.

   (b) Estimate, along with a 95% CI, the difference between the mean of *timelag* for men and the mean of *timelag* for women.

   (c) Why are the domain means for male and female Fellows uncorrelated for this data set?

6. Use the data in `nybight.csv` (see Exercise 22 of Chapter 3) for this problem. Using the 1974 data, estimate the coefficients in a straight-line regression model predicting weight of the catch from the number of fish caught. Give standard errors for your estimates. Be sure to plot the data.

7. Perform a model-based analysis for the setting in Exercise 6. Examine the residuals and postulate an appropriate variance structure for the model.

8. Repeat Exercises 6 and 7 for predicting number of species caught from the surface temperature.

9. Use the data in `teachers.csv` (described in Exercise 15 of Chapter 5) for this problem.

   (a) Estimate the coefficients in a straight-line regression model predicting *preprmin* from *size*. Give standard errors for your estimates. Is there evidence that the two variables are related? (Be sure to plot the data!)

   (b) Perform a model-based analysis of the same data. Examine the residuals and postulate an appropriate variance structure for the model.

10. Use the data in `books.csv` (described in Exercise 8 of Chapter 5) for this problem.

    (a) Plot *replace* vs. *purchase* for the raw data.

    (b) Plot *replace* vs. *purchase* using the sampling weights.

    (c) Using a design-based approach, estimate the regression equation for predicting *replace* from *purchase*, along with its standard error. How many df would you use in constructing a CI for the slope?

11. For the situation in Exercise 10, postulate a model for the variance structure. Using your model, estimate the slope of the regression line predicting *replace* from *purchase*. How do your estimate and its standard error compare with your answers in Exercise 10?

12. *Public health articles* (see Exercise 12 of Chapter 3; do not use an fpc for this exercise).

    (a) Investigate the relationship between use of a probability sample and number of authors by performing a logistic regression predicting *RandomSel* from *Numauthors*. Give a 95% CI for the regression slope.

    (b) Compare your results from (a) with those from the chi-square test using the binary variable *authors_cat* in Exercise 10(b) of Chapter 10.

13. *Mystery novels.* Exercise 13 of Chapter 3 described a stratified sample of mystery novels. Conduct a test for the null hypothesis that the average number of female detectives is equal for each type of mystery novel (variable *genre*). Use weight *p1weight*, and do not include an fpc.

14. Use your data set from Exercise 26 of Chapter 3 for this problem. Using the weights, fit a regression model predicting *acres92* from *largef92*. Give a standard error for the estimated slope. Now ignore the sampling design, and calculate the ordinary least squares estimate of the slope. Do your point estimates differ? Explain why or why not by examining plots of the data.

15. Using the data in `nhanes.csv`, fit a straight-line regression model predicting y = average sagittal abdominal diameter (variable *bmdavsad*) from x = body mass index (variable *bmxbmi*) for adults age 20 and over. You plotted these data in Exercise 14 of Chapter 7. Give a 95% CI for the slope, and calculate R^2 for these data. Draw your regression line on the plot.

16. Using the data in `nhanes.csv`, fit a straight-line regression model predicting y = waist circumference (variable *bmxwaist*) from x = upper arm circumference (variable *bmxarmc*) for adults age 20 and over. You plotted these data in Exercise 15 of Chapter 7. Give a 95% CI for the slope, and calculate R^2 for these data. Draw your regression line on the plot.

17. Using the data in `nhanes.csv`, calculate the estimated mean body mass index (variable *bmxbmi*) for the ten groups defined by the cross-classification of gender with the race/ethnicity variable used in Example 11.8. Conduct F tests for the main effects of gender and race/ethnicity, and for the interaction. Which demographic groups are significantly different? (Use a multiple comparison method if one is available in your software.)

18. For the situation in Example 11.12, fit a model that allows different regression slopes for each gender by including the term *bmxwaist*female* in the model. Give the parameter estimates and 95% CIs. Is there evidence that the slopes differ by gender?

19. Using the data in `schools.csv`, discussed in Example 5.7, fit a regression model predicting score on the reading test from the math test score. Give 95% CIs for the regression coefficients, using the with-replacement variance. Include a scatterplot of the data.

20. For the data in `schools.csv` (see Example 5.7) estimate the difference between the mean math test score for girls and the mean math test score for boys, along with a 95% CI. Do not use an fpc. Is there evidence that the mean test scores for girls and boys differ?

C. Working with Theory

21. *Comparison of domain means.* Suppose the population may be divided into two groups, with respective sizes N_1 and N_2 and population means $\bar{y}_{1\mathcal{U}}$ and $\bar{y}_{2\mathcal{U}}$. The overall population mean is $\bar{y}_{\mathcal{U}} = (N_1\bar{y}_{1\mathcal{U}} + N_2\bar{y}_{2\mathcal{U}})/N$, with $N = N_1 + N_2$. Let $x_i = 1$ if observation unit i is in group 1, and $x_i = 0$ if it is in group 2. The weight for observation unit i is w_i.

    Show that $B_1 = \bar{y}_{1\mathcal{U}} - \bar{y}_{2\mathcal{U}}$ and $B_0 = \bar{y}_{2\mathcal{U}}$. Also show that

$$\hat{B}_1 = \frac{\sum_{i\in\mathcal{S}} w_i x_i y_i}{\sum_{i\in\mathcal{S}} w_i x_i} - \frac{\sum_{i\in\mathcal{S}} w_i(1-x_i)y_i}{\sum_{i\in\mathcal{S}} w_i(1-x_i)} = \hat{\bar{y}}_1 - \hat{\bar{y}}_2$$

    and $\hat{B}_0 = \hat{\bar{y}}_2$.

22. *Weighting class adjustments for nonresponse.* Equation (8.11) of Section 8.5.2 presented a general model for nonresponse. We can express the model for weighting class adjustments as a linear regression, with

$$R_i = \sum_{c=1}^{C} x_{ic}\beta_c + \varepsilon_i,$$

where $R_i = 1$ if unit i responds to the survey and 0 otherwise, $x_{ic} = 1$ if unit i is in weighting class c and 0 otherwise, $E[\varepsilon_i] = 0$, and $V[\varepsilon_i] = \sigma^2$.

   (a) Show that

$$\hat{B}_c = \hat{\phi}_c = \frac{\sum_{i \in \mathcal{S}} x_{ic} w_i R_i}{\sum_{i \in \mathcal{S}} x_{ic} w_i}.$$

   (b) Show that the estimated response propensity for class c, $\hat{\phi}_c$, is always between 0 and 1.

   (c) Use (11.13) to find $\hat{V}(\hat{\mathbf{B}})$, for $\hat{\mathbf{B}} = (\hat{B}_1, \ldots, \hat{B}_C)^T$.

23. *Estimating response propensities using linear regression.* The estimated response propensities for the model in Exercise 22 are always between 0 and 1. Construct a data set for which a straight-line regression model, $R_i = \beta_0 + \beta_1 x_i + \varepsilon_i$, produces negative estimated response propensities.

24. Consider the data from an SRS in file `uneqvar.csv`.

   (a) Plot y vs. x.

   (b) Find the OLS regression equation under the assumption of equal variances.

   (c) Calculate $\hat{V}_M(\hat{\beta}_1)$ and $\hat{V}_L(\hat{B}_1)$. How do they compare?

25. *Informative sampling designs* (requires probability). This exercise is adapted from Example 1 of Pfeffermann (2011). Suppose that the population regression model is $Y_i|\mathbf{x}_i = \mathbf{x}_i^T\boldsymbol{\beta} + \varepsilon_i$, where $\varepsilon_i \sim N(0, \sigma^2)$. Let $Z_i = 1$ if unit i is included in the sample and 0 otherwise. Suppose that the inclusion probability can be described as $P(Z_i = 1|\mathbf{x}_i, Y_i = y) = \exp\left[\gamma_1 y + \gamma_2 y^2 + g(\mathbf{x}_i)\right]$, where γ_1 and $\gamma_2 \leq 0$ are constants and $g(\mathbf{x}_i)$ is a function of the covariates $\mathbf{x}_i$. Show that

$$Y_i|\mathbf{x}_i, Z_i = 1 \sim N\left[\frac{\gamma_1\sigma^2 + \mathbf{x}_i^T\boldsymbol{\beta}}{1 - 2\gamma_2\sigma^2}, \frac{\sigma^2}{1 - 2\gamma_2\sigma^2}\right].$$

   Thus, when $\gamma_1 \neq 0$ or $\gamma_2 \neq 0$ (that is, the design has information about the regression relationship), the distribution of Y_i in the sample differs from that in the population. HINT: Use Bayes' theorem.

26. Show that (11.12) is equivalent to (11.7) and (11.8) for straight-line regression.

27. (Requires linear algebra and calculus.) The linearization estimator of $V(\hat{\mathbf{B}})$ can be found by the method outlined in Section 9.1 (see Shah et al., 1977). However, the calculations are easier using the Demnati–Rao (2004) method discussed in Exercise 29 of Chapter 9. Show (11.13) using the Demnati–Rao method. HINT: If $\mathbf{F}$ is a nonsingular matrix whose entries are functions of u, then

$$\frac{\partial \mathbf{F}^{-1}}{\partial u} = -\mathbf{F}^{-1}\frac{\partial \mathbf{F}}{\partial u}\mathbf{F}^{-1}. \tag{11.29}$$

28. (Requires linear algebra and calculus.) Use the Demnati–Rao (2004) method discussed in Exercise 29 of Chapter 9 to estimate $V(\hat{t}_{y\text{GREG}})$ for $\hat{t}_{y\text{GREG}}$ defined in (11.24). This results in the variance estimator in (11.28). HINT: Use (11.29).

29. Example 4.7 used a straight-line regression model: $Y_i|x_i = \beta_0 + \beta_1 x_i + \varepsilon_i$, with $V(e_i) = \sigma^2$. Find the value of g_i for each observation in deadtrees.csv. Calculate $\sum_{i\in\mathcal{S}} w_i g_i y_i$ and the two variance estimates $\hat{V}_1(\hat{t}_{y\text{GREG}}) = \hat{V}\left(\sum_{i\in\mathcal{S}} w_i e_i\right)$ and $\hat{V}_2(\hat{t}_{y\text{GREG}}) = \hat{V}\left(\sum_{i\in\mathcal{S}} w_i g_i e_i\right)$.

30. *Plotting residuals from regression models with complex survey data.* In a design-based framework for inference, the regression coefficients $\hat{\mathbf{B}}$ estimate the population values $\mathbf{B}$. Inferences such as CIs depend on the inclusion probabilities in the sampling design and thus do not depend on model assumptions. Design-based inferences about the finite-population regression parameters of a poorly fitting model are valid. Nevertheless, as discussed in Section 11.4, we often are interested in an underlying theoretical model and want to assess how well the population regression model fits. We can plot residuals versus predicted values, incorporating the weights, using methods in Section 7.6.2. Plot the residuals versus predicted values for the regression model in Example 11.7. What do you see in your plot? Would you suggest any modifications to the model?

31. *Regression diagnostics for complex survey data.* Jenney (2005) and Li and Valliant (2009, 2011) developed regression diagnostics methods for complex survey data. The **leverages** for the population values are the diagonal elements of the matrix

$$\mathbf{H} = \mathbf{X}_{\mathcal{U}}(\mathbf{X}_{\mathcal{U}}^T\mathbf{X}_{\mathcal{U}})^{-1}\mathbf{X}_{\mathcal{U}}^T,$$

so that the leverage of unit i in the population is

$$h(\mathcal{U})_i = \mathbf{x}_i^T(\mathbf{X}_{\mathcal{U}}^T\mathbf{X}_{\mathcal{U}})^{-1}\mathbf{x}_i.$$

The leverage of an observation is a measure of the distance from $\mathbf{x}_i$ to the means of the set of explanatory variables (Kutner et al., 2005). Using the weights, define the leverage of unit i in the sample as

$$h(\mathcal{S})_i = w_i\mathbf{x}_i^T\left(\sum_{j\in\mathcal{S}} w_j\mathbf{x}_j\mathbf{x}_j^T\right)^{-1}\mathbf{x}_i.$$

(a) Show that $\sum_{i\in\mathcal{S}} h(\mathcal{S})_i = p$, the number of parameters in the regression model.

(b) Calculate the leverage, using the weights, for each observation in file anthuneq.csv, used in Section 11.2. Which points have the highest values of leverage? Does your assessment of the high-leverage points change if you do not use the weights?

32. The diagnostic statistic DFFITS (Belsley et al., 1980) is often used to assess influential observations in a data set. A complex survey version of DFFITS can be calculated using the survey-weighted leverage $h(\mathcal{S})_i$ from Exercise 31 along with the residual for observation i, $e_i = y_i - \hat{y}_i$:

$$\text{DFFITS} = \frac{h(\mathcal{S})_i e_i}{1 - h(\mathcal{S})_i}\frac{1}{\text{SE}(\hat{y}_i)},$$

where

$$\hat{V}(\hat{y}_i) = \hat{V}(\mathbf{x}_i^T\hat{\mathbf{B}}) = \mathbf{x}_i^T\hat{V}(\hat{\mathbf{B}})\mathbf{x}_i.$$

Calculate DFFITS for each observation in file anthuneq.csv. Which observation has the largest value of DFFITS? What is the regression equation for the jackknife replicate (see Example 11.6) where this observation is deleted?

33. DuMouchel and Duncan (1983) and Kott (1991) argued that using sampling weights in regression can provide robustness to model misspecification, because the weighted estimates are relatively unaffected if some x variables are left out of the model. Regression coefficients calculated with weights that differ substantially from coefficients calculated without weights can indicate that the sampling weights have information that should be included in the regression model.

    (a) Select a stratified random sample of size 400 from the data in `college.csv` (see Example 3.12) containing 80 public (*control* = 1) and 320 private (*control* = 2) institutions.

    (b) Using the sampling weights, fit a regression model predicting *avgfacsal* from x = *tuitionfee_in*.

    (c) Repeat (b) without the sampling weights. How do the estimates differ?

    (d) Now fit a model predicting *avgfacsal* from x, z, and $x * z$, where $z = 1$ if *control* = 2 and 0 otherwise. How do the weighted and unweighted estimates differ?

    (e) Explain the results in parts (a) through (d). Include scatterplots of the data with and without weights.

34. *Local polynomial regression with survey data* (requires theory of linear models). Section 7.6.2 discussed using smoothed trend lines in bivariate plots of survey data (see Korn and Graubard, 1998b; Bellhouse and Stafford, 2001). We posit a model $y_i = g(x_i) + \varepsilon_i$, where the second derivative of g is continuous, and estimate the underlying smooth function $g(x)$ by sliding a kernel window along the data and fitting a straight line (or higher-order polynomial) to the weighted data in that window. As with density estimation, briefly discussed in Section 7.6.1, the kernel function K is a symmetric density function such as the normal kernel function $K_N(t) = \exp(-t^2/2)/\sqrt{2\pi}$ or the quadratic kernel function $K_Q(t) = \frac{3}{4}(1-t^2)$ for $|t| < 1$. Since the data are from a complex survey, the weights used in fitting the local regression include the survey weights as well as the kernel weights. Let $x_1, \ldots, x_n$ and $y_1, \ldots, y_n$ denote the observations in the sample. For local linear regression, the function g at a point t is estimated by:

$$\hat{g}(t) = [1 \;\; 0]\,(\mathbf{X}_t^T \mathbf{W}_t \mathbf{X}_t)^{-1} \mathbf{X}_t^T \mathbf{W}_t \mathbf{y},$$

    where

$$\mathbf{X}_t = \begin{bmatrix} 1 & t - x_1 \\ \vdots & \vdots \\ 1 & t - x_n \end{bmatrix}, \quad \mathbf{y} = \begin{bmatrix} y_1 \\ \vdots \\ y_n \end{bmatrix},$$

    and

$$\mathbf{W}_t = \text{diag}\left[\frac{w_1}{b} K\left(\frac{t - x_1}{b}\right), \ldots, \frac{w_n}{b} K\left(\frac{t - x_n}{b}\right)\right].$$

    Calculate a local linear regression function for the NHANES data, using y = upper arm circumference (variable *bmxarmc*) and x = body mass index.

35. (Requires theory of linear models.) Suppose the "true" model describing the relation between x and y is

$$Y_i | x_i = \beta_0 + \beta_1 x_i + \varepsilon_i,$$

    where the ε_i are independently generated from a $N(0, \sigma_i^2)$ distribution. Let $\mathbf{\Sigma}$ be a matrix with diagonal entries $\sigma_1^2, \sigma_2^2, \ldots, \sigma_n^2$. What is the covariance matrix for the OLS parameter estimators? How does this relate to the discussion of different estimators of the variance in Section 11.2.2?

36. Assuming a model
$$y_i = \beta_0 + \beta_1 x_i + \varepsilon_i$$
with $V(\varepsilon_i) = \sigma^2$, what is the generalized regression estimator of t_y?

37. *Balanced sampling.* Stratification and poststratification both use information about the population sizes, N_h, of strata. The difference is that stratification uses the information in the design, forcing any sample that is selected to satisfy $\sum_{i\in\mathcal{S}} w_i x_{ih} = N_h$, where $x_{ih} = 1$ for unit i in stratum h and 0 otherwise, and w_i is the sampling weight. Poststratification uses the information at the estimation stage, so that the poststratified weights $\tilde{w}_i$ satisfy $\sum_{i\in\mathcal{S}} \tilde{w}_i x_{ih} = N_h$.

   Calibration, which is a generalization of poststratification, adjusts the sample weights so that (11.31) is satisfied. Calibration assumes the $\mathbf{x}$ variables are measured for each unit in the sample and adjusts the weights after the sample is collected. But if the sampling frame contains the values of $\mathbf{x}$ variables for every unit in the population and their population total $\mathbf{t}_\mathbf{x}$ is known, then **balanced sampling** (see, for example, Deville and Tillé, 2004; Tillé, 2006; Tillé and Wilhelm, 2017) may be used to guarantee that every sample that could be selected satisfies
$$\sum_{i\in\mathcal{S}} w_i \mathbf{x}_i = \mathbf{t}_\mathbf{x}, \qquad (11.30)$$
   where $w_i = 1/\pi_i$. Thus, balanced sampling bears the same relationship to calibration as stratification bears to poststratification. You can think of balanced sampling as pre-calibrating the survey.

   (a) Show that stratified random sampling with proportional allocation is a special case of balanced sampling. What are the $\mathbf{x}$ variables?

   (b) Suppose you want a sample that has approximately the same proportions as the population by gender, race (five categories), and education (five categories), and also has approximately the same average income. Explain why balanced sampling can achieve this goal, and how the balanced sample will differ from a sample that is stratified on gender, race, and education.

   (c) Suppose that you want to have a sample that will be approximately balanced on variables x and u so that $|\bar{x} - \bar{x}_\mathcal{U}| \le E_x$ and $|\bar{u} - \bar{u}_\mathcal{U}| \le E_u$, where E_x and E_u are small constants that allow the acceptable samples to be approximately, rather than exactly, balanced. Show that the following procedure results in a balanced sample of size n.

      i. Select an SRS of size n.
      ii. If $|\bar{x} - \bar{x}_\mathcal{U}| \le E_x$ and $|\bar{u} - \bar{u}_\mathcal{U}| \le E_u$, then keep the sample; otherwise, discard the sample.
      iii. Repeat Steps 1 and 2 until a sample is kept.

   (d) Explain why the sample selected using the procedure in (c) is a probability sample.

38. *Calibration estimators* (requires calculus). Deville and Särndal (1992) developed the theory of calibration as follows. Suppose a probability sample has weights w_i. A vector of auxiliary variables $\mathbf{x}_i$ is available for each unit in the sample $\mathcal{S}$, and the population totals $\mathbf{t}_\mathbf{x}$ are known from an external source. Assume that there is no nonresponse and the values of $\mathbf{t}_\mathbf{x}$ are known exactly, without error. It is desired to adjust the weights so that the calibrated weights $\tilde{w}_i$ satisfy
$$\sum_{i\in\mathcal{S}} \tilde{w}_i \mathbf{x}_i = \mathbf{t}_\mathbf{x} \qquad (11.31)$$

while still being close to the original sampling weights.

Show that the weights $\tilde{w}_i = w_i g_i$ in (11.26) minimize the distance function

$$h(\mathbf{w}, \tilde{\mathbf{w}}) = \frac{\sum_{i \in \mathcal{S}}(w_i - \tilde{w}_i)^2}{w_i}$$

subject to the constraint in (11.31). HINT: Use Lagrange multipliers.

39. *Nonresponse bias of calibration estimators.* Exercise 38 derived calibration weights when there is no nonresponse. When there is nonresponse, the calibrated weights are

$$w_i^* = w_i \left[1 + \left(\mathbf{t_x} - \sum_{j \in \mathcal{R}} w_j \mathbf{x}_j \right)^T \left(\sum_{j \in \mathcal{R}} w_j \mathbf{x}_j \mathbf{x}_j^T \right)^{-1} \mathbf{x}_i \right], \tag{11.32}$$

where w_i is the sampling weight for unit i and $\mathcal{R}$ denotes the set of respondents to the survey. Proposition 9.1 of Särndal and Lundström (2005) states that the bias of a calibration estimator of the population total, $\sum_{i \in \mathcal{R}} w_i^* y_i$, is approximately

$$-\sum_{i=1}^{N} (1 - \phi_i) \left[y_i - \mathbf{x}_i^T \left(\sum_{j=1}^{N} \phi_j \mathbf{x}_j \mathbf{x}_j^T \right)^{-1} \sum_{j=1}^{N} \phi_j \mathbf{x}_j y_j \right], \tag{11.33}$$

where ϕ_i is the response propensity for unit i (see Section 8.4.3).

(a) In the special case of poststratification, the H auxiliary variables indicate poststrata, that is, $x_{ih} = 1$ if unit i is in poststratum h and 0 otherwise, for $h = 1, \ldots, H$. The vector of population totals is $\mathbf{t_x} = [N_1, N_2, \ldots, N_H]^T$. Show that w_i^* in (11.32) is the poststratified weight of Equation (8.12).

(b) What is the expression in (11.33) when poststratification is used?

40. Table 11.7 gives the population counts for four cross-classified domains in a hypothetical population, along with the sample size observed in each domain in a sample drawn from the population. The sampling weight for each observation is $w_i = 20$.

TABLE 11.7
Population and sample sizes for Exercise 40.

Age Group	Gender	Population Size	Sample Size
25 or under	Female	50	15
25 or under	Male	250	26
Over 25	Female	100	5
Over 25	Male	900	14

(a) Find the poststratification weight adjustments g_i for observations in each of the four domains. Show that poststratification adjustments must always be positive, but they can be less than one.

(b) Now find weight adjustments g_i based on the model

$$y_i = \beta_0 + \beta_1 x_i + \beta_1 z_i + \varepsilon_i, \quad V_M(\varepsilon_i) - \sigma^2,$$

where $x_i = 1$ if observation i is female and 0 otherwise, and $z_i = 1$ if observation i is 25 or younger and 0 otherwise. Do the weight adjustments have to be positive in this model?

41. *Poststratification to control totals that are estimated from a survey.* The poststratified estimator in Example 11.14 can be written as $\hat{t}_{y\text{PS}} = \mathbf{t}_\mathbf{x}^T\hat{\mathbf{B}}$, where $\mathbf{t}_\mathbf{x} = (N_1, N_2, \ldots, N_C)^T$ is the vector of control totals for the C poststrata and $\hat{\mathbf{B}} = (\hat{t}_{y1}/\hat{N}_1, \hat{t}_{y2}/\hat{N}_2, \ldots, \hat{t}_{yC}/\hat{N}_C)^T$ is the vector of estimated poststratum means from the survey. The variance of $\hat{t}_{y\text{PS}}$ is

$$V(\hat{t}_{y\text{PS}}) \approx \mathbf{t}_\mathbf{x}^T V(\hat{\mathbf{B}})\mathbf{t}_\mathbf{x}. \tag{11.34}$$

Suppose that population poststratification totals are unknown, but estimates are available from another survey that is selected independently. Following Dever and Valliant (2010, 2016), let $\hat{\mathbf{t}}_{\mathbf{x},\text{aux}}$ be an unbiased estimator of $\mathbf{t}_\mathbf{x}$ from the auxiliary survey having sampling variance $V(\hat{\mathbf{t}}_{\mathbf{x},\text{aux}})$. Show that the variance of the estimator using the estimated poststratification totals, $\hat{t}_{y\text{EPS}} = \hat{\mathbf{t}}_{\mathbf{x},\text{aux}}^T\hat{\mathbf{B}}$, is

$$V(\hat{t}_{y\text{EPS}}) \approx \mathbf{t}_\mathbf{x}^T V(\hat{\mathbf{B}})\mathbf{t}_\mathbf{x} + \mathbf{B}^T V(\hat{\mathbf{t}}_{\mathbf{x},\text{aux}})\mathbf{B}, \tag{11.35}$$

where $\mathbf{B} = (t_{y1}/N_1, t_{y2}/N_2, \ldots, t_{yC}/N_C)^T$ is the vector of population poststratum means. Under what conditions will poststratifying to estimated totals reduce the variance of the estimator?

D. Projects and Activities

42. *Trucks.* Use the data in Exercise 49 of Chapter 3 and Exercise 30 of Chapter 7 for this exercise.

    (a) Fit a straight line model predicting y = *miles_annl* from x = model year (*adm_modelyear*). Give a 95% CI for the slope.

    (b) How well does this model fit the data?

    (c) What other variables in the data set might be useful for predicting y? Fit a multiple regression model predicting y using x variables of your choice.

43. *Baseball data.* Use the data from Exercise 32 in Chapter 7. What variables in the data do you think might be useful for predicting log(*salary*)? Fit a multiple regression model predicting log(*salary*) from these variables.

44. *IPUMS exercises.*

    (a) Regress *inctot* on covariates of your choice using your sample from Exercise 45 of Chapter 5. Write a paragraph interpreting the results of your analysis.

    (b) Perform a logistic regression predicting whether a person is in the labor force (variable *labforce*) from covariates of your choice.

45. *Activity for course project.* Using the survey you chose in Exercise 35 of Chapter 7, use regression methods to predict a response of interest from covariates in the data. If the survey has no continuous responses, use logistic regression to predict a binary response. Make sure you plot the data appropriately.

12

Two-Phase Sampling

Nearly the whole of the states have now returned their census. I send you the result, which as far as founded on actual returns is written in black ink, and the numbers not actually returned, yet pretty well known, are written in red ink. Making a very small allowance for omissions, we are upwards of four millions; and we know in fact that the omissions have been very great.

—Thomas Jefferson, letter to David Humphreys, August 23, 1791.

Sometimes you would like to use stratification, unequal-probability sampling, or ratio estimation to increase the precision of your estimator, but the sampling frame lacks information on useful auxiliary variables. For example, suppose you want to sample businesses with probability proportional to income but do not have income information in the sampling frame. Or you want to estimate the total timber volume that has been cut in the forest by measuring the total volume in a sample of truckloads of logs. Timber volume in a truck is related to the weight of the truckload, so you would expect to gain precision by using ratio estimation with y_i = timber volume in truck i and x_i = weight of truck i. But the ratio estimator $\hat{t}_{yr} = t_x \hat{t}_y / \hat{t}_x$ requires that the total weight for all truckloads be known, and weighing every truck in the population is impractical.

Two-phase sampling, also called **double sampling**, provides a solution. It is useful when the variable of interest y is relatively expensive to measure, but a correlated variable x can be measured fairly easily and used to improve the precision of the estimator of t_y. It may also be used to adjust for nonresponse, to sample rare populations, or to improve the sampling frame. We discuss some of these applications later in this chapter and in Chapter 14.

Suppose the population has N observation units. The sample is taken in two phases:

1. *Phase I sample.* Take a probability sample of $n^{(1)}$ units, called the phase I sample. Measure the auxiliary variables $\mathbf{x}$ for every unit in the phase I sample. In the survey of businesses, you could take a simple random sample (SRS) of tax records and record the reported income for each business in the sample. For measuring timber volume, you could weigh a sample of trucks selected either with an SRS or with probability proportional to estimated timber volume. The phase I sample is generally relatively large (and can be large because the auxiliary information is inexpensive to obtain), and should provide accurate information about the distribution of the $\mathbf{x}$'s.

2. *Phase II sample.* Now act as though the phase I sample is a population and select a probability sample of size $n^{(2)}$ from the phase I sample. Measure the variables of interest for each unit in the subsample, called the phase II sample. Since you are treating the phase I sample as the population from which the phase II sample is drawn, you may use the auxiliary information gathered in phase I when designing the phase II sample. You might select the businesses to be contacted with probability proportional to the income measured in the phase I sample. Alternatively, you might use the income information to stratify the businesses in the phase I sample and then contact a randomly selected subset

DOI: 10.1201/9780429298899-12

of the businesses in each income stratum to obtain the desired information on variables such as total expenses. You could select the truckloads on which timber volume is to be measured with probability proportional to weight, or you could use the information in the phase I sample to obtain a better estimate of total weight and use ratio estimation. In each case, the y variables are relatively expensive to measure, but y is related to $\mathbf{x}$.

Two-phase sampling can save time and money if the auxiliary information is relatively inexpensive to obtain and if having that auxiliary information can increase the precision of the estimates for quantities of interest.

Example 12.1. Stockford and Page (1984) used two-phase sampling to estimate the percentage of Vietnam-era veterans in U.S. Veterans Administration (VA) hospitals who actually served in Vietnam.

The 1982 VA Annual Patient Census (APC) included a random sample of 20% of the patients in VA hospitals. The following question was included: "If period of service is 'Vietnam era,' was service in Vietnam?" with answer categories "yes," "no," and "not available." The answers to the question were obtained from patients' medical records. But the response from medical records could be inaccurate for several reasons: (1) The medical record classification was largely self-reported, and the patient may not have been able to recall the location of service due to medical problems, or may have been confused about the definition of Vietnam service (some pilots whose duty station was officially recorded as Thailand flew missions over Vietnam); (2) a patient might misstate Vietnam service because he or she thought the answer might affect VA benefits; or (3) errors might be made in recording the response in the medical record. In addition, a large number of patients had "not available" for the answer. Thus, the answer to the question on Vietnam service in the APC survey was unsatisfactory for estimating the percentage of Vietnam-era veterans in VA hospitals who served in Vietnam.

Stockford and Page checked the military records for a stratified subsample of the hospitalized veterans to determine the true classification of Vietnam service. The information in the original survey was used for the stratification, as different percentages with Vietnam service were expected in the "yes", "no," and "not available" groups in the APC survey. Military records for all of the patients in the "not available" stratum were checked. It was expected that the within-stratum variances would be relatively low in the "yes" and "no" strata—even though the APC survey data are inaccurate, you would expect a higher percentage of "yes" respondents to have served in Vietnam than "no" respondents—and military records were checked for a subsample in each of those two strata.

TABLE 12.1
Number of persons who served in Vietnam for each stratum in Example 12.1.

APC Group	APC Survey Classification	Subsample Size	Vietnam Service in Subsample
Yes	755	67	49
No	804	72	11
Not available	505	505	211
Total	2064	644	271

The results for the question "Was service in Vietnam?" are given in Table 12.1. As expected, the percentage of veterans with Vietnam service differed for the three groups: Of the veterans with a "yes" response to the APC survey question, 73% actually served in Vietnam, compared with 15% for the "no" group and 42% for the veterans for which the information was not available. ■

Example 12.2. Two-phase sampling is often used in forestry surveys. Aerial photographs are available for the region of interest, and points are systematically distributed across the photographs. Areas around the points are inspected on the photographs and classified by land class: forest land, unproductive forest land, nonforest land, and water. A phase I sample of points is then drawn from the grid, with a higher sampling fraction for grid points classified as forest land than those classified as nonforest land. Areas in the phase I sample are examined more closely to classify them by stand size and density. Then, a subsample is taken of the points in the phase I sample, and ground measurements such as land use, volume, and mortality taken; the percentage of area that is forest from the phase II ground sample may differ somewhat from the photo estimate in phase I, and ratio estimation can be used in the phase II sample to increase the precision of the estimator. ■

Example 12.3. We have already seen two-phase sampling used for nonresponse adjustment in Section 8.3. A probability sample is taken from the population; the sampled units are then divided into the two strata of respondents and nonrespondents. Then a subsample is taken of the nonrespondents. The phase I sample is the original probability sample. The variable

$$x_i = \begin{cases} 1 \text{ if observation } i \text{ responds} \\ 0 \text{ if observation } i \text{ is a nonrespondent} \end{cases}$$

is observed for everyone in the phase I sample. Then the information about x_i is used in the phase II sample. The variable of interest y_i is observed for all observations with $x_i = 1$; a subsample is taken for observations with $x_i = 0$. ■

Two-phase sampling contrasted with two-stage sampling. What is the difference between a two-phase design, discussed in this chapter, and the two-stage designs discussed in Chapters 5 and 6? In two-stage sampling, different sizes of sampling units are collected at the two stages. Secondary sampling units (ssus) are sampled from all primary sampling units (psus) selected at stage 1, and the basic design for sampling ssus does not depend on which psus are selected.

In two-phase sampling, the phase I sample is used to collect inexpensive auxiliary information on the sampling units, and this information is used to modify or determine the phase II sample design.

Consider a sample of hospitals. A two-stage design might select hospitals with probability proportional to size at the first stage, and then subsample 20 patients from each hospital as ssus at the second stage. The sampling protocol for stage 2 does not depend on information gathered at the first stage of sampling, and patients are sampled from every psu selected at the first stage. A two-phase sample of hospitals might take a probability sample of hospitals in phase I, then divide the hospitals in the phase I sample into strata based on the number of cancer patients. In phase II, a stratified random subsample of the hospitals would be selected using the stratification information from phase I.

12.1 Theory for Two-Phase Sampling

A general framework for two-phase sampling is given in Särndal and Swensson (1987) and Legg and Fuller (2009). Let $\mathcal{S}^{(1)}$ denote the phase I sample; the units selected for the sample are determined by the random variables

$$Z_i = \begin{cases} 1 & \text{if unit } i \text{ is in the phase I sample} \\ 0 & \text{if unit } i \text{ is not in the phase I sample.} \end{cases}$$

Let $w_i^{(1)}$, for $i \in \mathcal{S}^{(1)}$, be the sampling weights for the phase I sample: $w_i^{(1)} = 1/P(Z_i = 1)$. We observe a vector of auxiliary characteristics $\mathbf{x}_i = (x_{i1}, x_{i2}, \ldots, x_{ik})^T$ for each observation unit in the phase I sample. Using the theory developed in earlier chapters, we can estimate the population total for auxiliary variable j as

$$\hat{t}_{x_j}^{(1)} = \sum_{i \in \mathcal{S}^{(1)}} w_i^{(1)} x_{ij} = \sum_{i=1}^{N} Z_i w_i^{(1)} x_{ij}.$$

Now, indicate membership in the phase II sample $\mathcal{S}^{(2)}$ by the random variable

$$D_i = \begin{cases} 1 & \text{if unit } i \text{ is in the phase II sample} \\ 0 & \text{if unit } i \text{ is not in the phase II sample} \end{cases}$$

The probability that a unit is in the phase II sample depends on whether it is in the phase I sample and also may depend on auxiliary information collected in the phase I sample; we denote this dependence by writing $P(D_i = 1 \mid \mathbf{Z})$, where $\mathbf{Z}$ is the vector $(Z_1, Z_2, \ldots, Z_N)$. Thus, when we find an expectation conditional on $\mathbf{Z}$, we are treating the information from the phase I sample as known. The subsampling weights for the final, phase II sample also depend on which units were selected to be in the phase I sample.

$$w_i^{(2)} = w_i^{(2)}(\mathbf{Z}) = \begin{cases} \dfrac{1}{P(D_i = 1|\mathbf{Z})} & \text{if } Z_i = 1 \\ 0 & \text{if } Z_i = 0. \end{cases}$$

An analog of the Horvitz-Thompson estimator for two-phase sampling is

$$\hat{t}_y^{(2)} = \sum_{i \in \mathcal{S}^{(2)}} w_i^{(1)} w_i^{(2)} y_i = \sum_{i=1}^{N} Z_i D_i w_i^{(1)} w_i^{(2)} y_i. \tag{12.1}$$

Kott and Stukel (1997) call (12.1) the double expansion estimator; it "expands" the weight on y_i by the product of the two sampling weights.

We use the following device to find properties of the estimator in (12.1). The phase II sample is selected by treating the phase I sample as the population, so we can find properties of the subsample relative to phase I using standard methods. Define

$$\hat{t}_y^{(1)} = \sum_{i \in \mathcal{S}^{(1)}} w_i^{(1)} y_i = \sum_{i=1}^{N} Z_i w_i^{(1)} y_i.$$

Now, we do not know what $\hat{t}_y^{(1)}$ is, because we only observe the y_is in the phase II sample. But $\hat{t}_y^{(1)}$ serves as the "population total" estimated in phase II— if we knew y_i for all units in the phase I sample, we would estimate t_y by $\hat{t}_y^{(1)}$. Treating the phase I sample as known, we have

$$E\left[\hat{t}_y^{(2)} | \mathbf{Z}\right] = \sum_{i=1}^{N} Z_i w_i^{(1)} w_i^{(2)} y_i E\left[D_i | \mathbf{Z}\right] = \sum_{i=1}^{N} Z_i w_i^{(1)} y_i = \hat{t}_y^{(1)}.$$

Then, using successive conditioning (see Section A.4),

$$E\left[\hat{t}_y^{(2)}\right] = E\left(E[\hat{t}_y^{(2)} | \mathbf{Z}]\right) = E\left[\sum_{i=1}^{N} Z_i w_i^{(1)} y_i\right] = t_y. \tag{12.2}$$

Also, from Property 5 of Table A.2,

$$V(\hat{t}_y^{(2)}) = V\left(E[\hat{t}_y^{(2)}|\mathbf{Z}]\right) + E\left(V[\hat{t}_y^{(2)}|\mathbf{Z}]\right) = V\left(\hat{t}_y^{(1)}\right) + E\left(V[\hat{t}_y^{(2)}|\mathbf{Z}]\right). \tag{12.3}$$

The first term is the variance that would be obtained if y_i had been observed for every observation in $\mathcal{S}^{(1)}$; the second term is the additional variance from subsampling in phase II. Consequently, the variance from two-phase sampling is always larger than if we measured y on every unit in the phase I sample of $n^{(1)}$ units. We hope, though, that if y is related to $\mathbf{x}$, the second term in (12.3) will be smaller than the variance of an estimator of t_y from a sample of size $n^{(2)}$ that does not use the auxiliary information in the design.

12.2 Two-Phase Sampling with Stratification

In two-phase sampling with stratification, information on a stratification variable is selected in phase I. That information is then used to select a stratified sample (the phase II sample) from the phase I sample. For simplicity, assume that an SRS is taken in phase I, and that stratified random sampling is used in phase II. Särndal et al. (1992, Chapter 9) give a more general treatment, allowing unequal-probability sampling for either phase. Define $\mathcal{S}^{(1)}$, $\mathcal{S}^{(2)}$, Z_i, and D_i as in Section 12.1. If an SRS of size n is taken in phase I, then $P(Z_i = 1) = n/N$.

The observation units are divided among H strata, but we do not know stratum membership for a unit until it is selected in phase I. In the population, however, stratum h has N_h units (assume N_h is unknown) and $N = \sum_{h=1}^{H} N_h$ (assume N is known). Let

$$x_{ih} = \begin{cases} 1 & \text{if unit } i \text{ is in stratum } h \\ 0 & \text{if unit } i \text{ is not in stratum } h. \end{cases}$$

Observe x_{ih}, $h = 1, \ldots, H$, for each unit in the phase I sample. The number of units in the phase I sample that belong to stratum h is a random variable:

$$n_h = \sum_{i=1}^{N} Z_i x_{ih}.$$

Now take a simple random subsample of size m_h in stratum h; m_h may depend on the first phase of the sampling. The subsamples in different strata are selected independently, given the information in the phase I sample. With random subsampling,

$$P(D_i = 1 \mid \mathbf{Z}) = Z_i \sum_{h=1}^{H} x_{ih} \frac{m_h}{n_h}.$$

Although $P(D_i = 1 \mid \mathbf{Z})$ is written as a sum, all but one of the x_{ih}'s ($h = 1, \ldots, H$) will equal zero because each unit belongs to exactly one stratum, so that $P(D_i = 1 \mid \mathbf{Z}) = Z_i m_h / n_h$ for unit i determined to be in stratum h. The sampling weight for a phase II unit in stratum h is $w_i^{(2)} = n_h/m_h$; in general, $w_i^{(2)} = Z_i \sum_{h=1}^{H} x_{ih} n_h / m_h$.

The two-phase stratified estimator of the population total is

$$\hat{t}_{\text{str}}^{(2)} = \sum_{i=1}^{N} Z_i D_i w_i^{(1)} w_i^{(2)} y_i = \sum_{i=1}^{N} Z_i D_i \frac{N}{n} \left(\sum_{h=1}^{H} \frac{n_h}{m_h} x_{ih} \right) y_i = N \sum_{h=1}^{H} \frac{n_h}{n} \bar{y}_h^{(2)}, \tag{12.4}$$

where $\bar{y}_h^{(2)} = \sum_{i \in \mathcal{S}^{(2)}} x_{ih} y_i / m_h$ is the average of the phase II units in stratum h. We showed in (12.2) that $E[\hat{t}_{\text{str}}^{(2)}] = t_y$. The corresponding estimator of the population mean is

$$\hat{\bar{y}}_{\text{str}}^{(2)} = \frac{1}{N} \sum_{i \in \mathcal{S}^{(2)}} w_i^{(1)} w_i^{(2)} y_i = \sum_{h=1}^{H} \frac{n_h}{n} \bar{y}_h^{(2)}. \tag{12.5}$$

Recall that a stratified random sampling estimator of the population total from (3.1) is

$$\hat{t}_{\text{str}} = N \sum_{h=1}^{H} \frac{N_h}{N} \bar{y}_h;$$

the two-phase estimator simply substitutes n_h/n for N_h/N.

The variance is also computed conditionally, using (12.3):

$$\begin{aligned} V\left(\hat{t}_{\text{str}}^{(2)}\right) &= V\left(E\left[\hat{t}_{\text{str}}^{(2)} \mid \mathbf{Z}\right]\right) + E\left(V\left[\hat{t}_{\text{str}}^{(2)} \mid \mathbf{Z}\right]\right) \\ &= V\left(\hat{t}_y^{(1)}\right) + N^2 E\left(V\left[\sum_{h=1}^{H} \frac{n_h}{n} \bar{y}_h^{(2)} \,\middle|\, \mathbf{Z}\right]\right) \\ &= N^2 \left(1 - \frac{n}{N}\right) \frac{S_y^2}{n} + N^2 E\left[\sum_{h=1}^{H} \left(\frac{n_h}{n}\right)^2 V\left(\bar{y}_h^{(2)} \mid \mathbf{Z}\right)\right]. \end{aligned} \tag{12.6}$$

The first term is the variance from the phase I SRS; the second term is the additional variance resulting from the subsampling in phase II. Here, $S_y^2 = \sum_{i=1}^{N} (y_i - \bar{y}_{\mathcal{U}})^2/(N-1)$ is the population variance of the y's. The second term in (12.6) is left as an expectation because n_h and m_h are random variables.

Rao (1973) estimated the variance in two-phase sampling with stratification as

$$\begin{aligned} \hat{V}\left(\hat{t}_{\text{str}}^{(2)}\right) = N(N-1) &\sum_{h=1}^{H} \left(\frac{n_h - 1}{n-1} - \frac{m_h - 1}{N-1}\right) \frac{n_h}{n} \frac{s_h^{2(2)}}{m_h} \\ &+ \frac{N^2}{n-1}\left(1 - \frac{n}{N}\right) \sum_{h=1}^{H} \frac{n_h}{n} \left(\bar{y}_h^{(2)} - \hat{\bar{y}}_{\text{str}}^{(2)}\right)^2, \end{aligned} \tag{12.7}$$

where

$$s_h^{2(2)} = \frac{1}{m_h - 1} \sum_{i \in \mathcal{S}^{(2)}} x_{ih} (y_i - \bar{y}_h^{(2)})^2$$

is the sample variance of the y_is in stratum h (see Exercise 12). If we can ignore the finite population corrections (fpcs),

$$\hat{V}(\hat{\bar{y}}_{\text{str}}^{(2)}) \approx \sum_{h=1}^{H} \frac{n_h - 1}{n-1} \frac{n_h}{n} \frac{s_h^{2(2)}}{m_h} + \frac{1}{n-1} \sum_{h=1}^{H} \frac{n_h}{n} \left(\bar{y}_h^{(2)} - \hat{\bar{y}}_{\text{str}}^{(2)}\right)^2. \tag{12.8}$$

Example 12.4. Table 12.2 contains the statistics from the phase II sample in Example 12.1. Because $\bar{y}_h^{(2)} = \hat{p}_h$ is a proportion, $s_h^{2(2)} = m_h \hat{p}_h (1 - \hat{p}_h)/(m_h - 1)$.

The estimated percentage of Vietnam-era VA hospital patients who served in Vietnam is, from (12.5),

$$\hat{\bar{y}}_{\text{str}}^{(2)} = \left(\frac{755}{2064}\right)(0.7313) + \left(\frac{804}{2064}\right)(0.1528) + \left(\frac{505}{2064}\right)(0.4178) = 0.4293.$$

TABLE 12.2
Statistics from phase II sample for Example 12.4.

Stratum	n_h	m_h	$\hat{p}_h$	$s_h^{2(2)}$
Yes	755	67	0.7313	0.1995
No	804	72	0.1528	0.1313
Not available	505	505	0.4178	0.2437
Total	2064	644		

The phase I sample is an SRS with $n/N = 0.2$, so the fpc should be included in the variance estimate. Calculating the terms in (12.7),

$$\sum_{h=1}^{H}\left(\frac{n_h-1}{n-1}-\frac{m_h-1}{N-1}\right)\frac{n_h}{n}\frac{s_h^{2(2)}}{m_h} = 0.000391 + 0.000271 + 0.0000231 = 0.000686,$$

and

$$\frac{1}{n-1}\left(1-\frac{n}{N}\right)\sum_{h=1}^{H}\frac{n_h}{n}(\bar{y}_h^{(2)}-\hat{\bar{y}}_{str}^{(2)})^2 = 1.29\times10^{-5}+1.16\times10^{-5}+1.24\times10^{-8} = 0.0000245.$$

Thus, $\hat{V}(\hat{\bar{y}}_{\text{str}}^{(2)}) = 0.000686 + 0.0000245 = 0.00071$, and $\text{SE}(\hat{\bar{y}}_{\text{str}}^{(2)}) = 0.027$.

Was two-phase sampling more efficient here? Had an SRS of size 644 been taken directly from the records, and had $\hat{p} = 0.429$ been observed, the standard error would have been $\text{SE}(\hat{p}) = 0.019$, which is actually smaller than the standard error from the two-phase sampling design. If you look at the individual terms in the variance estimates, you can see why two-phase sampling did not increase efficiency in this example. All of the phase I units in the "not available" stratum were subsampled, giving a very low value of $s_h^{2(2)}/m_h$ for that stratum. But the sample sizes in the other two strata were too small, leading to relatively large contributions to the overall variance from those two strata.

Suppose proportional allocation had been used in the phase II sample instead and that the same sample proportions had been observed. Then, you would subsample 236 records in the "yes" stratum, 251 records in the "no" stratum, and 157 records in the "not available" stratum. In that case, if the sample proportions remained the same, the standard error from the two-phase sample would have been 0.017, a modest decrease from the standard error of an SRS of size 644. But proportional allocation does not make the most efficient use of the phase I information. More savings would have been achieved if some sort of optimal allocation had been used (see Exercise 10). ■

Ideally, you would use the information about stratum membership from phase I to have a more efficient sampling design in phase II. This usually means using optimal allocation in the stratified phase II sample. For example, in a survey to study the total sales of manufacturing firms, you might obtain total revenue from the tax records (x) for a sample of manufacturing firms. Then you could use that tax information to stratify the phase I sample by the reported revenue, and take higher sampling fractions in phase II for the strata with higher revenue in the tax records.

Example 12.5. A screening survey is a special case of a two-phase sample using stratification. The U.S. National Immunization Child Survey (Wolter et al., 2017) collected a phase I sample using random digit dialing, including both landline and cellular telephones. The households in the phase I sample were divided into two strata: (1) households with children 19–35 months old, and (2) households with no children between 19 and 35 months old. Since the goal of the survey was to estimate vaccination rates for children in the 19–35 month age group, no households in stratum 2 were included in the phase II sample. The parent or guardian in stratum 1 households was asked to consent for information to be obtained from the child's vaccination providers, and those providers were asked about the child's immunizations. Nonresponse and other nonsampling errors in the phase II sample required weighting adjustments. ■

Example 12.6. McNamee (2003) discussed the use of two-phase sampling to estimate disease prevalence. In the first phase an inexpensive, but not completely accurate, method is used to classify persons as having the disease or not. The second phase is a more accurate test for the disease. For example, the phase I survey might ask people whether they have diabetes, and divide the respondents into stratum 1, persons who say they have diabetes, and stratum 2, persons who say they do not have diabetes. But some persons with diabetes are unaware that they have it. You therefore need to subsample both strata in the phase II sample, which evaluates persons through a medical examination, to guarantee that diabetics who are unaware they have diabetes can be included in the sample. Although we expect a smaller fraction of the persons in stratum 2 to have diabetes, compared with the fraction in stratum 1 who have diabetes, the characteristics of persons with diabetes in stratum 2 might be quite different from those in stratum 1. ■

12.3 Ratio and Regression Estimation in Two-Phase Samples

The stratified two-phase sampling design in Section 12.2 uses the auxiliary information collected from the phase I sample in the design of the phase II sample. Alternatively, or in addition, the information about the auxiliary variables x can be used in the estimator, through ratio and regression estimation.

12.3.1 Two-Phase Sampling with Ratio Estimation

Suppose that x, a variable thought to be highly correlated with y, can be measured inexpensively in the phase I sample. Define $\mathcal{S}^{(1)}$, $\mathcal{S}^{(2)}$, Z_i, and D_i as in Section 12.1. The auxiliary variable x_i is measured for each observation in the phase I sample; from that sample, we may estimate the population total $t_x = \sum_{i=1}^{N} x_i$ by

$$\hat{t}_x^{(1)} = \sum_{i \in \mathcal{S}^{(1)}} w_i^{(1)} x_i = \sum_{i=1}^{N} Z_i w_i^{(1)} x_i.$$

Now select the phase II subsample and measure y_i on units in the subsample. From the phase II sample $\mathcal{S}^{(2)}$, we can calculate $\hat{t}_y^{(2)}$ using (12.1) and

$$\hat{t}_x^{(2)} = \sum_{i \in \mathcal{S}^{(2)}} w_i^{(1)} w_i^{(2)} x_i = \sum_{i=1}^{N} Z_i D_i w_i^{(1)} w_i^{(2)} x_i.$$

Then,

$$\hat{t}_{yr}^{(2)} = \hat{t}_x^{(1)} \frac{\hat{t}_y^{(2)}}{\hat{t}_x^{(2)}} = \hat{t}_x^{(1)} \hat{B}^{(2)}. \tag{12.9}$$

Note that (12.9) is similar to the ratio estimator in (4.2); we substitute $\hat{t}_x^{(1)}$ from the phase I sample for the unknown quantity t_x.

Using linearization,

$$\hat{t}_{yr}^{(2)} \approx t_y + \frac{t_x}{t_x}(\hat{t}_y^{(2)} - t_y) + \frac{t_y}{t_x}(\hat{t}_x^{(1)} - t_x) - \frac{t_y t_x}{t_x^2}(\hat{t}_x^{(2)} - t_x).$$

Then,

$$\begin{aligned}
V(\hat{t}_{yr}^{(2)}) &\approx V\left[\hat{t}_y^{(2)} + \frac{t_y}{t_x}(\hat{t}_x^{(1)} - \hat{t}_x^{(2)})\right] \\
&= V\left\{E\left[\hat{t}_y^{(2)} + \frac{t_y}{t_x}(\hat{t}_x^{(1)} - \hat{t}_x^{(2)})\middle|\mathbf{Z}\right]\right\} + E\left\{V\left[\hat{t}_y^{(2)} + \frac{t_y}{t_x}(\hat{t}_x^{(1)} - \hat{t}_x^{(2)})\middle|\mathbf{Z}\right]\right\} \\
&= V[\hat{t}_y^{(1)}] + E\left[V\left(\hat{t}_y^{(2)} - \frac{t_y}{t_x}\hat{t}_x^{(2)}\middle|\mathbf{Z}\right)\right] \\
&= V[\hat{t}_y^{(1)}] + E\left[V(\hat{t}_d^{(2)}|\mathbf{Z})\right],
\end{aligned}$$

where $d_i = y_i - (t_y/t_x)x_i$. Thus, the variance of the two-phase ratio estimator is the variance that would be calculated for $\hat{t}_y^{(1)}$ if we observed y_i for every unit in the phase I sample, plus an extra term involving the variance of the residuals from the ratio model.

If an SRS of $n^{(1)}$ units is taken for phase I and an SRS of $n^{(2)}$ units is taken in phase II, then

$$V(\hat{t}_{yr}^{(2)}) \approx N^2\left(1 - \frac{n^{(1)}}{N}\right)\frac{S_y^2}{n^{(1)}} + N^2\left(1 - \frac{n^{(2)}}{n^{(1)}}\right)\frac{S_d^2}{n^{(2)}}, \tag{12.10}$$

where $d_i = y_i - Bx_i$ and $S_d^2 = \sum_{i=1}^N d_i^2/(N-1)$, and

$$\hat{V}(\hat{t}_{yr}^{(2)}) = N^2\left(1 - \frac{n^{(1)}}{N}\right)\frac{s_y^2}{n^{(1)}} + N^2\left(1 - \frac{n^{(2)}}{n^{(1)}}\right)\frac{s_e^2}{n^{(2)}}, \tag{12.11}$$

where $s_y^2 = \sum_{i\in\mathcal{S}^{(2)}}(y_i - \bar{y}^{(2)})^2/(n^{(2)}-1)$ and $s_e^2 = \sum_{i\in\mathcal{S}^{(2)}}(y_i - \hat{B}^{(2)}x_i)^2/(n^{(2)}-1)$, is an approximately unbiased estimator of $V(\hat{t}_{yr}^{(2)})$ (see Exercise 13). An alternative estimator of the variance is given in Exercise 14.

Example 12.7. Suppose, for the population in `agpop.csv` sampled in Examples 2.6 and 4.2, that we do not know the value of x_i = *acres87*, the acreage devoted to farms in 1987, or of t_x before sampling. To use x as auxiliary information through two-phase ratio estimation, we take an SRS of size 400, measure *acres87* on every unit in this phase I sample, and then take an SRS of size 30 from the phase I sample to serve as the phase II sample. We measure y = *acres92* on the units in the phase II sample. We then employ ratio estimation to estimate t_y, using $\hat{t}_x^{(1)}$ to estimate the unknown auxiliary population total t_x. We calculate

$$\hat{t}_{yr}^{(2)} = \hat{t}_x^{(1)}\frac{\hat{t}_y^{(2)}}{\hat{t}_x^{(2)}} = \hat{t}_x^{(1)}\frac{\bar{y}^{(2)}}{\bar{x}^{(2)}} = 1{,}002{,}814{,}347\left(\frac{322{,}385}{335{,}444}\right) = 963{,}774{,}784.$$

From (12.11),

$$\begin{aligned}
\hat{V}(\hat{t}_{yr}^{(2)}) &= N^2\left(1 - \frac{n^{(1)}}{N}\right)\frac{s_y^2}{n^{(1)}} + N^2\left(1 - \frac{n^{(2)}}{n^{(1)}}\right)\frac{s_e^2}{n^{(2)}} \\
&= 3000^2\left(1 - \frac{400}{3000}\right)\frac{112{,}160{,}218{,}976}{400} + 3000^2\left(1 - \frac{30}{400}\right)\frac{1{,}908{,}426{,}448}{30} \\
&= 2.7 \times 10^{15}.
\end{aligned}$$

Thus, an approximate 95% confidence interval (CI) for t_y is 963,774,784 $\pm\, 2.05\sqrt{2.7 \times 10^{15}}$ = [856,924,072, 1,070,625,496]. The corresponding CI for $\bar{y}_{\mathcal{U}}$ is [285,641, 356,875] with $\hat{\bar{y}}_r^{(2)}$ = 321,258. The widths of these CIs are comparable to those of the SRS of size 300 in Example 2.11, even though here y_i was measured only on the 30 units in the phase II sample. If measuring x is inexpensive relative to measuring y, the high correlation between x and y makes two-phase sampling with ratio estimation very efficient. ■

12.3.2 Generalized Regression Estimation in Two-Phase Sampling

We can also use a two-phase version of the generalized regression (GREG) estimator of Section 11.6. The two-phase GREG estimator takes the form

$$\hat{t}_{y\text{GREG}}^{(2)} = \hat{t}_y^{(2)} + (\hat{\mathbf{t}}_{\mathbf{x}}^{(1)} - \hat{\mathbf{t}}_{\mathbf{x}}^{(2)})^T \hat{\mathbf{B}}^{(2)}, \tag{12.12}$$

where

$$\hat{\mathbf{B}}^{(2)} = \left(\sum_{i \in \mathcal{S}^{(2)}} w_i^{(1)} w_i^{(2)} \frac{1}{\sigma_i^2} \mathbf{x}_i \mathbf{x}_i^T \right)^{-1} \sum_{i \in \mathcal{S}^{(2)}} w_i^{(1)} w_i^{(2)} \frac{1}{\sigma_i^2} \mathbf{x}_i y_i \tag{12.13}$$

and the constants σ_i^2 are determined by the analyst. In (12.12), the estimator is calibrated to the estimated population totals of $\mathbf{x}$ from phase I, $\hat{\mathbf{t}}_{\mathbf{x}}^{(1)}$ (see Exercise 16).

The estimator in (12.12) may be written using a modification of the weights. Analogously to (11.26), let

$$g_i = 1 + (\hat{\mathbf{t}}_{\mathbf{x}}^{(1)} - \hat{\mathbf{t}}_{\mathbf{x}}^{(2)})^T \left(\sum_{i \in \mathcal{S}^{(2)}} w_i^{(1)} w_i^{(2)} \frac{1}{\sigma_i^2} \mathbf{x}_i \mathbf{x}_i^T \right)^{-1} \frac{1}{\sigma_i^2} \mathbf{x}_i.$$

Then,

$$\hat{t}_{y\text{GREG}}^{(2)} = \sum_{i \in \mathcal{S}^{(2)}} w_i^{(1)} w_i^{(2)} g_i y_i.$$

We again use Property 5 in Table A.2 to find $V\left(\hat{t}_{y\text{GREG}}^{(2)}\right)$:

$$V\left(\hat{t}_{y\text{GREG}}^{(2)}\right) = V\left[E\left(\hat{t}_{y\text{GREG}}^{(2)} \,|\, \mathbf{Z}\right)\right] + E\left[V\left(\hat{t}_{y\text{GREG}}^{(2)} \,|\, \mathbf{Z}\right)\right]. \tag{12.14}$$

Since the GREG estimator is approximately unbiased, if the sizes of the phase I and phase II samples are sufficiently large, then

$$V\left[E\left(\hat{t}_{y\text{GREG}}^{(2)} \,|\, \mathbf{Z}\right)\right] \approx V\left(\hat{t}_y^{(1)}\right),$$

where $V(\hat{t}_y^{(1)})$ is the variance of the estimator $\hat{t}_y^{(1)} = \sum_{i \in \mathcal{S}^{(1)}} w_i^{(1)} y_i$ of the population total that we would have if we had been able to measure y on every unit in the phase I sample. By linearization, the conditional variance in the second term of (12.14) is

$$V\left(\hat{t}_{y\text{GREG}}^{(2)} | \mathbf{Z}\right) = V\left(\sum_{i \in \mathcal{S}^{(2)}} w_i^{(1)} w_i^{(2)} d_i^{(1)} \middle| \mathbf{Z} \right)$$

where $d_i^{(1)} = y_i - \mathbf{x}_i^T \hat{\mathbf{B}}^{(1)}$ and $\hat{\mathbf{B}}^{(1)} = \left(\sum_{i \in \mathcal{S}^{(1)}} (w_i^{(1)}/\sigma_i^2) \mathbf{x}_i \mathbf{x}_i^T\right)^{-1} \sum_{i \in \mathcal{S}^{(1)}} (w_i^{(1)}/\sigma_i^2) \mathbf{x}_i y_i$.

Särndal et al. (1992, Chapter 9) estimated the two terms in (12.14) separately. The conditional variance in the second term is unbiased for its expectation, so $E\left[V\left(\hat{t}^{(2)}_{y\text{GREG}} \mid \mathbf{Z}\right)\right]$ may be estimated by

$$\hat{V}\left(\hat{t}^{(2)}_{y\text{GREG}}|\mathbf{Z}\right) = \hat{V}\left(\sum_{i\in\mathcal{S}^{(2)}} w_i^{(1)} w_i^{(2)} e_i \middle| \mathbf{Z}\right),$$

with $e_i = y_i - x_i\hat{B}^{(2)}$ substituted for the unknown values $d_i^{(1)}$. If the phase I sampling design is an SRS of $n^{(1)}$ units, the first term in (12.14) may be estimated by

$$\hat{V}(\hat{t}_y^{(1)}) = N^2\left(1 - \frac{n^{(1)}}{N}\right)\frac{\hat{S}_y^2}{n^{(1)}},$$

with

$$\hat{S}_y^2 = \frac{1}{n^{(2)} - 1}\sum_{i\in\mathcal{S}^{(2)}}\left(y_i - \bar{y}^{(2)}\right)^2.$$

Estimating the first term in (12.14) is more challenging for complex designs; Exercise 17 presents an estimator for this situation.

Example 12.8. Barnett et al. (2001) described an application of two-phase sampling to an accounting problem. The auditor has access to a large phase I SRS of transactions. Each transaction in the phase I sample has been checked by internal auditors, who record errors they find. A small random subsample is taken of the phase I transactions; each transaction in the phase II SRS is examined by an external auditor. If the internal and external auditor disagree on a transaction, the external auditor is assumed to be correct. If the internal auditors are largely correct, then regression estimation can be used, either through ratio estimation or poststratification into classes based on types of errors, to greatly increase the precision of the amounts of errors in the population of transactions. ■

12.4 Jackknife Variance Estimation for Two-Phase Sampling

As we have seen, the formulas for variance estimators are complicated in two-phase sampling. Fortunately, in many cases we can use resampling methods to estimate variances. The jackknife method presented in Section 9.3.2 needs to be modified for two-phase sampling because y is observed only for units selected in phase II. The jackknife estimators from Kim et al. (2006) mimic the sampling design, including the two phases of sampling, in the resamples. The jackknife estimates the with-replacement variance; it is a good approximation to the without-replacement variance if the sampling fractions are small.

Suppose the phase I design is an SRS of size $n^{(1)}$ and the phase II design is an SRS of size $n^{(2)}$. Consider the ratio estimator of t_y in Section 12.3.1,

$$\hat{t}^{(2)}_{yr} = \hat{t}^{(1)}_x \frac{\hat{t}^{(2)}_y}{\hat{t}^{(2)}_x}.$$

When we delete unit j in the phase I sample, we obtain

$$\hat{t}^{(2)}_{yr(j)} = \hat{t}^{(1)}_{x(j)} \frac{\hat{t}^{(2)}_{y(j)}}{\hat{t}^{(2)}_{x(j)}},$$

where $\hat{t}^{(1)}_{x(j)}$, $\hat{t}^{(2)}_{y(j)}$, and $\hat{t}^{(2)}_{x(j)}$ are calculated using the jackknife weights as described in the next paragraph. Then,

$$\hat{V}_{\text{JK}}(\hat{t}^{(2)}_{yr}) = \sum_{j \in \mathcal{S}^{(1)}} \frac{n^{(1)}-1}{n^{(1)}} \left[\hat{t}^{(2)}_{yr(j)} - \hat{t}^{(2)}_{yr}\right]^2.$$

When both samples are SRSs, $w_i^{(1)} = N/n^{(1)}$ for the phase I sample and $w_i^{(2)} = n^{(1)}/n^{(2)}$ for the phase II sample. The jackknife weights for phase I are constructed exactly like those in Section 9.3.2. When unit j from the phase I sample is deleted, the modified weight for phase I is

$$w^{(1)}_{i(j)} = \begin{cases} 0 & \text{if } i = j \\ \dfrac{n^{(1)}}{n^{(1)}-1} w_i^{(1)} & \text{if } i \neq j. \end{cases}$$

The modified weight for phase II depends on whether the unit deleted from the phase I sample is in the phase II sample or not:

$$w^{(2)}_{i(j)} = \begin{cases} 0 & \text{if } i = j \text{ and } j \in \mathcal{S}^{(2)} \\ \dfrac{n^{(1)}-1}{n^{(2)}-1} & \text{if } i \neq j \text{ and } j \in \mathcal{S}^{(2)} \\ \dfrac{n^{(1)}-1}{n^{(2)}} & \text{if } j \notin \mathcal{S}^{(2)} \end{cases}$$

Using the jackknife weights,

$$\hat{t}^{(1)}_{x(j)} = \sum_{i \in \mathcal{S}^{(1)}} w^{(1)}_{i(j)} x_i = N \frac{n^{(1)}\bar{x}^{(1)} - x_j}{n^{(1)}-1}$$

$$\hat{t}^{(2)}_{x(j)} = \sum_{i \in \mathcal{S}^{(2)}} w^{(1)}_{i(j)} w^{(2)}_{i(j)} x_i = \begin{cases} N \dfrac{n^{(2)}\bar{x}^{(2)} - x_j}{n^{(2)}-1} & \text{if } j \in \mathcal{S}^{(2)} \\ \hat{t}^{(2)}_x & \text{if } j \notin \mathcal{S}^{(2)} \end{cases}$$

and

$$\hat{t}^{(2)}_{y(j)} = \sum_{i \in \mathcal{S}^{(2)}} w^{(1)}_{i(j)} w^{(2)}_{i(j)} y_i = \begin{cases} N \dfrac{n^{(2)}\bar{y}^{(2)} - y_j}{n^{(2)}-1} & \text{if } j \in \mathcal{S}^{(2)} \\ \hat{t}^{(2)}_y & \text{if } j \notin \mathcal{S}^{(2)} \end{cases} \quad ■$$

Jackknife variance estimation with stratification. For two-phase sampling with stratification, suppose that an SRS of size $n^{(1)}$ is taken in phase I and a stratified random sample is taken in phase II, where the sampling fractions for the strata are specified before the phase I sample is collected. Consider the estimator $\hat{t}^{(2)}_{\text{str}}$ in (12.4). Define the jackknife replicate, deleting unit j, by

$$\hat{t}^{(2)}_{\text{str}(j)} = \sum_{i \in \mathcal{S}^{(2)}} w^{(1)}_{i(j)} w^{(2)}_{i(j)} y_i,$$

where

$$w^{(1)}_{i(j)} = \begin{cases} 0 & \text{if } j = i \\ \dfrac{N}{n^{(1)}-1} & \text{if } j \neq i \end{cases} \qquad w^{(2)}_{i(j)} = \begin{cases} \dfrac{n_h}{m_h} & \text{if } x_{ih} = 1, x_{jh} \neq 1 \\ \dfrac{n_h - 1}{m_h} & \text{if } x_{ih} = 1, x_{jh} = 1, j \notin \mathcal{S}^{(2)} \\ \dfrac{n_h - 1}{m_h - 1} & \text{if } x_{ih} = 1, x_{jh} = 1, j \in \mathcal{S}^{(2)} \end{cases}$$

Then,

$$\hat{V}_{\text{JK}}(\hat{t}_{\text{str}}^{(2)}) = \sum_{j \in \mathcal{S}^{(1)}} \frac{n^{(1)} - 1}{n^{(1)}} \left[\hat{t}_{\text{str}(j)}^{(2)} - \hat{t}_{\text{str}}^{(2)}\right]^2.$$

As always, the jackknife estimates the with-replacement variance—in two-phase sampling, both phases are assumed to be sampled with replacement.

12.5 Designing a Two-Phase Sample

Two-phase sample designs require all the considerations of one-phase samples, plus the additional decision of how many resources to devote to each phase. A two-phase sample is more complicated than a one-phase sample; before you use one, study the relative costs and make sure a two-phase sample really will be more efficient. The two-phase design for the veterans survey in Examples 12.1 and 12.4 was actually less efficient than a one-phase design would have been. A two-phase sample uses resources to measure x on units that are not subsampled in phase II—resources that could alternatively be used to measure y on additional units. If x and y are strongly related, then the information in the phase I sample improves the efficiency of data collection. But if x and y are not related—for example, if x is the last digit of a student's telephone number and y is the student's grade point average—then the resources used to measure x are essentially wasted; you would be better off if you just sampled y directly.

Deming (1977) discussed issues to be considered when deciding whether to use a two-phase sample. A two-phase design adds complexity for both administering and analyzing the survey. It also can increase respondent burden, since respondents may need to be contacted twice. If a two-phase design is used to identify persons with a certain characteristic such as diabetes, persons ultimately selected for the phase II sample may first be asked to answer a questionnaire for phase I and then be asked to participate in a medical examination for phase II. However, the two-phase sample has the advantage that it can give useful information about the screening method used in phase I.

12.5.1 Two-Phase Sampling with Stratification

Consider the situation in which phase I is an SRS and phase II is a stratified random sample. Efficiency gains for two-phase sampling arise when more observations are subsampled in strata with large variance, large values of N_h, or low cost. Rao (1973) proposed letting $m_h = \nu_h n_h$ for stratum h, with ν_h, $h = 1, \ldots, H$, being constants to be determined before sampling. Let $c^{(1)}$ be the cost to sample a unit in the SRS taken for phase I and to determine its stratum membership. Let c_h be the cost of measuring y for a unit in stratum h in phase II. Assume the total cost will be a linear function, with

$$C = c^{(1)} n^{(1)} + \sum_{h=1}^{H} c_h m_h. \tag{12.15}$$

The total cost C varies from sample to sample, since the m_h's are only determined after the phase I sample is taken. The expected cost, however, is

$$E[C] = c^{(1)} n^{(1)} + n^{(1)} \sum_{h=1}^{H} c_h \nu_h W_h \tag{12.16}$$

where $W_h = N_h/N$. With ν_h fixed, we can write $V(\hat{\bar{y}}_{\text{str}}^{(2)})$ from (12.6) as:

$$V\left(\hat{\bar{y}}_{\text{str}}^{(2)}\right) = S_y^2\left(\frac{1}{n^{(1)}} - \frac{1}{N}\right) + \frac{1}{n^{(1)}}\sum_{h=1}^{H} W_h S_h^2\left(\frac{1}{\nu_h} - 1\right). \tag{12.17}$$

If $S^2 > \sum_{j=1}^H W_j S_j^2$, then $V(\hat{\bar{y}}_{\text{str}}^{(2)})$ is minimized, subject to the constraint in (12.16), when

$$\nu_{h,\text{opt}} = \sqrt{\frac{c^{(1)} S_h^2}{c_h\left(S^2 - \sum_{j=1}^H W_j S_j^2\right)}} \tag{12.18}$$

(see Exercise 18). If $\nu_{h,\text{opt}} > 1$ for a stratum h, then set $\nu_{h,\text{opt}} = 1$ and recalculate the other values. With a predetermined expected cost C^*, the phase I sample should have size

$$n_{\text{opt}}^{(1)} = \frac{C^*}{c^{(1)} + \sum_{h=1}^H c_h W_h \nu_{h,\text{opt}}}.$$

If $0 < \nu_{h,\text{opt}} < 1$ for $h = 1, \ldots, H$ and the optimal allocation is used, then, as shown in Exercise 19, the two-phase sample has variance

$$V_{\text{opt}}\left(\hat{\bar{y}}_{\text{str}}^{(2)}\right) = \frac{1}{C^*}\left[\sum_{h=1}^H W_h S_h\sqrt{c_h} + \sqrt{c^{(1)}}\sqrt{S_y^2 - \sum_{h=1}^H W_h S_h^2}\right]^2 - \frac{S_y^2}{N}. \tag{12.19}$$

Finding the optimal sample sizes for two-phase sampling with stratification requires estimates of the within-stratum variances S_h^2, similarly to the optimal allocation in stratified sampling discussed in Section 3.4.2. In addition, since the stratum sizes are unknown, the values of W_h and S^2 must also be estimated or guessed.

How much precision is gained by using two-phase sampling? We can compare the variance achieved by an optimally allocated two-phase stratified sample with the variance that we would get if we measured y on a one-phase sample with the same cost. For simplicity, assume $c_h = c^{(2)}$ for $h = 1, \ldots, H$. This is a reasonable cost structure for two-phase studies used to estimate disease prevalence, for example, in which all persons sampled in phase II are given the same medical examination. If, instead of taking a two-phase sample, we took an SRS in one phase with the same cost C^*, we could sample $n' = C^*/c^{(2)}$ units. Then, if S_y^2/N and $(1 - n'/N)$ are negligible, the ratio of the two-phase variance with optimal allocation to the one-phase variance with the same expected cost is approximately

$$\frac{V_{\text{opt}}(\hat{\bar{y}}_{\text{str}}^{(2)})}{V_{\text{SRS}}(\bar{y})} \approx \left[\sum_{h=1}^H W_h\frac{S_h}{S_y} + \sqrt{\frac{c^{(1)}}{c^{(2)}}}\sqrt{\frac{S_y^2 - \sum_{h=1}^H W_h S_h^2}{S_y^2}}\right]^2. \tag{12.20}$$

Thus, a two-phase sample with stratification is more efficient than an SRS for estimating $\bar{y}_{\mathcal{U}}$ if the within-strata variances S_h^2 are small relative to S_y^2 and if the cost to sample phase I units is smaller than the cost to sample phase II units.

For two-phase sampling with stratification, n_h, the number of units in the phase I sample that are in stratum h, is a random variable. If we select a different phase I sample from the population, it is likely that we will get a different value for n_h. It is possible that some phase I samples will have $n_h = 0$ for one or more strata. In that case, we cannot subsample that stratum in phase II, and y will not be measured on any unit in stratum h. The estimator in (12.4) is unbiased for the population total only if we assume $n_h > 0$ for all strata. Similarly, we need a subsample size of at least 2 in each stratum to estimate the variance within the stratum. We thus want to design a two-phase sample so that $P(n_h = 0)$ is extremely small.

12.5.2 Optimal Allocation for Ratio Estimation

How should the sample be allocated to phase I and phase II if ratio estimation is to be used? Suppose an SRS is taken in each of phase I and phase II and that the total cost of the two-phase sample is $C = c^{(1)}n^{(1)} + c^{(2)}n^{(2)}$. In Exercise 20, you will show that the variance of $\hat{t}_{yr}^{(2)}$, given in (12.10), is minimized subject to a fixed cost C when

$$\frac{n^{(2)}}{n^{(1)}} = \nu = \sqrt{\frac{c^{(1)}S_d^2}{c^{(2)}(S_y^2 - S_d^2)}}, \tag{12.21}$$

where S_d^2 is the population variance of the residuals $d_i = y_i - Bx_i$. Consequently, for a fixed cost C, the optimal phase I sample size is

$$n^{(1)} = \frac{C}{c^{(1)} + \nu c^{(2)}}$$

and the optimal phase II sample size is

$$n^{(2)} = \frac{C - n^{(1)}c^{(1)}}{c^{(2)}}.$$

The optimal sample sizes can often be estimated using results from a preliminary survey or prior work. If x and y are highly correlated, we expect S_d^2 to be small relative to S_y^2. In that situation, it makes sense to measure x on a large phase I sample, particularly if the cost of measuring x is small, and use a relatively small phase II sample to estimate B.

12.6 Chapter Summary

Two-phase sampling can increase precision of estimators of t_y for a fixed budget if there exist auxiliary variables $\mathbf{x}$ such that (1) the cost of measuring $\mathbf{x}$ is low compared with the cost of measuring y and (2) the auxiliary variables are correlated with y. The auxiliary information collected in the phase I sample can be used to improve the efficiency of the phase II sampling design, as when the auxiliary information $\mathbf{x}$ collected at phase I is used to stratify the phase II sample. Alternatively, or additionally, the information in the phase I sample can be used through ratio or regression estimation.

Key Terms

Phase I sample: A sample selected from a population on which auxiliary variables $\mathbf{x}$ are measured.

Phase II sample: A subsample selected from the phase I sample on which the variable of interest y is measured.

Two-phase sampling: A sampling design in which a preliminary (phase I) sample is selected from the population, and then a subsample (phase II sample) is selected from the phase I sample.

For Further Reading

Watson (1937) presented an early example of two-phase sample for regression, and Neyman (1938) developed theory for two-phase sampling with stratification. See Cochran (1977) for more discussion on two-phase sampling with simple random samples; Särndal et al. (1992, Chapter 9) and Legg and Fuller (2009) give a theoretical development for general probability sampling designs.

Hidiroglou et al. (2009) develop Sen–Yates–Grundy-type variance estimators for two-phase samples; Hidiroglou (2001) discusses non-nested two-phase sampling designs. Rao and Shao (1992), Rao and Sitter (1995, 1997), Kott and Stukel (1997), Sitter (1997), Kim et al. (2006), and Kim and Yu (2011) present jackknife variance estimators for two-phase sampling; also see the discussion in Beaumont et al. (2015).

12.7 Exercises

A. Introductory Exercises

1. A health official takes a two-phase sample to estimate the prevalence of diabetes in a population. In phase I, an SRS of size 1000 is taken from the population of size 100,000, and each individual is asked demographic information and whether he or she has diabetes. It is known that some demographic groups are more at risk for diabetes than others; in addition, the self report of diabetes may be inaccurate. Therefore, each individual in the phase II sample is given a medical exam to determine diabetes status. The phase I sample is divided into 4 strata. Stratum h has n_h observations in the phase I sample and m_h observations in the phase II sample. After the medical exam, r_h persons in stratum h of the phase II sample were determined to have diabetes.

Stratum	n_h	m_h	r_h
High-risk group and reports diabetes	241	96	86
High-risk group and does not report diabetes	113	45	17
Low-risk group and reports diabetes	174	35	29
Low-risk group and does not report diabetes	472	47	8

   Estimate the total number of persons with diabetes in the population, along with its standard error.

2. Data mining methods in statistics are used to discover relationships among variables in very large data sets (Hastie et al., 2009; James et al., 2013). A company, for example, has databases of all its financial transactions. A very small fraction of these transactions involve fraud, but fraudulent transactions are expensive for the company. Discovering whether a transaction is fraudulent requires an investigation, so the company can only determine whether transactions are fraudulent for a small sample. Discuss how two-phase sampling might be used to improve prediction of fraudulent transactions.

B. Working with Survey Data

3. Bart and Earnst (2002) described the use of two-phase sampling with ratio estimation to estimate the density of nesting birds. The phase I sample is conducted using a rapid search method involving bird sightings to obtain an approximate count of birds in each phase I plot. Then, a subsample of the phase I sample plots are surveyed using an intensive method to obtain a more accurate count of the number of nests in each plot. In the intensive method, a surveyor visits a plot for several hours over a period of days and searches for nests and other indications of territorial males in the plot. In this setting, x_i = number of birds counted in plot i using the rapid method, and y_i = number of territorial males counted in plot i using the intensive method. Using the data in file `shorebirds.csv`, which were generated using summary statistics from Bart and Earnst (2002), estimate the total number of nests using the two-phase ratio estimator.

4. Dunn et al. (1999) discussed issues in analyzing two-phase data to estimate prevalence of psychiatric disorders. Participants in a phase I sample were given the General Health Questionnaire (GHQ) and classified into three strata based on their GHQ score. The stratification was used to take a stratified random sample of 250 persons for the phase II sample; the Composite International Diagnostic Interview (CIDI), considered to be a more accurate diagnostic tool, was administered to each person in the phase II sample. The CIDI score was used to classify the phase II sample members as having at least one psychiatric disorder (case) or having no psychiatric disorder (non-case). The results are given in the following table. The counts of cases and non-cases are from the phase II sample.

Stratum	n_h	m_h	**Non-case**	**Case**
GHQ ≤ 3 (low)	1049	60	33	27
GHQ $= 4, 5$ (medium)	237	48	14	34
GHQ ≥ 6 (high)	272	142	23	119
Total	1558	250		

   (a) Calculate the phase II sampling weight $w_i^{(2)}$ for each stratum.

   (b) Use the two-phase sample to estimate the percentage of persons with at least one psychiatric disorder, along with its standard error. Since we do not know the population size N, use a relative phase I weight of $w_i^{(1)} = 1$.

5. Dunn et al. (1999) also classified the phase II sample by gender. In the following table, the entries in columns 3–7 are the counts from the phase II sample in the categories Male Non-Case (MNC), Male Case (MC), Female Non-Case (FNC), and Female Case (FC).

Stratum	n_h	m_h	**MNC**	**MC**	**FNC**	**FC**
GHQ ≤ 3 (low)	1049	60	16	8	17	19
GHQ $= 4, 5$ (medium)	237	48	9	8	5	26
GHQ ≥ 6 (high)	272	142	15	28	8	91

   (a) Estimate the percentages of persons in each cell of a 2×2 contingency table classified by gender and case/non-case. Find the standard error of each entry in the table.

(b) Find the design effect for each cell proportion $\hat{p}_{ij}$ and marginal proportion ($\hat{p}_{i+}$ and $\hat{p}_{+j}$) in the table.

(c) Use the Rao–Scott method (Section 10.3.2) to test $H_0 : p_{ij} = p_{i+}p_{+j}$.

6. Ismail et al. (2002) reported results of a two-phase survey to estimate prevalence of psychiatric disorders in Gulf War veterans. A random sample of the N = 53,462 Gulf War veterans was administered the SF-36 questionnaire, a 36-question survey on health. Respondents who scored below 72.2 on physical functioning subscale were defined as disabled (stratum 1, with n_1 = 406); respondents who scored above 72.2 were defined as not disabled (stratum 2, with n_2 = 3047). A random subsample of 111 veterans was taken from stratum 1, and a random subsample of 98 veterans was taken from stratum 2. The 209 veterans in the phase II sample were evaluated by psychiatrists; the counts in the table below give the number of veterans who are determined to have any alcohol-related disorder (Alcohol), any sleep disorder (Sleep), or any psychiatric-related disorder (Psych). A veteran can be in more than one of these categories.

Stratum	n_h	m_h	**Alcohol**	**Sleep**	**Psych**
Disabled	406	111	8	20	27
Not disabled	3047	98	10	17	12

Estimate the total number of Gulf War veterans with an alcohol-related disorder, and give a 95% CI.

7. Repeat Exercise 6 for sleep disorders.

8. Repeat Exercise 6 for any psychiatric disorder.

9. Some of the books in the sample of mystery novels in Exercise 13 of Chapter 3 were not available at the library—these were, in a sense, nonrespondents in the sample. The data collector gathered information on genre, authors, and detectives for all the nonrespondents from online descriptions of the books, so that those variables had complete information. She then selected a random subsample of about 30% of the nonrespondents within each stratum and bought those books. The phase II sample of books that were read thus consisted of the readily available books from the full sample plus the subsampled nonrespondents. The variable *p2weight* in data set `mysteries.csv` contains the final weight.

(a) The two-phase sample removes bias, but increases the variability of the weights. Find the coefficient of variation of the phase I sampling weights (variable *p1weight*) and the final sampling weights (variable *p2weight*).

(b) Use weight *p2weight* to estimate the total number of murder victims in the population of books, along with its standard error. In some of the strata, the random subsample of nonrespondents consisted of one book, so you may want to modify the variance estimator in (12.7).

(c) Estimate the difference between the mean number of victims in books written by women and the mean number of victims in books written by men, and give a 95% CI. Are they significantly different? Note: do not use an fpc for this comparison.

(d) Define *bfirearm* to equal 1 if at least one murder was committed using a firearm, and 0 otherwise. Estimate the population contingency table cross-classifying the variable *authorgender* by *bfirearm*, and carry out a chi-square test of independence.

10. Use the results of Section 12.5 to determine an optimal allocation for a follow-up survey similar to that in Example 12.1. Assume that the relative costs are $c^{(1)} = 1$ and $c_h = 20$ for $h = 1, 2, 3$. Use the data in Example 12.1 to estimate quantities such as W_h and S_h^2. How does your allocation differ from the one used? From proportional allocation?

C. Working with Theory

11. (Requires probability.) Suppose the phase I sample is an SRS of size $n^{(1)}$, and the phase II subsample is an SRS of size $n^{(2)}$, with $n^{(2)} < n^{(1)}$. Show that

$$V(\hat{t}_y^{(2)}) = N^2 \left(1 - \frac{n^{(2)}}{N}\right) \frac{S_y^2}{n^{(2)}}$$

the same variance that would result if an SRS of size $n^{(2)}$ were taken directly.

12. *Estimating the variance in two-phase sampling for stratification.* Show that (12.7) is an approximately unbiased estimator of $V\left(\hat{t}_{ystr}^{(2)}\right)$ in large samples. HINT: Use the result derived from Table 3.6 in Chapter 3 that

$$S^2 = \frac{1}{N-1}\left[\sum_{h=1}^{H}(N_h - 1)S_h^2 + \sum_{h=1}^{H} N_h(\bar{y}_{h\mathcal{U}} - \bar{y}_{\mathcal{U}})^2\right].$$

13. (Requires probability.) For two-phase sampling with ratio estimation (Section 12.3.1), suppose the phase I sample is an SRS of size $n^{(1)}$, and the phase II sample is an SRS of fixed size $n^{(2)}$.

    (a) Show that $P(Z_i = 1) = n^{(1)}/N$, and $P(D_i = 1 \mid \mathbf{Z}) = Z_i n^{(2)}/n^{(1)}$.

    (b) Show that (12.10) gives the approximate variance of $\hat{t}_{yr}^{(2)}$.

    (c) Let $e_i = y_i - \hat{B}^{(2)} x_i$ and let s_y^2 and s_e^2 be the sample variances of the y_is and the e_is from the phase II sample,

$$s_y^2 = \frac{1}{n^{(2)} - 1} \sum_{i \in \mathcal{S}^{(2)}} (y_i - \bar{y}^{(2)})^2 \quad \text{and} \quad s_e^2 = \frac{1}{n^{(2)} - 1} \sum_{i \in \mathcal{S}^{(2)}} e_i^2.$$

    Show that (12.11) is an approximately unbiased estimator of $V(\hat{t}_{yr}^{(2)})$.

14. Rao and Sitter (1995) proposed an alternative linearization variance estimator for the situation in Exercise 13,

    (a) Using part (b) of Exercise 13, show that

$$V(\hat{t}_{yr}^{(2)}) \approx N^2\left(1 - \frac{n^{(1)}}{N}\right)\frac{2BS_{xd} + B^2 S_x^2}{n^{(1)}} + N^2\left(1 - \frac{n^{(2)}}{N}\right)\frac{S_d^2}{n^{(2)}}$$

    where $S_{xd} = \sum_{i=1}^{N}(x_i - \bar{x}_{\mathcal{U}})d_i/(N-1)$. HINT: Write

$$S_y^2 = \frac{1}{N-1}\sum_{i=1}^{N}(y_i - \bar{y}_{\mathcal{U}})^2 = \frac{1}{N-1}\sum_{i=1}^{N}(y_i - Bx_i + Bx_i - B\bar{x}_{\mathcal{U}})^2.$$

(b) Show that

$$\hat{V}_2(\hat{t}_{yr}^{(2)}) = N^2\left(1-\frac{n^{(1)}}{N}\right)\frac{2\hat{B}^{(2)}s_{xe} + \hat{B}^{(2)}s_x^{2(1)}}{n^{(1)}} + N^2\left(1-\frac{n^{(2)}}{N}\right)\frac{s_e^2}{n^{(2)}},$$

is an approximately unbiased estimator of $V(\hat{t}_{yr}^{(2)})$, where

$$s_{xe} = \frac{1}{n^{(2)}-1}\sum_{i\in\mathcal{S}^{(2)}}(x_i - \bar{x}^{(2)})e_i, \quad s_x^{2(1)} = \frac{1}{n^{(1)}-1}\sum_{i\in\mathcal{S}^{(1)}}(x_i - \bar{x}^{(1)})^2.$$

15. *Demnati–Rao (2004) linearization variance estimator in two-phase sampling.* The linearization variance estimator presented in Exercise 29 of Chapter 9 can be extended to two-phase sampling. Let θ be the population quantity of interest, and define the estimator $\hat{\theta}$ to be a function of the vectors of sampling weights for the phase I and phase II samples and the population values:

$$\hat{\theta} = g(\mathbf{w}^{(1)}, \mathbf{w}, \mathbf{x}_1, \mathbf{x}_2, \ldots, \mathbf{x}_m, \mathbf{y}_1, \mathbf{y}_2, \ldots, \mathbf{y}_k),$$

where $\mathbf{w}^{(1)} = (w_1^{(1)}, \ldots, w_N^{(1)})^T$ with $w_i^{(1)}$ the phase I sampling weight of unit i ($w_i^{(1)} = 0$ if i is not in the phase I sample), $\mathbf{w} = (w_1, \ldots, w_N)^T$ with w_i the final sampling weight of unit i in the phase II sample ($w_i = w_i^{(1)}w_i^{(2)}$ if $i \in \mathcal{S}^{(2)}$ and $w_i = 0$ if $i \notin \mathcal{S}^{(2)}$), $\mathbf{x}_j$ is the vector of population values for the jth auxiliary variable (measured in phase I), and $\mathbf{y}_j$ is the vector of population values for the jth response variable (measured in phase II). Now let

$$z_i^{(1)} = \frac{\partial g(\mathbf{w}^{(1)}, \mathbf{w}, \mathbf{x}_1, \mathbf{x}_2, \ldots, \mathbf{x}_m, \mathbf{y}_1, \mathbf{y}_2, \ldots, \mathbf{y}_k)}{\partial w_i^{(1)}}$$

and

$$z_i^{(2)} = \frac{\partial g(\mathbf{w}^{(1)}, \mathbf{w}, \mathbf{x}_1, \mathbf{x}_2, \ldots, \mathbf{x}_m, \mathbf{y}_1, \mathbf{y}_2, \ldots, \mathbf{y}_k)}{\partial w_i}.$$

Then,

$$\hat{V}_{\text{DR}}(\hat{\theta}) = \hat{V}\left(\sum_{i\in\mathcal{S}^{(1)}} w_i^{(1)} z_i^{(1)} + \sum_{i\in\mathcal{S}^{(2)}} w_i z_i^{(2)}\right).$$

(a) Consider the two-phase ratio estimator in (12.9). We can write

$$\hat{t}_{yr}^{(2)} = \sum_{i\in\mathcal{S}^{(1)}} w_i^{(1)} x_i \frac{\sum_{i\in\mathcal{S}^{(2)}} w_i y_i}{\sum_{i\in\mathcal{S}^{(2)}} w_i x_i},$$

where $w_i = w_i^{(1)}w_i^{(2)}$. Show that the Demnati-Rao linearization variance estimator is

$$\hat{V}_{\text{DR}}\left(\hat{t}_{yr}^{(2)}\right) = \hat{V}\left[\sum_{i\in\mathcal{S}^{(1)}} w_i^{(1)} x_i \hat{B}^{(2)} + \frac{\hat{t}_x^{(1)}}{\hat{t}_x^{(2)}}\sum_{i\in\mathcal{S}^{(2)}} w_i(y_i - \hat{B}^{(2)}x_i)\right].$$

(b) Suppose that the phase I sample is an SRS of size $n^{(1)}$ and the phase II sample is an SRS of size $n^{(2)}$. What is $\hat{V}_{\text{DR}}\left(\hat{t}_{yr}^{(2)}\right)$ for this case?

16. Show that if the estimator in (12.12) is applied to any of the auxiliary variables in $\mathbf{x}$, then $\hat{t}^{(2)}_{\mathbf{x}\text{GREG}} = \hat{\mathbf{t}}^{(1)}_{\mathbf{x}}$.

17. (Requires probability.) Suppose the phase I sample is an unequal-probability sample of observations. If we observed y_i for every unit in $\mathcal{S}^{(1)}$, we could use the Horvitz–Thompson estimator of the variance of the Horvitz–Thompson estimator in (6.22) to estimate the first term in (12.14):

$$\hat{V}^{(1)}_{\text{HT}}(\hat{t}^{(1)}_y) = \sum_{i \in \mathcal{S}^{(1)}} \sum_{k \in \mathcal{S}^{(1)}} \frac{\pi^{(1)}_{ik} - \pi^{(1)}_i \pi^{(1)}_k}{\pi^{(1)}_{ik}} \frac{y_i}{\pi^{(1)}_i} \frac{y_k}{\pi^{(1)}_k},$$

where $\pi^{(1)}_i = P(Z_i = 1)$ and $\pi^{(1)}_{ik} = P(Z_i Z_k = 1)$ for $i \neq k$ and $\pi^{(1)}_{ii} = P(Z_i = 1)$. We need an estimator of $V(\hat{t}^{(1)}_y)$, however, that depends only on the y values in the phase II sample. Let $\pi^{(2)}_{ik} = P(D_i D_k = 1 \mid \mathbf{Z}) > 0$. Show that

$$\hat{V}_{\text{HT}}(\hat{t}^{(1)}_y) = \sum_{i \in \mathcal{S}^{(2)}} \sum_{k \in \mathcal{S}^{(2)}} \frac{\pi^{(1)}_{ik} - \pi^{(1)}_i \pi^{(1)}_k}{\pi^{(1)}_{ik} \pi^{(2)}_{ik}} \frac{y_i}{\pi^{(1)}_i} \frac{y_k}{\pi^{(1)}_k}$$

is an unbiased estimator of $V(\hat{t}^{(1)}_y)$.

18. *Optimal allocation for two-phase sampling with stratification* (requires calculus). Suppose phase I is an SRS and phase II is a stratified random sample, and that the total cost for the sample is given in (12.15), where $c^{(1)}$ is the cost to sample a unit in phase I and c_h is the cost to sample a unit in stratum h in phase II. Let $\nu_h = m_h/n_h$, $h = 1, \ldots, H$ be the proportion of phase I units in stratum h to be sampled in phase II.

    (a) Show that the expected cost is (12.16).

    (b) Prove (12.17).

    (c) Show that $V(\hat{\bar{y}}^{(2)}_{\text{str}})$ is minimized, subject to the constraint in (12.16), when ν_h is given in (12.18). HINT: Use Lagrange multipliers.

19. Show that when the optimal allocation is used, the variance of $\hat{\bar{y}}^{(2)}_{\text{str}}$ for two-phase sampling with stratification is given by (12.19).

20. (Requires calculus.) Show that if an SRS of size $n^{(1)}$ is taken in phase I, and an SRS of size $n^{(2)}$ is taken in phase II, then taking the ratio $n^{(2)}/n^{(1)}$ in (12.21) minimizes the variance in (12.10) for a fixed cost C.

21. This exercise is based on results in McNamee (2003) on the use of two-phase sampling to estimate disease prevalence. An inexpensive, but possibly inaccurate, screening test for the disease is given in the phase I sample, an SRS of size $n^{(1)}$. Let $x_i = 1$ if person i tests positive on the screening test and $x_i = 0$ if person i tests negative on the screening test. Persons are then classified into stratum 1 ($x_i = 0$) and stratum 2 ($x_i = 1$). The persons sampled in phase II are given a test for the presence of the disease that, for purposes of this exercise, is assumed to be 100% accurate: The phase II response is $y_i = 1$ if person i has the disease and 0 otherwise. We can display the population counts in a contingency table, shown in Table 12.3.

    We wish to estimate $p = \bar{y}_{\mathcal{U}} = C_{2+}/N$ from the two-phase sample; $p_1 = C_{21}/N_1$ and $p_2 = C_{22}/N_2$ are the proportions with the disease in strata 1 and 2, respectively.

TABLE 12.3
Contingency table for Exercise 21.

		Screening Test		
		Negative	Positive	
Disease	No	C_{11}	C_{12}	C_{1+}
Present?	Yes	C_{21}	C_{22}	C_{2+}
		$C_{+1} = N_1$	$C_{+2} = N_2$	N

(a) Epidemiologists often use the concepts of specificity and sensitivity to assess a test for a disease, with

$$S_1 = \text{ Specificity} = P(\text{test is negative} \mid \text{disease absent}) = \frac{C_{11}}{C_{1+}}$$

and

$$S_2 = \text{ Sensitivity} = P(\text{test is positive} \mid \text{disease present}) = \frac{C_{22}}{C_{2+}}.$$

Show that

$$\frac{N_1}{N}p_1 = (1-S_2)p, \; \frac{N_1}{N}(1-p_1) = (1-p)S_1, \; \frac{N_2}{N}p_2 = pS_2, \; \frac{N_2}{N}(1-p_2) = (1-p)(1-S_1).$$

(b) Suppose that the optimal allocation is used (see Section 12.5.1) and that $0 < \nu_{h,\text{opt}} < 1$ for $h = 1, 2$. Using (12.20) and part (a), show that

$$\frac{V_{\text{opt}}(\hat{p}^{(2)}_{\text{str}})}{V_{\text{SRS}}(\hat{p})} \approx \left[\sqrt{(1-S_2)S_1} + \sqrt{S_2(1-S_1)} + R\sqrt{\frac{c^{(1)}}{c^{(2)}}}\right]^2,$$

where R is the population Pearson correlation coefficient between x and y, given in (4.1). (HINT: For the second term, first show that $RS_y = p(S_2 - W_2)/\sqrt{W_1W_2}$.)

(c) Calculate the ratio of variances in (b) when $S_1 = S_2$ and $R = \min(S_1 + S_2 - 0.9, 0.95)$, for each combination of $S_1 \in \{0.5, 0.6, 0.7, 0.8, 0.9, 0.95\}$ and $c^{(1)}/c^{(2)} \in \{0.0001, 0.01, 0.1, 0.5, 1\}$. Display your results in a table. For which settings would you recommend two-phase sampling to estimate disease prevalence?

22. *Inverse Sampling.* Hinkins et al. (1997) (also see Rao et al., 2003) noted that in some situations one might want to apply a statistical procedure developed for an SRS to data from a complex survey, but the stratification and clustering in the survey make direct application inappropriate. They propose an inverse sampling algorithm to create a subsample from the complex survey that is an SRS from the population, essentially by inverting the procedure used to draw the complex sample. The procedure can be repeated multiple times.

    Suppose that the complex survey is a stratified random sample. The population stratum sizes are $N_1, \ldots, N_H$ and the sample sizes are $n_1, \ldots, n_H$. It would be possible for all the observations in an SRS from the population to be in one stratum, so the maximum possible size of the subsample is $m = \min\{n_1, \ldots, n_H\}$. Use the hypergeometric distribution to generate subsampling sizes $m_1, \ldots, m_H$ from the strata, with $\sum_{h=1}^{H} m_h = m$, where

$$P(M_1 = m_1, M_2 = m_2, \ldots, M_H = m_H) = \frac{1}{\binom{N}{m}} \binom{N_1}{m_1}\binom{N_2}{m_2}\cdots\binom{N_H}{m_H}.$$

In stratum h, select an SRS of m_h of the n_h sampled units to be in $\mathcal{S}^{(2)}$.

(a) Show that the probability that any subset of m units in the population is selected as the sample through this procedure (first taking a stratified random sample of size n, then using inverse sampling to select a subsample of size m) is $\binom{N}{m}^{-1}$.

(b) Use inverse sampling to select a subsample of size 21 from the stratified random sample in file `agstrat.csv`, first discussed in Example 3.2.

23. *Ranked set sampling.* McIntyre (1952) proposed ranked set sampling as a method of improving precision of estimates by a method related to two-phase sampling. In McIntyre's application y_i = pasture yield of field i. Measuring y_i requires mowing and weighing—a time-consuming process. An expert, however, can assess and rank a small number of fields from lowest to highest yield by visual inspection, which is much less effort. Wolfe (2012) and Patil et al. (2014) reviewed research on ranked set sampling and gave bibliographies.

To implement ranked set sampling, select k independent SRSs, each of size k. Rank each of the k samples from low to high, using either judgment or a correlated easy-to-measure variable. (This ranking must be done without knowing any values of y_i.) Then select the smallest unit of the first sample for measurement of y, the second smallest unit of the second sample, and so on until the largest unit of sample k is selected. Repeat this procedure until m replicates are obtained. At the end of the process mk^2 units have been ranked and y has been measured on a sample of $n = mk$ units. Let $\bar{y}_j$ be the mean of the y-values measured in replicate j, $j = 1, \ldots, m$ and let $\hat{\bar{y}}_{\text{RSS}} = \sum_{j=1}^{m} \bar{y}_j / m$.

We illustrate the method with a small example using the data in `agpop.csv` (Husby et al., 2005, have a similar example using NHANES data). We want to estimate the population total for y = *acres92*. We take mk SRSs, each of size k, using $m = 10$ and $k = 4$ (typically, k is relatively small to allow an expert to rank the elements). We rank each of the mk SRSs using the correlated variable *acres87*. Table 12.4 shows the values of *acres87* for the 4 samples in the first replicate.

TABLE 12.4
Values of *acres87* for first replicate in Exercise 23.

		Observation			
Sample 1:	x	119,956	144,986	108,861	302,659
	Rank	2	3	1	4
Sample 2:	x	351,106	294,551	80,104	226,954
	Rank	4	3	1	2
Sample 3:	x	241,276	253,421	702,173	412,225
	Rank	1	2	4	3
Sample 4:	x	529,964	823,729	355,973	121,119
	Rank	3	4	2	1

We then measure y on the third unit in Sample 1 (which has rank 1), the fourth unit in Sample 2 (rank 2), the fourth unit in Sample 3 (rank 3), and the second unit in Sample 4 (rank 4), obtaining the y values 106,206, 246,038, 379,044, and 783,715. Note that since y is highly correlated with x, these four y values are forced to be spread out.

The same procedure is repeated for the remaining nine replicates. We obtain

$\bar{y}_1$	$\bar{y}_2$	$\bar{y}_3$	$\bar{y}_4$	$\bar{y}_5$
378,750.75	51,841	280,658.5	187,791.5	175,436.75

$\bar{y}_6$	$\bar{y}_7$	$\bar{y}_8$	$\bar{y}_9$	$\bar{y}_{10}$
446,398.5	1,092,582.5	499,146	350,570	665,457.75

Consequently, $\hat{\bar{y}}_{\text{RSS}} = \sum_{j=1}^{m} \bar{y}_j / m = 412{,}863.33$.

(a) How is ranked set sampling similar to two-phase sampling? How does it differ?

(b) Argue that if the ranking is based on an auxiliary variable x, and the value of y itself is not used in the ranking, then $\hat{\bar{y}}_{\text{RSS}}$ is an unbiased estimator of the population mean.

(c) Show that if the m replicates are selected independently, then

$$\hat{V}(\hat{\bar{y}}_{\text{RSS}}) = \frac{1}{m(m-1)} \sum_{j=1}^{m} (\bar{y}_j - \hat{\bar{y}}_{\text{RSS}})^2$$

is an unbiased estimator of $V(\hat{\bar{y}}_{\text{RSS}})$. (HINT: See Section 9.2.) Using this method, we obtain $\text{SE}(\hat{\bar{y}}_{\text{RSS}}) = 93{,}943.21$ for our sample.

(d) The properties of ranked set sampling require that the ranking in each of the mk samples be done using the same method. What might go wrong in practice? How might a ranker who knows the sampling procedure produce bias in the estimates?

D. Projects and Activities

24. *Responsive design.* Section 8.3 described nonresponse follow-up as a two-phase sample, where the phase II sample contains the initial respondents plus the initial nonrespondents who responded after being subsampled. Groves and Heeringa (2006) proposed varying the follow-up modes and efforts based on information gathered in the first phase of sampling (see also Tourangeau et al., 2017; Schouten et al., 2017; Chun et al., 2018). For example, suppose the phase I sample is divided among H demographic strata, and the weighted response rate is calculated for each stratum as the sum of the weights for respondents in the stratum divided by the population count for the stratum from a census or population register. Then the phase II selection probabilities might be designed with one of several goals: (a) to equalize the response rates in the strata (assign higher probabilities to strata with low response rates), (b) to maximize the overall response rate (assign higher probabilities to strata with low response rates), (c) to obtain a desired minimum sample size in each stratum, or (d) to obtain minimum variance for a variable of interest y (use the phase I data to estimate the anticipated variance under different phase II allocations).

    Discuss the potential effects of options (a)–(d) on variance and bias.

25. Exercise 11 of Chapter 3 described the use of random selection to choose the city council in ancient Athens. Random selection for modern Citizens' Councils is typically done in two phases. First, a mailing is sent to a large randomly selected sample of citizens; for the United Kingdom Climate Assembly chosen in November 2019 (Harvey, 2019), 30,000 invitation letters were sent out. Then, from those who respond, a representative

subsample (usually between 20 and 150 persons) is selected as the Council. The subsample is selected in such a way that the Council has the same percentage of male/female members, members from different race/ethnicity groups, members from different subregions, members from different socioeconomic classes or education levels, and so on, as the region being represented.

(a) Discuss the advantages and disadvantages of using two-phase sampling to select a Citizens' Council.

(b) Describe how balanced sampling (see Exercise 37 of Chapter 11) might be used to obtain the desired representation at phase II.

(c) What are potential sources of selection bias in this procedure?

13

Estimating the Size of a Population

> I caught a large number of fishes in the neighbourhood of Suez. I passed a copper ring through their tails, and threw them back into the sea. Some months later, on the coast of Syria, I caught some of my fish ornamented with the ring.
>
> —Jules Verne, *Twenty Thousand Leagues Under the Sea*

13.1 Capture–Recapture Estimation

Example 13.1. Suppose we want to estimate N, the number of fish in a lake. One method is as follows: Catch and mark 200 fish in the lake, then release them. Allow the marked and released fish to mix with the other fish in the lake. Then, take a second, independent sample of 100 fish. Suppose that 20 of the fish in the second sample are marked. Then, assuming that the population of fish has not changed between the two samples and that each catch gives a simple random sample (SRS) of fish in the lake, estimate that 20% of the fish in the lake are marked, and, therefore, that the 200 fish tagged in the original sample represent approximately 20% of the population of fish. The population size N is then estimated to be approximately 1,000. ■

This method for estimating the size of a population is called two-sample **capture-recapture estimation**. Other names sometimes used are tag- or mark-recapture, multiple record system, the Petersen (1896) method or the Lincoln (1930) index. The method relies on the following assumptions:

1. The population is *closed*—no fish enter or leave the lake between the samples. This means that N is the same for each sample.

2. Each sample of fish is an SRS from the population. This means that each fish is equally likely to be chosen in a sample—it is not the case, for example, that smaller or less healthy fish are more likely to be caught. Also, there are no "hidden fish" in the population that are impossible to catch.

3. The two samples are independent. The marked fish from the first sample become remixed in the population, so that the marking status of a fish is unrelated to the probability that the fish is selected in the second sample. Also, fish included in the first sample do not become "trap-shy" or "trap-happy"—the probability that a fish will be caught in the second sample does not depend on its capture history.

4. Fish do not lose their markings, and marked fish can be identified as such. Water-soluble paint, for example, would not be a good choice for marking material.

In this simple form, capture-recapture is a special case of ratio estimation of a population total, and results from Chapter 4 may be used when the samples and population are large.

DOI: 10.1201/9780429298899-13

Let n_1 be the size of the first sample, n_2 the size of the second sample, and m the number of marked fish caught in the second sample. In Example 13.1, $n_1 = 200$, $n_2 = 100$, $m = 20$, and we used the estimator $\hat{N} = n_1 n_2/m$. To see how this estimator fits into the framework of Chapter 4, let

$$y_i = 1 \text{ for every fish in the lake}$$

and

$$x_i = \begin{cases} 1 & \text{if fish } i \text{ is marked} \\ 0 & \text{if fish } i \text{ is not marked} \end{cases}.$$

Then estimate $N = t_y = \sum_{i=1}^{N} y_i$ by $\hat{t}_{yr} = t_x \hat{B}$, where $t_x = \sum_{i=1}^{N} x_i = n_1$ and $\hat{B} = \bar{y}/\bar{x} = n_2/m$. This ratio estimator,

$$\hat{N} = \hat{t}_{yr} = \frac{n_1 n_2}{m}, \tag{13.1}$$

is also the maximum likelihood estimator (see Exercises 17 and 18). Applying (4.12) to the second SRS and ignoring the finite population correction (fpc), $s_e^2 = n_2(n_2 - m)/[m(n_2 - 1)]$ and

$$\hat{V}(\hat{N}) = t_x^2 \hat{V}(\hat{B}) = \left(\frac{n_1 n_2}{m}\right)^2 \frac{n_2 - m}{m(n_2 - 1)} \approx \frac{n_1^2 n_2 (n_2 - m)}{m^3}.$$

For the data in Example 13.1, $\hat{V}(\hat{N}) = 40{,}000$.

Being a ratio estimator, though, $\hat{N}$ is biased, and the bias can be large in wildlife applications with small sample sizes. Indeed, it is possible for the second sample to consist entirely of unmarked animals, making the estimate in (13.1) infinite. Chapman (1951) proposed the less biased estimator

$$\tilde{N} = \frac{(n_1 + 1)(n_2 + 1)}{m + 1} - 1. \tag{13.2}$$

A variance estimator for $\tilde{N}$ (Seber, 1970) is

$$\hat{V}(\tilde{N}) = \frac{(n_1 + 1)(n_2 + 1)(n_1 - m)(n_2 - m)}{(m + 1)^2 (m + 2)}. \tag{13.3}$$

The estimators in (13.2) and (13.3) are often used in wildlife applications. For the fish data, $\tilde{N} = (201)(101)/21 - 1 = 966$, and $\hat{V}(\tilde{N}) = 30{,}131$.

Many researchers have constructed confidence intervals (CIs) for the population size using either $\hat{N} \pm 1.96\sqrt{\hat{V}(\hat{N})}$ or $\tilde{N} \pm 1.96\sqrt{\hat{V}(\tilde{N})}$. These are not entirely satisfactory, however, because both require that $\hat{N}$ or $\tilde{N}$ be approximately normally distributed, and the normal distribution may not be a good approximation to the distribution of $\hat{N}$ or $\tilde{N}$ for small populations and samples. We'll discuss CIs in Section 13.1.2; first, however, let's look at another approach for these data that will be useful in developing CIs.

13.1.1 Contingency Tables for Capture–Recapture Experiments

Fienberg (1972) suggested viewing capture–recapture data in an incomplete contingency table. For the data in Example 13.1, the table is as follows:

		In Sample 2?		
		Yes	No	
In Sample 1?	Yes	20	180	200
	No	80	?	?
		100	?	N

In general, if x_{ij} is the observed count in cell (i, j), the contingency table looks as follows. An asterisk indicates that we do not observe that cell.

		In Sample 2?		
		Yes	No	
In Sample 1?	Yes	$x_{11}(= m)$	x_{12}	$x_{1+}(= n_1)$
	No	x_{21}	x^*_{22}	x^*_{2+}
		$x_{+1}(= n_2)$	x^*_{+2}	x^*_{++}

The expected counts are:

		In Sample 2?		
		Yes	No	
In Sample 1?	Yes	m_{11}	m_{12}	m_{1+}
	No	m_{21}	m^*_{22}	m^*_{2+}
		m_{+1}	m^*_{+2}	$m^*_{++} = N$

To estimate the expected counts, we use $\hat{m}_{11} = x_{11}$, $\hat{m}_{12} = x_{12}$, and $\hat{m}_{21} = x_{21}$. If presence in sample 1 is independent of presence in sample 2, then the odds of being in sample 2 are the same for marked fish as for unmarked fish: $m_{11}/m_{12} = m_{21}/m_{22}$. Consequently, under independence, the estimated count in the cell of fish not included in either sample is

$$\hat{m}_{22} = \frac{\hat{m}_{12}\hat{m}_{21}}{\hat{m}_{11}} = \frac{x_{12}x_{21}}{x_{11}},$$

and

$$\hat{N} = \hat{m}_{11} + \hat{m}_{12} + \hat{m}_{21} + \hat{m}_{22} = \frac{x_{+1}x_{1+}}{x_{11}} = \frac{n_1 n_2}{m}.$$

The estimator $\hat{N}$ is calculated based on the assumption that the two samples are independent; unfortunately, this assumption cannot be tested because only three of the four cells of the contingency table are observed.

13.1.2 Confidence Intervals for N

In many applications of capture-recapture estimation, CIs have been constructed using $\hat{N} \pm 1.96\sqrt{\hat{V}(\hat{N})}$ or $\tilde{N} \pm 1.96\sqrt{\hat{V}(\tilde{N})}$. If we use the first interval for the data in Example 13.1, $\hat{V}(\hat{N}) = 40{,}000$, and an asymptotic 95% CI would be $1000 \pm 1.96(200) = [608, 1392]$. The CI using the normal approximation and $\tilde{N}$ is [626, 1306]. Unfortunately, CIs based on the assumption that $\hat{N}$ or $\tilde{N}$ follow a normal distribution often have poor coverage probability in small samples because the distribution of $\hat{N}$ and $\tilde{N}$ is actually quite skewed, as you will see in Exercise 21. These CIs should only be used when the sample is large enough for $\hat{N}$ to be approximately normally distributed.

An additional shortcoming of CIs based on the normal distribution can occur in small samples. For example, suppose that $n_1 = 30$, $n_2 = 20$, and $m = 15$. Then $\hat{N} = (30)(20)/15 = 40$, and $\hat{V}(\hat{N}) = 26.7$. Using a normal approximation to the distribution of $\hat{N}$ results in the CI [30, 50]. The lower bound of 30 is silly, however; a total of 35 distinct animals were observed in the two samples, so we know that N must be at least 35.

Cormack (1992) discussed using the Pearson or likelihood ratio chi-square test for independence to construct a CI. Using this method, we fill in the missing observation x_{22} by some value u and perform a chi-square test for independence on the artificially completed data set. The 95% CI for m_{22} is then all values of u for which the null hypothesis of independence for the two samples would not be rejected at the 0.05 level. For the data in Example 13.1, let's try the value $u = 600$. With this value, the "completed" contingency table is

		In Sample 2?		
		Yes	No	
In Sample 1?	Yes	20	180	200
	No	80	600	680
		100	780	880

We can easily perform Pearson's chi-square test for independence on this table, obtaining a p-value of 0.49. As $0.49 > 0.05$, the value 600 would be inside the 95% CI for u, and the value 880 would be inside the 95% CI for N. Setting u equal to 1500, though, gives p-value 0.0043, so 1500 is outside the 95% CI for u, and 1780 is thus outside the 95% CI for N. Continuing in this manner, we find that values of u between 430 and 1198 are the only ones that result in p-value > 0.05, so [430, 1198] is a 95% CI for m_{22}. The corresponding CI for N is obtained by adding the number of observed animals in the other cells, 280, to the endpoints of the CI for m_{22}, resulting in the interval [710, 1478].

The likelihood ratio test may be used in similar manner, by including in the CI all values of u for which the p-value from the likelihood ratio test exceeds 0.05. We find that values of u between 436 and 1234 give a likelihood ratio p-value exceeding 0.05. The CI for N, using the likelihood ratio test, is then [716, 1514].

Another alternative for CIs is to use the bootstrap (Buckland, 1984). To apply the bootstrap here, take R resamples of size 100 with replacement from the 20 marked and 80 unmarked fish we observed in the second sample. Calculate $\tilde{N}_r$ (we use $\tilde{N}$ instead of $\hat{N}$ because some resamples might contain no marked fish) for each resample $r = 1, \ldots, R$, and find the 2.5th and 97.5th percentiles of $\{\tilde{N}_1, \ldots, \tilde{N}_R\}$. With $R = 2000$, this procedure gives a 95% CI for N of [699, 1561].

Note that all three of these CIs resulting from Pearson's chi-square test, the likelihood ratio chi-square test, and the bootstrap are similar, but all differ from the CIs based on the asymptotic normality of $\hat{N}$ or $\tilde{N}$.

13.1.3 Using Capture–Recapture on Lists

Capture–recapture estimation is not limited to estimating wildlife populations. It can also be used when the two samples are lists of individuals, provided that the assumptions for the method are met. Suppose you want to estimate the number of statisticians in the United States, and obtain membership lists from the American Statistical Association (ASA) and the Institute for Mathematical Statistics (IMS). Every statistician either is or is not a member of the ASA, and either is or is not a member of the IMS. (Of course, there are other worthy statistical organizations, but for simplicity let's limit the discussion here to these two.) Then n_1 is the number of ASA members, n_2 the number of IMS members, and m is the number of persons on both lists. We can estimate the number of statisticians using $\hat{N} = n_1 n_2 / m$, exactly as if statisticians were fish. The assumptions for this estimate are as above, but with slightly different implications than in wildlife settings:

(A1) The population is closed. In wildlife surveys, this assumption may not be met because animals often die or migrate between samples. When treating lists as the samples, though, we can usually act as though the population is closed if the lists are from the same time period.

(A2) Each list provides an SRS from the population of statisticians. This assumption is more of a problem; it implies that the probability of belonging to ASA is the same for all statisticians, and the probability of belonging to IMS is the same for all statisticians. It does not allow for the possibility that a group of statisticians may refuse to belong to either organization, or for the possibility that subgroups of statisticians may have different probabilities of belonging to an organization.

(A3) The two lists are independent. Here, this means that the probability that a statistician is in ASA does not depend on his or her membership in IMS. This assumption is also often not met—it may be that statisticians tend to belong to only one organization, and therefore that ASA members are less likely to belong to IMS than non-ASA members.

(A4) Individuals can be matched on the lists. This sounds easy, but often proves surprisingly difficult. Is J. Smith on List 1 the same person as Jonquil Smith on List 2? Christen (2012), Gilbert et al. (2018), and Han and Lahiri (2019) described some of the problems that can occur when you try to link records.

Example 13.2. Population censuses are often used to provide poststratification totals used to adjust for nonresponse in other surveys. But how does one know that the census counts themselves are accurate?

Capture-recapture estimation, called *dual-system estimation* in this context, is often used to evaluate census coverage (see Hogan, 1993; Mule, 2012; U.S. Census Bureau, 2020a; Chipperfield et al., 2017; Dolson, 2010; Brown et al., 2019, for how dual-system estimation has been used in the U.S., Australia, Canada, and England and Wales). Most countries that employ dual-system estimation use some variation of the following basic procedure.

Two surveys are used. A post-enumeration survey (PES) is taken directly from the population, independently of the census, and is used to estimate number of persons missed by the census. The second survey is the census enumeration itself (or a probability subsample taken from the census), and is used to estimate errors in the census such as nonexistent persons or duplicates.

In the simplest form of dual-system estimation, the population is divided among poststrata by region, race, ownership of dwelling unit, age, and other variables. The population count for each poststratum is then estimated separately from each survey. Poststrata are used because it is hoped that the assumption of equal recapture probabilities is approximately satisfied within each poststratum; the assumption is usually not satisfied for the population as a whole because undercount rates differ across subpopulations. The population table for a poststratum is as follows:

		In Census Enumeration?		
		Yes	No	
In PES?	Yes	N_{11}	N_{12}	N_{1+}
	No	N_{21}	N_{22}^*	N_{2+}^*
		N_{+1}	N_{+2}	N

Then the estimated count in the poststratum is

$$\hat{N} = \frac{\hat{N}_{+1}\hat{N}_{1+}}{\hat{N}_{11}}.$$

The estimates $\hat{N}_{1+}$ and $\hat{N}_{11}$ are from the post-enumeration survey: $\hat{N}_{1+}$ is the estimate of the poststratum total, using weights, from the PES, and $\hat{N}_{11}$ is a weighted estimate of matches between the PES and the census enumeration.

Assumptions (A1)–(A4) need to be met for dual-system estimation to give a better estimate of the population than the original census data. It is hoped that assumption (A2) holds within the poststrata. Assumption (A3) is also of some concern, though, as the PES also has nonresponse. Persons missed in the census may also be missed in the PES. Another concern is the ability to match persons in the PES to persons in the census. Because PES persons not matched are assumed to have been missed by the census, errors in matching persons in the two samples can lead to biases in the population estimates. ■

13.2 Multiple Recapture Estimation

The assumptions for the two-sample capture–recapture estimators described in Section 13.1 are strong: The population must be closed and the two random samples independent. Moreover, these assumptions cannot be tested, because we observe only three of the four cells in the contingency table—we need all four cells to be able to test independence of samples.

More complicated models may be fit if $K > 2$ random samples are taken, and especially if different markings are used for individuals caught in the different samples. With fish, for example, the left pectoral fin might be marked for fish caught in the first sample, the right pectoral fin marked for fish caught in the second sample, and a dorsal fin marked for fish caught in the third sample. A fish caught in Sample 4 that had markings on the left pectoral fin and dorsal fin, then, would be known to have been caught in Sample 1 and Sample 3, but not Sample 2.

Schnabel (1938) first discussed how to estimate N when K samples are taken. She found the maximum likelihood estimator of N to be the solution to

$$\sum_{i=1}^{K} \frac{(n_i - r_i)M_i}{N - M_i} = \sum_{i=1}^{K} r_i,$$

where n_i is the size of sample i, r_i is the number of recaptured fish in sample i, and M_i is the number of tagged fish in the lake when sample i is drawn.

If individual markings are used, we can also explore issues of immigration or emigration from the population, and test some of the assumptions of independence.

Example 13.3. Domingo-Salvany et al. (1995) used capture–recapture to estimate the prevalence of opiate addiction in Barcelona, Spain. One of their data sets consisted of three samples from 1989: (1) a list of opiate addicts from emergency rooms (E list), (2) a list of persons who started treatment for opiate addiction during 1989, reported to the Catalonia Information System on Drug Abuse (T list), and (3) a list of heroin overdose deaths registered by the forensic institute in 1989 (D list). A total of 2,864 distinct persons were on the three lists. Persons on the three lists were matched, with results in Table 13.1

It is unclear whether these data will fulfill the assumptions for the two-sample capture-recapture method. The assumption of independence among the samples may not be met—if treatment is useful, treated persons are less likely to appear in one of the other samples. In addition, persons on the death list are much less likely to subsequently appear on one of the other lists; the closed population assumption is also not met because one of the samples is a death list. Nevertheless, an analysis using the imperfectly met assumptions can

TABLE 13.1
Persons in each list combination, for Example 13.3.

		In D List?			
		Yes		No	
		In T List?		In T List?	
		Yes	No	Yes	No
In E List?	Yes	6	27	314	1728
	No	8	69	712	?

provide some information on the number of opiate addicts. Because there are more than two samples, we can assess the assumptions of independence among different samples by using loglinear models. There is one assumption, though, that we can *never* test: The missing cell follows the same model as the rest of the data. ■

If three samples are taken, the expected counts are:

		In Sample 3?			
		Yes		No	
		In Sample 2?		In Sample 2?	
		Yes	No	Yes	No
In Sample 1?	Yes	m_{111}	m_{121}	m_{112}	m_{122}
	No	m_{211}	m_{221}	m_{212}	m^*_{222}

Loglinear models were discussed in Section 10.4. The saturated model for three samples is:

$$\ln m_{ijk} = \mu + \alpha_i + \beta_j + \gamma_k + (\alpha\beta)_{ij} + (\alpha\gamma)_{ik} + (\beta\gamma)_{jk} + (\alpha\beta\gamma)_{ijk}.$$

This model cannot be fit, however, as it requires eight degrees of freedom (df) and we only have seven observations. The following models may be fit, with α referring to the E list, β referring to the T list, and γ referring to the D list.

1. *Complete independence.*

$$\ln m_{ijk} = \mu + \alpha_i + \beta_j + \gamma_k.$$

   This model implies that presence on any of the lists is independent of presence on any of the other lists. The independence model must always be adopted in two-sample capture-recapture.

2. *One list is independent of the other two.*

$$\ln m_{ijk} = \mu + \alpha_i + \beta_j + \gamma_k + (\alpha\beta)_{ij}.$$

   Presence on the E list is related to the probability that an individual is on the T list, but presence on the D list is independent of presence on the other lists. There are three versions of this model: the other two substitute $(\alpha\gamma)_{ik}$ or $(\beta\gamma)_{jk}$ for $(\alpha\beta)_{ij}$.

3. *Two samples are independent given the third.*

$$\ln m_{ijk} = \mu + \alpha_i + \beta_j + \gamma_k + (\alpha\beta)_{ij} + (\alpha\gamma)_{ik}.$$

   Three models of this type exist; the other two substitute either $(\alpha\beta)_{ij} + (\beta\gamma)_{ik}$ or $(\alpha\gamma)_{ij} + (\beta\gamma)_{ik}$ for $(\alpha\beta)_{ij} + (\alpha\gamma)_{ik}$. Presence on the death and treatment lists are conditionally independent given the E list status—once we know that a person is on the emergency room list, knowing that he or she is on the death list gives us no additional information about the probability that he or she will be on the treatment list.

4. *All two-way interactions.*

$$\ln m_{ijk} = \mu + \alpha_i + \beta_j + \gamma_k + (\alpha\beta)_{ij} + (\alpha\gamma)_{ik} + (\beta\gamma)_{jk}.$$

This model will always fit the data perfectly: It has the same number of parameters as there are cells in the contingency table.

Unfortunately, in none of these models can we test the hypothesis that the missing cell follows the model. But at least we can examine hypotheses of pairwise independence among the samples. For the addiction data, the loglinear models in Table 13.2 were fit to the data using maximum likelihood.

TABLE 13.2
Loglinear models for Example 13.3.

	Model	G^2	**df**	p**-value**	$\hat{m}_{222}$	$\hat{N}$	**95% CI**
1	Independence	1.80	3	0.62	3,967	6,831	[6,322, 7,407]
2a	E*T	1.09	2	0.58	4,634	7,499	[5,992, 9,706]
2b	E*D	1.79	2	0.41	3,959	6,823	[6,296, 7,425]
2c	T*D	1.21	2	0.55	3,929	6,793	[6,283, 7,373]
3a	E*T, E*D	0.19	1	0.67	6,141	9,005	[5,921,16,445]
3b	E*T, T*D	0.92	1	0.34	4,416	7,280	[5,687, 9,820]
3c	E*D, T*D	1.20	1	0.27	3,918	6,782	[6,253, 7,388]
4	E*T, E*D, T*D	—	0	—	7,510	10,374	[4,941, 25,964]

The value of G^2 is the likelihood ratio test statistic (deviance) for that model. Somewhat surprisingly, the model of independence fits the data well. The predicted cell counts under model 1, complete independence, are:

		In D List?			
		Yes		No	
		In T List?		**In T List?**	
		Yes	No	Yes	No
In E List?	Yes	5.1	28.3	310.8	1730.7
	No	11.7	64.9	712.4	3966.7

These predicted cell counts lead to the estimate

$$\hat{N} = 2864 + 3967 = 6831$$

if the model of independence is adopted. The values of $\hat{N}$ for the other models are calculated similarly, by estimating the value in the missing cell from the model and adding that estimate to the known total for the other cells, 2864.

We can use an inverted likelihood ratio test (Cormack, 1992) to construct a CI for N using any of the models. A 95% CI for the missing cell consists of those values u for which a 0.05-level hypothesis test of $H_0 : m_{222} = u$ would not be rejected for the loglinear model adopted. Let $G^2(u)$ be the likelihood-ratio test statistic (deviance) for the completed table with u substituted for the missing cell, let t be the total of the seven observed cells, and let

$\hat{u}$ be the estimate of the missing cell using that loglinear model. Cormack showed that the set

$$\left\{u : G^2(u) - G^2(\hat{u}) + \ln\left(\frac{u}{t+u}\right) - \ln\left(\frac{\hat{u}}{t+\hat{u}}\right) < q_1(\alpha)\right\},$$

where $q_1(\alpha)$ is the percentile of the χ^2_1 distribution with right-tail area α, is an approximate $100(1-\alpha)\%$ CI for m_{222}. This CI is conditional on the model selected and does not include uncertainty associated with the choice of model. Cormack also discussed extending the inverted Pearson chi-square test for goodness of fit, which produces a similar interval. Buckland and Garthwaite (1991) discussed using the bootstrap to find CIs for multiple recapture using loglinear models; they incorporated the model-selection procedure into each bootstrap iteration.

For these data, the point estimate and CI appear to rely heavily on the particular model fit, even though all seem to fit the observed cells. Note that the estimate $\hat{N}$ is larger and the CIs much wider for models including the E*T interaction, even though that interaction is not statistically significant. The good fit of the independence model is somewhat surprising because you would not expect the assumptions for independence to be satisfied. In addition, the population was not closed, but little information was available on migration in and out of the population.

13.3 Chapter Summary

Multiple samples from a population may be used to estimate its size. In the simplest form, two independent SRSs are taken and the number of population units found in both SRSs is used to estimate the population size. If the two samples are not independent—in particular, if individuals in the first sample are more likely to also appear in the second sample—then $\hat{N}$ calculated assuming independence is likely to underestimate the population size N.

Some forms of dependence can be assessed if three or more samples are taken. In that case, loglinear models can be fit to the data and used to predict the value of the missing cell.

Key Terms

Capture–recapture estimation: A method for estimating population size in which two independent samples are taken and the overlap used to estimate N.

Dual-system estimation: A form of capture-recapture estimation used to estimate errors in a population census.

Loglinear model: A model describing patterns of dependence in a contingency table.

For Further Reading

This chapter presented an introduction to estimating population size, under the assumption that the population is closed. Much other research has been done on models to estimate a population size, including models for populations with births, deaths, and migrations. The books authored or edited by Amstrup et al. (2010), McCrea and Morgan (2014), Böhning et al. (2017), and Seber and Schofield (2019) give good overviews with applications in medicine, social sciences, and biology. Lavallée and Rivest (2012) present a generalized version of capture-recapture estimation, in which units on the lists are related to, but not exactly the same as, the population of interest.

13.4 Exercises

A. Introductory Exercises

1. Suppose that an SRS of 500 fish is caught from a lake; each is marked and released, and the fish are allowed to remix with the other fish in the lake. A second sample of 300 fish has 120 marked fish. Estimate the total number of fish in the lake, along with a 95% CI.

B. Working with Survey Data

2. Investigators in the Wisconsin Department of Natural Resources (1993) used capture–recapture to estimate the number of fishers in the Monico Study Area in Wisconsin.
   (a) In the first study, 7 fishers were captured between August 11, 1981 and January 31, 1982. Twelve fishers were captured between February 1 and February 19, 1982; of those 12, 4 had also been captured in the first sample. Give an estimate of the total number of fishers in the area, along with a 95% CI.
   (b) In the second study, 16 fishers were captured between September 28, 1982 and October 31, 1982, and 19 fishers were captured between November 1 and November 17, 1982. Eleven of the 19 fishers in the second sample had also been caught in the first sample. Give an estimate of the total number of fishers in the area, along with a 95% CI.
   (c) What assumptions are you making to calculate these estimates? What do these assumptions mean in terms of fisher behavior and "catchability"?

3. Alexander et al. (1997) applied capture–recapture methods to estimate the number of Mead's milkweed plants in a tract in Kansas. In some years, Mead's milkweed plants do not produce aboveground parts; in addition, if they are in dense vegetation and are not flowering, they are difficult to observe. Thus, a census of observed plants in any given year is likely an undercount. From the first two years of observation, 15 plants were observed in year 1 but not in year 2, 12 plants were observed in year 2 but not in year 1, and 33 plants were observed in both years. Estimate the total number of plants in the tract, along with a 95% CI.

4. Example 1.2 described a sample that was recruited from persons (called Workers) registered on Mechanical Turk. Stewart et al. (2015) used capture-recapture methods to

estimate the number of Workers who would be willing to participate in an academic research study—that is, to estimate the size of the population from which samples are recruited.

One of their studies considered two surveys conducted by different researchers. 8,111 Workers participated in the first survey alone, 1,175 participated in the second survey alone, and 1,839 participated in both surveys. Use the methods in this chapter to estimate the number of Workers who would be willing to participate in a survey, along with a 95% CI. What assumptions are you making when calculating your estimates? Do you think these are reasonable for this data set?

5. Maritime accidents usually come to the attention of the public only when a large ship is involved or there is significant loss of life. Smaller accidents are often unreported. Hassel et al. (2011) described the important problem of estimating how many maritime accidents occur each year. They tallied the number of maritime accidents known to two maritime accident registries, called Sea-Web^{TM} and Flag-State, between 2005 and 2009 for each of seven countries. The columns in Table 13.3 show the number of accidents in the Sea-Web^{TM} database (including those that are also in the Flag-State data), the number of accidents in the Flag-State data (including those also in the Sea-Web^{TM} database), and the number of accidents found in both databases. Thus, the number of unique accidents recorded for Canada, between the two registries, is $608 + 722 - 454 = 876$, but that number does not include accidents that are missing from both databases.

TABLE 13.3
Number of maritime accidents between 2005 and 2009.

Country	Sea-Web^{TM}	Flag-State	Both
Canada	608	722	454
Denmark	189	220	45
Netherlands	304	342	71
Norway	529	596	203
Sweden	109	333	86
United Kingdom	401	1428	229
United States	632	2362	135

(a) For each country, estimate the total number of accidents assuming that presence on the two lists are independent. Give a 95% CI for each estimate.

(b) The Flag-State data contain the accidents that are registered in the individual countries. Calculate the percentage of total accidents captured in the Flag-State data for each country, along with a 95% CI for the percentage. Which countries capture the highest and lowest percentage of accidents in the Flag-State data? HINT: To calculate the CI, note that the percentage is a ratio.

6. The Centers for Disease Control and Prevention (CDC) publishes data on annual deaths in the U.S., categorized by cause of death (Murphy et al., 2017). In 2015, the CDC listed 523 deaths by legal intervention, that is, caused by law enforcement actions. A death is classified as legal intervention from the death certificate's description of how the injury occurred, but if the certificate does not mention police the death may be classified as something else (often as accident or homicide). Thus, it was thought that the CDC count underestimated the number of deaths caused by law enforcement actions. But how many deaths were missing from the CDC list?

The Guardian newspaper (Swaine and McCarthy, 2016), relying on crowdsourcing, news reports, and internet searches, compiled an independent database of 1,086 U.S. residents killed by law enforcement in 2015. Feldman et al. (2017), by matching names on the lists, estimated that 487 of the 1,086 deaths had been coded as legal intervention and were thus on both *The Guardian's* list and the CDC list.

Use this information to estimate, along with a 95% CI, the total number of persons killed by law enforcement actions in 2015. What assumptions must you make for your estimate? Do you think these are met? Explain.

7. Bellemain et al. (2005) relied on moose hunters in Norway to collect fecal samples from brown bears. Each sample was genotyped, and the number of distinct individuals was found in each of 2001 and 2002. In 2001, 311 unique genotypes were obtained (134 males and 177 females). In 2002, the procedure was repeated and 239 unique genotypes were obtained (106 males and 133 females). 165 of the individuals sampled in 2001 were also sampled in 2002.

   (a) Fifty-six bears in the area in 2001 had also been followed with radio transmitters; 36 of these bears were represented in the 311 genotypes from the 2001 feces samples. Estimate the number of bears in 2001, along with a 95% CI.

   (b) In 2002, 57 bears had radio transmitters, and 28 of them were among the 239 genotypes from the 2002 feces samples. Estimate the number of bears in 2002, along with a 95% CI.

   (c) Estimate the number of bears, along with a 95% CI, treating the samples from 2001 and 2002 as independent SRSs (and ignoring the radio transmitter data). What assumptions are needed to use capture-recapture estimators of population size?

8. Domingo-Salvany et al. (1995) also used capture-recapture on the emergency room survey by dividing the list into four samples according to trimester (TR). The following data are from Table 1 of their paper:

	TR1 yes TR2 yes	TR1 yes TR2 no	TR1 no TR2 yes	TR1 no TR2 no
TR3 yes, TR4 yes	29	35	35	96
TR3 yes, TR4 no	48	58	80	400
TR3 no, TR4 yes	25	77	50	376
TR3 no, TR4 no	97	357	312	?

   Fit loglinear models to these data. Which model do you think is best? Use your model to estimate the number of persons in the missing cell, and construct a 95% CI.

9. Chao et al. (2001) reported data on an outbreak of Hepatitis A virus among students of a college. Investigators wanted to estimate N, the total number of students with Hepatitis A. Cases were reported from three sources: (1) a serum test conducted by the Institute of Preventive Medicine (P list), (2) local hospital records from the National Quarantine Service (Q list), and (3) records collected by epidemiologists (E list). Table 13.4 gives the counts from the three sources.

   (a) Suppose that only the P list and Q list had been collected, with $n_1 = 135$, $n_2 = 122$ and $m = 49$. Calculate $\hat{N}$, Chapman's estimate $\tilde{N}$, and the standard error for each estimate.

   (b) Fit loglinear models to the data. Using the deviance, evaluate the fit of these models. Is there evidence that the lists are dependent?

TABLE 13.4
Data for Exercise 9.

P List?	Q List?	E List?	Count
no	no	yes	63
no	yes	no	55
no	yes	yes	18
yes	no	no	69
yes	no	yes	17
yes	yes	no	21
yes	yes	yes	28

10. Cochi et al. (1989) recorded data on congenital rubella syndrome from two sources. The National Congenital Rubella Syndrome Registry (NCRSR) obtained data through voluntary reports from state and local health departments. The Birth Defects Monitoring Program (BDMP) obtained data from hospital discharge records from a subset of hospitals. Table 13.5 gives data from 1970 to 1985, from the two systems.

TABLE 13.5
Data for Exercise 10.

Year	NCRSR	BDMP	Both
1970	45	15	2
1971	23	3	0
1972	20	6	2
1973	22	13	3
1974	12	6	1
1975	22	9	1
1976	15	7	2
1977	13	8	3
1978	18	9	2
1979	39	11	2
1980	12	4	1
1981	4	0	0
1982	11	2	0
1983	3	0	0
1984	3	0	0
1985	1	0	0

(a) The authors stated that the NCRSR and the BDMP are independent sources of information. Do you think that is plausible? What about the other assumptions for capture-recapture?

(b) Use Chapman's estimate in Equation (13.2) to find $\tilde{N}$ for each year for which you can calculate the estimate. What estimate will you use for the years in which Chapman's estimate cannot be calculated?

(c) Now aggregate the data for all the years, and estimate the total number of cases of congenital rubella syndrome between 1970 and 1985. How does your estimate from the aggregated data compare with the sum of the estimates from (b)? Which do you think is more reliable?

11. Frank (1978) reported on the following experiment to estimate the number of minnows in a tank. The first two samples used a minnow trap to catch fish, while the third used

a net to catch the fish. Minnows trapped in the first sample were marked by clipping their caudal fin, and minnows trapped in the second sample were marked by clipping the left pectoral fin. Data are given in Table 13.6.

TABLE 13.6
Data for Exercise 11.

Sample 1?	Sample 2?	Sample 3?	Number of Fish
yes	yes	yes	17
yes	no	yes	28
no	yes	yes	52
no	no	yes	234
yes	yes	no	80
yes	no	no	223
no	yes	no	400

Which loglinear model provides the best fit to these data? Using that model, estimate the total number of fish, and provide a 95% CI.

12. In the experiment in Exercise 11, what does it mean in terms of fish behavior if there is an interaction between presence in sample 1 and presence in sample 2? Between presence in sample 1 and presence in sample 3?

13. Egeland et al. (1995) used capture–recapture to estimate the total number of fetal alcohol syndrome cases among native Alaskans born between 1982 and 1989. Two sources of cases were used: 13 cases identified by private physicians, and 45 cases identified by the Indian Health Service (IHS). Eight cases were on both lists.

    (a) Estimate the total number of fetal alcohol syndrome cases. Give a 95% CI for your estimate, using either the inverted chi-square test or the bootstrap method.

    (b) The capture-recapture estimate relies on the assumption that the two sources of data are independent—that is, a child on the IHS list has the same probability of appearing on the private physicians list as a child not on the IHS list. Do you think this assumption will hold here? Why or why not? What advice would you give the investigators if they were concerned about independence?

    (c) Suppose that children who are seen by private physicians are less likely to be seen by the IHS. Is $\hat{N}$ then likely to underestimate or to overestimate the number of children with fetal alcohol syndrome? Explain.

14. Example 3.13 described New York City's sampling design for the annual point-in-time count of persons experiencing homelessness outside of shelters. But the volunteers may miss some locations in the sampled areas, or might not approach everyone encountered in the areas. The city conducts a "shadow count" to estimate how many unsheltered persons are missed by the volunteer teams, and to gather information for continuing to improve the survey procedures.

    Persons unknown to the volunteers are placed in sampled areas the night of the count. These "decoys" record whether they are surveyed during the point-in-time count. In the first shadow count in 2005, decoys were placed at 54 locations in the high-density areas; the decoys at 38 of those locations were approached by a volunteer team (Hopper et al., 2008).

(a) In Example 3.13, the high-density areas were treated as six separate strata. Explain why the high-density areas could equivalently be considered as belonging to one stratum, and thus that SRS formulas can be used to estimate the proportion of decoys counted by the volunteer teams.

(b) Estimate the proportion of decoys who were included in the count, along with a 95% CI.

(c) Hopper et al. (2008) did not list the number of persons counted by the volunteers in high-density areas. Suppose, though, that the point-in-time count in the high-density areas (excluding the decoys) was 3,000. Using the methods in Section 13.1, estimate the number of unsheltered persons in the area, along with a 95% CI.

(d) What assumptions are needed for the estimate in (c) to be approximately unbiased for the number of unsheltered persons in high-density areas of New York City?

C. Working with Theory

15. Note that in (13.1), $\hat{N} = n_1/\hat{p}$, where $\hat{p}$ is the sample proportion of individuals in the second sample that are tagged. Use the linearization method of Chapter 9 to find an estimator of $V(\hat{N})$.

16. The distribution of $\hat{N}$ in (13.1) is often not approximately normal. The distribution of $\hat{p} = m/n_2$, however, is often close to normality, and CIs for $\hat{p}$ are easily constructed. For the data in Example 13.1, find a 95% CI for $\hat{p}$. How can you use that interval to obtain a CI for $\hat{N}$? How does the resulting CI compare with others we calculated? Is the interval symmetric about $\hat{N}$?

17. (Requires mathematical statistics.) In a lake with N fish, n_1 of them tagged, the probability of obtaining m recaptured and $n_2 - m$ previously uncaught fish in an SRS of size n_2 is

$$\mathcal{L}(N \mid n_1, n_2) = \frac{\binom{n_1}{m}\binom{N-n_1}{n_2-m}}{\binom{N}{n_2}}.$$

The maximum likelihood estimator $\hat{N}$ of N is the value which maximizes $\mathcal{L}(N)$—it is the value that makes the observed value of m appear most probable if we know n_1 and n_2. Find the maximum likelihood estimator of N. HINT: When is $\mathcal{L}(N) \geq \mathcal{L}(N-1)$?

18. *Maximum likelihood estimation of N in large samples* (requires mathematical statistics). Suppose that n_1 of the N fish in a lake are marked. An SRS of n_2 fish is then taken, and m of those fish are found to be marked. Assume that N, n_1, and n_2 are all "large." Then the probability that m of the fish in the sample are marked is approximately:

$$\mathcal{L}(N) = \binom{n_2}{m}\left(\frac{n_1}{N}\right)^m\left(1-\frac{n_1}{N}\right)^{n_2-m}.$$

(a) Show that $\hat{N} = n_1 n_2/m$ is the maximum likelihood estimator of N.

(b) Using maximum likelihood theory, show that the asymptotic variance of $\hat{N}$ is approximately $N^2(N-n_1)/(n_1 n_2)$.

19. (Requires calculus.) For the situation in Exercise 18, suppose the cost of catching a fish is the same for each fish in the first and second samples, and you have enough resources to catch a total of $n_1 + n_2 = C$ fish altogether. If N and C are known and $C < N$, what should n_1 and n_2 be to minimize the variance in Exercise 18(b)?

20. (Requires probability.)

    (a) For Chapman's estimator $\tilde{N}$ in (13.2), let X be the random variable denoting the number of marked individuals in the second sample. What is the probability distribution of X?

    (b) Show that $E[\tilde{N}] = N$ if $n_2 \geq N - n_1$.

21. Suppose the lake has N fish, and n_1 of them are marked. A sample of size n_2 is then drawn from the lake. Choose three values of N, n_1, and n_2. Approximate the distribution of $\hat{N}$ by drawing 1,000 different samples of size n_2 from the population of N units and drawing a histogram of the $\hat{N}$ that result from the different samples. Repeat this for other values of N, n_1, and n_2. When does the histogram appear approximately normally distributed?

22. *Capture-recapture estimation with stratified sampling.* Suppose that capture-recapture estimation is carried out separately within each of H strata. In stratum h, a simple random sample of size n_{h1} is marked. Then a second, independent, sample in stratum h of size n_{h2} is taken. Let $x_{hi} = 1$ if item (hi) is marked and 0 otherwise, and let $y_{hi} = 1$ for every unit. The population size is $N = \sum_{h=1}^{H} \sum_{i=1}^{N_h} y_{hi}$.

    (a) What is the combined ratio estimator of the population size? What is the variance of the estimator?

    (b) What is the separate ratio estimator of the population size? What is the variance of the estimator?

    (c) Under what circumstances do you think the combined ratio estimator would be preferred? The separate ratio estimator?

D. Projects and Activities

23. Try out the two-sample capture–recapture method to estimate the total number of popcorn kernels or dried beans in a package, or to estimate the total number of coins in a jar. Describe fully what you did, and give the estimate of the population size along with a 95% CI for N. How did you select the sizes of the two samples?

24. Repeat the preceding exercise, using three samples and loglinear models. Would you expect the model of complete independence to fit well? Does it?

14

Rare Populations and Small Area Estimation

Housework can't kill you, but why take a chance?

—Phyllis Diller

Example 14.1. The bestselling book *The Millionaire Mind* (Stanley, 2000) used data from a survey of millionaire households in the United States. The population of millionaires is difficult to sample because no list of all millionaires exists. A simple random sample (SRS) from the U.S. population is likely to be inefficient because only a small fraction of American households have net worth over a million dollars; most returned surveys in an SRS will contain few members of the population of interest. Stanley estimated the proportion of millionaire households in each census block group of the United States. He then stratified the block groups by estimated proportion of millionaires, and took an SRS of block groups within each stratum having at least 30% estimated millionaire households. A total of 5,063 households were selected in those neighborhoods to receive the questionnaire; 1,001 questionnaires were returned, and 733 came from households reporting a net worth of at least one million dollars. The sample selected did not cover the entire population of millionaires in the United States, since households in block groups with fewer than 30% estimated millionaire households were not included in the sampling frame. ■

In this chapter, we discuss two situations for designing surveys. The first relates to Example 14.1: how to design a survey to sample units that belong to a rare population. A population can be rare in several ways. The number of individuals belonging to the rare population may be very small; snow leopards are a rare population simply because there are not very many of them. Or there may be a large number of individuals, but they form a small fraction of the population. Millionaires, for example, are reasonably plentiful but comprise a small percentage of the U.S. population. An SRS of persons in the United States, therefore, will yield few millionaires. Moreover, millionaires tend to be highly clustered, so that many geographic primary sampling units (psus) may have few, if any, millionaires. If we had a list of all millionaires in the United States, it would be quite easy to select a probability sample of them. For many rare populations, however, no such list exists; indeed, for some rare populations such as persons with Alzheimer's disease, it may be difficult to determine membership in the population because persons may be unaware they have the disease. The challenge is to obtain a sufficiently large probability sample of the rare population for the desired accuracy while controlling costs.

In many surveys, we want to estimate quantities for multiple subpopulations, for example, to estimate the unemployment rate for every county in the United States. If we only wanted to estimate the unemployment rate in one county, we could design a survey with large sample size in that county. But a survey with sufficiently large sample size in every county will have unacceptably large cost. Instead, we would like to estimate county unemployment rates using an existing national survey on unemployment. Such a survey will likely have small sample sizes (or perhaps even no observations) in some counties. A sample of 60,000 households may give accurate estimates of the national unemployment rate, but the sample might have only a few observations in Larimer County—so few that an estimate

DOI: 10.1201/9780429298899-14

of the unemployment rate in Larimer County that uses only the observations in the sample will have an unacceptably large margin of error. Larimer County, in this example, is a *small area* (also called a *small domain*); the population or land area of Larimer County may be large, but the sample size in the county is small. Section 14.2 explores models that may be used to improve accuracy of estimates for small areas.

14.1 Sampling Rare Populations

Sometimes you would like to investigate characteristics of a population that is difficult to find, or that is dispersed widely in the target population. For example, relatively few people are victims of violent crime in a given year, but you may want to obtain information about the population of violent crime victims. In an epidemiology survey, you may want to estimate the prevalence of a rare disease, and to make sure you have enough persons with the disease in your sample to analyze how the persons with the disease differ from persons without the disease.

One possibility, of course, is to take a very large sample. That is done in the American Community Survey (ACS), as discussed in Example 2.14. The sample is selected from all counties in the United States, and is designed so that annual estimates from areas with population size 65,000 have a predetermined precision. Even with the huge sample (more than 3.5 million in 2019), however, the sample sizes for some subpopulations may be small. In this section, we describe survey designs that have been proposed for estimating the prevalence of a rare characteristic or estimating quantities of interest for a rare populations; several are based on concepts we have already discussed in this book.

Nonresponse can be an especial hazard in surveys of rare populations. If population members with the rare characteristic are more likely to be nonrespondents than members without the rare characteristic, estimates of prevalence will be biased. In some health surveys, the characteristic itself can lead to nonresponse—a survey of cancer patients may have nonresponse because the illness prevents persons from responding. It is therefore important to try to minimize nonresponse for any survey of a rare population.

14.1.1 Stratified Sampling with Disproportional Allocation

Sometimes strata can be constructed so that the rare characteristic is much more prevalent in one of the strata (say, in stratum 1). Then a stratified sample in which the sampling fraction is higher in stratum 1 can give a more accurate estimate of the prevalence of the rare characteristic in the general population. The higher sampling fraction in stratum 1 also increases the domain sample size for population members with the rare characteristic. To sample millionaires, one might divide census block groups into strata by the estimated 90th percentile of income, and oversample the strata where the percentile is high.

Disproportional stratified sampling may work well when the allocation is efficient for all items of interest. Having higher sampling fractions in census block groups thought to have high proportions of millionaires is sensible for estimating characteristics of millionaires. But this disproportional allocation may be inefficient for estimating characteristics of persons who work at home, a rare population that is not necessarily concentrated in those block groups.

An additional challenge occurs if the sampling fraction in the concentrated stratum is much higher than that in the sparse strata thought to hold small numbers of the rare population. The sampling weights for persons in the sparse strata are then much higher than weights for persons in the concentrated stratum, and this can cause estimates to be influenced heavily by the data from a few individuals and have higher variance (see Exercise 17 of Chapter 7). With a small sampling fraction in the sparse strata, the sample might not capture any rare population members in those strata.

Example 14.2. Edwards et al. (2005) used models to construct strata for sampling rare lichen species in Washington, Oregon, and northern California. The rare species were uncommon in the pilot sample, so they fit classification tree models to predict presence of four common lichen species that frequently occur with the rare lichen species of interest. Each of the common lichen species was detected in at least 120 of the 840 sites sampled in a lichen air quality study, giving sufficient information to build models predicting presence of each species of common lichen from variables such as slope, aspect, precipitation, temperature, and relative humidity. Using data collected in a second sample, Edwards et al. (2005) estimated that a disproportional stratified sample based on the classification tree models would result in a 1.2- to 5-fold gain in sampling efficiency for four rare lichen species. ■

14.1.2 Two-Phase Sampling

Two-phase sampling methods were discussed in Chapter 12 as a way of using stratification when the information needed to form the strata is not available before sampling. To sample a rare population, we would like to stratify on the variable that indicates whether individuals belong to the population or not. Screen the phase I sample units to determine whether they have the rare characteristic. Then subsample all (or a high sampling fraction) of the units with the rare characteristic for the phase II sample. If the screening technique is completely accurate, use the phase I sample to estimate prevalence of the rare characteristic and the phase II sample to estimate other quantities for the rare population.

What if the screening technique is not completely accurate? If sampling arctic regions for presence of walruses, it is possible that you will not see walruses in some of the sectors from the air because the walruses are under the ice. Asking persons whether they have diabetes will not always produce an accurate response because persons do not always know whether they have it. As Deming (1977) pointed out, placing a person with diabetes in the "no-diabetes" stratum is more serious than placing a person without diabetes in the "diabetes" stratum: If only the "diabetes" stratum is subsampled, it is likely that the persons without diabetes who have been erroneously placed in that stratum will be discovered, while the errors from persons with diabetes misclassified into the "no-diabetes" stratum will not be found. One possible solution is to broaden the screening criterion so that it encompasses all units that might have the rare characteristic. Another solution is to subsample both strata in phase II, but to use a much higher sampling fraction in the "likely to have diabetes" stratum.

You may want to use a different two-phase design for estimating characteristics of rare population members than for estimating prevalence of the rare population. Exercise 21 of Chapter 12 presented optimal sampling strategies for using a two-phase sample to estimate prevalence of a disease.

14.1.3 Unequal-Probability Sampling

To oversample individuals with the rare characteristic, we can create a model for the inclusion probabilities based on related characteristics. This is similar to disproportional stratified sampling, except that the unequal probabilities may be used directly as well as in stratification. Hoeting et al. (2000) developed a model for predicting the presence or absence of a species from satellite data. The model gives a predicted probability that the species is present for each pixel in the satellite image. The predicted probabilities may then be used to form strata or to specify inclusion probabilities π_i.

The Mitofsky–Waksberg method for random digit dialing, discussed in Example 6.13, can be used to sample rare populations that are clustered. In a survey of millionaires, census block groups can be treated as clusters. A probability sample of block groups is drawn (the probability sampling design should rely on stratification as well). Select one household from each cluster; if it is a millionaire household, then sample additional households in that cluster. This procedure samples clusters with probability proportional to the number of millionaire households.

14.1.4 Multiple Frame Surveys

Even though you may not have a list of all of the members of the rare population, you may have some incomplete sampling frames that contain a high percentage of units with the rare characteristic. You can sometimes combine these incomplete frames, omitting duplicates, to construct a complete sampling frame for the population. Alternatively, you can select samples independently from the frames, then combine sample estimates from the incomplete frames (and, possibly, a complete frame) to obtain general population estimates. This **multiple frame survey** approach was pioneered by Hartley (1962).

Suppose you would like to estimate the prevalence of Alzheimer's disease, and characteristics of persons with the disease, in the noninstitutionalized population. Since many users of adult day care centers have Alzheimer's, you would expect that a sample of adult day care centers would yield a higher percentage of persons with Alzheimer's than a general population survey. But not all persons with Alzheimer's attend an adult day care center. Thus, you might have two sampling frames: frame A, which is the sampling frame for the general population survey, and frame B, which is the sampling frame for adult day care centers. All persons in frame B are presumed to also be in the frame for the general population survey, so the design in Figure 14.1(a) has two domains: ab, which consists of persons in frame A and also in frame B, and a, which consists of persons in frame A but not in frame B. In other situations, both frames are incomplete, leading to three domains as in Figure 14.1(b): domain a, consisting of persons in frame A but not frame B; domain b, with persons in frame B but not frame A; and domain ab, consisting of persons in both frames.

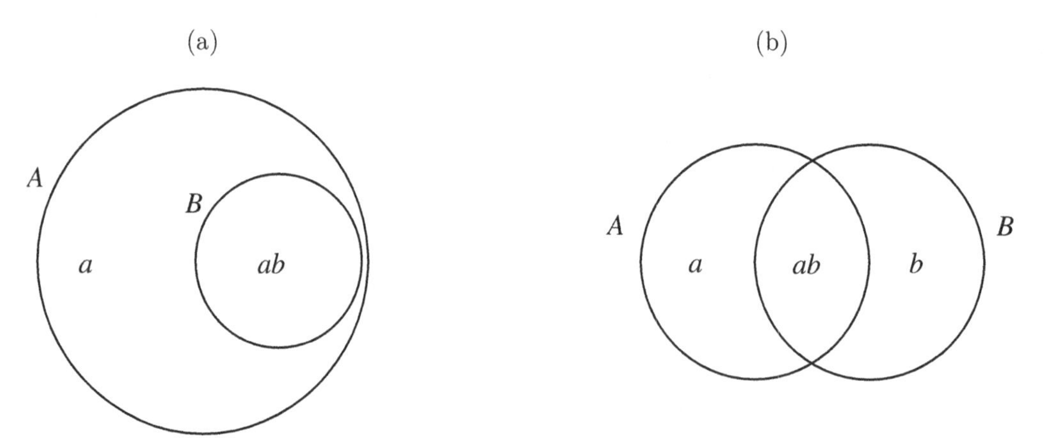

FIGURE 14.1
Examples of dual frame surveys. In (a), frame A is complete and frame B is incomplete. In (b), both frames are incomplete.

To estimate population quantities from the dual frame survey depicted in Figure 14.1(b), determine the domain membership of each sampled person. Estimate the population total $t = \sum_{i=1}^{N} y_i$ by $\hat{t}_a + \hat{t}_{ab} + \hat{t}_b$, where $\hat{t}_a$, $\hat{t}_{ab}$, and $\hat{t}_b$ estimate the population totals in domains a, ab, and b, respectively. Units in domain ab can be sampled from either survey, and Lohr (2011) reviews estimators that account for those units' multiple chances of selection. Exercise 3 gives Hartley's (1962) estimator for the survey depicted in Figure 14.1(b).

Example 14.3. One of the goals of the Survey of Consumer Finances, described in Example 8.12, is to calculate statistics about a wide range of asset classes, including classes that are owned by relatively few families in the population such as real estate investment trusts, tax-exempt bonds, and hedge funds. The survey employs a dual frame design of the type in Figure 14.1(a) to ensure that the sample contains an adequate sample size of families that might hold these assets. Frame A, an area frame used to select a stratified multistage sample of families, includes the entire population, and the sample from the frame can be used to provide estimates of population-wide characteristics such as median income and homeownership. Frame B is a list of wealthy families derived from tax returns. The frame-B sample increases the sample size of families holding assets such as hedge funds to allow calculation of statistics about those assets (Bhutta et al., 2020). ■

Example 14.4. Iachan and Dennis (1993) described the use of multiple frames to sample the population of persons experiencing homelessness in Washington, D.C. Four frames were used: (1) homeless shelters, (2) soup kitchens, (3) encampments such as vacant buildings and locations under bridges, and (4) streets, sampled by census blocks. Although the union of the frames should include more of the population than a single frame, it will not include all persons experiencing homelessness, as shown in Figure 14.2. Membership in more than one frame was estimated by asking survey respondents whether they had been or expected to be in soup kitchens, in shelters, or on the street in the 24-hour period of sampling. ■

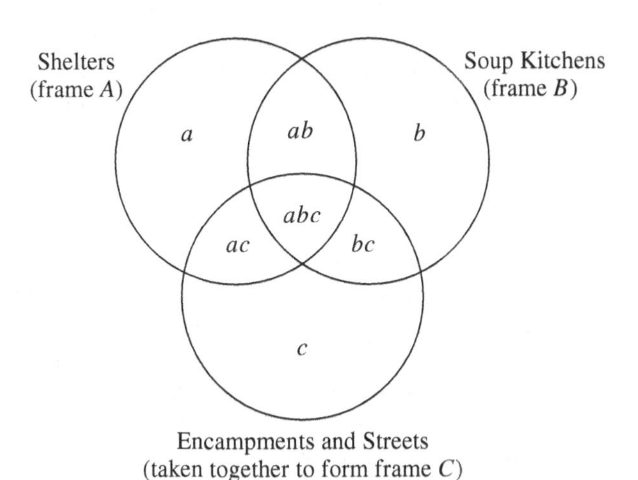

FIGURE 14.2
A multiple frame survey used to estimate the number of persons experiencing homelessness.

Combining surveys with other data sources. More generally, a probability sample can sometimes be combined with another data source to augment the sample size of a rare population, or to provide information on variables not measured in the probability sample.

The data sources can be thought of as coming from multiple sampling frames, with the difference that nonprobability samples may be taken from some of the frames. This necessitates a model-based approach, as discussed in Chapter 15. Lohr and Raghunathan (2017) reviewed methods that can be used to combine data from different surveys and other data sources, including multiple frame methods, linking individual records, mixed effects models, and imputation. Zhang and Chambers (2019) presented methods for analyzing data that have been integrated from multiple sources.

14.1.5 Network or Multiplicity Sampling

In a household survey on crime, each household provides information only on victimizations that have occurred to members of that household. In a network sample to study crime victimization, each household in the population is linked to other units the population; the sampled household can also provide information on units linked to it (called the network for that household). For example, the network of a household might be defined to be the adult siblings of adult household members.

Suppose a probability sample of households is taken. Define $\mathcal{G}_i$ to be the network for unit i in the probability sample. Suppose household 1 has adults John and Mary. Then, if networks are formed using the sibling rule, $\mathcal{G}_1$ consists of John, Mary, John's adult siblings (Suzy and Fred), and Mary's adult sibling (Mark). John and Mary are asked about crime incidents that occurred to either of them or to Suzy, Fred, or Mark. John's (or Suzy's or Fred's) response can be included up to three times in the sample: if John's household is selected, Suzy's household is selected, or Fred's household is selected. Mark's and Mary's information have only two chances of inclusion, if Mark's or Mary's household is chosen in the sample. An only child is included only if his or her household is selected in the probability sample.

The **multiplicity** of individual k is the number of links leading to that individual. Let $\omega_k = 1/(\text{multiplicity of person } k)$ be the multiplicity weight for person k in the population of interest. In our example, John, Suzy, and Fred each have multiplicity weight 1/3, and Mary and Mark each have multiplicity weight 1/2. Let y_k be an indicator for whether person k was a victim of crime. Estimate the total number of crime victims by

$$\hat{t}_{y,\text{net}} = \sum_{i \in \mathcal{S}} w_i \sum_{k \in \mathcal{G}_i} \omega_k y_k. \tag{14.1}$$

This estimator and its variance are derived in Exercise 5.

Network sampling can reduce the sampling variability of the estimated prevalence of a rare characteristic because it can provide more information per sampled individual. Czaja et al. (1986) found that network sampling provided greater precision for estimating prevalence of cancer cases. There are, however, additional possibilities for error in network sampling. If John is selected in the initial sample, he must report: (1) his value of the response y_i, (2) the response for each person in John's network (y_k for persons k linked to John), and (3) the number of population units linked to each person k in John's network (the multiplicity for person k in John's network).

John will probably give the correct multiplicity for his siblings. But with other linking rules, John's report of multiplicity for units in his network may be inaccurate—if John's network consists of students who are in class with him, John may not know the number of other classes taken by his classmates. Also, John might not report the correct value of the response for persons in his network. John might not be aware of criminal victimizations experienced by his or her siblings and give an inaccurate count. Social desirability of responses is also an issue. John may know which of his siblings have cancer, but may not know that one of them is a substance abuser.

14.1.6 Snowball Sampling

Snowball sampling is based on the premise that members of the rare population know one another. To take a snowball sample of jazz musicians, you would locate an initial sample of a few jazz musicians—these might be acquaintances or persons recruited from current performers in jazz clubs. Ask each of those persons to identify other jazz musicians who could be included in your sample, then ask the new persons in your sample to identify additional jazz musicians, and so on. As each wave of sampled musicians identifies more musicians, the sample grows like a snowball rolling down a hill. The chain-referral procedure continues until the desired sample size is achieved or no new musicians are identified.

As with network sampling, snowball sampling augments an initial sample through connections with persons who have already been sampled. In network sampling, however, the initial sample is a probability sample, and a person's network is determined by the researcher with well-defined criteria (for example, siblings of the sampled person). This makes network sampling a probability sampling method. In snowball sampling, the initial sample is usually chosen conveniently, and the sample accumulates through referrals from those who have already participated—in other words, sample members themselves choose who else can be in the sample. A snowball sample is thus a convenience sample, where the selection probabilities are unknown.

Snowball sampling can result in a large sample size for a rare population, but strong modeling assumptions are needed to generalize results to the population. Although the method can identify members of a rare population who would be difficult to find with other designs, the resulting sample can be far from an SRS. Persons with many connections in the population of interest are more likely to be included in the sample than persons with few connections. Isolated population members may not be reachable at all.

Example 14.5. Lee et al. (2021) used a form of snowball sampling known as respondent-driven sampling to obtain a sample of foreign-born Korean Americans. According to the American Community Survey (ACS), 1.9 million Korean Americans lived in the U.S. in 2017—a large number of persons, but a relatively small percentage of the U.S. population. In addition, although the population of interest was highly concentrated in specific regions and cities, some members lived in areas containing few other Korean Americans. Using a snowball sample allowed the investigators to collect data from a large sample of Korean Americans; they were also able to sample persons from geographic areas with low concentrations of Korean Americans, where a self-weighting probability sample would likely yield no members of the population of interest.

The researchers recruited an initial sample through referrals from organizations such as Korean churches and groceries, and through flyers and online advertisements. Eligible persons in the initial sample were asked to take an online survey; after completing the survey, they were sent a thank-you note along with two coupons to pass on to other foreign-born Korean Americans. When the coupon recipients visited the survey website, they were asked to take the online survey and were in turn sent coupons to distribute, and so on. Participants were paid \$20 for completing the main survey and \$5 for each coupon recipient who completed the survey, giving them a financial incentive to recruit others.

The selection probabilities are unknown, but Heckathorn (1997) proposed using weights that are inversely proportional to a respondent's social network size. The rationale is that persons with connections to many members of the target population are thought to have a higher chance of being recruited for the sample and thus given smaller weight. The weights are the inverses of estimated probabilities of selection, which may or may not be accurate representations of the true probabilities of participating in the sample; Gile et al. (2018) described other methods that might be used to estimate selection probabilities and calculate weights.

Lee et al. (2021) calculated unweighted estimates from the sample as well as two sets of weighted estimates, and found that the weighted and unweighted estimates were similar for most variables. They also compared survey estimates to independent estimates of the same variables from the ACS (considered to be approximately unbiased because they are from a high-response-rate probability sample). Survey estimates of household size, employment type, and health insurance coverage were similar to those from ACS. However, the survey estimates of the proportion of younger and more recent immigrants, and of persons with higher education and with disabilities, were significantly higher than the ACS estimates. In this comparison, weighted estimates sometimes performed worse than unweighted estimates. The bias of survey variables not measured in ACS (or another external probability sample) is unknown. ■

14.1.7 Sequential Sampling

In sequential sampling, observations or psus are sampled one or a few at a time, and information from previously drawn psus can be used to modify the sampling design for subsequently selected psus. In one method dating back to Stein (1945) and Cox (1952), an initial sample is taken, and results from that sample are used to estimate the additional sample size necessary to achieve a desired precision. If it is desired that the sample contain a certain number of members from the rare population, the initial sample could be used to obtain a preliminary estimate of prevalence, and that estimate of prevalence used to estimate the necessary size of the second sample. After the second sample is collected, it is combined with the initial sample to obtain estimates for the population. A sequential sampling scheme generally needs to be accounted for in the estimation; in Cox's method, for example, the sample variance obtained after combining the data from the initial and second samples is biased downward (Lohr, 1990). Lai (2001) reviewed history and uses of sequential methods.

Adaptive cluster sampling (Thompson, 1990, 2017) assumes that the rare population is clustered—caribou are in herds, an infectious disease is concentrated in certain areas of the country, or artifacts are clustered at specific sites of an archaeological dig. To conduct an adaptive cluster sample, select an initial probability sample of psus. For each psu in the initial sample, measure a response such as the number of caribou in the psu. If the number of caribou in psu i exceeds a predetermined value c, then add neighbors of psu i to the sample. Count the number of caribou in each of the neighboring units and add the neighbors of any of those units with more than c caribou to the sample. Continue the procedure until none of the neighbors has more than c caribou. The adaptive nature of the sampling scheme needs to be accounted for when estimating population quantities—if you estimate caribou density by (number of caribou observed)/(number of psus sampled) from an adaptive cluster sample, your estimate of caribou density will be far too high.

14.2 Small Area Estimation

In most surveys, estimates are desired not only for the population as a whole, but also for subpopulations (domains). We discussed estimation in domains in Section 4.3 for SRSs and showed that estimating domain means is a special case of ratio estimation because the sample size in the domain varies from sample to sample. But we noted that if the sample size for the domain in an SRS was large enough, we could essentially act as though the sample size was fixed for inference about the domain mean.

In complex surveys with many domains, estimation is not quite that simple. One worry is that the sample size for a given domain will be too small to provide a useful estimate. The National Health and Nutrition Examination Survey (NHANES), for example, gives an accurate estimate of the prevalence of hypertension for adults in the United States as a whole. However, if you are interested in estimating the prevalence of hypertension at the county level, the sample sizes for many counties are so small that direct estimates of the prevalence for those counties are of very little use—many counties have no sample members at all. You might conjecture, though, that prevalence might be similar among counties with similar characteristics, and use information from other counties to predict the prevalence for counties with small (or zero) sample sizes. You could also incorporate information from other sources, such as a population census or administrative health records, to improve your estimate.

Similarly, data collected in an educational survey in New York may be sufficient for estimating eighth grade mathematics achievement for students in the state, but not for a direct assessment of mathematics achievement in individual cities such as Rochester. The survey data from Rochester, though, can be combined with estimates from other cities and with school administrative data (scores on other standardized tests, for example, or information about mathematics instruction in the schools) to produce an estimate of eighth grade mathematics achievement for Rochester that we hope has smaller mean squared error.

Small area estimation techniques, in which estimates are obtained for domains with small sample sizes, have in recent years been the focus of intense research in statistics. Rao and Molina (2015) describe small area estimation methods and give a bibliography for further reading. Here, we summarize some of the proposed approaches. Let $a_{id} = 1$ if observation unit i is in domain d and 0 otherwise. In this section, the quantities of interest are the domain totals $t_{yd} = \sum_{i=1}^{N} a_{id} y_i$, the domain sizes $N_d = \sum_{i=1}^{N} a_{id}$, and the domain means $\bar{y}_{\mathcal{U}_d} = t_{yd}/N_d$, for domains $d = 1, \ldots, D$.

14.2.1 Direct Estimators

As we saw in Sections 4.3 and 11.3, a direct estimator of t_{yd} depends only upon the sampled observations in domain d:

$$\hat{t}_{yd}(\text{dir}) = \sum_{i \in \mathcal{S}} w_i a_{id} y_i. \tag{14.2}$$

The estimated domain totals in (14.2) satisfy the **benchmarking** property: The estimated totals for set of all domains sum to the estimated total for the entire population. This occurs because

$$\sum_{d=1}^{D} \hat{t}_{yd}(\text{dir}) = \sum_{d=1}^{D} \sum_{i \in \mathcal{S}} w_i a_{id} y_i = \sum_{i \in \mathcal{S}} w_i y_i = \hat{t}_y.$$

Benchmarking is desirable since we would like the estimated numbers of people with health insurance in the D domains to sum to the estimated number of people with health insurance in the population.

The domain mean is estimated by

$$\hat{\bar{y}}_d(\text{dir}) = \frac{\sum_{i \in \mathcal{S}} w_i a_{id} y_i}{\sum_{i \in \mathcal{S}} w_i a_{id}}. \tag{14.3}$$

Because $\hat{\bar{y}}_d(\text{dir})$ is a ratio, its variance is estimated using linearization (see Example 9.2) as

$$\hat{V}\left[\hat{\bar{y}}_d(\text{dir})\right] = \frac{1}{\hat{N}_d^2} \hat{V}\left[\sum_{i \in \mathcal{S}} w_i a_{id} \left\{ y_i - \hat{\bar{y}}_d(\text{dir}) \right\}\right]. \tag{14.4}$$

The approximation to the variance is valid if the expected sample size in the domain is sufficiently large. Section 11.3 discussed comparing domain means using regression.

Warning. In an SRS, if you create a new data set that consists solely of sampled observations in domain d and then apply the standard variance formula, your variance estimator is approximately unbiased. **Do not** adopt this approach for estimating the variance of domain means in complex samples. A sampled psu may contain no observations in domain d; if you eliminate such psus and then apply the standard variance formula, you may underestimate the variance (see Exercises 6 and 9).

In practice, the sample size in domain d may be so small that the variance of $\hat{\bar{y}}_d(\text{dir})$ is unacceptably large. Some domains of interest may have no observations at all so that a direct estimator cannot be calculated. The next sections describe methods that may be used to estimate domain mean and totals in these cases.

14.2.2 Synthetic and Composite Estimators

A **synthetic** estimator uses information from a reliable direct estimator of a large area under the assumption that small areas inside the large area have similar characteristics. Assume that we know some quantity associated with t_{yd} for each domain d. For estimating prevalence of hypertension, we might use t_{xd} = population in county d, or population scaled by a risk factor that is calculated from state health department information. A simple form of synthetic estimator is

$$\hat{t}_{yd}(\text{syn}) = \frac{\hat{t}_y}{\hat{t}_x} t_{xd} \tag{14.5}$$

The estimator does not use survey information about domain d directly—instead, it synthesizes information from a large-area direct estimator and small-area auxiliary information, hence the name *synthetic*.

If the ratios t_{yd}/t_{xd} are similar in different domains, and if each ratio is similar to the ratio of population totals t_y/t_x, the synthetic estimator may be more accurate than $\hat{t}_{yd}(\text{dir})$ in (14.2). Certainly the variance of $\hat{t}_{yd}(\text{syn})$ will be relatively small, since $(\hat{t}_y/\hat{t}_x)$ is estimated from the entire sample and is expected to be precise. If the ratios are not homogeneous, however, the synthetic estimator may have large bias.

You can also use synthetic estimation in subsets of the population, and then combine the synthetic estimators for each subset. For estimating prevalence of hypertension in small areas, you could divide the population into different age-race-gender classes. Then you could find a synthetic estimate of the total number of persons with hypertension in domain d for each age-race-gender class, and then sum the estimates for the age-race-gender classes to estimate the total number of persons with hypertension in small area d.

The direct estimator is unbiased but may have large variance; the synthetic estimator has smaller variance but may have large bias. They may be combined to form a **composite estimator**:

$$\hat{t}_{yd}(\text{comp}) = \alpha_d \hat{t}_{yd}(\text{dir}) + (1 - \alpha_d)\hat{t}_{yd}(\text{syn}), \tag{14.6}$$

where $0 \le \alpha_d \le 1$. How should one determine the relative weights α_d? One simple solution relates α_d to the sample size in domain d. If too few units are observed in domain d, α_d will be close to zero and more reliance will be placed on the synthetic estimator. The next section describes a model-based approach for determining α_d.

14.2.3 Model-Based Estimators

In a model-based approach, a superpopulation model is used to predict values in domain d. The model "borrows strength" from the data in related domains, or incorporates auxiliary information from administrative data or other surveys. The models can be used to determine the weights α_d in a composite estimator. Mixed models, described in Section 11.4.5, are often used in small area estimation.

Area-level models. In an area-level model, a vector of auxiliary information $\mathbf{x}_d$ is available for each domain $d = 1, \ldots, D$. If small areas are counties, $\mathbf{x}_d$ might contain information for county d such as the median home value, percentage of persons over age 65, number of births or deaths in the past year, or 75th percentile of 10th-grade student scores on a required mathematics test—quantities known from a population census or from county records.

The Fay–Herriot (1979) model is commonly used when the auxiliary information is available at the domain level. We wish to estimate the population domain mean $\theta_d = \bar{y}_{\mathcal{U}_d}$. Let $\hat{\bar{y}}_d = \hat{\bar{y}}_d(\text{dir})$ be a direct estimator of θ_d from the survey with sampling variance $V(\hat{\bar{y}}_d) = \psi_d$. Assume that

$$\hat{\bar{y}}_d = \theta_d + e_d,$$

where $e_d \sim N(0, \psi_d)$. Also assume that the population domain means θ_d are related to the covariates $\mathbf{x}_d$ through the model

$$\theta_d = \mathbf{x}_d^T \boldsymbol{\beta} + v_d,$$

where $v_d \sim N(0, \sigma_v^2)$ represents the domain effect. All random variables $e_1, \ldots, e_D$ and $v_1, \ldots, v_D$ are assumed to be independent. Combining the two models, we have

$$\hat{\bar{y}}_d = \mathbf{x}_d^T \boldsymbol{\beta} + v_d + e_d, \tag{14.7}$$

which includes the error term e_d from the direct estimator as well as the error v_d from the model that is assumed to hold for the population domain means. If ψ_d and σ_v^2 are known, then the best linear unbiased predictor of θ_d is

$$\tilde{\theta}_d = \alpha_d \hat{\bar{y}}_d + (1 - \alpha_d)\mathbf{x}_d^T \tilde{\boldsymbol{\beta}}, \tag{14.8}$$

where $\alpha_d = \sigma_v^2/(\sigma_v^2 + \psi_d)$ and $\tilde{\boldsymbol{\beta}}$ is the weighted least squares estimator of $\boldsymbol{\beta}$.

The estimator $\tilde{\theta}_d$ depends more heavily on the direct estimator $\hat{\bar{y}}_d$ when $\psi_d = V(\hat{\bar{y}}_d)$ is small; it depends more heavily on the predicted value from the regression model $\mathbf{x}_d^T \tilde{\boldsymbol{\beta}}$ when ψ_d is large. In practice, σ_v^2 must be estimated from the data and the estimator $\hat{\sigma}_v^2$ used to estimate α by $\hat{\alpha}$. Although the estimator in (14.8) does not satisfy the benchmarking property (because α_d varies for different y variables), Rao and Molina (2015) described modifications that may be used to ensure benchmarking.

Example 14.6. In the U.S. Small Area Income and Poverty Estimates program (U.S. Census Bureau, 2020c), the estimated poverty rate for a domain is a weighted average of (a) the direct estimate from the ACS and (b) the predicted value from a regression equation. Covariates include variables calculated from the domain population size, the number of food assistance recipients in the domain, and domain-level statistics compiled from tax records. In domains with large ACS sample size (with small value of ψ_d), $\tilde{\theta}_d$ is close to the ACS estimate. In domains where the ACS sample size is small, $\tilde{\theta}_d$ depends more heavily on the predicted value from the model. Estimates are benchmarked so that estimated totals are consistent over different units of geography.

Unit-level models. A *unit-level model* requires knowledge of covariate values for each individual record in the sample. In an educational survey, if Y_{dj} is the mathematics achievement of student j in domain d in the population, you might postulate a model such as

$$Y_{dj} = \beta_0 + x_{dj}\beta_1 + v_d + \varepsilon_{dj},$$

where $v_d \sim N(0, \sigma_v^2)$ and $\varepsilon_{dj} \sim N(0, \sigma_\varepsilon^2)$ for $d = 1, \ldots, D$ and student j in domain d, assuming all random variables v_d and ε_{dj} are independent (Battese et al., 1988). The student-level covariate x_{dj} (we included one covariate for simplicity, but of course multiple covariates could be included) could come from administrative records, for example the student's score on an achievement test given to all students in the state or the student's grades in mathematics classes. Assume that the population mean of x in domain d, $\bar{x}_{\mathcal{U}_d} = \frac{1}{N_d}\sum_{j=1}^{N_d} x_{dj}$, is known. Then, if the domain sample size n_d is small relative to the population size N_d, it can be shown that the best linear unbiased predictor of the modeled domain mean $\mu_d = \beta_0 + \bar{x}_{\mathcal{U}_d}\beta_1 + v_d$ is

$$\tilde{\mu}_d = \tilde{\beta}_0 + \bar{x}_{\mathcal{U}_d}\tilde{\beta}_1 + \gamma_d\left(\hat{\bar{y}}_d - \tilde{\beta}_0 - \hat{\bar{x}}_d\tilde{\beta}_1\right),$$

where $\gamma_d = \sigma_v^2/(\sigma_v^2 + \sigma_\varepsilon^2/n_d)$ and $\tilde{\beta}_0$ and $\tilde{\beta}_1$ are the best linear unbiased estimators of β_0 and β_1. The predictor depends more on the direct estimator $\hat{\bar{y}}_d$ if $V_M(\hat{\bar{y}}_d) = \sigma_\varepsilon^2/n_d$ is small; otherwise, it depends more on the predicted value from the regression at the population domain mean of x.

An indirect estimator, whether synthetic, composite, or model-based, is essentially an exercise in predicting missing data. Indirect estimators are thus highly dependent on the model used to predict the missing data—the synthetic estimator, for example, assumes that the ratios are homogeneous across domains. When possible, the model assumptions should be checked empirically; one method for exploring validity of the model assumptions is to pretend that some of the data you have is actually not available, and to compare the indirect estimator with the direct estimator computed with all the data.

14.3 Chapter Summary

Rare populations present special challenges for sampling, since many standard sampling designs yield few units in the rare population. Several designs discussed in previous chapters can be used to increase the number of rare population units in the sample. Auxiliary information associated with the rare characteristic can be used to design a stratified sample with disproportional allocation. If such auxiliary information is not known in advance, a two-phase sampling design can collect inexpensive screening information in the phase I sample, and then collect the detailed survey information in phase II.

Multiple frame surveys, in which independent probability samples are selected from sampling frames whose union is assumed to include the entire rare population, can greatly reduce the cost of a survey of a rare population. One frame might cover the entire population, while other frames might be incomplete yet inexpensive to sample.

Network, snowball, and adaptive cluster samples use connections among population members to increase the efficiency of the sampling design. In network sampling, persons in a probability sample are asked about themselves as well as persons defined to be in their network, for example, their adult siblings. A snowball sample often begins with a convenience sample of persons in the rare population, who are then asked to provide contact information for other persons in the rare population. In adaptive cluster sampling, the responses of an initial probability sample are used to select neighboring units for inclusion.

Small area estimation methods rely on auxiliary information and models to obtain estimators of population quantities in domains in which the sample size is too small for a direct estimator to be reliable.

Key Terms

Adaptive cluster sampling: A sequential probability sampling design in which estimates from an initial probability sample of clusters are used to determine inclusion probabilities for subsequent units.

Multiple frame survey: A survey in which independent samples are taken from two or more sampling frames that are thought to include the whole population.

Network sampling: A sampling method in which a probability sample is taken from a population and each sampled unit provides information on itself and on units in its network.

Rare population: A subpopulation that is uncommon relative to the whole population.

Small area: A subpopulation for which the sample size is small.

Snowball sampling: A sampling method in which members of a "seed" sample from a population suggest other population members for the sample.

For Further Reading

Kalton (2003), Christman (2009), and Tourangeau et al. (2014) review methods for sampling rare and hard-to-reach populations. The book edited by Thompson (2004) describes methods for sampling rare species in wildlife applications. See Thompson (2013) for theoretical developments and a bibliography for adaptive cluster sampling, and Lohr (2011) for a review of designs and estimators for multiple frame surveys.

Czaja and Blair (1986) and Sudman et al. (1988) describe the general method for network sampling. Frank (2010) presents general concepts and approaches that are used in network and snowball sampling, and describes applications of the methods. Heckathorn (1997), Salganik and Heckathorn (2004), Handcock and Gile (2010), Heckathorn and Cameron (2017), Pattison et al. (2013), Gile et al. (2018), and Lee et al. (2020) discuss snowball sampling and respondent-driven sampling, which use information about network connections in the sample to weight the sample units.

Thompson and Seber (1996), Thompson and Collins (2002), Turk and Borkowski (2005), and Seber and Salehi (2012) describe approaches for adaptive cluster sampling and give references to other work.

Rao and Molina (2015) describe models commonly used in small area estimation, with applications to estimating poverty, unemployment, disease prevalence, and census undercounts. Other useful references include Jiang and Lahiri (2006), Pratesi (2016), Pfeffermann (2013), and Ghosh (2020). Jiang and Rao (2020) review robust methods in small area estimation. Harter et al. (2019) give a practical step-by-step guide for how to carry out small area estimation.

14.4 Exercises

A. Introductory Exercises

1. What designs would you consider for obtaining a sample from each of the following rare populations?

   (a) Alumni of your university who are currently working as engineers.

   (b) Persons who are caregivers for a household member who has Alzheimer's disease.

   (c) Muslims in Canada.

   (d) Households with children aged 18–36 months.

   (e) Transgender students at California universities.

   (f) Urban agriculture operations (places in urban areas from which $1,000 or more of agricultural products were produced and sold). Examples include large gardens, nurseries, greenhouses, beekeeping or livestock operations, and egg producers.

C. Working with Theory

2. (Requires calculus.) Kalton and Anderson (1986) considered disproportional stratified random sampling for estimating the mean of a characteristic y_i in a rare population. Let $r_i = 1$ if person i is in the rare population and 0 otherwise. Stratum 1 contains N_1 persons, M_1 of whom are in the rare population; stratum 2 contains N_2 persons, with M_2 persons in the rare population. We wish to estimate the population mean $\bar{y}_{\mathcal{U}_d} = \sum_{i=1}^{N} r_i y_i/(M_1 + M_2)$ using a stratified random sample of n_1 persons in stratum 1 and n_2 persons in stratum 2.

   (a) Suppose $A = M_1/(M_1+M_2)$ is known. Let $\hat{\bar{y}}_d = A\bar{y}_1+(1-A)\bar{y}_2$, where $\bar{y}_1$ and $\bar{y}_2$ are the sample means of the rare population members in strata 1 and 2, respectively. Show that, if you ignore the finite population corrections (fpcs) and if the sampled number of persons in the rare population in each stratum is sufficiently large, then

   $$V(\hat{\bar{y}}_d) \approx \frac{A^2 S_1^2}{n_1 p_1} + \frac{(1-A)^2 S_2^2}{n_2 p_2},$$

   where S_j^2 is the variance of y for the rare population members in stratum j and $p_j = M_j/N_j$ for $j = 1, 2$.

   (b) Suppose that $S_1^2 = S_2^2$ and that the cost to sample each member of the population is the same. Let $f_2 = n_2/N_2$ be the sampling fraction in stratum 2, and write the sampling fraction in stratum 1 as $f_1 = kf_2$. Show that the variance in (a) is minimized when $k = \sqrt{p_1/p_2}$.

3. (Requires calculus.) Consider the dual frame survey in Figure 14.1(b) and suppose that all three domains are nonempty. Let $\mathcal{S}^A$ denote the sample from frame A, with inclusion probabilities $\pi_i^A = P(i \in \mathcal{S}^A)$ and sampling weights $w_i^A = 1/\pi_i^A$. Corresponding quantities for frame B are $\mathcal{S}^B$, π_i^B, and w_i^B. Let $\delta_i = 1$ if unit i is in domain ab and 0 otherwise. Then $\hat{t}_a^A = \sum_{i\in\mathcal{S}^A} w_i^A(1-\delta_i)y_i$ and $\hat{t}_b^B = \sum_{i\in\mathcal{S}^B} w_i^B(1-\delta_i)y_i$ estimate the domain totals t_a and t_b, respectively. There are two independent estimators of the population total in the intersection domain ab: $\hat{t}_{ab}^A = \sum_{i\in\mathcal{S}^A} w_i^A\delta_i y_i$ and $\hat{t}_{ab}^B = \sum_{i\in\mathcal{S}^B} w_i^B\delta_i y_i$.

(a) Let $\theta \in [0, 1]$. Show that

$$\hat{t}_{y,\theta} = \hat{t}_a^A + \theta \hat{t}_{ab}^A + (1-\theta)\hat{t}_{ab}^B + \hat{t}_b^B$$

is an unbiased estimator of $t_y = \sum_{i=1}^N y_i$ with

$$V(\hat{t}_{y,\theta}) = V\left[\hat{t}_a^A + \theta \hat{t}_{ab}^A\right] + V\left[(1-\theta)\hat{t}_{ab}^B + \hat{t}_b^B\right].$$

(b) Show that $V(\hat{t}_{y,\theta})$ is minimized when

$$\theta = \frac{V(\hat{t}_{ab}^B) + \text{Cov}\,(\hat{t}_b^B, \hat{t}_{ab}^B) - \text{Cov}\,(\hat{t}_a^A, \hat{t}_{ab}^A)}{V(\hat{t}_{ab}^A) + V(\hat{t}_{ab}^B)}.$$

4. (Requires calculus.) For the situation in Exercise 3, suppose that the total data-collection budget available for data collection is a fixed number C_0 and that an SRS is to be taken from each frame. Let c_A denote the cost of collecting data for one observation from frame A and let c_B denote the cost of collecting data for one observation from frame A, so that where $C_0 = c_A n_A + c_B n_B$, where n_A is the number of units to be sampled from frame A and n_B is the number of units to be sampled from frame B. Let α be a fixed number in [0,1], and suppose that

   (i) For an SRS of size n_A from frame A, $V\left[\hat{t}_a^A + \alpha \hat{t}_{ab}^A\right] = V_A/n_A$.

   (ii) For an SRS of size n_B from frame B, $V\left[\hat{t}_b^B + (1-\alpha)\hat{t}_{ab}^B\right] = V_B/n_B$.

   Show that $V\left[\hat{t}_a^A + \alpha \hat{t}_{ab}^A + (1-\alpha)\hat{t}_{ab}^B + \hat{t}_b^B\right]$ is minimized when

$$n_A = C_0 \frac{\sqrt{V_A/c_A}}{\sqrt{c_A V_A} + \sqrt{c_B V_B}}$$

   and

$$n_B = C_0 \frac{\sqrt{V_B/c_B}}{\sqrt{c_A V_A} + \sqrt{c_B V_B}}.$$

5. The estimator in Exercise 26 of Chapter 6 for indirect sampling can be applied to network sampling (Lavallée, 2007) to give the estimator in Sirken (1970). In the context of network sampling, $\mathcal{U}^A$ is the sampling frame population for the initial sample and $\mathcal{U}^B$ is the population of interest, with M elements. The links ℓ_{ik} define the networks: $\ell_{ik} = 1$ if person k in $\mathcal{U}^B$ is in the network of unit i in $\mathcal{U}^A$. Thus, $L_k = \sum_{i=1}^N \ell_{ik}$ is the multiplicity for person k.

   (a) Show that $\hat{t}_{y,\text{net}}$ in Equation (14.1) equals the estimator $\hat{t}_y$ given in Exercise 26(a) of Chapter 6. Consequently, $\hat{t}_{y,\text{net}}$ is an unbiased estimator of t_y.

   (b) Suppose that $\mathcal{U}^A = \mathcal{U}^B$ is a population of N persons, and the sample from $\mathcal{U}^A$, $\mathcal{S}^A$, is an SRS of size n. Let $y_k = 1$ if person k has the rare characteristic and 0 otherwise. Find $V(\hat{t}_{y,\text{net}})$.

   (c) How does the variance in (b) compare with the variance of $\hat{t}_y = \sum_{i \in \mathcal{S}^A} \frac{N}{n} y_i$, which uses only information from $\mathcal{S}^A$?

6. Consider a stratified sample in which an SRS of n_h psus is selected from the population of N_h psus in stratum h, for $h = 1, \ldots, H$. We wish to estimate the mean of domain d.

   (a) Find $\hat{V}(\hat{\bar{y}}_d)$ using linearization.

(b) Now suppose that a data analyst creates a new data set by deleting observations that are not in domain d. If you (incorrectly) act as though this is the full data set, what is the estimated variance of $\hat{\bar{y}}_d$?

(c) Show that the estimators of the variance in (a) and (b) are unequal if some sampled psus have no observations in domain d. The correct variance estimator is given in (a) and (14.4).

7. Estevao and Särndal (1999) and Hidiroglou and Patak (2004) studied the use of auxiliary information in domain estimation, which can reduce the variance of the direct domain estimator $\hat{t}_{yd}$ in Section 14.2.1.

(a) If the population total for an auxiliary variable x, t_x, is known, we may use the ratio estimator

$$\hat{t}_{ydr1} = \hat{t}_{yd}\frac{t_x}{\hat{t}_x}.$$

If the sample size in domain d is sufficiently large to use linearization, what is $\hat{V}(\hat{t}_{dr1})$? Does $\hat{t}_{ydr1}$ have the benchmarking property?

(b) If we know the population total of x for each domain d, with $t_{xd} = \sum_{i=1}^{N} a_{id}x_i$, then we can use a domain-specific ratio estimator

$$\hat{t}_{ydr2} = \hat{t}_{yd}\frac{t_{xd}}{\hat{t}_{xd}}.$$

What is $\hat{V}(\hat{t}_{ydr2})$? Does $\hat{t}_{ydr2}$ have the benchmarking property?

8. (Requires calculus.) Consider the Fay–Herriot model in (14.7). Suppose that ψ_d, σ_v^2, and $\boldsymbol{\beta}$ are known.

(a) Let

$$\tilde{\theta}_d(a) = a\hat{\bar{y}}_d + (1-a)\mathbf{x}_d^T\boldsymbol{\beta}$$

with $a \in [0,1]$. Show that, under the model in (14.7), $E_M[\tilde{\theta}_d(a) - \theta_d] = 0$ for any $a \in [0,1]$.

(b) Show that $V_M[\tilde{\theta}_d(a) - \theta_d]$ is minimized when $a = \alpha_d$ and that $V_M[\tilde{\theta}_d(\alpha_d) - \theta_d] = \alpha_d\psi_d$. Consequently, under the model, $V_M[\tilde{\theta}_d(\alpha_d) - \theta_d] \le V_M[\hat{\bar{y}}_d - \theta_d]$.

D. Projects and Activities

9. Construct a population with 20 strata. Each stratum has 8 psus and each psu has 3 secondary sampling units (ssus), so that the population has a total of 480 ssus. Observation j of psu i in stratum h has $y_{hij} = h$, for $h = 1, \ldots, 20$, so that all observations in stratum 1 have the value 1, all observations in stratum 2 have the value 2, and so on. Within each stratum, all observations in psus 1–4 are in domain 1, and all observations in psus 5–8 are in domain 2.

(a) Select a one-stage stratified sample from the population by selecting an SRS of two psus from each stratum and including all ssus within the selected psus in the sample. Your sample should have 120 observations. Estimate the population mean for each domain along with its standard error.

(b) Repeat (a) for a second stratified sample, selected independently (that is, use a different random seed). Compare the domain means from this sample with those from (a). Do the domain means vary from sample to sample?

(c) Now create a new data set for your sample in (a) that consists only of observations in domain 1, by deleting all the observations in domain 2. What is the estimated domain mean from this data set? What is the standard error using this data set, and why is it incorrect?

10. *Contact tracing.* Health departments often use a technique known as contact tracing to help curb the spread of a disease. Persons known to have the disease are asked to list other persons they have been in contact with. Those persons are then notified of their potential exposure to the disease and, if they test positive for the disease, asked to list persons they have been in contact with. The process is continued with each contact who tests positive for the disease.

    Which sampling method described in this chapter corresponds to contact tracing? What are the potential biases of the procedure for estimating the prevalence of the disease? For estimating characteristics of persons with the disease? Compare the advantages and disadvantages of the method with those from taking a probability sample of persons.

11. *Activity for course project.* Are there rare populations of interest for the survey you studied in Exercise 35 of Chapter 7? If so, what design features were used in the survey to sample members of the rare population?

15

Nonprobability Samples

'Let us understand one another, Captain Cunningham,' she said in her quiet voice. 'If you employ me, you will be employing me to discover facts. If I discover anything about these people, you will have the benefit of my discovery. It may be what you are expecting, or it may not. People are not always pleased to know the truth.' Miss Silver nodded her head in a gentle deprecating manner. 'You've no idea how often that happens. Very few people want to know the truth. They wish to be confirmed in their own opinions, which is a very different thing — very different indeed. I cannot promise that what I uncover will confirm you in your present opinion.'

—Patricia Wentworth, *The Case is Closed*

Probability sampling is a recent innovation in the history of statistics. For most of history, people—if they relied on data at all—collected a convenience sample to study the topic at hand. This is still the norm in many fields. Clinical trials in medicine often rely upon volunteers who may differ systematically from others in the population. Psychology is famous for its experiments and studies conducted on convenience samples of American undergraduate students taking Psychology 101.

For all of these data collections, however, investigators must make strong assumptions to apply the study results to population members who were not in the sample. In some cases, those assumptions may be true, or approximately true. It may be perfectly reasonable to assume that cancer patients who did not participate in a clinical trial will respond to radiation therapy similarly to the patients who did participate in the trial. In medicine, researchers continue to accrue data to monitor the performance of new therapies that have been validated in clinical experiments. But without additional data, one does not *know* that conclusions drawn from a convenience sample generalize to nonparticipants. When the only data available are from the convenience sample, there may be no way to assess the assumption that nonparticipants resemble the participants.

Probability sampling was developed to provide a mathematically rigorous way to make inferences about units that are not in the sample *without having to collect data from them or make assumptions about them.* It allows us to generalize from what we have seen to what we have not seen—with a measure of how accurately statistics from the sample describe the population.

But what about **nonprobability samples**—samples that are not collected by probability sampling? This chapter discusses the bias and mean squared error (MSE) of estimates from nonprobability samples, and presents methods that have been proposed to adjust (at least partially) for selection bias in samples that are not selected randomly. The methods are similar to those used in Chapter 8 to adjust for nonreponse in probability samples, and Section 15.4 compares low-response-rate probability samples with pure convenience samples.

First, though, let's look at some different types of nonprobability samples.

DOI: 10.1201/9780429298899-15

15.1 Types of Nonprobability Samples

In probability sampling, units are selected randomly from a population with a predetermined selection probability. This feature of the sample design ensures that a probability sample with full response has no selection bias. When accurate data are obtained from every unit selected for a probability sample, the only type of error is sampling error, which can be estimated using the methods in Chapter 9.

Nonprobability samples are not guaranteed to be free of selection bias, but some types may have less bias than others. In addition, even if they have selection bias, nonprobability samples may be useful for answering questions of interest about a population, and they are often easier and less expensive to collect than probability samples. A medical study may be conducted on volunteers, but if an experimental treatment reduces migraines in the study patients randomly assigned to that treatment, it may also be effective for persons not in the study.

15.1.1 Administrative Records

Example 15.1. Most police departments collect data on the number of crimes reported to the department. In the United States, the Uniform Crime Reports contain national and state estimates of the number of crimes of different types that are known to the police. But these records do not include all crimes. Some crimes are not reported to the police; others may be reported to the police but not included in the police department's statistics. Other crimes may be recorded incorrectly: for example, an attempted burglary may be misclassified as vandalism (Lohr, 2019a). In addition, some police agencies do not send their crime statistics to the Uniform Crime Reporting system, so the data from those agencies are missing and must be imputed.

Thus, while the Uniform Crime Reports contain information about a large number of crimes known to law enforcement agencies, they exclude crimes unknown to or unreported by those agencies. An increase in police-recorded crime from one year to the next does not necessarily mean that the total number of crimes in the area served by the law enforcement agency increased. The crime rate calculated from the agency's data may have increased because a higher percentage of the crimes in the area were reported to the police—that is, the undercoverage of the data set decreased—rather than because the actual amount of crime increased. ■

The Uniform Crime Reports are an example of **administrative records**: data that are collected by a government or a non-governmental organization by law or for operating a program. Other examples include tax records, electronic medical records, birth and death certificates, school records of children, and social service program data. Some countries keep population registers containing information on every person in the population; each person is assigned a unique identification number that allows information to be easily updated and to be linked across registers (Wallgren and Wallgren, 2014).

Administrative records can be treated as a *census* if the population of interest happens to be the population for which records are collected. If the target population contains units not in the administrative records, however, then the set of administrative records is a *sample* with undercoverage. In the U.S. the set of personal income tax records is a census of the population of persons who file taxes. If you want to know how many taxpayers reported self-employment income last year on their tax forms, the tax records are by far the best source of information. But the tax records undercover the population of all U.S. residents because persons with incomes below fixed thresholds are not required to file taxes. Because of this undercoverage, the number of persons with self-employment income calculated from tax records will underestimate the total number of U.S. residents with self-employment income.

Statistics calculated from administrative records may also be affected by measurement error. Crime misclassification affects the statistics in the Uniform Crime Reports. A death certificate may contain errors in the decedent's race, age, or cause of death. A medical clinic may enter an incorrect treatment code in a patient's electronic medical record. A tax filer may misreport the previous year's income to the taxing authority.

Statistics from administrative records are typically reported without margins of error. That does not mean that the statistics are without error, only that no measure of their likely accuracy is reported. Most of the errors in statistics calculated from administrative data are from measurement error or from under- or over-coverage of the population of interest. An external source of information is needed to study these errors.

Data from credit card transactions, property sales, retail store scanners, traffic sensors, surveillance cameras, locations recorded by mobile telephone service providers, visits to internet sites, and similar sources share features of administrative records. They too contain records of every transaction or activity monitored by the organization collecting the data. A credit card company has records of every purchase made by its customers. Editors of online publications know how many "clicks" each story received. But transaction and sensor data may have undercoverage of the target population (for example, the set of credit card transactions undercovers a target population of all purchases because it does not include purchases made by cash, cryptocurrencies, or other means) and measurement error. Moreover, for some types of transaction data, unlike for administrative data sets with clear rules for inclusion in the data, it may be difficult to specify the sampled population. What, for example, is the population corresponding to a convenience sample of persons who sell items on an online auction website?

15.1.2 Quota Samples

Many samples that masquerade as stratified random samples are actually quota samples. In **quota sampling**, the population is divided among subpopulations just as in stratified random sampling (in quota sampling, the subpopulations are called **quota classes** instead of strata), but with one important difference: Probability sampling is not used to choose quota class members for the sample.

In quota sampling, a specified number (quota) of observations from each quota class is required in the final sample. For example, to obtain a quota sample with $n = 4000$, you might specify that the sample should contain 1000 white males, 1000 white females, 1000 men of color, and 1000 women of color, but give no further instructions about how these quotas are to be filled. The sample could consist of the first 1000 persons who respond to the online survey invitation from each quota class.

In extreme versions of quota sampling, a convenience sample is taken within each quota class—the choice of units for the sample is entirely at the discretion of the interviewer, or the sample consists of volunteers. Thus, quota sampling is not a form of probability sampling—we do not know the probabilities with which each individual is included in the sample. It is often used when probability sampling is impractical, overly costly, or considered unnecessary, or when the persons designing the sample just do not know any better.

Example 15.2. The 1945 survey on reading habits taken for the Book Manufacturer's Institute (Link and Hopf, 1946), like many surveys in the 1940s and 1950s, used a quota sample. Some of the classifications used to define the quota classes were area, city size, age, sex, and socioeconomic status; a local supervising psychologist in each city determined the blocks of the city in which interviewers were to interview people from a specified socioeconomic group. The interviewers were then allowed to choose the specific households to be interviewed in the designated city blocks.

The quota procedure followed in the survey did not result in a sample that reflected demographic characteristics of the 1945 U.S. population. Table 15.1 compares the educational background of the survey respondents with figures from the 1940 U.S. Census, adjusted to reflect the wartime changes in population.

TABLE 15.1
Education level information for Example 15.2. Source: Link and Hopf (1946).

Distribution by Educational Levels	4,000 People Interviewed (%)	U.S. Census, Urban and Rural Nonfarm (%)
8th grade or less	28	48
1–3 years high school	18	19
4 years high school	25	21
1–3 years college	15	7
4 or more years college	13	5

The oversampling of better-educated persons casts doubt on many of the statistics given in Link and Hopf (1946). The study concluded that 31% of "active readers" (those who had read at least one book in the past month) had bought the last book they read, and that 25% of all last books read by active readers cost $1 or less. Who knows whether a stratified random sample would have given the same results? ■

The big drawback of quota sampling is the potential for selection bias. Interviewers with complete discretion are likely to choose the most accessible members of the population—for instance, persons who are easily reached by telephone, households without menacing dogs, or areas of the forest close to the road. The most accessible members of a population are likely to differ in a systematic way from less accessible members. In Example 15.2, the interviewers unwittingly oversampled better-educated persons. Now, a stratified random sample could be designed to oversample better-educated persons, but in stratified random sampling, the selection probabilities would be known and used in the estimation to compensate for the oversampling. With a quota sample, the degree of inadvertent oversampling of different subpopulations is unknown.

Thus, unlike in stratified random sampling, we cannot say that the estimator of the population total from quota sampling is unbiased over repeated sampling—one of our usual criteria of goodness in probability samples. In fact, in quota samples, we cannot measure sampling error over repeated samples and we have no way of estimating the bias from the sample data. Since selection of units is up to the individual interviewer, or depends on who happens to decide to participate, we cannot expect that repeating the sample procedure will give similar results.

Because we do not know the probabilities with which units were sampled, we must take a model-based approach, and make strong assumptions about the data structure, when analyzing data from a quota sample. The model generally adopted is that of Section 3.6—within each quota class, the random variables generating the subpopulation are assumed to be independent and identically distributed. Such a model implies that any sample of units from the quota class will be representative of the class. If the model holds, then quota sampling will likely give good estimates of the population quantity. If the model does not hold, then the estimates from quota sampling may be badly biased. The bias can sometimes be reduced by calibrating to known population totals, as we shall discuss in Section 15.3.1.

While quota sampling is not as good as probability sampling under ideal conditions, it may give better results than a completely haphazard sample because it at least forces the inclusion of members from different quota classes. Quota samples are often less expensive than probability samples. The quality of data from quota samples can be improved by allowing the interviewer less discretion in the choice of persons or households to be included in the sample.

Example 15.3. Sanzo et al. (1993) used a combination of stratified random sampling and quota sampling for estimating the prevalence of *Coxiella burnetii* infection within the Basque country in northern Spain. *Coxiella burnetii* can cause Q fever, which can lead to complications such as heart and nerve damage. Reviews of Q fever patient records from Basque hospitals showed that about three-fourths of the victims were male, about half were between 16 and 30 years old, and victims were disproportionately likely to be from areas with low population density.

The authors stratified the target population by population density and then randomly selected health care centers from the three strata. In selecting persons for blood testing, however, "a probabilistic approach was rejected as we considered that the refusal rate of blood testing would be high" (p. 1185). Instead, they used quota sampling to balance the sample by age and gender; physicians asked patients who needed laboratory tests whether they would participate in the study, and recruited subjects for the study until the desired sample sizes in the six quota classes were reached.

Because a quota sample was taken instead of a probability sample, persons analyzing the data must make strong assumptions about the representativeness of the sample in order to apply the results to the general population of the Basque country. First, the assumption must be made that persons attending a health clinic for laboratory tests (the sampled population of the study) are neither more nor less likely to be infected than persons who would not be visiting the clinic. Second, one must assume that persons who are requested and agree to do the study are similar (with respect to infection) to persons in the same quota class having laboratory tests that do not participate in the study. These are strong assumptions. The authors of the article argue that the assumptions are justified, but of course they cannot demonstrate that the assumptions hold unless follow-up investigations are done.

If they had taken a probability sample of persons instead of the quota sample, they would not have had to make these strong assumptions. A probability sample of persons, however, would have been exorbitantly expensive when compared with the quota sampling scheme used, and a probability sample would also have taken longer to design and implement. With the quota sample, the authors were able to collect information about the public health problem; it is unclear whether the results can be generalized to the entire population, but the data did provide a great deal of quick information on the prevalence of infection that could be used in future investigations of who is likely to be infected, and why. ■

A quota sample, while easier to collect than a probability sample, suffers from the same disadvantages as other nonprobability samples. Some survey organizations use quota sampling to recruit volunteers for online surveys; they accumulate respondents until they have specified sample sizes in the desired demographic quota classes. In such online surveys, the respondents in each quota class are self-selected—if, as argued by Couper (2000), internet users who volunteer for such surveys differ from members of the target population in those quota classes, results will be biased.

Other survey organizations recruit a large panel of potential respondents and draw a sample from the panel when it is desired to take a survey. Whether this method yields a probability sample or a quota sample depends on how the initial panel is recruited, as seen in the next example.

Example 15.4. *Online panel surveys.* An online panel is a group of persons who have been recruited by a survey organization for the purpose of taking surveys. Then, when a topic comes up for which polling is desired, the organization asks a sample of persons from the panel who meet the survey criteria to take the poll. Having a large group of potential respondents in the panel saves the expense of selecting and contacting a new sample for every poll.

Some organizations recruit panelists using probability sampling. They randomly select addresses (or, sometimes, telephone numbers) from a sampling frame and invite persons at the sampled addresses to join the panel. The sampling frame of addresses contains almost everyone, so undercoverage is low and unaffected by internet access. Panelists then complete self-administered surveys over the internet; those without internet access at home are provided with an internet connection and electronic device that can be used to take the surveys. This prevents, in theory at least, undercoverage of non-internet households. The panel is still subject to nonresponse, though, since not everyone invited to be on the panel does so. Panel members may also decline to respond to individual polls, causing additional nonresponse.

Nonprobability panels, sometimes called opt-in panels, are formed from volunteers who sign up on the survey organization's website. Then, when a survey is taken, the organization selects a subsample of panel members with the desired sample size in each quota class of interest. The original panel consists of volunteers, so any subsample of the panel also consists of volunteers—even if the subsample is selected randomly from the panel members within each quota class. The set of persons taking any individual survey is thus a quota sample. ■

15.1.3 Judgment Samples

A *judgment sample* is so called because persons conducting the survey use their judgment, or the judgment of others, to decide which population members should be in the sample. These are sometimes called purposive samples, because the sample is purposively selected.

Example 15.5. One of the most famous judgment samples in history was described by Neyman (1934), in his landmark article setting out the theory of stratified random sampling. In November 1926, after the summaries from the 1921 Italian General Census had been published, the original data sheets containing the information about individual families were to be destroyed in order to make room in the storage facility for new material. Statisticians Corrado Gini (famous for developing the Gini coefficient of inequality) and Luigi Galvani wanted to "obtain a representative sample of the whole country, with respect to its main demographic, social, economic, and geographic characteristics" in case information from the census was needed for research in the future (Gini and Galvani, 1929, p. 1, translated S. Lohr). They argued that a sample of about 15 percent of the records would be large enough to produce reliable results but not too large to store.

Because the data had already been tabulated, the records were sorted by districts (*circondarî*). Gini and Galvani decided to take a sample of districts, instead of sampling individual records, so that they could preserve the ability to make local comparisons within districts. They presented a table of 11 characteristics (reproduced in data file `gini.csv`) for each of the 214 districts.

Gini and Galvani knew that they could take an SRS or a systematic sample. But they calculated that the sampling error from an SRS would be too large for their purposes, and thought they could obtain better representation by choosing the sampled units themselves.

Their purposive selection process was quite onerous and time-consuming—much more difficult than taking any type of randomly selected sample would have been. In fact, they constructed an early example of a purposive balanced sample (see Langel and Tillé, 2011) of districts, which was designed to resemble the population with respect to seven "control" characteristics (birth, death, and marriage rates; percentage of males over ten years old who were engaged in agriculture; percentage of urban population; average income; and altitude).

The judgment sample worked beautifully for estimating the 1921 Census population values of the control variables. Neyman (1934), however, demonstrated that other statistics from the sample differed greatly from the population values. He proved mathematically that using random, rather than purposive, selection could have prevented this problem. ■

Judgment samples can exhibit strong selection bias, and in some cases can be deliberately engineered to yield a desired result. When resources permit only a small sample, however, or in preliminary stages of an investigation, a judgment sample often serves the purposes of the study as well as a probability sample would. Hansen and Madow (1978), who in general strongly advocated taking probability samples, made an exception when the sample size is small, writing: "We see losses and no benefit to be gained by altering the probability sampling model for sample surveys, except where quite small samples are used" (p. 142).

Example 15.6. Rivera (2016) selected a judgment sample of six New Jersey municipalities to investigate residents' experiences with applying for, and obtaining, federal disaster aid following Hurricane Sandy. Three of the municipalities were in the northern part of the state, and the other three were in the southern part. Random digit dialing, including both landline and cellular telephone numbers, was used to select a sample of residents within each municipality.

The municipalities were deliberately selected to represent the different types and extent of damage from the hurricane. Results from the sample, therefore, cannot be generalized to the population of New Jersey residents. For this study, however, the purposive selection met the research needs and allowed the researcher to investigate equity issues in aid distribution with a sample of six units. Moreover, because a probability sample was selected within each municipality, the data can be used to study contrasts among municipalities with different levels of hurricane damage; the data actually comprise six separate probability samples, each representing its municipality. ■

A *homogeneous sample* is a special case of a judgment sample. Homogeneous samples are purposely selected so that the individuals in the sample share similar traits. For example, the members of a homogeneous sample might all have the same age or live in the same city. Many psychology experiments have traditionally relied on homogeneous samples—undergraduate students in introductory psychology classes.

The rationale for using homogeneous samples is that they can be useful for experiments: the treatment comparison has small variance because all of the participants in the experiment are similar on the homogeneity factors. But that means that it is difficult to generalize from the sample to a larger population, since the sample reflects only a homogenous subset of the population.

15.1.4 Convenience Samples

In a sense, any nonprobability sample could be called a convenience sample, but, following Baker et al. (2013), we consider a convenience sample to be one in which the primary consideration for sample selection is the ease with which units can be recruited or located. A nonprobability sample that does not fit into one of the other categories in this section can be considered a pure convenience sample. Examples include:

- "Mall surveys," in which the survey conductor stands in a shopping mall, outside a library, or in another public place and asks passers-by to take a survey.

- Surveys for which the survey conductor asks friends and acquaintances to take the survey.
- Focus groups selected to provide responses to issues presented in a group setting. Focus groups are often used to pretest questionnaires or to obtain in-depth interviews from a small number of persons.
- Internet surveys that invite all website visitors to take the survey.
- Call-in, fax-in, or text-in polls, in which a survey invitation is broadcast to a wide audience and anyone who desires can respond.
- Snowball samples, described in Section 14.1.6, which obtain a sample of a rare population by having initial sample members recruit additional members.
- Samples of volunteers used in medical studies. The first Nurses' Health Study was launched in 1976 to investigate the potential long-term health consequences of oral contraceptives (Nurses' Health Study, 2019). Nurses were chosen for study because they were considered to be knowledgeable about health and able, because of their training, to answer questions about their health outcomes. Subsequent Nurses' Health Studies have asked participants about dietary patterns, lifestyle, and environment, and then explored the relationship between those attributes and later health outcomes.

Convenience samples are distinct from administrative or commercial records in that the researcher selects a convenient sample for the specific study. Administrative or commercial records are collected for an administrative or business purpose and often have a well-defined sampled population and criteria for inclusion.

Convenience samples can provide valuable information for science—the Nurses' Health Study, for example, has made tremendous contributions toward understanding relationships between diet, lifestyle, and health outcomes. But additional evidence is needed to show that the findings apply to persons who are not nurses, who might have different health knowledge and experiences. All that is known for certain is that statistics calculated from the Nurses' Health Study apply to the sample of nurses in the study.

15.2 Selection Bias and Mean Squared Error

Probability samples with full response allow the survey sampler to produce estimates of population means, quantiles, totals, and other population characteristics that are known to be approximately unbiased. The unbiasedness is a mathematical consequence of the procedure used to select the sample, as shown in Sections 2.9 and 6.6. We know the probability distribution of the random variables used for inference; these are the variables Z_i that indicate inclusion in the sample.

Recall from (2.6) in Section 2.2 that the MSE of a statistic T is:

$$\text{MSE}[T] = E\left[(T - E[T])^2\right] = V(T) + [\text{Bias}(T)]^2. \tag{15.1}$$

For most estimators used with a probability sample, the bias is small and the MSE can be approximated by the variance, which decreases as the sample size increases.

A statistic estimated from a large nonprobability sample has negligible variance because the sample size is so large. Almost all of the error from a set of administrative records or large convenience sample comes from the second, bias, term in (15.1). Selection bias occurs when the nonprobability sample systematically, and without the intention of the survey conductor, underrepresents parts of the population of interest. In general, we do not know how biased an individual statistic is, but we can study effects of bias in a general setting by considering what happens when we know the entire population.

15.2.1 Random Variables Describing Participation in a Sample

In this section, we study bias in the framework of Section 8.4 by viewing a nonprobability sample as a census with nonresponse from some population members. A random variable R_i—called the **participation indicator**—describes whether unit i from the population is in the sample: $R_i = 1$ if unit i is in the sample, and $R_i = 0$ if unit i is not in the sample. This structure assumes that a population member will appear at most once in the sample; if a population member can appear more than once in the sample, the random variables describing participation need to count the number of times unit i participates. As in Chapters 2 through 7, we assume that the characteristic of interest y_i is measured without error if unit i is in the sample.

The set of random variables Z_i, used in Chapters 2 through 14 to study properties of probability samples, can be considered to be a special case of the set of random variables R_i. There are two main differences between the general set of random variables R_i, used to study properties of nonprobability samples, and the variables Z_i, used to study properties of probability samples. The first is the probability mechanism for participation in the sample, which is well defined for a probability sample but may be unknown for a nonprobability sample. The second is that in a probability sample, every unit has a positive probability of being selected for the sample; in a nonprobability sample, some population members may have no chance of being in the sample.

Probability mechanism for sample participation. Recall the basic framework for probability sampling in Section 2.2. The probability that unit i is included in a probability sample is

$$P(Z_i = 1) = \pi_i = \sum_{\substack{\text{all possible samples} \\ \mathcal{S} \text{ containing unit } i}} P(\mathcal{S}).$$

Because the population is finite, there are a finite number of possible samples that could be selected. Each possible sample $\mathcal{S}$ is selected with known probability $P(\mathcal{S})$. The probability that unit i is in the sample is the sum of the probabilities of the samples that contain unit i. You can think of the inclusion probability π_i in the context of repeatedly drawing samples from the population under the sampling design. Randomly select a sample using the design. Then put the units back in the population and randomly select a second sample. Continue with a third sample, a fourth sample, and so on. Each time you select a sample, observe whether it contains unit i. The inclusion probability π_i is the limiting value (as the number of samples drawn goes to infinity) of the proportion of samples drawn that include unit i.

One can also conceive of participation probabilities $P(R_i = 1)$ for a nonprobability sample in a repeated sampling context. Think of repeating the procedure used to obtain the nonprobability sample over and over and over again. For each possible sample that results, observe whether unit i is in the sample. The participation probability $P(R_i = 1)$ is again the limiting proportion of samples obtained that include unit i.

But what does it mean to "repeat" the procedure? After all, the sample is collected once and, unlike in probability sampling, there is no random selection that allows one to think about alternative samples that could have been chosen. What, exactly, might change if the selection of a nonprobability sample were repeated? One framework is to consider the probability distribution of the participation variables R_i in a science fiction "alternative universes" sense.

In the episode "Parallels" of the television series *Star Trek: The Next Generation*, Lieutenant Worf shifts among parallel, alternative universes. Each differs in some small way from Worf's original universe. In one, the starship *Enterprise* has a different captain; in another, a character who left the ship in the "real" timeline is now back on board; in a third, Worf is married to counselor Deanna Troi. The producers of the show, however, did not make the alternative universes *too* different from the universe usually portrayed in the series. There were no universes in which the *Enterprise* did not exist, or in which the crew members were lizards.

If conditions in an alternative universe were different, would unit i be in the sample? Under other circumstances, person i might click on the web link to take the survey. Or person i may regularly visit the library where the survey is being taken, but just does not happen to visit at the time of the survey.

Of course, this requires one to think about how different the universes are allowed to be. In a probability sample, the alternative universes that we consider are those in which the population $\mathcal{U}$ and the characteristics y_i are the same; the only differences are that the dice are rolled differently, resulting in a different subset of the population as the sample. The probability distribution of the Z_is is known, since the sampler has determined it when establishing the probability sampling design.

The same population restrictions—of a fixed population $\mathcal{U}$ with characteristics y_i—can be assumed to hold in every alternative universe for a nonprobability sample. But within that context, a wide range of probability distributions might be adopted to describe participation in the survey, and the "true" probability distribution is often unknown.

Remember, the final sample consists of units for which the variable R_i took on the value 1. Units for which R_i took on the value 0 were not observed in the sample, and the sample contains no information about them. Knowledge or speculation about the distribution of the participation indicators R_i for a sample must come from information external to that provided in the sample. In a probability sample, that external knowledge comes from the procedure used to select the sample. For a nonprobability sample, there might be no or little external information about the participation probabilities.

Let's look at some examples of what might be known about the probability distribution of the R_is for different types of nonprobability samples.

Example 15.7. *Quota sampling.* A quota sampling procedure puts restrictions on the probability distribution of the R_is but does not fully determine it. The variables R_i for population members within a quota class must sum to the sample size taken within that quota class. Suppose a quota sample of 50 men and 100 women is taken from a population of 1,000 men and 2,000 women. Let $x_i = 1$ if person i is a man and $x_i = 0$ if person i is a woman. The quotas constrain the random variables to satisfy $\sum_{i=1}^{N} R_i x_i = 50$ and $\sum_{i=1}^{N} R_i(1 - x_i) = 100$, so that the participation indicators R_i must sum to 50 for men and 100 for women. This in turn implies that the participation probabilities $P(R_i = 1)$ for all men in the population must sum to 50, and the participation probabilities for all women in the population must sum to 100. As long as these restrictions are satisfied, however, the random variables in each quota class can have any distribution. Selecting the highest-income persons of each gender satisfies the quota sampling procedure but will give biased estimates of average income. ∎

Example 15.8. *Internet survey of volunteers.* An online survey where website visitors click on a link to take a survey may provide little information about the participation probabilities. Suppose that twice as many men as women have participated in a survey where the target population has approximately equal numbers of men and women. That implies that the sum of the participation probabilities for all men in the population is likely greater than the sum of the participation probabilities for all women in the population, but does not tell the probability that an individual man or woman in the population has participated in the survey.

Indeed, some surveys allow the same person to participate multiple times, as discussed in Section 1.3.3. Even if the survey conductors remove duplicate responses that come from the same IP address, a respondent can use different computers for each response, or a determined organization can encourage its members to flood a survey. When multiple responses are possible, properties of the nonprobability sample are determined by the distribution of random variables Q_i, the number of times unit i participates in the survey, rather than the distribution of random variables R_i (see Exercise 18). ■

Example 15.9. *Administrative records.* Participation in a set of administrative records often depends upon well-defined criteria, and those criteria provide information about the distribution of the random variables R_i for the population. The criteria also prevent the problem discussed in Example 15.8 where one person or organization can deliberately skew a statistic by responding multiple times. Most people have at most one driver's license and file at most one tax return. When a person appears more than once, as can occur in electronic medical records or social service administrative data, it may be possible to identify and link the set of records that belong to the same person.

The variables R_i do not have to be random: it might occur that in any conceivable universe, one set of units will always participate (have $P(R_i = 1) = 1$) and another set will always be excluded (have $P(R_i = 1) = 0$). Some administrative record data sets can be considered to be of this type. If $\mathbf{x}$ denotes the characteristics associated with requirement to file a tax return (in the U.S., these include marital status and income), and the law says that persons with characteristics $\mathbf{x}$ in the set $\mathcal{A}$ must file a return, then it might be reasonable to assume that $P(R_i = 1|\mathbf{x}_i \in \mathcal{A}) \approx 1$ (approximately equal because in reality, not all persons required by law to file a return actually do so). For some persons with $\mathbf{x}_i \notin \mathcal{A}$, the probability of being in the tax record data set can be considered to be zero. Other persons with $\mathbf{x}_i \notin \mathcal{A}$ may file a tax return anyway, and might have a participation probability between 0 and 1. ■

Participation probabilities can be zero. One of the characteristics of a probability sample with full response is that *every* unit in the population has a nonzero chance of being included in the sample. This is what enables us to say that the estimated population total $\hat{t} = \sum_{i=1}^{N} Z_i y_i / P(Z_i = 1)$ from the sample is an unbiased estimator of the population total $\sum_{i=1}^{N} y_i$ *for any* possible variable y that is studied, and that the estimated mean is approximately unbiased for the population mean.

In nonprobability samples, some population units may have a zero probability of participating in the sample. The same may be true in probability samples with nonresponse—some persons selected to be in the sample may never agree to participate, no matter what efforts are made to persuade them. That means that some units in the population may have absolutely no chance of appearing in the sample. In surveys or administrative data sets of people, these might be people who do not have internet access, or who do not file taxes, or who never go to the library where the survey is being taken, or who simply never participate in surveys. Wildlife surveys may exclude areas that are inaccessible by road. A convenience sample used for quality control that samples the first batch of pizzas made every day will never contain pizzas from later in the day.

For any sampling procedure in which some units have $P(R_i = 0) = 1$, and for any estimator, one can always conceive of a potential response variable y for which the estimator is biased (see Exercise 23). The only hope is that for the response variables of interest, the population units with zero chance of participating are similar enough to units in the sample that the bias is negligible.

15.2.2 Bias and Mean Squared Error of a Sample Mean

Many possible frameworks can be used to describe the random variables R_i that indicate participation in the sample. For some nonprobability samples, the R_is are deterministic (that is, for any particular unit i in the population either $P(R_i = 1) = 0$ or $P(R_i = 1) = 1$); for others, the probability distribution of the R_is is completely unknown. Even if the probability distribution is unknown, however, we can still explore properties of estimators under different scenarios. In this section, using the framework in Meng (2018), we examine properties of the sample mean under an assumed probability distribution for the participation indicators R_i.

Let $\mathcal{C}$ represent the set of units, out of the population $\mathcal{U}$ of size N, that are in the nonprobability sample (the $\mathcal{C}$ stands for "convenience"). Then the random variable R_i is observed to be 1 for every unit in $\mathcal{C}$, and is observed to be 0 for every population unit not in $\mathcal{C}$. The sample mean is:

$$\bar{y}_{\mathcal{C}} = \frac{1}{n}\sum_{i \in \mathcal{C}} y_i = \frac{\sum_{i=1}^{N} R_i y_i}{\sum_{i=1}^{N} R_i}. \tag{15.2}$$

How accurate is the mean of a sample $\bar{y}_{\mathcal{C}}$ as an estimator of the population mean $\bar{y}_{\mathcal{U}} = \sum_{i=1}^{N} y_i / N$? Let's look at the error as a function of the random variables R_i, without worrying about their probability distribution for the moment. Assume there are no measurement errors and that the sample size n is fixed by the investigator, so that

$$\sum_{i=1}^{N} R_i = n$$

for any possible sample. Let

$$S_y = \sqrt{\frac{1}{N-1}\sum_{i=1}^{N}(y_i - \bar{y}_{\mathcal{U}})^2} \quad \text{and} \quad S_R = \sqrt{\frac{1}{N-1}\sum_{i=1}^{N}(R_i - \bar{R}_{\mathcal{U}})^2}$$

denote the population standard deviations of y and R, respectively (see Equation (2.9)), and let

$$\text{Corr}\,(R, y) = \frac{\sum_{i=1}^{N}(R_i - \bar{R}_{\mathcal{U}})(y_i - \bar{y}_{\mathcal{U}})}{(N-1)S_y S_R}$$

denote the population correlation coefficient of R and y (see Equation (A.10)). When $S_y > 0$ and $n < N$, the error of the sample mean $\bar{y}_{\mathcal{C}}$ for estimating the population mean $\bar{y}_{\mathcal{U}}$ can be written as a product of three factors (see Exercise 14):

$$\bar{y}_{\mathcal{C}} - \bar{y}_{\mathcal{U}} = \text{Corr}\,(R, y) \times \sqrt{\frac{N-1}{n}\left(1 - \frac{n}{N}\right)} \times S_y. \tag{15.3}$$

The three factors are:

A. $\text{Corr}\,(R, y)$, the correlation between the sample participation mechanism and the variable of interest y. Suppose, for example, that $y_i = 1$ if person i has internet access and

$y_i = 0$ if person i does not have internet access. If having internet access is positively correlated with the participation indicator R_i, then persons with internet access are overrepresented in the sample and the estimated proportion of persons with internet access, $\bar{y}_C$, is too large. If having internet access is negatively correlated with the participation indicator, then persons with internet access are underrepresented in the sample and the estimated proportion of persons with internet access is too small.

B. A function of the sampling fraction n/N: $\sqrt{\frac{N-1}{n}\left(1-\frac{n}{N}\right)} = \sqrt{\frac{N-1}{N}\left(\frac{N}{n}-1\right)}$. The second factor in (15.3) is smaller when the sampling fraction n/N is larger. When the sampling fraction equals 1, the sample mean is identical to the population mean and there is no error.

C. The population standard deviation of y, S_y. If all y_i are equal (for the example above, this would occur if everyone in the population has internet access or if no one has internet access), then any sample gives the same value of the sample mean and there is no bias.

Equation (15.3) gives the error for the particular sample that was drawn, where $R_i = 1$ for units in the sample and $R_i = 0$ for units not in the sample. The bias of $\bar{y}_C$ is the expected value of the error in (15.3), where the expectation is over all possible samples that might be chosen:

$$\text{Bias}\,(\bar{y}_C) = E_R\left[\bar{y}_C - \bar{y}_{\mathcal{U}}\right] = E_R\left[\text{Corr}\,(R, y)\right] \times \sqrt{\frac{N-1}{n}\left(1-\frac{n}{N}\right)} \times S_y. \tag{15.4}$$

Similarly, the MSE of $\bar{y}_C$ is the expected value of the square of the error in (15.3):

$$\begin{aligned}\text{MSE}\,(\bar{y}_C) &= E_R\left[(\bar{y}_C - \bar{y}_{\mathcal{U}})^2\right] \\ &= E_R\left[\text{Corr}^2(R, y)\right] \times \frac{N-1}{n}\left(1-\frac{n}{N}\right) \times S_y^2 \end{aligned} \tag{15.5}$$

The first factor in (15.5)—the expected value of the squared correlation between the participation indicators R_i and the variable of interest y_i— is the key to the MSE of a non-probability sample.

Example 15.10. Let's look at the size of the first factor in (15.5) for an SRS of size n, a sample design for which we know the probability distribution of the participation indicators R_i ($= Z_i$ here since this is a probability sample). In Chapter 2, we proved that the sample mean $\bar{y}_{\text{SRS}}$ is unbiased and that its mean squared error is

$$\text{MSE}\,(\bar{y}_{\text{SRS}}) = V(\bar{y}_{\text{SRS}}) = \left(1-\frac{n}{N}\right)\frac{S_y^2}{n}.$$

Thus, for an SRS,

$$\text{MSE}\,(\bar{y}_{\text{SRS}}) = \left(1-\frac{n}{N}\right)\frac{S_y^2}{n} = E_Z\left[\text{Corr}^2(Z, y)\right] \times \frac{N-1}{n}\left(1-\frac{n}{N}\right) \times S_y^2,$$

where $Z_i = 1$ if unit i is in the SRS and 0 otherwise. Solving for the correlation term gives

$$E_Z\left[\text{Corr}^2(Z, y)\right] = \frac{1}{N-1}$$

(also see Exercise 15). Similarly, $E_Z\left[\text{Corr}\,(Z, y)\right] = 0$.

For an SRS, the expected correlation, and expected squared correlation, between the inclusion variables Z_i and y_i does not depend on the variable y at all. It does not matter whether y represents cholesterol level, or income, or political candidate preference, or number of *Star Trek* episodes watched in the past 10 years—the expected correlation of Z and y is *always* zero and the expected squared correlation of Z and y is *always* $1/(N-1)$ for an SRS. ■

Other types of probability samples have $E_Z\left[\text{Corr}^2(Z,y)\right]$ of the same order of magnitude—typically some constant divided by the population size. The expected value $E_Z\left[\text{Corr}\,(Z,y)\right]$ is approximately zero, making the estimator of the mean approximately unbiased. The low value of the expected squared correlation makes information from a full-response probability sample extremely valuable when compared with the amount of information from a typical nonprobability sample. A nonprobability sample can have a much larger value for the first factor in (15.5), and the expected squared correlation also usually depends on which variable y is being studied.

Example 15.11. *Comparing MSEs from SRSs and convenience samples.* In the social sciences, a correlation of 0.10 is often considered to be small. But what happens to the MSE of the sample mean from a nonprobability sample if the correlation between y and participating in the sample is of that size?

Suppose that $n = N/2$ and $E[\text{Corr}^2(R,y)] \geq 0.01$. Then (15.5) implies that

$$\text{MSE}\,(\bar{y}_{\mathcal{C}}) \geq 0.01 \times \frac{N-1}{n}\left(1-\frac{n}{N}\right) \times S_y^2 \approx \frac{S_y^2}{100}.$$

But $S_y^2/100$ is the MSE of the sample mean from an SRS of size 100 (when no fpc is used). Thus, with a correlation considered small in other contexts, the mean from a convenience sample of *half* the population has the same MSE as the mean of an SRS of size 100. This is true for any population size N, no matter how large. If N is one billion and $E[\text{Corr}^2(R,y)] \geq 0.01$, you will get a more accurate estimate of the population mean from an SRS of 101 people than from a convenience sample of *half a billion* people.

In an SRS, or any other type of probability sample, the MSE decreases as the sample size increases, regardless of population size. This occurs because in a probability sample, the random selection guarantees that $E[\text{Corr}^2(R,y)]$ is extremely small—it is $1/(N-1)$ for an SRS, and has a similar order of magnitude for other probability samples. In a nonprobability sample, there is no guarantee that $E[\text{Corr}^2(R,y)]$ decreases as N increases, and so increasing the sample size does not appreciably improve the accuracy until you sample nearly all of the population. ■

Reducing any of the three factors in Equation (15.5) (without increasing any of the others) will decrease the MSE of the estimated mean.

Let's start with the last factor, S_y^2. Typically, the sampler cannot change the population variance S_y^2—it is determined by the values of y_i in the population studied. But some subpopulations may have smaller variance than others. Stratified random sampling or quota sampling may reduce the third factor if the strata, or quota classes, are more homogeneous than the population as a whole. If, for example, the variance is zero within a particular quota class, then any sample will be representative of that class.

The second factor is roughly equal to $N/n - 1$, and the MSE is zero if the entire population is sampled. Some administrative data sets contain all or nearly all of the population of interest. It may be possible to increase the sampling fraction, or to take a supplemental sample of population members missing from $\mathcal{C}$ in order to reduce bias.

The final option is to reduce the expected squared correlation, $E\left[\text{Corr}^2(R,y)\right]$. Probability sampling allows the sampler to control the size of this expected value. But some nonprobability sampling schemes have smaller expected squared correlation than others. The factor may be further reduced by weighting the sample, as discussed in the next section.

15.3 Reducing Bias of Estimates from Nonprobability Samples

Nonprobability samples are not guaranteed to be representative. Nevertheless, it is often desired to use them to make statements about a population. Can that be done without the framework of probability sampling?

Many of the methods used to adjust for nonresponse, described in Chapter 8, can be adapted for nonprobability samples. As with nonresponse, there is no guarantee that the methods remove all selection bias. Two major methods are used in practice. The first, to estimate the probability that a unit self-selects to be in the sample, uses weighting. The second, to use statistical models to estimate the characteristics of units not in the sample, is related to imputation.

15.3.1 Weighting

In a probability sample, the random variable Z_i takes on the value 1 if unit i is in the sample and 0 otherwise. The inclusion probabilities $P(Z_i = 1)$ and joint inclusion probabilities $P(Z_i = 1, Z_j = 1)$ are known for all units in the sample, and these are used to construct the weights $w_i = 1/P(Z_i = 1)$ and calculate standard errors.

We cannot define the weight for unit i in a nonprobability sample to be the reciprocal of the inclusion probability, $1/P(R_i = 1)$, because the probability distribution of the participation indicators R_i is, in general, unknown and some units may have a participation probability of zero. But we may be able to make some reasonable conjectures about the probability distribution, or adjust the sample for observed discrepancies between the sample and population. Even if the same set of units have $R_i = 0$ in any conceivable alternative universe, these may be similar, with respect to y, to other units in the sample, so that weighting adjustments may be able to reduce undercoverage bias as well.

Weighting adjustments for a nonprobability sample set the weight for unit i in the sample equal to the reciprocal of an *estimate* of the unit's participation probability:

$$w_i = 1/P(\widehat{R_i = 1}).$$

Any of the weighting methods discussed in Chapter 8 can be applied to nonprobability samples to estimate the participation probability. Weights can be constructed using post-stratification, raking, generalized regression or calibration, propensity score estimation, or other statistical models.

If poststratification is used, for example, population counts N_h are known from an external source for the H poststrata, exactly as in Section 8.6. The sample $\mathcal{C}$ has n_h units in poststratum h. Then the estimated participation probability for sample unit i in poststratum h is $P(\widehat{R_i = 1}) = n_h/N_h$ and the weight for the unit is $w_i = N_h/n_h$. Poststratification requires the assumption that the participation probability is the same value, n_h/N_h, for every member of poststratum h. This is unlikely to be exactly true—in reality, some members in the poststratum will be more willing to participate than others, and some may have zero probability of participating. But the poststratification weights do capture disparities in participation across poststrata, and thus may result in estimates that have less bias than the unweighted mean $\bar{y}_{\mathcal{C}}$. The closer the estimate $\widehat{P(R_i = 1)}$ is to the true probability $P(R_i = 1)$, the better weighting will perform to reduce bias.

How does weighting affect the bias and MSE of sample statistics? After unit i in the sample is assigned weight w_i, the weighted mean is:

$$\bar{y}_{\mathcal{C}w} = \frac{\sum_{i=1}^{N} R_i w_i y_i}{\sum_{i=1}^{N} R_i w_i}. \tag{15.6}$$

This is of the same form as (15.2), but with $R_i w_i$ substituted for R_i. Unit i in the sample now counts w_i times toward any estimated population characteristic, so if unit i has characteristics that are underrepresented in the sample but a large weight, the weight can at least partially compensate for the underrepresentation.

An argument similar to that showing (15.5) implies that (see Exercise 20):

$$\bar{y}_{\mathcal{C}w} - \bar{y}_{\mathcal{U}} = \text{Corr}\,(Rw, y) \times \sqrt{\frac{N-1}{n}\left\{\frac{n-1}{n}\text{CV}_w^2 + \left(1 - \frac{n}{N}\right)\right\}} \times S_y, \tag{15.7}$$

where

$$\text{Corr}\,(Rw, y) = \frac{\sum_{i=1}^{N}(R_i w_i - \sum_{j=1}^{N} R_j w_j/N)(y_i - \bar{y}_{\mathcal{U}})}{\sqrt{\sum_{i=1}^{N}(y_i - \bar{y}_{\mathcal{U}})^2 \sum_{k=1}^{N}(R_k w_k - \sum_{j=1}^{N} R_j w_j/N)^2}},$$

$\bar{w}_{\mathcal{C}} = (\sum_{i=1}^{N} R_i w_i)/n$ is the average weight of the sample units, and

$$\text{CV}_w = \frac{1}{\bar{w}_{\mathcal{C}}}\sqrt{\frac{1}{n-1}\sum_{i=1}^{N} R_i \left(w_i - \bar{w}_{\mathcal{C}}\right)^2} \tag{15.8}$$

is the coefficient of variation of the weights for the sample units.

The bias of the weighted estimator is the expected value of (15.7):

$$\text{Bias}\,(\bar{y}_{\mathcal{C}w}) = E\left[\text{Corr}\,(Rw, y)\sqrt{\frac{n-1}{n}\text{CV}_w^2 + \left(1 - \frac{n}{N}\right)}\right] \times \sqrt{\frac{N-1}{n}} \times S_y. \tag{15.9}$$

The bias of the weighted estimator would be approximately zero if we knew the true participation probability $P(R_i = 1)$ for all units (see Exercise 22). In general, the closer the weights are to the "true" inverses of the participation probabilities, the more that we expect weighting to reduce the bias.

The MSE of the weighted estimator is the expected value of the squared error in (15.7):

$$\text{MSE}\,(\bar{y}_{\mathcal{C}w}) = E\left[\text{Corr}^2(Rw, y)\left\{\frac{n-1}{n}\text{CV}_w^2 + \left(1 - \frac{n}{N}\right)\right\}\right] \times \frac{N-1}{n} \times S_y^2. \tag{15.10}$$

Equation (15.10) has an extra component that is not found in (15.5): the squared coefficient of variation of the weights. This is related to the result in Exercise 17 of Chapter 7 that having variable weights increases the variance of an estimator. Thus, if $E\left[\text{Corr}^2(Rw, y)\right]$ is the same as $E\left[\text{Corr}^2(R, y)\right]$, using weights *increases* the mean squared error. The hope, though, is that the correlation between $R_i w_i$ and y_i for the population members is much smaller than the correlation between R_i and y_i, and that the reduction in the correlation more than compensates for the variance inflation caused by unequal weights.

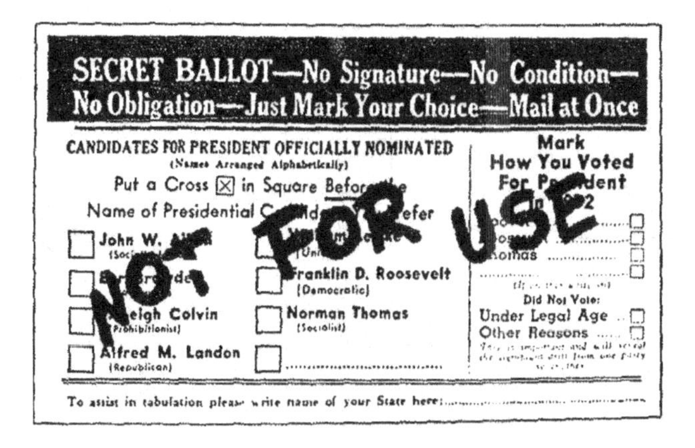

SECRET BALLOT—No Signature—No Condition—
No Obligation—Just Mark Your Choice—Mail at Once

CANDIDATES FOR PRESIDENT OFFICIALLY NOMINATED
(Names Arranged Alphabetically)
Put a Cross ☒ in Square Before the
Name of Presidential Candidate You Prefer

☐ John W. Aiken
☐ Franklin D. Roosevelt (Democratic)
☐ D. Leigh Colvin (Prohibitionist)
☐ Norman Thomas (Socialist)
☐ Alfred M. Landon (Republican)
☐ ....

Mark How You Voted For President In 1932

Thomas .... ☐

Did Not Vote:
Under Legal Age .. ☐
Other Reasons ..... ☐

To assist in tabulation please write name of your State here: ....

FIGURE 15.1
Sample ballot from the 1936 *Literary Digest* poll.
Source: Literary Digest (1936b).

Example 15.12. Let's revisit the *Literary Digest* poll from Example 1.1. The magazine editors, using the unweighted poll results, predicted that Republican candidate Alf Landon would handily win the 1936 U.S. presidential election but in the election, Democratic candidate Franklin Roosevelt received 61% of the popular vote and 523 out of the total possible 531 electoral votes. (In U.S. presidential elections, voters in each state choose electors who have pledged to support one of the candidates; the number of electors allocated to a state equals the size of its Congressional delegation. In 1936, a candidate who won a state received all of its electoral votes.)

The sample—intended to be a census—had several sources of selection bias. First, the sampling frame was constructed from subscription lists, motor vehicle registrations, telephone directories, and similar sources. This undercoverage meant that voters not listed in the frame had no chance of being selected for the sample—they had $P(R_i = 1) = 0$.

An additional problem was the low response rate of less than 25%, and weighting methods might help correct for some of the nonresponse bias. The survey did not collect demographic information from the participants, but the ballot (Figure 15.1) collected two important pieces of information by asking participants to list the state of residence and to "Mark How You Voted for President in 1932."

The proportion of the vote in state s for candidate j, p_{sj}, is known from the official election results for 1932, and Lohr and Brick (2017) used this information to weight the poll data. For a voter in state s who supported candidate j in 1932, they defined relative weight w_{sj} as $p_{sj}/\hat{p}_{sj}$, where $\hat{p}_{sj}$ was the proportion of poll respondents from state s who supported candidate j in 1932 (among those who reported voting in 1932). A respondent who supported candidate c in 1936 but who did not vote in 1932, or who did not indicate their 1932 candidate on the ballot, was assigned weight $\sum_j w_{sj} a_{csj}$, where a_{csj} is the proportion of *Literary Digest* 1936 poll respondents from state s (among those who reported a choice for 1932) who supported candidate c in 1936 and candidate j in 1932.

In 43 of the 48 states, the actual percentage vote for the Democratic candidate in the 1932 election, p_{sD}, was greater than the percentage of poll respondents who said they had voted for the Democratic candidate in 1932, $\hat{p}_{sD}$, so that w_{sD} was greater than one. Correspondingly, the weights in those states for respondents who said they had voted for the Republican candidate in 1932, w_{sR}, were less than one. The weights thus partially corrected for the underrepresentation of persons who had voted for the Democratic candidate in 1932.

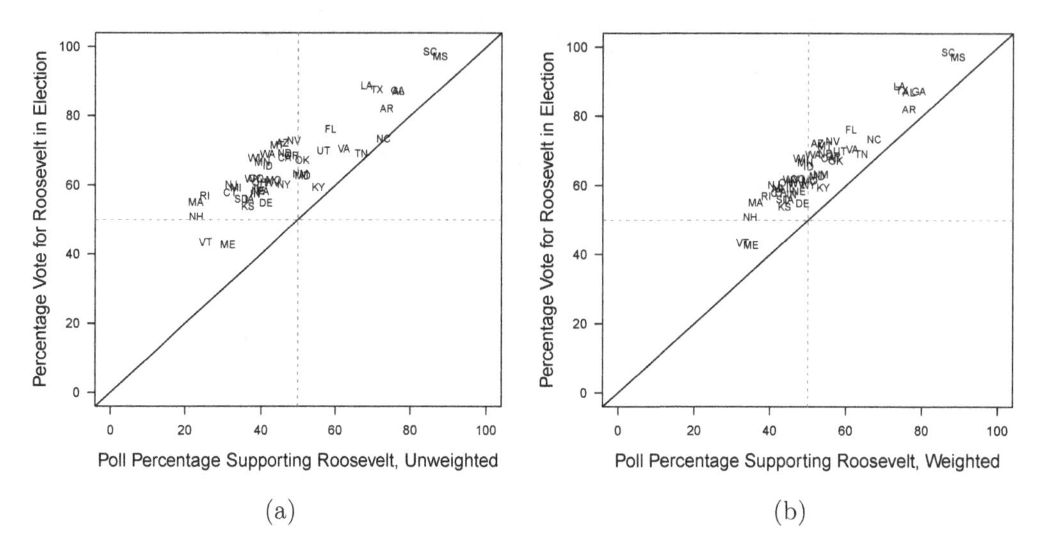

FIGURE 15.2
Results from 1936 *Literary Digest* poll (a) without weighting and (b) with weighting.

Roosevelt was the Democratic candidate in both 1932 and 1936. Overall, about 70 percent of the poll respondents who reported voting for Roosevelt in 1932 also supported him in 1936, and about 85 percent of the poll respondents who reported voting for the Republican candidate in 1932 (Herbert Hoover) supported Landon in 1936. Calibrating the weights to the state-by-state 1932 election results thus increases the estimated percentages supporting Roosevelt in the 1936 election.

Figure 15.2(a) compares the raw, unweighted, percentage supporting Roosevelt from the poll with the percentage that voted for him in the election for each state. The two-letter symbols are the state abbreviations. A state such as North Carolina (NC), whose election result is accurately predicted, falls approximately on the 45-degree line: the predicted percentage support for Roosevelt from the poll is approximately equal to Roosevelt's percentage in the election. Votes for candidates other than Roosevelt and Landon were excluded from the calculations, so the graph shows that the *Literary Digest* unweighted poll results predicted the correct winner for a state if either (i) its predicted support for Roosevelt exceeded 50 percent and Roosevelt won the state in the election (the 16 states in the upper right quadrant of the graph) or (ii) its predicted support for Roosevelt was less than 50 percent and Roosevelt lost the state in the election (the 2 states in the lower left quadrant of the graph). Without weights, the *Literary Digest* predicted that Roosevelt would win 16 states and 161 electoral votes, less than the 262 electoral votes needed to be declared the winner of the election.

Figure 15.2(b) displays the estimated support for Roosevelt from the *Literary Digest* poll when each poll respondent from state s who supported candidate j in 1932 receives weight w_{sj}. With the weights, the estimated percentage of voters supporting Roosevelt is still lower in most states than his percentage vote in the 1936 election. But most states have moved closer to the 45-degree line. And, crucially, ten states shifted from the upper left quadrant (incorrect prediction) to the upper right quadrant (correct prediction). With the weights, the poll would have predicted that Roosevelt would win 26 states and 276 electoral votes, and thus would have predicted Roosevelt to be the winner of the election.

Even with the weighted estimates, the *Literary Digest* editors would have still severely underestimated Roosevelt's support in the popular vote and the electoral college. But if they had used weights they would have predicted that Roosevelt would win in 1936 (and perhaps continued to flourish as a news magazine). ∎

For the *Literary Digest* poll, we can conclude that weights improved the estimates because we know the outcome of the election. For most surveys, however, the true population values of the characteristics of interest are unknown. One hopes that the weighting reduces bias, but it is usually too much to hope that it completely removes bias for all key survey outcomes.

Poststratification removes bias for the poststratification variables if the external population totals are accurate. If you poststratify the convenience sample to external population totals for age groups, then the weighted estimates of percentages in each age group from the convenience sample will equal the external population percentages. To the extent that the characteristic of interest y is associated with the weighting variables, the bias in the estimated mean of y will be reduced too.

But weighting does not necessarily remove bias in other characteristics. Poststratifying to population totals for age/race/gender groups assumes that the nonparticipants in each group have similar values of y to the participants in that group. But members of any demographic group—say Black women over age 65—have diverse views, life experiences, and opinions. A model that weights the data solely on demographic characteristics does not capture differences between participants and nonparticipants within a demographic group.

In general, the amount of bias adjustment that can be afforded by weighting adjustments depends on how much information is available for use in the weighting. Some nonprobability samples have a great deal of information that can be used to construct weights; others have very little. The weights constructed for the *Literary Digest* poll used the only information available for that purpose: state and vote in previous election. The bias in the estimates might have been further reduced if the poll had collected information such as age/race/gender or urban/rural status from the respondents, so that information could have been used in the weighting too.

Many nonprobability samples have large numbers of observations. The concern with these is not about the sample size, because sampling error is negligible, but about bias. How well does the sample represent the population? If the sample contains high-quality auxiliary information, and a probability sample is available that measures the same auxiliary information, the probability sample can be used to provide weights for units in the nonprobability sample.

Example 15.13. *Sample Matching.* Suppose a large nonprobability sample $\mathcal{C}$ measures the variable you want to study, y, as well as a rich set of auxiliary variables $\mathbf{x}$. You also have a high-quality probability sample $\mathcal{S}$ that measures $\mathbf{x}$ but not y. Thus, $\mathcal{S}$ is representative of the population but does not contain the variable you are interested in; $\mathcal{C}$ contains the variable you are interested in but is not representative.

Sample matching (Bethlehem, 2016) combines the two samples to take advantage of the desirable feature of each. In a simple form of matching, suppose that $\mathcal{C}$ is much larger than the probability sample $\mathcal{S}$. For each member of $\mathcal{S}$ find the member of $\mathcal{C}$ that is most similar to it with respect to the auxiliary variables $\mathbf{x}$. Then assign the probability sampling weight from that unit in $\mathcal{S}$ to the unit in $\mathcal{C}$. At the end of the process, you have a subsample from $\mathcal{C}$ of the same size as $\mathcal{S}$, with weights transferred from $\mathcal{S}$. Then use that subsample and set of weights to estimate $\bar{y}_{\mathcal{U}}$ and other characteristics.

Sample matching is thus a method for weighting units in $\mathcal{C}$. In the simple form described in the preceding paragraph, each unit in $\mathcal{C}$ that is matched to a unit in $\mathcal{S}$ receives the weight from the unit in $\mathcal{S}$; unmatched units receive a weight of zero. Many other methods can be used for matching—for example, the weight from a unit in $\mathcal{S}$ can be divided among a set of closely matching units from $\mathcal{C}$. In another variation, all units from $\mathcal{C}$ are kept, but are assigned weights according to the distribution of the $\mathbf{x}$ variables in $\mathcal{S}$. Stuart (2010) provided a comprehensive review of methods that may be used to match units across surveys.

Note, though, that having a matched sample is not the same as having a probability sample to begin with. Even if the matched units from $\mathcal{S}$ and $\mathcal{C}$ agree perfectly with respect to the $\mathbf{x}$ variables, their y values may have systematic differences. For example, suppose $\mathcal{C}$ is from the membership list of an advocacy organization. You can match $\mathcal{C}$ to $\mathcal{S}$ with respect to demographics, education, income, and a host of other variables, but after the matching, the estimated percentage of people who support the cause of the advocacy organization is likely to be much higher than the population percentage.

Rivers and Bailey (2009) studied the performance of matched samples for predicting the 2008 U.S. presidential election. They started with an opt-in nonprobability panel of more than 1.5 million persons who had been recruited through internet advertising. They then located the 3,000 persons in the nonprobability panel who were most similar to those in a stratified random subsample of size 3,000 from the American Community Survey (ACS). Similarity was measured using auxiliary variables that included age, race, gender, education, marital status, employment status, and state of residence. Because some of the matches were not perfect, estimated propensity scores were used to produce weights for the matched sample.

Sometimes a richer set of auxiliary information can be used with sample matching than with other weighting methods. Information available for postratification from a census or population register may be limited to demographic, education, and similar variables. But a small probability sample can include questions that might be more closely related to the response propensity. Schonlau et al. (2007), for example, suggested that including questions about privacy views, reading habits, and out-of-town travel might result in better matches between probability and nonprobability sample units. ■

Mercer et al. (2018) argued that the information available for weighting matters more than the specific method used to construct the weights. If the sample has rich auxiliary information—for example, demographics, education, home ownership, marital status, or other information for which reliable external population statistics exist—then poststratification, raking, propensity score estimation, and sample matching are all likely to result in similar bias reduction. On the other hand, a volunteer web poll that asks only the question of interest has no auxiliary information that can be used to weight the sample.

15.3.2 Estimate the Values of the Missing Units

The second method to reduce bias from a nonprobability sample is to estimate data for the population members who are not in the sample. To do this, you must have some information about the missing units in order to construct a model to predict their values of y.

In the model-based prediction approach introduced in Section 2.10, the population total for y is written as the sum of the values in the sample and the values not in the sample (see Equation (2.38)):

$$t_y = \sum_{i=1}^{N} y_i = \sum_{i \in \mathcal{C}} y_i + \sum_{i \notin \mathcal{C}} y_i.$$

Then a model is used to predict the values of y_i for the units not in $\mathcal{C}$:

$$\hat{t}_y = \sum_{i=1}^{N} y_i = \sum_{i \in \mathcal{C}} y_i + \sum_{i \notin \mathcal{C}} \hat{y}_i.$$

How are the predictions $\hat{y}_i$ made? Suppose auxiliary variables $\mathbf{x}_i$ are measured for each unit in $\mathcal{C}$, and the population totals $t_{\mathbf{x}}$ are known. One might posit a regression model

$$y_i = \mathbf{x}_i^T \boldsymbol{\beta} + \varepsilon_i$$

relating y to $\mathbf{x}$, where the error term ε_i is assumed to have mean 0. Then, if $\hat{\mathbf{B}}$ is an estimator of $\boldsymbol{\beta}$, the values of y_i for unsampled units can be estimated by

$$\hat{y}_i = \mathbf{x}_i^T \hat{\mathbf{B}}$$

and thus

$$\sum_{i \notin \mathcal{C}} \hat{y}_i = \left(\sum_{i \notin \mathcal{C}} \mathbf{x}_i \right)^T \hat{\mathbf{B}} = \left(t_{\mathbf{x}} - \sum_{i \in \mathcal{C}} \mathbf{x}_i \right)^T \hat{\mathbf{B}}. \tag{15.11}$$

The model is used, as in Section 4.6, to predict the values of y_i for units not in the sample.

The estimator $\hat{\mathbf{B}}$ might come from:

- The sample $\mathcal{C}$, with $\hat{\mathbf{B}}_{\mathcal{C}} = \left(\sum_{i \in \mathcal{C}} \mathbf{x}_i \mathbf{x}_i^T \right)^{-1} \sum_{i \in \mathcal{C}} \mathbf{x}_i y_i$. This requires the assumption that the regression relationship between y and $\mathbf{x}$ in $\mathcal{C}$ is the same as that between y and $\mathbf{x}$ among the nonsampled units. The validity of the predictions depends on how well the regression model predicts the values of y for units not in the sample, and one cannot assess that without additional information about $\bar{\mathcal{C}}$.

  In a nonprobability sample, there is no assurance that the regression relationship between y and $\mathbf{x}$ in $\mathcal{C}$ is the same as that between y and $\mathbf{x}$ among the nonsampled units. Just as $\bar{y}_{\mathcal{C}}$ may be a biased estimator of $\bar{y}_{\mathcal{U}}$, $\hat{\mathbf{B}}_{\mathcal{C}}$ may be a biased estimator of the finite population regression coefficient $\mathbf{B}$ given in (11.11). Consider an extreme situation in which y takes on the values 0 or 1, and $\mathcal{C}$ contains all of the population units with $y_i = 0$ and none of the units with $y_i = 1$. Then, $\hat{\mathbf{B}}_{\mathcal{C}} = 0$ for any set of explanatory variables $\mathbf{x}$, and the regression relationship developed on $\mathcal{C}$ will predict that all of the units not in $\mathcal{C}$ have $y_i = 0$.

- A high-quality probability sample $\mathcal{S}$ that measures both $\mathbf{x}$ and y. In this case, $\hat{\mathbf{B}}_{\mathcal{S}}$ is computed from $\mathcal{S}$ using (11.12). If the probability sample has full response, then $\hat{\mathbf{B}}_{\mathcal{S}}$ is an approximately unbiased estimator of the finite population quantity $\mathbf{B}$. The MSE of the predicted total depends on the size of the probability sample and the closeness of the relationship between $\mathbf{x}$ and y, as well as the sampling fraction for the nonprobability sample.

The regression model imputes a value of y for the observations not in the sample. Any method of imputation may be used to predict the values; it is not limited to regression imputation. But note that the same strong assumption—that the imputation model holds for the unsampled units—is needed for any imputation model.

Some researchers recommend taking both approaches described in Sections 15.3.1 and 15.3.2: weighting the data for nonresponse *and* using a model to predict the values of the nonparticipants. In that way, the estimate is approximately unbiased if *either* the response mechanism or the model predicting y is approximately correct. See, for example, Kim and Haziza (2014) and Chen et al. (2020).

15.3.3 Measures of Uncertainty for Nonprobability Samples

If there is no measurement error, the sample mean $\bar{y}_{\mathcal{C}}$ from a nonprobability sample accurately describes the set of units in the sample. The MSE of $\bar{y}_{\mathcal{C}}$ equals zero when the target population $\mathcal{U}$ is defined to equal $\mathcal{C}$. Then the nonprobability sample is a census and has no selection bias or sampling error. Some sets of administrative data fall in this category. If you are interested in the median amount of property taxes paid on houses in different neighborhoods, the property tax records would have the desired information on every house in every neighborhood; the median amount of property taxes is known exactly and there is no need to use sampling.

Inference for nonprobability samples that do not coincide with the target population is necessarily model-based. You must adopt a model relating the unsampled observations in the population to information from the sample and from other data sources about the population.

Model-based inference for different types of samples was discussed in Sections 2.10, 3.6, 4.6, 5.6, and 6.7. The standard errors in these sections describe the uncertainty about the estimate when the model assumptions are met; when the model assumptions are not met, a standard error calculated under the model usually underestimates the uncertainty.

For example, quota samples are often treated using the same model as for stratified sampling (Section 3.6). The model assumes that the observations sampled from quota class h are independent with mean μ_h and variance σ_h^2. If the assumptions from the model are met, then the formula in Equation (3.6) for the estimated variance of a mean from a stratified random sample can be applied to estimate the variance of the mean from a quota sample. If those assumptions are not met, however, the estimated variance in (3.6) can severely understate the true mean squared error of the statistic.

Many of the methods that have been proposed for estimating uncertainty in nonprobability samples (see, for example, Valliant, 2020) assume that the variables used in the weighting or matching or imputation have removed all of the bias. This assumption cannot be tested without external knowledge of the population (Mercer et al., 2017).

Many samples are likely to have bias remaining after the weighting or imputation—even though those techniques often reduce it—so that confidence intervals reported for statistics from nonprobability samples generally are too small. A survey may report a 95% confidence interval for the mean, but, if the estimator of the mean is biased, the true coverage probability will be much less than 0.95 (see Exercise 16).

Example 15.14. *Unequal weighting adjustments for margins of error.* Example 8.15 described a telephone poll taken by the New York Times (2018) in Illinois' 6th Congressional District in September 2018. Each respondent received an initial weight of 1, and the final weights were constructed such that the weighted sample matched voter turnout model projections for the composition of the likely electorate by age, political party, gender, region, race, and education. The final weights have mean 1 and standard deviation 0.44553745.

Many pollsters—whether taking a probability sample such as this one or taking a nonprobability sample—use the weighting design effect in Equation (7.12) to calculate the margin of error. For this survey,

$$\hat{V}(\bar{y}_{\mathcal{C}w}) = \frac{s_y^2}{n}\left(1 + \text{CV}_w^2\right) = \frac{s_y^2}{n}\left(1 + 0.44553745^2\right) = 1.1985\frac{s_y^2}{n}.$$

This expression multiplies the variance of the sample mean under an SRS by the variance inflation due to having unequal weights (here, 1.1985). The margin of error for an estimated proportion of 0.5 is thus

$$1.96\sqrt{\frac{(0.5)(0.5)}{512} \times 1.1985} = 0.047.$$

This margin of error will give a 95% CI with approximately correct coverage probability if (1) All observations in $\mathcal{C}$ are independent, and (2) $E[\text{Corr}^2(Rw, y)] = 1/(N-1)$, that is, the expected squared correlation between the weighted participation indicators and the response variable is the same as that from an SRS with full response. Assumption (2), which implies that the weighting has approximately removed the bias, cannot be tested with the survey data. If these assumptions are not met, the margin of error understates the true uncertainty about the population value. ■

When imputation is used to predict the values of missing data, the uncertainty about the estimates depends on the particular imputation model used (Chipperfield et al., 2012). Elliott and Valliant (2017) discussed general issues for assessing uncertainty about estimates from nonprobability samples.

15.4 Nonprobability versus Low-Response Probability Samples

Some probability samples have such low response rates that it has been argued that they are, essentially, convenience samples. When fewer than 10 percent of the persons randomly selected for a survey participate in it, weighting or imputation methods used to account for the missing data have strong assumptions: that the more than 90 percent of sample members who cannot be reached or decline to respond can be predicted from the 10 percent who agree to participate. If the weighting or imputation model fails to capture differences between respondents and nonrespondents, then estimates from the sample will be biased.

When the response rate is expected to be small, is it still worth the effort of drawing a probability sample and then weighting for nonresponse? Or would a weighted nonprobability sample, which might be less expensive and easier to collect, be just as accurate?

The answer depends, in the framework of Section 15.2, on the correlation between the response of interest y and the participation indicator R for the two surveys. Unfortunately, in general this correlation is unknown. However, sometimes one can evaluate the two types of samples on characteristics that might be related to y and that are known in the population.

Some researchers have empirically compared estimates from nonprobability and low-response-rate probability samples with estimates from high-response-rate surveys (see Example 8.13 and the nonprobability sample investigations by Dutwin and Buskirk, 2017, MacInnis et al., 2018, and Pennay et al., 2018). In some of the comparisons, the weighted nonprobability and low-response-rate probability samples are both close to the estimates from the high-response-rate survey (assumed to be approximately unbiased); in others, estimates from the low-response-rate probability samples are more accurate than those from the nonprobability samples.

These empirical investigations are important, and provide guidance for surveys that may be similar to those studied empirically. But the results of empirical investigations do not generalize beyond the surveys in those investigations, and do not substitute for theory. The theory says that the only type of sample for which we can provide estimates for the population with a known margin of error is a probability sample with full response. All others, including nonprobability samples and low-response-rate probability samples, require modeling assumptions. Even if those assumptions have been validated in the past, there is no guarantee that they will hold in the future, or for other outcome variables. A model that weights on demographic variables, past voting behavior, and education may produce accurate results for an election poll, but it might completely fail to adjust for selection bias in a survey about volunteering in the community. It might also fail for election polls in the future if the nonresponse mechanism changes, as occurred with the *Literary Digest* poll in Example 15.12.

In the following example, there is no high-response-rate survey measuring the quantity of interest. The decision of which survey results to trust must be made by looking at the statistical procedures used and by comparing results for other variables.

Example 15.15. In 2006, the state of North Carolina was considering whether to allow hunting on Sundays. The prohibition on hunting with firearms on Sunday had been in place since 1868, and there were vocal supporters on each side of the issue. But what did the general public think?

Duda and Nobile (2010) described two polls taken in North Carolina to assess attitudes toward legalizing Sunday hunting. The first was an online poll; more than 10,000 responses were received at the survey website. The second poll was conducted by telephone and received 1,212 responses. More than half (55 percent) of the respondents to the online poll supported Sunday hunting, compared with the estimate of 25 percent computed from the telephone survey respondents.

When statistics from two surveys are so disparate, how can you tell which (if either) to trust? The true percentage of North Carolina residents supporting Sunday hunting was unknown (which is why the surveys were conducted in the first place), so one cannot examine the differences between the two survey statistics and the true value.

The statistics must be judged by the procedures used to produce them. As stated in Section 1.1, reliable statistics result from using statistically sound sampling and estimation procedures. What were the procedures for the two surveys?

The online poll was conducted on the website of the North Carolina Wildlife Resources Commission, and anyone could participate by visiting the website. Although the survey asked about the respondent's gender, county of residence, and hunting participation, nothing prevented a female non-hunter from Florida from claiming to be a male hunter residing in a North Carolina county. Some responses were discarded from the analysis because of "strong evidence of multiple submissions by the same respondent," but it is possible that there were additional multiple submissions that were not detected. The report on the online survey acknowledged that its results "are not representative of the views of the general population or hunters because only those who were aware of the hunting on Sunday issue submitted their comments through the web portal" (Virginia Polytechnic and State University/Responsive Management, 2006). The sample was entirely self-selected, and likely attracted responses from persons with strong opinions on the issue.

The telephone survey used random digit dialing of landline telephone numbers. It had a response rate of about 20 percent and did not include cellular telephones (in June 2006, about 12 percent of adults nationwide lived in a household without landline telephone service; see Blumberg and Luke, 2007). The respondents were weighted to account for non-response and undercoverage. After accounting for the weight variation, the margin of error for estimated percentages from the survey was 3.6 percentage points (see Exercise 11).

We cannot *know* for this example which survey better represented public opinion. But the telephone survey, even though not perfect, at least started with a randomly selected sample. This prevented the sample from being dominated by advocates for one position or the other. It also prevented persons or organizations from flooding the poll with responses in a particular direction. In addition, weighting was used with the telephone survey in order to reduce bias. It is possible that the weighting did not correct for all bias in the telephone poll, but it is likely that the residual bias in the telephone poll is far less than in the online poll, which was entirely self-selected.

One can also check to see how closely other characteristics estimated from the telephone poll correspond with known population characteristics. On most measures, there was high agreement. For example, the weighted telephone survey percentages of persons with different education levels were close to estimates for North Carolina from the ACS.

A higher percentage of telephone survey respondents had no opinion about Sunday hunting (10 percent vs. 2 percent), suggesting that the online respondents may have been unusually interested and engaged with the issue. For this example, the evidence indicates that the telephone survey, despite its smaller sample size, likely gave a more accurate estimate of the support for Sunday hunting. ■

Advantages of using a nonprobability sample. The primary advantage associated with using a nonprobability sample, as opposed to a low-response-rate probability sample, is that nonprobability samples often have much lower costs. This allows a larger sample to be collected, for the same budget, with a nonprobability sample than with a probability sample.

While a low response rate for a probability sample is not always indicative of bias, it does increase the costs for conducting the survey. A larger initial sample is required to obtain the desired sample size, and much of the effort spent trying to recruit sample members is unproductive. The cost per completed interview increases as the response rate decreases.

For some types of nonprobability samples, such as administrative records or traffic sensor data, the data collection is a by-product of a program or another activity, and large amounts of data are thus available. A smaller probability sample may have few or no members of some subpopulations, while a large nonprobability sample may contain many members of such a subpopulation. For example, the National Crime Victimization Survey, an annual probability survey asking U.S. residents about their experiences as crime victims, typically has fewer than 300 respondents who report they were victims of robbery. In 2018, the Uniform Crime Reports, discussed in Example 15.1, contained detailed information on more than 78,000 robbery incidents. The Uniform Crime Reports provided a much larger sample size for studying characteristics of robbery victims, even though it excluded robberies from non-participating jurisdictions and those unknown to the police.

Advantages of starting with a probability sample. A low-response-rate probability sample lacks some of the sources of error from certain nonprobability samples. Many sampling frames used for probability sampling have high coverage. Area frames, lists of all known addresses in a region, random-digit-dialing telephone frames, and databases of businesses in a particular industry often include all, or a high percentage, of the units in the population of interest. With high-coverage frames, almost every member of the population has a positive probability of being selected for the sample and the selection bias comes primarily from nonresponse.

By contrast, coverage for a convenience sample may be low or unknown. A survey taken of volunteers to a website is limited to persons with internet access who would visit that website. The survey described in Example 1.2 was limited to persons who register with Mechanical Turk. For these surveys, the majority of population units have $P(R_i = 0) = 1$ simply because they never have an opportunity to be solicited for the survey.

Even when nonresponse to a probability sample is high, the survey conductor still has control over the initial list of units that are allowed to participate the survey. This is in contrast to anyone-can-respond-by-clicking-the-button internet surveys, where there is no control over the sample and, in some cases, the same person or organization can respond repeatedly.

When people can volunteer for a survey, or for a panel from which samples are drawn, some persons may join multiple panels or take a multitude of surveys, particularly when paid for their survey-taking efforts (Hillygus et al., 2014; Matthijsse et al., 2015; Toepoel, 2016). Persons who take surveys "professionally" may have different characteristics and attitudes than persons who are randomly selected and agree to participate in a survey.

A probability sample often has more information available for nonresponse modeling than a nonprobability sample has. Even if a high percentage of units selected for the probability sample fail to provide data, the sampler knows which units are nonrespondents and can use information from the sampling frame and external information to construct weights or impute missing values. A nonprobability sample, which is not randomly selected from a sampling frame, often does not have as much information to use in bias adjustments.

Advice for either type of sample. For either type of sample, it is desirable to collect as much auxiliary information as possible for assessing bias and constructing a weighting or imputation model. Plan in advance which data source(s) will be used for weighting and comparison. For example, if you plan to weight your survey estimates to statistics from the ACS, then you should use that survey's questions on race, ethnicity, and other variables, so that your survey measures the weighting variables the same way. As discussed in Section 8.5.1, it is desirable to have auxiliary variables that are related both to the participation probabilities and to the key variables of the survey. Many surveys use demographic information for weighting, but you can also include survey questions on other topics that are (a) measured in high-quality surveys, and (b) thought to be related to survey participation or responses.

The next chapter discusses steps you can take to improve the data quality, considering errors from all sources, while designing a survey.

15.5 Chapter Summary

Nonprobability samples are in common use for many purposes. Administrative records are used for program administration; they represent a census of the population in the program but may have undercoverage for studying other populations. Judgment, quota, and convenience samples are used in a wide variety of scientific fields—much experimental work in the medical and social sciences is conducted on samples of volunteers, and quota and convenience samples are sometimes employed to measure public opinion.

Unlike estimates of the mean from a probability sample with full response, estimates from a nonprobability sample may be biased. The mean squared error of an estimate of a population mean from a nonprobability sample depends on three factors: the correlation between participating in the sample and the characteristic of interest, the percentage of the population that is included in the sample, and the population variance of the characteristic being studied. Of these factors, the correlation is the most important as well as the most difficult to estimate. Even a small correlation between sample participation and y can lead to a large bias for the estimated population mean of y.

Methods similar to those used to adjust for nonresponse may be used to reduce bias for estimates from nonprobability samples. These include poststratification, sample matching, and imputation. But there is no guarantee that any of the methods completely remove the bias from estimates computed from nonprobability samples, even though most measures of uncertainty reported for such estimates are calculated under the assumption that no bias remains after the adjustments.

Some probability samples have very low response rates, which require the same types of assumptions and adjustments as used for nonprobability samples. The decision between using a low-response-rate probability sample and a nonprobability sample then depends on the relative costs, bias, and information available for nonresponse adjustments from the two types of samples.

Key Terms

Administrative records: Data that are collected by a government or a non-governmental organization by law or for operating a program.

Convenience sample: A sample in which the primary consideration for sample selection is the ease with which units can be recruited or located.

Focus group: A small group of persons selected to provide responses to issues presented in a group setting.

Judgment sample: A sample selected based on the judgment of the survey conductor or persons thought to be experts.

Nonprobability sample: A sample selected by a method other than probability sampling.

Quota sampling: A nonprobability sampling method in which population members are divided among quota classes (analogous to strata). A nonprobability sampling method such as convenience sampling is used to select survey participants and attain the desired sample size in each quota class.

For Further Reading

In 2013, the American Association of Public Opinion Research published a report of its task force on nonprobability samples (Baker et al., 2013), which provides an overview of the topic. Hand (2018) gives a thorough discussion of issues with administrative data. Citro (2014) and Thompson (2019) discuss issues involved with using administrative data and other, non-survey, sources of data for official statistics. Lohr and Raghunathan (2017) review statistical methods for combining data from surveys with other data sources. The book edited by Chun et al. (2021) contains several articles on using administrative data together with, or in place of, survey data. Callegaro and Yang (2018) discuss the role of surveys in the "big data" era.

Many researchers, including Dutwin and Buskirk (2017), MacInnis et al. (2018), and Tourangeau (2020), have contrasted probability and nonprobability samples. Keiding and Louis (2018) discuss issues of self-selection in online surveys. Smith and Dawber (2019) examine differences between quota and random probability samples. Sometimes a probability sample cannot be taken; Rivera (2019) evaluates the suitability of different types of nonprobability samples for meeting specific research needs. Langer (2018) and Link (2018) discuss advantages and disadvantages of nonprobability sampling methods.

The results, examples, and discussion in Section 15.2 draw heavily from articles by Meng (2018) and Rao (2021), and these papers, and the references they contain, are good sources for learning more about issues of bias and mean squared error in nonprobability samples.

Elliott and Valliant (2017), Valliant (2020), and Rao (2021) describe the methods for making inferences from nonprobability samples that are presented in Section 15.3. Chapter 18 of Valliant et al. (2018) discusses model-based estimates of variance for nonprobability samples and gives examples for different surveys. West et al. (2018) describe software available for analyzing data from complex probability and nonprobability samples.

A number of authors have proposed Bayesian approaches for inference in probability and nonprobability samples. This topic is beyond the scope of the present book; helpful references include Ghosh (2009) and Rao (2011). Si et al. (2015) present a Bayesian model for sampling when weights are used. Credible intervals, which reflect a survey conductor's belief about the value of a population characteristic as informed by the sample data, can be derived through a Bayesian approach; Chen et al. (2012) and Zhao et al. (2020) provide more information about their use with sample surveys.

15.6 Exercises

A. Introductory Exercises

1. For each data source summarized below, describe the type of nonprobability sample that was taken. Explain why the nonprobability sample is, or is not, appropriate for its purpose.

   (a) A sample of 584 adults in Flanders, Belgium was recruited for an insomnia study by students taking a research methodology class. The goal was to obtain a sample that had the same percentages as a recent census with respect to three variables: gender, education, and age. About one out of eight respondents reported having consulted a doctor about sleep difficulties (Exelmans et al., 2018).

   (b) Upon being told by a sales clerk that girls' shoes are manufactured to be narrower because boys have wider feet, Meyer (2006) decided to investigate. She measured the length and width of 20 boys' and 19 girls' feet in two fourth-grade classes at her daughter's school, and estimated that the average width of the boys' feet was 2.3 mm larger than the average width of the girls' feet.

   (c) Erba et al. (2018) evaluated the sampling methods used in all 1,173 quantitative mass communication studies published between 2000 and 2014 in six major communication journals, including only studies in which the data collection was done in the U.S. They found that 83% of the papers used nonprobability samples; in more than half, the samples consisted of college students. (The sample to evaluate for this exercise is that of the 1,173 studies.)

   (d) A sample of 267 acute-care hospitals and laboratories from about 70 counties in 10 states provides information on laboratory-confirmed cases of influenza among hospitalized patients, to allow monitoring of the disease burden of influenza in the U.S. The sample is not randomly selected, but is deliberately selected to be geographically diverse (Chaves et al., 2015).

   (e) Researchers studying postmenopausal women wanted to obtain a sample containing women from all race/ethnicity groups. They put up posters advertising the study in hospitals and clinics in different neighborhoods, and stopped recruiting women when they had achieved the target sample size in each demographic group.

   (f) To measure concern about the environment over time, researchers obtained the frequencies of internet searches for 19 environment-related terms for each year in a 9-year period. They concluded from the declining frequencies of the searches that the public was growing less interested in the environment.

(g) As of February 2016, 55 million persons in England had an electronic Summary Care Record, containing information about medicines taken, allergies to medication, and, optionally, additional information about medical history, immunizations, and communications needs (House of Commons, 2016).

(h) River sampling, is sometimes used by persons wanting to take a sample from a "stream" of visitors to websites. Visitors who click on banners or advertisements at participating websites are asked screening questions, and, if the screening questions indicate they are eligible for the survey, are invited to participate.

2. How do books attain bestseller status? The *New York Times* (2019) described its methods for selecting and ranking books.

> Rankings reflect unit sales reported on a confidential basis by vendors offering a wide range of general interest titles published in the United States. Every week, thousands of diverse selling locations report their actual sales on hundreds of thousands of individual titles. The panel of reporting retailers is comprehensive and reflects sales in tens of thousands of stores of all sizes and demographics across the United States .... Institutional, special interest, group or bulk purchases, if and when they are included, are at the discretion of The New York Times Best-Seller List Desk editors based on standards for inclusion that encompass proprietary vetting and audit protocols, corroborative reporting and other statistical determinations .... Sales are statistically weighted to represent and accurately reflect all outlets proportionally nationwide.

What conditions are needed for the list to be an accurate reflection of books' sales? Of readership?

3. Consider the population of objects in Exercises 2.13 and 3.9. The units that you can access for a convenience sample have $conv = 1$.

(a) Since the population is known, we can calculate the bias that will result from taking a convenience sample as the difference between the mean area of population units having $conv = 1$ and the mean area of all population units. What is this bias?

(b) Select an SRS of 200 items from the set of units having $conv = 1$. Calculate $\bar{y}$ from your sample. Now weight the data by the known population characteristics of shape. Does that reduce the bias? Why, or why not?

4. The following statistic from a survey taken in summer 2019 was widely reported: 64 percent of workers worldwide "would trust a robot manager more than their current human one." The company taking the survey described it as follows:

> In total, 8,370 completed the survey. The study was administered online and fielded in 10 different countries (and in six languages). Permanent full-time employees between the ages 18-74 years old were eligible to participate. The survey targeted HR Leaders, Managers and Employees. Respondents are recruited through a number of different mechanisms, via different sources to join the panels and participate in market research surveys. All panelists have passed a double opt-in process and complete on average 300 profiling data points prior to taking part in surveys. Respondents are invited to take part via email and are provided with a small monetary incentive for doing so. (Oracle, 2019)

What type of nonprobability sample was taken? What are some possible biases in the survey? Are the results representative of workers worldwide?

B. Working with Survey Data

5. The Lake Wobegon effect, named after a fictional town where "all the women are strong, all the men are good-looking, and all the children are above average," states that "for nearly any subjective and socially desirable dimension ... most people see themselves as better than average" (Myers, 1998). When the distribution of that dimension is symmetric about the average, however, some of those people must be wrong.

   Heck et al. (2018) took two surveys to estimate the percentage of Americans who feel that they have above-average intelligence. Participants in each survey were asked to choose one of the response options (Strongly Agree; Mostly Agree; Mostly Disagree; Strongly Disagree; Don't Know) for the statement "I am more intelligent than the average person."

   In Survey 1, conducted in June 2009, 79,014 landline telephone numbers were selected using random digit dialing; 1,838 of the numbers called yielded a completed interview. No cell phone numbers were called in Survey 1; according to Blumberg and Luke (2010), 24.5% of U.S. households in 2009 had cell phone service only and an additional 2% had no telephone service.

   Survey 2 recruited 983 respondents through Mechanical Turk (described in Example 1.2) in July–August 2011, soliciting Workers in the U.S. to take a "short survey of your beliefs about psychology" and offering $0.25 payment upon completion of the survey.

   Separately for each survey, a set of weights was calculated that adjusted for discrepancies between the percentages of survey respondents and the 2010 U.S. Census percentages for eight age/race/gender categories. Files `intelltel.csv` and `intellonline.csv` contain data from the two surveys.

   (a) Data file `intellwts.csv` summarizes the weighting adjustments for the eight demographic groups. Which groups were least likely to respond to the telephone survey? To the online survey?

   (b) Calculate the coefficient of variation for the weights from Survey 1.

   (c) Calculate the coefficient of variation for the weights from Survey 2. Which survey has more weight variability?

   (d) For each survey, define variable *agree* to equal 1 if the person chooses either "strongly agree" or "agree" for the statement "I am more intelligent than the average person" and 0 otherwise. Using Survey 1 with the weights, estimate the percentage of persons with $agree = 1$. Repeat for Survey 2. Which survey indicates that the population has a higher percentage of persons who agree they are "above average"?

   (e) What are the sources of selection bias in these surveys? Do you think that the survey weighting has compensated for these?

6. Demographic information was available about the population of all American Statistical Association (ASA) members for the survey described in Exercise 13 of Chapter 1. Table 15.2 compares the percentages of the sample having certain demographic characteristics with the corresponding percentages for the full population.

TABLE 15.2
Comparison of ASA survey participants with full membership.

Characteristic	Survey Participants (%)	Full Membership (%)
Gender		
Female	39.1	36.8
Male	60.9	63.2
Race/Ethnicity		
White	77.7	59.9
Black/African American	2.4	3.4
Asian	13.8	30.5
Hispanic/Latino/Latina	1.8	3.1
Other or multiple races	4.3	2.0
Membership Type		
Regular	55.1	46.7
Senior	10.6	8.1
Student	16.0	29.4
Other	18.3	15.8

Source: Langer Research Associates (2019, p. 18).

(a) Fill in the cell counts (marked by ?) in the following 2 × 2 contingency table of gender by respondent status.

	Respondent	Nonrespondent	Total
Female	?	?	?
Male	?	?	?
	3,191	?	15,769

Perform a chi-square test of association between gender and respondent status.

(b) Construct a 5 × 2 contingency table of race/ethnicity by respondent status and perform a χ^2 test of association.

(c) Construct a 4 × 2 contingency table of membership type by respondent status and perform a χ^2 test of association.

(d) Which variable tested in parts (a), (b), and (c) has the highest association with responding to the survey?

(e) How would you use the information in Table 15.2 to construct weights for the respondents? Describe the steps you would take.

7. Population percentages for gender, race/ethnicity, and membership type are known for the survey in Exercise 6. We can thus calculate the correlation between survey participation and each of these characteristics. Calculate Corr (R, y) for each of the following (HINT: Use (15.12) in Exercise 13):

(a) $y_i = 1$ if person i is female and 0 otherwise.

(b) $y_i = 1$ if person i is Asian and 0 otherwise.

(c) $y_i = 1$ is person i is a student and 0 otherwise.

8. For the *Literary Digest* poll discussed in Example 15.12, the election results are known. Roosevelt obtained 27.753 million votes and Landon obtained 16.675 million votes. The *Literary Digest* obtained a total of 966,352 ballots for Roosevelt and 1,286,511 ballots for Landon.

(a) Assume that everyone who returned the *Literary Digest* ballot voted the same way in the election. Calculate Corr (R, y), where $y_i = 1$ if person i voted for Landon and $y_i = 0$ if person i voted for Roosevelt. HINT: Use (15.12) in Exercise 13.

(b) Using the correlation in (a), what sample size in an SRS would have been needed to attain the same MSE as the *Literary Digest* poll?

9. Zhang et al. (2020) studied "professional respondents" to online surveys—people who take many online surveys, often in order to receive remuneration. File `profresp.csv` contains data from an online panel of survey volunteers used for the study.

   (a) Construct a contingency table cross-classifying respondents by gender and "professional respondent" status (variable *prof_cat*). Are men more likely to be professional respondents than women?

   (b) Construct a contingency table cross-classifying respondents by age category and "professional respondent" status (variable *prof_cat*). Which age groups have the highest percentage of professional respondents in this survey?

   (c) Calculate the percentage of persons motivated by money (variable *motive*) for each level of variable *prof_cat*. Conduct a test for the null hypothesis that these three percentages are equal.

   (d) Discuss potential sources of bias in this survey.

10. Use the data in Exercise 9 for this exercise.

   (a) Compare the percentage of survey respondents who are women, college graduates, and who are age 65 and over with the corresponding percentages from the 2011 ACS in file `profrespacs.csv`. What does this tell you about persons who tend to participate in the survey?

   (b) Create weights that poststratify the weights of the survey respondents to the population numbers from the ACS, using the 18 poststrata defined in `profrespacs.csv`. What is the coefficient of variation of the poststratified weights for the survey?

   (c) Calculate estimates of the mean values of variables *freq_q1* to *freq_q5* without weights, and then with the poststratification weights. For which variables does the poststratification make a difference for the estimated mean? For the standard error?

   (d) How can you assess the accuracy of the poststratified estimates of variables *freq_q1* to *freq_q5* for estimating the corresponding population quantities?

11. The telephone survey of residents of North Carolina, discussed in Example 15.15, weighted the respondents to account for nonresponse. File `hunting.csv` gives the population and sample sizes for each poststratum used. Use the method in Example 15.14 to calculate the variance inflation due to having unequal weights.

12. *Calculating positive predictive value for a diagnostic test.* Consider a disease or medical condition called X. Some people in the population have X, and the remaining people do not have X. There exists a medical diagnostic test for X, where some people will test positive (the test says they have X) and others will test negative (the test says they do not have X). But the test is not completely accurate: some of the people who have X test negative (a false negative), and some of the people who do not have X test positive (a false positive).

    False positive and false negative rates are usually estimated from two convenience samples. Sample A is of persons known to have X, and Sample B is of persons known not to have X. Each sample is administered the diagnostic test. The false negative rate is estimated by the percentage of persons in Sample A for whom the test gives a negative result; the false positive rate is estimated by the percentage of persons in Sample B for whom the test gives a positive result.

Suppose Sample A consists of 108 persons known to have had X, of whom 95 persons tested positive and the remaining 13 tested negative. Of the 433 persons in Sample B, known to have not had X, 428 tested negative and 5 tested positive.

(a) Assume that the population contains 100,000 persons, and that in reality 5,000 of them have X and 95,000 do not have X. If Samples A and B are both self-weighting, what weight should be assigned to each member of Sample A so that the weights sum to the number of population persons who do not have X? What weight should be assigned to each member of Sample B so that the weights for the sample sum to the number of persons in the population who have X?

(b) Use the results from Samples A and B, with the weights in part (a), to fill in estimates for the entries marked by ?? in the following contingency table. Give a 95% CI for each entry. You may want to use Clopper-Pearson CIs (see Exercise 27 of Chapter 7) for this exercise if they are available in your software because some of the estimated percentages are close to 100%.

	Test Negative for X	**Test Positive for X**	**Total**
Don't Have X	??	??	95,000
Have X	??	??	5,000
Total	??	??	100,000

(c) The false negative rate for a test is 100 × P(person tests negative for X | person has X) and the false positive rate is 100 × P(person tests positive for X | person does not have X). But a person taking the test is actually interested in the positive predictive value, which is P(person has X | person tests positive for X), or the negative predictive value, which is P(person does not have X | person tests negative for X). Let's look at using survey data to estimate the positive predictive value.

Use the data from Samples A and B, together with the weights, to estimate the positive predictive value for the test along with a 95% CI. HINT: Treat the combined samples as a stratified sample from the population, and then estimate the ratio.

(d) What assumptions are made for this analysis? How might those assumptions be violated in the data collection?

C. Working with Theory
All exercises in this section require mathematical statistics

13. Suppose that x and y are both binary variables and the population counts are given in the following contingency table:

	$y = 0$	$y = 1$	**Total**
$x = 0$	N_{00}	N_{01}	N_{0+}
$x = 1$	N_{10}	N_{11}	N_{1+}
Total	N_{+0}	N_{+1}	N

Show that the population correlation coefficient of x and y, defined in (A.10), is

$$R = \frac{N_{11}N_{00} - N_{10}N_{01}}{\sqrt{N_{0+}\, N_{1+}\, N_{+0}\, N_{+1}}}. \tag{15.12}$$

14. Prove the identity in (15.3). HINT: First show that

$$\bar{y}_{\mathcal{C}} - \bar{y}_{\mathcal{U}} = \frac{\sum_{i=1}^{N} (R_i - \bar{R}_{\mathcal{U}})\,(y_i - \bar{y}_{\mathcal{U}})}{\sum_{i=1}^{N} R_i}.$$

Then use an argument similar to that in Equation (4.6) of Section 4.1.2.

15. In an SRS, $R_i = Z_i$, $P(Z_i = 1) = n/N$, and $P(Z_iZ_j = 1) = n(n-1)/[N(N-1)]$ for $i \neq j$. Assume that $n < N$ and $S_y^2 > 0$.

    (a) Show that $E\left[\text{Corr}\,(Z, y)\right] = 0$. Thus the first factor in the expression for the bias in (15.4) is zero.

    (b) Show, using the random variables directly (not relying on the variance shown in Chapter 2), that the first factor in (15.5) is

$$E\left[\text{Corr}^{\,2}(Z, y)\right] = \frac{1}{N-1}.$$

16. *Coverage probability of a confidence interval when there is bias* (requires computing). Many people using a nonprobability sample report a confidence interval for the population mean $\bar{y}_{\mathcal{U}}$ of the form

$$\bar{y}_{\mathcal{C}} \pm 1.96\sqrt{\hat{v}}, \tag{15.13}$$

where $\hat{v}$ is a consistent estimator of $v = V(\bar{y}_{\mathcal{C}}) = E\left[\{\bar{y}_{\mathcal{C}} - E(\bar{y}_{\mathcal{C}})\}^2\right]$. Suppose that

$$T = \frac{\bar{y}_{\mathcal{C}} - E(\bar{y}_{\mathcal{C}})}{\sqrt{v}}$$

follows a standard normal distribution with mean 0 and variance 1. Let

$$b = E(\bar{y}_{\mathcal{C}}) - \bar{y}_{\mathcal{U}}$$

denote the bias of $\bar{y}_{\mathcal{C}}$. When there is no bias (that is, $b = 0$), then (15.13) is a 95% confidence interval because the coverage probability is

$$P\left(|\bar{y}_{\mathcal{C}} - \bar{y}_{\mathcal{U}}| \leq 1.96\sqrt{v}\right) = 0.95.$$

In this exercise, you will investigate the effects of bias on the coverage probability.

   (a) Show that the coverage probability is

$$P\left(|\bar{y}_{\mathcal{C}} - \bar{y}_{\mathcal{U}}| \leq 1.96\sqrt{v}\right) = 1 - \frac{1}{\sqrt{2\pi}}\int_{\frac{-b}{\sqrt{v}}+1.96}^{\infty} e^{-t^2/2}dt - \frac{1}{\sqrt{2\pi}}\int_{-\infty}^{\frac{-b}{\sqrt{v}}-1.96} e^{-t^2/2}dt.$$

   Thus the coverage probability depends on the size of the bias b relative to $\sqrt{v}$.

   (b) Compute the coverage probability in (a) for values of the relative bias $b/\sqrt{v}$ in {0, 0.01, 0.02, ..., 5}. Draw a graph of the coverage probability (on the y axis) as a function of $b/\sqrt{v}$ (on the x axis).

(c) Relate this result to ratio estimation in an SRS. What is the difference between the bias caused by using ratio estimation and that resulting from a nonprobability sample?

17. *"Design effect" for a nonprobability sample.* In Chapter 7, we defined the *design effect* for a probability sampling design to be the ratio of the variance of a statistic under the sampling design to the variance of a statistic calculated from an SRS of the same size. A similar quantity can be calculated for a nonprobability sample (even though a convenience sample is not, strictly speaking, "designed"). Show that the ratio of the mean squared error of $\bar{y}_C$ for a nonprobability sample of size n to the mean squared error of $\bar{y}_{\text{SRS}}$ for an SRS of size n is

$$\frac{\text{MSE}\,(\bar{y}_C)}{\text{MSE}\,(\bar{y}_{\text{SRS}})} = (N-1)E\left[\text{Corr}^{\,2}(R, y)\right].$$

18. The results in Section 15.2 assumed that a population member appears at most once in a sample. In some convenience samples, however, the same person can participate multiple times. A person may participate in an opt-in internet poll as many times as desired, and may even write a computer program to automate participation. Let Q_i denote the number of times population unit i appears in the sample, and assume that unit i provides the same answer y_i for each of its Q_i appearances.

   (a) Show that for a sample of fixed size n, each variable Q_i can take on values in the set $\{0, 1, \ldots, n\}$ subject to the constraint $\sum_{i=1}^{N} Q_i = n$.

   (b) Let

   $$\bar{y}_Q = \frac{\sum_{i=1}^{N} Q_i y_i}{\sum_{i=1}^{N} Q_i}$$

   denote the sample mean of the n responses to the survey, counting each vote equally. Show that

   $$\bar{y}_Q - \bar{y}_{\mathcal{U}} = (N-1)\,\text{Corr}\,(Q, y)\, S_Q S_y,$$

   where $S_Q^2 = \sum_{i=1}^{N}(Q_i - \bar{Q}_{\mathcal{U}})^2/(N-1)$.

   (c) Suppose that you have a choice between two samples of size n. In sample 1, each unit can participate at most once; in sample 2, a unit can participate multiple times. Thus, the error for the mean of sample 1, $\bar{y}_C$, is given by (15.2) and the error for the mean of sample 2, $\bar{y}_Q$, is given in part (b) of this exercise. Show that $S_Q^2 \geq S_R^2$ and consequently, if $\text{Corr}\,(R, y) = \text{Corr}\,(Q, y)$, then

   $$|\bar{y}_Q - \bar{y}_{\mathcal{U}}| \geq |\bar{y}_C - \bar{y}_{\mathcal{U}}|\,.$$

19. Example 3.13 described New York City's sampling plan for estimating the number of persons experiencing homelessness. Some other cities conduct their surveys by having volunteers visit a convenience sample of areas thought to contain persons experiencing homelessness. The estimated count can be thought of as follows. In the population of size N, $y_i = 1$ if person i is experiencing homelessness and $y_i = 0$ otherwise. The total number of persons experiencing homelessness is $t_y = \sum_{i=1}^{N} y_i$; the city's estimated count is $\hat{t}_y = \sum_{i=1}^{N} R_i y_i$, where $R_i = 1$ if person i is observed in the sample and $R_i = 0$ if person i is not observed in the sample.

(a) Show that when $S_R > 0$ and $S_y > 0$, the error in the estimated population count is

$$\sum_{i=1}^{N} R_i y_i - t_y = N\text{Corr}(R, y)\sqrt{\bar{R}_{\mathcal{U}}(1 - \bar{R}_{\mathcal{U}})\bar{y}_{\mathcal{U}}(1 - \bar{y}_{\mathcal{U}})} - N(1 - \bar{R}_{\mathcal{U}})\bar{y}_{\mathcal{U}}.$$

(b) Show that the error in the estimated population count is zero if any of the following are true:

i. $\bar{R}_{\mathcal{U}} = 1$

ii. $\bar{y}_{\mathcal{U}} = 0$

iii. $\text{Corr}(R, y) = \sqrt{\dfrac{(1 - \bar{R}_{\mathcal{U}})\,\bar{y}_{\mathcal{U}}}{(1 - \bar{y}_{\mathcal{U}})\,\bar{R}_{\mathcal{U}}}}$.

The third condition holds if $R_i = 1$ whenever $y_i = 1$, which includes the case when $\text{Corr}(R, y) = 1$. (By contrast, the error for estimating a population *mean* is zero if $\text{Corr}(R, y) = 0$.)

20. Prove Equation (15.7).

21. *Ratio estimation.* Suppose the population total of an auxiliary variable x is known to be t_x. In ratio estimation, as shown in Example 11.13, the ratio weight for each unit in $\mathcal{C}$ is $w_i = t_x / \sum_{i=1}^{N} R_i x_i$. Show that when the ratio relationship between y and x is exactly the same for the sampled and unsampled units—that is, if $B_{\mathcal{C}} = B_{\bar{\mathcal{C}}}$, where

$$B_{\mathcal{C}} = \frac{\sum_{i \in \mathcal{C}} y_i}{\sum_{i \in \mathcal{C}} x_i} \quad \text{and} \quad B_{\bar{\mathcal{C}}} = \frac{\sum_{i \notin \mathcal{C}} y_i}{\sum_{i \notin \mathcal{C}} x_i},$$

then $\sum_{i \in \mathcal{C}} w_i y_i = t_y$.

22. Suppose that $w_i = 1/P(R_i = 1)$ for every unit in $\mathcal{C}$, for a sample of size n. Also suppose that there exists a constant $k > 0$ such that $P(R_i = 1) \geq kn/N$ for $i = 1, \ldots, N$, which ensures that no unit has negligible probability of participation relative to the other units. Prove that

$$E\left[\bar{y}_{\mathcal{C}w} - \bar{y}_{\mathcal{U}}\right] \approx 0$$

for any response variable y satisfying $|y_i| \leq M$ for some fixed value M.

23. Consider the situation in Exercise 22, but with one difference. While $P(R_i = 1) \geq kn/N$ for 90 percent of the units in $\mathcal{U}$, the remaining 10 percent of units have $P(R_i = 0) = 1$.

Show that for any fixed value of the bias b, it is possible to construct a response variable y that achieves that bias, that is, $E\left[\bar{y}_{\mathcal{C}w} - \bar{y}_{\mathcal{U}}\right] > b$.

D. Projects and Activities

24. In 2008 the online service provider AOL conducted a straw poll about the U.S. presidential election. Visitors to the website were asked to confirm that they were not robots and then to express their preference for candidate John McCain or candidate Barack Obama. As of September 25, 2008, the poll had received 154,732 responses, with 97,050 (63%) supporting McCain and 57,682 (37%) supporting Obama. Various online commentators argued that the AOL poll was less biased than contemporaneous polls that relied on random digit dialing, because:

- Most of the other polls sampled only 1,000 or 2,000 people, while the AOL poll had an enormous sample size.
- Participants in other polls consist of people that are chosen by the pollsters, and thus may reflect the pollster's bias.
- The AOL poll was available to anyone with internet service, and persons who care enough to respond to a poll are also more likely to vote.
- Other public opinion polls are unfair because they do not have equal numbers of Republicans and Democrats.

How would you respond to each of these arguments?

25. Henrich et al. (2010) wrote: "Behavioral scientists routinely publish broad claims about human psychology and behavior in the world's top journals based on samples drawn entirely from Western, Educated, Industrialized, Rich, and Democratic (WEIRD) societies. Researchers — often implicitly — assume that either there is little variation across human populations, or that these 'standard subjects' are as representative of the species as any other population. Are these assumptions justified?" They were describing the common practice of performing experiments using undergraduate students in psychology lectures as test subjects.

    Find a recent journal article reporting on results of an experiment with humans as test subjects. How were the participants selected? How do you think the sample selection process might affect conclusions from the study?

26. The Consumer Bankruptcy Project of 2001 (Warren and Tyagi, 2003) surveyed 2,220 households who filed for Chapter 7 or Chapter 13 bankruptcy, with the goal of studying why families file for bankruptcy. Questionnaires were given to debtors attending the mandatory meeting with the bankruptcy trustee assigned to their case in the five districts selected by the investigators for the study (these districts included the cities of Nashville, Chicago, Dallas, Philadelphia, and Los Angeles) on specified target dates. Additional samples were taken from two rural districts in Tennessee and Iowa. Quota sampling was used in each district, with the goal of collecting 250 questionnaires from each district that had the same proportions of Chapter 7 and Chapter 13 bankruptcies as were filed in the district. Discuss the relative merits and disadvantages of using quota sampling for this study.

27. To estimate how many New York state residents had antibodies to the virus causing COVID–19, Rosenberg et al. (2020) selected 99 grocery stores in 26 counties; the stores were chosen to "increase sample coverage of the racial and ethnic diversity of the statewide population." Testing was done on 6 separate days in April, 2020. Adults were recruited through a flyer posted at each store on the day of testing and by approaching customers as they entered. A total of 15,101 adults participated and provided sufficient information to be assigned to one of the 160 poststrata defined by sex, race and ethnicity, age group, and region of state.

    The weight for an adult in poststratum h was calculated by dividing an estimate of the state population in poststratum h by the number of sample members in poststratum h. Using the weights, the researchers estimated that 14.0% (95% CI: 13.3–14.7%) of the adults in New York would test positive for antibodies.

    Discuss the relative merits of using this nonprobability sampling design, contrasted with using a probability sample such as that described in Example 6.12, to estimate the percentage of adults in the state with antibodies. What is the design effect from weighting (you can estimate this by comparing the published CI with the CI you would get from an SRS)? What are the possible sources of bias, and do you think the weighting reduces these? How would you have designed the study?

28. Kreuter et al. (2020) described a global rapid-response survey that collected more than one million responses every week to questions about symptoms, mental health issues, social distancing behavior, and financial impacts from the COVID–19 pandemic. Read the three-page article. How were participants selected? What type of sampling was done? Can the results be considered representative of the world's population? Discuss the strengths and weaknesses of the sampling design and weighting scheme.

29. The movie review website `www.rottentomatoes.com` presents four ratings for each movie: (a) Percentage of critics (from a list of critics approved by the site) who gave the movie a favorable review, (b) Percentage of "top critics" who gave the movie a favorable review, (c) Percentage, of all website visitors who reviewed the movie, who rated the movie 3.5 stars or higher (out of 5), and (d) Percentage of "verified ticket purchaser" reviewers who rated the movie 3.5 stars or higher. To have a review appear in category (d), a user checks the box "Verify I bought my ticket" underneath the review text and selects the online ticket vendor who sold the ticket. The review is verified if the e-mail address of the user's Rotten Tomatoes® account matches that used to buy the ticket.

    (a) What type of sample is represented by each set of reviews? What are some potential sources of bias?

    (b) Take an SRS or systematic sample of 30 movies released in the past 12 months. Describe how you selected the sample. For the sample, record the percentage of each type of review. Which of the four types of review has the highest ratings, on average? The lowest? Relate your findings to the sources of bias you identified in part (a).

30. The Evidence-Based Medicine Pyramid (Straus et al., 2018) summarizes the quality of evidence about treatment effectiveness, with the most compelling evidence considered to be a systematic review of multiple high-quality studies and the least compelling evidence considered to be opinion uninformed by data. In decreasing order, the levels of evidence are:

    (a) Systematic reviews of studies, which compare and draw conclusions from multiple randomized controlled trials.

    (b) Randomized controlled trials. Study participants are randomly assigned (individually or in groups) to one of the medical treatments (one of the treatments is usually a control). The randomization ensures that, on average, other characteristics of the participants that might affect the outcomes are equally likely to be found in each treatment. In a randomized study, the treatments may be said to have *caused* the differences in outcomes.

    (c) Cohort studies, which follow people over time to see how receiving the treatment is related to later health outcomes. The health outcomes of persons who have received the treatment are compared with those of persons who have not received the treatment. Unlike randomized studies, the study participants have themselves chosen their treatment group or dosage. For example, a nutrition study may follow people over time and note that persons who eat the most vegetables have a lower incidence of a certain type of cancer. The study, however, does not *prove* that eating vegetables prevents cancer; perhaps people who eat the most vegetables tend to have other characteristics (for example, exercising more, having cancer-protective genetic factors, or being exposed to fewer environmental carcinogens) that might reduce their risk of cancer.

(d) Case studies, which describe how one, or a few, patients responded to a medical treatment. Case studies typically have no formal comparison group.

(e) Opinion. Experts (or people who think they are experts) speculate about treatment effects, often based on anecdotal or unsystematically collected evidence, or even on personal beliefs.

Construct one or more evidence-based pyramids for the amount of evidence about a population afforded by different types of samples, including probability samples and the types of samples discussed in Section 15.1. What types of studies provide the highest-quality information about population characteristics? You may want to consider characteristics other than accuracy, such as the ability to generate detailed information about small subsets of the population.

31. File `gini.csv` contains the data from the entire population of districts and the purposive sample taken by Gini and Galvani (described in Example 15.5). The list is ordered from highest to lowest birth rate.

(a) Because the data for all 214 districts are available, population characteristics can be calculated. Suppose that the goal is to obtain a sample that accurately estimates the average of the 214 district statistics for each characteristic. For each characteristic, find the margin of error for the estimate from an SRS of 29 districts, using the population variances S^2.

(b) Gini and Galvani said that a sample would be "excellent" if an estimate was within 1.5% of its population value (that is, |estimate minus population value|/(population value) ≤ 0.015), "satisfactory" if an estimate was between 1.5% and 5% of its population value, and "sufficient" if an estimate was between 5% and 10% of its population value. For an SRS of size 29, which characteristics have margins of error that will place them in the "excellent" category for 95% of possible samples? Satisfactory? Sufficient? Insufficient?

(c) Given your answer to (b), which variable(s) would you use to stratify the population? Within strata, would you want to sample with equal or unequal probabilities?

(d) Design and take a stratified sample of 29 districts from the list of 214 districts. What is the margin of error for each characteristic? How does the set of units you selected compare with that selected by Gini and Galvani?

32. Repeat Exercise 31, assuming that the weighted average values across all districts are the population characteristics of interest. As an example, the population birth rate is calculated as $\sum_{i=1}^{214} a_i y_i / \sum_{i=1}^{214} a_i$, where a_i is the population for district i and y_i is the birth rate for district i.

33. Find a nonprobability sample used in a recent journal article. What method(s) did the authors use to try to obtain a more representative sample or to reduce the selection bias? How well do you think the methods worked?

16

Survey Quality

DUDLEY. Do you think you've learned from your mistakes?

PETER. Oh, yes, I've learned from my mistakes and I'm sure I could repeat them exactly.

—Peter Cook and Dudley Moore, *Good Evening: A Comedy-revue in Two Acts*

Example 16.1. The American Community Survey (ACS) is the largest demographic sample survey in the United States. Each year, about 3.5 million addresses across the United States are included in the ACS sample. Of course, a survey of this scale requires a great deal of planning and development, and potential inaccuracies in the data need to be resolved before the survey is launched. For national estimates of quantities such as unemployment and household size, the sampling error of the survey will be very small. But other sources of error such as nonresponse, undercoverage, and measurement error are important. It is thus crucial in planning such a survey that all errors be considered in the design. ■

Throughout this book, we have concentrated on designing surveys that will produce accurate and timely statistics. Chapters 2–7 discussed survey designs that could be used to control sampling error for estimating population means and totals. Chapters 9–14 outlined other methods for analyzing data from complex surveys. Chapters 1, 8, and 15 discussed nonsampling errors that can arise in surveys.

In Chapters 2–7, we assumed that there are no nonsampling errors, and the only reason that survey estimates differ from population quantities is that a sample was taken rather than a census. In many surveys, the margin of error reported is based entirely on the sampling error; nonsampling errors are sometimes acknowledged in the survey documentation, but generally are not included in confidence intervals or other measures of uncertainty.

Dalenius (1977, p. 21) referred to the practice of reporting only sampling error and ignoring other sources of error as "'strain at a gnat and swallow a camel'; this characterization applies especially to the practice with respect to the accuracy: the sampling error plays the role of the gnat, sometimes malformed, while the non-sampling error plays the role of the camel, often of unknown size and always of unwieldy shape."

In this chapter, we explore approaches to survey design and analysis that consider the whole camel. Much of the early inspiration for this approach came from W. Edwards Deming, who in addition to writing one of the first books on survey sampling (Deming, 1950) was also one of the leaders in developing quality improvement methods for industry after World War II (Boardman, 1994). Not surprisingly, Deming was one of the earliest writers to consider factors that might affect the quality of survey estimates. Deming (1944) discussed survey errors due to interviewer variability, survey mode, questionnaire design, sampling variability, nonresponse, and other sources now considered to be part of total survey error.

Quality in surveys draws on many ideas from Deming's work on quality improvement (Deming, 1986). Biemer and Lyberg (2003), recognizing the multiple purposes of survey data, defined survey quality as "fitness for use." Table 16.1, adapted from Eurostat (2020), lists some of the dimensions of survey quality.

DOI: 10.1201/9780429298899-16

TABLE 16.1
Some dimensions of survey quality, adapted from Eurostat (2020).

1. *Relevance.* The statistics produced should meet user needs.
2. *Accuracy and reliability.* Estimates should be close to the true values of population quantities.
3. *Timeliness.* Results may need to be disseminated quickly to be useful. Indeed, as argued in Chapter 1, one reason for taking a survey rather than conducting a census is that the survey can sometimes be completed in a shorter time period.
4. *Accessibility and clarity.* Data and data products must be accessible to users, and sufficient documentation should be provided to enable users to interpret the results.
5. *Coherence and comparability.* It is often desired to compare estimates of quantities such as unemployment rate over time. Surveys measuring such quantities must be conducted consistently so that these comparisons are meaningful. When survey results are to be compared for different countries, care must be taken to ensure the concepts being measured are interpreted the same way across countries and that appropriate methodologies are used to ensure comparable results (Harkness et al., 2003). Common definitions and standards should be used when data come from multiple sources.
6. *Cost and burden.* Data should be collected efficiently and so as to impose minimal burden on respondents who supply the data.
7. *Confidentiality.* Most organizations taking household and economic surveys assure respondents that their data will be protected and will be used for "statistical purposes" only—that the data will not be used in a way that could cause harm to them.

While all of the quality dimensions in Table 16.1 may be important in different contexts, we argue that accuracy is crucial. Timely and coherent statistics are of little use if they are wildly inaccurate. As defined in Chapter 2, an estimator $\hat{\theta}$ of a population quantity θ is **accurate** if it is close to the true value of the quantity being estimated, that is, if the mean squared error $\text{MSE}[\hat{\theta}] = E[(\hat{\theta} - \theta)^2]$ is small.

We can consider the **total survey error** to be the sum of five main sources of error, shown in Figure 16.1:

$$\begin{aligned}\text{Total survey error} = {} & \text{sampling error} + \text{coverage error} + \text{nonresponse error} \\ & + \text{measurement error} + \text{processing error}.\end{aligned}$$

Chapters 2 to 7 treated sampling error from a probability sample, telling how to measure it and how to design a survey whose margin of error for key survey quantities are within prespecified tolerances. The other four components of total survey error—error from coverage, nonresponse, measurement, and processing—are all part of nonsampling error. The next sections of this chapter discuss methods for assessing and controlling these sources of error.

Total survey design. The survey designer needs to consider all types of errors in the survey design, not just sampling error. Of course, to design a survey that has high accuracy, you need to know what the major error components are. If you know that most of the error in survey estimates is caused by coverage problems, then you can devote resources to improving the coverage. If you know that the coding is highly accurate, then you do not need to devote as many resources to improving the quality of the coding procedures.

Total survey design calls for an interdisciplinary approach. The areas of expertise needed to study and reduce sources of error include theory of complex surveys including weighting

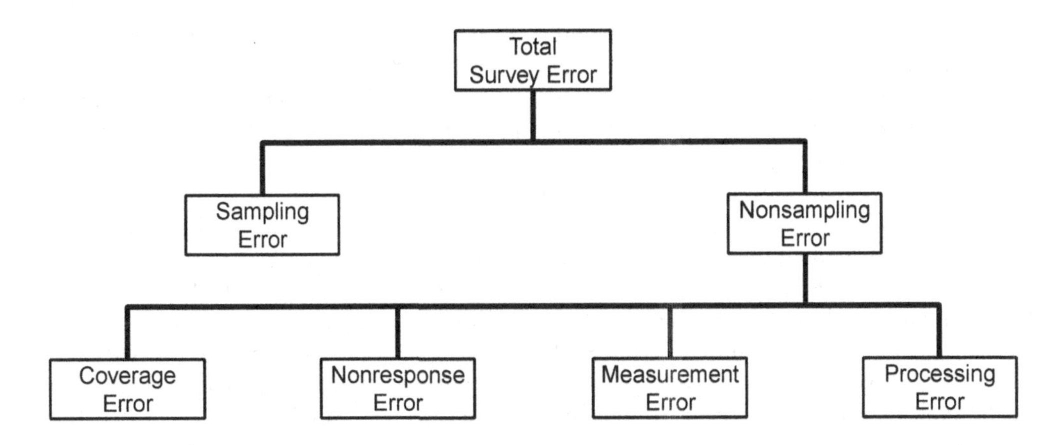

FIGURE 16.1
Components of total survey error.

and variance estimation, design of experiments, statistical process control, mixed models, cognitive psychology, management, and ethnography.

16.1 Coverage Error

As discussed in Chapter 1, coverage is the percentage of the population of interest that is included in the sampling frame. A mismatch between the target population and the sampling frame can cause coverage bias. Most common is undercoverage, where the sampling frame misses part of the population. Consider the target population mean of variable y for all N units in the population, $\bar{y}_{\mathcal{U}}$. Let $\bar{y}_{\mathcal{U}F}$ be the mean for the N_F units in the sampling frame, and let $\bar{y}_{\mathcal{U}N}$ be the mean for the $N - N_F$ units that are in the target population but not in the sampling frame. The bias due to undercoverage is then

$$\bar{y}_{\mathcal{U}F} - \bar{y}_{\mathcal{U}} = \frac{N - N_F}{N}(\bar{y}_{\mathcal{U}F} - \bar{y}_{\mathcal{U}N}). \tag{16.1}$$

The bias is thus low if (1) the population means are approximately the same for the covered and noncovered units in the population, that is, $\bar{y}_{\mathcal{U}F} \approx \bar{y}_{\mathcal{U}N}$, or (2) the coverage rate, N_F/N, is high.

16.1.1 Measuring Coverage and Coverage Bias

Estimating undercoverage or bias caused by undercoverage is, in general, difficult. If it were easy to identify and reach units missed by the sampling frame, those units would have been included when the frame was constructed. By definition, undercoverage is external to the survey and thus information external to the survey must be used to assess it. Example 13.2 described one method that is used to assess undercoverage in population censuses.

Comparing survey estimates of the number of persons in demographic groups to counts from a census or large survey such as the ACS can identify undercoverage or nonresponse among demographic subgroups, and suggest variables that might be useful for calibration. If your estimated number of 18- to 24-year-old males from the survey is much lower than the total number of 18- to 24-year-old males from an accurate census, then there is likely undercoverage or nonresponse (or both) for that subpopulation.

The coverage rate can sometimes be estimated using information from other studies or external records. For example, undercoverage in a survey of households with infants might be assessed by taking a sample of recent birth certificates and checking whether those households are in the sampling frame. The coverage of a survey using a voter registration list as a sampling frame could be estimated by comparing the number of persons on the list with the target population size from a census or population register. As discussed in Example 8.6, one of the measures used to assess coverage in the ACS itself is to compare ACS estimates with independent estimates of numbers of housing units.

Network or snowball sampling (see Section 14.1) can sometimes be used to estimate a coverage rate, particularly in surveys of rare populations. These can identify members of the population that are missing from your sampling frame. Rothbart et al. (1982) found that a network sample of Vietnam veterans gave improved coverage of Vietnam veterans from minority groups.

Election polls present many challenges for constructing sampling frames and assessing coverage. The target population for an election poll is persons who will vote in the election, but no one knows in advance exactly who those persons will be. A list of registered voters will include many persons who do not vote on election day, and exclude persons who register to vote after the list is obtained. A sampling frame of persons who voted in the last election will miss new voters. Many election polls in the United States use models to predict who is likely to vote, taking into account voting history and other information. Polls conducted by telephone exclude nontelephone households; online polls may exclude households without internet access.

The methods discussed so far involve estimating the coverage rate, N_F/N, in (16.1). The second factor of the undercoverage bias in (16.1), $\bar{y}_{\mathcal{U}F} - \bar{y}_{\mathcal{U}N}$, depends on the mean of the units not in the sampling frame. Estimating $\bar{y}_{\mathcal{U}N}$ requires data from the uncovered part of the population, which in general must be obtained from an external source.

Large, high-quality surveys can sometimes be used to estimate coverage bias on characteristics related to variables of interest. The ACS, for example, includes telephone and nontelephone households. It could thus be used to estimate the bias for some responses from an education survey that is conducted by telephone. If the telephone and nontelephone households in the ACS have significantly different proportions of college graduates, then you would expect an education survey that excludes nontelephone households to have bias for estimating the proportion of college graduates and related items.

16.1.2 Coverage and Survey Mode

The mode of survey administration (such as in-person, telephone, mail, e-mail, or internet) exerts great influence on the coverage properties, and choice of mode should depend in part on the coverage that can be obtained. Other considerations for choice of mode, such as response rate and accuracy of responses for various modes, are discussed in Chapter 8.

Area frames usually have the highest coverage. An area frame is constructed by selecting a sample of geographical primary sampling units (psus) from the region of interest. A list of housing units is constructed for the psus selected for the sample, and a probability sample of housing units is selected from the list in each psu.

The sampling frame for a mail or e-mail survey is a list of physical or e-mail addresses. The coverage of the frame depends on the completeness and accuracy of the list. Even if the frame contains everyone in the target population, the addresses may be incorrect or out of date. E-mail surveys may work well for populations in a university or organization in which everyone uses e-mail; for other populations, they exclude persons who do not have e-mail or who never check their e-mail accounts. Mail and e-mail surveys carry risks that the questionnaire will not reach the intended respondent. A mail survey might be discarded by another household member; an e-mail survey invitation might be deleted by a spam filter.

Telephone surveys may use list frames constructed from directories, random digit dialing (see Example 6.13), or a combination. List frames for telephone surveys may be incomplete or have incorrect telephone numbers. Frames that consist exclusively of listed landline telephone numbers often have serious undercoverage of the population (Sala and Lillini, 2017). A dual frame survey, with one sample from a frame of landline telephones and another sample from a frame of cellular telephones, will miss nontelephone households, but will have better coverage than a sample of landline telephones that excludes persons who exclusively use a cellular telephone. In some countries where almost everyone has access to a cellular telephone, a single frame of cellular telephone numbers is used—this simplifies the survey operations but excludes landline-only households.

Internet surveys are appealing because of their low cost, but obtaining good coverage of the target population is challenging. Persons with internet access may differ from those lacking internet access (Mohorko et al., 2013; Sterrett et al., 2017). At this writing, the most reliable surveys that use the internet to collect data start with a probability sample selected from a mail, telephone, or area frame. They contact and recruit the sampled individuals through the initial probability sample and then ask them to submit survey responses through the internet. Some survey organizations provide computers and internet access for persons in the sample who do not have them. These surveys thus include members of the population who do not have internet access.

It is challenging to assess coverage in other internet surveys (Couper, 2000). Internet surveys in which website visitors volunteer to participate are untrustworthy and cannot be used to estimate characteristics of a population. The coverage of such surveys is unknown because the sample consists of volunteers. Some of the volunteers may take the survey many times in an attempt to influence the results. Even if the sample matches the target population on demographic characteristics, or is weighted to match demographic characteristics through poststratification, as discussed in Section 15.3, other characteristics may differ.

Example 16.2. In April 2007 the city of Tempe, Arizona, arranged for a market research company to conduct a survey of Tempe residents to solicit opinions about a proposed neighborhood shuttle bus service. An announcement of the upcoming survey and a map of the proposed bus route was mailed to every address in the neighborhood in March, 2007. The market research company took a telephone survey of approximately 700 Tempe residents. Because the survey was done by telephone, the questionnaire started with screening questions to ensure the respondent lived in the area of interest. The respondent was then asked to refer to the map that had been mailed earlier to answer questions about proposed bus routes. Neighborhood residents who were not selected to be in the telephone sample were given the opportunity to respond to the survey over the internet.

Telephone was a poor choice for the survey mode. The telephone survey required screening questions to exclude persons not in the neighborhood. Cellular phones were not sampled, resulting in undercoverage in this neighborhood close to Arizona State University. The survey required respondents to refer to a map that had been mailed earlier, and it is likely that many respondents would not have ready access to that map when they were called, resulting in measurement error. In addition, city planners and neighborhood activists were interested in whether residents adjacent to the proposed bus route had different opinions than other residents. The telephone survey could not guarantee a sufficient sample size of residents along the route.

A mail survey would have been a much better choice. The city had already gone to the expense of mailing the map to all neighborhood residents. It could have easily selected a stratified sample (stratified by proximity to proposed route, with a higher sampling fraction for addresses close to the proposed route) of those addresses and included the survey in the envelopes mailed to the households in the stratified sample. The money saved by not taking a telephone sample could have been used for nonresponse follow-up. ■

16.1.3 Improving Coverage

As with nonresponse, the best way to deal with undercoverage and overcoverage is to prevent it. Some options for improving coverage in the survey design are:

- Check the sampling frame to remove duplicates and ensure the information is accurate.
- Compare the sampling frame with external sources to check for members of the target population that are missing in the frame.
- Choose a survey mode or modes that give frames with high coverage of the target population.
- Use a multiple frame survey. An area frame, though expensive to sample, often has good coverage of the population. Data from the area frame sample can then be combined with data from incomplete frames that are inexpensive to sample, as discussed in Section 14.1.4. Coverage can often be improved by combining samples from several incomplete frames; even if the union of the frames does not include the entire target population, the multiple frame survey will have better coverage than any of the frames taken singly.

Poststratification, discussed in Section 8.6, can partially alleviate coverage bias, but, as with all after-the-fact adjustments for nonresponse or coverage errors, you do not know whether the adjustment truly compensates for coverage bias unless you obtain data on the persons not covered by the sampling frame.

16.2 Nonresponse Error

In Chapter 8, we looked at weighting and imputation as remedies for nonresponse that has already occurred. It is far better, of course, to be able to prevent or reduce nonresponse before it occurs. The methods outlined in Section 8.2 can be used to reduce nonresponse at the survey design stage.

As with undercoverage, it is often challenging to assess the bias due to nonresponse. Section 8.8.3 described some methods that can be used. In some cases, you can obtain accurate data for nonrespondents from an external source such as a population register and use the external records to evaluate the bias due to nonresponse. In a health survey, you might be able to access the medical records of a subsample of nonrespondents and a subsample of respondents. You can then compare the respondents and nonrespondents on quantities given in the medical records. If the nonrespondents have significantly higher blood pressure than the respondents, they may differ from the respondents on key survey items as well.

Similar comparisons can be done if your sampling frame has substantial auxiliary information about each individual in the frame. A university administrator taking a survey of students can compare the grade point averages and majors of survey respondents with those of the survey nonrespondents. In addition to identifying potential nonresponse bias, the frame information can be used to construct weighting classes or impute values for nonrespondents.

Under a response propensity framework, nonresponse bias depends on the relation between the (unknown) response propensity ϕ_i of each unit and the variable of interest y_i (see Section 8.5.1). If ϕ_i and y_i are uncorrelated, then the bias incurred by estimating the population mean from the respondents alone is approximately zero. A model adopted for nonresponse that estimates the response propensities accurately will also reduce the nonresponse bias. Unfortunately, we do not know the correlation between the propensity to respond and the variables of interest, or whether the nonresponse-adjustment model compensates for nonresponse, because we do not observe y_i for the nonrespondents. We can, however, fit multiple models for nonresponse and investigate the sensitivity of the results to the modeling assumptions, as described in Section 8.8.3.

Groves (2006) concluded from a review of 30 empirical studies that nonresponse bias occurs but is not necessarily correlated with the response rate. Some studies have relatively high response rates and yet still have high bias, while other studies with lower response rates have low bias. In general, higher response rates are better and complete response is best of all, but a low response rate does not invalidate a survey.

Paradoxically, sometimes efforts to increase response rates can also increase nonresponse bias. This occurs, for example, when the measures taken to increase response rates also increase the correlation between ϕ_i and y_i. For example, an incentive given in a survey might increase the propensity of low-income persons to respond and thus result in more bias for estimates of income.

Much research (see, for example, Groves et al., 2002; Keusch, 2015; Dillman, 2020) has been done on why persons choose to respond to a survey and how surveys can be designed to increase cooperation. Cialdini (1984, 2009) identified factors associated with willingness to respond to a survey:

1. *Reciprocation.* Will the potential respondent gain something by participating in the survey? Singer (2011, 2016), summarizing experimental and survey research on survey participation, argued that people choose to respond to a survey "when in their subjective calculus the benefits of doing so outweigh the costs." The benefits can be personal or societal. Some persons who participate in the National Health and Nutrition Examination Survey (NHANES), for example, may do so in part because they are contributing to knowledge about health.

   Incentives can increase survey cooperation in some instances; Singer (2002) concluded that monetary incentives are most effective in surveys for which persons have few other motivations to participate. Singer and Ye (2013) summarized results from experiments that have been conducted about the effects of incentives on response rates.

2. *Authority.* Persons are often more likely to provide responses to a survey if it is issued by a recognized authority. University faculty members may be more likely to respond to a survey sent by the university president than a survey distributed by a graduate student. The U.S. Census Bureau (Griffin et al., 2003) sent one group of potential respondents a "mandatory" letter saying that participation in the ACS is "required by law. We are conducting this survey under the authority of Title 13, United States Code, sections 141–193, and 221." Another group was sent a "voluntary" letter saying, "Your participation in the survey is important; however, you may decline to answer any or all questions." The mail response rate was more than 20 percentage points higher with the "mandatory" letter than with the "voluntary" letter.

3. *Consistency.* Once someone is persuaded to participate in a survey, that person is likely to continue and perhaps participate in other surveys. Online panel surveys take samples from a set of panel members who have already been recruited, and the panel members may be asked to participate in multiple surveys.

4. *Scarcity.* The scarcity heuristic is related to reciprocation: A potential respondent who believes that the opportunity to participate in the survey is reserved for the select few may be more likely to respond.

5. *Social validation.* Potential respondents may be more likely to participate if they believe others do so.

6. *Liking.* Potential respondents may be more amenable to participation if they like the interviewer.

Persons choose to participate in a survey for many different reasons, so a flexible approach to soliciting responses is helpful. Different survey introductions may work better with some subsets of the population. Skilled interviewers use a variety of approaches to persuade persons to respond to a survey. We know that some nonresponse will occur despite the best efforts of the survey designer. Thus, it is valuable for the sampling frame to contain as much information as possible—not just for adjusting the estimates for nonresponse after the data are collected, but for giving interviewers additional information to use when recruiting respondents.

16.3 Measurement Error

In Chapters 2–7, we assumed that y_i, a characteristic of interest on unit i, is a fixed quantity measured without error. When there is measurement error, however, y_i is not the true characteristic of interest for unit i. Instead, there is some underlying "true" value μ_i, and y_i is a measurement of μ_i taken from the survey. For example, suppose that you want to find out $\mu_i =$ the true amount that household i spent on medical care between March and June of last year. The response provided by the household, y_i, is not necessarily equal to μ_i. The question may be worded confusingly so that the respondent omits some medical expenses (perhaps omitting over-the-counter medications); the respondent may forget some expenses or include expenses from July; characteristics of the interviewer may lead the respondent to give different answers (perhaps excluding expenses they are embarrassed by); or other circumstances may lead y_i to differ from μ_i. Measurement error is the difference between the value y_i provided by a survey respondent and the true value of the response, μ_i. Estimating measurement error, like estimating coverage and nonresponse bias, requires additional information and modeling.

Often, as for measuring the amount spent on medical care between March and June, μ_i is a fixed value that could be found exactly using specific definitions of "medical care" and "household." Demographic characteristics such as age or ethnicity, physical measurements such as body mass index, behavioral variables such as number of visits to doctors, and monetary variables can be thought of as having a true underlying value μ_i that could be determined if the definitions and measuring instruments were precise enough. In other cases, the true characteristic of interest may not have a precise physical meaning, as when a consumer confidence survey asks you whether you think you will be better off, worse off, or about the same financially a year from now. Although it would be possible to compare your financial status 12 months from now with your financial status now, that is not the point of the survey—the survey researchers want to know how optimistic or confident you are about your short-term financial future.

Psychometricians call a possibly unobservable underlying characteristic, in this case consumer confidence, a **construct**, and attempt to approximate the construct through items that can be measured. It is rarely possible for survey questions to correspond exactly to certain underlying constructs, however, which is the reason for the advice in Section 1.5 to report the actual questions asked when summarizing results from a survey.

The survey instrument, the interviewer, and the respondent all can contribute to measurement error. To reduce measurement error due to the survey instrument, follow Bradburn's (2004) Law for Questionnaires: "Ask what you want to know, not something else." Bradburn reviewed research by linguists, psychologists, and statisticians on how to reduce measurement error when constructing a questionnaire. Presser et al. (2004) described methods that can be used to test and evaluate questionnaires.

Interviewers can contribute to response variability and bias, and interviewer effects vary across modes of data collection. Interviewers can often increase response rates and improve accuracy by explaining questions to respondents. But some respondents may give a more socially desirable response to a survey conducted by an interviewer than to a self-administered survey, and may report that they exercise more, and gamble less, than they actually do. Some interviewers may prompt a respondent toward a particular response.

A particular interviewer may also affect the accuracy of the response by misreading questions, recording responses inaccurately, or antagonizing the respondent. In a survey about abortion, a poorly trained interviewer with strong feelings on the issue may encourage the respondent to provide one answer rather than another. In extreme cases, an interviewer may change the answers given by the respondent, or simply make up data and not contact the respondent at all (Murphy et al., 2016). Proper interviewer training and quality improvement methods can avoid many of these problems.

Respondents may deliberately or inadvertently provide inaccurate information to a survey. Respondents to a victimization survey may forget about criminal victimizations that have occurred to them or, if reporting for another person, be unaware of that person's experiences with criminal victimization. A respondent may choose to omit reporting an incident of domestic violence to the survey, particularly if the perpetrator is present during the interview.

16.3.1 Measuring and Modeling Measurement Error

Suppose that T replications of the measurement of unit i could be taken, and let y_{it} be the value of the tth replicate measurement on unit i. A simple additive model for the measurement error is:

$$y_{it} = \mu_i + \beta_i + \varepsilon_{it}, \tag{16.2}$$

where β_i is a fixed bias for respondent i and ε_{it} is a random variable representing unexplained sources of measurement error. In the simplest model, the ε_{it}'s are assumed to be independent random variables with mean 0 and variance σ_i^2. The assumptions of this model imply that all conditions remain the same for replicate measurements, and that there are no carryover effects for multiple responses of the same person.

Validity and reliability. If μ_i is the true characteristic of interest, the survey measurement y_{it} should be as close to μ_i as possible. For the model in (16.2), β_i and σ_i^2 should both be close to zero. In psychometrics, two concepts called **validity** and **reliability** are used to assess this closeness.

Validity deals with how well the survey item measures the true value μ_i. Many types of validity have been proposed (Groves, 1989); we define theoretical validity to be the correlation between the true score and its observed value: theoretical validity $=$ Corr (y_i, μ_i).

Sometimes you can find the true value μ_i by checking external records. In a survey asking the question "Did you vote in the election on November 3, 2020?" you may be able to check the voting records to determine whether the person actually voted (of course, you must have an accurate way to link persons from the survey to the voting records for this to work). Then $\mu_i = 1$ if the person is listed as voting in the voting records and 0 otherwise; $y_i = 1$ if the person responds that he or she voted. The validity of the question is estimated by the correlation between y_i and μ_i. If there is no external source of the true value, you can sometimes estimate validity by other methods such as looking at the correlations among answers to closely related questions.

Note that theoretical validity is not the same thing as unbiasedness or accuracy. Suppose μ_i is weight of person i, and the scale has negligible variability but erroneously adds 5 kg to every measurement. Then Corr $(y_i, \mu_i) \approx 1$ but $E_M[y_i|\mu_i] = \mu_i + 5$; in a simple random sample (SRS), $\bar{y}$ will overestimate the true mean weight of the population. In general, you need an external source of information to be able to evaluate measurement bias.

Reliability deals with variability of responses under repeated measurements. If all the values of σ_i^2 are equal to σ^2 in the model in (16.2),

$$\text{Reliability} = \frac{V_M(\mu)}{\sigma^2 + V_M(\mu)} = \frac{\text{variance of true values}}{\text{variance of values reported to the survey}}, \tag{16.3}$$

where $V_M(\mu) = \sum_{i=1}^{N}(\mu_i - \bar{\mu})^2/(N-1)$, $\bar{\mu} = \sum_{i=1}^{N}\mu_i/N$, and the subscript M denotes inference under the model. If the reliability is 1, then $\sigma^2 = 0$, that is, respondent i gives exactly the same answer over repeated trials. If the answers of respondent i are highly variable over repeated trials, then the reliability is low.

Cronbach's alpha (Cronbach, 1951) is often used to estimate reliability when multiple questions are used to assess the same construct:

$$\alpha = \frac{k\bar{r}}{1 + (k-1)\bar{r}},$$

where k is the number of items, and $\bar{r}$ is the average of the pairwise correlations of the items. If α is close to one, then there is high reliability. High reliability can occur, however, when the questionnaire is constructed so that answers to one question affect answers to another. It is possible for all questions to be highly consistent yet for none of them to measure the true construct of interest.

Measurement error from interviewers. The model in (16.2) can be expanded with terms to represent effects of interviewers on the response given. Let y_{ijt} be the response given by respondent i to interviewer j on replicate t:

$$y_{ijt} = \mu_i + \beta_i + b_j + \varepsilon_{ijt} \tag{16.4}$$

where b_j is the systematic effect of interviewer j, with $E_M(b_j) = 0$, $V_M(b_j) = \sigma_b^2$, $E_M(\varepsilon_{ijt}) = 0$, $V_M(\varepsilon_{ijt}) = \sigma_i^2$, and that all of the b_j's and ε_{ijt}'s are uncorrelated. In this model, replicate measurements must be taken from the respondents in order to distinguish the error component ε_{ijt} from the sampling error.

In model (16.4), it is assumed that any respondent asked a question by interviewer j is likely to deviate from the true value by an amount b_j that is intrinsic to interviewer j. For example, in a health survey, perhaps Fred has a tendency to take blood pressure readings just a little below the true value. Then every person examined by Fred will have a blood pressure reading that is slightly too low, and respondents examined by Fred will tend to be more similar to each other than respondents selected at random. Or, in a victimization survey, respondents may tend to find an interviewer more sympathetic and be more likely to tell him or her about victimizations. That interviewer would tend to have more reported victimizations than other interviewers. The variability due to interviewers can be estimated using standard methods for mixed models (Demidenko, 2013).

The model in (16.4) can be expanded by including interaction effects between interviewers and respondents; for example, it might be thought that female respondents will report a different number of criminal victimizations to a female interviewer than to a male interviewer. Terms for mode effects can be added in a mixed-mode survey.

Mahalanobis (1946) proposed interpenetrating subsampling for estimating interviewer effects. The basic idea is the same as for estimating the variance of systematic sampling (Section 5.5): Assign each interviewer a random subsample of the interviews. Often in surveys, interviewers are assigned according to convenience; for example, an interviewer might be assigned to all households in a psu, which confounds the effect of the psu with the effect of the interviewer. In interpenetrating subsampling with an SRS, interviewers are assigned households at random.

16.3.2 Reducing Measurement Error

The first step in reducing measurement error is to estimate its prevalence and identify the main sources. If the largest component of measurement error is interviewer variability, then more standardized interview procedures may reduce the variance component. If respondents misinterpret questions, then better questions should be written and tested.

We recommend collecting data using randomized experiments to estimate components of variability and likely sources of bias. Hartley and Rao (1978), Hartley and Biemer (1981), and O'Muircheartaigh and Campanelli (1998) presented experimental designs that can be used to estimate interviewer variability from surveys. Randomized experiments, conducted before a survey is implemented, can compare alternative versions of questionnaires, field procedures, methods of interviewer training, and almost any other factor affecting survey quality. Reducing measurement error, like other aspects of improving survey quality, is an ongoing process.

Fowler (1991) provided advice for reducing interviewer-related measurement error:

- *Test your questions.* Interview potential respondents to see if they interpret the questions as you intend.

- *Write clear questions.* If a respondent does not know how to answer a question, the interviewer is likely to have more influence on the response. In a self-administered survey, unclear questions can lead to more variability or bias in the responses. Open-ended questions may be more susceptible to interviewer effects than closed questions.

- *Institute procedures for administering the survey that will reduce errors.*

- *Provide training and supervision for interviewers so they act consistently.* Interviewers should read the questions exactly as written, and should not indicate that one response is preferred over another. An interviewer should have a professional and neutral demeanor.

- *Give interviewers a reasonable workload.* Deming (1986) argued that assigning numerical quotas to workers decreases quality: An industrial worker who is required to make 130 parts per hour cannot pay attention to the quality of the part. Cannell et al. (1977) found similar effects for survey interviewers: Interviewers with high assignments had more errors in responses.

- *Apply quality improvement principles to the interviewing process.* Montgomery (2012) described quality improvement methods for many settings. Reducing measurement error in surveys fits nicely into this framework.

16.3.3 Sensitive Questions

Many surveys involve questions that persons might view as sensitive. The ACS asks respondents to report their income in the past 12 months from eight different possible sources of income (wages, alimony and child support, interest income, and other sources). The National Survey on Drug Use and Health asks respondents about their use of tobacco, alcohol, and drugs such as marijuana and cocaine (Substance Abuse and Mental Health Services Administration, 2020). Some respondents view such questions as intrusive; others may fear that providing accurate information may expose them to penalties (for example, they may fear that reporting their true income on a survey may lead to penalties for underpayment of income taxes). Some persons may protect their personal information by refusing to respond to the survey or to specific items, while others may give inaccurate answers to sensitive questions.

Reputable survey organizations promise respondents that their answers will be kept confidential, and keep that promise. Singer (2003) reported that persons who said they were concerned about confidentiality were less likely to return their census forms by mail, although other factors such as age had a higher association with nonresponse than did confidentiality concerns. Even if promises of confidentiality do not influence the response rate, they are an ethical obligation to the respondents.

There is much evidence that many people simply do not provide accurate answers to sensitive questions. Tourangeau et al. (2000, Chapter 9) summarized studies in which record checks indicate underreporting of certain behaviors. Urine samples may contradict persons who say they do not use illegal drugs; counts of abortions from abortion clinics may exceed estimates of the total number of abortions from surveys. There is also overreporting of behaviors that are deemed to be socially desirable: Kreuter et al. (2008) found that college alumni who had experienced academic problems as undergraduates were more likely to misreport their academic records (and also less likely to respond to the survey).

As with coverage, mode of administration can have a great effect on responses to sensitive questions (see Tourangeau and Smith, 1996; Kreuter et al., 2008). Numerous studies report that higher percentages of people say they have used illegal drugs when they fill out the questionnaire themselves than when the questionnaire is administered by an interviewer (Tourangeau et al., 2000, p. 295). One option for sensitive topics is to use computer-assisted self-administration, where the respondent types answers directly onto the computer. An interviewer may be in the room to answer questions, but the interviewer does not see the responses typed into the computer.

16.3.4 Randomized Response

Suppose that the main goal of the survey is to learn about sensitive questions such as "Do you use cocaine?" or "Have you ever shoplifted?" or "Did you understate your income on your tax return?"

These are all questions that "yes" respondents could be expected to lie about. A question form that encourages truthful answers but makes people comfortable is desired. Horvitz et al. (1967), in a variation of Warner's (1965) original idea, suggested using two questions: the sensitive question and an innocuous question, and using a randomizing device (such as a coin flip) to determine which question the respondent should answer. If a coin flip is used as the randomizing device, the respondent might be instructed to answer the question "Did you use cocaine in the past week?" if the coin is heads, and "Is the second hand on your watch between 0 and 30?" if the coin is tails. The interviewer does not know whether the coin was heads or tails, and hence does not know which question is being answered. It is hoped that the randomization, and the knowledge that the interviewer does not know which question is being answered, will encourage respondents to tell the truth if they have used cocaine in the past week.

The randomizing device can be anything, but it must have known probability P that the person is asked the sensitive question and probability $1-P$ that the person is asked the innocuous question. Other forms of randomized response are described in Blair et al. (2015). The key to randomized response is that the probability that the person responds yes to the innocuous question, p_I, is known. We want to estimate p_S, the proportion responding yes to the sensitive question. If everyone answers the questions truthfully, then

$$\begin{aligned}
\phi &= P(\text{respondent replies yes}) \\
&= P(\text{yes} \mid \text{asked sensitive question})P(\text{asked sensitive question}) \\
&\quad + P(\text{yes} \mid \text{asked innocuous question})P(\text{asked innocuous question}) \\
&= p_S P + p_I(1-P).
\end{aligned}$$

Let $\hat{\phi}$ be the estimated proportion of "yesses" from the sample. Since both P and p_i are known, p_S may be estimated by

$$\hat{p}_S = \frac{\hat{\phi} - (1-P)p_I}{P}. \tag{16.5}$$

Then the estimated variance of $\hat{p}_S$ is

$$\hat{V}(\hat{p}_S) = \frac{\hat{V}(\hat{\phi})}{P^2}.$$

The penalty for randomized response appears in the factor $1/P^2$ in the estimated variance. If $P = 1/3$, for example, the variance is nine times as great as it would have been had everyone in the sample been asked the sensitive question and responded truthfully. The larger P is, the smaller the variance of $\hat{p}_S$. But if P is too large, respondents may think that the interviewer will know which question is being answered. Some respondents may think that only a P of 0.5 is "fair" and that no other probabilities exist when choosing among two items.

Example 16.3. An SRS of high school seniors is selected. Each senior in the sample is presented with a card containing the following two questions:

Question 1: Have you ever cheated on an exam?
Question 2: Were you born in July?

We know from birth records that $p_I = 0.085$. Suppose the randomizing device is a spinner, with $P = 1/5$. Of the 800 people surveyed, 175 say yes to whichever question the spinner indicated they should answer. Then $\hat{\phi} = 175/800$. Because this is an SRS,

$$\hat{V}(\hat{\phi}) = \hat{\phi}(1-\hat{\phi})/(n-1) = 0.0002139.$$

Thus,

$$\hat{p}_S = \frac{175/800 - (4/5).085}{1/5} = .75375,$$

and $\hat{V}(\hat{p}_S) = (0.0002139)/(1/5)^2 = 0.0053.$ ■

Before using randomized response methods in your survey, you should test the method to see if the extra complication does indeed increase compliance and reduce bias. Danermark and Swensson (1987) found that randomized response methods worked well for estimating drug use in schools and appeared to reduce response bias. Duffy and Waterton (1988), however, found that randomized response methods were not helpful in their survey to estimate incidence of various alcohol-related problems in Edinburgh, Scotland. They compared response rates and responses for a randomized response group with those for a group asked the questions directly, and found that the randomized response group had a lower response rate and lower estimated proportion of persons who had drunk more than the legal limit immediately before driving a car. Randomized response did, however, increase the complexity of the interviews, and interviewers reported that many persons were confused by the method.

16.4 Processing Error

Measurement error occurs when an answer given by a respondent differs from the truth. Sometimes the answer given (or intended) by the respondent differs from the value that ends up in the data set; that is a data processing error. Data processing errors include errors from data entry, coding, editing, calculation of survey weights, and tabulation of statistics. Statistical quality improvement methods can be used to measure and reduce them.

Data entry errors are less common with modern surveys than in the past, when all data were entered manually by clerks. Surveys in which respondents enter their own data directly into the computer often have few data entry errors. For surveys conducted by other modes, errors can occur when data are transcribed or scanned into the data set. Character recognition software may misread answers or pick up stray marks.

Some surveys require an additional step in data entry, where the answers given by the respondent are coded as one of the allowable response categories. Coding error occurs when the coded response misrepresents the original answer. Open-ended questions are particularly prone to coding error, since someone must make a decision about how to classify a response. For example, some crime victimization surveys ask the respondent to describe each incident in words. Using that description, the coder classifies the crime as burglary, assault, robbery, theft, or another crime in the allowable list of types. But the description may be ambiguous: was the broken window the result of vandalism (malicious destruction of property) or burglary (forced entry with intent to commit a felony or theft)?

Most survey organizations edit data files to remove internal inconsistencies and correct obvious errors. Some organizations also impute values for missing data, or for items that are internally inconsistent. Public-use data files may be edited to protect confidentiality of respondents' data—observations may be swapped from one location to another or noise may be added to the answers provided by the respondents.

Example 16.4. Data provided by respondents to the ACS may contain internal inconsistencies. This occurs when, for example, a person with age 103 is listed as the child of another household member, or a person born in Arizona is not listed as a citizen. Other combinations of responses may be possible but implausible, for example, when property taxes for a housing unit are much higher than would be expected from the property value and other variables. The Census Bureau edits the data to resolve these types of inconsistencies, using a lengthy list of editing rules that have been developed for each topic (U.S. Census Bureau, 2014). ■

Editing, in general, is thought to remove errors introduced in other stages of data collection. Some edits, however, can introduce errors. Granquist and Kovar (1997) reported that "...it was not until a demographer 'discovered' that wives are on average two years younger than their husbands that the edit rule which performed this exact imputation was removed from the Canadian Census system!"

16.5 Total Survey Quality

A survey is a system of procedures intended to produce data and statistics that meet the needs of the organization sponsoring the survey and, in the case of government surveys, for the public good. Consequently, a systems-based approach, considering all aspects of the system, is needed to ensure the quality of the statistics that are produced.

It can happen that reducing one source of error actually increases another. For example, heroic efforts by interviewers might result in some of the die-hard nonrespondents providing answers for the survey. But there is no guarantee that those responses are accurate—reluctant respondents may make up data to stop the calls (Olson, 2013). Similarly, a monetary incentive intended to increase the survey response rate may affect other aspects of accuracy. West and Olson (2010) and West et al. (2018) noted that interviewers play a dual role in recruitment and asking questions, and thus can contribute to both nonresponse and measurement error. Särndal et al. (2018) considered the interaction between the sampling design and nonresponse error.

A systems-based approach also recognizes that accuracy of estimates, which is the focus of total survey error (and of most of this book), is important but is not the only dimension of quality.

Example 16.5. During the spring of 2020, the U.S. Census Bureau initiated the Household Pulse Survey to measure the social and economic effects of the COVID-19 pandemic on American households. Surveys conducted by the Census Bureau usually undergo years of planning and testing before being launched but, because of the need for timely data, data collection for this survey began within a month of its conception. Households in the Census Bureau's master address file were contacted by e-mail or text message. Fields et al. (2020) explained the trade-off: "These modes are expected to yield response rates much lower than traditional in-person or mail surveys usually conducted by the U.S. Census Bureau. The benefits to this collection plan are implementation efficiency, cost, and timeliness of responses."

The weighted response rates for the weekly surveys conducted between April and July of 2020 were less than 5%, and the samples did not include households lacking e-mail and cell phone contact information. The data were weighted to attempt to reduce errors from undercoverage and nonresponse, but if respondent households within a weighting class differed systematically from the nonrespondents or non-covered households in that class, the weighted estimates would still have bias (see U.S. Census Bureau, 2020d, for a description of the weighting procedures and Peterson et al., 2021, for nonresponse bias analyses). However, it would have taken months or years to implement a survey with the high coverage and response rates enjoyed by the ACS—too late for the data to be of use. The Household Pulse Survey provided rapid data needed at a time of crisis. ■

Much research is still needed on how to estimate and improve survey quality. Biemer and Lyberg (2003) recommended a holistic approach to survey design, considering all possible sources of errors at the design stage. Most of this book has focused on methods for reducing and estimating sampling error using a design-based approach. The study of error sources involves proposing and fitting stochastic models for the sources or error; commonly, mixed models are used that incorporate terms describing bias and different sources of variability. A multivariate approach is needed since most surveys have multiple responses and errors may be correlated among different responses.

Survey quality must be an ongoing effort that involves every part of the organization. Populations, response propensities, and measurement methods change, and a survey organization must continually seek to improve procedures for collecting and analyzing data (Biemer and Caspar, 1994).

Example 16.6. *Continual quality improvement for the ACS.* The Census Bureau has an ongoing research program to improve the quality and utility of data from the ACS. Questions considered for inclusion on the questionnaire undergo extensive field testing (U.S. Census Bureau, 2017). Reports on the research are regularly posted on the ACS web page at `www.census.gov`. Recent examples include research on possibilities for reducing the length of the questionnaire by substituting information from property tax records (Dillon, 2019), exploring effects of messages on the envelope and wording about the mandatory nature of the ACS on self-response rate (Risley and Berkley, 2020), and using administrative records to edit and impute data for the ACS (Clark, 2020). ■

The quality of the survey should be communicated to data users. Kasprzyk and Giesbrecht (2003) recommended that even abbreviated reports on survey results should contain the following:

- Information about the data set, including whether it was based on a probability sample
- Sources of sampling and nonsampling error
- Total in-scope sample size
- Unit nonresponse rates
- A reference to more detailed information about data collection
- A contact for more information

The Eurostat (2020) guidelines specify that a full survey report should include 18 topics that address the quality dimensions listed in Table 16.1. These include, in addition to measures of sampling and nonsampling errors, descriptions of steps taken to protect data confidentiality and security, and descriptions of the procedures used to promote data quality.

Some organizations, claiming that sampling error is a small part of total survey error, have returned to taking convenience samples, using opt-in internet panels of respondents. As discussed in Chapter 15, models must be used for inference in these surveys, and there is no guarantee that the model assumptions are met. Some of them create potential biases by offering financial rewards to persons who volunteer to take surveys, and all of them live outside the framework of design-based inference.

Probability samples are not perfect, and are subject to nonresponse and other nonsampling errors. But they remove many possible sources of bias, including the possibility that advocacy groups will bias a poll by encouraging their members to participate. The arguments put forward by some proponents of inexpensive convenience samples that their results agree with those from organizations that take probability samples do not prove the quality of their surveys. After all, the procedures used by the *Literary Digest* poll discussed in Section 1.1 resulted in accurate election predictions for 1920, 1924, 1928, and 1932—before spectacularly misfiring in 1936.

Groves (2006, p. 670) wrote:

> Probability sampling offers measurable sampling errors and unbiased estimates when 100 percent response rates are obtained. There is no such guarantee with low response rate surveys. Thus, within the probability sampling paradigm, high response rates are valued. Unfortunately, the alternative research designs for descriptive statistics, most notably volunteer panels, quota samples from large compilations of personal data records, and so forth, require even more heroic assumptions to derive the unbiased survey estimates.

Sampling and the future. Statistical sampling is a relatively young field, with many dating its origin as a modern discipline to Anders Kiaer's 1895 proposal to take "representative samples" (Kiaer, 1896). The discipline has been spurred by societal needs as well as technological developments. Over the past 125 years, survey researchers have developed methods for probability sampling, nonresponse and undercoverage adjustment, measurement error models, designed experiments for improving sample design and reducing nonsampling errors, computer-intensive inference, small area estimation, sampling rare populations, and many other applications.

You can now solve some of the challenges of the next 125 years. The fundamental problem of survey sampling remains the same as it was in 1895: How can data from a sample be used to infer properties of the population? The increasing availability of massive amounts of data from the internet or automated data collections does not change the problem when those data come from only part of the population—the statistician must still generalize from what was seen to what was not seen.

To date, probability sampling is the only method that has been developed that allows the statistician to make inferences about unsampled parts of the population without making assumptions about their structure. But the properties of probability sampling depend on there being full coverage, full response, and accurate measurements. With the increasing availability of large data sets from nonprobability samples, research is needed to develop ever-improving methods for using such information (National Academies of Sciences, Engineering, and Medicine, 2017, 2018). At the same time, methods from sampling theory can be used to improve data collection in other arenas, through improving coverage, taking probability samples to assess data quality, and developing models for missing data.

As Gertrude Cox (1957) said in her 1956 presidential address to the American Statistical Association, "We are surrounded with ever widening horizons of thought, which demand that we find better ways of analytical thinking. We must recognize that the observer is part of what he observes and that the thinker is part of what he thinks. We cannot passively observe the statistical universe as outsiders, for we are all in it."

16.6 Chapter Summary

Total survey error is the sum of five components: coverage error, nonresponse error, measurement error, processing error, and sampling error. The main concern with undercoverage and nonresponse is bias. Sampling error produces variability in the estimates. Measurement and processing error have both bias and variance aspects.

Reports of survey results should include an assessment of errors. Quality improvement methods can be used to control errors throughout the survey process. Designed experiments are useful for improving survey quality.

Key Terms

Processing error: Error that arises during data processing, including errors from data entry, coding, and editing. This also includes errors in calculating survey weights or imputed values.

Reliability: The overall consistency of a measure; it has high reliability if it produces similar results when administered under the same conditions.

Total survey design: A system for designing a survey that meets its objectives, through considering all features of the survey collection system.

Total survey error: All errors from the survey process, including sampling, coverage, nonresponse, measurement, and processing error.

Validity: How well an assessment measures what it is intended to measure.

For Further Reading

The book edited by Biemer et al. (2017) reviews the history of the concepts of total survey error and total survey design, and discusses how to improve survey quality through careful survey design and processing. The authors apply the ideas to surveys on topics ranging from immunization to crime, with consideration of online, smartphone, and social media surveys as well as surveys conducted in-person, by telephone, and through mail.

The collection of essays in Engel et al. (2015) give research-based guidance for improving response rates, treating nonresponse, obtaining accurate answers to sensitive questions, conducting online panel surveys, measuring effect of survey mode, reducing variation from interviewers, and more. Lessler and Kalsbeek (1992), Linacre and Trewin (1993), Groves and Lyberg (2010), Biemer (2010), Lyberg (2012), and Weisberg (2018) discuss components of total survey error and design. Lyberg and Weisberg (2016) describes sources of survey error as well as how to measure and reduce errors. The books edited by Lyberg et al. (1997), de Leeuw et al. (2008), and Pfeffermann and Rao (2009a) each contain several chapters on improving quality in survey data collection. Biemer and Lyberg (2003), Groves et al. (2011), and Dillman et al. (2014) present methods for improving survey quality at the design stage. Dillman (2020) discusses a systems-based approach to nonresponse in surveys and outlines how multiple response-rate-improving techniques can interact. Yung et al. (2018) discuss areas where machine learning methods might improve the quality of official statistics.

Marker and Morganstein (2004) describe the use of statistical quality improvement methods in survey organizations. Quality improvement programs require that every person in the organization, starting with leadership, be committed to survey quality. Designed experiments and methods used for quality improvement in industry are also useful for improving survey quality. The books on experimental design and quality improvement by Deming (1986), Oehlert (2000), and Montgomery (2012), while not specific for survey operations, give principles that can be applied to survey design and processing. The book edited by Lavrakas et al. (2020) describes methods for embedding designed experiments in a survey.

The guidelines from Statistics Canada (2019) and Eurostat (2019, 2020) describe procedures for assuring quality in all steps of survey development and implementation, and for reporting the quality of statistics. Daas et al. (2009) provide a checklist for assessing the quality of administrative data. Manski (2015) stresses the importance of communicating uncertainty about statistics. Numerous organizations (see, for example, American Association of Public Opinion Research, 2016a; United Nations Statistics Division, 2014; National Academies of Sciences, Engineering, and Medicine, 2021) have outlined principles and practices that can help an organization obtain high-quality statistics through a survey.

For more information on measurement error models, see Alwin (2010) and Bohrnstedt (2010). Schaeffer et al. (2010) discuss interviewer effects. Geisen and Bergstrom (2017) tell how to conduct usability tests on surveys—to help ensure that interviewers and respondents can record responses easily and accurately. Having great questions will not help you if survey respondents cannot figure out how to enter their answers.

For the very important topic of protecting the confidentiality of data provided by respondents, see Rinott et al. (2018) and Reiter (2019). Two reports by the National Academies of Sciences, Engineering, and Medicine (2017, 2018) review issues and methods for protecting confidentiality of respondents' data. The *Journal of Privacy and Confidentiality* publishes many articles on the topic.

See Brick (2011), Singer (2016), Rao and Fuller (2017), and Beaumont (2020) for thoughts about the future of sampling. Citro (2014), Lohr and Raghunathan (2017), Zhang and Chambers (2019), Thompson (2019), and Rao (2021) discuss options for integrating data sources to improve quality of estimates.

16.7 Exercises

A. Introductory Exercises

1. The National Do Not Call Registry allows U.S. residents to prohibit telemarketers from calling the registered numbers. Conkey (2005) reported on a telephone survey conducted by the Customer Care Alliance. Respondents who had signed up for the registry were asked about their satisfaction with the list; 51% said they are still getting calls they thought the registry was supposed to block. Discuss possible sources of measurement error in this survey.

2. When the Current Population Survey asked the adults in its sample of 50,000 households if they voted in the 2000 presidential election, 55% said they had. The margin of error was less than 1%. But only 51% of the adult population actually voted in that election. Why do you think the difference between the CPS estimate and the election records was so much larger than the margin of error? What might be done to study the source(s) of the error?

3. A university wishes to estimate the proportion of its students who have used cocaine. Students were classified into one of three groups: undergraduate, graduate, or professional school (medical or law school) and were sampled randomly within the groups. Since there was some concern that students might be unwilling to disclose their use of cocaine to a university official, the following method was used. Thirty red balls, sixteen blue balls, and four white balls were placed in a box and mixed well. The student was then asked to draw one ball from the box. If the ball drawn was red, the person answered question (a). Otherwise question (b) was answered.

   Question (a): Have you ever used cocaine?
   Question (b): Is the ball you drew white?

The results are as follows:

Group	Undergraduates	Graduates	Professional
Number of students in group	8972	1548	860
Number of students sampled	900	150	80
Number answering yes	123	27	27

Assuming that all responses were truthful, estimate the proportion of students who have used cocaine and report the standard error of your estimate. Compare this standard error with the standard error you would expect to have if you asked the sample students question (a) directly and if all answered truthfully.

Now suppose that all respondents answer truthfully with the randomized response method but 25% of those who have used cocaine deny the fact when asked directly. Which method gives an estimate of the overall proportion of students who have used cocaine with the smallest MSE?

B. Working with Survey Data

4. Ulrich et al. (2018) used randomized response techniques in surveys to estimate the prevalence of doping in sports. They asked athletes participating in two international competitions to take a survey on a tablet computer (the survey was available in multiple languages). The athletes were asked to think of a person close to them (for example, a parent, sibling, or spouse) whose birthday they knew. If the birthday was in the first 10 days of the month, the athlete was asked to answer an innocuous question: "Is the person's date of birth in the first half of the year (January through June inclusive)?" If the birthday was between the 11th and 31st day of the month, the athlete was asked to answer the question of interest: "Have you knowingly violated anti-doping regulations by using a prohibited substance or method in the last 12 months?" Both questions were on the same screen, and the athlete was assured, "Note that only you can know which of the two questions you are answering!"

   (a) In the first survey, 551 of the 1202 athletes answered yes. Give an estimate and 95% CI for p_S.

   (b) In the second survey, 528 of the 965 athletes answered yes. Give an estimate and 95% CI for p_S.

   (c) What assumptions are made when calculating these estimates?

C. Working with Theory

5. Kuk (1990) proposed the following randomized response method. Ask the respondent to generate two independent binary variables X_1 and X_2 with $P(X_1 = 1) = \theta_1$ and $P(X_2 = 1) = \theta_2$. The probabilities θ_1 and θ_2 are known. Now ask the respondent to tell you the value of X_1 if the respondent is in the sensitive class, and X_2 if not in the sensitive class. Suppose the true proportion of persons in the sensitive class is p_S.

   (a) What is the probability that the respondent reports 1?

   (b) Using your answer to (a), give an estimator $\hat{p}_S$ of p_S. What conditions must θ_1 and θ_2 satisfy to allow estimation of p_S?

   (c) What is $V(\hat{p}_S)$ if an SRS is taken?

6. Randomized response methods can be applied retroactively to data that have already been collected in order to protect the confidentiality of responses (Dwork and Roth, 2014, give general theoretical results for using randomization mechanisms to protect privacy). Let μ_i denote the response of person i to a question with possible answers 0 (no) and 1 (yes). Then flip a coin where the probability of heads is P. If the coin is heads, then $y_i = \mu_i$. If the coin is tails, flip a second coin where the probability of heads is p_I; set $y_i = 1$ if the second coin is heads and $y_i = 0$ if the second coin is tails.

   (a) Find $P(y_i = j|\mu_i = k)$ for each combination $(j, k) \in \{(0,0), (0,1), (1,0), (1,1)\}$.

   (b) Find the smallest value of M satisfying $P(y_i = 1|\mu_i = 1) \leq MP(y_i = 1|\mu_i = 0)$ and $P(y_i = 0|\mu_i = 0) \leq MP(y_i = 0|\mu_i = 1)$.

   (c) Let $p_S = \sum_{i=1}^{N} \mu_i/N$. Suppose a simple random sample of size n is taken, and let $\hat{\phi} = \bar{y}$ and $\hat{p}_S = [\hat{\phi} - (1-P)p_I]/P$ as in (16.5). Explore the values of $V(\hat{p}_S)$ and M for all 125 combinations where p_S, p_I, and P are in the set $\{0.1, 0.3, 0.5, 0.7, 0.9\}$. What values would you choose for p_I and P if you want both $V(\hat{p}_S)$ and M to be small?

7. (Requires linear models.) In Section 16.3.1, we discussed the reliability of a survey instrument. Suppose we measure each individual twice, under the same conditions. Let X_i be the score of person i on the first survey administration, and let Y_i be the score of person i on the second survey administration. Consider the following model: Assume $U_i \sim N(\mu, \sigma_S^2)$. Now let $X_i = U_i + R_{i1}$ and $Y_i = U_i + R_{i2}$, where R_{i1} and R_{i2} are independent $N(0, \sigma_R^2)$ random variables (and are also independent of U_i).

   (a) Using matrices, find the distribution of $\begin{bmatrix} X_i \\ Y_i \end{bmatrix}$.

   (b) What is the reliability of the test under this model?

   (c) Find $E[Y \mid X = x]$ and $V[Y \mid X = x]$.

8. Find $V(y_{ijt})$ and $\text{Cov}(y_{ijt}, y_{ijs})$ (for $s \neq t$) under the model in (16.4). Using these quantities, define an intraclass correlation coefficient for interviewers analogous to that for clustering given in (5.39).

D. Projects and Activities

9. Read the article "On errors in surveys" (Deming, 1944). What sources of error identified by Deming are still considered part of total survey error? Did Deming discuss any errors that are no longer relevant? What new sources of error have arisen since Deming wrote the article?

10. Return to the survey you critiqued in Exercise 36 of Chapter 1 or Exercise 33 of Chapter 7. What sources of error were reported? What steps were taken to implement survey quality? What suggestions do you have for improvement?

11. Read the guidelines on statistical ethics by the American Statistical Association (2018) or the International Statistical Institute (2010). What specific recommendations in the guidelines apply to survey research?

12. Read the Code of Ethics from the American Association of Public Opinion Research (2015). Give examples of how adhering to the standards in this code might improve the quality of survey data.

13. *Activity for course project.* What methods were used to improve the quality of the survey you studied in Exercise 35 of Chapter 7? In your opinion, how effective were these methods? How could the survey have been improved?

A

Probability Concepts Used in Sampling

I recollect nothing that passed that day, except Johnson's quickness, who, when Dr. Beattie observed, as something remarkable which had happened to him, that he had chanced to see both No. 1, and No. 1000, of the hackney-coaches, the first and the last; "Why, Sir, (said Johnson,) there is an equal chance for one's seeing those two numbers as any other two." He was clearly right; yet the seeing of the two extremes, each of which is in some degree more conspicuous than the rest, could not but strike one in a stronger manner than the sight of any other two numbers.

—James Boswell, *The Life of Samuel Johnson*

The essence of probability sampling is that we can calculate the probability with which any subset of observations in the population will be selected as the sample. Most of the randomization theory results used in this book depend on probability concepts for their proof. In this appendix, we present a brief review of some of the basic ideas used. The reader should consult a more comprehensive reference on probability, such as Ross (2019) or Blitzstein and Hwang (2019), for more detail and for derivations and proofs.

Because all work in randomization theory concerns discrete random variables, only results for discrete random variables are given in this section. We use the results from Sections A.1–A.3 in Chapters 2–4, and the results from Sections A.3–A.4 in Chapters 5 and 6.

A.1 Probability

Consider performing an experiment in which you can write out all of the outcomes that could possibly happen, but you do not know exactly which one of those outcomes will occur. You might flip a coin, or draw a card from a deck, or pick three names out of a hat containing 20 names. Probabilities are assigned to the different outcomes and to sets composed of outcomes (called **events**), in accordance with the likelihood that the events will occur. Let Ω be the **sample space**, the list of all possible outcomes. For flipping a coin, $\Omega = \{\text{heads, tails}\}$. Probabilities in finite sample spaces have three basic properties:

1. $\text{P}(\Omega) = 1$.
2. For any event A, $0 \le P(A) \le 1$.
3. If the events $A_1, \ldots, A_k$ are disjoint, then $P\left(\bigcup_{i=1}^{k} A_i\right) = \sum_{i=1}^{k} P(A_i)$.

In sampling, we have a population of N units and use a probability sampling scheme to select n of those units. We can think of those N units as balls in a box labeled 1 through N, and we draw n balls from the box. For illustration, suppose $N = 5$ and $n = 2$. Then we draw two labeled balls out of the box in Figure A.1.

DOI: 10.1201/9780429298899-A

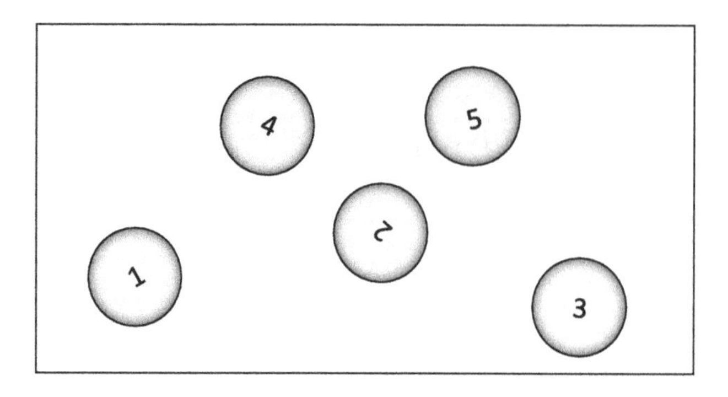

FIGURE A.1
Box containing labeled balls.

If we take a simple random sample (SRS) of one ball, each ball has an equal probability $1/N$ of being chosen as the sample.

A.1.1 Simple Random Sampling with Replacement

In a simple random sample with replacement (SRSWR), we put a ball back after it is chosen, so the same population is used on successive draws from the population. For the box in Figure A.1, there are 25 possible samples (a, b) in Ω, where a represents the first ball chosen and b represents the second ball chosen:

$$\begin{array}{ccccc} (1,1) & (2,1) & (3,1) & (4,1) & (5,1) \\ (1,2) & (2,2) & (3,2) & (4,2) & (5,2) \\ (1,3) & (2,3) & (3,3) & (4,3) & (5,3) \\ (1,4) & (2,4) & (3,4) & (4,4) & (5,4) \\ (1,5) & (2,5) & (3,5) & (4,5) & (5,5) \end{array}$$

Since we are taking a random sample, each of the possible samples has the same probability, 1/25, of being the one chosen. When we take a sample, though, we usually do not care whether we chose unit 4 first and unit 5 second, or the other way around. Instead, we are interested in the probability that our sample consists of units 4 and 5 in either order, which we write as $\mathcal{S} = \{4, 5\}$. By the third property in the definition of a probability,

$$P(\{4,5\}) = P[(4,5) \cup (5,4)] = P[(4,5)] + P[(5,4)] = \frac{2}{25}.$$

Suppose we want to find P(unit 2 is in the sample). We can either count that nine of the outcomes above contain 2, so the probability is 9/25, or we can use the **addition formula:**

$$P(A \cup B) = P(A) + P(B) - P(A \cap B). \tag{A.1}$$

Here, let $A = \{$unit 2 is chosen on the first draw$\}$ and let $B = \{$unit 2 is chosen on the second draw$\}$. Then,

$$P(\text{unit 2 is in the sample}) = P(A) + P(B) - P(A \cap B) = 1/5 + 1/5 - 1/25 = 9/25.$$

Note that, for this example,

$$P(A \cap B) = P(A) \times P(B).$$

That occurs in this situation because events A and B are **independent**, that is, whatever happens on the first draw has no effect on the probabilities of what will happen on the second draw. In finite population sampling, the draws are independent when we sample with replacement.

A.1.2 Simple Random Sampling without Replacement

Most of the time, we sample without replacement because it is more efficient—if Heather is already in the sample, why should we use resources by sampling her again? If we plan to take an SRS (recall that SRS refers to a simple random sample without replacement) of our population with N balls, the ten possible samples (ignoring the ordering) are

$$\begin{array}{ccccc} \{1, 2\} & \{1, 3\} & \{1, 4\} & \{1, 5\} & \{2, 3\} \\ \{2, 4\} & \{2, 5\} & \{3, 4\} & \{3, 5\} & \{4, 5\} \end{array}$$

Since there are ten possible samples and we are sampling with equal probabilities, the probability that a given sample will be chosen is $1/10$.

In general, there are

$$\binom{N}{n} = \frac{N!}{n!(N-n)!} \tag{A.2}$$

possible samples of size n that can be drawn without replacement and with equal probabilities from a population of size N, where

$$k! = k(k-1)(k-2)\cdots 1 \text{ and } 0! = 1.$$

For our example, there are

$$\binom{5}{2} = \frac{5!}{2!(5-2)!} = \frac{5 \times 4 \times 3 \times 2 \times 1}{(2 \times 1)(3 \times 2 \times 1)} = 10$$

possible samples of size 2, as we found when we listed them.

Note that in sampling without replacement, successive draws are *not* independent. The probability that the sample consists of units 4 and 5 equals $1/10$, which differs from the probability of $2/25$ for with-replacement sampling.

Example A.1. Players of the lottery game "Fantasy 5" choose 5 numbers without replacement from the numbers 1 through 35. If the 5 numbers you choose match the 5 official winning numbers, you win \$50,000. What is the probability you win \$50,000?

You could select a total of

$$\binom{35}{5} = \frac{35!}{5!\,30!} = 324{,}632$$

possible sets of 5 numbers. But only

$$\binom{5}{5} = 1$$

of those sets will match the official winning numbers, so your probability of winning \$50,000 is $1/324{,}632$.

Cash prizes are also given if you match three or four of the numbers. To match four, you must select four numbers out of the set of five winning numbers, and the remaining number out of the set of 30 nonwinning numbers, so the probability is

$$P(\text{match exactly 4 balls}) = \frac{\binom{5}{4}\binom{30}{1}}{\binom{35}{5}} = \frac{150}{324{,}632}. \quad \blacksquare$$

A.2 Random Variables and Expected Value

A **random variable** is a function that assigns a number to each outcome in the sample space. Which number the random variable will actually assume is only determined after we conduct the experiment and depends on a random process: Before we conduct the experiment, we only know probabilities with which the different outcomes can occur. The set of possible values of a random variable, along with the probability with which each value occurs, is called the **probability distribution** of the random variable. Random variables are denoted by capital letters in this book to distinguish them from the fixed values y_i. If X is a random variable, then $P(X = x)$ is the probability that the random variable X takes on the value x. The quantity x is sometimes called a **realization** of the random variable X; x is one of the values that could occur if we performed the experiment.

Example A.2. In the game "Fantasy 5," let X be the amount of money you will win from your selection of numbers. You win \$50,000 if you match all 5 winning numbers, \$500 if you match 4, \$5 if you match 3, and nothing if you match fewer than 3. Then the probability distribution of X is given in the following table:

x	0	5	500	50,000
$P(X = x)$	$\frac{320{,}131}{324{,}632}$	$\frac{4350}{324{,}632}$	$\frac{150}{324{,}632}$	$\frac{1}{324{,}632}$

$\blacksquare$

If you played "Fantasy 5" many, many times, what would you expect your average winnings per game to be? The answer is the **expected value** of X, defined by

$$E(X) = \sum_x xP(X = x). \tag{A.3}$$

For "Fantasy 5,"

$$\begin{aligned} E(X) &= \left(0 \times \frac{320{,}131}{324{,}632}\right) + \left(5 \times \frac{4350}{324{,}632}\right) + \left(500 \times \frac{150}{324{,}632}\right) + \left(50{,}000 \times \frac{1}{324{,}632}\right) \\ &= \frac{176{,}750}{324{,}632} = 0.45. \end{aligned}$$

Think of a box containing 324,632 balls, in which 1 ball contains the number 50,000, 150 balls contain the number 500, 4350 balls contain the number 5, and the remaining 320,131 balls contain the number 0. The expected value is simply the average of the numbers written inside all the balls in the box. One way to think about expected value is to imagine repeating the experiment over and over again and calculating the long-run average of the results. If

TABLE A.1
Properties of expected value.

1. If g is a function, then $E[g(X)] = \sum_x g(x)P(X = x)$.
2. If a and b are constants, then $E[aX + b] = aE[X] + b$.
3. If X and Y are independent, then $E[XY] = [E(X)][E(Y)]$.
4. $\text{Cov}(X, Y) = E[XY] - [E(X)][E(Y)]$.
5. $\text{Cov}\left[\sum_{i=1}^{n}(a_iX_i + b_i), \sum_{j=1}^{m}(c_jY_j + d_j)\right] = \sum_{i=1}^{n}\sum_{j=1}^{m} a_ic_j \,\text{Cov}(X_i, Y_j)$.
6. $V(X) = E\left[X^2\right] - [E(X)]^2$.
7. $V(X + Y) = V(X) + V(Y) + 2\,\text{Cov}(X, Y)$.
8. If X and Y are independent, then $V(X + Y) = V(X) + V(Y)$.
9. $-1 \leq \text{Corr}(X, Y) \leq 1$.

you play "Fantasy 5" many, many times, you would expect to win about 45 cents per game, even though 45 cents is not one of the possible realizations of X.

The variance, covariance, correlation, and coefficient of variation are defined directly in terms of the expected value:

$$V(X) = E\left[\{X - E(X)\}^2\right] = \text{Cov}(X, X) \tag{A.4}$$

$$\text{Cov}(X, Y) = E[\{X - E(X)\}\{Y - E(Y)\}] \tag{A.5}$$

$$\text{Corr}(X, Y) = \frac{\text{Cov}(X, Y)}{\sqrt{V(X)V(Y)}} \tag{A.6}$$

$$\text{CV}(X) = \frac{\sqrt{V(X)}}{E(X)}, \text{ for } E(X) \neq 0. \tag{A.7}$$

Table A.1 lists properties of expected value and variance. These follow directly from the definitions in (A.4) through (A.7).

In sampling, we often use estimators that are ratios of two random variables. But $E[Y/X]$ usually does not equal $E[Y]/E[X]$. To illustrate this, consider the following probability distribution for X and Y:

x	y	y/x	$P(X = x, Y = y)$
1	2	2	$\frac{1}{4}$
2	8	4	$\frac{1}{4}$
3	6	2	$\frac{1}{4}$
4	8	2	$\frac{1}{4}$

Then E[Y]/E[X] = 6/2.5 = 2.4, but E[Y/X] = 2.5. In this example, the values are close but not equal.

The random variable we use most frequently in this book is

$$Z_i = \begin{cases} 1 \text{ if unit } i \text{ is in the sample} \\ 0 \text{ if unit } i \text{ is not in the sample.} \end{cases} \tag{A.8}$$

This indicator variable tells us whether the ith unit is in the sample or not. In an SRS, n of the random variables $Z_1, Z_2, \ldots, Z_N$ will take on the value 1, and the remaining $N - n$ will be 0. For Z_i to equal 1, one of the units in the sample must be unit i, and the other $n - 1$ units must come from the remaining $N - 1$ units in the population, so

$$\begin{aligned} P(Z_i = 1) &= P(i\text{th unit is in the sample}) \\ &= \frac{\binom{1}{1}\binom{N-1}{n-1}}{\binom{N}{n}} \\ &= \frac{n}{N}. \end{aligned} \tag{A.9}$$

Thus,

$$\begin{aligned} E[Z_i] &= 0 \times P(Z_i = 0) + 1 \times P(Z_i = 1) \\ &= P(Z_i = 1) = \frac{n}{N}. \end{aligned}$$

Similarly, for $i \neq j$,

$$\begin{aligned} P(Z_i Z_j = 1) &= P(Z_i = 1 \text{ and } Z_j = 1) \\ &= P(i\text{th unit is in the sample and } j\text{th unit is in the sample}) \\ &= \frac{\binom{2}{2}\binom{N-2}{n-2}}{\binom{N}{n}} \\ &= \frac{n(n-1)}{N(N-1)}. \end{aligned}$$

Thus for $i \neq j$,

$$\begin{aligned} E[Z_i Z_j] &= 0 \times P(Z_i Z_j = 0) + 1 \times P(Z_i Z_j = 1) \\ &= P(Z_i Z_j = 1) = \frac{n(n-1)}{N(N-1)}. \end{aligned}$$

The properties of expectation and covariance may be used to prove many results in finite population sampling. In Chapter 4, we use the covariance of $\bar{x}$ and $\bar{y}$ from an SRS. Let

$$\bar{x}_{\mathcal{U}} = \frac{1}{N}\sum_{i=1}^{N} x_i, \qquad \bar{y}_{\mathcal{U}} = \frac{1}{N}\sum_{i=1}^{N} y_i,$$

$$\bar{x} = \frac{1}{n}\sum_{i=1}^{N} Z_i x_i, \qquad \bar{y} = \frac{1}{n}\sum_{i=1}^{N} Z_i y_i,$$

and

$$R = \frac{\sum_{i=1}^{N}(x_i - \bar{x}_{\mathcal{U}})(y_i - \bar{y}_{\mathcal{U}})}{(N-1)S_x S_y}. \tag{A.10}$$

Then,

$$\text{Cov}\,(\bar{x}, \bar{y}) = \left(1 - \frac{n}{N}\right)\frac{RS_x S_y}{n}. \tag{A.11}$$

We use properties 5 and 6 in Table A.1, along with the results of Exercise 4, to show (A.11):

$$\begin{aligned}
\text{Cov}\,(\bar{x}, \bar{y}) &= \frac{1}{n^2}\text{Cov}\left(\sum_{i=1}^{N} Z_i x_i, \sum_{j=1}^{N} Z_j y_j\right) \\
&= \frac{1}{n^2}\sum_{i=1}^{N}\sum_{j=1}^{N} x_i y_j \text{Cov}\,(Z_i, Z_j) \\
&= \frac{1}{n^2}\sum_{i=1}^{N} x_i y_i\ \text{V}(Z_i) + \frac{1}{n^2}\sum_{i=1}^{N}\sum_{j\neq i}^{N} x_i y_j \text{Cov}\,(Z_i, Z_j) \\
&= \frac{1}{n}\frac{N-n}{N^2}\sum_{i=1}^{N} x_i y_i - \frac{1}{n}\frac{N-n}{N^2(N-1)}\sum_{i=1}^{N}\sum_{j\neq i}^{N} x_i y_j \\
&= \frac{1}{n}\left[\frac{N-n}{N^2} + \frac{N-n}{N^2(N-1)}\right]\sum_{i=1}^{N} x_i y_i - \frac{1}{n}\frac{N-n}{N^2(N-1)}\sum_{i=1}^{N}\sum_{j=1}^{N} x_i y_j \\
&= \frac{1}{n}\frac{N-n}{N(N-1)}\sum_{i=1}^{N} x_i y_i - \frac{1}{n}\frac{N-n}{N-1}\bar{x}_{\mathcal{U}}\bar{y}_{\mathcal{U}} \\
&= \frac{1}{n}\frac{N-n}{N(N-1)}\sum_{i=1}^{N}(x_i - \bar{x}_{\mathcal{U}})(y_i - \bar{y}_{\mathcal{U}}) \\
&= \frac{1}{n}\left(1 - \frac{n}{N}\right) RS_x S_y.
\end{aligned}$$

A.3 Conditional Probability

In sampling without replacement, successive draws from the population are **dependent**: The unit we choose on the first draw changes the probabilities of selecting the other units on subsequent draws. When taking an SRS from the box in Figure A.1, each ball has probability 1/5 of being chosen on the first draw. If we choose ball 2 on the first draw and sample without replacement, then

$$P(\text{select ball 3 on second draw} \mid \text{select ball 2 on first draw}) = \frac{1}{4}.$$

(Read as "the conditional probability that ball 3 is selected on the second draw given that ball 2 is selected on the first draw equals 1/4.") Conditional probability allows us to adjust the probability of an event if we know that a related event occurred.

The **conditional probability** of A given B is defined to be

$$P(A|B) = \frac{P(A \cap B)}{P(B)}. \tag{A.12}$$

In sampling, we usually use this definition the other way around:

$$P(A \cap B) = P(A|B)P(B). \tag{A.13}$$

If events A and B are independent—that is, knowing whether A occurred gives us absolutely no information about whether B occurred—then $P(A|B) = P(A)$ and $P(B|A) = P(B)$.

Suppose we have a population with 8 households and 15 persons living in the households, as follows:

Household	Persons
1	1, 2, 3
2	4
3	5
4	6, 7
5	8
6	9, 10
7	11, 12, 13, 14
8	15

In a one-stage cluster sample, as discussed in Chapter 5, we might take an SRS of two households, then interview each person in the selected households. Then,

$$\begin{aligned} P(\text{select person 10}) &= P(\text{select household 6})\, P(\text{select person 10} \mid \text{select household 6}) \\ &= \left(\frac{2}{8}\right)\left(\frac{2}{2}\right) = \frac{2}{8}. \end{aligned}$$

In fact, for this example, the probability that any individual in the population is interviewed is the same value, 2/8, because each household is equally likely to be chosen and the probability a person is selected is the same as the probability that the household is selected.

Suppose now that we take a two-stage cluster sample instead of a one-stage cluster sample, and we interview only one randomly selected person in each selected household. Then, in this example, we are more likely to interview persons living alone than those living with others:

$$\begin{aligned} P(\text{select person 4}) &= P(\text{select household 2})\, P(\text{select person 4} \mid \text{select household 2}) \\ &= \left(\frac{2}{8}\right)\left(\frac{1}{1}\right) = \frac{2}{8}, \end{aligned}$$

but

$$\begin{aligned} P(\text{select person 12}) &= P(\text{select household 7})\, P(\text{select person 12} \mid \text{select household 7}) \\ &= \left(\frac{2}{8}\right)\left(\frac{1}{4}\right) = \frac{2}{32}. \end{aligned}$$

These calculations extend to multistage cluster sampling because of the general result

$$P(A_1 \cap A_2 \cap \cdots \cap A_k) = P(A_1 \mid A_2, \cdots, A_k)P(A_2 \mid A_3 \ldots, A_k) \ldots P(A_k). \tag{A.14}$$

Suppose we take a three-stage cluster sample of grade school students. First, we take an SRS of schools, then an SRS of classes within schools, then an SRS of students within classes. Then the event {Joe is selected in the sample} is the same as {Joe's school is selected ∩ Joe's class is selected ∩ Joe is selected} and we can find Joe's probability of inclusion by

$$\begin{aligned} P(\text{Joe in sample}) = P&(\text{Joe's school is selected}) \\ &\times P(\text{Joe's class is selected} \mid \text{Joe's school is selected}) \\ &\times P(\text{Joe is selected} \mid \text{Joe's school and class are selected}). \end{aligned}$$

If we sample 10% of schools, 20% of classes within selected schools, and 50% of students within selected classes, then

$$P(\text{Joe in sample}) = (0.10)(0.20)(0.50) = 0.01.$$

A.4 Conditional Expectation

Conditional expectation is used extensively in the theory of cluster sampling. Let X and Y be random variables. Then, using the definition of conditional probability,

$$P(Y = y|X = x) = \frac{P(Y = y \cap X = x)}{P(X = x)}. \tag{A.15}$$

This gives the **conditional distribution** of Y given that $X = x$. The **conditional expectation** of Y given that $X = x$ simply follows the definition of expectation using the conditional distribution:

$$E(Y|X = x) = \sum_y yP(Y = y|X = x). \tag{A.16}$$

The **conditional variance** of Y given that $X = x$ is defined similarly:

$$V(Y|X = x) = \sum_y [y - E(Y|X = x)]^2 P(Y = y|X = x). \tag{A.17}$$

Example A.3. Consider a box with two balls, A and B:

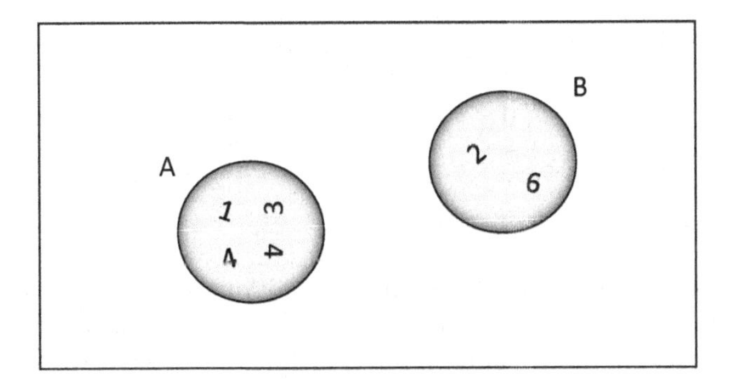

Choose one of the balls at random, then randomly select one of the numbers inside that ball. Let Y = the number that we choose and let

$$Z = \begin{cases} 1 \text{ if we choose ball A} \\ 0 \text{ if we choose ball B.} \end{cases}$$

Then,

$$P(Y = 1|Z = 1) = \frac{1}{4},$$
$$P(Y = 3|Z = 1) = \frac{1}{4},$$
$$P(Y = 4|Z = 1) = \frac{1}{2},$$

and

$$E(Y|Z = 1) = \left(1 \times \frac{1}{4}\right) + \left(3 \times \frac{1}{4}\right) + \left(4 \times \frac{1}{2}\right) = 3.$$

Similarly,

$$P(Y = 2|Z = 0) = \frac{1}{2}$$

and

$$P(Y = 6|Z = 0) = \frac{1}{2},$$

so

$$E(Y|Z = 0) = \left(2 \times \frac{1}{2}\right) + \left(6 \times \frac{1}{2}\right) = 4.$$

In short, if we know that ball A is picked, then the conditional expectation of Y is the average of numbers in ball A since an SRS of size 1 is taken from the ball; the conditional expectation of Y given that ball B is picked is the average of the numbers in ball B. ■

Note that $E(Y|X = x)$ is a function of x; call it $g(x)$. Define the **conditional expectation** of Y given X, $E(Y|X)$, to be $g(X)$, the same function but of the random variable instead. $E(Y|X)$ is a random variable and gives us the conditional expected value of Y for the general random variable X: for each possible value of x, the value $E(Y|X = x)$ occurs with probability $P(X = x)$.

Example A.4. In Example A.3, we know the probability distribution of Z and can thus use the conditional expectations calculated to write the probability distribution of $E(Y|Z)$:

z	$E(Y\|Z = z)$	Probability
0	4	$\frac{1}{2}$
1	3	$\frac{1}{2}$

In sampling, we need this general concept of conditional expectation primarily so we can use the properties in Table A.2 to find expected values and variances for cluster samples. Conditional expectation can be confusing, so let's talk about what these properties mean.

TABLE A.2
Properties of conditional expectation.

1.	$E(X\|X) = X$.
2.	For any function g, $E[g(X)Y\|X] = g(X)E(Y\|X)$.
3.	If X and Y are independent, then $E[Y\|X] = E[Y]$.
4.	$E(Y) = E[E(Y\|X)]$.
5.	$V[Y] = V[E(Y\|X)] + E[V(Y\|X)]$.

1. $E(X|X) = X$. If we know what X is already, then we expect X to be X. The probability distribution of $E(X|X)$ is the same as the probability distribution of X.

2. $E[g(X)Y|X] = g(X)E(Y|X)$. If we know what X is, then we know X^2, or $\ln X$, or any function $g(X)$ of X.

3. If X and Y are independent, then $E(Y|X) = E(Y)$. If X and Y are independent, then knowing X gives us no information about Y. Thus, the expected value of Y, the average of all the possible outcomes of Y in the experiment, is the same no matter what X is.

4. $E(Y) = E[E(Y|X)]$. This property, called **successive conditioning**, and property 5 are the ones we use the most in sampling; we use them to find the bias and variance of estimators in cluster sampling. Successive conditioning simply says that if we take the weighted average of the conditional expected value of Y given that $X = x$, with weights $P(X = x)$, the result is the expected value of Y. You use successive conditioning every time you take a weighted average of a quantity over subpopulations: If a population has 60 women and 40 men, and if the average height of the women is 64 inches and the average height of the men is 69 inches, then the average height for the class is

$$(64 \times 0.6) + (69 \times 0.4) = 66 \text{ inches.}$$

   In this example, 64 is the conditional expected value of height given that the person is a woman, 69 is the conditional expected value of height given that the person is a man, and 66 is the expected value of height for all persons in the population.

5. $V[Y] = V[E(Y|X)] + E[V(Y|X)]$. This property gives an easy way of calculating variances in two-stage cluster samples. It says that the total variability has two parts: (a) the variability that arises because $E(Y|X = x)$ varies with different values of x, and (b) the variability that arises because there can be different values of y associated with the same value of x. Note that, using property 6 in Table A.1,

$$V(Y|X) = E\{[Y - E(Y|X)]^2|X\} = E[Y^2|X] - [E(Y|X)]^2 \tag{A.18}$$

   and

$$\begin{aligned} V[E(Y|X)] &= E\left(\{E(Y|X) - E[E(Y|X)]\}^2\right) \\ &= E\left(\{E(Y|X) - E(Y)\}^2\right) \\ &= E\{[E(Y|X)]^2\} - [E(Y)]^2. \end{aligned} \tag{A.19}$$

Example A.5. Here's how conditional expectation properties work in Example A.3. Successive conditioning implies that

$$
\begin{aligned}
E(Y) &= E(Y|Z=0)P(Z=0) + E(Y|Z=1)P(Z=1) \\
&= \left(4 \times \frac{1}{2}\right) + \left(3 \times \frac{1}{2}\right) = 3.5.
\end{aligned}
$$

We can find the distribution of $V(Y|Z)$ using (A.18):

$$
\begin{aligned}
V(Y|Z=0) &= E(Y^2|Z=0) - [E(Y|Z=0)]^2 \\
&= \left(2^2 \times \frac{1}{2}\right) + \left(6^2 \times \frac{1}{2}\right) - (4)^2 = 4, \\
V(Y|Z=1) &= E(Y^2|Z=1) - [E(Y|Z=1)]^2 \\
&= \left(1^2 \times \frac{1}{4}\right) + \left(3^2 \times \frac{1}{4}\right) + \left(4^2 \times \frac{1}{2}\right) - (3)^2 = 1.5.
\end{aligned}
$$

These calculations give the following probability distribution for $V(Y|Z)$:

z	$V(Y\|Z=z)$	Probability
0	4	$\frac{1}{2}$
1	1.5	$\frac{1}{2}$

Thus, using (A.19),

$$
\begin{aligned}
V[E(Y|Z)] &= E\left\{[E(Y|Z) - E(Y)]^2\right\} \\
&= [E(Y|Z=0) - E(Y)]^2 P(Z=0) + [E(Y|Z=1) - E(Y)]^2 P(Z=1) \\
&= \left[(4-3.5)^2 \times \frac{1}{2}\right] + \left[(3-3.5)^2 \times \frac{1}{2}\right] \\
&= 0.25.
\end{aligned}
$$

Using the probability distribution of $V(Y|Z)$,

$$
E[V(Y|Z)] = \left(4 \times \frac{1}{2}\right) + \left(1.5 \times \frac{1}{2}\right) = 2.75.
$$

Consequently,

$$
V(Y) = V[E(Y|Z)] + E[V(Y|Z)] = 0.25 + 2.75 = 3.00. \quad \blacksquare
$$

If we did not have the properties of conditional expectation, would need to find the unconditional probability distribution of Y to calculate its expectation and variance—a relatively easy task for the small number of options in Example A.3 but cumbersome to do for general multistage cluster sampling.

A.5 Exercises

A. Introductory Exercises

1. What is the probability you match exactly 3 of the numbers in the game "Fantasy 5"? That you match at least one of the numbers?

2. *Calculating the sampling distribution in Example 2.4.* A box has eight balls; three of the balls contain the number 7. You select an SRS (without replacement) of size 4. What is the probability that your sample contains no 7s? Exactly one 7? Exactly two 7s?

3. Prove the nine Properties of Expected Value in Table A.1, using the definitions in (A.3) through (A.7).

4. Show that

$$V(Z_i) = \text{Cov}\,(Z_i, Z_i) = \frac{n(N-n)}{N^2}$$

and that, for $i \neq j$,

$$\text{Cov}\,(Z_i, Z_j) = -\frac{n(N-n)}{N^2(N-1)}.$$

5. Show that in an SRS,

$$\text{Corr}\,(\bar{x}, \bar{y}) = R. \tag{A.20}$$

6. Consider the box below, with three balls labeled A, B, and C:

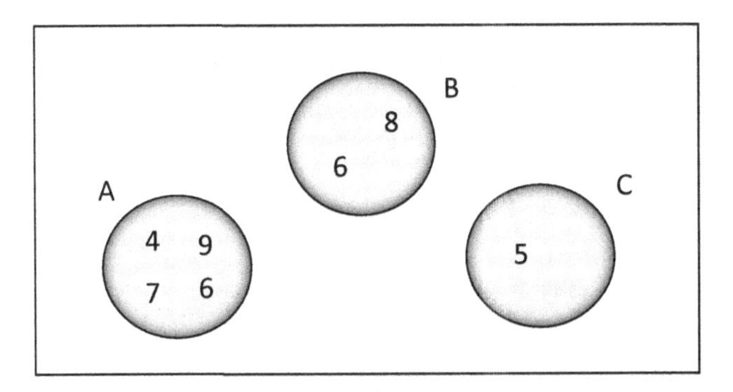

Suppose we take an SRS of one ball, then subsample an SRS of one number from the selected ball. Let Z represent the ball chosen, and let Y represent the number we choose from the ball. Use the properties of conditional expectation to find $E(Y)$ and $V(Y)$.

Bibliography

Agresti, A. (2013). *Categorical Data Analysis, 3rd ed.* Hoboken, NJ: Wiley.

Alba, S., F. Wong, and Y. Bråten (2019). Gender matters in household surveys. *Significance 16*(6), 38–41.

Albrizio, A. (2007). Biometry and anthropometry: From Galton to constitutional medicine. *Journal of Anthropological Sciences 85*, 101–123.

Alexander, H. M., N. A. Slade, and W. D. Kettle (1997). Application of mark-recapture models to estimation of the population size of plants. *Ecology 78*, 1230–1237.

Alf, C. and S. L. Lohr (2007). Sampling assumptions in introductory statistics classes. *The American Statistician 61*, 71–77.

Alonso-Revenga, J., N. Martín, and L. Pardo (2017). New improved estimators for overdispersion in models with clustered multinomial data and unequal cluster sizes. *Statistics and Computing 27*(1), 193–217.

Altham, P. M. E. (1976). Discrete variable analysis for individuals grouped into families. *Biometrika 63*, 263–269.

Alwin, D. F. (2010). How good is survey measurement? Assessing the reliability and validity of survey measures. In P. Marsden and J. D. Wright (Eds.), *Handbook of Survey Research, 2nd ed.*, pp. 405–434. Bingley, UK: Emerald Publishing.

American Association of Public Opinion Research (2015). *AAPOR Code of Ethics.* Oakbrook Terrace, IL: AAPOR. `https://www.aapor.org/Standards-Ethics/AAPOR-Code-of-Ethics.aspx` (accessed January 16, 2020).

American Association of Public Opinion Research (2016a). *Best Practices for Survey Research.* Oakbrook Terrace, IL: AAPOR. `https://www.aapor.org/Standards-Ethics/Best-Practices.aspx` (accessed January 16, 2020).

American Association of Public Opinion Research (2016b). *Standard Definitions: Final Dispositions of Case Codes and Outcome Rates for Surveys, 9th ed.* Oakbrook Terrace, IL: AAPOR. `www.aapor.org/AAPOR_Main/media/publications/Standard-Definitions20169theditionfinal.pdf` (accessed September 27, 2020).

American Society of Composers, Authors and Publishers (2015). ASCAP's survey and distribution system: Rules & policies. `https://www.ascap.com/-/media/files/pdf/members/payment/drd.pdf` (accessed August 15, 2020).

American Statistical Association (2018). *Ethical Guidelines for Statistical Practice.* Alexandria, VA: American Statistical Association.

Amstrup, S. C., T. L. McDonald, and B. F. Manly (2010). *Handbook of Capture–Recapture Analysis.* Princeton: Princeton University Press.

Anderson, M. J. (2015). *The American Census: A Social History, 2nd ed.* New Haven, CT: Yale University Press.

Andridge, R. R. and R. J. Little (2010). A review of hot deck imputation for survey non-response. *International Statistical Review 78*(1), 40–64.

Ardilly, P. and Y. Tillé (2006). *Sampling Methods: Exercises and Solutions.* New York: Springer.

Aristotle (350 BCE). Atheniensium respublica. In W. D. Ross (Ed.), *The Works of Aristotle*, pp. 369–452. Translated by F. G. Kenyon. Oxford: Clarendon Press.

Arnold, T. W. (1991). Intraclutch variation in egg size of American coots. *The Condor 93*, 19–27.

Aye Maung, N. (1995). Survey design and interpretation of the British Crime Survey. In M. Walker (Ed.), *Interpreting Crime Statistics*, pp. 207–227. Oxford: Oxford University Press.

Azur, M. J., E. A. Stuart, C. Frangakis, and P. J. Leaf (2011). Multiple imputation by chained equations: What is it and how does it work? *International Journal of Methods in Psychiatric Research 20*(1), 40–49.

Babbie, E. R. (2007). *The Practice of Social Research, 11th ed.* Belmont, CA: Wadsworth.

Baker, R., J. M. Brick, N. A. Bates, M. Battaglia, M. P. Couper, J. A. Dever, K. J. Gile, and R. Tourangeau (2013). Summary report of the AAPOR task force on non-probability sampling. *Journal of Survey Statistics and Methodology 1*(2), 90–143.

Bankier, M. D. (1988). Power allocations: Determining sample sizes for subnational areas. *The American Statistician 42*(3), 174–177.

Barnett, V., J. Haworth, and T. M. F. Smith (2001). A two-phase sampling scheme with applications to auditing or *sed quis custodiet ipsos custodes*? *Journal of the Royal Statistical Society, Series A 164*, 407–422.

Bart, J. and S. Earnst (2002). Double-sampling to estimate density and population trends in birds. *The Auk 119*, 36–45.

Basow, S. A. and J. L. Martin (2012). Bias in student evaluations. In M. E. Kite (Ed.), *Effective Evaluation of Teaching: A Guide for Faculty and Administrators*, pp. 40–49. Washington, DC: Society for the Teaching of Psychology.

Basow, S. A. and N. T. Silberg (1987). Student evaluations of college professors: Are female and male professors rated differently? *Journal of Educational Psychology 87*, 656–665.

Basu, D. (1971). An essay on the logical foundations of survey sampling, part 1. In V. P. Godambe and D. A. Sprott (Eds.), *Foundations of Statistical Inference*, pp. 203–242. Toronto: Holt, Rinehart & Winston.

Battese, G. E., R. M. Harter, and W. A. Fuller (1988). An error-components model for prediction of county crop areas using survey and satellite data. *Journal of the American Statistical Association 83*, 28–36.

Beatty, P. and D. Herrmann (2002). To answer or not to answer: Decision processes related to survey item nonresponse. In R. M. Groves, D. Dillman, J. Eltinge, and R. Little (Eds.), *Survey Nonresponse*, pp. 71–85. New York: Wiley.

Beaumont, J.-F. (2008). A new approach to weighting and inference in sample surveys. *Biometrika 95*(3), 539–553.

Beaumont, J.-F. (2020). Are probability surveys bound to disappear for the production of official statistics? *Survey Methodology 46*(1), 1–28.

Beaumont, J.-F. and A. Alavi (2004). Robust generalized regression estimation. *Survey Methodology 30*, 195–208.

Beaumont, J.-F., A. BÉliveau, and D. Haziza (2015). Clarifying some aspects of variance estimation in two-phase sampling. *Journal of Survey Statistics and Methodology 3*(4), 524–542.

Beck, A. J., S. A. Kline, and L. A. Greenfeld (1988). Survey of Youth in Custody. Technical Report NCJ-113365, Bureau of Justice Statistics, Washington, DC.

Bedrick, E. J. (1983). Adjusted chi-squared tests for cross-classified tables of survey data. *Biometrika 70*, 591–595.

Beebe, T. J., J. Y. Ziegenfuss, J. L. St. Sauver, S. M. Jenkins, L. Haas, M. E. Davern, and N. J. Talley (2011). HIPAA authorization and survey nonresponse bias. *Medical Care 49*(4), 365–370.

Bellemain, E., J. E. Swenson, D. Tallmon, S. Brunberg, and P. Taberlet (2005). Estimating population size of elusive animals with DNA from hunter-collected feces: Four methods for brown bears. *Conservation Biology 19*, 150–161.

Bellhouse, D. R. (1984). A review of optimal designs in survey sampling. *The Canadian Journal of Statistics 12*, 53–65.

Bellhouse, D. R. and J. E. Stafford (1999). Density estimation from complex surveys. *Statistica Sinica 9*, 407–424.

Bellhouse, D. R. and J. E. Stafford (2001). Local polynomial regression in complex surveys. *Survey Methodology 27*(2), 197–203.

Belsley, D. A., E. Kuh, and R. E. Welsch (1980). *Regression Diagnostics.* New York: Wiley.

Belson, W. A. (1981). *The Design and Understanding of Survey Questions.* Aldershot, UK: Gower Publishing.

Berg, E., J. K. Kim, and C. Skinner (2016). Imputation under informative sampling. *Journal of Survey Statistics and Methodology 4*(4), 436–462.

Berger, Y. G. (2004). A simple variance estimator for unequal probability sampling without replacement. *Journal of Applied Statistics 31*, 305–315.

Berger, Y. G. (2007). A jackknife variance estimator for unistage stratified samples with unequal probabilities. *Biometrika 94*(4), 953–964.

Berger, Y. G. and C. J. Skinner (2005). A jackknife variance estimator for unequal probability sampling. *Journal of the Royal Statistical Society: Series B 67*(1), 79–89.

Bethel, J. (1989). Sample allocation in multivariate surveys. *Survey Methodology 15*, 47–57.

Bethlehem, J. (2002). Weighting nonresponse adjustments based on auxiliary information. In R. M. Groves, D. Dillman, J. Eltinge, and R. Little (Eds.), *Survey Nonresponse*, pp. 275–287. New York: Wiley.

Bethlehem, J. (2016). Solving the nonresponse problem with sample matching? *Social Science Computer Review 34*(1), 59–77.

Bethlehem, J., F. Cobben, and B. Schouten (2011). *Handbook of Nonresponse in Household Surveys.* Hoboken, NJ: Wiley.

Bhaskaran, K. and L. Smeeth (2014). What is the difference between missing completely at random and missing at random? *International Journal of Epidemiology 43*(4), 1336–1339.

Bhutta, N., J. Bricker, A. C. Chang, L. J. Dettling, S. Goodman, J. W. Hsu, K. B. Moore, S. Reber, A. H. Volz, and R. A. Windle (2020). Changes in U.S. family finances from 2016 to 2019: Evidence from the Survey of Consumer Finances. *Federal Reserve Bulletin 106*(5), 1–42.

Biemer, P. P. (2010). Total survey error: Design, implementation, and evaluation. *Public Opinion Quarterly 74*(5), 817–848.

Biemer, P. P. and R. Caspar (1994). Continuous quality improvement for survey operations: Some general principles and applications. *Journal of Official Statistics 10*, 307–326.

Biemer, P. P., E. D. de Leeuw, S. Eckman, B. Edwards, F. Kreuter, L. E. Lyberg, N. C. Tucker, and B. T. West (Eds.) (2017). *Total Survey Error in Practice.* Hoboken, NJ: Wiley.

Biemer, P. P. and L. E. Lyberg (2003). *Introduction to Survey Quality.* New York: Wiley.

Biemer, P. P., J. Murphy, S. Zimmer, C. Berry, G. Deng, and K. Lewis (2018). Using bonus monetary incentives to encourage web response in mixed-mode household surveys. *Journal of Survey Statistics and Methodology 6*(2), 240–261.

Bilder, C. R. and T. M. Loughin (2014). *Analysis of Categorical Data with R.* Boca Raton, FL: CRC Press.

Binder, D. A. (1983). On the variances of asymptotically normal estimators from complex surveys. *International Statistical Review 51*, 279–292.

Binder, D. A. (1996). Linearization methods for single phase and two-phase samples: A cookbook approach. *Survey Methodology 22*, 17–22.

Binder, D. A. and G. R. Roberts (2003). Design-based and model-based methods for estimating model parameters. In R. L. Chambers and C. J. Skinner (Eds.), *Analysis of Survey Data*, pp. 29–48. New York: Wiley.

Blair, G., K. Imai, and Y.-Y. Zhou (2015). Design and analysis of the randomized response technique. *Journal of the American Statistical Association 110*(511), 1304–1319.

Blitzstein, J. K. and J. Hwang (2019). *Introduction to Probability, 2nd ed.* Boca Raton, FL: CRC Press.

Blumberg, S. J. and J. V. Luke (2007). *Wireless Substitution: Early Release of Estimates from the National Health Interview Survey, July–December 2006.* Hyattsville, MD: National Center for Health Statistics.

Blumberg, S. J. and J. V. Luke (2010). *Wireless Substitution: Early Release of Estimates from the National Health Interview Survey, July–December 2009.* Hyattsville, MD: National Center for Health Statistics.

Boardman, T. J. (1994). The statistician who changed the world: W. Edwards Deming, 1900–1993. *The American Statistician 48*, 179–187.

Böhning, D., P. G. Van der Heijden, and J. Bunge (2017). *Capture-Recapture Methods for the Social and Medical Sciences.* Boca Raton, FL: CRC Press.

Bohrnstedt, G. W. (2010). Measurement models for survey research. In P. Marsden and J. D. Wright (Eds.), *Handbook of Survey Research, 2nd ed.*, pp. 347–404. Bingley, UK: Emerald Publishing.

Box, G. E. P. (1979). Robustness in the strategy of scientific model building. In G. N. Wilkinson and R. L. Launer (Eds.), *Robustness in Statistics*, pp. 201–236. New York: Academic Press.

Box, G. E. P., W. G. Hunter, and J. S. Hunter (1978). *Statistics for Experimenters: An Introduction to Design, Data Analysis, and Model Building.* New York: Wiley.

Bradburn, N. M. (1978). Respondent burden. In *Proceedings of the Survey Research Methods Section*, pp. 35–40. Alexandria, VA: American Statistical Association.

Bradburn, N. M. (2004). Understanding the question-answer process. *Survey Methodology 30*, 5–15.

Bradburn, N. M. and S. Sudman (1979). *Improving Interview Method and Questionnaire Design: Response Effects to Threatening Questions in Survey Research.* San Francisco: Jossey-Bass.

Bradburn, N. M., S. Sudman, and B. Wansink (2004). *Asking Questions: The Definitive Guide to Questionnaire Design—for Market Research, Political Polls, and Social and Health Questionnaires.* Hoboken, NJ: Wiley.

Breidt, F. J. and J. D. Opsomer (2000). Local polynomial regression estimators in survey sampling. *The Annals of Statistics 28*, 1026–1053.

Brewer, K. R. W. (1963). Ratio estimation and finite populations: Some results deducible from the assumption of an underlying stochastic process. *The Australian Journal of Statistics 5*(3), 93–105.

Brewer, K. R. W. (1975). A simple procedure for sampling πpswor. *The Australian Journal of Statistics 17*, 166–172.

Brewer, K. R. W. (2002). *Combined Survey Sampling Inference: Weighing Basu's Elephants.* London: Arnold.

Brewer, K. R. W. (2013). Three controversies in the history of survey sampling. *Survey Methodology 39*(2), 249–262.

Brewer, K. R. W. and M. E. Donadio (2003). The high entropy variance of the Horvitz-Thompson estimator. *Survey Methodology 29*, 189–196.

Brewer, K. R. W. and M. Hanif (1983). *Sampling with Unequal Probabilities.* New York: Springer-Verlag.

Brewer, K. R. W. and R. W. Mellor (1973). The effect of sample structure on analytical surveys. *The Australian Journal of Statistics 15*, 145–152.

Brick, J. M. (2011). The future of survey sampling. *Public Opinion Quarterly 75*(5), 872–888.

Brick, J. M. (2013). Unit nonresponse and weighting adjustments: A critical review. *Journal of Official Statistics 29*, 329–353.

Brick, J. M. and S. L. Lohr (2019). Experimental evaluation of mail questionnaires in a probability sample on victimization. *Journal of the Royal Statistical Society: Series A 182*(2), 669–687.

Brick, J. M. and C. Tucker (2007). Mitofsky–Waksberg: Learning from the past. *Public Opinion Quarterly 71*, 703–716.

Bricker, J., L. J. Dettling, A. Henriques, J. W. Hsu, L. Jacobs, K. B. Moore, S. Pack, J. Sabelhaus, J. Thompson, and R. A. Windle (2017). Changes in U.S. family finances from 2013 to 2016: Evidence from the Survey of Consumer Finances. *Federal Reserve Bulletin 103*(3), 1–42.

Brier, S. S. (1980). Analysis of contingency tables under cluster sampling. *Biometrika 67*, 591–596.

Brito, J. A. M., P. L. N. Silva, G. S. Semaan, and N. Maculan (2015). Integer programming formulations applied to optimal allocation in stratified sampling. *Survey Methodology 41*(2), 427–442.

Brogan, D. (2015). Analysis of complex survey data, misuse of standard statistical procedures. In *Wiley StatsRef: Statistics Reference Online*. Hoboken, NJ: Wiley Online Library.

Brown, J., C. Sexton, O. Abbott, and P. Smith (2019). The framework for estimating coverage in the 2011 Census of England and Wales: Combining dual-system estimation with ratio estimation. *Statistical Journal of the IAOS 35*, 481–499.

Bruch, C., R. Münnich, and S. Zins (2011). *Variance Estimation for Complex Surveys*. Luxembourg: Advanced Methodology for European Laeken Indicators. `http://ameli.surveystatistics.net` (accessed December 10, 2020).

Buckland, S. T. (1984). Monte Carlo confidence intervals. *Biometrics 40*, 811–817.

Buckland, S. T., D. R. Anderson, K. P. Burnham, J. L. Laake, D. L. Borchers, and L. Thomas (2001). *Introduction to Distance Sampling: Estimating Abundance of Biological Populations*. Oxford: Oxford University Press.

Buckland, S. T. and P. H. Garthwaite (1991). Quantifying precision of mark-recapture estimates using the bootstrap and related methods. *Biometrics 47*, 255–268.

Burnard, P. (1992). Learning from experience: Nurse tutors' and student nurses' perceptions of experiential learning in nurse education: Some initial findings. *International Journal of Nursing Studies 29*, 151–161.

Buskirk, T. D. and S. L. Lohr (2005). Asymptotic properties of kernel density estimation with complex survey data. *Journal of Statistical Planning and Inference 128*, 165–190.

Cai, S. and J. N. K. Rao (2019). Empirical likelihood inference for missing survey data under unequal probability sampling. *Survey Methodology 45-1 45*(1), 145–164.

Calahan, D. (1989). The *Digest* poll rides again. *Public Opinion Quarterly 53*, 129–133.

Callegaro, M. and Y. Yang (2018). The role of surveys in the era of "big data". In D. L. Vannette and J. A. Krosnick (Eds.), *The Palgrave Handbook of Survey Research*, pp. 175–192. Cham, Switzerland: Springer.

Campbell, C. (1980). A different view of finite population estimation. In *Proceedings of the Survey Research Methods Section*, pp. 319–324. Alexandria, VA: American Statistical Association.

Cannell, C. F., K. H. Marquis, and A. Laurent (1977). A summary of studies of interviewing methodology. In *Vital and Health Statistics, Series 2, 69.* Washington, DC: U.S. Government Printing Office.

Cantor, D., B. Fisher, S. H. Chibnall, R. Townsend, H. Lee, G. Thomas, and C. Bruce (2015). *Report on the AAU Campus Climate Survey on Sexual Assault and Sexual Misconduct.* Washington, DC: Association of American Universities.

Carpenter, J. and M. Kenward (2012). *Multiple Imputation and its Application.* Hoboken, NJ: Wiley.

Chambers, R. L. and R. Clark (2012). *An Introduction to Model-Based Survey Sampling with Applications.* Oxford, UK: Oxford University Press.

Chambers, R. L. and C. J. Skinner (Eds.) (2003). *Analysis of Survey Data.* New York: Wiley.

Chambers, R. L., D. G. Steel, S. Wang, and A. Welsh (2012). *Maximum Likelihood Estimation for Sample Surveys.* Boca Raton, FL: CRC Press.

Chambless, L. E. and K. E. Boyle (1985). Maximum likelihood methods for complex sample data: Logistic regression and discrete proportional hazards models. *Communications in Statistics: Theory and Methods 14*, 1377–1392.

Chao, A., P. K. Tsay, S.-H. Lin, W.-Y. Shau, and D.-Y. Chao (2001). The applications of capture-recapture models to epidemiological data. *Statistics in Medicine 20*, 3123–3157.

Chapman, D. G. (1951). Some properties of the hypergeometric distribution with applications to zoological sample censuses. *University of California Publications in Statistics 1*, 131–160.

Chapman, D. W. (2005). Sample design for the FDIC's asset loss reserve project. In *Proceedings of the Survey Research Methods Section*, pp. 2839–2843. Alexandria, VA: American Statistical Association.

Chatterjee, S. and A. S. Hadi (2012). *Regression Analysis by Example, 5th ed.* Hoboken, NJ: Wiley.

Chaves, S. S., R. Lynfield, M. L. Lindegren, J. Bresee, and L. Finelli (2015). The US Influenza Hospitalization Surveillance Network. *Emerging Infectious Diseases 21*(9), 1543–1550.

Chen, Q., M. R. Elliott, and R. J. Little (2012). Bayesian inference for finite population quantiles from unequal probability samples. *Survey Methodology 38*(2), 203–214.

Chen, S. and D. Haziza (2019). Recent developments in dealing with item non-response in surveys: A critical review. *International Statistical Review 87*, S192–S218.

Chen, S., D. Haziza, C. Léger, and Z. Mashreghi (2019). Pseudo-population bootstrap methods for imputed survey data. *Biometrika 106*(2), 369–384.

Chen, T. C., J. Clark, M. K. Riddles, L. K. Mohadjer, and T. H. I. Fakhouri (2020). *National Health and Nutrition Examination Survey, 2015–2018: Sample Design and Estimation Procedures.* Vital and Health Statistics 2(184). Hyattsville, MD: National Center for Health Statistics.

Chen, Y., P. Li, and C. Wu (2020). Doubly robust inference with non-probability survey samples. *Journal of the American Statistical Association 115*(532), 2011–2021.

Chipperfield, J., J. Brown, and P. Bell (2017). Estimating the count error in the Australian census. *Journal of Official Statistics 33*(1), 43–59.

Chipperfield, J., J. Chessman, and R. Lim (2012). Combining household surveys using mass imputation to estimate population totals. *Australian & New Zealand Journal of Statistics 54*(2), 223–238.

Christen, P. (2012). *Data Matching: Concepts and Techniques for Record Linkage, Entity Resolution, and Duplicate Detection.* New York: Springer Science & Business Media.

Christensen, R. (2020). *Plane Answers to Complex Questions, Fifth Edition.* New York: Springer.

Christman, M. C. (2000). A review of quadrat-based sampling of rare, geographically clustered populations. *Journal of Agricultural, Biological, and Environmental Statistics 5*(2), 168–201.

Christman, M. C. (2009). Sampling of rare populations. In D. Pfeffermann and C. R. Rao (Eds.), *Handbook of Statistics, Vol. 29A, Sample Surveys: Design, Methods and Applications*, pp. 109–124. Amsterdam: North Holland.

Chun, A. Y., S. G. Heeringa, and B. Schouten (2018). Responsive and adaptive design for survey optimization. *Journal of Official Statistics 34*(3), 581–597.

Chun, A. Y., M. D. Larsen, G. Durrant, and J. P. Reiter (Eds.) (2021). *Administrative Records For Survey Methodology.* Hoboken, NJ: Wiley.

Cialdini, R. B. (1984). *Influence: The New Psychology of Modern Persuasion.* New York: Quill.

Cialdini, R. B. (2009). *Influence: Science and Practice.* Boston: Pearson.

Citro, C. F. (2014). From multiple modes for surveys to multiple data sources for estimates. *Survey Methodology 40*, 137–161.

Clark, S. L. (2020). Using alternative data sources to fill-in missing values for demographic characteristics on the American Community Survey. Paper presented at the 2020 Joint Statistical Meetings.

Clopper, C. J. and E. S. Pearson (1934). The use of confidence or fiducial limits illustrated in the case of the Binomial. *Biometrika 26*(4), 404–413.

Cochi, S. L., L. E. Edmonds, K. Dyer, W. L. Greaves, J. S. Marks, E. Z. Rovira, S. R. Preblud, and W. A. Orenstein (1989). Congenital rubella syndrome in the United States: 1970–1985. *American Journal of Epidemiology 129*, 349–361.

Cochran, W. G. (1939). The use of analysis of variance in enumeration by sampling. *Journal of the American Statistical Association 34*, 492–510.

Cochran, W. G. (1977). *Sampling Techniques, 3rd ed.* New York: Wiley.

Cochran, W. G. (1978). Laplace's ratio estimator. In H. A. David (Ed.), *Contributions to Survey Sampling and Applied Statistics, in Honor of H. O. Hartley*, pp. 3–10. New York: Academic Press.

Cohen, J. E. (1976). The distribution of the chi-squared statistic under clustered sampling from contingency tables. *Journal of the American Statistical Association 71*, 665–670.

Cohn, N. (2018, September 6). Our polling methodology: Underneath the hood of the Upshot/Siena survey. *The New York Times.* `https://www.nytimes.com/2018/09/06/upshot/live-poll-method.html`. (accessed March 29, 2019).

Collins, D. and Y. Lu (2019). A stratified reservoir sampling algorithm in streams and large datasets. *Communications in Statistics—Simulation and Computation*, 1–16. DOI: 10.1080/03610918.2019.1682159.

Conkey, C. (2005). Do-not-call lists under fire. *The Wall Street Journal September 28*, D1.

Conti, P. L. and D. Marella (2015). Inference for quantiles of a finite population: Asymptotic versus resampling results. *Scandinavian Journal of Statistics 42*(2), 545–561.

Conti, P. L., D. Marella, F. Mecatti, and F. Andreis (2020). A unified principled framework for resampling based on pseudo-populations: Asymptotic theory. *Bernoulli 26*(2), 1044–1069.

Converse, J. M. (1987). *Survey Research in the United States: Roots and Emergence, 1890–1960.* Berkeley: University of California Press.

Converse, J. M. and S. Presser (1986). *Survey Questions: Handcrafting the Standardized Questionnaire.* Beverly Hills, CA: Sage.

Cormack, R. M. (1992). Interval estimation for mark-recapture studies of closed populations. *Biometrics 48*, 567–576.

Cornfield, J. (1944). On samples from finite populations. *Journal of the American Statistical Association 39*, 236–239.

Cornfield, J. (1951). Modern methods in the sampling of human populations. *American Journal of Public Health 41*, 654–661.

Couper, M. P. (2000). Web surveys: A review of issues and approaches. *Public Opinion Quarterly 64*, 464–494.

Couper, M. P. (2008). *Designing Effective Web Surveys.* New York: Cambridge University Press.

Couzens, G. L., B. Shook-Sa, P. Lee, and M. Berzofsky (2015). *Users' Guide to the National Crime Victimization Survey (NCVS) Generalized Variance Functions.* Washington, DC: U.S. Bureau of Justice Statistics.

Cox, D. R. (1952). Estimation by double sampling. *Biometrika 39*, 217–227.

Cox, G. M. (1957). Statistical frontiers. *Journal of the American Statistical Association 52*, 1–12.

Cronbach, L. J. (1951). Coefficient alpha and the internal structure of tests. *Psychometrika 16*, 297–334.

Cullen, R. (1994). Sample survey methods as a quality assurance tool in a general practice immunisation audit. *New Zealand Medical Journal 107*, 152–153.

Czaja, R. F. and J. Blair (1986). Using network sampling in crime victimization surveys. *Journal of Quantitative Criminology 6*, 185–206.

Czaja, R. F., C. B. Snowden, and R. J. Casady (1986). Reporting bias and sampling errors in a survey of a rare population using multiplicity counting rules. *Journal of the American Statistical Association 81*, 411–419.

Da Silva, D. N. and J. D. Opsomer (2009). Nonparametric propensity weighting for survey nonresponse through local polynomial regression. *Survey Methodology 35*(2), 165–176.

Daas, P., S. Ossen, R. Vis-Visschers, and J. Arends-Tóth (2009). *Checklist for the Quality Evaluation of Administrative Data Sources.* The Hague: Statistics Netherlands.

Dalenius, T. (1977). Strain at a gnat and swallow a camel: Or, the problem of measuring sampling and non-sampling errors. In *Proceedings of the Social Statistics Section*, pp. 21–25. Alexandria, VA: American Statistical Association.

D'Alessandro, U., M. K. Aikins, P. Langerock, S. Bennett, and B. M. Greenwood (1994). Nationwide survey of bednet use in rural Gambia. *Bulletin of the World Health Organization 72*, 391–394.

Danermark, B. and B. Swensson (1987). Measuring drug use among Swedish adolescents: Randomized response versus anonymous questionnaires. *Journal of Official Statistics 3*, 439–448.

David, I. P. and B. V. Sukhatme (1974). On the bias and mean squared error of the ratio estimator. *Journal of the American Statistical Association 69*, 464–466.

Davison, A. C. and D. V. Hinkley (1997). *Bootstrap Methods and Their Application.* Cambridge: Cambridge University Press.

de Jonge, T., R. Veenhoven, and W. Kalmijn (2017). *Diversity in Survey Questions on the Same Topic: Techniques for Improving Comparability.* Cham, Switzerland: Springer.

de Leeuw, E. D. (2008). Choosing the method of data collection. In E. D. de Leeuw, J. J. Hox, and D. A. Dillman (Eds.), *International Handbook of Survey Methodology*, pp. 113–135. New York: Erlbaum.

de Leeuw, E. D., J. J. Hox, and D. A. Dillman (Eds.) (2008). *International Handbook of Survey Methodology.* New York: Erlbaum.

Deal, R. L. and D. H. Olson (2015). *The Smart Stepfamily Marriage.* Minneapolis, MN: Bethany House Publishers.

Dean, N. and M. Pagano (2015). Evaluating confidence interval methods for binomial proportions in clustered surveys. *Journal of Survey Statistics and Methodology 3*(4), 484–503.

Dell-Kuster, S., R. Droeser, J. Schäfer, V. Gloy, H. Ewald, S. Schandelmaier, L. Hemkens, H. Bucher, J. Young, and R. Rosenthal (2018). Systematic review and simulation study of ignoring clustered data in surgical trials. *British Journal of Surgery 105*(3), 182–191.

Demidenko, E. (2013). *Mixed Models: Theory and Applications with R, 2nd ed.* Hoboken, NJ: Wiley.

Deming, W. E. (1944). On errors in surveys. *American Sociological Review 9*, 359–369.

Deming, W. E. (1950). *Some Theory of Sampling.* New York: Dover.

Deming, W. E. (1977). An essay on screening, or on two-phase sampling, applied to surveys of a community. *International Statistical Review 45*, 29–37.

Deming, W. E. (1986). *Out of the Crisis.* Cambridge, MA: Massachusetts Institute of Technology.

Deming, W. E. and F. F. Stephan (1940). On a least squares adjustment of a sampled frequency table when the expected marginal totals are known. *Annals of Mathematical Statistics 11*, 427–444.

Demnati, A. and J. N. K. Rao (2004). Linearization variance estimators for survey data (with discussion). *Survey Methodology 30*, 17–34.

Devaud, D. and Y. Tillé (2019). Deville and Särndal's calibration: Revisiting a 25-years-old successful optimization problem. *Test 28*(4), 1033–1065.

Dever, J. A. and R. Valliant (2010). A comparison of variance estimators for poststratification to estimated control totals. *Survey Methodology 36*(1), 45–56.

Dever, J. A. and R. Valliant (2016). General regression estimation adjusted for undercoverage and estimated control totals. *Journal of Survey Statistics and Methodology 4*, 289–318.

Deville, J.-C. and C.-E. Särndal (1992). Calibration estimators in survey sampling. *Journal of the American Statistical Association 87*, 376–382.

Deville, J.-C. and Y. Tillé (2004). Efficient balanced sampling: The cube method. *Biometrika 91*(4), 893–912.

Dillman, D. A. (2006). Why choice of survey mode makes a difference. *Public Health Reports 121*(1), 11–13.

Dillman, D. A. (2008). The logic and psychology of constructing questionnaires. In E. D. deLeeuw, J. J. Hox, and D. A. Dillman (Eds.), *International Handbook of Survey Methodology*, pp. 161–175. New York: Erlbaum.

Dillman, D. A. (2020). Towards survey response rate theories that no longer pass each other like strangers in the night. In P. S. Brenner (Ed.), *Understanding Survey Methodology*, pp. 15–44. Cham, Switzerland: Springer.

Dillman, D. A. and L. M. Christian (2005). Survey mode as a source of instability in responses across surveys. *Field Methods 17*(1), 30–52.

Dillman, D. A., J. D. Smyth, and L. M. Christian (2014). *Internet, Phone, Mail, and Mixed-Mode Surveys: The Tailored Design Method, 4th ed.* Hoboken, NJ: Wiley.

Dillon, M. (2019). Use of administrative records to replace or enhance questions about housing characteristics on the American Community Survey. Technical Report ACS18-RER-02, U.S. Census Bureau, Washington, DC.

Dippo, C. S., R. E. Fay, and D. H. Morganstein (1984). Computing variances from complex samples with replicate weights. In *Proceedings of the Survey Research Methods Section*, pp. 489–494. Alexandria, VA: American Statistical Association.

Dolson, D. (2010). Census coverage studies in Canada: A history with emphasis on the 2011 census. In *Proceedings of the Survey Research Methods Section*, pp. 441–455. Alexandria, VA: American Statistical Association.

Domingo-Salvany, A., R. L. Hartnoll, A. Maquire, J. M. Suelves, and J. M. Anto (1995). Use of capture-recapture to estimate the prevalence of opiate addiction in Barcelona, Spain, 1989. *American Journal of Epidemiology 141*, 567–574.

Duda, M. D. and J. L. Nobile (2010). The fallacy of online surveys: No data are better than bad data. *Human Dimensions of Wildlife 15*, 55–64.

Duffy, J. C. and J. J. Waterton (1988). Randomised response vs. direct questioning: Estimating the prevalence of alcohol related problems in a field survey. *The Australian Journal of Statistics 30*, 1–14.

Dumitrescu, L., W. Qian, and J. N. K. Rao (2021). A weighted composite likelihood approach to inference from clustered survey data under a two-level model. *Sankhyā, Series A*, 1–30.

DuMouchel, W. H. and G. J. Duncan (1983). Using sample survey weights in multiple regression analyses of stratified samples. *Journal of the American Statistical Association 78*, 535–543.

Dunn, G., A. Pickles, M. Tansella, and J. Vazquez-Barquero (1999). Two-phase epidemiological surveys in psychiatric research. *British Journal of Psychiatry 174*, 95–100.

Durbin, J. (1953). Some results in sampling theory when the units are sampled with unequal probabilities. *Journal of the Royal Statistical Society, Series B 15*, 262–269.

Dutwin, D. and T. D. Buskirk (2017). Apples to oranges or gala versus golden delicious? Comparing data quality of nonprobability internet samples to low response rate probability samples. *Public Opinion Quarterly 81*(S1), 213–239.

Dwork, C. and A. Roth (2014). The algorithmic foundations of differential privacy. *Foundations and Trends in Theoretical Computer Science 9*(3–4), 211–407.

Ebersole, S. (2000). Uses and gratifications of the web among students. *Journal of Computer-Mediated Communication 6*(1).

Edwards, T. C., D. R. Cutler, N. E. Zimmerman, L. Geiser, and J. Alegria (2005). Model-based stratifications for enhancing the detection of rare ecological events. *Ecology 86*, 1081–1090.

Efron, B. and R. Tibshirani (1993). *An Introduction to the Bootstrap.* London: Chapman & Hall.

Egeland, G. M., K. A. Perham-Hester, and E. B. Hook (1995). Use of capture-recapture analyses in fetal alcohol syndrome surveillance in Alaska. *American Journal of Epidemiology 141*, 335–341.

Eilers, P. H. C. and B. D. Marx (2021). *Practical Smoothing: The Joys of P-splines.* Cambridge: Cambridge University Press.

Einarsen, S., S. Matthiessen, and A. Skogstad (1998). Bullying, burnout, and well-being among assistant nurses. *Journal of Occupational Health and Safety Australia and New Zealand 14*, 563–568.

Eisinga, R., M. te Grotenhuis, J. K. Larsen, B. Pelzer, and T. van Strien (2011). BMI of interviewer effects. *International Journal of Public Opinion Research 23*(4), 530–543.

Elliott, M. R. and R. J. A. Little (2000). Model-based alternatives to trimming survey weights. *Journal of Official Statistics 16*(3), 191–209.

Elliott, M. R. and R. Valliant (2017). Inference for nonprobability samples. *Statistical Science 32*(2), 249–264.

Ellis, J. M., H. F. Zickgraf, A. T. Galloway, J. H. Essayli, and M. C. Whited (2018). A functional description of adult picky eating using latent profile analysis. *International Journal of Behavioral Nutrition and Physical Activity 15*(109), 1–12.

Eltinge, J. L. (1994). Sufficient conditions for moment approximations for a sample ratio or regression coefficient under simple random sampling. *Sankhyā, Series B 53*(4), 400–414.

Eltinge, J. L. and I. S. Yansaneh (1997). Diagnostics for formation of nonresponse adjustment cells, with an application to income nonresponse in the U.S. Consumer Expenditure Survey. *Survey Methodology 23*, 33–40.

Engel, U., B. Jann, P. Lynn, A. Scherpenzeel, and P. Sturgis (2015). *Improving Survey Methods: Lessons from Recent Research.* New York: Routledge.

Erba, J., B. Ternes, P. Bobkowski, T. Logan, and Y. Liu (2018). Sampling methods and sample populations in quantitative mass communication research studies: A 15-year census of six journals. *Communication Research Reports 35*(1), 42–47.

Erlich, Y., T. Shor, I. Pe'er, and S. Carmi (2018). Identity inference of genomic data using long-range familial searches. *Science 362*(6415), 690–694.

Estevao, V. M. and C.-E. Särndal (1999). The use of auxiliary information in design-based estimation for domains. *Survey Methodology 25*, 213–231.

Eurostat (2019). *Quality Assurance Framework of the European Statistical System, V2.0.* Luxembourg: Publications Office of the European Union.

Eurostat (2020). *European Statistical System Handbook for Quality and Metadata Reports.* Luxembourg: Publications Office of the European Union.

Exelmans, L., M. Gradisar, and J. Van den Bulck (2018). Sleep latency versus shuteye latency: Prevalence, predictors and relation to insomnia symptoms in a representative sample of adults. *Journal of Sleep Research 27*(6), e12737.

Fay, R. E. (1984). Some properties of estimates of variance based on replication methods. In *Proceedings of the Survey Research Methods Section*, pp. 495–500. Alexandria, VA: American Statistical Association.

Fay, R. E. (1985). A jackknifed chi-squared test for complex samples. *Journal of the American Statistical Association 80*, 148–157.

Fay, R. E. (1989). Theory and application of replicate weighting for variance calculations. In *Proceedings of the Survey Research Methods Section*, pp. 212–217. Alexandria, VA: American Statistical Association.

Fay, R. E. and R. A. Herriot (1979). Estimates of income for small places: An empirical Bayes application of James-Stein procedures to census data. *Journal of the American Statistical Association 74*, 269–277.

Feldman, J. M., S. Gruskin, B. A. Coull, and N. Krieger (2017). Quantifying underreporting of law-enforcement-related deaths in United States vital statistics and news-media-based data sources: A capture–recapture analysis. *PLoS Medicine 14*(10), e1002399.

Fields, J., J. Hunter-Childs, A. Tersine, J. Sisson, E. Parker, V. Velkoff, C. Logan, and H. Shin (2020). *Design and Operation of the 2020 Household Pulse Survey.* Washington, DC: U.S. Census Bureau.

Fienberg, S. E. (1972). The multiple recapture census for closed populations and incomplete 2^k contingency tables. *Biometrika 59*, 591–603.

Fienberg, S. E. (1979). Use of chi-squared statistics for categorical data problems. *Journal of the Royal Statistical Society, Series B 41*, 54–64.

Fienberg, S. E. and J. M. Tanur (2018). The interlocking world of surveys and experiments. *The Annals of Applied Statistics 12*(2), 1157–1179.

Fisher, N. (2019). A comprehensive approach to problems of performance measurement. *Journal of the Royal Statistical Society: Series A 182*(3), 755–803.

Fisher, R. A. (1938). Presidential address. In *Proceedings of the Indian Statistical Conference.* Calcutta: Statistical Publishing Society.

Forman, S. L. (2004). Baseball-reference.com—Major league statistics and information. `www.baseball-reference.com` (accessed November 2004).

Fowler, F. J. (1991). Reducing interviewer-related error through interviewer training, supervision, and other means. In P. P. Biemer, R. M. Groves, L. E. Lyberg, N. A. Mathiowetz, and S. Sudman (Eds.), *Measurement Error in Surveys*, pp. 259–278. New York: Wiley.

Fowler, F. J. (1995). *Improving Survey Questions: Design and Evaluation.* Thousand Oaks, CA: Sage.

Fowler, F. J. (2014). *Survey Research Methods, 5th ed.* Thousand Oaks, CA: Sage Publications Inc.

Fowler, F. J. and C. Cosenza (2009). Design and evaluation of survey questions. In L. Bickman and D. J. Rog (Eds.), *The SAGE Handbook of Applied Social Research Methods, 2nd ed.*, pp. 375–412. Thousand Oaks, CA: Sage Publications.

Francisco, C. A. and W. A. Fuller (1991). Quantile estimation with a complex survey design. *The Annals of Statistics 19*, 454–469.

Frank, A. (1978). The contingency table approach to mark-recapture population estimation. Department of Applied Statistics Plan B Paper. St. Paul, MN: University of Minnesota.

Frank, O. (2010). Network sampling. In M. Carlson, H. Nyquist, and M. Villani (Eds.), *Official Statistics—Methodology and Applications in Honour of Daniel Thorburn*, pp. 51–60. Stockholm: Stockholm University.

Fuller, W. A. (2002). Regression estimation for survey samples. *Survey Methodology 28*, 5–23.

Fuller, W. A. (2009). *Sampling Statistics.* Hoboken, NJ: Wiley.

Gabler, S. (1984). On unequal probability sampling: Sufficient conditions for the superiority of sampling without replacement. *Biometrika 71*, 171–175.

Gabler, S., S. Häder, and P. Lahiri (1999). A model based justification of Kish's formula for design effects for weighting and clustering. *Survey Methodology 25*, 105–106.

Gabler, S., S. Häder, and P. Lynn (2006). Design effects for multiple design samples. *Survey Methodology 32*, 115–120.

Garbarski, D., N. C. Schaeffer, and J. Dykema (2015). The effects of response option order and question order on self-rated health. *Quality of Life Research 24*(6), 1443–1453.

Geisen, E. and J. R. Bergstrom (2017). *Usability Testing for Survey Research.* Cambridge, MA: Morgan Kaufmann.

Gelman, A. (2007). Struggles with survey weighting and regression modeling (with discussion). *Statistical Science 22*, 153–164.

Gelman, A. and J. Hill (2007). *Data Analysis using Regression and Multilevel/Hierarchical Models.* New York: Cambridge University Press.

Gentle, J., C. Perry, and W. Wigton (2006). Modeling nonsampling errors in agricultural surveys. In *Proceedings of the Survey Research Methods Section*, pp. 3035–3041. Alexandria, VA: American Statistical Association.

Ghosh, M. (2009). Bayesian developments in survey sampling. In D. Pfeffermann and C. R. Rao (Eds.), *Handbook of Statistics, Volume 29B, Sample Surveys: Inference and Analysis*, pp. 153–187. Amsterdam: Elsevier.

Ghosh, M. (2020). Small area estimation: Its evolution in five decades (with discussion). *Statistics in Transition 21*(4), 1–22.

Gilbert, R., R. Lafferty, G. Hagger-Johnson, K. Harron, L.-C. Zhang, P. Smith, C. Dibben, and H. Goldstein (2018). GUILD: Guidance for information about linking data sets. *Journal of Public Health 40*(1), 191–198.

Gile, K. J., I. S. Beaudry, M. S. Handcock, and M. Q. Ott (2018). Methods for inference from respondent-driven sampling data. *Annual Review of Statistics and Its Application 5*, 65–93.

Giles, J. (2005). Internet encyclopaedias go head to head. *Nature 438*, 900–901.

Gini, C. and L. Galvani (1929). Di una applicazione del metodo rappresentativo all'ultimo censimento italiano della popolazione. *Annali di Statistica 6*(4), 1–105.

Gnanadesikan, M., R. L. Scheaffer, A. E. Watkins, and J. A. Witmer (1997). An activity-based statistics course. *Journal of Statistics Education 5*(2), 1–16.

Gnap, R. (1995). *Teacher Load in Arizona Elementary School Districts in Maricopa County.* Ph.D. dissertation. Tempe, AZ: Arizona State University.

Goga, C. (2005). Variance reduction in surveys with auxiliary information: A nonparametric approach involving regression splines. *The Canadian Journal of Statistics 33*(2), 163–180.

Golbeck, A. L., T. H. Barr, and C. A. Rose (2019). Fall 2017 departmental profile report. *Notices of the American Mathematical Society 66*(10), 1721–1730.

Goldstein, H. (2011). *Multilevel Statistical Models.* Hoboken, NJ: Wiley.

Gonzalez, J. M. and J. L. Eltinge (2010). Optimal survey design: A review. In *Proceedings of the Survey Research Methods Section*, pp. 4970–4983. Alexandria, VA: American Statistical Association.

Goren, S., L. Silverstein, and N. Gonzales (1993). A survey of food service managers of Washington State boarding homes for the elderly. *Journal of Nutrition for the Elderly 12*, 27–36.

Granquist, L. and J. Kovar (1997). Editing of survey data: How much is enough? In L. Lyberg, P. Biemer, M. Collins, E. de Leeuw, C. Dippo, N. Schwarz, and D. Trewin (Eds.), *Survey Measurement and Process Quality*, pp. 415–435. New York: Wiley.

Griffin, D. H., D. A. Raglin, T. F. Leslie, P. D. McGovern, and J. K. Broadwater (2003). *Meeting 21st Century Demographic Data Needs—Implementing the American Community Survey.* Washington, DC: U.S. Census Bureau.

Gross, S. T. (1980). Median estimation in sample surveys. In *Proceedings of the Survey Research Methods Section*, pp. 181–184. Alexandria, VA: American Statistical Association.

Groves, R. M. (1989). *Survey Errors and Survey Costs.* New York: Wiley.

Groves, R. M. (2006). Nonresponse rates and nonresponse bias in household surveys. *Public Opinion Quarterly 70*, 646–675.

Groves, R. M. and M. P. Couper (2002). Designing surveys acknowledging nonresponse. In *Studies of Welfare Populations: Data Collection and Research Issues*, pp. 13–54. Washington DC: The National Academies Press.

Groves, R. M., D. Dillman, J. Eltinge, and R. Little (Eds.) (2002). *Survey Nonresponse.* New York: Wiley.

Groves, R. M., F. J. Fowler, M. P. Couper, J. M. Lepkowski, E. Singer, and R. Tourangeau (2011). *Survey Methodology, 2nd ed.* Hoboken, NJ: Wiley.

Groves, R. M. and S. G. Heeringa (2006). Responsive design for household surveys: Tools for actively controlling survey errors and costs. *Journal of the Royal Statistical Society: Series A 169*(3), 439–457.

Groves, R. M. and L. Lyberg (2010). Total survey error: Past, present, and future. *Public Opinion Quarterly 74*(5), 849–879.

Groves, R. M. and E. Peytcheva (2008). The impact of nonresponse rates on nonresponse bias: A meta-analysis. *Public Opinion Quarterly 72*(2), 167–189.

Guo, Y., J. A. Kopec, J. Cibere, L. C. Li, and C. H. Goldsmith (2016). Population survey features and response rates: A randomized experiment. *American Journal of Public Health 106*(8), 1422–1426.

Hájek, J. (1960). Limiting distributions in simple random sampling from a finite population. *Publications of the Mathematical Institute of the Hungarian Academy of Sciences 5*, 361–371.

Hájek, J. (1964). Asymptotic theory of rejection sampling with varying probabilities from a finite population. *Annals of Mathematical Statistics 35*, 1491–1523.

Halbesleben, J. R. and M. V. Whitman (2013). Evaluating survey quality in health services research: A decision framework for assessing nonresponse bias. *Health Services Research 48*(3), 913–930.

Han, D. and R. Valliant (2021). Effects of outcome and response models on single-step calibration estimators. *Journal of Survey Statistics and Methodology 9*(3), 574–597.

Han, Y. and P. Lahiri (2019). Statistical analysis with linked data. *International Statistical Review 87*, S139–S157.

Hand, D. J. (2018). Statistical challenges of administrative and transaction data. *Journal of the Royal Statistical Society: Series A 181*(3), 555–605.

Hand, D. J., F. Daly, A. D. Lunn, K. J. McConway, and E. Ostrowski (1994). *A Handbook of Small Data Sets.* London: Chapman and Hall.

Handcock, M. S. and K. J. Gile (2010). Modeling social networks from sampled data. *The Annals of Applied Statistics 4*(1), 5–25.

Hansen, M. H. and W. N. Hurwitz (1943). On the theory of sampling from a finite population. *Annals of Mathematical Statistics 14*, 333–362.

Hansen, M. H. and W. N. Hurwitz (1946). The problem of non-response in sample surveys. *Journal of the American Statistical Association 41*, 517–529.

Hansen, M. H., W. N. Hurwitz, and W. G. Madow (1953a). *Sample Survey Methods and Theory. Volume 1: Methods and Applications.* New York: Wiley.

Hansen, M. H., W. N. Hurwitz, and W. G. Madow (1953b). *Sample Survey Methods and Theory. Volume 2: Theory.* New York: Wiley.

Hansen, M. H. and W. G. Madow (1978). Estimation and inferences from sample surveys: Some comments on recent developments. In *Survey Sampling and Measurement*, pp. 341–357. New York: Academic Press.

Hansen, M. H., W. G. Madow, and B. J. Tepping (1983). An evaluation of model-dependent and probability-sampling inferences in sample surveys. *Journal of the American Statistical Association 78*, 776–793.

Hanurav, T. V. (1967). Optimum utilization of auxiliary information: πps sampling of two units from a stratum. *Journal of the Royal Statistical Society, Series B 29*, 374–391.

Harkness, J., F. J. R. van de Vijver, and P. P. Mohler (2003). *Cross-Cultural Survey Methods.* Hoboken, NJ: Wiley.

Harms, T. and P. Duchesne (2010). On kernel nonparametric regression designed for complex survey data. *Metrika 72*(1), 111–138.

Harter, R., A. Vaish, A. Sukasih, J. Iriondo-Perez, K. Jones, and B. Lu (2019). A practical guide to small area estimation, illustrated using the Ohio Medicaid Assessment Survey. In *Proceedings of the Survey Research Methods Section*, pp. 1992–2004. Alexandria, VA: American Statistical Association.

Hartley, H. O. (1946). Discussion of paper by F. Yates. *Journal of the Royal Statistical Society 109*, 38–39.

Hartley, H. O. (1962). Multiple frame surveys. In *Proceedings of the Social Statistics Section*, pp. 203–206. Alexandria, VA: American Statistical Association.

Hartley, H. O. and P. Biemer (1981). The estimation of nonsampling variance in current surveys. In *Proceedings of the Survey Research Methods Section*, pp. 257–262. Alexandria, VA: American Statistical Association.

Hartley, H. O. and J. N. K. Rao (1962). Sampling with unequal probabilities and without replacement. *Annals of Mathematical Statistics 33*, 350–374.

Hartley, H. O. and J. N. K. Rao (1978). Estimation of nonsampling variance components in sample surveys. In N. K. Namboodiri (Ed.), *Survey Sampling and Measurement*, pp. 35–44. New York: Academic Press.

Hartley, H. O. and A. Ross (1954). Unbiased ratio estimators. *Nature 174*, 270–271.

Harvey, F. (2019). Thousands of Britons invited to climate crisis citizens' assembly. *The Guardian*. `https://www.theguardian.com/environment/2019/nov/02/thousands-britons-invited-take-part-climate-crisis-citizens-assembly` (accessed November 27, 2019).

Hassel, M., B. E. Asbjørnslett, and L. P. Hole (2011). Underreporting of maritime accidents to vessel accident databases. *Accident Analysis & Prevention 43*(6), 2053–2063.

Hastie, T., R. Tibshirani, and J. Friedman (2009). *The Elements of Statistical Learning, 2nd ed.* New York: Springer.

Hayat, M. and T. Knapp (2017). Randomness and inference in medical and public health research. *Journal of the Georgia Public Health Association 7*(1), 7–11.

Hayes, L. R. (2000). Are prices higher for the poor in New York City? *Journal of Consumer Policy 23*, 127–152.

Haziza, D. (2009). Imputation and inference in the presence of missing data. In D. Pfeffermann and C. R. Rao (Eds.), *Sample Surveys: Design, Methods, and Applications. Handbook of Statistics, Volume 29A*, pp. 215–246. Amsterdam: North-Holland.

Haziza, D. and J.-F. Beaumont (2017). Construction of weights in surveys: A review. *Statistical Science 32*(2), 206–226.

Haziza, D. and É. Lesage (2016). A discussion of weighting procedures for unit nonresponse. *Journal of Official Statistics 32*(1), 129–145.

Haziza, D. and J. N. K. Rao (2006). A nonresponse model approach to inference under imputation for missing survey data. *Survey Methodology 32*, 53–64.

Haziza, D. and A.-A. Vallée (2020). Variance estimation procedures in the presence of singly imputed survey data: A critical review. *Japanese Journal of Statistics and Data Science 3*(2), 583–623.

He, Y., A. M. Zaslavsky, D. Harrington, P. Catalano, and M. Landrum (2010). Multiple imputation in a large-scale complex survey: A practical guide. *Statistical Methods in Medical Research 19*(6), 653–670.

Heck, P. R., D. J. Simons, and C. F. Chabris (2018). 65% of Americans believe they are above average in intelligence: Results of two nationally representative surveys. *PloS One 13*(7), 1–11.

Heckathorn, D. D. (1997). Respondent driven sampling: A new approach to the study of hidden populations. *Social Problems 44*, 174–199.

Heckathorn, D. D. and C. J. Cameron (2017). Network sampling: From snowball and multiplicity to respondent-driven sampling. *Annual Review of Sociology 43*, 101–119.

Hedges, L. V. and E. C. Hedberg (2007). Intraclass correlation values for planning group-randomized trials in education. *Educational Evaluation and Policy Analysis 29*(1), 60–87.

Heeringa, S. G., B. T. West, and P. A. Berglund (2017). *Applied Survey Data Analysis, 2nd ed.* Boca Raton, FL: CRC Press.

Henrich, J., S. J. Heine, and A. Norenzayan (2010). Most people are not WEIRD. *Nature 466*(7302), 29.

Henry, K. A. and R. Valliant (2009). Comparing sampling and estimation strategies in establishment populations. *Survey Research Methods 3*(1), 27–44.

Herzog, R. (2020, January 19). Little Rock poll: Plastic bags out of favor. *Arkansas Democrat Gazette.* `https://www.arkansasonline.com/news/2020/jan/19/lr-poll-plastic-bags-out-of-favor-20200/` (accessed January 22, 2020).

Hetzel, A. M. (1997). *U.S. Vital Statistics System: Major Activities and Developments, 1950–95.* Hyattsville, MD: National Center for Health Statistics.

Hidiroglou, M. A. (2001). Double sampling. *Survey Methodology 27*, 143–154.

Hidiroglou, M. A., J. D. Drew, and G. B. Gray (1993). A framework for measuring and reducing nonresponse in surveys. *Survey Methodology 19*, 81–94.

Hidiroglou, M. A. and Z. Patak (2004). Domain estimation using linear regression. *Survey Methodology 30*, 67–78.

Hidiroglou, M. A., J. N. K. Rao, and D. Haziza (2009). Variance estimation in two-phase sampling. *Australian and New Zealand Journal of Statistics 51*, 127–141.

Hillygus, D. S., N. Jackson, and M. Young (2014). Professional respondents in nonprobability online panels. In M. Callegaro, R. Baker, J. Bethlehem, A. S. Göritz, J. A. Krosnick, and P. J. Lavrakas (Eds.), *Online Panel Research*, pp. 219–237. Hoboken, NJ: Wiley.

Hinkins, S., H. L. Oh, and F. Scheuren (1997). Inverse sampling design algorithms. *Survey Methodology 23*, 11–21.

Hitlin, P. (2016). *Research in the Crowdsourcing Age, A Case Study.* Washington, DC: Pew Research.

Hoeting, J. A., M. Leecaster, and D. Bowden (2000). An improved model for spatially correlated binary responses. *Journal of Agricultural, Biological, and Environmental Statistics 5*, 102–114.

Hogan, H. (1993). The 1990 Post-Enumeration Survey: Operations and results. *Journal of the American Statistical Association 88*, 1047–1060.

Hogan, H., P. J. Cantwell, J. Devine, V. T. Mule, and V. Velkoff (2013). Quality and the 2010 census. *Population Research and Policy Review 32*(5), 637–662.

Holt, D., A. J. Scott, and P. D. Ewings (1980). Chi-squared tests with survey data. *Journal of the Royal Statistical Society, Series A 143*, 303–320.

Hopper, K., M. Shinn, E. Laska, M. Meisner, and J. Wanderling (2008). Estimating numbers of unsheltered homeless people through plant-capture and postcount survey methods. *American Journal of Public Health 98*(8), 1438–1442.

Horvitz, D. G., B. V. Shah, and W. R. Simmons (1967). The unrelated question randomized response model. In *Proceedings of the Social Statistics Section*, pp. 65–72. Alexandria, VA: American Statistical Association.

Horvitz, D. G. and D. J. Thompson (1952). A generalization of sampling without replacement from a finite universe. *Journal of the American Statistical Association 47*, 663–685.

Hosmer, D. W., S. Lemeshow, and R. X. Sturdivant (2013). *Applied Logistic Regression, 3rd ed.* Hoboken, NJ: Wiley.

House of Commons (2016, April 25). A paperless NHS: Electronic health records. `researchbriefings.parliament.uk/ResearchBriefing/Summary/CBP-7572` (accessed January 6, 2020).

Hox, J., E. de Leeuw, and T. Klausch (2017). Mixed-mode research: Issues in design and analysis. In P. P. Biemer, E. de Leeuw, S. Eckman, B. Edwards, F. Kreuter, L. E. Lyberg, N. C. Tucker, and B. T. West (Eds.), *Total Survey Error in Practice*, pp. 511–530. New York: Wiley.

Hurlbert, S. H. (1984). Pseudoreplication and the design of ecological field experiments. *Ecological Monographs 54*, 187–211.

Hurlbert, S. H. (2009). The ancient black art and transdisciplinary extent of pseudoreplication. *Journal of Comparative Psychology 123*(4), 434–443.

Husby, C. E., E. A. Stasny, and D. A. Wolfe (2005). An application of ranked set sampling for mean and median estimation using USDA crop production data. *Journal of Agricultural, Biological, and Environmental Statistics 10*, 354–373.

Hyndman, R. J. and Y. Fan (1996). Sample quantiles in statistical packages. *The American Statistician 50*(4), 361–365.

Iachan, R. and M. L. Dennis (1993). A multiple frame approach to sampling the homeless and transient population. *Journal of Official Statistics 9*, 747–764.

International Statistical Institute (2010). *ISI Declaration on Professional Ethics.* The Hague, Netherlands: International Statistical Institute. `https://isi-web.org/images/about/Declaration-EN2010.pdf` (accessed January 14, 2020).

Ismail, K., K. Kent, T. Brugha, M. Hotopf, L. Hull, P. Seed, I. Palmer, S. Reid, C. Unwin, A. S. David, and S. Wessely (2002). The mental health of UK gulf war veterans: Phase 2 of a two phase cohort study. *British Medical Journal 325*, 576–579.

Jackson, K. W., I. W. Eastwood, and M. S. Wild (1987). Stratified sampling protocol for monitoring trace metal concentrations in soil. *Soil Science 143*, 436–443.

Jacoby, J. and A. H. Handlin (1991). Non-probability sampling designs for litigation surveys. *Trademark Reporter 81*, 169–179.

James, G., D. Witten, T. Hastie, and R. Tibshirani (2013). *An Introduction to Statistical Learning.* New York: Springer.

Jenney, B. (2005). *Regression Diagnostics with Complex Survey Data.* Master's thesis. Tempe, AZ: Arizona State University.

Jiang, J. and P. Lahiri (2006). Mixed model prediction and small area estimation. *Test 15*(1), 1–96.

Jiang, J. and J. S. Rao (2020). Robust small area estimation: An overview. *Annual Review of Statistics and Its Application 7*, 337–360.

Judkins, D. (1990). Fay's method for variance estimation. *Journal of Official Statistics 6*, 223–240.

Kalton, G. (2003). Practical methods for sampling rare and mobile populations. *Statistics in Transition 6*, 491–501.

Kalton, G. and D. W. Anderson (1986). Sampling rare populations. *Journal of the Royal Statistical Society, Series A 149*, 65–82.

Kan, I. P. and A. B. Drummey (2018). Do imposters threaten data quality? An examination of worker misrepresentation and downstream consequences in Amazon's Mechanical Turk workforce. *Computers in Human Behavior 83*, 243–253.

Karras, T. (2008). The disorder next door. *SELF 30* (May), 248–253.

Kasprzyk, D. and L. Giesbrecht (2003). Reporting sources of error in U.S. federal government surveys. *Journal of Official Statistics 19*(4), 343–363.

Keeter, S. (2015). *From Telephone to the Web: The Challenge of Mode of Interview Effects in Public Opinion Polls.* Washington, DC: Pew Research.

Keiding, N. and T. A. Louis (2018). Web-based enrollment and other types of self-selection in surveys and studies: Consequences for generalizability. *Annual Review of Statistics and Its Application 5*, 25–47.

Kempthorne, O. (1952). *The Design and Analysis of Experiments.* New York: Wiley.

Kennedy, C., K. McGeeney, S. Keeter, E. Patten, A. Perrin, A. Lee, and J. Best (2018). Implications of moving public opinion surveys to a single-frame cell-phone random-digit-dial design. *Public Opinion Quarterly 82*(2), 279–299.

Kennickell, A. B. (2017). Multiple imputation in the Survey of Consumer Finances. *Statistical Journal of the IAOS 33*(1), 143–151.

Keusch, F. (2015). Why do people participate in web surveys? Applying survey participation theory to internet survey data collection. *Management Review Quarterly 65*(3), 183–216.

Keyes, K. M., C. Rutherford, F. Popham, S. S. Martins, and L. Gray (2018). How healthy are survey respondents compared with the general population?: Using survey-linked death records to compare mortality outcomes. *Epidemiology 29*(2), 299–307.

Kiaer, A. N. (1896). Observations et expériences concernant des dénombrements représentatifs. *Bulletin de l'Institut International de Statistique 9*(42), 176–183.

Kim, J. K. and D. Haziza (2014). Doubly robust inference with missing data in survey sampling. *Statistica Sinica 24*, 375–394.

Kim, J. K. and J. J. Kim (2007). Nonresponse weighting adjustment using estimated response probability. *Canadian Journal of Statistics 35*(4), 501–514.

Kim, J. K., A. Navarro, and W. A. Fuller (2006). Replication variance estimation for two-phase stratified sampling. *Journal of the American Statistical Association 101*, 312–320.

Kim, J. K. and M. K. Riddles (2012). Some theory for propensity-score-adjustment estimators in survey sampling. *Survey Methodology 38*(2), 157–165.

Kim, J. K. and J. Shao (2013). *Statistical Methods for Handling Incomplete Data.* Boca Raton, FL: CRC Press.

Kim, J. K. and C. L. Yu (2011). Replication variance estimation under two-phase sampling. *Survey Methodology 37*(1), 67.

Kish, L. (1965). *Survey Sampling.* New York: Wiley.

Kish, L. (1992). Weighting for unequal P_i. *Journal of Official Statistics 8*, 183–200.

Kish, L. (1995). Methods for design effects. *Journal of Official Statistics 11*, 55–77.

Kish, L. and M. R. Frankel (1974). Inference from complex samples (with discussion). *Journal of the Royal Statistical Society, Series B 36*, 1–37.

Kleppel, G. S., S. A. Madewell, and S. E. Hazzard (2004). Responses of emergent marsh wetlands in upstate New York to variations in urban typology. *Ecology and Society 5*(1), 1–18.

Kloke, J. and J. W. McKean (2014). *Nonparametric Statistical Methods Using R.* Boca Raton, FL: CRC Press.

Koch, G. G., D. H. Freeman, and J. L. Freeman (1975). Strategies in the multivariate analysis of data from complex surveys. *International Statistical Review 43*, 59–78.

Kohut, A., S. Keeter, C. Doherty, M. Dimock, and L. Christian (2012). *Assessing the Representativeness of Public Opinion Surveys.* Washington DC: Pew Research Center.

Kolenikov, S. (2010). Resampling variance estimation for complex survey data. *The Stata Journal 10*(2), 165–199.

Korn, E. L. and B. I. Graubard (1995). Analysis of large health surveys: Accounting for the sampling design. *Journal of the Royal Statistical Society, Series A 158*, 263–295.

Korn, E. L. and B. I. Graubard (1998a). Confidence intervals for proportions with small expected number of positive counts estimated from survey data. *Survey Methodology 24*, 193–201.

Korn, E. L. and B. I. Graubard (1998b). Scatterplots with survey data. *The American Statistician 52*, 58–69.

Korn, E. L. and B. I. Graubard (1999). *Analysis of Health Surveys.* New York: Wiley.

Kosmin, B. A. and S. P. Lachman (1993). *One Nation Under God: Religion in Contemporary American Society.* New York: Harmony Books.

Kott, P. S. (1991). A model-based look at linear regression with survey data. *The American Statistician 45*, 107–112.

Kott, P. S. (2001). The delete-a-group jackknife. *Journal of Official Statistics 17*(4), 521–526.

Kott, P. S. (2016). Calibration weighting in survey sampling. *Wiley Interdisciplinary Reviews: Computational Statistics 8*(1), 39–53.

Kott, P. S. and D. M. Stukel (1997). Can the jackknife be used with a two-phase sample? *Survey Methodology 23*, 81–89.

Kovar, J. G., J. N. K. Rao, and C. F. J. Wu (1988). Bootstrap and other methods to measure errors in survey estimates. *Canadian Journal of Statistics 16*(S), 25–45.

Krenzke, T. (1995). Reevaluating generalized variance model parameters for the National Crime Victimization Survey. In *Proceedings of the Survey Research Methods Section*, pp. 327–332. Alexandria, VA: American Statistical Association.

Kreuter, F. (2013). Facing the nonresponse challenge. *The Annals of the American Academy of Political and Social Science 645*(1), 23–35.

Kreuter, F., N. Barkay, A. Bilinski, A. Bradford, S. Chiu, R. Eliat, J. Fan, T. Galili, D. Haimovich, B. Kim, S. LaRocca, Y. Li, K. Morris, S. Presser, T. Sarig, J. A. Salomon, K. Steward, E. A. Stuart, and R. Tibshirani (2020). Partnering with a global platform to inform research and public policy making. *Survey Research Methods 14*(2), 159–163.

Kreuter, F., S. Presser, and R. Tourangeau (2008). Social desirability bias in CATI, IVR, and web surveys: The effects of mode and question sensitivity. *Public Opinion Quarterly 72*(5), 847–865.

Krewski, D. and J. N. K. Rao (1981). Inference from stratified samples: Properties of the linearization, jackknife and balanced repeated replication methods. *The Annals of Statistics 9*, 1010–1019.

Kripke, D. F., L. Garfinkel, D. L. Wingard, M. R. Klauber, and M. R. Marler (2002). Mortality associated with sleep duration and insomnia. *Archives of General Psychiatry 59*, 131–136.

Krosnick, J. A. and S. Presser (2010). Question and questionnaire design. In P. Marsden and J. D. Wright (Eds.), *Handbook of Survey Research, 2nd ed.*, pp. 263–313. Bingley, UK: Emerald Publishing.

Krumpal, I. (2013). Determinants of social desirability bias in sensitive surveys: A literature review. *Quality & Quantity 47*(4), 2025–2047.

Kruuk, H., A. Moorhouse, J. W. H. Conroy, L. Durbin, and S. Frears (1989). An estimate of numbers and habitat preferences of otters *Lutra lutra* in Shetland, UK. *Biological Conservation 49*, 241–254.

Kuk, A. Y. C. (1990). Asking sensitive questions indirectly. *Biometrika 77*, 436–438.

Kutner, M. H., C. J. Nachtsheim, J. Neter, and W. Li (2005). *Applied Linear Statistical Models, 5th ed.* Boston: McGraw-Hill/Irwin.

Lahiri, D. B. (1951). A method of sample selection providing unbiased ratio estimates. *Bulletin of the International Statistical Institute 33*, 133–140.

Lai, T. L. (2001). Sequential analysis: Some classical problems and new challenges (with discussion). *Statistica Sinica 11*, 303–408.

Landers, A. (1976). If you had it to do over again, would you have children? *Good Housekeeping 182* (June), 100–101, 215–216, 223–224.

Langel, M. and Y. Tillé (2011). Corrado Gini, a pioneer in balanced sampling and inequality theory. *Metron 69*(1), 45–65.

Langer, G. (2018). Probability versus non-probability methods. In D. L. Vannette and J. A. Krosnick (Eds.), *The Palgrave Handbook of Survey Research*, pp. 393–403. Cham, Switzerland: Springer.

Langer Research Associates (2019). Sexual and gender-based misconduct at American Statistical Association events. `www.amstat.org/asa/files/pdfs/LangerResearchFinalReport.pdf` (accessed July 15, 2019).

Langkjær-Bain, R. (2019). The troubling legacy of Francis Galton. *Significance 16*(3), 16–21.

Laplace, P. S. (1814). *Essai Philosophique sur les Probabilités.* Paris: Courcier.

Lavallée, P. (2007). *Indirect Sampling.* New York: Springer.

Lavallée, P. and J.-F. Beaumont (2015). Why we should put some weight on weights. *Survey Insights: Methods from the Field*, 1–18. `https://surveyinsights.org/?p=6255` (accessed February 15, 2018).

Lavallée, P. and L.-P. Rivest (2012). Capture-recapture sampling and indirect sampling. *Journal of Official Statistics 28*(1), 1–27.

Lavrakas, P., M. Traugott, C. Kennedy, A. Holbrook, E. de Leeuw, and B. West (Eds.) (2020). *Experimental Methods in Survey Research: Techniques that Combine Random Sampling with Random Assignment.* Hoboken, NJ: Wiley.

League of Women Voters of Texas (2019). *2019 County Election Website Review.* Austin, TX: League of Women Voters of Texas. `https://my.lwv.org/texas/article/league-women-voters-texas-finds-only-20-texas-counties-following-website-security-best` (accessed November 25, 2019).

Lee, S., A. R. Ong, C. Chen, and M. Elliott (2021). Respondent driven sampling for immigrant populations: A health survey of foreign-born Korean Americans. *Journal of Immigrant and Minority Health 23*, 784–792.

Lee, S., A. R. Ong, and M. Elliott (2020). Exploring mechanisms of recruitment and recruitment cooperation in respondent driven sampling. *Journal of Official Statistics 36*(2), 339–360.

Leemis, L. M. and K. S. Trivedi (1996). A comparison of approximate interval estimators for the Bernoulli parameter. *The American Statistician 50*(1), 63–68.

Legg, J. C. and W. A. Fuller (2009). Two-phase sampling. In D. Pfeffermann and C. R. Rao (Eds.), *Handbook of Statistics: Vol. 29A. Sample Surveys: Design, Methods and Applications*, pp. 55–70. Amsterdam: North Holland.

Lehmann, E. L. (1999). *Elements of Large-Sample Theory.* New York: Springer-Verlag.

Lehtonen, R. and E. Pahkinen (2004). *Practical Methods for Design and Analysis of Complex Surveys, 2nd ed.* Hoboken, NJ: Wiley.

Leitch, S., S. M. Dovey, A. Samaranayaka, D. M. Reith, K. A. Wallis, K. S. Eggleton, A. W. McMenamin, W. K. Cunningham, M. I. Williamson, S. Lillis, et al. (2018). Characteristics of a stratified random sample of New Zealand general practices. *Journal of Primary Health Care 10*(2), 114–124.

Lennert-Cody, C. E., S. T. Buckland, T. Gerrodette, A. Webb, J. Barlow, P. T. Fretwell, M. N. Maunder, T. Kitakado, J. E. Moore, M. D. Scott, and H. J. Skaug (2018). Review of potential line-transect methodologies for estimating abundance of dolphin stocks in the eastern tropical Pacific. *Journal of Cetacean Research and Management 19*, 9–21.

Lenski, G. and J. Leggett (1960). Caste, class, and deference in the research interview. *American Journal of Sociology 65*, 463–467.

Lenzner, T. and N. Menold (2016). *Question Wording (Version 2.0, GESIS Survey Guidelines).* Mannheim, Germany: GESIS — Leibniz-Institut für Sozialwissenschaften.

Lesser, V. M. and W. D. Kalsbeek (1999). Nonsampling errors in environmental surveys. *Journal of Agricultural, Biological, and Environmental Statistics 4*, 473–488.

Lessler, J. T. and W. Kalsbeek (1992). *Nonsampling Errors in Surveys.* New York: Wiley.

Levy, P. S. and S. Lemeshow (2008). *Sampling of Populations: Methods and Applications, 4th ed.* Hoboken, NJ: Wiley.

Lewis, N., K. McCaffrey, K. Sage, et al. (2019). E-cigarette use, or vaping, practices and characteristics among persons with associated lung injury—Utah, April–October 2019. *Morbidity and Mortality Weekly Report 68*, 953–956.

Li, J. and R. Valliant (2009). Survey weighted hat matrix and leverages. *Survey Methodology 35*(1), 15–24.

Li, J. and R. Valliant (2011). Linear regression influence diagnostics for unclustered survey data. *Journal of Official Statistics 27*(1), 99–119.

Linacre, S. J. and D. J. Trewin (1993). Total survey design—application to a collection of the construction industry. *Journal of Official Statistics 9*, 611–621.

Lincoln, F. C. (1930). Calculating waterfowl abundance on the basis of banding returns. *Circular of the U.S. Department of Agriculture 118*, 1–4.

Lind, L. H., M. F. Schober, F. G. Conrad, and H. Reichert (2013). Why do survey respondents disclose more when computers ask the questions? *Public Opinion Quarterly 77*(4), 888–935.

Lineback, J. F. and K. J. Thompson (2010). Conducting nonresponse bias analysis for business surveys. In *Proceedings of the Government Statistics Section*, pp. 317–331. Alexandria, VA: American Statistical Association.

Link, H. C. and H. A. Hopf (1946). *People and Books: A Study of Reading and Book-Buying Habits.* New York: Book Manufacturer's Institute.

Link, M. (2018). New data strategies: Nonprobability sampling, mobile, big data. *Quality Assurance in Education 26*(2), 303–314.

Literary Digest (1916, September 16). A nation-wide "straw vote" on the presidency. *Literary Digest 53*, 659.

Literary Digest (1936a, October 31). Landon, 1,293,669: Roosevelt, 972,897. *Literary Digest 122*, 5–6.

Literary Digest (1936b, August 22). "The *Digest*" presidential poll is on: Famous forecasting machine is thrown into gear for 1936. *Literary Digest 122*, 3–4.

Literary Digest (1936c, November 14). What went wrong with the polls? *Literary Digest 122*, 7–8.

Little, B. (2016, November 7). Four of history's worst political predictions. *National Geographic [online]*. `http://news.nationalgeographic.com/2016/11/presidential-election-predictions-history/` (accessed March 16, 2017).

Little, R. J. A. (1986). Survey nonresponse adjustments for estimates of means. *International Statistical Review 54*, 139–157.

Little, R. J. A. (1991). Inference with survey weights. *Journal of Official Statistics 7*, 405–424.

Little, R. J. A. (2004). To model or not to model: Competing modes of inference for finite population sampling. *Journal of the American Statistical Association 99*, 546–556.

Little, R. J. A. and D. B. Rubin (2019). *Statistical Analysis with Missing Data, 3rd ed.* Hoboken, NJ: Wiley.

Lohr, S. L. (1990). Accurate multivariate estimation using triple sampling. *The Annals of Statistics 18*, 1615–1633.

Lohr, S. L. (2001). Sample surveys: Model-based approaches. In N. J. Smelser and P. B. Baltes (Eds.), *International Encyclopedia of the Social & Behavioral Sciences*, pp. 13462–13467. New York: Elsevier.

Lohr, S. L. (2011). Alternative survey sample designs: Sampling with multiple overlapping frames. *Survey Methodology 37*, 197–213.

Lohr, S. L. (2012). Using SAS® for the design, analysis, and visualization of complex surveys. In *Proceedings of SAS Global Forum*, Number 343–2012. Cary, NC: SAS Institute, Inc. `https://support.sas.com/resources/papers/proceedings12/343-2012.pdf` (accessed August 22, 2020).

Lohr, S. L. (2014). Design effects for a regression slope in a cluster sample. *Journal of Survey Statistics and Methodology 2*(2), 97–125.

Lohr, S. L. (2019a). *Measuring Crime: Behind the Statistics.* Boca Raton, FL: CRC Press.

Lohr, S. L. (2019b). Scamming their way into college—but why? `https://www.sharonlohr.com/blog/2019/3/16/scamming-their-way-into-college-but-why` (accessed August 25, 2020).

Lohr, S. L. (2022). *SAS® Software Companion for* Sampling: Design and Analysis, 3rd ed. Boca Raton, FL: CRC Press.

Lohr, S. L. and J. M. Brick (2017). Roosevelt predicted to win: Revisiting the 1936 *Literary Digest* poll. *Statistics, Politics and Policy 8*(1), 65–84.

Lohr, S. L., V. Hsu, and J. Montaquila (2015). Using classification and regression trees to model survey nonresponse. In *Proceedings of the Survey Research Methods Section*, pp. 2071–2085. Alexandria, VA: American Statistical Association.

Lohr, S. L. and T. E. Raghunathan (2017). Combining survey data with other data sources. *Statistical Science 32*(2), 293–312.

Lorant, V., S. Demarest, P.-J. Miermans, and H. Van Oyen (2007). Survey error in measuring socio-economic risk factors of health status: A comparison of a survey and a census. *International Journal of Epidemiology 36*(6), 1292–1299.

Loy, A., H. Hofmann, and D. Cook (2017). Model choice and diagnostics for linear mixed-effects models using statistics on street corners. *Journal of Computational and Graphical Statistics 26*(3), 478–492.

Lu, Y. and S. L. Lohr (2021). *SDAResources: Datasets and Functions for "Sampling: Design and Analysis."* R package version 0.1.0. `https://CRAN.R-project.org/package=SDAResources` (accessed May 17, 2021).

Lu, Y. and S. L. Lohr (2022). *R Companion for* Sampling: Design and Analysis, 3rd ed. Boca Raton, FL: CRC Press.

Lumley, T. (2020). *survey: Analysis of Complex Survey Samples.* R package version 4.0. `https://CRAN.R-project.org/package=survey` (accessed September 20, 2020).

Lumley, T. and A. Scott (2014). Tests for regression models fitted to survey data. *Australian & New Zealand Journal of Statistics 56*(1), 1–14.

Lumley, T. and A. Scott (2017). Fitting regression models to survey data. *Statistical Science 32*(2), 265–278.

Lundquist, P. and C.-E. Särndal (2013). Aspects of responsive design with applications to the Swedish Living Conditions Survey. *Journal of Official Statistics 29*(4), 557–582.

Lusinchi, D. (2012). "President" Roosevelt and the 1936 Literary Digest poll. *Social Science History 36*(1), 23–54.

Lyberg, L. (2012). Survey quality. *Survey Methodology 38*(2), 107–130.

Lyberg, L., P. Biemer, M. Collins, E. deLeeuw, C. Dippo, N. Schwarz, and D. Trewin (1997). *Survey Measurement and Process Quality.* New York: Wiley.

Lyberg, L. and H. Weisberg (2016). Total survey error: A paradigm for survey methodology. In C. Wolf, D. Joye, T. W. Smith, and Y.-C. Fu (Eds.), *The SAGE Handbook of Survey Methodology*, pp. 27–41. Thousand Oaks, CA: SAGE.

Lydersen, C. and M. Ryg (1991). Evaluating breeding habitat and populations of ringed seals *Phoca hispida* in Svalbard fjords. *Polar Record 27*, 223–228.

Lynn, P. (2019). The advantage and disadvantage of implicitly stratified sampling. *Methods, Data, Analyses 13*(2), 253–266.

Macdonell, W. R. (1901). On criminal anthropometry and the identification of criminals. *Biometrika 1*, 177–227.

MacInnis, B., J. A. Krosnick, A. S. Ho, and M.-J. Cho (2018). The accuracy of measurements with probability and nonprobability survey samples: Replication and extension. *Public Opinion Quarterly 82*(4), 707–744.

Madans, J., K. Miller, A. Maitland, and G. Willis (Eds.) (2011). *Question Evaluation Methods: Contributing to the Science of Data Quality.* Hoboken, NJ: Wiley.

Mahalanobis, P. C. (1939). A sample survey of the acreage under jute in Bengal. *Sankhyā 4*, 511–531.

Mahalanobis, P. C. (1946). Recent experiments in statistical sampling in the Indian Statistical Institute. *Journal of the Royal Statistical Society 109*, 325–378.

Maher, B. (2008). Poll results: Look who's doping. *Nature 452*(7188), 674–675.

Manly, B. F. J. and J. A. Navarro Alberto (2020). *Randomization, Bootstrap and Monte Carlo Methods in Biology, 4th ed.* Boca Raton, FL: Chapman & Hall/CRC Press.

Manski, C. F. (2015). Communicating uncertainty in official economic statistics: An appraisal fifty years after Morgenstern. *Journal of Economic Literature 53*(3), 631–653.

Marker, D. A. and D. Morganstein (2004). Keys to successful implementation of continuous quality improvement in a statistical agency. *Journal of Official Statistics 20*, 125–136.

Mashreghi, Z., D. Haziza, and C. Léger (2016). A survey of bootstrap methods in finite population sampling. *Statistics Surveys 10*, 1–52.

Massarsky, B. M. (2013). The operating dynamics behind ASCAP, BMI and SESAC, the U.S. performing rights societies. `www.cni.org` (accessed August 15, 2020).

Matthijsse, S. M., E. D. de Leeuw, and J. J. Hox (2015). Internet panels, professional respondents, and data quality. *Methodology 11*(3), 81–88.

Mayr, J., M. Gaisl, K. Purtscher, H. Noeres, G. Schimpl, and G. Fasching (1994). Baby walkers—an underestimated hazard for our children? *European Journal of Pediatrics 153*, 531–534.

McAuley, R. G., W. M. Paul, G. H. Morrison, R. F. Beckett, and C. H. Goldsmith (1990). Five-year results of the peer assessment program of the College of Physicians and Surgeons of Ontario. *Canadian Medical Association Journal 143*, 1193–1199.

McCarthy, P. J. (1966). *Replication: An Approach to the Analysis of Data from Complex Surveys.* Washington, DC: National Center for Health Statistics.

McCarthy, P. J. (1969). Pseudo-replication: Half-samples. *Review of the International Statistical Institute 37*, 239–264.

McCarthy, P. J. (1993). Standard error and confidence interval estimation for the median. *Journal of Official Statistics 9*, 673–689.

McConville, K. and F. Breidt (2013). Survey design asymptotics for the model-assisted penalised spline regression estimator. *Journal of Nonparametric Statistics 25*(3), 745–763.

McConville, K. S., F. J. Breidt, T. C. Lee, and G. G. Moisen (2017). Model-assisted survey regression estimation with the lasso. *Journal of Survey Statistics and Methodology 5*(2), 131–158.

McConville, K. S. and D. Toth (2019). Automated selection of post-strata using a model-assisted regression tree estimator. *Scandinavian Journal of Statistics 46*(2), 389–413.

McCrea, R. S. and B. J. Morgan (2014). *Analysis of Capture-Recapture Data.* Boca Raton, FL: CRC Press.

McFarland, S. G. (1981). Effects of question order on survey responses. *Public Opinion Quarterly 45*, 208–215.

McGeeney, K. and C. Kennedy (2017). A next step in the evolution of landline sampling: Evaluating the assignment-based frame. *Journal of Survey Statistics and Methodology 5*(1), 37–47.

McIlwee, J. S. and J. G. Robinson (1992). *Women in Engineering: Gender, Power, and Workplace Culture*. Albany, NY: State University of New York Press.

McIntyre, G. A. (1952). A method of unbiased selective sampling, using ranked sets. *Australian Journal of Agricultural Research 3*, 385–390.

McNamee, R. (2003). Efficiency of two-phase designs for prevalence estimation. *International Journal of Epidemiology 32*, 1072–1078.

Meng, X.-L. (2018). Statistical paradises and paradoxes in big data (I): Law of large populations, big data paradox, and the 2016 US presidential election. *The Annals of Applied Statistics 12*(2), 685–726.

Mercer, A., A. Caporaso, D. Cantor, and R. Townsend (2015). How much gets you how much? Monetary incentives and response rates in household surveys. *Public Opinion Quarterly 79*, 105–129.

Mercer, A., F. Kreuter, S. Keeter, and E. Stuart (2017). Theory and practice in nonprobability surveys: Parallels between causal inference and survey inference. *Public Opinion Quarterly 81*(S1), 250–271.

Mercer, A., A. Lau, and C. Kennedy (2018). *For Weighting Online Opt-In Samples, What Matters Most?* Washington, DC: Pew Research.

Meyer, M. C. (2006). Wider shoes for wider feet? *Journal of Statistics Education 14*(1), 1–6.

Mitofsky, W. (1970). Sampling of telephone households. Unpublished CBS News memorandum.

Mohadjer, L. and G. H. Choudhry (2002). Adjusting for missing data in low-income surveys. In *Studies of Welfare Populations: Data Collection and Research Issues*, pp. 129–156. Washington DC: The National Academies Press.

Mohorko, A., E. de Leeuw, and J. Hox (2013). Internet coverage and coverage bias in Europe: Developments across countries and over time. *Journal of Official Statistics 29*(4), 609–622.

Montanari, G. E. (1987). Post-sampling efficient QR-prediction in large-scale surveys. *International Statistical Review 55*, 191–202.

Montanari, G. E. and M. G. Ranalli (2005). Nonparametric model calibration estimation in survey sampling. *Journal of the American Statistical Association 100*, 1429–1442.

Montaquila, J. and K. M. Olson (2012). Practical tools for nonresponse bias studies. Webinar, Survey Research Methods Section, American Statistical Association, `https://community.amstat.org/surveyresearchmethodssection/programs/new-item2` (accessed May 16, 2020).

Montgomery, D. C. (2012). *Introduction to Statistical Quality Control, 7th ed.* Hoboken, NJ: Wiley.

Montgomery, D. C., E. A. Peck, and G. G. Vining (2012). *Introduction to Linear Regression Analysis, 5th ed.* Hoboken, NJ: Wiley.

Moran, B. (2017, July 26). CTE found in 99 percent of former NFL players studied. *The Brink.* `https://www.bu.edu/articles/2017/cte-former-nfl-players/` (accessed January 17, 2020).

Morel, J. G. and N. Neerchal (2012). *Overdispersion Models in SAS®.* Cary, NC: SAS Institute, Inc.

Moritz, E. D., L. B. Zapata, A. Lekiachvili, et al. (2019). Update: Characteristics of patients in a national outbreak of e-cigarette, or vaping, product use-associated lung injuries—United States, October 2019. *Morbidity and Mortality Weekly Report 68*(43), 985–989.

Mortensen, K., M. G. Alcalá, M. T. French, and T. Hu (2018). Self-reported health status differs for Amazon's Mechanical Turk respondents compared with nationally representative surveys. *Medical care 56*(3), 211–215.

Morton, H. C. and A. J. Price (1989). *The ACLS Survey of Scholars: Final Report of Views on Publications, Computers, and Libraries.* Washington, DC: University Press of America.

Mule, T. (2012). Census coverage measurement estimation report: Summary of estimates of coverage for persons in the United States. DSSD 2010 census coverage measurement memorandum series #2010-G-01. Washington, DC: U.S. Census Bureau.

Murphy, J., P. Biemer, C. Stringer, R. Thissen, O. Day, and Y. P. Hsieh (2016). Interviewer falsification: Current and best practices for prevention, detection, and mitigation. *Statistical Journal of the IAOS 32*(3), 313–326.

Murphy, S. L., J. Xu, K. D. Kochanek, S. C. Curtin, and E. Arias (2017). Deaths: Final data for 2015. *National Vital Statistics Reports 66*(6), 1–75.

Murray, J. S. (2018). Multiple imputation: A review of practical and theoretical findings. *Statistical Science 33*(2), 142–159.

Muthén, L. K. and B. O. Muthén (2017). *Mplus User's Guide: Statistical Analysis with Latent Variables.* Los Angeles: Muthén & Muthén.

Myers, D. G. (1998). *Psychology.* New York: Macmillan.

Narain, R. D. (1951). On sampling without replacement with varying probabilities. *Journal of the Indian Society of Agricultural Statistics 3*, 169–174.

Nathan, G. (2005). More advanced approaches to the analysis of survey data. In *Household Sample Surveys in Developing and Transition Countries*, pp. 419–445. New York: United Nations Statistics Division.

National Academies of Sciences, Engineering, and Medicine (2017). *Innovations in Federal Statistics: Combining Data Sources While Protecting Privacy.* Washington, DC: National Academies Press.

National Academies of Sciences, Engineering, and Medicine (2018). *Federal Statistics, Multiple Data Sources, and Privacy Protection: Next Steps.* Washington, DC: National Academies Press.

National Academies of Sciences, Engineering, and Medicine (2021). *Principles and Practices for a Federal Statistical Agency, 7th ed.* Washington, DC: National Academies Press.

National Research Council (2004). *On Evaluating Curricular Effectiveness: Judging the Quality of K–12 Mathematics Evaluations.* Washington, DC: National Academies Press.

National Research Council (2013). *Nonresponse in Social Science Surveys: A Research Agenda.* R. Tourangeau and T. J. Plewes, (Eds). Panel on a Research Agenda for the Future of Social Science Data Collection, Committee on National Statistics. Division of Behavioral and Social Sciences and Education. Washington, DC: The National Academies Press.

National Science Foundation (2019). Survey of Doctorate Recipients. `https://www.nsf.gov/statistics/srvydoctoratework` (accessed November 19, 2019).

Nelson, G. A. (2014). Cluster sampling: A pervasive, yet little recognized survey design in fisheries research. *Transactions of the American Fisheries Society 143*(4), 926–938.

Neter, J., R. A. Leitch, and S. E. Fienberg (1978). Dollar unit sampling: Multinomial bounds for total overstatement and understatement errors. *Accounting Review 53*, 77–93.

New York City Department of Homeless Services (2019). HOPE 2019: NYC HOPE 2019 results. `https://www1.nyc.gov/assets/dhs/downloads/pdf/hope-2019-results.pdf` (accessed January 9, 2020).

New York Times (2018, November 5). Polling in real time: The 2018 midterm elections. *The New York Times.* `https://www.nytimes.com/interactive/2018/upshot/elections-polls.html` (accessed March 21, 2019).

New York Times (2019). About the best sellers. `www.nytimes.com/books/best-sellers/methodology/` (accessed November 18, 2019).

Newsday (1976, June 13). 91% would have children (Take that, Ann Landers). *Newsday*, 6.

Neyman, J. (1934). On the two different aspects of the representative method: The method of stratified sampling and the method of purposive selection. *Journal of the Royal Statistical Society 97*, 558–606.

Neyman, J. (1938). Contribution to the theory of sampling human populations. *Journal of the American Statistical Association 33*, 101–116.

Nieves, K. E., L. A. Sacerdote, A. Ginder, Y. Liu, and R. M. Figazzotto (2018). 2017 caseload statistics of the Unified Judicial System of Pennsylvania. `http://www.pacourts.us/news-and-statistics/research-and-statistics/caseload-statistics` (accessed November 9, 2019).

Nurses' Health Study (2019). Nurses' Health Study. `https://www.nurseshealthstudy.org` (accessed December 30, 2019).

Oberski, D. (2018). Questionnaire science. In L. Atkeson and R. Alvarez (Eds.), *The Oxford Handbook of Polling and Survey Methods*, pp. 114–138. Oxford, UK: Oxford University Press.

O'Brien, L. A., J. A. Grisso, G. Maislin, K. LaPann, K. P. Krotki, P. J. Greco, E. A. Siegert, and L. K. Evans (1995). Nursing home residents' preferences for life-sustaining treatments. *Journal of the American Medical Association 274*, 1775–1779.

Oehlert, G. W. (2000). *A First Course in Design and Analysis of Experiments.* New York: Freeman.

O'Hare, W. P. (2014). Assessing net coverage error for young children in the 2010 U.S. Decennial Census. Survey Measurement Report #2014-02. Washington DC: U.S. Census Bureau.

Olson, K. (2013). Do non-response follow-ups improve or reduce data quality?: A review of the existing literature. *Journal of the Royal Statistical Society: Series A 176*(1), 129–145.

O'Muircheartaigh, C. and P. Campanelli (1998). The relative impact of interviewer effects and sample design effects on survey precision. *Journal of the Royal Statistical Society, Series A 161*(1), 63–77.

Opsomer, J. and C. Miller (2005). Selecting the amount of smoothing in nonparametric regression estimation for complex surveys. *Nonparametric Statistics 17*(5), 593–611.

Oracle (2019, October 15). New study: 64% of people trust a robot more than their manager. `https://www.oracle.com/corporate/pressrelease/robots-at-work-101519.html` (accessed December 28, 2019).

Orlandini, A. (2018). Simulation of the allotment of *Dikastai.* `https://www.academia.edu/36510282/KLEROTERION_simulation_of_the_allotment_of_dikastai` (accessed October 12, 2019).

Overton, W. S. and S. V. Stehman (1995). The Horvitz-Thompson theorem as a unifying perspective for probability sampling: With examples from natural resource sampling. *The American Statistician 49*, 261–268.

Park, I. and H. Lee (2004). Design effects for the weighted mean and total estimators under complex survey sampling. *Survey Methodology 30*(2), 183–193.

Parten, M. (1950). *Surveys, Polls, and Samples.* New York: Harper & Brothers.

Parzen, E. (2004). Quantile probability and statistical data modeling. *Statistical Science 19*(4), 652–662.

Patil, G. P., B. Surucu, and D. Egemen (2014). Ranked set sampling. In *Wiley StatsRef: Statistics Reference Online.* Hoboken, NJ: Wiley Online Library.

Pattison, P. E., G. L. Robins, T. A. Snijders, and P. Wang (2013). Conditional estimation of exponential random graph models from snowball sampling designs. *Journal of Mathematical Psychology 57*(6), 284–296.

Pearl, J. (2009). Causal inference in statistics: An overview. *Statistics Surveys 3*, 96–146.

Peart, D. (1994). *Impacts of Feral Pig Activity on Vegetation Patterns Associated with Quercus agrifolia on Santa Cruz Island, California.* Ph.D. dissertation. Tempe, AZ: Arizona State University.

Pennay, D., D. Neiger, P. Lavrakas, and K. Borg (2018). *The Online Panels Benchmarking Study: A Total Survey Error Comparison of Findings from Probability-Based Surveys and Nonprobability Online Panel Surveys in Australia.* Canberra: Australian National University.

Petersen, C. G. J. (1896). The yearly immigration of young plaice into the Limfjord from the German Sea. *Reports of the Danish Biological Station 6*, 5–84.

Peterson, S., N. Toribio, J. Farber, and D. Hornick (2021). *Nonresponse Bias Report for the 2020 Household Pulse Survey, Version 1.0.* Washington, DC: U.S. Census Bureau.

Peytcheva, E. and R. M. Groves (2009). Using variation in response rates of demographic subgroups as evidence of nonresponse bias in survey estimates. *Journal of Official Statistics 25*(2), 193.

Pfeffermann, D. (1993). The role of sampling weights when modeling survey data. *International Statistical Review 61*, 317–337.

Pfeffermann, D. (1996). The use of sampling weights for survey data analysis. *Statistical Methods in Medical Research 5*, 239–261.

Pfeffermann, D. (2011). Modelling of complex survey data: Why is it a problem? How should we approach it? *Survey Methodology 37*(2), 115–136.

Pfeffermann, D. (2013). New important developments in small area estimation. *Statistical Science 28*(1), 40–68.

Pfeffermann, D. and D. J. Holmes (1985). Robustness considerations in the choice of a method of inference for regression analysis of survey data. *Journal of the Royal Statistical Society, Series A 148*, 268–278.

Pfeffermann, D. and C. R. Rao (2009a). *Handbook of Statistics: Vol. 29A. Sample Surveys: Design, Methods and Applications.* Amsterdam: North Holland.

Pfeffermann, D. and C. R. Rao (2009b). *Handbook of Statistics: Vol. 29B. Sample Surveys: Inference and Analysis.* Amsterdam: North Holland.

Pfeffermann, D., C. J. Skinner, D. J. Holmes, H. Goldstein, and J. Rabash (1998). Weighting for unequal selection probabilities in multilevel models. *Journal of the Royal Statistical Society, Series B 60*, 23–40.

Pfeffermann, D. and M. Sverchkov (1999). Parametric and semi-parametric estimation of regression models fitted to survey data. *Sankhyā, Series B 61*, 166–186.

Phipps, P. and D. Toth (2012). Analyzing establishment nonresponse using an interpretable regression tree model with linked administrative data. *The Annals of Applied Statistics 6*(2), 772–794.

Pincus, T. (1993). Arthritis and rheumatic diseases: What doctors can learn from their patients. In D. Goleman and J. Gurin (Eds.), *Mind/Body Medicine: How to Use Your Mind for Better Health*, pp. 177–192. Yonkers, NY: Consumer Reports Books.

Platek, R. (1977). Some factors affecting non-response. *Survey Methodology 3*, 191–214.

Politz, A. and W. Simmons (1949). An attempt to get the "not at homes" into the sample without callbacks. *Journal of the American Statistical Association 44*, 9–31.

Pollán, M., B. Pérez-Gómez, R. Pastor-Barriuso, J. Oteo, M. A. Hernán, M. Pérez-Olmeda, J. L. Sanmartín, A. Fernández-García, I. Cruz, N. Fernández de Larrea, et al. (2020). Prevalence of SARS-CoV-2 in Spain (ENE-COVID): A nationwide, population-based seroepidemiological study. *The Lancet 396*(10250), 535–544.

Pratesi, M. (Ed.) (2016). *Analysis of Poverty Data by Small Area Estimation.* Hoboken, NJ: Wiley.

Presser, S., J. M. Rothgeb, M. P. Couper, J. T. Lessler, E. Martin, J. Martin, and E. Singer (2004). *Methods for Testing and Evaluating Survey Questionnaires.* New York: Wiley.

Quenouille, M. H. (1956). Notes on bias in estimation. *Biometrika 43*, 353–360.

Rabe-Hesketh, S. and A. Skrondal (2006). Multilevel modelling of complex survey data. *Journal of the Royal Statistical Society, Series A 169*, 805–827.

Raghunathan, T., P. Solenberger, P. Berglund, and J. van Hoewyk (2016). *IVEware: Imputation and Variance Estimation Software (Version 0.3).* Ann Arbor, MI: University of Michigan.

Raghunathan, T. E. and J. E. Grizzle (1995). A split questionnaire survey design. *Journal of the American Statistical Association 90*, 54–63.

Raj, D. (1968). *Sampling Theory.* New York: McGraw-Hill.

Ranalli, M. G. and F. Mecatti (2012). Comparing recent approaches for bootstrapping sample survey data: A first step towards a unified approach. In *Proceedings of the Survey Research Methods Section*, pp. 4088–4099. Alexandria, VA: American Statistical Association.

Rao, J. N. K. (1963). On three procedures of unequal probability sampling without replacement. *Journal of the American Statistical Association 58*, 202–215.

Rao, J. N. K. (1973). On double sampling for stratification and analytical surveys. *Biometrika 60*, 125–133.

Rao, J. N. K. (1979a). On deriving mean square errors and their non-negative unbiased estimators in finite population sampling. *Journal of the Indian Statistical Association 17*, 125–136.

Rao, J. N. K. (1979b). Optimization in the design of sample surveys. In J. S. Rustagi (Ed.), *Optimizing Methods in Statistics: Proceedings of an International Conference*, pp. 419–434. New York: Academic Press.

Rao, J. N. K. (1988). Variance estimation in sample surveys. In P. R. Krishnaiah and C. R. Rao (Eds.), *Handbook of Statistics Volume 6: Sampling*, pp. 427–447. New York: Elsevier Science.

Rao, J. N. K. (1994). Estimating totals and distribution functions using auxiliary information at the estimation stage. *Journal of Official Statistics 10*, 153–165.

Rao, J. N. K. (1997). Developments in sample survey theory: An appraisal. *Canadian Journal of Statistics 25*, 1–21.

Rao, J. N. K. (2005). Interplay between sample survey theory and practice: An appraisal. *Survey Methodology 31*, 117–338.

Rao, J. N. K. (2011). Impact of frequentist and Bayesian methods on survey sampling practice: A selective appraisal. *Statistical Science 26*(2), 240–256.

Rao, J. N. K. (2021). On making valid inferences by combining data from surveys and other sources. *Sankhyā, Series B 83-B*(1), 242–272.

Rao, J. N. K. and W. A. Fuller (2017). Sample survey theory and methods: Past, present, and future directions. *Survey Methodology 43*(2), 145–160.

Rao, J. N. K., H. O. Hartley, and W. G. Cochran (1962). A simple procedure for unequal probability sampling without replacement. *Journal of the Royal Statistical Society, Series B 24*, 482–491.

Rao, J. N. K. and I. Molina (2015). *Small Area Estimation, 2nd ed.* Hoboken, NJ: Wiley.

Rao, J. N. K. and A. J. Scott (1981). The analysis of categorical data from complex sample surveys: Chi-squared tests for goodness of fit and independence in two-way tables. *Journal of the American Statistical Association 76*, 221–230.

Rao, J. N. K. and A. J. Scott (1984). On chi-squared tests for multiway contingency tables with cell proportions estimated from survey data. *The Annals of Statistics 12*, 46–60.

Rao, J. N. K., A. J. Scott, and E. Benhin (2003). Undoing complex survey data structures: Some theory and applications of inverse sampling. *Survey Methodology 29*(2), 107–121.

Rao, J. N. K. and J. Shao (1992). Jackknife variance estimation with survey data under hot deck imputation. *Biometrika 79*, 811–822.

Rao, J. N. K. and J. Shao (1999). Modified balanced repeated replication for complex survey data. *Biometrika 86*, 403–415.

Rao, J. N. K. and R. R. Sitter (1995). Variance estimation under two-phase sampling with application to imputation for missing data. *Biometrika 82*, 453–460.

Rao, J. N. K. and R. R. Sitter (1997). Variance estimation under stratified two-phase sampling with applications to measurement bias. In L. Lyberg, P. Biemer, M. Collins, E. de Leeuw, C. Dippo, N. Schwarz, and D. Trewin (Eds.), *Survey Measurement and Process Quality*, pp. 753–768. New York: Wiley.

Rao, J. N. K. and D. R. Thomas (1988). The analysis of cross-classified categorical data from complex sample surveys. *Sociological Methodology 18*, 213–269.

Rao, J. N. K. and D. R. Thomas (1989). Chi-squared tests for contingency tables. In C. J. Skinner, D. Holt, and T. M. F. Smith (Eds.), *Analysis of Complex Surveys*, pp. 89–114. New York: Wiley.

Rao, J. N. K. and D. R. Thomas (2003). Analysis of categorical response data from complex surveys: An appraisal and update. In R. L. Chambers and C. J. Skinner (Eds.), *Analysis of Survey Data*, pp. 85–108. New York: Wiley.

Rao, J. N. K., F. Verret, and M. A. Hidiroglou (2013). A weighted composite likelihood approach to inference for two-level models from survey data. *Survey Methodology 39*(2), 263–282.

Rao, J. N. K. and C. F. J. Wu (1985). Inference from stratified samples: Second-order analysis of three methods for nonlinear statistics. *Journal of the American Statistical Association 80*, 620–630.

Rao, J. N. K. and C. F. J. Wu (1987). Methods for standard errors and confidence intervals from sample survey data: Some recent work. *Bulletin of the International Statistical Institute 52*, 5–21.

Rao, J. N. K. and C. F. J. Wu (1988). Resampling inference with complex survey data. *Journal of the American Statistical Association 83*, 231–241.

Rao, J. N. K., C. F. J. Wu, and K. Yue (1992). Some recent work on resampling methods for complex surveys. *Survey Methodology 18*, 209–217.

Rässler, S., D. B. Rubin, and N. Schenker (2008). Incomplete data: Diagnosis, imputation, and estimation. In E. D. deLeeuw, J. J. Hox, and D. A. Dillman (Eds.), *International Handbook of Survey Methodology*, pp. 370–386. New York: Erlbaum.

Reiter, J. P. (2019). Differential privacy and federal data releases. *Annual Review of Statistics and Its Application 6*, 85–101.

Reiter, J. P., T. E. Raghunathan, and S. K. Kinney (2006). The importance of modeling the sampling design in multiple imputation for missing data. *Survey Methodology 32*(2), 143–149.

Rinott, Y., C. M. O'Keefe, N. Shlomo, and C. Skinner (2018). Confidentiality and differential privacy in the dissemination of frequency tables. *Statistical Science 33*(3), 358–385.

Risley, M. and J. Berkley (2020). 2018 ACS mail materials test. Technical Report ACS21-RER-01, U.S. Census Bureau, Washington, DC.

Rivera, J. D. (2016). *Acquiring Federal Disaster Assistance: Investigating Equitable Resource Distribution Within FEMA's Home Assistance Program.* Ph.D. dissertation. Camden, NJ: Rutgers University.

Rivera, J. D. (2019). When attaining the best sample is out of reach: Nonprobability alternatives when engaging in public administration research. *Journal of Public Affairs Education 25*(3), 314–342.

Rivers, D. and D. Bailey (2009). Inference from matched samples in the 2008 U.S. national elections. In *Proceedings of the Joint Statistical Meetings*, pp. 627–639. Alexandria, VA: American Statistical Association.

Roberts, G., J. N. K. Rao, and S. Kumar (1987). Logistic regression analysis of sample survey data. *Biometrika 74*, 1–12.

Roberts, R. J., Q. D. Sandifer, M. R. Evans, M. Z. Nolan-Ferrell, and P. M. Davis (1995). Reasons for non-uptake of measles, mumps, and rubella catch up immunisation in a measles epidemic and side effects of the vaccine. *British Medical Journal 310*, 1629–1632.

Rosenbaum, P. R. and D. B. Rubin (1983). The central role of the propensity score in observational studies for causal effects. *Biometrika 70*, 41–55.

Rosenberg, E., J. Tesoriero, E. Rosenthal, R. Chung, M. Barranco, L. Styer, M. Parker, S. Y. J. Leung, J. E. Morne, D. Greene, D. R. Holtgrave, D. Hoefer, J. Kumar, T. Udo, B. Hutton, and H. A. Zucker (2020). Cumulative incidence and diagnosis of SARS–CoV–2 infection in New York. *Annals of Epidemiology 48*, 23–29.

Ross, S. M. (2019). *A First Course in Probability, 10th ed.* Boston: Pearson.

Rothbart, G. S., M. Fine, and S. Sudman (1982). On finding and interviewing the needles in the haystack: The use of multiplicity sampling. *Public Opinion Quarterly 46*, 408–421.

Rothenberg, R. B., A. Lobanov, K. B. Singh, and G. Stroh (1985). Observations on the application of EPI cluster survey methods for estimating disease incidence. *Bulletin of the World Health Organization 63*, 93–99.

Royall, R. M. (1970). On finite population sampling theory under certain linear regression models. *Biometrika 57*, 377–387.

Royall, R. M. (1976). The linear least-squares prediction approach to two-stage sampling. *Journal of the American Statistical Association 71*, 657–664.

Royall, R. M. (1992). Robustness and optimal design under prediction models for finite populations. *Survey Methodology 18*, 179–185.

Royall, R. M. and K. R. Eberhardt (1975). Variance estimates for the ratio estimator. *Sankhyā, Series C 37*, 43–52.

RTI International (2012). *SUDAAN Language Manual, Release 11.0.* Research Triangle Park, NC: RTI International.

Rubin, D. B. (1985). The use of propensity scores in applied Bayesian inference. In J. M. Bernardo, M. H. DeGroot, D. V. Lindley, and A. F. M. Smith (Eds.), *Bayesian Statistics 2*, pp. 463–472. Amsterdam: Elsevier.

Rubin, D. B. (1987). *Multiple Imputation for Nonresponse in Surveys.* New York: Wiley.

Rubin-Bleuer, S. and I. S. Kratina (2005). On the two-phase framework for joint model and design-based inference. *Annals of Statistics 33*, 2789–2810.

Ruggles, S. (1995). Sampling designs and sampling errors. *Historical Methods 28*, 40–46.

Ruggles, S., M. Sobek, T. Alexander, C. A. Fitch, R. Goeken, P. K. Hall, M. King, and C. Ronnander (2004). Integrated Public Use Microdata Series: Version 3.0 [machine-readable database]. `www.ipums/org/usa` (accessed September 17, 2008).

Russell, H. J. (1972). Use of a commercial dredge to estimate a hardshell clam population by stratified random sampling. *Journal of the Fisheries Research Board of Canada 29*, 1731–1735.

Rust, K. F. and J. N. K. Rao (1996). Variance estimation for complex surveys using replication techniques. *Statistical Methods in Medical Research 5*, 283–310.

Sala, E. and R. Lillini (2017). Undercoverage bias in telephone surveys in Europe: The Italian case. *International Journal of Public Opinion Research 29*(1), 133–156.

Salganik, M. J. and D. D. Heckathorn (2004). Sampling and estimation in hidden populations using respondent-driven sampling. *Sociological Methodology 34*, 193–239.

Samuels, C. (1996). Full-time vs. part-time instructors. *Arizona AAUP Advocate 45*, 1–3.

Santacatterina, M. and M. Bottai (2018). Optimal probability weights for inference with constrained precision. *Journal of the American Statistical Association 113*(523), 983–991.

Sanzo, J. M., M. A. Garcia-Calabuig, A. Audicana, and V. Dehesa (1993). Q fever: Prevalence of antibodies to *Coxiella burnetii* in the Basque country. *International Journal of Epidemiology 22*, 1183–1188.

Saris, W. E. and I. N. Gallhofer (2014). *Design, Evaluation, and Analysis of Questionnaires for Survey Research, 2nd ed.* Hoboken, NJ: Wiley.

Saris, W. E., M. Revilla, J. A. Krosnick, and E. M. Shaeffer (2010). Comparing questions with agree/disagree response options to questions with item-specific response options. *Survey Research Methods 4*(1), 61–79.

Särndal, C.-E. (1996). Efficient estimators with simple variance in unequal probability sampling. *Journal of the American Statistical Association 91*, 1289–1300.

Särndal, C.-E. (2007). The calibration approach in survey theory and practice. *Survey Methodology 33*, 99–119.

Särndal, C.-E. and S. Lundström (2005). *Estimation in Surveys with Nonresponse.* Hoboken, NJ: Wiley.

Särndal, C.-E. and B. Swensson (1987). A general view of estimation for two phases of selection with applications to two-phase sampling and nonresponse. *International Statistical Review 55*, 279–294.

Särndal, C.-E., B. Swensson, and J. Wretman (1992). *Model Assisted Survey Sampling.* New York: Springer-Verlag.

Särndal, C.-E., I. Traat, and K. Lumiste (2018). Interaction between data collection and estimation phases in surveys with nonresponse. *Statistics in Transition 183*(2), 183–200.

SAS Institute Inc. (2020). *The Quality Imperative: SAS Institute's Commitment to Quality.* Cary, NC: SAS Institute Inc. `https://www.sas.com/content/dam/SAS/en_us/doc/whitepaper1/quality-imperative-commitment-to-quality-106810.pdf` (accessed September 15, 2020).

Satterthwaite, F. E. (1946). An approximate distribution of estimates of variance components. *Biometrics 2*, 110–114.

Sauer, J. R., K. L. Pardieck, D. J. Ziolkowski Jr, A. C. Smith, M.-A. R. Hudson, V. Rodriguez, H. Berlanga, D. K. Niven, and W. A. Link (2017). The first 50 years of the North American Breeding Bird Survey. *The Condor 119*(3), 576–593.

Schaeffer, N. C., J. Dykema, and D. W. Maynard (2010). Interviewers and interviewing. In P. V. Marsden and J. D. Wright (Eds.), *Handbook of Survey Research, 2nd ed.*, pp. 437–470. Bingley, UK: Emerald Group Publishing Limited.

Schei, B. and L. S. Bakketeig (1989). Gynaecological impact of sexual and physical abuse by spouse: A study of a random sample of Norwegian women. *British Journal of Obstetrics and Gynaecology 96*, 1379–1383.

Schnabel, Z. E. (1938). The estimation of the total fish population of a lake. *American Mathematical Monthly 45*, 348–352.

Schneider, M., D. Brisson, and D. Burnes (2016). Do we really know how many are homeless?: An analysis of the point-in-time homelessness count. *Families in Society 97*(4), 321–329.

Schonlau, M., A. Van Soest, and A. Kapteyn (2007). Are "webographic" or attitudinal questions useful for adjusting estimates from web surveys using propensity scoring? *Survey Research Methods 1*(3), 155–163.

Schouten, B., F. Cobben, J. Bethlehem, et al. (2009). Indicators for the representativeness of survey response. *Survey Methodology 35*(1), 101–113.

Schouten, B., A. Peytchev, and J. Wagner (2017). *Adaptive Survey Design.* Boca Raton, FL: CRC Press.

Schreuder, H. T., J. Sedransk, and K. D. Ware (1968). 3-P sampling and some alternatives, I. *Forest Science 14*, 429–453.

Schuitemaker, L. (2019). *Childless Living: The Joys and Challenges of Life without Children.* Rochester, VT: Findhorn Press.

Schuman, H. and S. Presser (1981). *Questions and Answers in Attitude Surveys: Experiments on Question Form, Wording, and Context.* New York: Academic Press.

Schuman, H. and J. Scott (1987). Problems in the use of survey questions to measure public opinion. *Science 280*, 957–959.

Scott, A. J. (2007). Rao–Scott corrections and their impact. In *Proceedings of the Survey Research Methods Section*, pp. 3514–3518. Alexandria, VA: American Statistical Association.

Scott, A. J. and T. M. F. Smith (1969). Estimation in multi-stage surveys. *Journal of the American Statistical Association 64*, 830–840.

Scott, D. W. (2015). *Multivariate Density Estimation: Theory, Practice, and Visualization, 2nd ed.* Hoboken, NJ: Wiley.

Seber, G. A. F. (1970). The effects of trap response on tag recapture estimates. *Biometrics 26*, 13–22.

Seber, G. A. F. and M. M. Salehi (2012). *Adaptive Sampling Designs: Inference for Sparse and Clustered Populations.* New York: Springer.

Seber, G. A. F. and M. R. Schofield (2019). *Capture-Recapture: Parameter Estimation for Open Animal Populations.* Cham, Switzerland: Springer.

Sen, A. R. (1953). On the estimate of the variance in sampling with varying probabilities. *Journal of the Indian Society of Agricultural Statistics 5*, 119–127.

Senturia, Y. D., K. K. Christoffel, and M. Donovan (1994). Children's household exposure to guns: A pediatric practice-based survey. *Pediatrics 93*, 469–475.

Shah, B. V., M. M. Holt, and R. E. Folsom (1977). Inference about regression models from sample survey data. *Bulletin of the International Statistical Institute 47*, 43–57.

Shao, J. (1994). L–statistics in complex survey problems. *The Annals of Statistics 22*(2), 946–967.

Shao, J. (2003). Impact of the bootstrap on sample surveys. *Statistical Science 18*, 191–198.

Shao, J. and Y. Chen (1998). Bootstrapping sample quantiles based on complex survey data under hot deck imputation. *Statistica Sinica 8*, 1071–1085.

Shao, J. and R. R. Sitter (1996). Bootstrap for imputed survey data. *Journal of the American Statistical Association 91*(435), 1278–1288.

Shao, J. and P. Steel (1999). Variance estimation for survey data with composite imputation and nonnegligible sampling fractions. *Journal of the American Statistical Association 94*(445), 254–265.

Shao, J. and D. Tu (1995). *The Jackknife and Bootstrap.* New York: Springer.

Shao, J. and C. F. J. Wu (1989). A general theory for jackknife variance estimation. *The Annals of Statistics*, 1176–1197.

Shao, J. and C. F. J. Wu (1992). Asymptotic properties of the balanced repeated replication method for sample quantiles. *The Annals of Statistics 20*, 1571–1593.

Shen, J. and C.-L. Hsieh (1999). Improving the professional status of teaching: Perspectives of future teachers, current teachers, and education professors. *Teaching and Teacher Education 15*, 315–323.

Si, Y., N. S. Pillai, and A. Gelman (2015). Bayesian nonparametric weighted sampling inference. *Bayesian Analysis 10*(3), 605–625.

Silva, P. L. D. N. and C. J. Skinner (1997). Variable selection for regression estimation in finite populations. *Survey Methodology 23*, 23–32.

Simonoff, J. S. (2006). *Analyzing Categorical Data.* New York: Springer.

Singer, E. (2002). The use of incentives to reduce nonresponse in household surveys. In R. M. Groves, D. Dillman, J. Eltinge, and R. Little (Eds.), *Survey Nonresponse*, pp. 163–177. New York: Wiley.

Singer, E. (2003). The Eleventh Morris Hansen Lecture: Public perceptions of confidentiality. *Journal of Official Statistics 19*(4), 333–341.

Singer, E. (2011). Toward a benefit-cost theory of survey participation: Evidence, further tests, and implications. *Journal of Official Statistics 27*(2), 379.

Singer, E. (2016). Reflections on surveys' past and future. *Journal of Survey Statistics and Methodology 4*(4), 463–475.

Singer, E. and C. Ye (2013). The use and effects of incentives in surveys. *The ANNALS of the American Academy of Political and Social Science 645*(1), 112–141.

Siniff, D. and R. O. Skoog (1964). Aerial censusing of caribou using stratified random sampling. *Journal of Wildlife Management 28*, 391–401.

Sirken, M. (1970). Household surveys with multiplicity. *Journal of the American Statistical Association 65*, 257–266.

Sitter, R. R. (1992). Comparing three bootstrap methods for survey data. *The Canadian Journal of Statistics 20*, 135–154.

Sitter, R. R. (1997). Variance estimation for the regression estimator in two-phase sampling. *Journal of the American Statistical Association 92*, 780–787.

Sitter, R. R. and C. F. J. Wu (2001). A note on Woodruff confidence intervals for quantiles. *Statistics and Probability Letters 52*, 353–358.

Skinner, C. J., D. Holt, and T. M. F. Smith (Eds.) (1989a). *Analysis of Complex Surveys.* New York: Wiley.

Skinner, C. J., D. Holt, and T. M. F. Smith (1989b). General introduction. In C. J. Skinner, D. Holt, and T. M. F. Smith (Eds.), *Analysis of Complex Surveys*, pp. 1–20. New York: Wiley.

Smith, P. A. and J. Dawber (2019). Random probability vs quota sampling. https://eprints.soton.ac.uk/435300/1/WP5_Random_probability_vs_quota_sampling.pdf (accessed February 5, 2020).

Smith, T. M. F. (1988). To weight or not to weight, that is the question. In J. M. Bernardo, M. H. DeGroot, D. V. Lindley, and A. F. M. Smith (Eds.), *Bayesian Statistics 3*, pp. 437–451. Oxford: Clarendon Press [Oxford University Press].

Snijders, T. A. and R. J. Bosker (2011). *Multilevel Analysis: An Introduction to Basic and Advanced Multilevel Modeling, 2nd ed.* Thousand Oaks, CA: SAGE.

Squire, P. (1988). Why the 1936 Literary Digest poll failed. *Public Opinion Quarterly 52*, 125–133.

Stanley, T. J. (2000). *The Millionaire Mind.* Kansas City, MO: Andrews McMeel.

Statistics Canada (2019). *Statistics Canada Quality Guidelines, 6th ed.* Ottawa, ON: Statistics Canada.

Statistics Canada (2020). *Guide to the Survey of Employment, Payrolls and Hours.* Ottawa, ON: Statistics Canada. https://www150.statcan.gc.ca/n1/pub/72-203-g/72-203-g2020001-eng.htm (accessed May 18, 2020).

Steffey, D. L., S. E. Fienberg, and R. H. Sturgess (2006). Statistical assessment of damages in breach of contract litigation. *Jurimetrics 46*, 129–138.

Stein, C. (1945). A two-sample test for a linear hypothesis whose power is independent of the variance. *Annals of Mathematical Statistics 37*, 36–50.

Sterrett, D., D. Malato, J. Benz, T. Tompson, and N. English (2017). Assessing changes in coverage bias of web surveys in the United States. *Public Opinion Quarterly 81*(S1), 338–356.

Stewart, N., C. Ungemach, A. J. Harris, D. M. Bartels, B. R. Newell, G. Paolacci, and J. Chandler (2015). The average laboratory samples a population of 7,300 Amazon Mechanical Turk workers. *Judgment and Decision Making 10*(5), 479–491.

Stockford, D. D. and W. F. Page (1984). Double sampling and the misclassification of Vietnam service. In *Proceedings of the Social Statistics Section*, pp. 261–264. Alexandria, VA: American Statistical Association.

Stokes, S. L. and J. Plummer (2004). Using spreadsheet solvers in sample design. *Computational Statistics and Data Analysis 44*(1), 527–546.

Straus, S. E., P. Glasziou, W. S. Richardson, and R. B. Haynes (2018). *Evidence-Based Medicine: How to Practice and Teach EBM, 5th ed.* Edinburgh: Elsevier Health Sciences.

Strunk, W. and E. B. White (1959). *The Elements of Style.* New York: Macmillan.

Stuart, A. (1984). *The Ideas of Sampling, 3rd ed.* New York: Macmillan.

Stuart, E. A. (2010). Matching methods for causal inference: A review and a look forward. *Statistical Science 25*(1), 1–21.

Student (1908). On the probable error of the mean. *Biometrika 6*, 1–25.

Substance Abuse and Mental Health Services Administration (2020). *Key Substance Use and Mental Health Indicators in the United States: Results from the 2019 National Survey on Drug Use and Health.* HHS Publication No. PEP20-07-01-001, NSDUH Series H-55. Rockville, MD: Center for Behavioral Health Statistics and Quality, Substance Abuse and Mental Health Services Administration.

Sudman, S., M. G. Sirken, and C. D. Cowan (1988). Sampling rare and elusive populations. *Science 240*, 991–996.

Suessbrick, A., M. F. Schober, and F. G. Conrad (2000). Different respondents interpret ordinary questions quite differently. In *Proceedings of the Survey Research Methods Section*, pp. 907–912. Alexandria, VA: American Statistical Association.

Sugden, R. A., T. M. F. Smith, and R. P. Jones (2000). Cochran's rule for simple random sampling. *Journal of the Royal Statistical Society, Series B 62*, 787–793.

Sukhatme, P. V., B. V. Sukhatme, S. Sukhatme, and C. Asok (1984). *Sampling Theory of Surveys with Applications, 3rd ed.* Ames, IA: Iowa State University Press.

Swaine, J. and C. McCarthy (2016, December 15). Killings by US police logged at twice the previous rate under new federal program. *The Guardian.* `https://www.theguardian.com/us-news/2016/dec/15/us-police-killings-department-of-justice-program` (accessed March 26, 2018).

Tate, E. E. (1988). *Survey of Radon in Minnesota Homes.* Minneapolis: Minnesota Department of Health.

The Physicians Foundation (2018). 2018 survey of America's physicians. `https://physiciansfoundation.org/wp-content/uploads/2018/09/physicians-survey-results-final-2018.pdf` (accessed October 26, 2019).

Theoharakis, V. and M. Skordia (2003). How do statisticians perceive statistics journals? *The American Statistician 54*, 115–123.

Thomas, D. R. (1989). Simultaneous confidence intervals for proportions under cluster sampling. *Survey Methodology 15*, 187–201.

Thomas, D. R., A. C. Singh, and G. R. Roberts (1996). Tests of independence on two-way tables under cluster sampling: An evaluation. *International Statistical Review 64*, 295–311.

Thompson, M. E. (1997). *Theory of Sample Surveys.* London: Chapman & Hall.

Thompson, M. E. (2019). Combining data from new and traditional sources in population surveys. *International Statistical Review 87*, S79–S89.

Thompson, M. E. and C. Wu (2008). Simulation-based randomized systematic PPS sampling under substitution of units. *Survey Methodology 34*(1), 3.

Thompson, S. K. (1990). Adaptive cluster sampling. *Journal of the American Statistical Association 85*, 1050–1059.

Thompson, S. K. (2013). Adaptive web sampling in ecology. *Statistical Methods & Applications 22*(1), 33–43.

Thompson, S. K. (2017). Adaptive and network sampling for inference and interventions in changing populations. *Journal of Survey Statistics and Methodology 5*, 1–21.

Thompson, S. K. and L. M. Collins (2002). Adaptive sampling in research on risk-related behaviors. *Drug and Alcohol Dependence 68*, S57–S67.

Thompson, S. K. and G. A. F. Seber (1996). *Adaptive Sampling.* New York: Wiley.

Thompson, W. L. (Ed.) (2004). *Sampling Rare or Elusive Species: Concepts, Designs, and Techniques for Estimating Population Parameters.* Washington DC: Island Press.

Thomsen, I. and E. Siring (1983). On the causes and effects of nonresponse: Norwegian experience. In W. G. Madow and I. Olkin (Eds.), *Incomplete Data in Sample Surveys* Vol. 3, pp. 25–59. New York: Academic Press.

Tillé, Y. (2006). *Sampling Algorithms.* New York: Springer.

Tillé, Y. and A. Matei (2021). *sampling: Survey Sampling.* R package version 2.9, `https://CRAN.R-project.org/package=sampling` (accessed March 12, 2021).

Tillé, Y. and M. Wilhelm (2017). Probability sampling designs: Principles for choice of design and balancing. *Statistical Science 32*(2), 176–189.

Toepoel, V. (2016). *Doing Surveys Online.* Los Angeles: SAGE.

Tourangeau, R. (2020). How errors cumulate: Two examples. *Journal of Survey Statistics and Methodology 8*(3), 413–432.

Tourangeau, R. and N. Bradburn (2010). The psychology of survey response. In P. V. Marsden and J. D. Wright (Eds.), *Handbook of Survey Research, 2nd ed.*, pp. 315–346. Bingley, UK: Emerald.

Tourangeau, R., B. Edwards, T. P. Johnson, K. M. Wolter, and N. Bates (Eds.) (2014). *Hard-to-Survey Populations.* Cambridge: Cambridge University Press.

Tourangeau, R., J. Michael Brick, S. Lohr, and J. Li (2017). Adaptive and responsive survey designs: A review and assessment. *Journal of the Royal Statistical Society: Series A 180*(1), 203–223.

Tourangeau, R., L. J. Rips, and K. Rasinski (2000). *The Psychology of Survey Responses.* Cambridge: Cambridge University Press.

Tourangeau, R. and T. W. Smith (1996). Asking sensitive questions: The impact of data collection mode, question format, and question context. *Public Opinion Quarterly 60*, 275–304.

Traat, I., L. Bondesson, and K. Meister (2004). Sampling design and sample selection through distribution theory. *Journal of Statistical Planning and Inference 123*(2), 395–413.

Tremblay, V. (1986). Practical criteria for definition of weighting classes. *Survey Methodology 12*, 85–97.

Tukey, J. W. (1958). Bias and confidence in not-quite large samples. *Annals of Mathematical Statistics 29*, 614.

Turk, P. and J. J. Borkowski (2005). A review of adaptive cluster sampling: 1990–2003. *Environmental and Ecological Statistics 12*, 55–94.

Ulrich, R., H. G. Pope, L. Cléret, A. Petróczi, T. Nepusz, J. Schaffer, G. Kanayama, R. D. Comstock, and P. Simon (2018). Doping in two elite athletics competitions assessed by randomized-response surveys. *Sports Medicine 48*(1), 211–219.

United Nations Statistics Division (2014). *Fundamental Principles of Official Statistics.* New York: United Nations.

United Press International (2006, April 18). Parade magazine says Cruise poll skewed. https://www.upi.com/Entertainment_News/2006/04/18/Parade-magazine-says-Cruise-poll-skewed/59541145374281/ (accessed November 25, 2019).

U.S. Bureau of the Census (1995). *1992 Census of Agriculture, Volume 1: Geographic Area Series.* Washington, DC: U.S. Bureau of the Census.

U.S. Census Bureau (1994). *County and City Data Book: 1994.* Washington, DC: U.S. Census Bureau.

U.S. Census Bureau (2006). *Vehicle Inventory and Use Survey—Methods.* Washington, DC: U.S. Census Bureau.

U.S. Census Bureau (2014). *American Community Survey Design and Methodology.* Washington, DC: U.S. Census Bureau.

U.S. Census Bureau (2017). How a question becomes part of the American Community Survey. www.census.gov/content/dam/Census/library/visualizations/2017/comm/acs-questions.pdf (accessed December 3, 2019).

U.S. Census Bureau (2019a). American Community Survey questionnaire. www2.census.gov/programs-surveys/acs/methodology/questionnaires/2019/ (accessed November 30, 2019).

U.S. Census Bureau (2019b). *Current Population Survey Design and Methodology Technical Paper 77.* Washington, DC: U.S. Census Bureau.

U.S. Census Bureau (2020a). *2020 Census Detailed Operational Plan for: Post-Enumeration Survey (PES) Operations.* Washington, DC: U.S. Census Bureau.

U.S. Census Bureau (2020b). American Community Survey demographic and housing one-year estimates, 2017. Table B15001 from https://data.census.gov (accessed January 28, 2020).

U.S. Census Bureau (2020c). Small area income and poverty estimates (SAIPE) program. https://www.census.gov/programs-surveys/saipe.html (accessed January 15, 2021).

U.S. Census Bureau (2020d). *Source of the Data and Accuracy of the Estimates for the 2020 Household Pulse Survey, May 21–May 26.* Washington, DC: U.S. Census Bureau. https://www.census.gov/programs-surveys/household-pulse-survey/technical-documentation.html (accessed April 7, 2021).

U.S. Census Bureau (2020e). *Understanding and Using American Community Survey Data: What Users Need to Know.* Washington, DC: U.S. Census Bureau. https://www.census.gov/content/dam/Census/library/publications/2020/acs/acs_researchers_handbook_2020.pdf (accessed December 22, 2020).

U.S. Census Bureau (2020f). *Understanding and Using the American Community Survey Public Use Microdata Sample Files: What Data Users Need to Know.* Washington, DC: U.S. Census Bureau.

U.S. Department of Education (2020). College scorecard data. `https://collegescorecard.ed.gov/data/` (accessed August 25, 2020).

U.S. Department of Health and Human Services, Centers for Disease Control and Prevention (2014, June). *National Intimate Partner and Sexual Violence Survey (NISVS): General Population Survey Raw Data, 2010: User's Guide.* Ann Arbor, MI: Inter-University Consortium for Political and Social Research [distributor].

U.S. Department of Housing and Urban Development (2014). *Point-in-Time Count Methodology Guide.* Washington, DC: U.S. Department of Housing and Urban Development.

U.S. Department of Justice (1989). *Survey of Youth in Custody, 1987, United States computer file, Conducted by Department of Commerce, Bureau of the Census, 2nd ICPSR ed.* Ann Arbor, MI: Inter-University Consortium for Political and Social Research.

U.S. Federal Reserve Board (2017). *Codebook for 2016 Survey of Consumer Finances.* Washington, DC: Federal Reserve Board.

Valliant, R. (1987). Generalized variance functions in stratified two-stage sampling. *Journal of the American Statistical Association 82*, 499–508.

Valliant, R. (2002). Variance estimation for the general regression estimator. *Survey Methodology 28*, 103–114.

Valliant, R. (2009). Model-based prediction of finite population totals. In D. Pfeffermann and C. R. Rao (Eds.), *Handbook of Statistics: Vol. 29B. Sample Surveys: Inference and Analysis*, pp. 11–31. Amsterdam: North Holland.

Valliant, R. (2020). Comparing alternatives for estimation from nonprobability samples. *Journal of Survey Statistics and Methodology 8*(2), 231–263.

Valliant, R., J. M. Brick, and J. A. Dever (2008). Weight adjustments for the grouped jackknife variance estimator. *Journal of Official Statistics 24*(3), 469.

Valliant, R., J. A. Dever, and F. Kreuter (2018). *Practical Tools for Designing and Weighting Survey Samples.* New York: Springer.

Valliant, R., A. H. Dorfman, and R. M. Royall (2000). *Finite Population Sampling and Inference: A Prediction Approach.* New York: Wiley.

Valliant, R. and K. F. Rust (2010). Degrees of freedom approximations and rules-of-thumb. *Journal of Official Statistics 26*(4), 585–602.

van Buuren, S. (2018). *Flexible Imputation of Missing Data.* Boca Raton, FL: CRC Press.

Van Patter, L., T. Flockhart, J. Coe, O. Berke, R. Goller, A. Hovorka, and S. Bateman (2019). Perceptions of community cats and preferences for their management in Guelph, Ontario. Part II: A qualitative analysis. *The Canadian Veterinary Journal 60*(1), 48–54.

Vijayan, K. (1968). An exact πps sampling scheme: Generalization of a method of Hanurav. *Journal of the Royal Statistical Society, Series B 30*, 556–566.

Virginia Polytechnic and State University/Responsive Management (2006). *An Assessment of Public and Hunter Opinions and the Costs and Benefits to North Carolina of Sunday Hunting.* Blacksburg, VA: Virginia Polytechnic and State University.

Waksberg, J. (1978). Sampling methods for random digit dialing. *Journal of the American Statistical Association 73*, 40–46.

Wald, A. (1943). Tests of statistical hypotheses concerning several parameters when the number of observations is large. *Transactions of the American Mathematical Society 54*, 426–482.

Wallgren, A. and B. Wallgren (2014). *Register-Based Statistics: Statistical Methods for Administrative Data.* Hoboken, NJ: Wiley.

Wang, J. (2021). The pseudo maximum likelihood estimator for quantiles of survey variables. *Journal of Survey Statistics and Methodology 9*(1), 185–201.

Wang, J. C. and J. D. Opsomer (2011). On asymptotic normality and variance estimation for nondifferentiable survey estimators. *Biometrika 98*(1), 91–106.

Warner, S. L. (1965). Randomized response: A survey technique for eliminating evasive answer bias. *Journal of the American Statistical Association 60*, 63–69.

Warren, E. and A. W. Tyagi (2003). *The Two-Income Trap.* New York: Basic Books.

Watson, D. J. (1937). The estimation of leaf area in field crops. *Journal of Agricultural Science 27*, 474–483.

Webb, W. B. (1955). The illusive phenomena in accident proneness. *Public Health Reports 70*, 951.

Weisberg, H. F. (2018). Total survey error. In L. R. Atkeson and R. M. Alvarez (Eds.), *The Oxford Handbook of Polling and Survey Methods*, pp. 1–16. Oxford, UK: Oxford University Press.

Weisberg, S. (2014). *Applied Linear Regression, 4th ed.* Hoboken, NJ: Wiley.

Welford, P. B. (1962). Note on a method for calculating corrected sums of squares and products. *Technometrics 4*(3), 419–420.

Welsh, A. H. and D. P. Wiens (2013). Robust model-based sampling designs. *Statistics and Computing 23*(6), 689–701.

West, B. T. and A. G. Blom (2017). Explaining interviewer effects: A research synthesis. *Journal of Survey Statistics and Methodology 5*(2), 175–211.

West, B. T., F. G. Conrad, F. Kreuter, and F. Mittereder (2018). Nonresponse and measurement error variance among interviewers in standardized and conversational interviewing. *Journal of Survey Statistics and Methodology 6*(3), 335–359.

West, B. T. and K. Olson (2010). How much of interviewer variance is really nonresponse error variance? *Public Opinion Quarterly 74*(5), 1004–1026.

West, B. T., J. W. Sakshaug, and G. A. S. Aurelien (2018). Accounting for complex sampling in survey estimation: A review of current software tools. *Journal of Official Statistics 34*(3), 721–752.

Westat (2015). *WesVar® 5.1 User's Guide.* Rockville, MD: Westat.

Wilk, S. J., W. W. Morse, D. E. Ralph, and T. R. Azarovitz (1977). *Fishes and Associated Environmental Data Collected in New York Bight, June 1974–June 1975.* NOAA Tech. Rep. No. NMFS SSRF-716. Washington, DC: U.S. Government Printing Office.

Williams, D. and J. M. Brick (2018). Trends in U.S. face-to-face household survey nonresponse and level of effort. *Journal of Survey Statistics and Methodology 6*(2), 186–211.

Wisconsin Department of Natural Resources (1993). *The Fisher in Wisconsin.* Technical Bulletin no. 183. Madison, WI: Department of Natural Resources.

Wolfe, D. A. (2012). Ranked set sampling: Its relevance and impact on statistical inference. *International Scholarly Research Notices 2012*, 1–32.

Wolter, K. M. (2007). *Introduction to Variance Estimation, 2nd ed.* New York: Springer.

Wolter, K. M., P. J. Smith, M. Khare, B. Welch, K. R. Copeland, V. J. Pineau, and N. Davis (2017). Statistical methodology of the National Immunization Survey, 2005–2014. *Vital and Health Statistics 1*(61), 1–96.

Woodruff, R. S. (1952). Confidence intervals for medians and other position measures. *Journal of the American Statistical Association 47*, 636–646.

Woodruff, R. S. (1971). A simple method for approximating the variance of a complicated estimate. *Journal of the American Statistical Association 66*, 411–414.

Wright, J. (1988). The mentally ill homeless: What is myth and what is fact? *Social Problems 35*, 182–191.

Wright, T. (2012). The equivalence of Neyman optimum allocation for sampling and equal proportions for apportioning the U.S. House of Representatives. *The American Statistician 66*(4), 217–224.

Wu, C. and W. W. Lu (2016). Calibration weighting methods for complex surveys. *International Statistical Review 84*(1), 79–98.

Wu, C. and M. E. Thompson (2020). *Sampling Theory and Practice.* Cham, Switzerland: Springer.

Wynia, W., A. Sudar, and G. Jones (1993). Recycling human waste: Composting toilets as a remedial action plan option for Hamilton Harbour. *Water Pollution Research Journal of Canada 28*, 355–368.

Yan, T., S. Fricker, and S. Tsai (2020). Response burden: What is it and what predicts it? In P. Beatty, D. Collins, L. Kaye, J. L. Padilla, G. Willis, and A. Wilmot (Eds.), *Advances in Questionnaire Design, Development, Evaluation and Testing*, pp. 193–212. Hoboken, NJ: Wiley.

Yang, S. and J. K. Kim (2016). Fractional imputation in survey sampling: A comparative review. *Statistical Science 31*(3), 415–432.

Yates, F. (1981). *Sampling Methods for Censuses and Surveys, 4th ed.* New York: Macmillan.

Yates, F. and P. M. Grundy (1953). Selection without replacement from within strata with probability proportional to size. *Journal of the Royal Statistical Society, Series B 109*, 12–30.

Yi, G. Y., J. N. K. Rao, and H. Li (2016). A weighted composite likelihood approach for analysis of survey data under two-level models. *Statistica Sinica 26*(2), 569–587.

Yung, W., J. Karkimaa, M. Scannapieco, G. Barcarolli, D. Zardetto, J. A. R. Sanchez, B. Braaksma, B. Buelens, and J. Burger (2018). *The Use of Machine Learning in Official Statistics.* Geneva: UNECE.

Yung, W. and J. N. K. Rao (2000). Jackknife variance estimation under imputation for estimators using poststratification information. *Journal of the American Statistical Association 95*(451), 903–915.

Zanutto, E. L. (2006). A comparison of propensity score and linear regression analysis of complex survey data. *Journal of Data Science 4*(1), 67–91.

Zehnder, G. W., D. M. Kolodny-Hirsch, and J. J. Linduska (1990). Evaluation of various potato plant sample units for cost-effective sampling of Colorado potato beetle (*Coleoptera chrysomelidae*). *Journal of Economic Entomology 83*, 428–433.

Zhang, C., C. Antoun, H. Y. Yan, and F. G. Conrad (2020). Professional respondents in opt-in online panels: What do we really know? *Social Science Computer Review 38*(6), 703–719.

Zhang, G., F. Christensen, and W. Zheng (2015). Nonparametric regression estimators in complex surveys. *Journal of Statistical Computation and Simulation 85*(5), 1026–1034.

Zhang, L.-C. and R. L. Chambers (Eds.) (2019). *Analysis of Integrated Data.* Boca Raton, FL: CRC Press.

Zhao, P., M. Ghosh, J. N. K. Rao, and C. Wu (2020). Bayesian empirical likelihood inference with complex survey data. *Journal of the Royal Statistical Society: Series B 82*, 155–174.

Zimmerman, D. L. (2020). *Linear Model Theory: With Examples and Exercises.* New York: Springer.

Ziolkowski, D., K. Pardieck, and J. R. Sauer (2010). On the road again for a bird survey that counts. *American Birding 42*(4), 32–40.

Zou, D., J. E. Lloyd, and J. L. Baumbusch (2020). Using SPSS to analyze complex survey data: A primer. *Journal of Modern Applied Statistical Methods 18*(1), 1–22.

Index

Printed in the United States
by Baker & Taylor Publisher Services